TK 7868 .P6 B343 2008

Basso, Christophe P.

Switch-mode power supplies

SWITCH-MODE
POWER SUPPLIES

ABOUT THE AUTHOR

Christophe P. Basso is currently a technical engineer at ON Semiconductor in France. He is a frequent contributor to various power electronics magazines and the author of McGraw-Hill's *Switch-Mode Power Supply SPICE Cookbook*.

SWITCH-MODE POWER SUPPLIES

SPICE Simulations and Practical Designs

Christophe P. Basso

New York Chicago San Francisco Lisbon London Madrid
Mexico City Milan New Delhi San Juan Seoul
Singapore Sydney Toronto

The McGraw·Hill Companies

Cataloging-in-Publication Data is on file with the Library of Congress

Copyright © 2008 by The McGraw-Hill Companies, Inc. All rights reserved. Printed in the United States of America. Except as permitted under the United States Copyright Act of 1976, no part of this publication may be reproduced or distributed in any form or by any means, or stored in a data base or retrieval system, without the prior written permission of the publisher.

2 3 4 5 6 7 8 9 0 DOC/DOC 0 1 3 2 1 0 9 8

ISBN: P/N 978-0-07-150859-9 of set 978-0-07-150858-2
MHID: P/N 0-07-150859-7 of set 0-07-150858-9

ISBN: P/N 978-0-07-159769-2 of set 978-0-07-159751-7 (ON Semiconductor Edition)
MHID: P/N 0-07-159769-7 of set 0-07-159751-4 (ON Semiconductor Edition)

Printed and bound by RR Donnelley.

This book is printed on acid-free paper.

McGraw-Hill books are available at special quantity discounts to use as premiums and sales promotions, or for use in corporate training programs. For more information, please write to the Director of Special Sales, McGraw-Hill Professional, Two Penn Plaza, New York, NY 10121-2298. Or contact your local bookstore.

Sponsoring Editor
Stephen S. Chapman

Production Supervisor
Richard C. Ruzycka

Editing Supervisor
Stephen M. Smith

Project Manager
Madhu Bhardwaj

Copy Editor
Patti Scott

Proofreader
Sanjukta Chandra Chowdhury

Indexer
Kevin Broccoli

Art Director, Cover
Jeff Weeks

Composition
International Typesetting and Composition

Information contained in this work has been obtained by The McGraw-Hill Companies, Inc. ("McGraw-Hill") from sources believed to be reliable. However, neither McGraw-Hill nor its authors guarantee the accuracy or completeness of any information published herein, and neither McGraw-Hill nor its authors shall be responsible for any errors, omissions, or damages arising out of use of this information. This work is published with the understanding that McGraw-Hill and its authors are supplying information but are not attempting to render engineering or other professional services. If such services are required, the assistance of an appropriate professional should be sought.

CONTENTS

Foreword xiii
Preface xv
Nomenclature xvii

Chapter 1. Introduction to Power Conversion   1

1.1. "Do You Really Need to Simulate?" / *1*
1.2. What You Will Find in the Following Pages / *2*
1.3. What You Will Not Find in This Book / *3*
1.4. Converting Power with Resistors / *3*
    1.4.1. Associating Resistors / *3*
    1.4.2. A Closed-Loop System / *5*
    1.4.3. Deriving Useful Equations with the Linear Regulator / *7*
    1.4.4. A Practical Working Example / *10*
    1.4.5. Building a Simple Generic Linear Regulator / *14*
    1.4.6. Conclusion on Linear Regulators / *17*
1.5. Converting Power with Switches / *18*
    1.5.1. A Filter Is Needed / *19*
    1.5.2. Current in the Inductance, Continuous or Discontinuous? / *21*
    1.5.3. Charge and Flux Balance / *25*
    1.5.4. Energy Storage / *27*
1.6. The Duty Cycle Factory / *27*
    1.6.1. Voltage-Mode Operation / *27*
    1.6.2. Current-Mode Operation / *29*
1.7. The Buck Converter / *30*
    1.7.1. On-Time Event / *30*
    1.7.2. Off-Time Event / *31*
    1.7.3. Buck Waveforms—CCM / *31*
    1.7.4. Buck Waveforms—DCM / *34*
    1.7.5. Buck Transition Point DCM—CCM / *37*
    1.7.6. Buck CCM Output Ripple Voltage Calculation / *39*
    1.7.7. Now with the ESR / *41*
    1.7.8. Buck Ripple, the Numerical Application / *41*
1.8. The Boost Converter / *42*
    1.8.1. On-Time Event / *43*
    1.8.2. Off-Time Event / *44*
    1.8.3. Boost Waveforms—CCM / *44*
    1.8.4. Boost Waveforms—DCM / *47*
    1.8.5. Boost Transition Point DCM—CCM / *50*
    1.8.6. Boost CCM Output Ripple Voltage Calculations / *51*
    1.8.7. Now with the ESR / *54*
    1.8.8. Boost Ripple, the Numerical Application / *54*
1.9. The Buck-Boost Converter / *55*
    1.9.1. On-Time Event / *56*
    1.9.2. Off-Time Event / *56*

1.9.3. Buck-Boost Waveforms—CCM / 57
1.9.4. Buck-Boost Waveforms—DCM / 59
1.9.5. Buck-Boost Transition Point DCM—CCM / 63
1.9.6. Buck-Boost CCM Output Ripple Voltage Calculation / 64
1.9.7. Now with the ESR / 65
1.9.8. Buck-Boost Ripple, the Numerical Application / 65
1.10. Input Filtering / 66
    1.10.1. The *RLC* Filter / 67
    1.10.2. A More Comprehensive Representation / 70
    1.10.3. Creating a Simple Closed-Loop Current Source with SPICE / 71
    1.10.4. Understanding Overlapping Impedances / 72
    1.10.5 Damping the Filter / 76
    1.10.6 Calculating the Required Attenuation / 79
    1.10.7 Fundamental Frequency Evaluation / 80
    1.10.8 Selecting the Right Cutoff Frequency / 82
What I Should Retain from Chap. 1 / 85
References / 85
Appendix 1A A RLC Transfer Function / 86
Appendix 1B The Capacitor Equivalent Model / 89
Appendix 1C Power Supply Classification by Topologies / 93

Chapter 2. Small-Signal Modeling   95

2.1. State-Space Averaging / 98
    2.1.1. SSA at Work for the Buck Converter—First Step / 100
    2.1.2. The DC Transformer / 102
    2.1.3. Large-Signal Simulations / 105
    2.1.4. SSA at Work for the Buck Converter, the Linearization—Second Step / 106
    2.1.5. SSA at Work for the Buck Converter, the Small-Signal Model—Final Step / 108
2.2. The PWM Switch Model—the Voltage-Mode Case / 111
    2.2.1. Back to the Good Old Bipolars / 112
    2.2.2. An Invariant Internal Architecture / 113
    2.2.3. Waveform Averaging / 114
    2.2.4. Terminal Currents / 116
    2.2.5. Terminal Voltages / 117
    2.2.6. A Transformer Representation / 117
    2.2.7. Large-Signal Simulations / 118
    2.2.8. A More Complex Representation / 121
    2.2.9. A Small-Signal Model / 123
    2.2.10. Helping with Simulation / 128
    2.2.11. Discontinuous Mode Model / 129
    2.2.12. Deriving the d_2 Variable / 132
    2.2.13. Clamping Sources / 132
    2.2.14. Encapsulating the Model / 134
    2.2.15. The PWM Modulator Gain / 138
    2.2.16. Testing the Model / 142
    2.2.17. Mode Transition / 143
2.3. The PWM Switch Model—the Current-Mode Case / 145
    2.3.1. Current-Mode Instabilities / 146
    2.3.2. Preventing Instabilities / 151
    2.3.3. The Current-Mode Model in CCM / 153
    2.3.4. Upgrading the Model / 158
    2.3.5. The Current-Mode Model in DCM / 161
    2.3.6. Deriving the Duty Cycles d_1 and d_2 / 163
    2.3.7. Building the DCM Model / 165
    2.3.8. Testing the Model / 168
    2.3.9. Buck DCM, Instability in DC / 172
    2.3.10. Checking the Model in CCM / 172

2.4. The PWM Switch Model—Parasitic Elements Effects / 175
  2.4.1. A Variable Resistor / 179
  2.4.2. Ohmic Losses, Voltage Drops: The VM Case / 180
  2.4.3. Ohmic Losses, Voltage Drops: The CM Case / 182
  2.4.4. Testing the Lossy Model in Current Mode / 183
  2.4.5. Convergence Issues with the CM Model / 186
2.5. PWM Switch Model in Borderline Conduction / 187
  2.5.1. Borderline Conduction—the Voltage-Mode Case / 187
  2.5.2. Testing the Voltage-Mode BCM Model / 191
  2.5.3. Borderline Conduction—the Current-Mode Case / 194
  2.5.4. Testing the Current-Mode BCM Model / 198
2.6. The PWM Switch Model—a Collection of Circuits / 202
  2.6.1. The Buck / 203
  2.6.2. The Tapped Buck / 204
  2.6.3. The Forward / 205
  2.6.4. The Buck-Boost / 206
  2.6.5. The Flyback / 207
  2.6.6. The Boost / 208
  2.6.7. The Tapped Boost / 208
  2.6.8. The Nonisolated SEPIC / 209
  2.6.9. The Isolated SEPIC / 210
  2.6.10. The Nonisolated Ćuk Converter / 211
  2.6.11. The Isolated Ćuk Converter / 212
2.7. Other Averaged Models / 213
  2.7.1. Ridley Models / 213
  2.7.2. Small-Signal Current-Mode Models / 213
  2.7.3. Ridley Models at Work / 214
  2.7.4. CoPEC Models / 216
  2.7.5. CoPEC Models at Work / 218
  2.7.6. Ben-Yaakov Models / 220
What I Should Retain from Chap. 2 / 224
References / 224
Appendix 2A Basic Transfer Functions for Converters / 225
  2A.1. Buck / 226
  2A.2. Boost / 229
  2A.3. Buck-Boost / 231
    References / 235
Appendix 2B Poles, Zeros, and Complex Plane—a Simple Introduction / 235
    References / 240

Chapter 3. Feedback and Control Loops   241

3.1. Observation Points / 243
3.2. Stability Criteria / 247
3.3. Phase Margin and Transient Response / 248
3.4. Choosing the Crossover Frequency / 249
3.5. Shaping the Compensation Loop / 250
  3.5.1. The Passive Pole / 250
  3.5.2. The Passive Zero / 251
  3.5.3. Right Half-Plane Zero / 253
  3.5.4. Type 1 Amplifier—Active Integrator / 255
  3.5.5. Type 2 Amplifier—Zero-Pole Pair / 256
  3.5.6. Type 2a—Origin Pole Plus a Zero / 258
  3.5.7. Type 2b—Proportional Plus a Pole / 259
  3.5.8. Type 3—Origin Pole Plus Two Coincident Zero-Pole Pairs / 261
  3.5.9. Selecting the Right Amplifier Type / 262
3.6. An Easy Stabilization Tool—the k Factor / 263
  3.6.1. Type 1 Derivation / 264
  3.6.2. Type 2 Derivation / 264

3.6.3. Type 3 Derivation / 266
3.6.4. Stabilizing a Voltage-Mode Buck Converter with the k Factor / 267
3.6.5. Conditional Stability / 270
3.6.6. Independent Pole-Zero Placement / 272
3.6.7. Crossing Over Right at the Selected Frequency / 273
3.6.8. The k Factor Versus Manual Pole-Zero Placement / 275
3.6.9. Stabilizing a Current-Mode Buck Converter with the k Factor / 280
3.6.10. The Current-Mode Model and Transient Steps / 286
3.7. Feedback with the TL431 / 286
   3.7.1. A Type 2 Amplifier Design Example with the TL431 / 291
   3.7.2. A Type 3 Amplifier with the TL431 / 292
   3.7.3. Biasing the TL431 / 298
   3.7.4. The Resistive Divider / 303
3.8. The Optocoupler / 304
   3.8.1. A Simplified Model / 305
   3.8.2. Extracting the Pole / 306
   3.8.3. Accounting for the Pole / 308
3.9. Shunt Regulators / 312
   3.9.1. SPICE Model of the Shunt Regulator / 313
   3.9.2. Quickly Stabilizing a Converter Using the Shunt Regulator / 314
3.10. Small-Signal Responses with PSIM and SIMPLIS / 316
What I Should Retain from Chap. 3 / 322
References / 322
Appendix 3A Automated Pole-Zero Placement / 323
Appendix 3B A TL431 Spice Model / 326
   3B.1. A Behavioral TL431 Spice Model / 326
   3B.2. Cathode Current Versus Cathode Voltage / 328
   3B.3. Output Impedance / 329
   3B.4. Open-Loop Gain / 330
   3B.5. Transient Test / 331
   3B.6. Model Netlist / 331
Appendix 3C Type 2 Manual Pole-Zero Placement / 332
Appendix 3D Understanding the Virtual Ground in Closed-Loop Systems / 335
   3D.1. Numerical Example / 336
   3D.2. Loop Gain Is Unchanged / 337

Chapter 4. Basic Blocks and Generic Switched Models   341

4.1. Generic Models for Faster Simulations / 341
   4.1.1. In-Line Equations / 341
4.2. Operational Amplifiers / 343
   4.2.1. A More Realistic Model / 344
   4.2.2. A UC384X Error Amplifier / 345
4.3. Sources with a Given Fan-Out / 348
4.4. Voltage-Adjustable Passive Elements / 349
   4.4.1. The Resistor / 350
   4.4.2. The Capacitor / 351
   4.4.3. The Inductor / 353
4.5. A Hysteresis Switch / 355
4.6. An Undervoltage Lockout Block / 358
4.7. Leading Edge Blanking / 359
4.8. Comparator with Hysteresis / 361
4.9. Logic Gates / 362
4.10. Transformers / 364
   4.10.1. A Simple Saturable Core Model / 366
   4.10.2. Multioutput Transformers / 372
4.11. Astable Generator / 372
   4.11.1. A Voltage-Controlled Oscillator / 374
   4.11.2. A Voltage-Controlled Oscillator Featuring Dead Time Control / 377

CONTENTS ix

4.12. Generic Controllers / 377
    4.12.1. Current-Mode Controllers / 378
    4.12.2. Current-Mode Model with a Buck / 380
    4.12.3. Current-Mode Instabilities / 381
    4.12.4. The Voltage-Mode Model / 382
    4.12.5. The Duty Cycle Generation / 382
    4.12.6. A Quick Example with a Forward Converter / 384
4.13. Dead Time Generation / 387
4.14. List of Generic Models / 387
4.15. Convergence Options / 388
What I Should Retain from Chap. 4 / 391
References / 392
Appendix 4A An Incomplete Review of the Terminology Used in Magnetic Designs / 392
    4A.1. Introduction / 392
    4A.2. Field Definition / 393
    4A.3. Permeability / 393
    4A.4. Founding Laws / 396
    4A.5. Inductance / 396
    4A.6. Avoiding Saturation / 397
        References / 398
Appendix 4B Feeding Transformer Models with Physical Values / 398
    4B.1. Understanding the Equivalent Inductor Model / 398
    4B.2. Determining the Physical Values of the Two-Winding T Model / 400
    4B.3. The Three-Winding T Model / 401
        References / 405

Chapter 5. Simulations and Practical Designs of Nonisolated Converters   407

5.1. The Buck Converter / 407
    5.1.1. A 12 V, 4 A Voltage-Mode Buck from a 28 V Source / 407
    5.1.2. Ac Analysis / 410
    5.1.3. Transient Analysis / 413
    5.1.4. The Power Switch / 417
    5.1.5. The Diode / 418
    5.1.6. Output Ripple and Transient Response / 419
    5.1.7. Input Ripple / 421
    5.1.8. A 5 V, 10 A Current-Mode Buck from a Car Battery / 425
    5.1.9. Ac Analysis / 426
    5.1.10. Transient Analysis / 429
    5.1.11. A Synchronous Buck Converter / 433
    5.1.12. A Low-Cost Floating Buck Converter / 434
    5.1.13. Component Constraints for the Buck Converter / 439
5.2. The Boost Converter / 441
    5.2.1. A Voltage-Mode 48 V, 2 A Boost from a Car Battery / 441
    5.2.2. Ac Analysis / 444
    5.2.3. Transient Analysis / 449
    5.2.4. A Current-Mode 5 V, 1 A Boost from a Li-Ion Battery / 452
    5.2.5. Ac Analysis / 454
    5.2.6. Transient Analysis / 459
    5.2.7. Input Filter / 460
    5.2.8. Component Constraints for the Boost Converter / 465
5.3. The Buck-Boost Converter / 465
    5.3.1. A Voltage-Mode 12 V, 2 A Buck-Boost Converter Powered
        from a Car Battery / 465
    5.3.2. Ac Analysis / 468
    5.3.3. Transient Analysis / 474
    5.3.4. A Discontinuous Current-Mode 12 V, 2 A Buck-Boost Converter Operating
        from a Car Battery / 476
    5.3.5. Ac Analysis / 479

5.3.6. Transient Analysis / 483
5.3.7. Component Constraints for the Buck-Boost Converter / 486
References / 486
Appendix 5A The Boost in Discontinuous Mode, Design Equations / 487
5A.1. Input Current / 487
5A.2. Output Ripple Voltage / 489

Chapter 6. Simulations and Practical Designs of Off-Line Converters—The Front End   491

6.1. The Rectifier Bridge / 491
   6.1.1. Capacitor Selection / 493
   6.1.2. Diode Conduction Time / 495
   6.1.3. Rms Current in the Capacitor / 496
   6.1.4. Current in the Diodes / 498
   6.1.5. Input Power Factor / 498
   6.1.6. A 100 W Rectifier Operated on Universal Mains / 499
   6.1.7. Hold-Up Time / 501
   6.1.8. Waveforms and Line Impedance / 502
   6.1.9. In-Rush Current / 506
   6.1.10. Voltage Doubler / 508
6.2. Power Factor Correction / 510
   6.2.1. Definition of Power Factor / 512
   6.2.2. Nonsinusoidal Signals / 512
   6.2.3. A Link to the Distortion / 514
   6.2.4. Why Power Factor Correction? / 515
   6.2.5. Harmonic Limits / 517
   6.2.6. A Need for Storage / 518
   6.2.7. Passive PFC / 520
   6.2.8. Improving the Harmonic Content / 524
   6.2.9. The Valley-Fill Passive Corrector / 526
   6.2.10. Active Power Factor Correction / 527
   6.2.11. Different Techniques / 528
   6.2.12. Constant On-Time Borderline Operation / 529
   6.2.13. Frequency Variations in BCM / 531
   6.2.14. Averaged Modeling of the BCM Boost / 532
   6.2.15. Fixed-Frequency Average Current-Mode Control / 535
   6.2.16. Shaping the Current / 540
   6.2.17. Fixed-Frequency Peak Current-Mode Control / 543
   6.2.18. Compensating the Peak Current-Mode Control PFC / 544
   6.2.19. Average Modeling of the Peak Current-Mode PFC / 546
   6.2.20. Hysteretic Power Factor Correction / 549
   6.2.21. Fixed-Frequency DCM Boost / 550
   6.2.22. Flyback Converter / 555
   6.2.23. Testing the Flyback PFC / 559
6.3. Designing a BCM Boost PFC / 559
   6.3.1. Average Simulations / 567
   6.3.2. Reducing the Simulation Time / 570
   6.3.3. Cycle-by-Cycle Simulation / 571
   6.3.4. The Follow-Boost Technique / 574
What I Should Retain from Chap. 6 / 575
References / 576

Chapter 7. Simulations and Practical Designs of Flyback Converters   579

7.1. An Isolated Buck-Boost / 579
7.2. Flyback Waveforms, No Parasitic Elements / 583

- 7.3. Flyback Waveforms with Parasitic Elements / *586*
- 7.4. Observing the Drain Signal, No Clamping Action / *588*
- 7.5. Clamping the Drain Excursion / *591*
- 7.6. DCM, Looking for Valleys / *597*
- 7.7. Designing the Clamping Network / *599*
  - 7.7.1. The *RCD* Configuration / *601*
  - 7.7.2. Selecting k_c / *604*
  - 7.7.3. Curing the Leakage Ringing / *605*
  - 7.7.4. Which Diode to Select? / *609*
  - 7.7.5. Beware of Voltage Variations / *610*
  - 7.7.6. TVS Clamp / *612*
- 7.8. Two-Switch Flyback / *614*
- 7.9. Active Clamp / *616*
  - 7.9.1. Design Example / *622*
  - 7.9.2. Simulation Circuit / *625*
- 7.10. Small-Signal Response of the Flyback Topology / *628*
  - 7.10.1. DCM Voltage Mode / *628*
  - 7.10.2. CCM Voltage Mode / *635*
  - 7.10.3. DCM Current Mode / *636*
  - 7.10.4. CCM Current Mode / *638*
- 7.11. Practical Considerations about the Flyback / *642*
  - 7.11.1. Start-Up of the Controller / *642*
  - 7.11.2. Start-Up Resistor Design Example / *644*
  - 7.11.3. Half-Wave Connection / *646*
  - 7.11.4. Good Riddance, Start-up Resistor! / *648*
  - 7.11.5. High-Voltage Current Source / *649*
  - 7.11.6. The Auxiliary Winding / *651*
  - 7.11.7. Short-Circuit Protection / *653*
  - 7.11.8. Observing the Feedback Pin / *654*
  - 7.11.9. Compensating the Propagation Delay / *655*
  - 7.11.10. Sensing the Secondary Side Current / *660*
  - 7.11.11. Improving the Drive Capability / *662*
  - 7.11.12. Overvoltage Protection / *663*
- 7.12. Standby Power of Converters / *665*
  - 7.12.1. What Is Standby Power? / *666*
  - 7.12.2. The Origins of Losses / *666*
  - 7.12.3. Skipping Unwanted Cycles / *667*
  - 7.12.4. Skipping Cycles with a UC384X / *669*
  - 7.12.5. Frequency Foldback / *670*
- 7.13. A 20 W, Single-Output Power Supply / *670*
- 7.14. A 90 W, Single-Output Power Supply / *687*
- 7.15. A 35 W, Multioutput Power Supply / *706*
- 7.16. Component Constraints for the Flyback Converter / *725*

What I Should Retain from Chap. 7 / *726*
References / *727*
Appendix 7A Reading the Waveforms to Extract the Transformer Parameters / *727*
Appendix 7B The Stress / *729*
- 7B.1. Voltage / *730*
- 7B.2. Current / *731*

Appendix 7C Transformer Design for the 90 W Adapter / *732*
- 7C.1. Core Selection / *732*
- 7C.2. Determining the Primary and Secondary Turns / *733*
- 7C.3. Choosing the Primary and Secondary Wire Sizes / *734*
- 7C.4. Choosing the Material, Based on the Desired Inductance, or Gapping the Core If Necessary / *735*
- 7C.5. Designs Using Intusoft Magnetic Designer / *735*

Chapter 8. Simulations and Practical Designs of Forward Converters 739

8.1. An Isolated Buck Converter / 739
    8.1.1. Need for a Complete Core Reset / 742
8.2. Reset Solution 1, a Third Winding / 746
    8.2.1. Leakage Inductance and Overlap / 752
8.3. Reset Solution 2, a Two-Switch Configuration / 756
    8.3.1. Two-Switch Forward and Half-Bridge Driver / 760
8.4. Reset Solution 3, the Resonant Demagnetization / 762
8.5. Reset Solution 4, the RCD Clamp / 767
8.6. Reset Solution 5, the Active Clamp / 778
8.7. Synchronous Rectification / 796
8.8. Multioutput Forward Converters / 799
    8.8.1. Magnetic Amplifiers / 799
    8.8.2. Synchronous Postregulation / 804
    8.8.3. Coupled Inductors / 806
8.9. Small-Signal Response of the Forward Converter / 817
    8.9.1. Voltage Mode / 817
    8.9.2. Current Mode / 821
    8.9.3. Multioutput Forward / 825
8.10. A Single-Output 12 V, 250 W Forward Design Example / 828
    8.10.1. MOSFET Selection / 833
    8.10.2. Installing a Snubber / 835
    8.10.3. Diode Selection / 838
    8.10.4. Small-Signal Analysis / 839
    8.10.5. Transient Results / 841
    8.10.6. Short-Circuit Protection / 846
8.11. Component Constraints for the Forward Converter / 849
What I Should Retain from Chap. 8 / 849
References / 850
Appendix 8A Half-Bridge Drivers Using the Bootstrap Technique / 851
Appendix 8B Impedance Reflections / 855
Appendix 8C Transformer and Inductor Designs for the 250 W Adapter / 859
    8C.1. Transformer Variables / 859
    8C.2. Transformer Core Selection / 859
    8C.3. Determining the Primary and Secondary Turns / 860
    8C.4. Choosing the Primary and Secondary Wire Sizes / 861
    8C.5. Gapping the Core / 861
    8C.6. Designs Using Intusoft Magnetic Designer / 862
    8C.7. Inductor Design / 865
    8C.8. Core Selection / 866
    8C.9. Choosing the Wire Size and Checking the DC Resistive Loss / 867
    8C.10. Checking the Core Loss / 867
    8C.11. Estimating the Temperature Rise / 867
Appendix 8D CD-ROM Content / 868

Conclusion 869
Index 871

FOREWORD

First, a word about switching power supply design. It is a field that seldom receives the budget, expertise, cooling, testing, or schedule that it deserves. Many high-profile commercial product failures are the result of inadequate power system design. These product failures plague manufacturers and consumers at an ever-increasing rate today.

While the circuits for switching power supplies are quite simple in schematic form, successful power supply design for manufacturing is never simple or straightforward, regardless of what people may tell you. Mr. Basso understands this precept very well. There are layers upon layers of design detail which, if overlooked, will cause a power supply to fail. This book devotes over 800 pages to explaining many of those layers—essential information for all power supply designers.

Electrical engineers tasked with designing their first power supply will often turn to familiar simulation tools for guidance. Granted, there is a comforting familiarity in seeing a simulated waveform before building a production power supply. However, this can be fraught with problems. You cannot simply toss a schematic in SPICE or other software and expect it to tell you something useful about your design. The production schematic must be modified before you attempt any simulation.

Mr. Basso clearly explains the different ways that power supplies can be simulated and the level of modeling needed to achieve the desired results. If you have never worked with SPICE before for switching power supplies, you should not attempt to do so without absorbing the valuable experience contained in the models of this book. Mr. Basso, while being perhaps the foremost expert in SPICE simulations of power supplies in the world, is also very quick to point out that testing on the lab bench is essential, and that SPICE simulations without lab verification are not to be trusted alone. They should be used as a tool to aid the design process.

Power supply design continues to be a field where experience is paramount. There is no single book that explains from A to Z the complete design process for every power application, nor will there ever be. Mr. Basso takes the disciplined road in writing specifically about converters and topics that he knows in great detail through personal experience. If you are building power factor correctors, forward and flyback converters, the most common power supply topologies, you should heed Mr. Basso's wisdom before you embark on your design. Pay close attention to his comments on semiconductor deratings and advice for safe operating regions. If just this part of the book were absorbed into our industry, the number of power supply failures would drop dramatically.

I highly recommend this book to all levels of designers, from the newcomer to the advanced. *Switch-Mode Power Supplies: SPICE Simulations and Practical Designs* should be on every lab bench. Read the pages of this book, run the simulations to fully understand the points that Mr. Basso makes, and absorb his valuable advice to start testing on the bench as soon as possible.

Dr. Ray Ridley
Ridley Engineering

PREFACE

This book corresponds to a fully rewritten *Switch-Mode Power Supply SPICE Cookbook* in many aspects, hence the change in title now including "practical designs." When I started to write again, I realized that there were many books on the market dedicated to switching power supplies. However, these books either were too academic without a connection to the industrial world or were simply too practical, lacking theoretical foundations for most of the formulas used in the design examples. This book tries to fill the gap, by offering a technical content balanced between analytical descriptions and practical designs. For example, in Chap. 3, compensation techniques are described not only with an operational amplifier as found in many textbooks but also with the industry-standard TL431 found in today's power supplies. Therefore, both students and design engineers should find relevant data in their respective fields. Concerning the analytical content, the format of equations now respects IEEE notations, and I recognize in all honesty that this was not the case in *Switch-Mode Power Supply SPICE Cookbook*! The proposed models are now better and more thoroughly described than before: Remarks from some readers concerning this point have been taken into account. As with my earlier book, chapter examples will be available in some of the popular SPICE editors' versions, offering an easy means to learn and progress from these examples. Numerous files appear on the book CD whereas some more complex circuits are distributed separately. For those of you who would like to have a taste of the current market simulation on offer, the CD includes a variety of demonstration versions of PSIM, Transim, Tina, Multisim, Intusoft, and other interesting packages.

My reader friends, this book represents a huge amount of time spent writing, usually at night, for almost three years. Despite all the efforts put into the descriptions, equation definitions, and other little details, there may be mistakes or typos left uncorrected. Please be kind enough to report them via my website where updates, new models, and information on the full library file will also appear: http://pagesperso-orange.fr/cbasso/Spice.htm. Thank you in advance.

As I have implied above, my warmest thanks and love go, first, to my dear family: Anne, my wife, and my two beloved children, Lucile and Paul, who all endured my mood swings when facing nonconvergence issues and a lack of inspiration starting a blank page. I also express my gratitude to my beloved parents, Michele and Paul Basso, who bought me my first power supply when I was 14 and let me develop my passion for electronic circuits, at the expense of numerous breaker trips. As we have returned to my youth, "merci" to teachers such as René Vinci and Bernard Métral from the "Clos-Banet Lycée," who instilled their passion and knowledge into the restless student that I was. At the same time, I published my first article in *Radio-Plans* (1982), thanks to my friends Claude Ducros and Christian Duchemin, last editors-in-chief of the now defunct magazine. Finally, Claude Duchemin from the Montpellier University added the finishing touches and plugged my fingers into the switching power supply world!

A book such as this could never have been printed without the help and involvement of many people. Among them, I had endless discussions and brainstorming sessions with my colleague and friend Joël Turchi from ON Semiconductor. I wish to thank him for his dedication in discussing and reviewing all my work. These discussions also took place with the application team with which I am lucky enough to be able to work: Thierry Sutto, Nicolas Cyr, Stéphanie Conseil, and François Lhermite. I must also thank my friend Christophe Warin from the marketing team, for pointing out with obvious pleasure all the small typos in several chapters!

My peer review team members count among the most prestigious people I have been honored to work with. Their names appear below and I wish to thank them warmly for the amount of time they spent reviewing the book content:

- Dr. Vatché Vorpérian, from the Jet Pulpulsory Laboratory (JPL) in California (United States), who really sparked my interest in the PWM switch approach and spent time reviewing my work.
- Dr. Richard Redl, from Elfi (Switzerland), through his various tutorial sessions that I attended over the years and his incisive comments on my book chapters.
- Ed Bloom, from e/j BLOOM associates Inc. (United States), for his thorough review of the dc-dc design chapter.
- Dr. Raymond Ridley, from Ridley Engineering (United States), first for kindly writing the Foreword of this book and then for providing me with comments and suggestions on the chapters.
- Dr. Ivo Barbi, from the Power Electronics Institute of the Federal University of Santa Catarina (Brazil). Thank you for your guidance on chapter organization and various corrections.
- Jeff Hall, Dhaval Dalal, and not forgetting Monsieur Mullett, for the two appendices kindly contributed on magnetic designs! All three are with ON Semiconductor in the United States.
- Christian Zardini, "professeur émérite" from the ENSEIRB engineering school (France), who carefully reviewed portions of my work.
- Dr. Franki Poon and Dr. S. C. Tan, from PowerElab Ltd. and the Hong-Kong Polytechnic University (China), respectively, for a thorough review of some of the chapters. I appreciate your useful comments on the text.
- Dr. Dylan Lu, from Sydney University (Australia), for his review of one chapter.
- Arnaud Obin, from the company Lord Engineering (France), for his comments on the flyback section.
- Dr. V. Ramanarayanan, from the Electrical Engineering Department of the Indian Institute of Science in Bangalore (India), who reviewed and offered interesting suggestions on the book content.
- Dr. Jean-Paul Ferrieux, from the Laboratoire d'Electrotechnique de Grenoble (France), who made suggestions on my power factor correction chapter.
- Steve Sandler, consultant with AEi Systems (United States), who carefully checked and edited several chapters while traveling.
- Dr. Didier Balocco, from Saft Power Systems (France), for his kind review of the forward converter chapter.
- Pierre Aloisi, former MOTOROLA application engineer (France) and specialist of power components, for his interesting suggestions on the electrical and thermal stress.

I would also like to thank the people at Intusoft, Larry and Lise Meares and all their great support team (George, Farhad, Everett, Tim), who helped me during the testing phase of the numerous book examples. I want to also thank the editors of the demonstration versions included in the accompanying CD, who have kindly granted me the authorization to distribute their great products.

Thank you to Steve Chapman, at McGraw-Hill, for giving me the opportunity to publish this new book.

Finally, this work would have been difficult to achieve without the tremendous help from the Miller family, Noreen and Steve, who painfully corrected my poor English, full of Gallicisms!

Christophe Basso

NOMENCLATURE

A_e	the cross section area of a magnetic material
BV_{DSS}	the MOSFET drain-source breakdown voltage
B	the induction flux density in a magnetic medium
BCM	borderline conduction mode (same as CRM) or boundary conduction mode
B_r	the remanent induction flux level when the magnetizing field is zero
B_{sat}	the induction flux density at which μ_r drops to 1
CCM	continuous conduction mode
CL	closed loop
C_{lump}	the total capacitance seen on a particular point of the circuit
CRM	critical conduction mode
CTR	current transfer ratio for an optocoupler
D or d	the converter duty cycle; also noted d_1 in DCM analysis
D' or d'	the duty cycle off time ($d' = 1 - d$)
d_2, d_3	the duty cycle off times in DCM: $1 = d_1 + d_2 + d_3$
DT	the dead time between switching events
D_0	the converter static duty cycle during a bias-point analysis
ΔI_L	the peak to peak ripple current in the inductor
ESR	equivalent series resistor
ESL	equivalent series inductance
η	the converter efficiency, eta
f_c	the crossover frequency, where $|T(f_c)| = 0$ dB
F_{sw}	the switching frequency
F_{line}	the mains frequency
$G(s)$	the compensator frequency response
Gf_c	the gain deficit (or excess) at the selected crossover frequency
φ	the flux in a magnetic medium
φ_m	the phase margin read at f_c
gm	the transconductance of an operational transconductance amplifier (OTA)
H	the magnetizing force
H_c	the coercive field which brings the flux density back to zero
I_a, I_p and I_c	the average currents flowing in or out of the PWM switch terminals
I_C	the current inside a capacitor
I_d	the diode current
I_D	the MOSFET drain current
I_{in}	the input current of a given converter
$I_{in,rms}$ or I_{ac}	the input rms current in a mains powered converter

NOMENCLATURE

I_L	the current inside an inductor
I_{mag}	the magnetizing inductor current in a forward converter
I_{out}	the output current of a given converter
I_p	the primary current in a transformer-based converter
I_{peak}	the peak current in a given element
I_{sec}	the secondary current in a transformer-based converter
I_{valley}	the valley current in a given element
k_D	the derating factor for the MOSFET BV_{DSS}
k_d	the derating factor for the diode V_{RRM}
l, l_e, l_m	the mean magnetic path length
l_g	the gap length in a transformer
L_p	the primary inductor of a transformer (usually in a flyback converter)
LHP	a zero (LHPZ) or a pole (LHPP) placed in the left portion in an s-plane plot
L_{leak}	the transformer total leakage inductance seen from the primary (all outputs shorted)
L_{mag}	the magnetizing inductance of a transformer (usually in a forward converter)
L_{sec}	the secondary-side inductor of a transformer
M	the converter conversion ratio, V_{out}/V_{in}
M_c	the ramp compensation level in a current mode converter (per Dr. Ridley's definition)
M_r	the ramp coefficient in current-mode designs (as a percentage of the off slope)
μ_r	the permeability of a material relative to that of free space
μ_i	the initial permeability describes the slope of the magnetization curve at the origin
μ_0	the permeability of the air
N	the turns ratio of a transformer normalized to its primary winding. For instance, if $N_p = 10$ and $N_s = 3$, then $N = 0.3$
OL	open-loop; for instance a gain, a phase, or an output impedance
P_{cond}	the conduction losses of an element implying a resistive path and a rms current squared
PF	power factor
PFC	power factor correction
P_{out}	the converter output power
PIV	the peak inverse voltage a diode has to sustain
P_{SW}	switching losses of an element implying an overlap area between a current and voltage
Q	the quality coefficient of a filter or the quantity of electricity (coulombs)
Q_r	the charge the diode needs to evacuate before recovering its blocking capabilities
Q_{rr}	the total diode recovery charge
Q_G	the amount of coulombs you need to bring to the MOSFET for its full enhancement
r_{Cf}	the series resistor of the capacitor; also noted the ESR
r_{Lf}	the series resistor of the inductor; also noted the ESL
$R_{DS(on)}$	the MOSFET drain-source resistance when turned on
rms	root mean square

R_{sense} or R_i	the sense resistor in a current-mode converter; sometimes called the burden resistor
RHP	right half plane zero (RHPZ) or pole (RHPP); a zero or pole located in the right portion in an s-plane plot
S_a or S_e	the external compensation ramp
S_{on} or S_1	the inductor slope during the on time
SEPIC	single-ended primary inductance converter
SMPS	switch-mode power supply
SPICE	Simulation Program with Integrated Circuit Emphasis
S_{off} or S_2	the inductor slope during the off time
S_r	the externally imposed blocking slope when blocking a diode
t_c	the rectifying diode conduction time
t_d	the bulk capacitor discharge time
t_{on}	the time during which the power switch is turned on
t_{off}	the time during which the power switch is turned off
t_{prop}	the propagation delay of the logic blocks in a controller
t_{rr}	the reverse recovery time of a diode
THD	the total harmonic distortion
TVS	transient voltage suppressor
$T(s)$	the compensated loop gain
T_j	the junction temperature
T_{sw}	the switching period
V_{ac}, V_{cp}	the average voltages across the PWM switch terminals
V_{bulk}	the bulk voltage
$V_{bulk,max}$ or V_{peak}	the bulk voltage at the highest line (the ripple is neglected in this case)
V_C	the voltage across a capacitor
$V_{ce(sat)}$	the saturation voltage of a bipolar transistor
V_{clamp}	the clamping voltage level
V_{DS}	the MOSFET drain-source voltage
V_f	the diode forward drop
V_{GS}	the MOSFET gate-source voltage
V_{in}	the input voltage of the converter
$V_{in,rms}$ or V_{ac}	the mains rms voltage
V_L	the voltage across an inductor
V_{leak}	the voltage across the leakage inductance
V_{min} or $V_{bulk,min}$	the bulk valley voltage, low-line only
V_{OS}	the voltage overshoot on the RCD clamp
V_{out}	the output voltage
V_{peak}	the peak amplitude of sawtooth ramp in a voltage-mode PWM
V_p	the peak undershoot voltage in response to a load step
V_r	the secondary side voltage reflected on the primary side in a transformer-based converter
V_{sense}	the voltage developed across the sense resistor in a current-mode converter
V_{ripple}	the peak-to-peak ripple voltage
V_{RRM}	the diode maximum repetitive reverse voltage
ζ	the Greek letter zeta, representative of the damping factor [often mixed with ξ (Xi)]

SWITCH-MODE POWER SUPPLIES

CHAPTER 1
INTRODUCTION TO POWER CONVERSION

User friendliness is a key factor for the commercial success of any simulation program. The growing complexity of integrated circuits and equipment makes this aspect more and more important. Despite numerous publications devoted to the Simulation Program with Integrated Circuit Emphasis (SPICE), it still scares the novice when its name is mentioned.

Developed in the mid-1970s by the University of California, Berkeley, the SPICE program's main aim was to fulfill the needs of the electronic industry—mainly integrated circuit makers. However, with the support and funds from private editors, the SPICE program has evolved over a number of years into many practical and affordable packages, with emphasis on providing both low-priced and friendly access to beginners.

The performance of SPICE can significantly help you speed up the design phase of the equipment you are currently working on, even if SPICE is not able to generate an electronic schematic by itself! SPICE is inherently efficient because if you start working with an unfamiliar concept, it will quickly enable you to grasp the full meaning of any particular architecture by unveiling its peculiar waveforms. You can thus use the simulator to gain insight into the circuit you have to build and also ensure all parameters are taken into account before the breadboard phase.

This book has been written for power supply designers, experts in their fields, but also for beginners who would like to understand the secrets of switch-mode power conversion. Manipulating virtual components on a computer screen, without the hazard of high voltage, offers an interesting and safe way to learn the technique. Furthermore, the "experience" gained in simulation, and it is also true for experts simulating a novel concept, will let you feel more comfortable when breadboarding on the bench.

1.1 "DO YOU REALLY NEED TO SIMULATE?"

How many times have you heard this question when asking for a simulation package or a new computer? The following statements do not represent an exhaustive list of pros about computer simulation, but they can certainly be thought as a "help list" available during the negotiations:

1. Here is an argument: simulation can avoid waste of time and money. With its inherent iterative power, SPICE covers numerous application cases in which you could easily detect any design flaw or product weakness. The stability of a closed-loop SMPS represents a typical application when some key feedback elements are moving (i.e., the variable load that affects a pole) or start to degrade with temperature and aging (as the electrolytic equivalent series resistor). Moreover, design ideas can also be tested or assessed in a snapshot through a computer and, if they are worth trying, further refined in the lab.

2. You can start to work on a project by downloading components models and become familiar with the key elements, before going to the bench or waiting for the samples to be delivered. Once they arrive, you will have already gained insight and the debug phase on the bench will clearly have been beneficial!
3. Simulate test measurements whenever you do not own the adequate equipment: bandwidth measurements represent a good example. If you cannot afford a network analyzer, then a proven small-signal model can start to help you refine your feedback loop. When run on the final prototype, stability assessments will be faster and more efficient.
4. Power libraries are safe: they let you experiment "what if" when amperes and kilovolts are flowing in the circuit without blowing up in the case of a wrong connection! Also, they let you see how your design reacts to a short-circuit of the optocoupler, or the opening of a resistor. SPICE can begin to give you the answer.

1.2 WHAT YOU WILL FIND IN THE FOLLOWING PAGES

This book thoroughly details the advantages of the SPICE power to let you understand, simulate, test, and finally improve the switch-mode power supply (SMPS) you want to design. By providing you with specific simulation recipes, this work intends to facilitate as much as possible your SMPS design. Unlike in other books, the author has striven to balance the theoretical content, necessary to understand and question simulation results, and the need for practical design examples. This is developed throughout the eight chapters of the book.

Chapter 1 explains switch-mode power supply techniques and types of converters, and it introduces a few important results to help you better understand averaging techniques. Chapter 2 explains how *average* models were derived, and different types are described. A good comprehension of this chapter is fundamental: it will help you question certain weird SPICE data, resulting from a bad model implementation. If you do not understand the way the model has been derived, you will obviously face some difficulties in resolving these issues. In Chap. 2, you will also learn the way to wire an average model and run basic simulations. Closing the loop is obviously an important aspect of converter design and is often overlooked. This is not the case here, and Chap. 3 will guide you through control loop design, again using practical examples with a TL431 and not op amps only, as often seen in the literature. Since not every integrated circuit always comes with a SPICE model, Chap. 4 describes how the generic *switched* models were derived. The reading of this chapter will interest the person who wants to strengthen her or his knowledge in SPICE model writing. Chapter 5 describes practical designs of the three basic nonisolated topologies, including the front-end filter. Before analyzing off-line converters, Chap. 6 shows how to design the rectifying section and spends time on the various power factor correction techniques. Chapter 7 is entirely dedicated to the flyback converter, with specific design examples at the end. Finally, the forward converter appears in Chap. 8, again associated with a design example.

When one is discussing SPICE simulations, one of the greatest issues finds its root in the version syntax. Most SPICE editors deal with a proprietary syntax, sometimes SPICE3 conformant, that makes translation from one platform to another a difficult and painful exercise. To allow the use of different simulators, the standard models presented throughout the pages are compatible with Intusoft *IsSpice* (San Pedro, Calif.) and CADENCE's *PSpice* (Irvine, Calif.).

To help you quickly copy and paste the examples, we have included a CD-ROM in this book. Some selected simulation examples are offered in *IsSpice* and *PSpice* syntax, and you can easily load them on your computer if you are equipped with one of these software programs.

For students or newcomers to the SPICE world, we add the demonstration versions of these editors. They let you open the aforementioned files and simulate some of them (those demos are size-limited) to give you a taste of what the full version can do. Other demonstration versions also find a place in the CD-ROM, such as PSIM, Transim, Tina, Multisim, Power 4-5-6 B2 Spice, Micro-Cap, 5Spice and TopSPICE.

For professional power supply designers, another disk is separately distributed. This disk contains the design examples presented in the book plus numerous other industrial applications using real controllers. Please visit the author's website for distribution details (http://perso.orange.fr/cbasso/Spice.htm).

1.3 WHAT YOU WILL NOT FIND IN THIS BOOK

This book does not describe the way SPICE operates, nor does it solve typical electric circuits. It assumes that the reader is already familiar with the basics of SPICE simulations. Numerous books and papers are available on the subject as the given References section details [1, 2]. Whenever possible, the extended bibliography will guide your choice if you wish to strengthen your knowledge in a particular domain, such as some topologies that you are unfamiliar with. If some theoretical results are sometimes delivered just "as is," we strongly encourage the reader to dig further into the appropriate literature and acquire the theory that precedes the result.

The book also only focuses on a system approach. No SPICE description of typical discrete power elements such as diodes, MOSFETs, etc., is proposed.

Finally, here is the important statement, probably the most interesting one! SPICE does *not* replace the breadboard phase, nor does it shield you from writing equations or understanding electronics. It looks like a simple sentence, but the author has often been confronted by designers showing boards in the trash and claiming, "But SPICE said it would work!!" Yes, all ideas work on paper until they face the soldering iron condemnation.... Use SPICE as a design companion, a circuit insider that can reveal waveforms difficult to observe. But always question the delivered data: is this the real behavior, have I been misled somewhere, does a simple calculation more or less confirm what I see?

After this brief introduction, it is time to plunge into the intricacy of the Switch-Mode Power Supply design and simulation with SPICE.

1.4 CONVERTING POWER WITH RESISTORS

In the electronics world, different types of circuitries must cohabit: logic devices, analog circuits, microprocessors, and so on. Unfortunately for the designer, these circuits do not cope with a single, fixed, power supply rail: a microprocessor or a digital signal processor (DSP) will need a stable 3.3 V source or less, a front-end acquisition board will require ± 15 V and perhaps some logic glue around a standard 5 V. For the final board being supplied from a single power point, e.g., the mains outlet or a battery, how is one to adapt and distribute all these different voltages to the concerned portions? The solution consists of inserting a so-called converter to adapt the voltage distribution to the circuit needs.

1.4.1 Associating Resistors

Figure 1-1 portrays the simplest option a designer can think of: resistive dividers! If our DSP consumes 66 mA over 3.3 V, then it can be replaced by a 50 Ω resistor, the same as for our

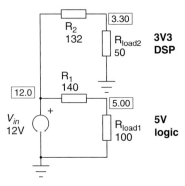

FIGURE 1-1 The simplest voltage distribution via resistors.

50 mA, 5 V logic circuit via the 100 Ω resistor. From a 12 V source, we can then calculate the dropping resistors:

$$R_1 = \frac{12 - 5}{50m} = 140\,\Omega \tag{1-1}$$

$$R_2 = \frac{12 - 3.3}{66m} = 132\,\Omega \tag{1-2}$$

Before going further, please note that 0.066 A or 0.05 A was, respectively, noted in the calculation as 66m or 50m. This is done to retain the SPICE notation for units, without any space. It adheres to the following rules and will be extensively used in the remaining portions of the book:

$$p = pico = 10^{-12}$$
$$n = nano = 10^{-9}$$
$$u = micro = 10^{-6}$$
$$m = milli = 10^{-3}$$
$$k = kilo = 10^{3}$$
$$Meg = mega = 10^{6}$$

- Beware not to mix mega and milli, a very common mistake: 10 mΩ = 10m, 1 MΩ = 1Meg.

Unfortunately, these resistors will be the seat of a permanent voltage drop, and power dissipation (in heat) will occur. The dissipated power for each resistor is

$$P_1 = \frac{(12 - 5)^2}{140} = 350\,mW \tag{1-3a}$$

$$P_2 = \frac{(12 - 3.3)^2}{132} = 573.4\,mW \tag{1-3b}$$

From these values, we can now evaluate the system efficiency obtained by dividing the delivered output power P_{out} by the power taken away from the source P_{in}:

$$P_{out} = \frac{5^2}{100} + \frac{3.3^2}{50} = 250m + 218m = 468\,mW \tag{1-4a}$$

$$P_{in} = \frac{12^2}{100 + 140} + \frac{12^2}{50 + 132} = 600m + 791m = 1.39\,W \tag{1-4b}$$

INTRODUCTION TO POWER CONVERSION

The efficiency, represented by the Greek letter η or "eta," can be computed by dividing P_{out} by P_{in}, or

$$\eta = \frac{P_{out}}{P_{in}} = \frac{0.468}{1.39} \times 100 = 33.6\% \quad (1\text{-}5)$$

which is an extremely poor performance!

The loss, dissipated in heat, is simply the difference between the power delivered by the source and the power, converted as the real work, P_{out}.

$$P_{loss} = P_{in} - P_{out} = \frac{P_{out}}{\eta} - P_{out} = P_{out}\left(\frac{1}{\eta} - 1\right) \quad (1\text{-}6)$$

In our example, the loss is $1.39 - 0.468 = 922$ mW.

1.4.2 A Closed-Loop System

If the load changes, or if the input voltage drifts, what is going to happen? Well, since our input-to-output transfer ratio, denoted M, is fixed, the output voltage will also vary. Therefore, we need to think of a kind of regulated system that permanently observes the output power demand and adjusts the series resistor to maintain a constant output voltage, if the output voltage represents the variable of interest. For a well-designed system, the converter must also ensure a proper regulation independently from input voltage variations. To reach this goal, we need to use several particular components such as

- A reference voltage V_{ref}: This voltage is by definition extremely stable in temperature and precise in value (e.g., $\pm 1\%$). A programmable shunt regulator, such as a TL431 adjustable zener, could do the job.
- An operational amplifier (op amp): This device will observe a portion of the output voltage (αV_{out}) and compare it to the reference V_{ref}. It will actually "amplify" the error, the difference between αV_{out} and V_{ref}, to drive a series-pass element. The error monitored by the op amp is usually denominated by the Greek letter ε or "epsilon": $\varepsilon = \alpha V_{out} - V_{ref}$.
- A series-pass element: It can be a MOSFET or a bipolar transistor but working in a linear mode, playing the role of the necessary variable resistor. If it is a MOSFET, the static driving power is null. For a bipolar, there is a need to supply a sufficient amount of base current to deliver the right collector or emitter current. This is called the bias current.

Figure 1-2 finally shows how our resistive converter could be improved, let's say for the 5 V section. The error amplifier is made via a voltage-controlled voltage source (E primitive) and features a gain of 10k (or 80 dB). One of its input receives the voltage reference whereas the other one, the inverting input, is biased by a portion of the output voltage. This is actually a linear regulator, however, limited in the input voltage range since V_{in} shall be above V_{out} by a V_{be}, at least, to guarantee a proper drive for Q_1. If V_{out} is below the target (5 V in our example), E_1 output increases and strengthens Q_1 bias current: V_{out} goes up. On the other hand, suppose the load has suddenly been reduced, therefore V_{out} exceeds 5 V. Thanks to E_1, Q_1 bias current goes down, reducing the output voltage until regulation is met again.

The output voltage observation, which delivers αV_{out} (a fraction of the output voltage), is obtained through a resistive divider made of R_{upper} and R_{lower}. Calculating their values is straightforward:

1. Let us fix a current circulating in the divider bridge. Since there is no biasing current for E_1 in the example (this is the case for most MOS-based technologies), we could take $I_b = 250$ µA, for example. A lower value is acceptable, but degrades the noise immunity in a noisy environment.

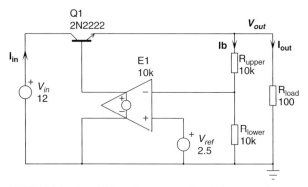

FIGURE 1-2 The addition of an error amplifier brings regulation to the circuit: we have built a linear regulator.

2. I_b equals 250 μA and entirely circulates in R_{lower}. Thanks to the control loop, 2.5 V is "seen" across R_{lower}. Therefore, $R_{lower} = \dfrac{2.5}{250u} = 10\,\text{k}\Omega$.

3. The voltage drop across R_{upper} is $V_{out} - V_{ref}$. Thus, $R_{upper} = \dfrac{V_{out} - V_{ref}}{I_b} = \dfrac{5 - 2.5}{250u} = 10\,\text{k}\Omega$.

If we neglect the power needed to drive Q_1, then all the source current I_{in} flows into the load as I_{out}. Therefore, applying Eq. (1-5), we can derive the efficiency for this linear regulator:

$$\eta = \frac{P_{out}}{P_{in}} = \frac{V_{out} I_{out}}{V_{in} I_{in}} \approx \frac{V_{out}}{V_{in}} = M \tag{1-7}$$

If we now plot the efficiency versus the input voltage, we can see how difficult the situation becomes in the presence of small M ratios (Fig. 1-3). For these reasons, resistive divider type of converters, that is to say series-pass regulators, are limited to application where M does not fall below 0.3. Otherwise the heat dissipation burden shall become a real handicap. On the

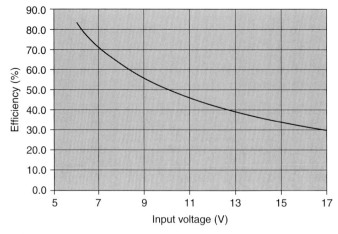

FIGURE 1-3 As soon as M diminishes, the efficiency dramatically drops ($V_{out} = 5$ V).

other hand, when the user really needs to operate his regulator to ratios M closer to 1 (V_{in} very close to V_{out}), the low-dropout (LDO) regulator made with a PNP becomes a good choice. The input low limit is now linked to the transistor $V_{ce(sat)}$ (a few hundred millivolts, or less) rather than its V_{be} (around 650 mV at room temperature, 25 °C).

To close the study on regulator efficiency, we can take three different output examples with linear regulators, where output and input conditions vary:

1. $V_{in} = 14$ V $V_{out} = 5$ V $\Delta V = V_{in} - V_{out} = 9$ V $\eta = \frac{5}{14} 100 = 35.7\%$

2. $V_{in} = 14$ V $V_{out} = 12$ V $\Delta V = V_{in} - V_{out} = 2$ V $\eta = \frac{12}{14} 100 = 85.7\%$

3. $V_{in} = 5$ V $V_{out} = 3$ V $\Delta V = V_{in} - V_{out} = 2$ V $\eta = \frac{3}{5} 100 = 60\%$

As a result, one can see that a high efficiency can be obtained with a linear regulator if ΔV is small (as plotted in Fig. 1-3), but also if $V_{out} \gg \Delta V$.

1.4.3 Deriving Useful Equations with the Linear Regulator

Figure 1-2 is interesting because we can use it to derive general statements, pertinent to the closed-loop world we are going to enter, linear, or switched. Suppose that we remove the error amplifier and replace it by a fixed voltage source of 5.77 V, our actual op amp output (look at Fig. 1-2 values) as Fig. 1-4a shows. The regulator becomes a simple emitter follower circuit, affected by an output impedance and an output voltage. As such, it can be described with its equivalent Thévenin generator, what Fig. 1-4b suggests. $R_{s,OL}$ represents the open-loop output impedance and V_{th}, the voltage delivered when biased by a control voltage V_c, our fixed 5.77 V in the application. Let us now use this representation and redraw our closed-loop regulator around it, ignoring, for now, the input voltage contribution:

In Fig. 1-5, $V_{out}(s)$ is compared to $V_{ref}(s)$ via a resistive divider affected by a transfer ratio of α. $H(0)$ illustrates the *static* or dc relationship between the output voltage and the control voltage V_c, e.g., $V_c = 5.77$ V to obtain $V_{out} = 5$ V in this example. The *theoretical* dc voltage ($s = 0$, but we purposely avoided this subscript below for the sake of clarity) you would expect from such a configuration is

$$V_{out} = \frac{V_{ref}}{\alpha} \tag{1-8}$$

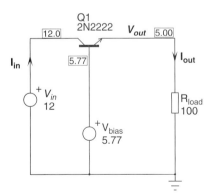

FIGURE 1-4a If the feedback is suppressed, there is no output voltage observation to adjust Q_1 bias point: we are running open-loop.

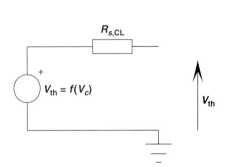

FIGURE 1-4b A Thévenin generator portrays the regulator when run in closed loop.

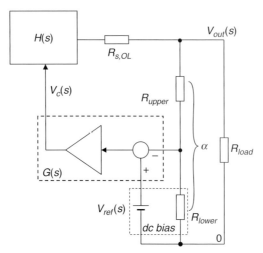

FIGURE 1-5 When closing the loop, our Thévenin generator undergoes a transformation in its dynamic behavior. Here, the input perturbation is purposely omitted.

Unfortunately, the whole gain chain and various impedances will affect this value. With a few lines of algebra, we can write the *static* output voltage definition (again, $s = 0$) simply by following the meshes:

$$V_{out} = (V_{ref} - \alpha V_{out})HG - R_{s,OL}\frac{V_{out}}{R_{load}} \tag{1-9}$$

$$V_{out} = \frac{V_{ref}HG}{1 + \alpha HG + \dfrac{R_{s,OL}}{R_{load}}} \tag{1-10}$$

The static error on the output, actually the deviation between what we really want and what we finally obtain, is derived by subtracting the V_{out} expression [Eq. (1-10)] from Eq. (1-8):

$$V_{error} = \frac{V_{ref}}{\alpha} - \frac{V_{ref}HG}{1 + \alpha HG + \dfrac{R_{s,OL}}{R_{load}}} = V_{ref}\left(\frac{1}{\alpha} - \frac{1}{\dfrac{1}{HG} + \alpha + \left(\dfrac{R_{s,OL}}{R_{load}}\dfrac{1}{HG}\right)}\right) \tag{1-11}$$

If we consider $R_{s,OL} \ll R_{load}$, then Eq. (1-11) simplifies to

$$V_{error} = V_{ref}\left(\frac{1}{\alpha} - \frac{1}{\alpha + \dfrac{1}{HG}}\right) \tag{1-12a}$$

which equals zero if

$$\alpha = \alpha + \frac{1}{HG} \tag{1-12b}$$

From this equation, we can see that increasing the dc gain, $G(0)$, helps diminish the static error which finally affects our output voltage precision.

Another important parameter influenced by the loop gain is the closed-loop output impedance. The output impedance of a system can be derived in different manners. As Fig. 1-4b has shown, our closed-loop generator can now be reduced to its Thévenin equivalent, that is, a voltage source V_{th} [V_{out} measured without any load, or $R_{load} = \infty$ in Eq. (1-10)], followed by an output impedance $R_{s,CL}$, which we actually look for. One option consists of calculating a resistor R_{LX} which, once wired between the output and ground, will reduce $V_{out} = V_{th}$ to $V_{out} = \dfrac{V_{th}}{2}$. When this occurs, R_{LX} simply equals $R_{s,CL}$ (we have built a simple resistive divider with equal resistors). We can quickly manipulate Eq. (1-10), assuming $R_{load} = \infty$:

$$\frac{V_{th}}{2} = V_{out}(R_{LX}) \quad \text{or} \quad \text{"What value of } R_{LX} \text{ will divide the Thévenin voltage by 2?"}$$

$$\frac{V_{ref}HG}{2(1+\alpha HG)} = \frac{V_{ref}HG}{1+\alpha HG + \dfrac{R_{s,OL}}{R_{LX}}} \tag{1-13}$$

If we call αHG the *static* loop gain T, then the closed-loop output impedance is

$$R_{LX} = R_{s,CL} = \frac{R_{s,OL}}{1+\alpha HG} = \frac{R_{s,OL}}{1+T} \tag{1-14}$$

Equation (1-14) teaches us different things:

1. If we have a large dc loop gain $T(0)$, then $R_{s,CL}$ is close to zero.
2. Because we have compensated the feedback return path $G(s)$ for stability purposes, when the loop gain $T(s)$ reduces as the frequency increases, $R_{s,CL}$ starts to rise: an impedance whose module grows with frequency looks like an inductance! We will come back to this result later.
3. When the loop gain $T(s)$ has dropped to zero, then the system exhibits an output impedance that is the same as in the lack of feedback, $R_{s,OL}$: the system runs open-loop.

Why do we talk about a static (dc) and a frequency-dependent gain? Well, this is so simply because Figs. 1-5 and 1-6 do not represent genuine regulators. In reality, $G(s)$ is made via a

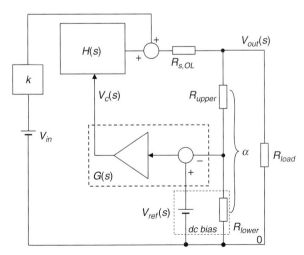

FIGURE 1-6 Our previous regulator mode is now upgraded with an input perturbation kV_{in}.

real operational amplifier, imposing a virtual ground on its inverting pin as soon as local feedback exists. In other words, R_{lower} simply goes off the picture in the small-signal model, and α no longer plays a role. This is described in App. 3D.

In this example, we purposely did not account for an input voltage perturbation. This assumption is valid for bipolar transistors as the weak influence of the Early effect makes them good current generators, almost independent from their V_{ce} variations. However, when V_{out} and V_{in} are close to each other, the transistor becomes a closed switch rather than a current source! Therefore, the input voltage starts to play a role. Let us redraw the Fig. 1-5 sketch, including the input voltage contribution. As drawn in Fig. 1-6, the term k represents the open-loop audio susceptibility, denoted $A_{s,OL}$. It represents the input voltage contribution to the output, also called input ripple rejection.

Let's now write the mesh equations as we did previously:

$$V_{out} = (V_{ref} - \alpha V_{out})HG + (kV_{in}) - R_{s,OL}\frac{V_{out}}{R_{load}} \tag{1-15}$$

$$V_{out}\left(1 + \alpha HG + \frac{R_{s,OL}}{R_{load}}\right) = V_{ref}HG + kV_{in} \tag{1-16}$$

$$V_{out} = \frac{V_{ref}HG}{\left(1 + \alpha HG + \frac{R_{s,OL}}{R_{load}}\right)} + \frac{kV_{in}}{\left(1 + \alpha HG + \frac{R_{s,OL}}{R_{load}}\right)} \tag{1-17}$$

Again, if we consider $R_{s,OL} \ll R_{load}$, then

$$V_{out} = \frac{V_{ref}}{\left(\frac{1}{HG} + \alpha\right)} + \frac{kV_{in}}{(1 + \alpha HG)} \tag{1-18}$$

As Eq. (1-18) shows, V_{out} is made of two terms:

1. The theoretical output voltage, similar to what Eq. (1-10), simplified, gave ($R_{load} = \infty$)
2. The input voltage contribution whose new term is $\frac{k}{(1 + \alpha HG)}$ or, sticking to the previous definition,

$$A_{s,CL} = \frac{A_{s,OL}}{1 + \alpha HG} = \frac{A_{s,OL}}{1 + T} \tag{1-19}$$

Again, operating with a large dc gain ensures an excellent rejection of the input voltage ripple (100 or 120 Hz for full-wave rectification). When $T(s)$ reduces in the high-frequency domain, the system runs open-loop. Please note that we purposely selected a positive polarity for k, but a negative value could also have been chosen. It actually depends on the topology under study.

1.4.4 A Practical Working Example

Thanks to SPICE, we can simulate a completely theoretical regulator by associating blocks. Figure 1-7 depicts the circuit where you will recognize the block discussed above. The operating parameters are the following:

$$R_{s,OL} = 1\ \Omega$$
$$A_{s,OL} = 50\ m$$
$$V_{in} = 15\ V$$
$$V_{out} = 5\ V\ (\text{target})$$

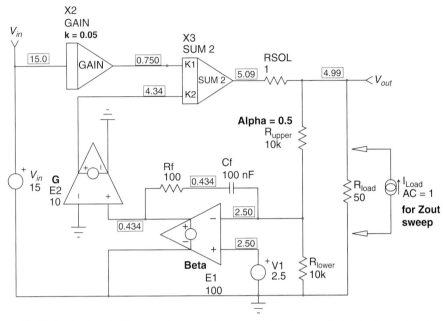

FIGURE 1-7 Our theoretical linear regulator, including the output impedance and the input voltage perturbation.

$$V_{ref} = 2.5 \text{ V}$$
$$\alpha = 0.5$$

Please ignore, for now, the presence of the compensation network R_f-C_f. Applying our above input numbers reveals these closed-loop values:

Equation (1-14), $R_{s,CL}$ = 1.996 mΩ, or in dBΩ: $20 \log_{10}(R_{s,CL}) = 20 \log_{10}(1.996m) = -54$ dBΩ

Equation (1-17), V_{out} = 4.991318 V

Equation (1-19), $A_{s,CL}$ = 99.8u, or in decibels: $20 \log_{10}(A_{s,CL}) = 20 \log_{10}(99.8u) = -80$ dB

Now, let us compare to what the SPICE simulator will give. We have several options. The first one uses a .TF statement, which performs a $\frac{dV_{out}}{dV_{in}}$ calculation as well as an output impedance measurement. The SPICE code is the following: .TF V(vout) vin. Once it has run, we obtain the results in the .OUT file:

```
***** SMALL SIGNAL DC TRANSFER FUNCTION
output_impedance_at_V(vout)          1.995928e-003
vin#Input_impedance                  1.000000e+020
Transfer_function                    9.979642e-005
```

As one can see, we are very close to our theoretical numbers. The input impedance value does not make sense here since we do not take any current from the source. Now, we can transient step the input and observe the output. Transient step the input means replacing the fixed

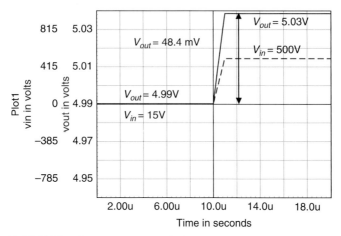

FIGURE 1-8a The output response to the input step.

V_{in} source by a *pulsewise linear* (PWL) statement. This SPICE function builds the curve you wish by (time, amplitude) couples. Here, we start for $t = 0$ at $V_{in} = 15$ V and suddenly increase V_{in} when $t = 10$ μs with a 1 μs slope to 500 V:

```
Vin 7 0     PWL 0 15 10u 15 11u 500
```

Figure 1-8a reveals the output response when the input is suddenly increased from 15 up to 500 V. We can observe a deviation of 48.4 mV engendered by a $500 - 15 = 485$ V input variation. Therefore, the dc audio susceptibility is $0.0484/485 = 99.8u$, which is what we found before.

By stepping the circuit output with a current source, we will be able to extract the static output impedance. To do so, we remove R_{load} and connect a current source affected by a PWL statement:

```
ILoad  vout 0   PWL 0 0.1 10u 0.1 11u 1end
```

Here, the output will be pulsed from 100 mA to 1 A in 1 μs. The output impedance will simply be $\dfrac{\Delta V_{out}}{\Delta I_{out}}$. Figure 1-8b displays the simulation results. For the 900 mA step, we can see a

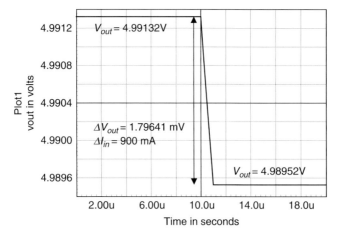

FIGURE 1-8b The response to an output step.

deviation of 1.79641 mV. The dc output resistance is therefore $1.79641m / 900m = 1.996$ mΩ, again exactly what our calculations predicted.

Now, we reconnect the compensation network made of R_f and C_f. This network will make $T(s)$, the total loop gain, depend on the sweep frequency. By connecting an ac source of 1 A in place of our static load resistor, we can sweep the ac output resistance/impedance, called Z_{out}, of our converter.

- *The current source is connected with its positive terminal to ground and negative terminal to the output.*

Then, using the graphical interface, plotting V_{out} alone will reveal Z_{out} since $I_{out} = 1$ A, as Fig. 1-8c portrays.

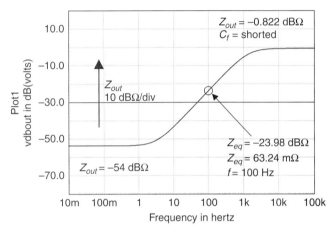

FIGURE 1-8c The output impedance sweep with a compensation network.

On these graphics, we can observe three areas:

1. **dc region, $f < 1$ Hz**: The loop gain $T(0)$ is extremely high, C_f can be considered as open. Therefore, Z_{out} is defined by Eq. (1-14) and is extremely small.
2. **Above 1 Hz, C_f starts to play a role**: The loop gain $T(s)$ starts to diminish and the denominator of Eq. (1-14) goes down. As a result, Z_{out} increases. An impedance growing with frequency reproduces an inductive behavior. If we take a point, let us select $f = 100$ Hz, we can calculate the equivalent inductance L_{eq}. From Fig. 1-8c, the equivalent inductance L_{eq} is equal to $\frac{63.24m}{2\pi 100} = 100.6$ μH at $f = 100$ Hz.
3. C_f becomes a complete short, and the loop gain is fixed by R_{upper} and R_f. Now Z_{out} is close to its open-loop value, which is 0 dBΩ (1 Ω).

If for filtering reasons we connect a 100 μF capacitor on our regulator output, we create an LC filter that is going to resonate. The resonating frequency is defined by $\frac{1}{2\pi \sqrt{L_{eq}C}}$, which is approximately 1.586 kHz. If we restart a Z_{out} sweep with the 100 μF capacitor, we obtain Fig. 1-8d: a resonance occurs as predicted. This behavior is typical of low-dropout (LDO) regulators where one can sometimes observe on a spectrum analyzer an increase of the noise density when output capacitors of different values are connected to the output [3].

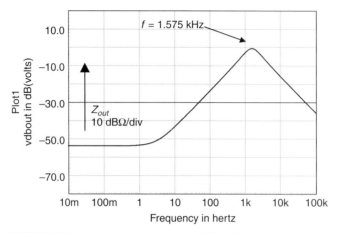

FIGURE 1-8d A resonance takes place at 1.5 kHz if a 100 μF decoupling capacitor is connected!

Since our converter output impedance looks like an inductance, we might see an inductive behavior if we step its output. Well, if you try to interrupt the current $I(t)$ flowing in an inductance, you will verify that the inductance "fights" to keep its ampere-turns product constant by suddenly reversing its terminals voltage $V_L(t)$. This is defined by the well-known Lenz formula

$$V_L(t) = -L\frac{di(t)}{dt} \qquad (1\text{-}20)$$

Stepping the Fig. 1-7 circuit will reveal exactly the same behavior! The feedback loop tries to fight against the voltage variations induced by the applied current step, and overshoots and undershoots appear as in Fig. 1-9a. The overshoot amplitude is quite severe and reaches 750 mV peak. In some applications, it might be incompatible with the initial specifications. How can we improve the situation? A solution consists of removing the integral term made of C_f in the compensation network. If we short C_f and increase R_f to 470 kΩ, we might degrade the static error, but see in Fig. 1-9b how the overshoots are gone: 3.8 mV peak! This is so because $T(s)$ is now independent of the frequency and Fig. 1-8c totally flattens as the output impedance no longer looks inductive. Of course, this is purely theoretical and a small capacitor will always straddle R_f, but the overshoot effects will be seriously reduced. This technique appears in Ref. 4 where a supply for microprocessors is described.

1.4.5 Building a Simple Generic Linear Regulator

If Fig. 1-2 works properly for the sake of the example, we need a more elaborate regulator to meet our future needs of power supply simulations. Very often, multioutput switching converters are associated with linear regulators to (1) reduce the switching ripple and (2) downsize the output voltage to the right value, difficult to obtain from a single winding.

Figure 1-10a shows how we can build a simple positive regulator, further encapsulated in a subcircuit, as Fig. 1-10b. The regulator capitalizes on a TL431 device, diverting to ground the biasing current delivered by R_1. Resistor R_2 is calculated as a parameter passed to the subcircuit.

Thanks to a simple integral compensation, the regulator exhibits a stable behavior as Fig. 1-10c indicates. The current was pulsed up to 250 mA but starting from two different values. The first

INTRODUCTION TO POWER CONVERSION 15

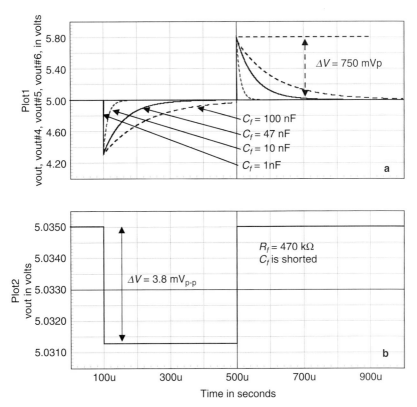

FIGURE 1-9a, b   An integral compensation brings severe overshoots in (a), compared to a pure proportional gain in (b).

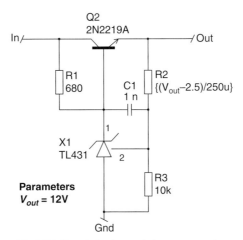

FIGURE 1-10a   A simple generic low-power regulator.

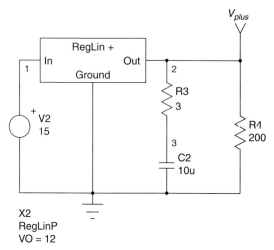

FIGURE 1-10b The same regulator, once encapsulated into a graphical symbol.

start was a no-load condition, $I_{out} = 0$, whereas the second one started from 150 mA. As one can see, the regulator response drastically differs, depending on where we start: in no-load conditions, Q_2 is just slightly biased to maintain 12 V on the output since no current needs to be delivered. Therefore, when the 250 mA pulse appears, the TL431 requires a longer time to react as C_1 charges up to the right value. This implies a deep undershoot as shown in Fig. 1-10c. To the opposite, if Q_2 bias is already established to deliver 150 mA, then the small jump to reach the 250 mA bias requires less time and thus reduces the undershoot.

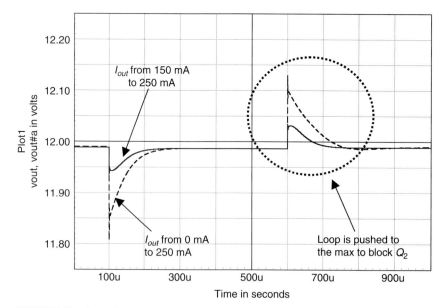

FIGURE 1-10c Depending on the bias point from which the regulator is excited, its output response varies.

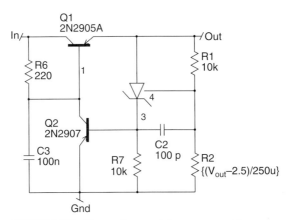

FIGURE 1-10d The negative generic low-power regulator.

When the load is suddenly released, the output voltage goes up. There is nothing the feedback loop can do to fight against this situation, except to block Q_2. If the load were reapplied during this time, the undershoot would be even more severe than previously observed.

A negative version of our generic regulator appears in Fig. 1-10d, whereas Fig. 1-10e shows how to wire it.

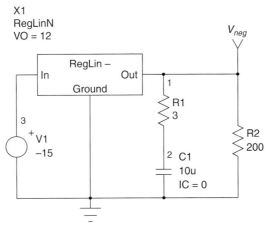

FIGURE 1-10e The same regulator, once encapsulated into a graphical symbol.

This negative regulator differs from the previous implementation: we still use our TL431 but wired in a different manner, given the negative output. The inclusion of Q_2 brings additional gain to the loop and therefore gives the regulator a larger bandwidth than its positive counterpart. It is stable, however.

1.4.6 Conclusion on Linear Regulators

As we have seen, linear regulators do not lend themselves very well to high-efficiency converters, unless the headroom between V_{out} and V_{in} reduces to a few hundred of millivolts. However, they are very often able to reject the input ripple and can serve as additional filtering regulators on noisy output lines. They are a safe value if you need to power sensitive circuitries such as A/D converters.

Reference 5 offers a good introduction to the regulator world for those of you who would like to strengthen your knowledge in this domain.

1.5 CONVERTING POWER WITH SWITCHES

Section 1.4 has shown that resistive drops imply losses. In that case, why not use power switches that, once closed, feature an almost 0 Ω resistance and, once open, behave as an open circuit? Good suggestion, indeed. A possible arrangement would be as what Fig. 1-11a depicts: one single switch routing V_{in} to V_{out} when activated by a clock signal. When the clock is high, the upper switch is closed: this is the on time, denoted by t_{on}. When the clock is low, the upper switch opens and V_{out} drops to zero: this is the off time, denoted by t_{off}. The output then looks like a low-impedance switching pattern, toggling between V_{in} and 0 V at a quick pace, here 100 kHz in the example:

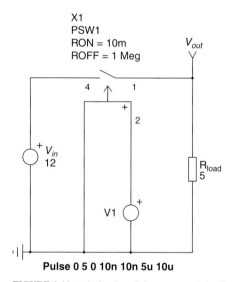

FIGURE 1-11a A simple switch arrangement drastically reduces the power loss.

If we apply an old-fashioned needle-based voltmeter across R_{load}, the inertia of the measuring mechanism will remain insensitive to the alternating signal: it will measure the average value, naturally extracting the dc component. In our case, the instrument would measure 6 V. If we now try to estimate the transmitted power, we take a true-rms voltmeter (more recent than a hot-wire wattmeter!) and evaluate the root mean square (rms) voltage across R_{load}: 8.47 Vrms. We could also calculate it since the rms value of a square wave signal is

$$V_{out,rms} = V_{in}\sqrt{D} \qquad (1\text{-}21)$$

where D is the duty cycle, defined by

$$D = \frac{t_{on}}{T_{sw}} \qquad (1\text{-}22a)$$

INTRODUCTION TO POWER CONVERSION

The off-time duration is often expressed by D' via the equation

$$D' = \frac{t_{off}}{T_{sw}} \qquad (1\text{-}22b)$$

If we consider the sum $t_{on} + t_{off}$ to equal the switching period T_{sw}, then Eq. (1-22b) can be rearranged as

$$D' = \frac{T_{sw} - t_{on}}{T_{sw}} = \frac{T_{sw}}{T_{sw}} - \frac{t_{on}}{T_{sw}} = 1 - D \qquad (1\text{-}22c)$$

D equals 50% in our example. Equation (1-21) helps to evaluate the output power delivered to R_{load}:

$$P_{out} = \frac{V^2_{out,rms}}{R_{load}} = 14.37\,W \qquad (1\text{-}23)$$

For the input power, we measure the *average* current flowing through the V_{in} source and multiply it by V_{in}. We find 1.2 A, leading to an input power of $P_{in} = 1.2 \times 12 = 14.4\,W$. Dividing P_{in} by P_{out} gives an efficiency of . . . 99.8%!

Well, that's not too bad for a simple switch activated at a high-frequency pace. Why is the efficiency so high? Because the only resistive path is X_1 which features an on-state resistance R_{ON} of 10 mΩ. When crossed by the input rms current (1.7 A, measured), subcircuit X_1 will dissipate $I^2_{rms} R_{ON} = 29\,mW$. It is the only source of loss here.

Instead of using the good old needle-based voltmeter, we could calculate the average voltage across R_{load}. To obtain the average value of a waveform, we need to integrate it over one switching period. Mathematically, it can be written as

$$V_{out,avg} = \frac{1}{T_{sw}} \int_0^{T_{sw}} f(t)dt \qquad (1\text{-}24)$$

In our case $f(t)$ is a rather simple function. It is V_{in} between 0 and t_{on}, then 0 during t_{off} (see Fig. 1-11b). Therefore, Eq. (1-24) can be rewritten as

$$V_{out,avg} = \frac{1}{T_{sw}} \int_0^{t_{on}} V_{in} dt = \frac{t_{on}}{T_{sw}} V_{in} = DV_{in} \qquad (1\text{-}25)$$

Thus, by adjusting D, the duty cycle, we have a means to adjust the average output voltage. Usually, the duty cycle adjustment is controlled by a feedback loop which instructs a pulse width modulator (PWM) to properly generate the on-time duration. We will come back later to that important block.

1.5.1 A Filter Is Needed

You would think we would be happy with our single switch converter; however, there is something bothering us: what we want is a dc generator, and Fig. 1-11b does not really look like a clean continuous signal. Or actually, it is a dc signal but together with a high harmonic content. Evaluating the rms voltage applied over R_{load}, reveals a portion linked to the dc value (that we actually want), plus harmonic terms gathered into an ac component that we do not need:

$$V_{out,rms} = \sqrt{V^2_{out,dc} + V^2_{out,ac}} \qquad (1\text{-}26)$$

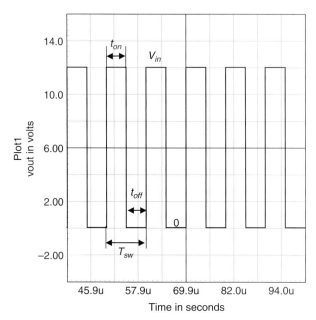

FIGURE 1-11b   The resulting voltage across R_{load}.

Equation (1-26) is nothing else than the result of a Fourier series decomposition, where the dc portion is the A_0 term, with the rest of the harmonic constituting the lump ac terms.

To remove all the unwanted harmonics, or get rid of the ac portion, we must install a filter. We could put an *RC* filter to remove the unwanted components. However, all the efforts we put in the switching circuits to minimize ohmic losses would fade away in heat through any additional resistive path, such as in the *RC* filter. A filter whose losses are limited, consists in the combination of an inductance and a capacitor, the *LC* filter. Let's install a (perfect) *LC* filter in series with our power switch. Figure 1-11c shows the result.

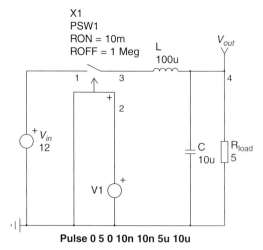

FIGURE 1-11c   Installing an *LC* filter removes unwanted harmonics.

This converter looks really simple now, but if we plug it into a dc source, the power switch (most likely to be a MOSFET) will immediately be destroyed at the opening. Why? Because when it closes during t_{on}, it magnetizes the inductance L. Since V_{in} appears on the inductance left terminal whereas V_{out} is present on its right end terminal, the voltage difference imposes a slope (a di/dt in ampere-seconds) obtained from rearranging Eq. (1-20):

$$S_{on} = \frac{V_{in} - V_{out}}{L} \qquad (1\text{-}27)$$

As a result, if X_1 closes during $t_{on} = 5$ μs, then the peak current excursion $\Delta I_{L,on}$ in L will be, assuming there is no initial current,

$$\Delta I_{L,on} = \frac{V_{in} - V_{out}}{L} t_{on} = \frac{12 - 5}{100u} \times 5u = 350 \text{ mA} \qquad (1\text{-}28)$$

At the end of the on time, the current in the inductance reaches its peak I_{peak}. However, as Eq. (1-20) states, there cannot be a current discontinuity in an inductance. That is, if a current circulates in a coil, you have to provide a way to keep it circulating at the switch opening, in the same direction, to ensure the ampere-turns continuity. In our case, switching X_1 off will generate a negative peak of a few kilovolts [Eq. (1-20)], immediately destroying the power switch. The solution consists of adding a second switch which will offer a path to the current at the switch-opening event: this is the so-called freewheel diode D_1 in Fig. 1-11d. We have built a buck converter.

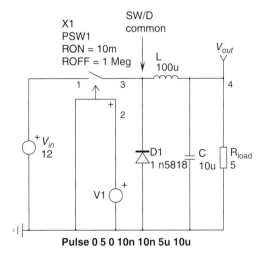

FIGURE 1-11d . . . But a diode is needed to ensure the ampere-turns continuity in the coil!

1.5.2 Current in the Inductance, Continuous or Discontinuous?

As Eq. (1-27) depicts, the inductor undergoes a magnetization cycle during the on time: its magnetic flux builds up to a point where the switch opens. The switch opens because the controller that activates it makes the decision to do so. This decision is usually linked to the observation of a state variable such as the output voltage or the inductor current via a feedback loop.

Now that our inductor L is magnetized, we have seen that the diode D_1 will provide a path to let it discharge and supply the output during the off time, the time where SW is open. If our diode features a small forward voltage compared to V_{out}, then the left terminal of L swings down to nearly 0 V. The voltage seen by the inductance is now $-V_{out}$. Therefore, the off-slope S_{off} can be defined as follows:

$$S_{off} = -\frac{V_{out}}{L} \tag{1-29}$$

Similarly, we can evaluate the current excursion during the off time $\Delta I_{L,off}$

$$\Delta I_{L,off} = \frac{V_{out}}{L} t_{off} = \frac{5}{100u} \times 5u = 250 \text{ mA} \tag{1-30}$$

At the end of the off time, the current in the inductance reaches its valley I_{valley}.

At power-on, the switch closes and the current in the inductance starts from zero to rise up to 350 mA in 5 μs. Then, during the next 5 μs time frame, it will drop down to $350 - 250 = 100$ mA. At this time, since our switching period is 10 μs, the switch will close again with current still circulating in the inductor L. We are in what is called continuous conduction mode or CCM. It means that, from one cycle to another cycle, there is always current flowing in the inductor. We can therefore think about various cases occurring in the inductor operating mode:

- **CCM**, continuous conduction mode: the current never goes back to zero within a switching cycle. Otherwise stated, the inductor is never "reset," meaning that its flux never returns to zero within the considered switching cycle. The switch always closes with current circulating in the coil.
- **DCM**, discontinuous conduction mode: the current always goes back to zero within a switching cycle, also meaning that the inductor is properly "reset." The switch always closes in zero-current conditions.
- **BCM**, boundary or borderline conduction mode: a controller observes the current in the inductance, and when it is detected to be zero, the switch immediately closes. The controller always waits for the inductance "reset" to reactivate the switch. Yes, if the peak current is high and the off-slope rather flat, the switching period expands: a BCM converter is a variable-frequency system. The BCM converter can also be found under the name of *critical conduction mode* converter or CRM.

Figure 1-12a portrays all three different situations by plotting the inductor current. The difference between I_{peak} and I_{valley} is called the ripple current. Its amplitude depends on the inductance value and directly reflects the ac portion of Eq. (1-26). If we now look at Fig. 1-12b, it becomes possible to derive a simple analytical expression for this ripple current. The peak-to-peak amplitude is defined by

$$\Delta I_L = I_{peak} - I_{valley} \tag{1-31}$$

We can see that the continuous dc portion $I_{L,avg}$ is located halfway between I_{peak} and I_{valley}. Therefore, $I_{L,avg}$ lying in the middle of S_{on} and S_{off}, we can update Eq. (1-31) quite quickly:

$$\Delta I_L = I_{L,avg} + \frac{S_{on}}{2} t_{on} - \left(I_{L,avg} - \frac{S_{off}}{2} t_{off}\right) = \frac{V_1 t_{on} + V_2 t_{off}}{2L} \tag{1-32}$$

For a buck, $V_1 = V_{in} - V_{out}$, $V_2 = V_{out}$.

Figure 1-12b portrays a curve at the equilibrium: when the current starts from I_{valley} to I_{peak}, the flux in the inductor builds up during the on time. Then, during the off time, I_{peak} goes down

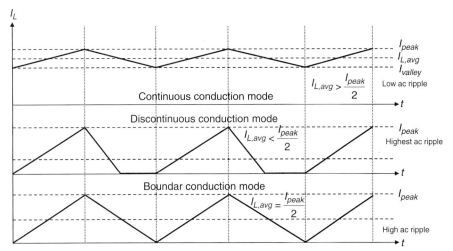

FIGURE 1-12a All three inductor operating modes.

to I_{valley}, meaning that the flux goes back to its starting point. Figure 1-13 pictures what we would observe in the inductor if we could monitor its flux activity:

φ represents the flux in the inductor

N depicts the turns number the inductor is made of

I is the current flowing in the inductor

Fortunately, without having a direct information on the inductor flux, we can apply Faraday's law which states that the voltage $V_L(t)$ measured across the inductor is

$$V_L(t) = N \frac{d\varphi(t)}{dt} \tag{1-33}$$

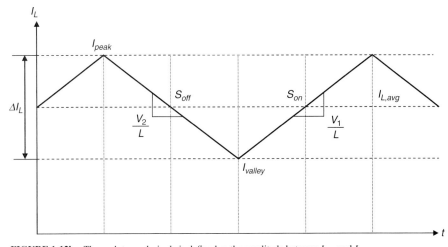

FIGURE 1-12b The peak-to-peak ripple is defined as the amplitude between I_{peak} and I_{valley}.

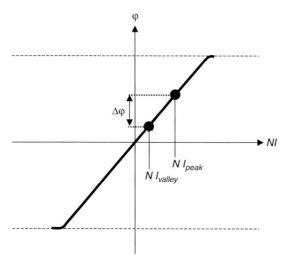

FIGURE 1-13 The flux goes up and down during the switching events. At the equilibrium, it always comes back to its starting position on the graph.

If we now integrate both parties, we obtain

$$\int V_L(t) \cdot dt = \int N \frac{d\varphi(t)}{dt} dt \qquad (1\text{-}34)$$

In this equation, N is obviously constant, it is the turns ratio of the inductor. $V_L(t)$ can be a variable, but in our case, it is either V_1 or V_2 [respectively ($V_{in} - V_{out}$) and V_{out} for a buck], constant values during the on or off times. Equation (1-34) can therefore be tweaked as

$$Vt = N\varphi \qquad (1\text{-}35a)$$

or, to account for initial conditions,

$$V \Delta t = N \Delta \varphi \qquad (1\text{-}35b)$$

As we can see, the Eq. (1-35b) unit is volt-seconds and directly reflects the flux activity in the inductor. This formula will come back when we are studying the forward converter.

In Eq. (1-32), the products $V_1 t_{on}$ and $V_2 t_{off}$, respectively, represent the flux excursion $\Delta\varphi$ during the on and off times. Since at the equilibrium or steady state (it means the converter is stabilized to its nominal operating point) the flux comes back to its starting point,

$$V_1 t_{on} = V_2 t_{off} \qquad (1\text{-}36)$$

As a result, Eq. (1-32) can be rewritten as

$$\Delta I_L = \frac{N \Delta\varphi + N \Delta\varphi}{2L} = \frac{N \Delta\varphi}{L} \qquad (1\text{-}37)$$

From Eq. (1-37), we can deduce simple observations:

- A large inductor induces a low current ripple. But as S_{off} and S_{on} depend on L, a large inductor selection will naturally slow down the system response as L will oppose any wanted current variations.

- As stated by Eq. (1-26), an output current affected by a low ripple amplitude features a lower rms value compared to the same current, but on which are superimposed large ac variations (Fig. 1-12a, middle and bottom traces): CCM operation engenders lower conduction losses compared to DCM or BCM.
- On the contrary, a small inductor implies a larger ripple and perhaps a DCM operation. The system now reacts quicker since a smaller inductor offers less opposition to current changes. However, the ripple current being important, you pay for it through higher conduction losses (all resistive paths, $R_{DS(on)}$ etc.): DCM operation brings larger conduction losses compared to CCM.
- As depicted by Fig. 1-12a, observing the inductor peak current and comparing it to the average value reveal the operating mode:

CCM: $$I_{L,avg} > \frac{I_{peak}}{2} \tag{1-38}$$

DCM: $$I_{L,avg} < \frac{I_{peak}}{2} \tag{1-39}$$

BCM: $$I_{L,avg} = \frac{I_{peak}}{2} \tag{1-40}$$

This observation is often mathematically implemented in some average models to determine the operating mode of the converter. It can also be used to derive the border at which the converters enter or leave a given mode.

1.5.3 Charge and Flux Balance

Now, if we take a closer look at Eq. (1-35a), we can reveal an interesting result. As portrayed by Fig. 1-12b, we apply V_1 during t_{on} over the inductor to magnetize it, and during the off time, we apply V_2 to reset it. If we plot the inductor voltage excursion at the equilibrium, we obtain something like Fig. 1-14. If we recall Eq. (1-35a), the on and off volt-second products must be equal to satisfy the Fig. 1-13 drawing: the flux starts from a given point and comes back to the same point at the end of the switching period. If we apply this rule to Fig. 1-14, $V_1 t_{on} = V_2 t_{off}$ then we conclude that the *average* voltage across L, at the equilibrium, is *null*:

$$\frac{1}{T_{sw}} \int_0^{T_{sw}} V_L(t) \cdot dt = \langle V_L \rangle_{T_{sw}} = 0 \tag{1-41}$$

The law describing Eq. (1-41) is called the inductor volt-second balance.

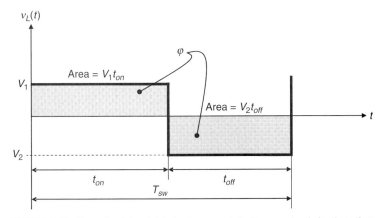

FIGURE 1-14 The portion integrated during t_{on} represents the flux excursion during the on time.

Since the capacitor is the dual element of the inductor, a similar law exists and is now called the capacitor charge balance. A current $I_C(t)$ circulates inside a capacitor C if external conditions force a voltage variation across its terminals. Equation (1-42) describes this current evolution with time:

$$I_C(t) = \frac{dQ(t)}{dt} = C \frac{dV_C(t)}{dt} \qquad (1\text{-}42)$$

If we transfer the capacitor C on the left side of the equation, and integrate both parties, we unveil Eq. (1-43):

$$\frac{1}{C} \int_0^t I_C(t) \cdot dt = \int_0^t \frac{dV_C(t)}{dt} dt \qquad (1\text{-}43)$$

which transforms to

$$\frac{1}{C} \int_0^t I_C(t) \cdot dt = \Delta V_C(t) \qquad (1\text{-}44)$$

The current integration of Eq. (1-44) represents a charge Q expressed in coulombs (or ampere-seconds) and leads to the well-known equation $\frac{1}{C} \Delta Q_C(t) = \Delta V_C(t)$. As with the inductor flux, there is some amount of charge pushed in the capacitor for $I_C(t)$ positive, whereas the same amount of charge is extracted during the negative portion of $I_C(t)$. Figure 1-15 shows the current evolution in C:

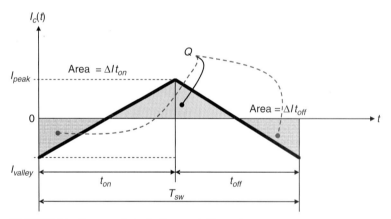

FIGURE 1-15 Current evolution in C over switching cycles.

At the equilibrium, if one calculates the area of the positive portion and compares it to the area of the negative portion, they are equal. As a result, the average current flowing inside the capacitor is *null*. This is the second law, called the charge-balance law, formulated via Eq. (1-45):

$$\int_0^{T_{sw}} I_C(t) \cdot dt = \langle I_C \rangle_{T_{sw}} = 0 \qquad (1\text{-}45)$$

1.5.4 Energy Storage

As everyone knows, capacitors and inductors are energy storage devices. Let us start with the inductor L, when magnetized. From Eq. (1-20), which describes Lenz' law, we need to find the amount of needed work to force a current variation inside the inductor. First, the instantaneous inductor power $P(t)$, when a bias voltage $V(t)$ is applied to its terminals, can easily be described by

$$P(t) = V_L(t) I_L(t) = L \frac{dI_L(t)}{dt} I_L(t) \qquad (1\text{-}46)$$

To calculate the amount of energy W necessary to bring the inductive current from 0 ($t = 0$) to I_L ($t = t_1$), we need to integrate Eq. (1-46):

$$W(t) = \int_0^{t_1} P(t) \cdot dt = L \int_0^{I_L} \frac{dI_L(t)}{dt} I_L(t) \cdot dt = \frac{1}{2} L I_L^2 \qquad (1\text{-}47)$$

Equation (1-47) defines the energy W, in joules, stored in an inductor when magnetized by a current I_L. If we multiply W by a switching frequency F_{sw}, we obtain watts.

For the capacitor, the methodology is very close. From Eq. (1-47), we can calculate the amount of energy necessary to bring the capacitor voltage from 0 ($t = 0$) to a voltage V_C reached for $t = t_1$:

$$W(t) = \int_0^{t_1} P(t) \cdot dt = \int_0^{t_1} V_C(t) I_C(t) \cdot dt \qquad (1\text{-}48)$$

from Eq. (1-42), we can rearrange this equation:

$$W(t) = C \int_0^{V_C} V_C(t) \frac{dV_C(t)}{dt} dt \qquad (1\text{-}49)$$

which leads to the well-known equation

$$W = \frac{1}{2} C V_C^2 \qquad (1\text{-}50)$$

The above equations, charge balance, volt-second balance and storage results play an important role in the analysis of switching converters.

1.6 THE DUTY CYCLE FACTORY

We have seen previously that our converter output characteristics can be modified by playing on the duty cycle D. Actually, the feedback loop will control D, for instance, to keep V_{out} constant, whatever the input/output conditions are. But how is D elaborated?

1.6.1 Voltage-Mode Operation

Voltage-mode control is probably the most common way to control a power supply. In essence, an error voltage (remember, the Greek letter epsilon ε) obtained from the difference between a reference voltage and a portion of the output voltage is permanently compared to a fixed frequency and amplitude sawtooth. The crossing point between these two signals generates a transition on the comparator's output. When the output voltage deviates from its natural target,

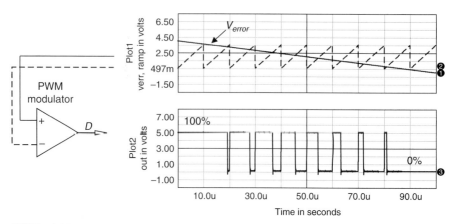

FIGURE 1-16a The voltage-mode duty cycle factory: a sawtooth compared to a dc level, the error voltage.

the error voltage ε increases. As a result, the point at which both the error and sawtooth signals cross, naturally expands the distance between the toggling points: D is increased. Figure 1-16a shows how the pulse width modulator (the PWM, our duty cycle factory) is usually built:

In Fig. 1-16a, the error signal is purposely decreased to show the duty cycle reduction. One can see the two extreme positions: V_{error} is greater than the sawtooth peak, and the comparator's output is permanently high. This is a 100% duty cycle situation.

On the contrary, when the error voltage passes below the sawtooth signal offset, then the comparator permanently stays low, keeping the power switch off: this is a 0% duty cycle situation. Please note that some integrated controllers accept this behavior (so-called skip-cycle operation), while others do not and are stuck in a minimum duty cycle situation. Figure 1-16b depicts a simple converter operated by a voltage-mode PWM controller.

Voltage mode is also called direct duty cycle control as the error voltage directly drives the duty cycle.

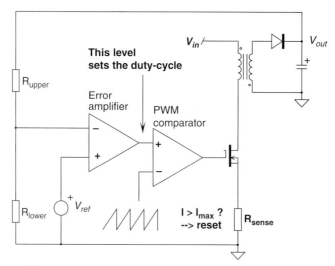

FIGURE 1-16b A practical implementation of a voltage-mode control circuitry where the error voltage sets the on-time duration independently from the inductor current.

INTRODUCTION TO POWER CONVERSION 29

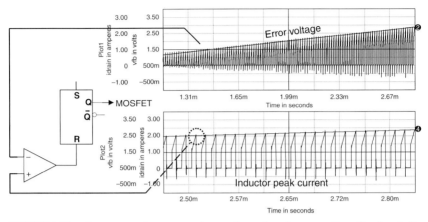

FIGURE 1-17a A current-mode power supply waveform where the error voltage directly sets the peak current limit.

1.6.2 Current-Mode Operation

In the previous example, the PWM does not rely upon the circulating inductor current to make its toggling decision. Some systems, however, include a current limit circuitry which resets the switch, activated in the presence of a fault condition. Current-mode modulators, however, do rely upon the instantaneous inductor current. A clock pulse sets a latch which closes the power switch. The current ramps up in the inductor, following a $\frac{V}{L}$ slope. When the current reaches a given set point value, a comparator detects it and resets the latch. SW now opens and waits for the next clock cycle to close again. You understood that the feedback loop now controls the peak current set point and *indirectly* the duty cycle. We have a current-mode power supply, as depicted by Fig. 1-17a.

Figure 1-17b portrays a standard implementation of a current-mode controller.

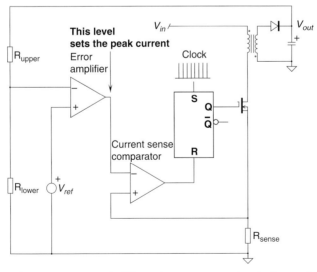

FIGURE 1-17b The main difference between current mode and voltage mode lies in the duty cycle generation: direct control with voltage mode, indirect control via the inductor peak current with current mode.

Current-mode control (denoted CM control) and voltage-mode control (denoted VM control) offer drastically different dynamic behaviors. However, looking at a typical converter waveform on the oscilloscope would not tell you which PWM technique was implemented.

1.7 THE BUCK CONVERTER

Figure 1-11d described a buck converter, basically a square wave generator followed by an LC filter. To refine our understanding of the converter, we can split the on and off events as proposed by Figs. 1-18a and 1-18b. We assume that the converter implements the voltage-mode technique. In Fig. 1-18a, we can see the current flow during the switch closing time.

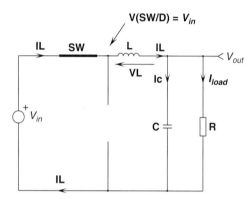

FIGURE 1-18a On-time current path during the switch closing period.

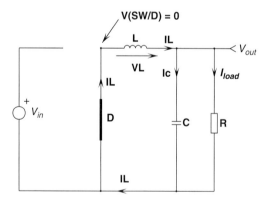

FIGURE 1-18b During the off time, the inductor current keeps circulating in the same direction.

1.7.1 On-Time Event

As soon as the SW closes, a current rise takes place in L. Applying a voltage V to an inductor L affected by a series resistor r_{Lf} normally gives rise to an exponential current whose shape follows the equation

$$I_L(t) = \frac{V_L}{r_{Lf}}\left(1 - e^{-t\frac{r_{Lf}}{L}}\right) \qquad (1\text{-}51)$$

where V_L is the voltage applied across the inductor and r_{Lf} is the inductor series resistor. However, as the series resistor r_{Lf} is obviously kept small, we can try to develop Eq. (1-51) in the vicinity of zero. We obtain the well-known equation

$$I_L(t) \approx \frac{V_L}{r_{Lf}}\left(1 - 1 + t\frac{r_{Lf}}{L}\right) = \frac{V_L}{L}t \qquad (1\text{-}52)$$

If we look at the buck configuration during the on event, one of the inductor terminals "sees" V_{in} (assuming SW drop is null) whereas the other one is directly connected to V_{out}. Applying Eq. (1-52) to this situation implies a current reaching a peak value imposed by the SW closing duration, the on time, denoted t_{on}. This is Eq. (1-28), already presented but updated with initial conditions

$$I_{peak} = I_{valley} + \frac{V_{in} - V_{out}}{L}t_{on} \qquad (1\text{-}53)$$

As current I_L flows in the inductor, it also crosses the capacitor C (I_C) and the load connected to the output (I_{load}). As we have already shown, at the equilibrium, all the alternating current (the ac portion) will go in C, whereas the dc portion will flow in R_{load}. Understanding this fact will help us to evaluate the output voltage ripple amplitude.

1.7.2 Off-Time Event

The current having reached a value imposed by t_{on}, the PWM modulator now instructs the switch to open. To fight against its collapsing magnetic field, the inductor reverses its voltage as described by Eq. (1-20). Since the inductive current still needs to flow somewhere, in the same direction, the diode gets activated: we obtain Fig. 1.18b. If we neglect the diode voltage drop, the inductor left terminal is grounded, whereas the right one "sees" V_{out}. The inductor valley current I_{valley} can be described by Eq. (1-30) already derived, again updated with the initial condition

$$I_{valley} = I_{peak} - \frac{V_{out}}{L}t_{off} \qquad (1\text{-}54)$$

When the new clock cycle occurs, SW closes again and 1.53 comes in play again: we end up with a square wave signal present on the common point SW/D, swinging between V_{in} and 0 and further integrated by the LC filter. Let us now reveal all waveforms with a simulation.

1.7.3 Buck Waveforms—CCM

To gain a comprehensive understanding of the buck operation, let us observe all individual waveforms brought by Fig. 1-11d. They appear in Fig. 1-19.

Waveform 1 (plot 1) represents the PWM pattern, activating the switch on and off. When the switch SW is on, V_{in} appears on the common point SW/D. On the contrary, when SW opens, we would expect this node to swing negative. However, thanks to the presence of diode D, the inductor current biases it, and a negative drop appears: this is the freewheel action.

Waveform 3 describes how the voltage evolves across the inductor. If you remember Eq. (1-41), the average voltage across L is null at equilibrium, meaning that $S_1 + S_2 = 0$. S_1

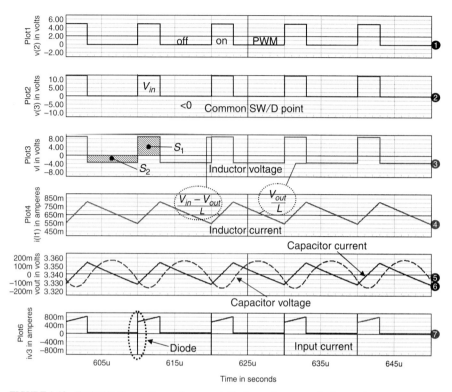

FIGURE 1-19 Typical buck waveforms when operated in the continuous conduction mode.

corresponds to the on volts-seconds area, whereas S_2 represents the off volts-seconds area. S_1 is simply the rectangle height $V_{in} - V_{out}$ multiplied by the duration DT_{sw}, whereas S_2 is also the rectangle height $-V_{out}$ multiplied by the duration $(1 - D)T_{sw}$. If we sum up both S_1 and S_2 and average that over T_{sw}, we have

$$(D(V_{in} - V_{out})T_{sw} - V_{out}(1 - D)T_{sw})\frac{1}{T_{sw}} = 0 \qquad (1\text{-}55a)$$

If we rearrange the equation, we obtain the well-known buck dc transfer function M in CCM:

$$V_{out} = DV_{in} \quad \text{or} \quad M = \frac{V_{out}}{V_{in}} = D \qquad (1\text{-}55b)$$

If we plot this function, to see how V_{out} evolves with D, we can see a linear variation portrayed by Fig. 1-20. As Eq. (1-55b) ideally states, the transfer characteristic is independent of the output load. We will see later that it is not exactly true.

Another simple trick, based on the null average inductor voltage, can help to speed up the transfer ratio determination. First, let us write the instantaneous inductor voltage V_L

$$V_L = V_{SW/D} - V_{out} \qquad (1\text{-}56a)$$

To benefit from Eq. (1-41), let us reformulate Eq. (1-56a) via averaged values:

$$\langle V_L \rangle_{T_{sw}} = \langle V_{SW/D} \rangle_{T_{sw}} - \langle V_{out} \rangle_{T_{sw}} \qquad (1\text{-}56b)$$

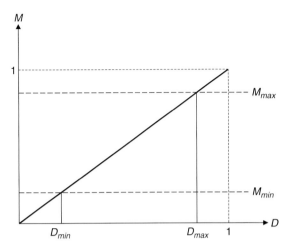

FIGURE 1-20 Dc conversion ratio of the buck operated in CCM.

By definition, V_{out} is kept constant by the feedback loop. However, the common point SW/D swings between V_{in} (during the on time) and 0 during the off time. Equation (1-56b) can thus be rearranged as:

$$V_{in}D - V_{out} = 0 \text{ or } V_{in}D = V_{out} \text{ which leads to}$$

$$M = \frac{V_{out}}{V_{in}} = D \qquad (1\text{-}56c)$$

A closer observation of the bottom waveform shows several things. First, there is a large spike when SW is turned on. This is so because the switch brutally interrupts the diode conduction period by applying V_{in} to its cathode. If we use a PN diode, then we need to bring the crystal back to its electrical neutrality to block it. This is done by removing all the minority carriers: the diode sweeps away all injected charges and needs a certain time to recover its blocking action. This period is called the t_{rr}. Before the diode has fully recovered, it behaves as a short-circuit. For a Schottky diode, we have a metal-silicon junction and there is no recovery effect. However, there can be a large parasitic capacitance. Once discharged (when the diode conducts), SW quickly applies V_{in} over this discharged capacitor and a current spike appears. Striving to slow down the turn-on SW transition will help lower the spike amplitude.

The second point relates to the current shape. You can observe a nice ripple on the output. We say that the output ripple is smooth, "nonpulsating." It implies a better acceptance by the downstream electronic circuits, meaning less pollution on the line. On the other hand, the input current not only features the spike, but also looks like a square wave. If L grew to infinity, the shape would be a true square wave. It is a "pulsating" current, carrying a large polluting spectrum, more difficult to filter than an almost sinusoidal shape.

As a summary, we can write some comments related to the CCM buck converter:

- As D is constrained below 1, the output voltage of the buck is always smaller than the input voltage ($M < 1$).
- If we neglect the various ohmic losses, the conversion ratio M is independent of the load current.
- By varying the duty cycle D, we can control the output voltage, as already mentioned.

- Operating the buck in CCM brings additional losses as the diode reverse recovery charge needs to be evacuated (t_{rr}). This is seen as an additional loss burden on the power switch SW.
- There is nonpulsating output ripple, but pulsating input current.
- Unless a P channel or a PNP transistor is used, the control of an N channel or an NPN requires a special circuitry as its source or emitter floats.
- Short-circuit protection or in-rush limitation is feasible as the power switch can interrupt the input current flow.

Let us now reduce the load and see how the waveforms evolve in discontinuous conduction mode.

1.7.4 Buck Waveforms—DCM

We have now reduced the output load to 40 (instead of 5 Ω. Therefore, during the off period, the inductor is fully depleted to zero, waiting for SW to turn on again. Since this inductor current also flows in the freewheel diode, we expect the diode to naturally block. As SW is opened and the diode is blocked, we have a third state, of duration $D_3 T_{sw}$, where the capacitor supplies the load on its own (Fig. 1-21a). The simulation waveforms of the DCM buck are available in Fig. 1-21b.

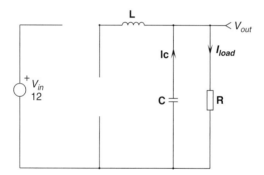

FIGURE 1-21a In DCM, a third state exists where all switches are open.

Observing plot number 4, we can now clearly see the inductor current going down to zero, leading to the diode blocking. When it happens, the inductor left end terminal becomes open. In theory, the voltage on this node should go back to V_{out}, without oscillation, as L no longer carries current. Because of all the parasitic capacitance around (for instance, the diode and SW parasitic capacitors), a resonating tank is formed. As observed on waveforms 2 and 3, a sinusoidal signal appears and dies out after a few cycles, depending on the resistive damping. Figure 1-23 graphs the evolution of both $V_L(t)$ and $I_L(t)$. Thanks to this drawing, it becomes possible to calculate the new transfer function of the buck DCM, by still applying Eq. (1-41), where D_2 represents the demagnetization phase (Fig. 1-23):

$$(V_{in} - V_{out})DT_{sw} - V_{out}D_2 T_{sw} = 0 \tag{1-57}$$

In the buck configuration, the output current is the dc, or the average value of the current flowing in L. Therefore, a second equation averaging $I_L(t)$ will help us to evaluate the transfer function

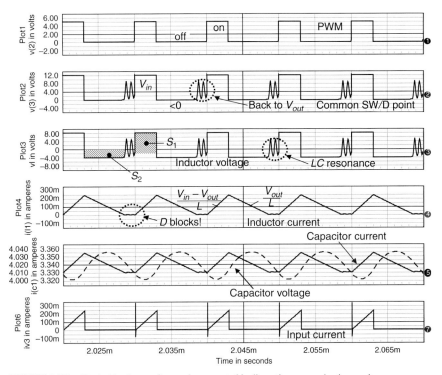

FIGURE 1-21b   Typical buck waveforms when operated in discontinuous conduction mode.

of our buck operated in DCM:

$$\frac{DT_{sw}I_{peak}}{2T_{sw}} + \frac{D_2 T_{sw} I_{peak}}{2T_{sw}} = I_{out} \tag{1-58}$$

From Eq. (1-53), we extract the off duty cycle D_2:

$$D_2 = \frac{2I_{out} - DI_{peak}}{I_{peak}} \tag{1-59}$$

into which we can substitute the peak current definition I_{peak} involving D_2:

$$I_{peak} = \frac{V_{out}}{L} D_2 T_{sw} \tag{1-60}$$

$$D_2 = \frac{2I_{out} - DD_2 \frac{V_{out} T_{sw}}{L}}{V_{out}} \frac{L}{D_2 T_{sw}} \tag{1-61}$$

Solving Eq. (1-61) for D_2 and substituting the result into Eq. (1-57) give a V_{out} definition:

$$V_{out} = \frac{V_{in}}{1 + \frac{2I_{out}L}{D^2 T_{sw} V_{in}}} = \frac{V_{in}}{1 + \frac{2V_{out}L}{D^2 T_{sw} V_{in} R}} \tag{1-62}$$

In Eq. (1-62), let us multiply both terms by $\frac{1}{V_{in}}$ to reveal our ratio $M = \frac{V_{out}}{V_{in}}$:

$$M = \frac{1}{1 + \frac{2ML}{D^2 T_{sw} R}} \quad (1\text{-}63)$$

Solving for M in this second-order equation gives us

$$M = \frac{RT_{sw}D^2}{4L}\left[\sqrt{1 + \frac{8L}{RT_{sw}D^2}} - 1\right] = \frac{D^2}{\tau_L 4}\left[\sqrt{1 + \frac{8\tau_L}{D^2}} - 1\right] \quad (1\text{-}64)$$

where $\tau_L = \frac{L}{RT_{sw}}$ is the normalized inductor time constant.

To rearrange Eq. (1-64), we will multiply the numerator and the denominator by $\sqrt{1 + \frac{8\tau_L}{D^2}} + 1$. Thanks to this operation, Eq. (1-65) describes the final conversion ratio for the buck operated in DCM:

$$M = \frac{2}{1 + \sqrt{1 + \frac{8\tau_L}{D^2}}} \quad (1\text{-}65)$$

If we now plot the transfer ratio M versus the duty cycle D, we obtain the curve family of Fig. 1-22 where τ_L takes different values. For small values, where the output current is low, we are in deep DCM and we can easily reach an M ratio of 1. However, as we increase the load current, still keeping the DCM operation, the curve finishes away from the value $M = 1$.

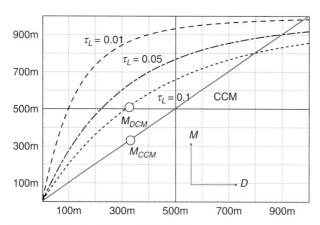

FIGURE 1-22 The buck operated in DCM.

As a summary, some comments can be made for the buck converter operating in DCM:

- M is now dependent on the load current.
- For an identical duty cycle, the ratio M is larger for a DCM operation than for a CCM operation.

In particular because of the first bullet, a buck—or a buck-derived circuit such as the forward converter—is never designed for a full-load operation in DCM.

INTRODUCTION TO POWER CONVERSION   37

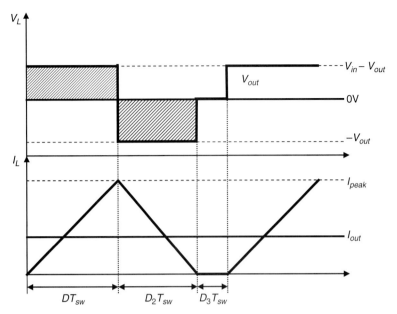

FIGURE 1-23   The inductor voltage and current signals for the buck working in DCM.

1.7.5   Buck Transition Point DCM—CCM

When the inductor current vanishes to zero during a switching period, the converter is said to be operated in DCM. On the contrary, when it never reaches zero during the switching period, we are operating in CCM. As the load goes down, so does the average current in the inductor. The point at which the ripple touches zero to immediately restart is called the *boundary* point, the *borderline* point, or the *critical* point. This is the place where the transition from CCM to DCM (or the other way round) occurs: there is no dead time as in Fig. 1-23. Let us calculate the value of the load resistor or the inductor value that determines the occurrence of this point. Figure 1-24 portrays the

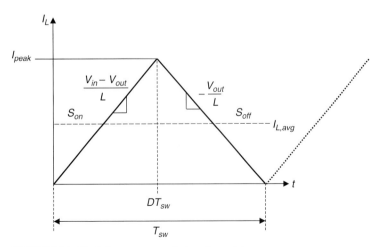

FIGURE 1-24   The inductor current at the boundary point.

current activity in the inductor at the boundary point. As highlighted previously, in boundary mode, the average value of the symmetric triangle is its peak divided by 2. Otherwise stated,

$$\langle I_L \rangle_{T_{sw}} = \frac{I_{peak}}{2} \tag{1-66}$$

From Fig. 1-24 we can define the various slopes:

$$S_{on} = \frac{V_{in} - V_{out}}{L} \tag{1-67}$$

$$S_{off} = -\frac{V_{out}}{L} \tag{1-68}$$

From Eq. (1-67), the peak value I_{peak} is found to be

$$I_{peak} = \frac{(V_{in} - V_{out})DT_{sw}}{L} \tag{1-69}$$

Thanks to Eq. (1-66),

$$\langle I_L \rangle_{T_{sw}} = \frac{(V_{in} - V_{out})DT_{sw}}{2L} \tag{1-70}$$

The inductor dc component is the output current I_{out}, thus

$$\langle I_L \rangle_{T_{sw}} = \frac{V_{out}}{R} \tag{1-71}$$

$$\frac{(V_{in} - V_{out})DT_{sw}}{2L} = \frac{V_{out}}{R} = \frac{DV_{in}}{R} \tag{1-72}$$

Simplifying by elimination of D yields

$$(V_{in} - V_{out})T_{sw}R = V_{in}2L \tag{1-73}$$

Factoring V_{in} and knowing that $V_{out} = DV_{in}$, we obtain

$$\mathbf{R}_{critical} = \frac{2LF_{sw}}{1 - D} \tag{1-74a}$$

$$\mathbf{L}_{critical} = \frac{(1 - D)R}{2F_{sw}} \tag{1-75a}$$

Rearranging these above equations by replacing D by the buck dc transfer function yields

$$\mathbf{R}_{critical} = 2F_{sw}L\frac{V_{in}}{V_{in} - V_{out}} \tag{1-74b}$$

$$\mathbf{L}_{critical} = \frac{R(V_{in} - V_{out})}{2F_{sw}V_{in}} \tag{1-75b}$$

When you are designing a buck converter, using Eqs. (1-74a) or (1-74b) or (1-75a) or (1-75b) will give indications related to the loss of the CCM operation. The designer will thus strive to keep the operating current above the critical point. Increasing L also represents a viable solution.

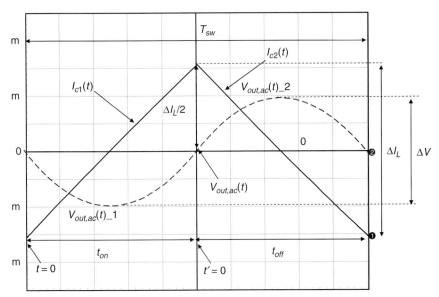

FIGURE 1-25   The linear inductor current induces a quasi-sinusoidal voltage over C.

1.7.6   Buck CCM Output Ripple Voltage Calculation

Figure 1-25 shows the linearly ramping up and down current flowing in the output capacitor. From Eq. (1-44), we know that the voltage across the capacitor, which is V_{out} in our application, is obtained by integrating the capacitor current $I_C(t)$. The current is a ramp, described by a simple straight-line equation in the form of ax. Its integral will therefore look like a parabolic expression, satisfying the form ax^2. Since the signal is discontinuous, we need to bound it between known limits. The first one will be between 0 and t_{on} (I_{C1}), whereas the second one, I_{C2}, could be between t_{on} and T_{sw}. However, to simplify the analysis, we will bound I_{C2} between 0 and t_{off}.

Please note that we only care about the ac portion of V_{out} (the actual ripple), hence the notation $V_{out,ac}(t)$ which is $V_{out}(t)$ minus its dc portion.

Let us first express the capacitor current expression for I_{C1} and I_{C2}:

$$I_{C1}(t) = -\frac{\Delta I_L}{2} + \Delta I_L \frac{t}{t_{on}} \qquad t \in [0, t_{on}] \tag{1-76}$$

$$I_{C2}(t) = \frac{\Delta I_L}{2} - \Delta I_L \frac{t'}{t_{off}} \qquad t' \in [0, t_{off}] \tag{1-77}$$

where $\frac{\Delta I_L}{2}$ or $-\frac{\Delta I_L}{2}$ represents the starting point for $t = t' = 0$, actually one-half of the total current ripple ΔI_L. To obtain an image of the output ripple, we need to satisfy Eq. (1-44) by integrating I_{C1} and I_{C2} from 0 to t.

$$\frac{1}{C}\int_0^t I_{C1}(t) \cdot dt = -\frac{\Delta I_L t}{2C} + \frac{\Delta I_L t^2}{2 t_{on} C} = \frac{\Delta I_L}{2C}\left(\frac{t^2}{t_{on}} - t\right) \tag{1-78a}$$

$$\frac{1}{C}\int_0^t I_{C2}(t) \cdot dt = \frac{\Delta I_L t}{2C} - \frac{\Delta I_L t^2}{2 t_{off} C} = \frac{\Delta I_L}{2C}\left(t - \frac{t^2}{t_{off}}\right) \tag{1-79a}$$

Since we are interested in the amplitude of the above functions, we know that a peak affects them. By nulling their respective derivative terms [Eqs. (1-76) and (1-77)], we will find the exact moment at which these functions peak:

$$-\frac{\Delta I_L}{2} + \Delta I_L \frac{t}{t_{on}} = 0 \quad \text{which gives} \quad t = \frac{t_{on}}{2} \quad (1\text{-}78b)$$

$$\frac{\Delta I_L}{2} - \Delta I_L \frac{t}{t_{off}} = 0 \quad \text{which gives} \quad t = \frac{t_{off}}{2} \quad (1\text{-}79b)$$

If we replace these results in Eqs. (1-78a) and (1-79a), we obtain a generic definition for the ripple voltage across a capacitor in which a linear current flows:

$$\text{For } t \in [0, t_{on}] \quad V_{out,ac}(t)\_1 = -\frac{\Delta I_L t_{on}}{8C} \quad (1\text{-}80)$$

$$\text{For } t \in [0, t_{off}] \quad V_{out,ac}(t)\_2 = \frac{\Delta I_L t_{off}}{8C} \quad (1\text{-}81)$$

The total peak-to-peak ripple ΔV can now be found by subtracting Eq. (1-80) from Eq. (1-81):

$$\Delta V = \frac{\Delta I_L t_{off}}{8C} - \left(-\frac{\Delta I_L t_{on}}{8C}\right) = \frac{\Delta I_L}{8C}\left[t_{on} + t_{off}\right] = \frac{\Delta I_L T_{sw}}{8C} \quad (1\text{-}82)$$

This equation describes the peak-to-peak ripple definition for the CCM buck converter and will be useful for calculating the output capacitor in relation to the required ripple. Please note that this definition does not include the equivalent series resistor (ESR) effects.

The exercise now lies in calculating ΔI_L for the buck converter, the considered subject. From Fig. 1-25, at equilibrium, the current starts from $-\frac{\Delta I_L}{2}$, rises up to $+\frac{\Delta I_L}{2}$, and goes back again to $-\frac{\Delta I_L}{2}$. A simple equality can thus be derived from this expression:

$$-\frac{\Delta I_L}{2} + \frac{(V_{in} - V_{out})}{L} t_{on} = \frac{\Delta I_L}{2} - \frac{V_{out}}{L} t_{off} \quad (1\text{-}83)$$

Rearranging to extract ΔI_L leads to

$$\Delta I_L = \frac{1}{L}\left[(V_{in} - V_{out})t_{on} + V_{out} t_{off}\right] \quad (1\text{-}84)$$

Substituting Eq. (1-84) into Eq. (1-82) gives

$$\Delta V = \frac{1}{8L}\left(t_{on} V_{in} - t_{on} V_{out} + V_{out} t_{off}\right) \frac{T_{sw}}{C} \quad (1\text{-}85)$$

Knowing that $t_{on} = DT_{sw}$ and $t_{off} = (1 - D)T_{sw}$ gives us a chance to massage Eq. (1-85):

$$\Delta V = \frac{T_{sw}^2}{8LC}\left(DV_{in} - 2DV_{out} + V_{out}\right) \quad (1\text{-}86)$$

Since we consider our buck converter as a square wave generator followed by an *LC* filter, we know that the cutoff frequency of such a filter is

$$f_0 = \frac{1}{2\pi\sqrt{LC}} \quad (1\text{-}87)$$

Then

$$LC = \frac{1}{4\pi^2 f_0^2} \quad (1\text{-}88)$$

INTRODUCTION TO POWER CONVERSION   41

Replacing LC in Eq. (1-86) and introducing F_{sw}, the switching frequency, yield

$$\Delta V = \frac{\pi^2}{2}\left(\frac{f_0}{F_{sw}}\right)^2 (DV_{in} - 2DV_{out} + V_{out}) \quad (1\text{-}89)$$

To simplify this expression, a possibility exists to normalize the ripple to the output voltage V_{out}. Equation (1-89) can thus be updated as

$$\frac{\Delta V}{V_{out}} = \frac{\pi^2}{2}\left(\frac{f_0}{F_{sw}}\right)^2 \left(D\frac{V_{in}}{V_{out}} - 2D + 1\right) \quad (1\text{-}90)$$

Knowing that $D = \dfrac{V_{out}}{V_{in}}$, we have

$$\frac{\Delta V}{V_{out}} = \frac{\pi^2}{2}\left(\frac{f_0}{F_{sw}}\right)^2 (1 - D) \quad (1\text{-}91)$$

Well, it looks pretty complicated to obtain the ripple expression, after numerous equations. Actually, a more elegant way exists, without using brute force. Reference 6, pages 372–373, details it for you. The result is derived in ... four lines of algebra! Sorry, I cannot compete.

1.7.7 Now with the ESR

The ESR appears like a resistor in series with the capacitor (Fig. 1-26a). The ac inductor current ripple no longer crosses C alone, but a combination of C plus a resistor in series R_{ESR}. Actually, an equivalent inductor also exists, but we will ignore it in this case.
The drop incurred to this element is simply

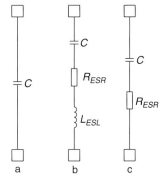

FIGURE 1-26a A capacitor can be represented in series with a resistor, its ESR alone, when the parasitic inductor is neglected.

$$\Delta V_{ESR} = \Delta I_L R_{ESR} \quad (1\text{-}92)$$

From Eq. (1-68), we know that the inductor ripple current flows in R_{ESR}. Therefore,

$$\Delta V_{ESR} = \frac{V_{in} - V_{out}}{L} t_{on} R_{ESR} = \frac{V_{in} - V_{out}}{L} DT_{sw} R_{ESR} \quad (1\text{-}93a)$$

As usual, knowing that $D = \dfrac{V_{out}}{V_{in}}$, we deduce, after normalization to V_{out},

$$\frac{\Delta V_{ESR}}{V_{out}} = \frac{\dfrac{V_{out}}{D} - \dfrac{DV_{out}}{D}}{L} DT_{sw} R_{ESR} = \frac{1 - D}{LF_{sw}} R_{ESR} \quad (1\text{-}94)$$

Depending on the domination of either the capacitive or the resistive term, either the final ripple curve will look almost sinusoidal for a negligible ESR effect, or it will transform to a triangular waveform in the case of a large ESR contribution.

1.7.8 Buck Ripple, the Numerical Application

In our example, Fig. 1-11d, we have the following element values, upgraded with the ESR value placed in series with the capacitor C:

$$L = 100 \text{ μH}$$
$$C = 10 \text{ μF}$$

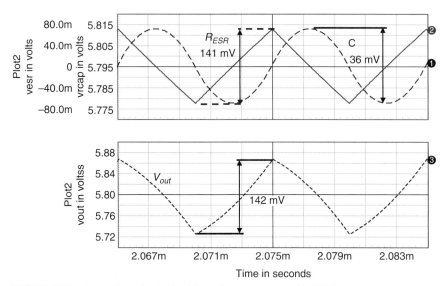

FIGURE 1-26b Output simulation ripple of the buck converter operated in CCM.

$$R_{ESR} = 500 \text{ m}\Omega$$
$$V_{in} = 12 \text{ V}$$
$$V_{out} = 5.8 \text{ V}$$
$$F_{sw} = 100 \text{ kHz}$$
$$D = 0.5$$

The LC cutoff frequency is found to be 5.03 kHz. Applying Eq. (1-91), we should obtain the peak-to-peak capacitive ripple value

$$\Delta V = \frac{9.87}{2}\left(\frac{5.03}{100}\right)^2 0.5 \times 5.8 = 36\,mV \qquad (1\text{-}93b)$$

Applying Eq. (1-94) gives the ESR addition

$$\Delta V_{ESR} = \frac{1-D}{LF_{sw}} R_{ESR} V_{out} = \frac{0.5}{100u \times 100k} \times 0.5 \times 5.8 = 145\,mV \qquad (1\text{-}93c)$$

Figure 1-26b gives the simulation result and confirms our approach. In this example, we can see the contribution of each term. However, as the ESR dominates the ripple expression, the final waveform looks almost triangular (lower curve in Fig. 1-26b).

1.8 THE BOOST CONVERTER

The boost converter belongs to the family of indirect energy transfer converters. The power process involves an energy-storing phase and an energy-release phase. During the on time, the inductor stores energy, and the output capacitor alone powers the load. At the switch opening, the stored inductive energy appears in series with the input source and contributes to supply the output. If the buck converter draws current from the source during the on time, the boost does during both the on- and off-time cycles, where the inductor discharges into the output network for the latter. As we will see shortly, this architecture confers a nonpulsating input current signature to the CCM boost converter and makes it a low-input ripple topology.

Figure 1-27 represents the boost converter, still involving a switch (MOSFET or bipolar), an inductor, and a capacitor.

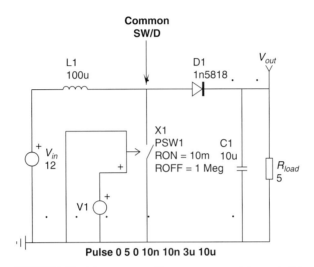

FIGURE 1-27 A boost converter first stores energy and then routes it to the output capacitor.

From inspection of Fig. 1-27 we can make several comments. First, the inductor lies in series with the input source. We already know that an inductor opposes fast current variations. Therefore, when the inductor is placed where it is, its action will naturally smooth the input current signature. Second, the power switch is now driven by a source V_1 referenced to the ground. Compared to the buck architecture, where the reference control switch terminal (e.g., the emitter or the source) connects to V_{out}, the boost configuration offers a simpler driving situation for the controller. Let us now study the situation during the *on* and *off* events:

1.8.1 On-Time Event

When the switch SW closes, V_{in} immediately appears across L (neglecting the switch voltage drop). As a result, the inductor current reaches a peak value defined by

$$I_{peak} = I_{valley} + \frac{V_{in}}{L} t_{on} \tag{1-95}$$

Actually I_{valley} represents the condition for $t = 0$. It can be zero, for the DCM case, or a nonzero value for CCM.

Figure 1-28a portrays the situation where the load and capacitor are left alone while the power switch SW closes: as you can imagine, the voltage over C is going to decrease during the on-time.

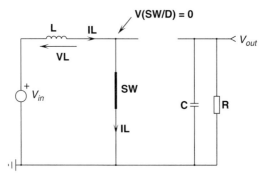

FIGURE 1-28a During the on time, V_{in} is applied across the main inductor L.

1.8.2 Off-Time Event

At the switch opening, the inductive current must find a circulating path. The inductor voltage reverses, trying to keep the ampere-turns constant. This voltage now comes in series with the input voltage V_{in}. The diode routes the inductive current to the output capacitor and helps dump the stored energy into the capacitor C (current I_C) and the load (current I_{load}). Figure 1-28b shows

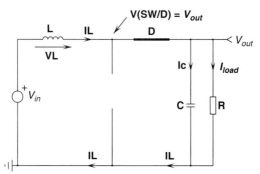

FIGURE 1-28b During the off time, the current circulates in the same direction and the coil voltage reverses.

the situation. During the off time, the energy stored in L during the on time depletes to a rate described by Eq. (1-96):

$$I_{valley} = I_{peak} - \frac{V_{out} - V_{in}}{L} t_{off} \qquad (1\text{-}96)$$

When the new clock cycle occurs, SW closes again and Eq. (1-95) comes back into play. It is time to reveal the boost characteristic waveforms in CCM.

1.8.3 Boost Waveforms—CCM

To have a comprehensive understanding of the boost operation, let us observe all individual waveforms brought by Fig. 1-29.

Waveform 1 (in plot 1) represents the PWM pattern, activating the switch *on* and *off*. When the switch SW is *on*, the common point SW/D drops to almost zero. To the opposite, when SW opens, its right terminal increases to a value imposed by the output voltage plus the forward drop of the diode, V_f.

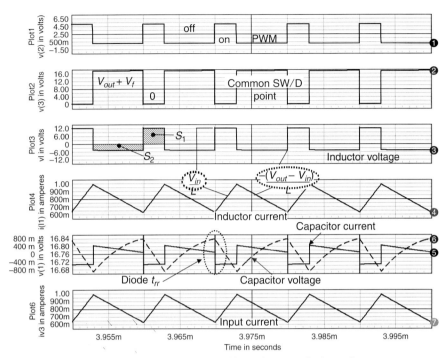

FIGURE 1-29 Typical boost waveforms when operated in continuous conduction mode.

Waveform 3 describes how the voltage evolves across the inductor. If we observe it as a positive value during the on time, the inductor voltage reverses at the switch opening. The inductance left terminal stays at V_{in} whereas the right terminal jumps to V_{out} as diode D conducts. Therefore, observing the inductor voltage with a + plus at the V_{in} connections implies a negative swing during the off time. This negative swing is simply $-(V_{out} - V_{in})$, implying that V_{out} is greater than V_{in}.

As highlighted by Eq. (1-41), the average voltage across L is null at equilibrium, meaning that $S_1 + S_2 = 0$. S_1 corresponds to the on volts-seconds area, whereas S_2 represents the off volts-seconds area. S_1 is simply the rectangle height V_{in} multiplied by the duration DT_{sw}, whereas S_2 is also the rectangle height $-(V_{out} - V_{in})$ multiplied by the duration $(1 - D)T_{sw}$, or $D'T_{sw}$. If we sum up both S_1 and S_2 and average it over T_{sw}, we have

$$(DV_{in}T_{sw} - (V_{out} - V_{in})D'T_{sw})\frac{1}{T_{sw}} = 0 \qquad (1\text{-}97)$$

If we rearrange the equation, we obtain the well-known boost dc transfer function M in CCM:

$$V_{in}(D + D') = V_{out}D' \quad \text{or} \quad M = \frac{V_{out}}{V_{in}} = \frac{1}{D'} = \frac{1}{1 - D} \qquad (1\text{-}98)$$

If we plot this function, to see how V_{out} evolves with D, we can observe a nonlinear variation portrayed by Fig. 1-30. As Eq. (1-98) ideally states, the transfer characteristic is independent of the output load and could increase to infinity as D approaches 1. We will see later that boost conversion ratios above 4 to 5 are difficult to obtain as soon as some output current is needed.

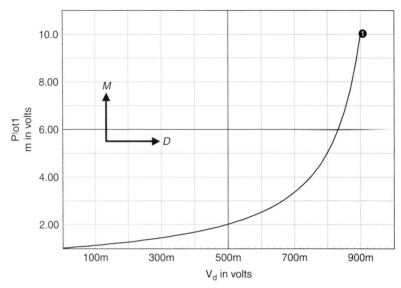

FIGURE 1-30 Dc conversion ratio of the boost operated in CCM.

As we have already done in the buck analysis, we can quickly derive the transfer function of the boost by using averaged values. If we assume that the instantaneous voltage across the inductor V_L is

$$V_L = V_{in} - V_{SW/D} \qquad (1\text{-}99)$$

Applying Eq. (1-41) yields

$$\langle V_L \rangle_{T_{sw}} = \langle V_{in} \rangle_{T_{sw}} - \langle V_{SW/D} \rangle_{T_{sw}} = 0 \qquad (1\text{-}100\text{a})$$

By definition V_{in} is constant, and the common point SW/D drops to zero during t_{on} but jumps to V_{out} during the off time. Therefore, Eq. (1-100a) can be reformulated as

$$V_{in} = V_{out} D' \quad \text{or} \quad \frac{V_{out}}{V_{in}} = \frac{1}{D'} = M \qquad (1\text{-}100\text{b})$$

Almost the same buck comments apply to the boost. Since we operate the converter in CCM, the diode blocking point is solely dictated by the activation of SW. As a result, brutally blocking a diode makes it behave as a short-circuit until it fully recovers its neutral state. A "shorted" diode means V_{out} is applied over SW when it is closing. A current spike occurs in the power switch. We do not see it in Fig. 1-29 as SW current does not show up. However, we clearly see the recovery spike on the capacitor waveform. Slowing down SW will help to reduce the recovery current. This is the major problem in power factor correction (PFC) circuits operated in CCM.

Unlike the buck, the boost output current is made of sharp transitions, typical of indirect energy transfer converters (we will see these shapes again in the flyback and buck-boost topologies). The output ripple is "pulsating" whereas the input current, the final bottom waveform, is "nonpulsating." This ensures the boost converter a soft input signature but a noisy output one.

In summary, these comments pertain to the CCM boost converter:

- The output voltage is always greater than the input voltage.
- If we neglect the various ohmic losses, the conversion ratio M is independent of the load current. However, we will discover later that these losses limit M excursion to values around 4 to 5.

- By varying the duty cycle D, we can control the output voltage, as mentioned.
- Operating the boost in CCM brings additional losses as the diode reverse recovery charge needs to be evacuated (t_{rr}). This is seen as an additional loss burden on the power switch SW.
- There is nonpulsating input ripple, but pulsating output current.
- Easy-to-control power switch as the ground is common to one of its terminals.
- The boost converter cannot be protected against short-circuits: a direct path exists between V_{in} and V_{out} via L and D.

Let us now reduce the load and see how the waveforms evolve in discontinuous conduction mode.

1.8.4 Boost Waveforms—DCM

We have now reduced the output load to 200 Ω instead of 5 Ω. Therefore, during the *off* period, the inductor is fully depleted to zero, waiting for SW to turn on again. Since this inductor current also flows in the freewheel diode, we expect the diode to naturally block. As SW is opened and the diode is blocked, we have a third state, of $D_3 T_{sw}$ width, depicted by Fig. 1-31, typical of a DCM operation. The energy stored in the capacitor C solely supplies the load R.

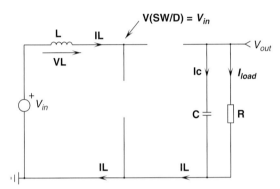

FIGURE 1-31 In DCM, a third state exists where all switches are open.

In Fig. 1-32, we can now clearly see the inductor current going from I_{peak} down to zero, leading to the diode blocking. The diode stays blocked for a duration called the dead time. This is our third state, of $D_3 T_{sw}$ duration. When the diode blocks, the inductor right end terminal becomes open. Because of all the parasitic capacitance around (for instance, the diode and SW parasitic capacitors), a resonating tank is formed. The parasitic capacitor being initally charged to V_{out}, the inductor right end terminal goes back to V_{in} through an oscillation sequence between L and the parasitic capacitor. As observed in waveforms 2 and 3, a sinusoidal waveform appears whose damping depends on the various ohmic losses in the considered mesh.

It is now time to evaluate the transfer ratio M of the boost converter operated in DCM. Still using the volt-inductor balance, but this time, describing the inductor voltage as already done in Eqs. (1-99) and (1-100a):

$$\langle V_L \rangle_{T_{sw}} = \langle V_{in} \rangle_{T_{sw}} - \langle V_{SW/D} \rangle_{T_{sw}} = 0 \tag{1-101}$$

As the input voltage does not change across a switching cycle, we need to evaluate the voltage swing across the common point SW/D which appears in Fig. 1-33. The exercise now consists in deriving the average value of this common point. From Fig. 1-33, we have

$$\langle V_{SW/D} \rangle_{T_{sw}} = D_2 V_{out} + D_3 V_{in} \tag{1-102}$$

48 CHAPTER ONE

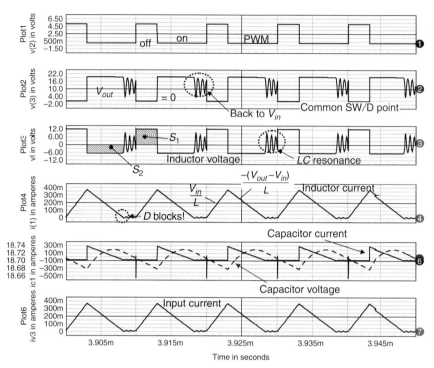

FIGURE 1-32 Typical boost waveforms when operated in discontinuous conduction mode.

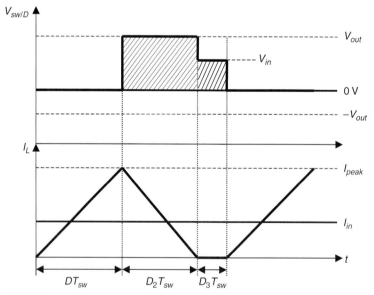

FIGURE 1-33 The common point SW/D and current signals for the boost DCM.

substituting $D_3 = 1 - D - D_2$ in Eq. (1-102), implies that

$$\langle V_{SW/D} \rangle_{T_{sw}} = D_2 V_{out} + (1 - D - D_2) V_{in} \tag{1-103}$$

Since $\langle V_{in} \rangle_{T_{sw}} = \langle V_{SW/D} \rangle_{T_{sw}}$ according to Eq. (1-100a),

$$\frac{V_{out}}{V_{in}} = \frac{D}{D_2} + 1 \tag{1-104}$$

In a boost converter, the average inductor current $\langle I_L \rangle$ represents the average input current I_{in}. Therefore, assuming 100% efficiency, we can write

$$V_{in} I_{in} = V_{out} I_{out} \tag{1-105}$$

Knowing that $I_{out} = \frac{V_{out}}{R}$, we can rewrite Eq. (1-105) by averaging the inductor current:

$$\langle I_L \rangle_{T_{sw}} = \frac{D I_{peak}}{2} + \frac{D_2 I_{peak}}{2} = \frac{V_{out}}{V_{in}} \frac{V_{out}}{R} \tag{1-106}$$

From Eq. (1-106), we extract the off duty cycle D_2:

$$D_2 = \frac{2 V_{out}^2}{V_{in} R I_{peak}} - D \tag{1-107}$$

into which we can substitute the peak current definition I_{peak} involving D_2:

$$I_{peak} = \frac{V_{out} - V_{in}}{L} D_2 T_{sw} \tag{1-108}$$

We plug Eq. (1-108) into (1-107):

$$D_2 = \frac{2 V_{out}^2}{V_{in} R (V_{out} - V_{in}) D_2 T_{sw}} - D \tag{1-109}$$

Solving Eq. (1-109) for D_2 and substituting the result into Eq. (1-104) yield (I confess a program like Mathcad helps a little!)

$$M = \frac{1}{2} \left(1 + \sqrt{1 + \frac{2 T_{sw} R D^2}{L}} \right) \tag{1-110}$$

As for the buck converter, if we define a normalized inductor time constant $\tau_L = \frac{L}{R T_{sw}}$, we finally obtain

$$M = \frac{1 + \sqrt{1 + \frac{2 D^2}{\tau_L}}}{2} \tag{1-111}$$

We can now plot the transfer ratio M versus the duty cycle D, to obtain the curve family of Fig. 1-34 where τ_L takes different values.

For small τ_L values, where the output current is low, for instance (R is large), we are in deep DCM and we can easily reach rather high M ratios, above 5. However, as we increase the load current (τ_L goes up), still keeping the DCM operation, it becomes more difficult to obtain decent M ratios, e.g., in the range of 2 or 3.

In summary, for the boost converter operating in DCM:

- M is now dependent on the load current.
- A certain linearity exists between M and D, as on a buck in CCM.
- For an identical duty cycle, the ratio M can be larger for DCM than for CCM.

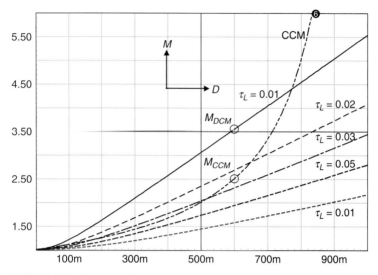

FIGURE 1-34 Various transfer ratios as τ_L is swept.

1.8.5 Boost Transition Point DCM—CCM

In the buck study, we defined the point at which the converter enters or leaves one of the two operating modes, CCM or DCM: when the inductor current ripple "touches" zero, the power switch SW is immediately reactivated and the current ramps up again. Figure 1-35 portrays the current activity in the inductor at the boundary point for the boost converter. In boundary mode, the average value of the symmetric triangle is its peak divided by 2 [Eq. (1-66)].

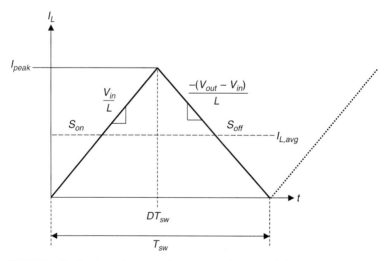

FIGURE 1-35 In a boost, the average *input* current is the average *inductor* current.

From Fig. 1-35 we can define the various slopes:

$$S_{on} = \frac{V_{in}}{L} \quad (1\text{-}112)$$

$$S_{off} = -\frac{V_{out} - V_{in}}{L} \quad (1\text{-}113)$$

From Eq. (1-112), the peak value I_{peak} is found to be

$$I_{peak} = \frac{V_{in}DT_{sw}}{L} \quad (1\text{-}114)$$

Since we are in critical conduction mode, $\langle I_L \rangle_{T_{sw}} = \frac{I_{peak}}{2}$, thus we can rearrange Eq. (1-114):

$$\langle I_L \rangle_{T_{sw}} = \frac{V_{in}DT_{sw}}{2L} \quad (1\text{-}115)$$

Assuming a 100% efficiency for the boost converter gives

$$\langle I_L \rangle_{T_{sw}} = \frac{V_{out}^2}{RV_{in}} \quad (1\text{-}116)$$

Equation (1-100b) teaches us that $\frac{V_{out}}{V_{in}} = \frac{1}{1-D}$. Therefore,

$$\frac{V_{in}DT_{sw}}{2L} = \frac{V_{out}V_{out}}{RV_{in}} = \frac{V_{out}}{R(1-D)} \quad (1\text{-}117)$$

Multiplying both sides by $\frac{1}{V_{in}}$ leads to

$$\frac{DT_{sw}}{2L} = \frac{1}{R(1-D)} \frac{1}{(1-D)} \quad (1\text{-}118)$$

Extracting R and L as critical values gives

$$\mathbf{R}_{critical} = \frac{2F_{sw}L}{D(1-D)^2} \quad (1\text{-}119a)$$

$$\mathbf{L}_{critical} = \frac{RD(1-D)^2}{2F_{sw}} \quad (1\text{-}120a)$$

Rearranging the above equations by replacing D by the boost dc transfer function gives

$$\mathbf{R}_{critical} = \frac{2F_{sw}LV_{out}^2}{\left(1 - \frac{V_{in}}{V_{out}}\right)V_{in}^2} \quad (1\text{-}119b)$$

$$\mathbf{L}_{critical} = \frac{\left(1 - \frac{V_{in}}{V_{out}}\right)V_{in}^2 R}{2F_{sw}V_{out}^2} \quad (1\text{-}120b)$$

1.8.6 Boost CCM Output Ripple Voltage Calculations

Unfortunately, the current circulating in the capacitor no longer looks like a linearly ramping up and down current. Here, energy is first stored in the inductor and then brutally dumped into the capacitor. This event creates the current discontinuity seen in Fig. 1-29. To properly derive the

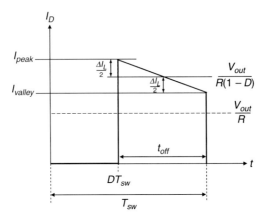

FIGURE 1-36a The current evolution in the output diode.

output ripple, we need to understand the expression of the current flowing into the capacitor. It is no more than the ac portion of the total current flowing through the diode, otherwise expressed by

$$I_C(t) = I_D(t) - \frac{V_{out}}{R_{load}} \quad (1\text{-}121)$$

Figure 1-36a shows the current evolution in the diode, highlighting the dc portion $\frac{V_{out}}{R_{load}}$, which is the output current. The capacitive current is depicted by Fig. 1-36b.

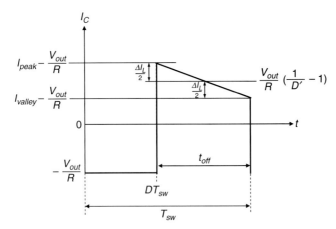

FIGURE 1-36b The diode current minus the dc output.

The simulated ripple appears in Fig. 1-37a. We can see two signal portions, one taking place during the on-time, actually the capacitor depletion, and the other one, evolving as the switch opens, refueling the capacitor. We will write the time equation for this later in the section. As Fig. 1-36a indicates, the middle of the triangle (the inductor ripple) represents the inductor average current $\langle I_L \rangle$. But this average current is routed to the output during the off time only. Therefore, we can write

$$I_{out} = \langle I_L \rangle_{T_{sw}} D' \quad (1\text{-}122)$$

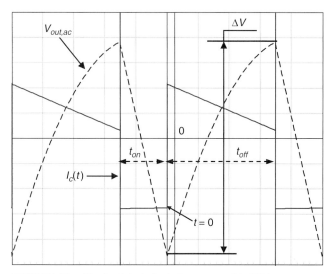

FIGURE 1-37a Simulated ripple for the boost operated in CCM.

Knowing that $I_{out} = \dfrac{V_{out}}{R_{load}}$, we can update and rearrange Eq. (1-122) to reveal the inductor dc component

$$\frac{V_{out}}{R_{load}} \frac{1}{D'} = \langle I_L \rangle_{T_{sw}} \tag{1-123}$$

This point corresponds to $\dfrac{I_{peak} + I_{valley}}{2}$ in Fig. 1-36a. From Eq. (1-121), we know that the capacitive current is actually the diode current minus its dc portion, the output current $I_{out} = \dfrac{V_{out}}{R_{load}}$. As a result, if we subtract I_{out} from all points in Fig. 1-36a, we obtain Fig. 1-36b where the middle point $\dfrac{I_{peak} + I_{valley}}{2}$ becomes

$$\frac{V_{out}}{R_{load}} \frac{1}{D'} - \frac{V_{out}}{R_{load}} = \frac{V_{out}}{R_{load}} \left(\frac{1}{D'} - 1 \right) \tag{1-124}$$

If we now define the function $I_C(t)$ as a function of time bounded between 0 and t_{off}, we can express it as

$$I_C(t) = \Delta I_L \left[\frac{t_{off} - t}{t_{off}} \right] + \frac{V_{out}}{R} \left[\frac{1}{D'} - 1 \right] - \frac{\Delta I_L}{2} \quad \text{for } t \in [0, t_{off}] \tag{1-125}$$

As we did for the buck CCM, we will integrate $I_C(t)$ between 0 and t and divide by C to obtain a voltage expression:

$$\frac{1}{C} \int_0^t I_C(t) \cdot dt = \frac{1}{C} \left[\frac{\Delta I_L}{t_{off}} \left(t_{off} t - \frac{t^2}{2} \right) + \frac{V_{out}}{R_{load}} \left(\frac{1}{D'} - 1 \right) - \frac{\Delta I_L}{2} t \right] \tag{1-126}$$

The peak-to-peak ripple is actually obtained for $t = t_{off}$; therefore, substituting t_{off} into Eq. (1-126) gives

$$\Delta V = \left(\frac{V_{out}}{R_{load}}\left(\frac{1}{D'} - 1\right)t_{off}\right)\frac{1}{C} \tag{1-127}$$

Replacing t_{off} by $D'T_{sw}$ lets us update Eq. (1-127) to

$$\Delta V = \frac{T_{sw} D V_{out}}{R_{load} C} \tag{1-128a}$$

or, normalized to V_{out},

$$\frac{\Delta V}{V_{out}} = \frac{T_{sw} D}{R_{load} C} \tag{1-128b}$$

Well, I know, another example where many lines of brute-force algebra leads to a simple result! For these ripple derivation exercises, it is easier to use the charge Q stored and released by the capacitor. We will show how in several other examples.

1.8.7 Now with the ESR

The ESR appears as a resistor in series with the capacitor, as Fig. 1-26 has already shown. The ac diode current ripple no longer crosses C alone, but a combination of C plus a resistor in series R_{ESR}.
The drop incurred to this element is simply

$$\Delta V_{ESR} = \Delta I_C R_{ESR} \tag{1-129}$$

From Fig. 1-36a and b, ΔI_C corresponds to I_{peak}. Therefore, applying Eqs. (1-95) and (1-113), we have

$$I_{peak} = \frac{V_{out}}{R_{load} D'} - \frac{S_{off}}{2} D'T_{sw} = \frac{I_{out}}{D'} - \frac{(V_{in} - V_{out}) D'T_{sw}}{2L} \tag{1-130}$$

If we use Eq. (1-98) $V_{out} = \frac{1}{1-D} V_{in}$, we obtain the final ESR voltage contribution:

$$\Delta V_{ESR} = \left(\frac{I_{out}}{D'} + \frac{V_{in} DT_{sw}}{2L}\right) R_{ESR} \tag{1-131a}$$

If we divide all terms by V_{out} for the normalization to the output voltage, it becomes

$$\frac{\Delta V_{ESR}}{V_{out}} = \left(\frac{I_{out}}{V_{out} D'} + \frac{V_{in} DT_{sw}}{V_{out} 2L}\right) R_{ESR} = \left(\frac{1}{R_{load} D'} + \frac{D'DT_{sw}}{2L}\right) R_{ESR} \tag{1-132}$$

Depending on the domination of one of the above contributors, either the final ripple curve will look like Fig. 1-37a (almost triangular) for a negligible ESR effect, or the final ripple will transform into a square wave signal, in case of a large ESR contribution.

1.8.8 Boost Ripple, the Numerical Application

In our example, Fig. 1-27, we have the following element values, upgraded with the ESR value placed in series with the capacitor C:

$L = 100\ \mu H$
$C = 10\ \mu F$

INTRODUCTION TO POWER CONVERSION 55

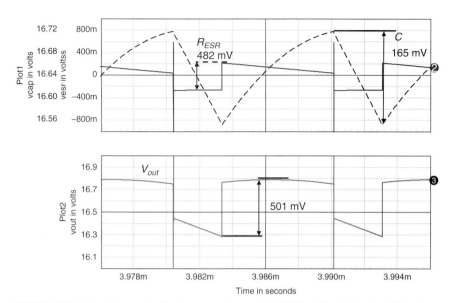

FIGURE 1-37b Simulation results of the boost converter operated in CCM with the addition of an ESR resistor.

$R_{ESR} = 500$ mΩ
$V_{in} = 12$ V
$V_{out} = 16.6$ V
$F_{sw} = 100$ kHz
$D = 0.3$
$R_{load} = 30$ Ω

Applying Eq. (1-128a), we should obtain the peak-to-peak capacitive ripple value

$$\Delta V = \frac{10u \times 0.3 \times 16.6}{30 \times 10u} = 166\, mV \quad (1\text{-}131\text{b})$$

Applying Eq. (1-132) gives the ESR addition

$$\Delta V_{ESR} = \left(\frac{1}{30 \times 0.7} + \frac{0.7 \times 0.3 \times 10u}{2 \times 100u}\right) \times 0.5 \times 16.6$$

$$= (47.6m + 105m) \times 0.5 \times 16.6 = 482\, mV \quad (1\text{-}131\text{c})$$

Figure 1-37b delivers the simulation result and confirms our approach. In this example, the ESR dominates the ripple expression, and the final waveform looks almost rectangular.

1.9 THE BUCK-BOOST CONVERTER

As the boost topology, the buck-boost converter belongs to the family of indirect energy transfer converters. A cycle is thus necessary to first store energy in a coil, then release it to the output capacitor. The buck-boost structure slightly changes the component arrangement, compared to the boost. The switch is now floating, like the buck, and one of the inductor terminals connects to ground. Please note that the diode has changed its orientation, now implying a negative output voltage. Figure 1-38 portrays the buck-boost arrangement:

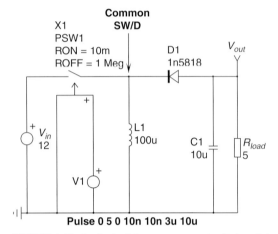

FIGURE 1-38 In the buck-boost converter, the diode-switch junction voltages swing negative at the switch opening.

From inspection of Fig. 1-38, we can again offer several comments. The inductor does not lie in series with the input source or with the output. Its connection to either V_{in} or V_{out} is brutal, via the respective switches SW and D: the buck-boost converter features a very poor input/output current signature! Also, since the power switch reference control point (emitter or source) is floating, it means you need either a P-type MOSFET (or a PNP transistor), or an N-channel (or a NPN bipolar) with a kind of bootstrap circuit to generate the floating V_{GS} signal. This increases the converter complexity and cost.

1.9.1 On-Time Event

When the switch SW closes, V_{in} immediately appears across L, as the for the boost (neglecting the switch voltage drop). As a result, the inductor current increases to a peak value defined by

$$I_{peak} = I_{valley} + \frac{V_{in}}{L} t_{on} \qquad (1\text{-}133a)$$

I_{valley} actually represents the condition for $t = 0$. It can be zero, for the DCM case, or a nonzero value for CCM. Figure 1-39a portrays the situation where the load and capacitor are left alone during the inductor magnetization time. As on the boost, the capacitor supplies the load during this time and will deplete according to the time constant RC.

1.9.2 Off-Time Event

At the switch opening, the inductive current must find a circulating path to keep the ampere-turns constant. The inductor voltage reverses, hence a negative output voltage, and the diode D now conducts to dump the stored energy into the capacitor C (current I_C) and the load (current I_{load}). Figure 1-39b shows the situation. During the off time, the energy stored in L during the

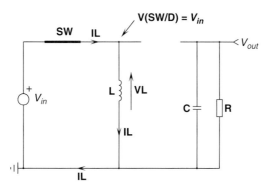

FIGURE 1-39a During the on time, V_{in} is applied across the main inductor L.

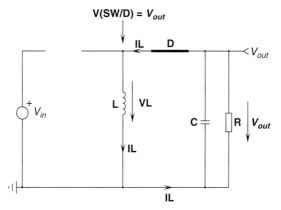

FIGURE 1-39b During the off time, the current circulates in the same direction and the coil voltage reverses, negatively charging C.

on time depletes to a rate described by Eq. (1-133b):

$$I_{valley} = I_{peak} - \frac{V_{out}}{L} t_{off} \qquad (1\text{-}133\text{b})$$

When the new clock cycle occurs, SW closes again and Eq. (1-133a) rules again. Let's unveil the buck-boost characteristic waveforms in CCM.

1.9.3 Buck-Boost Waveforms—CCM

To gain a comprehensive understanding of the *buck*-boost operation, we can observe all individual waveforms brought by Fig. 1-40.

Waveform 1 (in plot 1) represents the PWM pattern, driving the switch *on* and *off*. When the switch SW is *on*, the common point SW/D reaches the input voltage V_{in}. On the contrary, when SW opens, its right terminal swings down to a value imposed by the output voltage plus the forward drop of the diode. On all drawings and analysis, this V_f is ignored. We will depict its role, as well as $R_{DS(on)}$ in Chap. 2.

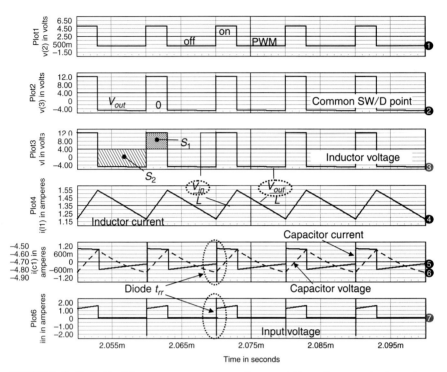

FIGURE 1-40 Typical buck-boost converter waveforms when operated in continuous conduction mode.

Waveform 3 describes how the voltage evolves across the inductor. Please note that waveforms 2 and 3 are identical since the SW/D point connects right to the inductor and is observed referenced to ground, as is the inductor. Sticking to our previous method, we can write that this particular node, SW/D, shall be null in average:

$$\langle V_L \rangle_{T_{sw}} = \langle V_{SW/D} \rangle_{T_{sw}} = 0 \tag{1-134a}$$

which implies that

$$V_{in} D - V_{out} D' = 0 \tag{1-134b}$$

Rearranging this equation gives M, the transfer ratio, where the negative V_{out} sign appears:

$$\frac{V_{out}}{V_{in}} = -\frac{D}{1-D} = M \tag{1-135}$$

Figure 1-41 plots the absolute value of the ratio $|M|$ versus the control duty cycle D.

Almost the same buck comments apply for the buck-boost. Since we operate the converter in CCM, the diode blocking point is exclusively dictated by the *on* activation of SW. As a result, brutally blocking a diode makes it behave as a short-circuit until it fully recovers. A current spike takes place in the power switch. We can see it in Fig. 1-40, in the bottom waveform depicting the input current. Slowing down the SW control voltage will reduce the recovery spike.

Unlike the buck and the boost, the buck-boost output *and* input currents are made of sharp transitions, again typical of indirect energy transfer converters. The output and input ripples

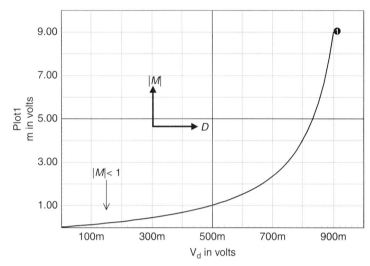

FIGURE 1-41 The transfer ratio excursion for a buck-boost operated in CCM.

are pulsating, naturally requiring adequate filtering to avoid polluting the source and the downstream circuitries powered by the converter.

In summary, for the CCM buck-boost converter

- The output voltage can be either greater or lower than the input voltage.
- As in the other topologies, if we neglect the various ohmic losses, the conversion ratio M appears to be independent of the load current. By varying the duty cycle D, we can control the output voltage, as mentioned.
- Operating the boost in CCM brings additional losses as the diode reverse recovery charge needs to be evacuated (t_{rr}). This is seen as an additional loss burden on the power switch SW.
- There is pulsating input ripple and pulsating output current.
- As for the buck, the power switch "floats," meaning that its control looks more difficult.
- The buck-boost can implement short-circuit protection as SW appears in series with the input source.

Let us now reduce the load and see how the waveforms evolve in discontinuous conduction mode.

1.9.4 Buck-Boost Waveforms—DCM

We have now reduced the output load to 100 Ω instead of 5 Ω. Therefore, during the off period, the inductor is now fully discharged to zero, waiting for SW to energize it again. Since this inductor current also flows in the freewheel diode, the diode naturally blocks. Since SW is opened and the diode blocked, we reach a third state depicted by Fig. 1-42, typical of the DCM operation. Its duration is always denoted D_3T_{sw}. During this dead time, the energy stored in the capacitor C solely supplies the load R.

Observing Fig. 1-43, as for other DCM cases already seen, we can see the inductor current building up during the on time to I_{peak} then going down to zero during the off time, engendering the diode blocking. The upper inductor terminal becomes open and, thanks to the parasitic

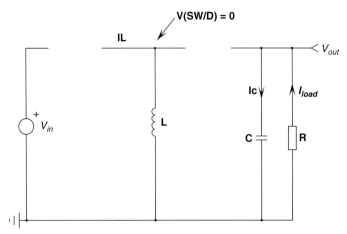

FIGURE 1-42 In DCM, a third state exists where all switches are open.

capacitance around (for instance, the diode and SW parasitic capacitors), a resonating signal starts to appear (waveforms 2 and 3). In theory, the voltage should go back to 0, without oscillation, as L no longer carries current.

It is now time to evaluate the transfer ratio M of the buck-boost converter operated in DCM. Still using the volt-inductor balance, but this time describing the inductor voltage as already

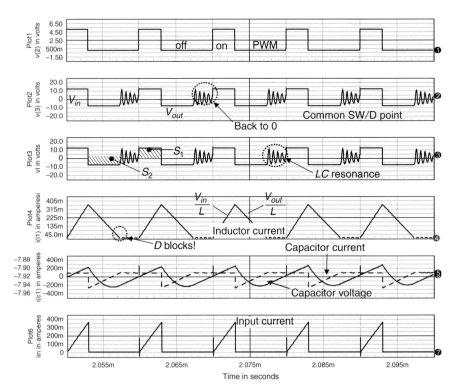

FIGURE 1-43 Typical buck-boost waveforms when operated in discontinuous conduction mode.

done in Eq. (1-134a):

$$\langle V_L \rangle_{T_{sw}} = \langle V_{SW/D} \rangle_{T_{sw}} = 0 \tag{1-136a}$$

The voltage swing across the common point SW/D appears in Fig. 1-44 for a more detailed view. The exercise now consists of deriving the average value of this common point. From Fig. 1-44, we have

$$0 = DV_{in} - D_2 V_{out} \tag{1-136b}$$

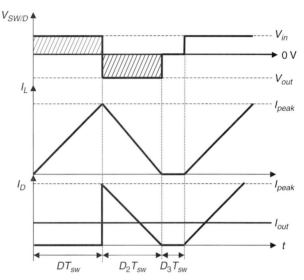

FIGURE 1-44 The common point SW/D and current signals for the buck-boost DCM.

D_2 can be derived when you are looking at the diode current, which only circulates during D_2, that is, during the inductor demagnetization time. The diode average current is nothing else than the dc output

$$\langle I_D \rangle_{T_{sw}} = \frac{I_{peak} D_2 T_{sw}}{2} \frac{1}{T_{sw}} = I_{out} \tag{1-137}$$

Knowing that $I_{out} = \dfrac{V_{out}}{R}$, we can rewrite Eq. (1-137) and unveil D_2:

$$D_2 = \frac{2V_{out}}{RI_{peak}} \tag{1-138}$$

During the on time, the inductor current increases from 0 (we are in DCM) to its peak value:

$$I_{peak} = DT_{sw}\frac{V_{in}}{L} \tag{1-139}$$

We can now substitute Eq. (1-139) into Eq. (1-138) to refine the D_2 definition:

$$D_2 = \frac{2V_{out}}{R}\frac{L}{DV_{in}T_{sw}} = \frac{2V_{out}\tau_L}{DV_{in}} \tag{1-140}$$

with $\tau_L = \dfrac{L}{RT_{sw}}$. From Eq. (1-136b), let us update Eq. (1-140):

$$\frac{2V_{out}\tau_L}{DV_{in}} V_{out} = DV_{in} \tag{1-141}$$

Rearranging the above equation leads to

$$D^2 V_{in}^2 = 2V_{out}^2 \tau_L \tag{1-142}$$

from which we can extract our transfer ratio M, giving back to V_{out} its negative sign:

$$M = -D\sqrt{\frac{1}{2\tau_L}} \tag{1-143}$$

We can now plot the transfer ratio M versus the duty cycle D, to obtain the curve family of Fig. 1-45 where τ_L takes different values.

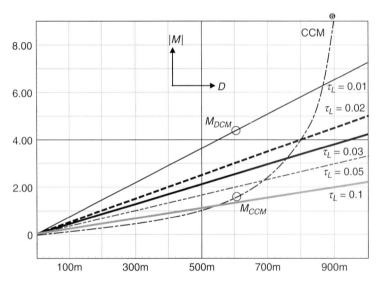

FIGURE 1-45 Various transfer ratios as τ_L is swept.

In Eq. (1-143), despite the presence of the square root, we are dealing with a linear equation, simply because the square root relates to a constant term. This is well illustrated in Fig. 1-45.

For small τ_L values, where the output current is low, for instance (R is large), we are in deep DCM and we can easily reach rather high M ratios. However, as we increase the load current (τ_L goes up), still keeping the DCM operation, it becomes more difficult to obtain large M ratios.

In summary for the buck-boost converter operating in DCM:

- M is now dependent on the load current.
- Linearity exists between M and D, as on a buck in CCM.
- For an identical duty cycle, the ratio M can be larger for DCM than for CCM.

1.9.5 Buck-Boost Transition Point DCM—CCM

In this section, we are going to derive the transition point, where the converter toggles from CCM to DCM, and vice versa. We need to define the point at which the inductor current ripple touches zero, and as the power switch SW is immediately reactivated, the current ramps up again. Figure 1-46 portrays the input current activity of the buck-boost converter.

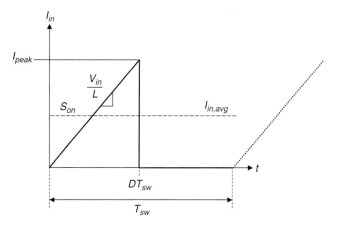

FIGURE 1-46 The buck-boost input current signature in BCM.

From Fig. 1-46 we can define the on current slope:

$$S_{on} = \frac{V_{in}}{L} \qquad (1\text{-}144)$$

From Eq. (1-133a), the peak value I_{peak} is found to be

$$I_{peak} = \frac{V_{in} DT_{sw}}{L} \qquad (1\text{-}145)$$

Observing Fig. 1-46, we write the average input current as

$$\langle I_{in} \rangle_{T_{sw}} = \frac{I_{peak} D}{2} \qquad (1\text{-}146)$$

Substituting Eq. (1-145) into (1-146) reveals

$$\langle I_{in} \rangle_{T_{sw}} = \frac{V_{in} D^2 T_{sw}}{2L} \qquad (1\text{-}147)$$

Assuming a 100% efficiency, then

$$V_{in} \langle I_{in} \rangle_{T_{sw}} = V_{out} I_{out} \rightarrow \frac{V_{in}^2 D^2 T_{sw}}{2L} = \frac{V_{out}^2}{R} \qquad (1\text{-}148)$$

Knowing that $V_{out} = \frac{D}{1-D} V_{in}$, we obtain

$$V_{in}^2 D^2 T_{sw} R = 2L V_{out}^2 \qquad (1\text{-}149)$$

Extracting R and L as critical values gives

$$R = \frac{2LV_{out}^2}{V_{in}^2 D^2 T_{sw}} = \frac{2L}{D^2 T_{sw}}\left(\frac{V_{out}}{V_{in}}\right)^2 = \frac{2LF_{sw}}{(1-D)^2} \tag{1-150}$$

$$\mathbf{R}_{critical} = \frac{2LF_{sw}}{(1-D)^2} \tag{1-151a}$$

$$\mathbf{L}_{critical} = \frac{(1-D)^2 R}{2F_{sw}} \tag{1-151b}$$

We can also rearrange the above equations by replacing D by the buck-boost dc transfer function. Using Eqs. (1-135) gives

$$\mathbf{R}_{critical} = 2LF_{sw}\left(\frac{V_{in}+V_{out}}{V_{in}}\right)^2 \tag{1-151c}$$

$$\mathbf{L}_{critical} = \frac{R}{2F_{sw}}\left(\frac{V_{in}}{V_{in}+V_{out}}\right)^2 \tag{1-151d}$$

1.9.6 Buck-Boost CCM Output Ripple Voltage Calculation

Fortunately, the buck-boost capacitor current is similar to that of the boost converter operating in CCM. Therefore, we do not need to go through all the derivations as we did before. Figure 1-47 represents the buck-boost capacitor current and the output voltage, but multiplied by -1 to look like the boost voltage (the boost output is positive whereas the buck-boost delivers a negative output). Therefore, we can directly reuse equation Eq. (1-128b) seen in the boost section:

$$\frac{\Delta V}{V_{out}} = \frac{T_{sw} D}{R_{load} C} \tag{1-152}$$

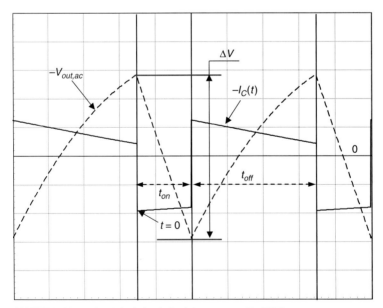

FIGURE 1-47 Simulated ripple for the buck-boost operated in CCM.

We can tweak it to replace the variable D by the buck-boost transfer ratio, as defined in Eq. (1-135),

$$D = \frac{M}{M+1} \quad (1\text{-}153)$$

and obtain the ripple voltage normalized to the output voltage

$$\frac{\Delta V}{V_{out}} = \frac{T_{sw}}{R_{load}C} \frac{M}{M+1} \quad (1\text{-}154)$$

1.9.7 Now with the ESR

The ESR appears like a resistor in series with the capacitor, as Fig. 1-26a has already shown. The ac diode current ripple no longer crosses C alone, but a combination of C plus a resistor in series R_{ESR}. The drop incurred to this element, again, is simply

$$\Delta V_{ESR} = \Delta I_C R_{ESR} \quad (1\text{-}155a)$$

Waveforms from Fig. 1-36a and b are still valid, since the diode current shape is similar to that of the boost. We just need to update the original Eq. (1-130) with the buck-boost S_{off} definition:

$$I_{peak} = \frac{V_{out}}{R_{load}D'} + \frac{S_{off}}{2}D'T_{sw} = \frac{I_{out}}{D'} + \frac{V_{out}D'T_{sw}}{2L} \quad (1\text{-}155b)$$

If we plug in the result of Eq. (1-135), $V_{out} = \frac{D}{1-D}V_{in}$, we obtain the final ESR voltage contribution

$$\Delta V_{ESR} = \left(\frac{I_{out}}{D'} + \frac{V_{in}DT_{sw}}{2L}\right)R_{ESR} \quad (1\text{-}155c)$$

If we divide all terms by V_{out} for the normalization to the output voltage, it becomes

$$\frac{\Delta V_{ESR}}{V_{out}} = \left(\frac{I_{out}}{V_{out}D'} + \frac{V_{in}}{V_{out}}\frac{DT_{sw}}{2L}\right)R_{ESR} = \left(\frac{1}{R_{load}D'} + \frac{D'T_{sw}}{2L}\right)R_{ESR} \quad (1\text{-}155d)$$

Depending on the domination of one of the above contributors, either the final ripple curve will look like Fig. 1-47 (almost triangular) for a negligible ESR effect, or the final ripple will transform into a square wave signal, in the case of a large ESR contribution.

1.9.8 Buck-Boost Ripple, the Numerical Application

In our example, Fig. 1-38, we have the following element values, upgraded with the ESR value placed in series with the capacitor C:

$$L = 100\ \mu H$$
$$C = 10\ \mu F$$
$$R_{ESR} = 500\ m\Omega$$
$$V_{in} = 12\ V$$
$$V_{out} = 4.68\ V$$
$$F_{sw} = 100\ kHz$$

$$D = 0.3$$
$$R_{load} = 10 \, \Omega$$

Applying Eq. (1-128a), we should obtain the peak-to-peak capacitive ripple value

$$\Delta V = \frac{10u \times 0.3 \times 4.68}{10 \times 10u} = 140 \, mV \tag{1-156}$$

Applying Eq. (1-155d) gives the ESR addition

$$\Delta V_{ESR} = \left(\frac{1}{10 \times 0.7} + \frac{0.7 \times 10u}{2 \times 100u} \right) \times 0.5 \times 4.68$$
$$= (142.8m + 35m) \times 0.5 \times 4.68 = 416 \, mV \tag{1-157}$$

Figure 1-48 gives the simulation result and confirms our approach. In this example, the ESR dominates the ripple expression, and the final waveform looks almost rectangular.

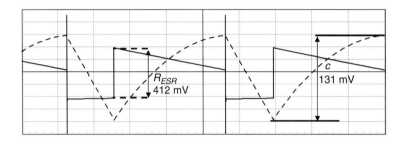

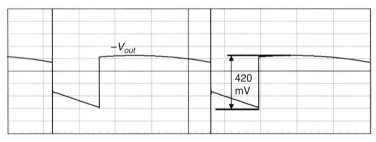

FIGURE 1-48 Simulation results of the buck-boost operated in CCM with the addition of an ESR resistor.

1.10 INPUT FILTERING

Switched-mode converters are inherently noisy and can interfere with equipment sharing the same supply lines. This is particularly true in automotive applications, telecommunication equipment, measurement crates, and so on. We have seen from our previous studies related to simple topologies such as buck, buck-boost, and boost that their respective input current signature (or the way the input current evolves with time) can be either smooth or pulsating. Trying to supply a sensitive circuitry like a sample and hold from a line where a buck converter is already connected can be seen as a difficult exercise if no precaution is taken. A filtering circuit must be inserted between the supply output and the converter input. Most of the time, this

INTRODUCTION TO POWER CONVERSION 67

electromagnetic interference (EMI) filter will be reactive. The problem comes from the nature of the load applied to the filter output: the converter's input. Let's take a closer look at it.

In a closed-loop system, the loop strives to maintain a constant output power regardless of the operating conditions. If you insert an ammeter in series with your converter and modulate the input line, you will see that if V_{in} grows, the input current diminishes (this makes sense, the supply strives to keep P_{out} constant) while a drop of V_{in} is followed by an increase of I_{in}. From this input side, the converter looks like a negative resistor! Please keep in mind that the same converter operated in open-loop loses its negative input impedance.

If we assume 100% efficiency, we can describe this behavior via a few lines of algebra:

$$P_{in} = P_{out} = I_{in}V_{in} = I_{out}V_{out} \qquad (1\text{-}158)$$

Instead of considering M, the dc transfer ratio, let's consider μ, the inverse of M, obtained by rearranging Eq. (1-158):

$$\frac{V_{in}}{V_{out}} = \frac{I_{out}}{I_{in}} = \mu \qquad (1\text{-}159)$$

The static input resistance of the converter can simply be calculated from

$$R_{in} = \frac{V_{in}}{I_{in}} \qquad (1\text{-}160)$$

What interests us is the *incremental* resistance

$$R_{in,inc} = \frac{dV_{in}}{dI_{in}} \qquad (1\text{-}161)$$

This expression represents the way this resistance moves with perturbations. Via simple manipulations, the incremental resistance can be formulated:

$$V_{in} = \frac{P_{in}}{I_{in}} \qquad (1\text{-}162)$$

$$P_{in} = P_{out} = R_{load}I_{out}^2 \qquad (1\text{-}163)$$

Plugging Eq. (1-163) into (1-162) and deriving as Eq. (1-161) suggests yield

$$\frac{dV_{in}}{dI_{in}} = \frac{d}{dI_{in}}\frac{R_{load}I_{out}^2}{I_{in}} \qquad (1\text{-}164)$$

$$R_{in,inc} = -R_{load}\frac{I_{out}^2}{I_{in}^2} = -R_{load}\mu^2 \qquad (1\text{-}165)$$

Please note that R_{load} can represent another switching regulator, a simple resistor, or a linear regulator. Therefore, Eq. (1-165) can be reformulated as

$$R_{in,inc} = -\frac{V_{out}}{I_{out}}\mu^2 = -\frac{V_{out}^2}{P_{out}}\mu^2 \qquad (1\text{-}166)$$

1.10.1 The *RLC* Filter

Equation (1-166) teaches us that our *closed-loop* converter input resistance looks negative. What are the implications of this result? Well, since we need to connect a low-insertion-loss filter, we will probably choose a filter made of L and C elements, rather than a filter made of R and C elements. A filter we could think of is depicted in Fig. 1-49. We can see the converter and the filter placed between the supply and the converter input.

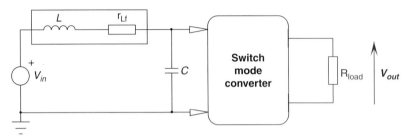

FIGURE 1-49 The *LC* filter is inserted in series with the input line.

The inductance L appears in series with a resistor r_{Lf}, testifying for its dc ohmic losses. This *RLC* combination forms a resonating tank. From the theory, we know that when an *RLC* network is excited on its input by voltage pulses, a decaying sinusoidal waveform appears on the output. The transfer function of the *RLC* network can be described by the well-known second-order function

$$T(s) = \frac{1}{LCs^2 + RCs + 1} \quad (1\text{-}167)$$

By introducing the resonant frequency ω_0, we can rearrange Eq. (1-167) in a more familiar format:

$$T(s) = \frac{1}{\frac{s^2}{\omega_0^2} + 2\zeta\frac{s}{\omega_0} + 1} \quad (1\text{-}168a)$$

where

$$\omega_0 = \frac{1}{\sqrt{LC}} \text{ is the nondamped natural oscillation} \quad (1\text{-}168b)$$

$$\zeta = R\sqrt{\frac{C}{4L}} \text{ represents the damping factor, pronounced "zeta"} \quad (1\text{-}168c)$$

The characteristic impedance Z_0 can be introduced as

$$Z_0 = \sqrt{\frac{L}{C}} \quad (1\text{-}168d)$$

Again, if we decide to deal with the quality coefficient Q, rather than the damping factor, Eq. (1-168a) can be reformulated as

$$T(s) = \frac{1}{\frac{s^2}{\omega_0^2} + \frac{s}{\omega_0 Q} + 1} \quad (1\text{-}169a)$$

where

$$Q = \frac{1}{2\zeta} \quad (1\text{-}169b)$$

If we now excite the input with a unit-step function, we obtain the following Laplace equation:

$$Y(s) = \frac{1}{\frac{s^2}{\omega_0^2} + 2\zeta\frac{s}{\omega_0} + 1} \cdot \frac{1}{s} \quad (1\text{-}170)$$

INTRODUCTION TO POWER CONVERSION

To obtain the response in the time domain, we need to use the inverse Laplace transform applied to Eq. (1-170). This has been derived numerous times, and the result appears below [7]:

$$Y(t) = 1 - \frac{e^{-\zeta \omega_0 t}}{\sqrt{1 - \zeta^2}} \sin(\omega_d t + \theta) \quad \text{for } \zeta < 1 \quad (1\text{-}171\text{a})$$

where

$$\omega_d = \omega_0 \sqrt{1 - \zeta^2} \quad (1\text{-}171\text{b})$$

$$\theta = \cos^{-1}(\zeta) \quad (1\text{-}171\text{c})$$

In Eq. (1-168a), the damping factor zeta plays an important role, as Fig. 1-50 shows. It plots the output voltage of an *RLC* filter ($L = 1$ μH, $C = 1$ μF), excited by a 1 V input step. The damping factor is swept from 0.1 up to 0.9. As zeta increases, the oscillation amplitude decays faster to stabilize around the 1 V input step. For zeta exhibiting lower values, we have an oscillating answer.

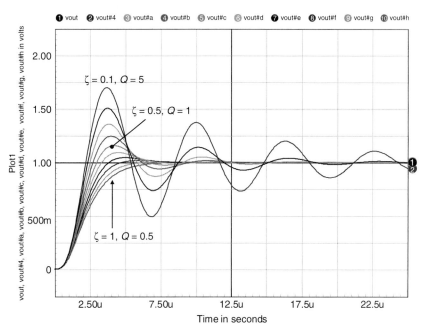

FIGURE 1-50 The damping factor is swept between 0.1 and 0.9.

The poles affecting Eq. (1-168a), denoted p_1 and p_2, represent the denominator roots for which $s^2 + 2\zeta\omega_0 s + \omega_0^2 = 0$. Depending on zeta value, these poles affect the stability of a system described by Eq. (1-170):

- $\zeta < 0$—In that case, poles affecting Eq. (1-170) feature a positive real portion. Whatever the excitation level, the transient response diverges.
- $\zeta = 0$—This particular case implies two pure imaginary poles $p_{1,2} = \pm j\omega_0$, making the system output permanently oscillating (no decay).
- $\zeta > 0$—The two poles now have a real portion (ohmic losses), and the system exhibits different responses depending on whether $\zeta > 1$ (overdamping), $\zeta = 1$ (critical damping), or $0 < \zeta < 1$ for which we obtain a decaying oscillating response.

1.10.2 A More Comprehensive Representation

Since our capacitors and inductors feature ohmic losses, Fig. 1-49 needs to be updated to reflect the reality (Fig. 1-51). However, Eq. (1-167) is no longer valid since elements have been added. The new transfer function can be derived by using matrix analysis, as detailed at the end of this chapter in Appendix 1A. We obtain the following equation:

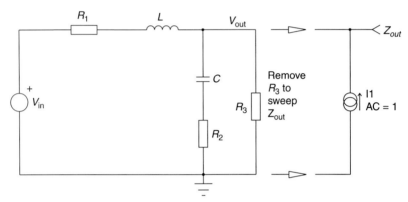

FIGURE 1-51 The complete *RLC* filter featuring parasitic elements. If R_3 is replaced by a 1 A ac source, it is possible to obtain $Z_{out}(s)$, the output impedance.

$$T(s) = \frac{R_3}{R_1 + R_3} \frac{1 + sR_2C}{s^2LC\left(\frac{R_3 + R_2}{R_1 + R_3}\right) + s\frac{L + C(R_2R_3 + R_1R_3 + R_2R_1)}{R_1 + R_3} + 1} \quad (1\text{-}172)$$

The exercise now lies in identifying the Eq. (1-168a) elements to unveil the new damping factor definition and the resonant frequency value. By looking at the original equation, Eq. (1-168a), we have

$$\omega_0 = \frac{1}{\sqrt{LC}}\sqrt{\frac{R_1 + R_3}{R_2 + R_3}} \quad (1\text{-}173)$$

$$\zeta = \frac{L + C(R_2R_3 + R_1R_3 + R_2R_1)}{2(R_1 + R_3)} \omega_0 \quad (1\text{-}174)$$

and, according to Eq. (1-169b):

$$Q = \frac{R_1 + R_3}{L + C(R_2R_3 + R_1R_3 + R_2R_1)} \frac{1}{\omega_0} \quad (1\text{-}175)$$

Figure 1-51 represents a practical implementation of Fig. 1-49 where resistor R_3 illustrates the loading of this filter: this is the input impedance of the switch-mode power converter. The instability exists if a particular combination of R_3 cancels the damping factor or makes it negative, as we have seen in the bullets. To see what R_3 value brings this particular result, let us null the numerator of Eq. (1-174). To simplify the equation, we can consider $R_1 \ll R_3$ and $R_2 \ll R_3$, making ω_0 a constant term equal to $\frac{1}{\sqrt{LC}}$. Hence,

$$L + C(R_2R_3 + R_1R_3 + R_2R_1) = 0 \quad (1\text{-}176)$$

which leads to

$$R_3 = -\frac{R_1 R_2 C + L}{C(R_1 + R_2)} \qquad (1\text{-}177)$$

As a result, if R_3 becomes negative, as will occur thanks to Eq. (1-165), oscillations can take place and jeopardize the full conversion chain. Loading an LC filter with a negative resistor is a well-known technique to create oscillators! Techniques include negative impedance converters (NICs), tunnel effect diodes, Gunn diodes, etc. (I know, who remembers tunnel diodes these days?)

Let's run a quick simulation based on a numerical example where $R_1 = 100$ mΩ, $R_2 = 500$ mΩ, $C = 1$ µF, and $L = 100$ µH. Applying Eq. (1-177), we obtain zeta cancellation for R_3 equal to -166.75 Ω. If we plug these values in Fig. 1-51 and run the simulator, Fig. 1-52 appears, showing diverging oscillations ($R_3 = -150$ Ω, $\zeta < 0$), steady-state oscillations ($R_3 = -166$ Ω, $\zeta = 0$), or decaying oscillations ($R_3 = -175$ Ω, $\zeta > 0$).

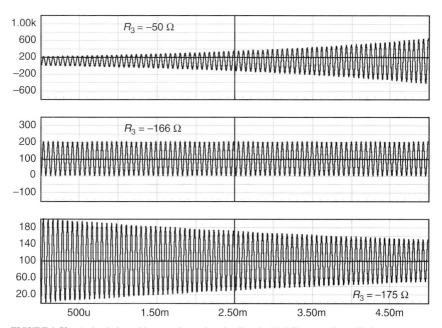

FIGURE 1-52 A simulation with a negative resistor loading the RLC filter unveils oscillating responses.

1.10.3 Creating a Simple Closed-Loop Current Source with SPICE

Using an in-line equation is a rather easy way to mimic a closed-loop converter. In our converter, the output and input power are kept constant and equal, assuming 100% efficiency. The input current is thus permanently adjusted to satisfy this fact: that is the origin of the negative incremental resistance. A voltage-controlled current source can do the job if the current value equals the power divided by the current source terminal voltage: $I = P/V$. If P is a fixed number, and V changes, I automatically adjusts to satisfy P equals constant. Figure 1-53 shows how to wire the current source.

SPICE hosts an analysis primitive called .TF which performs a transfer function calculation from one point to a source reference, V_1 in our case. Here, the 60 W converter (hence

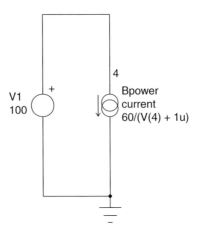

FIGURE 1-53 An in-line equation quickly mimics a closed-loop converter.

the number 60 on the B_{power} source), is supplied from a 100 V input. If we run the simulator, *IsSpice* in this example, we obtain the following netlist extracted from the output file (*.OUT extension):

```
.TF V(4) V1 ; transfer function analysis
***** SMALL SIGNAL DC TRANSFER FUNCTION
output_impedance_at_V(4)    0.000000e+000
v1#Input_impedance         -1.66667e+002
Transfer_function           1.000000e+000
```

As the second line details, we really have a negative impedance ($-166\ \Omega$) seen from the input source V_1. The same simulation could be run with *PSpice*, but the B_{power} controlled current source syntax would change to

```
G1 4 0 value = { 60 / (V(4)+1) }
```

Please note the presence of the "1" value, placed here to avoid any divide-by-zero error in case the node 4 level is null.

The other example consists of connecting the constant power load to the *RLC* filter, as Fig. 1-54 portrays. The input represents a rectified ac voltage moving from 150 V down to 100 V in successive steps. This situation can arise if a perturbation or a severe loading occurs on a distribution network. Unfortunately, the negative loading will affect the damping ratio, and oscillations will occur as Fig. 1-55 shows. When we decrease the input voltage, the feedback loop adjusts the loading current which affects the damping factor. As long as ζ stays positive, oscillations decay. However, for a null ζ, oscillations are permanent. The situation becomes worse as ζ becomes negative. A diverging situation takes place, bound by clamping effects brought by protection elements. If no precaution is taken, a heavy smoke ends the oscillating process!

1.10.4 Understanding Overlapping Impedances

One of the design criteria, as demonstrated by Dr. Middlebrook [8], is that the output impedance of the filter $Z_{outFILTER}$ must be much smaller than the input impedance of the filtered converter Z_{inSMPS}:

$$Z_{outFILTER} \ll Z_{inSMPS} \qquad (1\text{-}178)$$

INTRODUCTION TO POWER CONVERSION 73

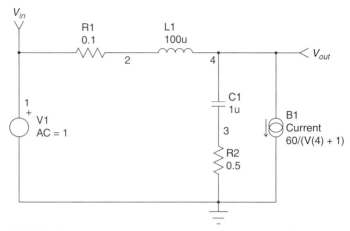

FIGURE 1-54 Connecting the negative impedance load on the *RLC* filter output.

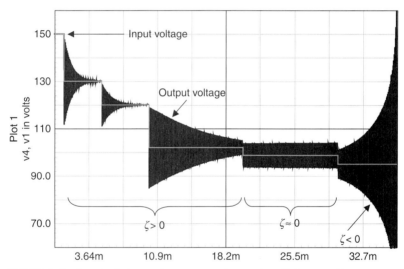

FIGURE 1-55 Stepping the input voltage can engender several oscillating cases if no precautions are taken.

In dc conditions, we can already check whether this condition is met. The dc condition in Fig. 1-54 implies that L is shorted and C open. Therefore, R_1 (or r_{Lf}, the inductive ohmic losses) is left in series with the input source, and this is our *RLC* filter dc output impedance:

$$Z_{outFILTER,dc} = r_{Lf} \tag{1-179a}$$

The converter static input impedance can be found, given the input power and the operating voltage:

$$P_{in} = \frac{P_{out}}{\eta} = \frac{V_{in}^2}{Z_{inSMPS}} \tag{1-180}$$

From the above equation, we can extract the dc input impedance:

$$Z_{inSMPS,dc} = \frac{V_{in}^2 \eta}{P_{out}} \tag{1-181a}$$

If we go back our example, let's see what dc impedances we were talking about:

$$Z_{outFILTER,dc} = 100\,m\Omega \quad \text{or} \quad 20\log_{10}(100m) = -20\,dB\Omega \tag{1-179b}$$

$$Z_{inSMPS,dc} = \frac{100^2 \times 1}{60} = 166\,\Omega \quad \text{or} \quad 20\log_{10}(166) = 44.4\,dB\Omega \tag{1-181b}$$

Apparently, the Eq. (1-178) criterion is perfectly respected given the difference between the equation results. However, as with any tuned filter, the ω_0 peaking affects the output impedance, through the quality coefficient Q. The exercise now consists of calculating the output impedance of the Fig. 1-51 network. Thanks to our matrix approach (detailed in Appendix 1, we can immediately extract it from the final T expression: $Z_{outFILTER} = T_{2,2}$, otherwise stated as

$$Z_{outFILTER} = \frac{(R_2 sC + 1)(sL + R_1)R_3}{s^2 LC\,(R_3 + R_2) + s[(R_2 R_3 + R_3 R_1 + R_1 R_2)C + L] + R_1 + R_3} \tag{1-182}$$

This term, however, includes R_3, representative of the load. As we need the output impedance, without R_3, we can update the impedance in a no-load condition:

$$\lim_{R_3 \to \infty} Z_{outFILTER} = \frac{s(L + R_1 R_2 C) + s^2 LCR_2 + R_1}{s^2 LC + sC(R_1 + R_2) + 1} \tag{1-183}$$

As we have Laplace terms, let's extract the Eq. (1-183) module, replacing s by $j\omega$:

$$\left\| Z_{outFILTER} \right\| = \sqrt{\frac{[(CR_2\omega)^2 + 1][(\omega L)^2 + R_1^2]}{\omega^2\left(L^2 C^2 \omega^2 - 2LC + R_2^2 C^2 + 2R_2 R_1 C^2 + R_1^2 C^2\right) + 1}} \tag{1-184}$$

Fortunately, this expression reduces to R_1 for $\omega = 0$.

What is of interest to us is the peaking point at which the output impedance is maximum. This point is reached for $\omega_0 = \dfrac{1}{\sqrt{LC}}$. Therefore, we can update Eq. (1-184) by replacing ω with ω_0. After rearranging all terms, we obtain a much simpler expression:

$$\left\| Z_{outFILTER} \right\|_{max} = \sqrt{\frac{(CR_2^2 + L)(CR_1^2 + L)}{C^2(R_2 + R_1)^2}} \tag{1-185}$$

For $R_2 = 0$ (e.g., almost no ESR in series with the capacitor), Eq. (1-185) simplifies to

$$\left\| Z_{outFILTER} \right\|_{max} = \sqrt{\frac{L(CR_1^2 + L)}{C^2 R_1^2}} \tag{1-186}$$

If we express the characteristic impedance of the RLC network by

$$Z_0 = \sqrt{\frac{L}{C}} \tag{1-187}$$

and we substitute it into Eq. (1-186), we obtain Eq. (1-188a):

$$\left\| Z_{outFILTER} \right\|_{max} = \frac{Z_0^2}{R_1}\sqrt{1 + \left(\frac{R_1}{Z_0}\right)^2} \tag{1-188a}$$

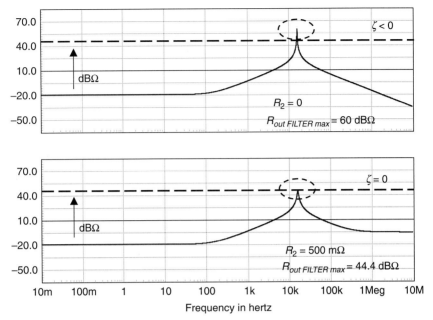

FIGURE 1-56 Connecting an ac source of 1 A on the *RLC* filter output unveils its output impedance.

This equation gives the maximum output impedance value reached at the resonance. To fulfill Eq. (1-188a), we now must be sure that despite the peaking region, we always have a large difference between $Z_{outFILTER}$ and Z_{inSMPS} over the frequency of interest. Figure 1-56 plots the output impedance of the Fig. 1-54 circuit. As we have seen, to get the output impedance of a network, we connect an ac source I_{ac} of 1 A amplitude on the filter output. The output impedance is thus directly V_{out} as I_{ac} equals 1. If we use our previous numerical values with Eq. (1-188a), we reach the following peak:

$R_1 = 100$ mΩ, $L = 100$ μH, $C = 1$ μF   and $R_2 = 0$ or $R_2 = 500$ mΩ

According to Eqs. (1-168d) and (1-168b),

- $Z_0 = \sqrt{\dfrac{100u}{1u}} = 10\,\Omega$ \hfill (1-188b)

- $f_0 = \dfrac{1}{2\pi\sqrt{LC}} = 15.9\,kHz$ \hfill (1-188c)

- $R_2 = 0$, then $\left\|Z_{outFILTER}\right\|_{max} = \dfrac{10^2}{100m}\sqrt{1 + \left(\dfrac{100m}{10}\right)^2} = 1000\,\Omega = 60\,dB\Omega$ \hfill (1-188d)

- $R_2 = 500$ mΩ, then

$$\left\|Z_{outFILTER}\right\|_{max} = \sqrt{\dfrac{(1u \times 0.5^2 + 100u) \times (1u \times 0.1^2 + 100u)}{1u^2 \times (0.5 + 0.1)^2}}$$

$$= \sqrt{\dfrac{100.25u \times 100u}{0.36p}} = 166.8\,\Omega = 44.4\,dB\Omega \tag{1-188e}$$

One can see that the above numbers are confirmed by the simulation. Just a precision: as the resonant peak is rather sharp, make sure enough data points per decade are collected by the simulator during the ac analysis; e.g., put this variable to 1000 to gain in finesse.

An instability is likely to occur if both curves, SMPS input impedance and filter output impedance, cross at some point. If no intersection ever occurs, then the configuration is said to be stable. On the upper curve, where R_2 has been reduced to zero, the peaking is maximum. When the resistor R_2 starts to increase, the peaking lowers and contributes to increase the stability. Should R_2 grow further, the instability would completely go away: we would "damp" the RLC filter. Damping is a technique used to crush the quality coefficient in such a way that both impedance curves never intersect. However, care must be taken not to compromise the efficiency. Inserting a resistor in series with C_1 is a possible way, but at the expense of power dissipation. We will see shortly how to damp the filter in a more appropriate way. Once it is damped, new ac sweeps have to be run to confirm that stability is unconditional.

In our example, the input impedance is a static number. However, in a closed-loop system, a gain affects the transmission loop. As long as we have gain in the chain, a negative incremental resistance exists. When the gain collapses below 0 dB, we run open-loop and the negative effect disappears. The static, flat, SMPS input impedance will actually not be horizontal in reality as a compensation network affects the gain loop over frequency. The SMPS input impedance is therefore likely to be the site of peaks and valleys.

1.10.5 Damping the Filter

To damp the filter means to reduce the quality coefficient Q in such a way that the peaking no longer jeopardizes the stability. A good solution consists in inserting a resistor R_{damp} in parallel with the output load; this technique is called *parallel damping* (series damping also exists but is beyond the scope of this chapter). We actually come back to our previous RLC filter, as depicted by Fig. 1-57, where R_3 portrays the damping resistor R_{damp} in parallel with the SMPS input resistance.

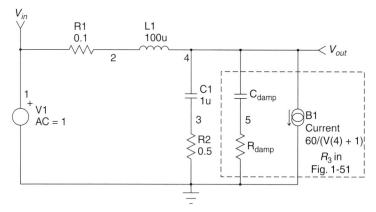

FIGURE 1-57 A resistor is added as a filter load to help reduce the quality coefficient, hence the peaking.

However, to avoid increasing the dc consumption (V_{out} is actually a continuous voltage), a capacitor C_{damp} is placed in series with R_{damp} and stops the dc component. Usually C_{damp} is chosen as

$$C_{damp} = 10 C_1 \tag{1-189}$$

The quality coefficient will be selected to be around 1, meaning that in Eq. (1-175) the numerator equals the denominator. Otherwise stated,

$$L + C(R_2R_3 + R_1R_3 + R_2R_1) = (R_1 + R_3)\frac{1}{\omega_0} \quad (1\text{-}190)$$

We can now extract the R_3 value which will lead to $Q \approx 1$, assuming that $\omega_0 = \frac{1}{\sqrt{LC}}$ ($R_1 \ll R_3$ and $R_2 \ll R_3$):

$$R_3 = \frac{R_1 - \omega_0(L + R_1R_2C)}{2R_1C\omega_0 - 1} \quad (1\text{-}191)$$

Now, knowing that the R_3 resistor actually corresponds to $R_{damp} \parallel Z_{inSMPS,dc}$, we need to extract the final value of our damping resistor R_{damp}:

$$R_{damp} \parallel Z_{inSMPS,dc} = \frac{R_1 - \omega_0(L + R_1R_2C)}{2R_1C\omega_0 - 1} \quad (1\text{-}192)$$

$$R_{damp} = Z_{inSMPS,dc} \frac{L + CR_1R_2 - \frac{R_1}{\omega_0}}{\frac{Z_{inSMPS,dc}}{\omega_0} + L + CR_2R_1 - \frac{R_1}{\omega_0} - 2Z_{inSMPS}CR_1} \quad (1\text{-}193)$$

Back to the numerical values of Fig. 1-54, we have $R_1 = 100$ mΩ, $R_2 = 500$ mΩ, $C = 1$ μF, $L = 100$ μH, and $Z_{inSMPS,dc} = 166$ Ω. Therefore, Eq. (1-193) indicates that the damping resistor must be

$$R_{damp} = 9.52 \text{ Ω}$$

$$C_{damp} = 10 \text{ μF}$$

Let us plug these values into the original *RLC* filter and have a transmittance sweep, as shown in Fig. 1-58.

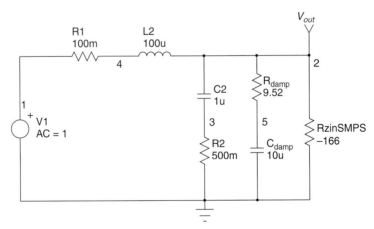

FIGURE 1-58 The updated *RLC* filter featuring the negative loading and its damping elements.

Figure 1-59 unveils the results and confirms the absence of peaking. A damping capacitor of 10 times the original value can sometimes be unacceptably large. As you can see in Fig. 1-59,

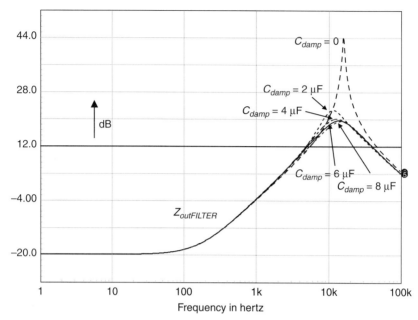

FIGURE 1-59 The output impedance sweep confirms the absence of resonance peak when a capacitor of the right value is placed in series with R_{damp}.

capacitors values down to four times the filter capacitor can also provide good damping. We can assume the filter is properly damped and instabilities are cured. To confirm this hypothesis, we can rerun the Fig. 1-54 input steps excitation and see how the output behaves (Fig. 1-60):

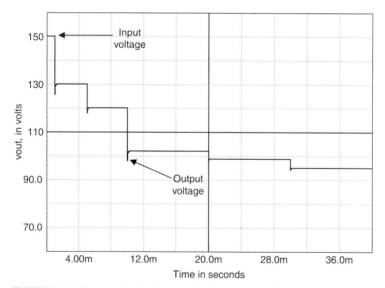

FIGURE 1-60 Thanks to the damping network, all unwanted oscillations have disappeared!

INTRODUCTION TO POWER CONVERSION

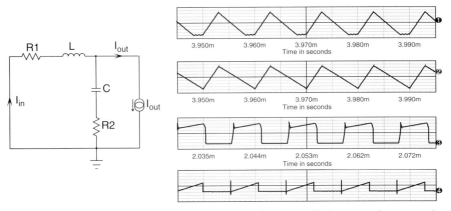

FIGURE 1-61 The load is replaced by a current source whose shape and amplitude represent the converter signature. The right side gathers several possible converter input current shapes such as DCM or CCM.

1.10.6 Calculating the Required Attenuation

Now that we have seen the possible interactions between the filter and the negative loading, let us take a look at how to calculate the corner frequency, given an input current ripple specification. For instance, how should we position f_0 to reach a 5 mA rms current in the main source when a noisy converter generating 1 A pulses loads the filter output? We first need to find an equivalent model of the *RLC* filter and the power supply. A simplified version is presented in Fig. 1-61, where the input source is actually shorted, as what matters is the alternating current flowing into it. The load is replaced by a pulsed current source, whose shape and frequency depend on the converter. It can be pulsating, nonpulsating, discontinuous, critical, etc. The damping filter has been removed for the sake of clarity and simplification.

Please note that all parasitic elements are kept in this model as they clearly affect the attenuation ability of the filter. Since we have two impedances actually in parallel, we can redraw the model in a more convenient way, as Fig. 1-62 shows.

Our task consists of finding the relationship between the inductive current I_{in} and the converter current I_{out}. Applying the current division law in Fig. 1-62, we can write

$$I_{in} = I_{out} \frac{R_2 + \frac{1}{sC}}{R_2 + \frac{1}{sC} + R_1 + sL} \quad (1\text{-}194)$$

Rearranging this equation unveils a familiar format:

$$\frac{I_{in}}{I_{out}} = \frac{1 + R_2 Cs}{s^2 LC + sC(R_1 + R_2) + 1} \quad (1\text{-}195)$$

Equation (1-195) reveals a classical second-order low-pass filter affected by a zero, due to the capacitor ESR R_2. As a result, despite the square wave shape of the converter signature, we can assume that all harmonics will be blocked by the filter and only the attenuated fundamental will pass. This approximation is called the *first harmonic approximation* (FHA) and consists of analyzing the circuit by using

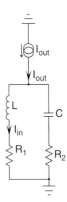

FIGURE 1-62 Another way to represent the model.

sinusoidal signals. We can therefore replace s by $j\omega$, were ω will represent the fundamental pulsation. We will see later on how we obtain the fundamental from the current signature.

Introducing imaginary notation in Eq. (1-195) requires that we extract the module, to finally manipulate the result:

$$\left\|\frac{I_{in}}{I_{out}}\right\| = \sqrt{\frac{R_2^2 + \frac{1}{(\omega C)^2}}{(R_1 + R_2)^2 + \frac{1}{(\omega C)^2} - \frac{2L}{C} + (\omega L)^2}} \quad (1\text{-}196)$$

If we neglect R_1 and R_2, we can simplify this result:

$$\left\|\frac{I_{in}}{I_{out}}\right\| = \sqrt{\frac{1}{1 - 2\left(\frac{\omega}{\omega_0}\right)^2 + \left(\frac{\omega}{\omega_0}\right)^4}} = \sqrt{\frac{1}{\left[1 - \left(\frac{\omega}{\omega_0}\right)^2\right]^2}} = \frac{1}{\left|1 - \left(\frac{\omega}{\omega_0}\right)^2\right|} \approx \left(\frac{\omega_0}{\omega}\right)^2 \quad (1\text{-}197)$$

From this equation, we need to extract ω_0 in order to properly position the cutoff frequency in relationship with the needed attenuation $\frac{I_{in}}{I_{out}}$. This is what Eq. (1-198) describes with A_{filter} being the required attenuation:

$$\omega_0 = \sqrt{A_{filter}} \cdot \omega \quad (1\text{-}198)$$

Equation (1-198) will help us position the LC filter cutoff frequency in relation to the input ripple design constraint.

1.10.7 Fundamental Frequency Evaluation

In the FHA technique, since all harmonics are removed, what matters is the fundamental amplitude of the converter current signature. This amplitude, of course, depends on several parameters such as rise time and duty cycle. The Fourier series for a periodic function $f(t)$ of period T_{sw} can be written in the following form:

$$f(t) = a_0 + \sum_{n=1}^{\infty}(a_n \cos n\omega t + b_n \sin n\omega t) \quad (1\text{-}199)$$

where the coefficients of this series are defined by

$$a_0 = \frac{1}{T_{sw}}\int_0^{T_{sw}} f(t) \cdot dt \qquad \text{this is the } \textit{average} \text{ value or the dc term} \quad (1\text{-}200)$$

$$a_n = \frac{2}{T_{sw}}\int_0^{T_{sw}} f(t)\cos n\omega t \cdot dt \quad (1\text{-}201)$$

$$b_n = \frac{2}{T_{sw}}\int_0^{T_{sw}} f(t)\sin n\omega t \cdot dt \quad (1\text{-}202)$$

and correspond to the Fourier coefficients. The designer task thus consists in deriving the fundamental term ($n = 1$) for the considered waveform, i.e., the input current signature $i(t)$:

$$i_{fund} = a_1 \cos \omega t + b_1 \sin \omega t \quad (1\text{-}203)$$

$$i_{fund,rms} = \sqrt{\frac{a_1^2 + b_1^2}{2}} \qquad (1\text{-}204)$$

Suppose that the converter operates in the discontinuous conduction mode, with a duty cycle D of 66.6% and a frequency of 66 kHz and exhibits a signature as portrayed by Fig. 1-63. This

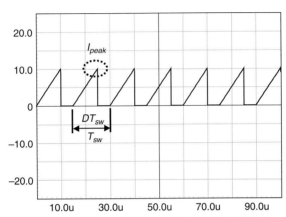

FIGURE 1-63   The converter input current signature: a DCM signal pulsing up to 10 A peak.

is a ramping up signal, growing from 0 up to the peak value $I_{peak} = 10$ A. A ramping up signal can be described by the following function:

$$i(t) = I_{peak}\frac{t}{DT_{sw}} \qquad \text{for } t \in [0, DT_{sw}] \qquad (1\text{-}205)$$

$$i(t) = 0 \qquad \text{for } t \in [DT_{sw}, T_{sw}] \qquad (1\text{-}206)$$

Plugging Eq. (1-205) into Eqs. (1-201) and (1-202) gives

$$a_1 = 2F_{sw}\int_0^{\frac{D}{F_{sw}}} \frac{I_{peak}F_{sw}}{D} \cdot t \cdot \cos(2\pi F_{sw}t)dt \qquad (1\text{-}207)$$

$$b_1 = 2F_{sw}\int_0^{\frac{D}{F_{sw}}} \frac{I_{peak}F_{sw}}{D} \cdot t \cdot \sin(2\pi F_{sw}t)dt \qquad (1\text{-}208)$$

Thanks to a Mathcad spreadsheet, we obtain the following results:

$$a_1 = I_{peak}\frac{\cos^2(D\pi) - 1 + 2D\pi \sin(D\pi)\cos(D\pi)}{D\pi^2} \qquad (1\text{-}209)$$

$$b_1 = I_{peak}\frac{\sin(D\pi)\cos(D\pi) - 2D\pi \cos^2(D\pi) + D\pi}{D\pi^2} \qquad (1\text{-}210)$$

For a 50% duty cycle, or $D = 0.5$, the above equations would greatly simplify, as $\cos\left(\frac{\pi}{2}\right) = 0$:

$$a_1 = -\frac{I_{peak}}{D\pi^2} \qquad (1\text{-}211)$$

$$b_1 = \frac{I_{peak}}{\pi} \qquad (1\text{-}212)$$

Applying Eq. (1-204) to a_1 (−3.89) and b_1 (0.946) gives a fundamental value of 2.83 A rms for this particular waveform. To check this result, we can use SPICE's ability to generate a Fourier analysis via the .FOUR keyword. If the node of observation is 1, then the statement looks like

```
.FOUR 66.6kHz V(1)
```

Please note that an observation frequency must be passed as the main parameter. If we run the simulator, we obtain a result up to the ninth harmonic, available in the .OUT file:

```
Fourier analysis for v(1):
No. Harmonics: 10, THD: 56.4316%, Gridsize: 200, Interpolation
Degree: 1

Harmonic  Frequency  Magnitude   Phase     Norm. Mag   Norm. Phase
--------  ---------  ---------   -----     ---------   -----------
0         0          3.35495     0         0           0
1         66600      4.02125     162.732   1           0
2         133200     1.4641      166.822   0.36409     4.08998
3         199800     1.06248     177.307   0.264217    14.5752
4         266400     0.844345    171.698   0.209971    8.96609
5         333000     0.616013    171.129   0.153189    8.39686
6         399600     0.531826    174.614   0.132254    11.8823
7         466200     0.47188     171.028   0.117347    8.29572
8         532800     0.39115     170.079   0.0972707   7.34702
9         599400     0.355201    171.921   0.0883309   9.1894
```

For harmonic 1, the peak amplitude is 4.0 A which converts to 2.83 A rms as we have calculated. We could derive a lot of Fourier series as input signatures can be of various types. Fortunately, SPICE can do the calculation for us in a snapshot, whatever the signal shapes are! Therefore, let's use it for our design example.

1.10.8 Selecting the Right Cutoff Frequency

Suppose we want to filter a 100 kHz CCM boost converter exhibiting a signature as Fig. 1-64 depicts. The input ripple on the source shall be less than 1 mA rms, as the design specification states. Below are the steps to follow in order to get the right values:

1. Evaluate the fundamental of the input current via a .FOUR statement. From Fig. 1.64, we obtain

```
Harmonic  Frequency  Magnitude  Phase     Norm. mag   Norm. phase
--------  ---------  ---------  -----     ---------   -----------
1         100000     0.141061   -54.262   1           0
```

2. Calculate the fundamental rms value and the necessary attenuation:

$$i_{fund,rms} = \frac{141m}{\sqrt{2}} = 99\,\text{mA}$$

$$A_{filter} < \frac{1m}{99m} < 10m \quad \text{or better than a 40 dB attenuation}$$

3. Position the cutoff frequency of the LC filter using Eq. (1-198):

$$f_0 < \sqrt{0.01 \cdot F_{sw}} < 10\,kHz$$

Let us select $f_0 = 9$ kHz.

INTRODUCTION TO POWER CONVERSION

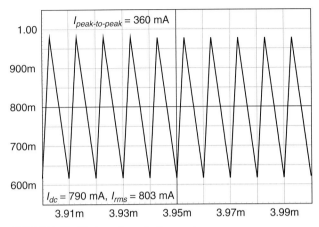

FIGURE 1-64 The boost input ripple example.

4. From Fig. 1-64 measurements, we can evaluate the rms current circulating in the filtering capacitor via the simulator measurement tools. The result represents one of the key selection criteria for filtering capacitors:

$$I_{ac} = \sqrt{I_{rms}^2 - I_{dc}^2} = 13\,\text{mA rms} \quad \text{a rather low value in this case}$$

5. From the step 3 result, we have two variables to play with: L and C. Most of the time, what matters is the total combined volume brought by the two elements. Once completed, a check must be made to see whether the important variables are compatible with the component maximum ratings: rms current for the capacitor, peak current for the inductor, and so on. Let us choose a 100 µH inductor; we can quickly determine the capacitor value via

$$C = \frac{1}{4\pi^2 f_0^2 L} = 3.16\,\mu\text{F}$$

or 4.7 µF for the normalized value. We checked the data sheet and found that a 13 mA rms current was below the maximum rating for this particular type of capacitor.

6. From the vendor catalog, the data sheet reveals the parasitic components of L and C: $C_{esr} = 150\,\text{m}\Omega$ (R_2) and $r_{Lf} = 10\,\text{m}\Omega$ (R_1).

7. We now plug these parasitic resistors into Eq. (1-196) to check if the final attenuation remains within the limit, e.g., below $10m$, despite the addition of the parasitic elements:

$$L = 100\,\mu\text{H} - R_1 = 10\,\text{m}\Omega - C = 4.7\,\mu\text{F} - R_2 = 150\,\text{m}\Omega - \omega = 628\,\text{krd/s}$$

$$\left\|\frac{I_{in}}{I_{out}}\right\| = \sqrt{\frac{0.15^2 + \frac{1}{(2.95)^2}}{(160m)^2 + \frac{1}{(2.95)^2} - \frac{200u}{4.7u} + (62.8)^2}} = 5.9m$$

which is below our $10m$ specification.

Therefore, our initial 99 mA rms input current shall be seen as an input ripple whose amplitude is now $99m \times 5.9m = 584\,\mu\text{A rms}$, or 826 µA peak.

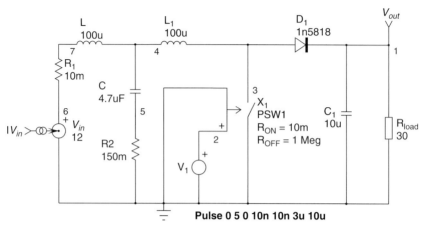

FIGURE 1-65 The open-loop boost converter with its filtering section.

The performance check is made through a complete simulation of the converter, to which the front-end filtering elements are added. In this simple case, we are running open-loop; therefore, no damping elements shall be installed. See Fig. 1-65.

The results appear in Fig. 1-66 where a fast Fourier transform (FFT) analysis has been carried out on both source and boost input currents. We can clearly see the good relationship between the expected result and the measured numbers. Here, we have used FFT command to display the spectral content of the signals of interest. A .*FOUR* statement on the input current

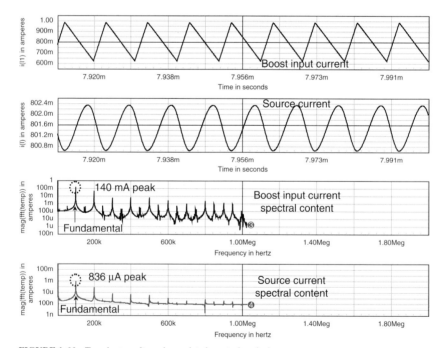

FIGURE 1-66 Transient results and associated spectral analysis.

would have given us the same number. However, this simple measurement would not reveal any annoying resonance, whereas the FFT would.

```
Harmonic   Frequency   Magnitude    Phase     Norm. Mag   Norm. Phase
--------   ---------   ---------    -----     ---------   -----------
1          100000      0.000839704  -30.437   1           0
```

WHAT I SHOULD RETAIN FROM CHAP. 1

1. Linear regulators do not lend themselves very well to build efficient converters. If the designer maintains a low voltage difference between the output and the input of the regulator, a decent efficiency can, however, be obtained.
2. The simplest form of a switching regulator associates one single switch. The converter then delivers a square pattern. However, as we want a continuous dc voltage, we need to associate a filter made of an inductor L and a filtering capacitor C to get rid of harmonics contained in the square signal. As we have used an inductor, a second switching device (a diode) is added to the circuit to ensure current continuity during the period where the switch is open: we have built a buck converter.
3. The current in the inductor can either be continuous or discontinuous. That is, if the inductor current is zero at the beginning of the next switching cycle, the converter is told to operate in discontinuous conduction mode (DCM). On the contrary, if the current in the inductor is greater than zero at the beginning of the next cycle, the converter is told to operate in continuous conduction mode (CCM). Static and dynamic behavior of a converter operating in DCM differs from that operating in CCM. When the load is increased or decreased, the converter changes its operating mode, DCM or CCM. A converter operating at the border of these two regions is told to operate in boundary conduction mode (BCM) or in *cr*itical conduction *m*ode (CRM).
4. Various types of converters exist. The three basic structures are
   - The buck: reduces the input voltage
   - The boost: increases the input voltage
   - The buck-boost: reduces or increases the input voltage but with an opposite polarity

   Other topologies are derived from the above converters:
   - Buck → single-switch, multiswitch forward converters (brings isolated primary/secondary grounds)
   - Buck-boost → flyback converter (also brings isolation)
5. If linear regulators are quiet by nature (no switching elements), switch-mode converters are inherently noisy. A filter must be placed in front of the regulator to reduce the harmonic content conducted over the power cord. However, the filter resonates and its presence can conflict with the converter. This converter actually presents a negative input resistance which can cancel the filter damping factor. To avoid this situation, damping elements such as *RC* components must be installed to calm possible oscillations.

REFERENCES

1. R. Kielkowski, *Inside SPICE,* McGraw-Hill, 1998
2. A. Vladimirescu, *The SPICE Book*, John Wiley & Sons, New York, 1993
3. R. Pease, *Analog Troubleshooting,* Newnes, 1991

4. R. Mamano, "Fueling the Megaprocessors—Empowering Dynamic Energy Management," Unitrode, SEM-1100

5. C. Basso et al., "Get the Best from Your LDO Regulator," *EDN Magazine*, February 18, 1999

6. V. Vorperian, "Fast Analytical Techniques for Electrical and Electronic Circuits," Cambridge University Press, 2002

7. H. Ozbay, *Introduction to Feedback Control Theory*, CRC Press, 1999

8. D. Middlebrook, "Design Techniques for Preventing Input-Filter Oscillations in Switched-Mode Regulator," *Advances in Switched-Mode Power Conversion*, vols. 1 and 2, TESLAco, 1983

APPENDIX 1A A RLC TRANSFER FUNCTION

This appendix details how the *RLC* transfer function and output impedance were derived. Since we want to obtain various parameters affecting this passive circuit, a good method is to use matrix algebra. Matrix algebra is well suited for numerical computations on a computer, and SPICE makes extensive use of it. It is true that the symbolic answer given by a transfer matrix does not give the designer much insight into the circuit's operation. However, one remarkable point is that once you have found the matrix coefficients, the resulting transfer matrix $T(s)$ contains, in one shot, all the parameters of interest:

$$T(s) = \begin{bmatrix} \dfrac{Y_1(s)}{U_1(s)} & \dfrac{Y_1(s)}{U_2(s)} \\ \dfrac{Y_2(s)}{U_1(s)} & \dfrac{Y_2(s)}{U_2(s)} \end{bmatrix} \qquad (1A\text{-}1)$$

where Y_1 and Y_2, respectively, represent the input current and output voltage and U_1 and U_2, represent the input voltage and output current (Fig. 1A-1). Thanks to this arrangement, we obtain from Eq. (1A-1) all our variables of interest:

$T_{1,1}$ = input admittance

$T_{1,2}$ = output current susceptibility

$T_{2,1}$ = audiosusceptibility or transfer function

$T_{2,2}$ = output impedance

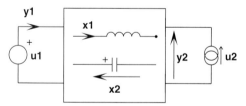

FIGURE 1A-1 Arrangement of various variables to fit the matrix analysis format.

In Fig. 1A-1, we can identify the so-called source variables (u_k), state variables (x_k), and output variables (y_k). State variables can be seen as the necessary initial conditions for $t = 0$. Without these elements, we cannot solve the differential equations describing our network. For instance, in a simple *LR* circuit, if we know the initial current in the inductance, then we know the "state" of that system for $t > t_0$. As we can see, this is x_1 for the inductor current and x_2 for the initial charging voltage describing a capacitor.

INTRODUCTION TO POWER CONVERSION

The exercise consists of writing all mesh and node equations as functions of variables x and u to further arrange the state and output variables as the following format shows:

$$\dot{x}_1 = \mathbf{A}x_1 + \mathbf{B}u_1 \quad \text{general form of state equation} \quad (1\text{A-}2)$$

$$y_1 = \mathbf{M}x_1 + \mathbf{N}u_1 \quad \text{general form of output equation} \quad (1\text{A-}3)$$

where

$$\dot{x}_1 = \frac{dx_1}{dt}$$

A = state equation state coefficient matrix
u = source vector coefficient matrix
B = state equation source coefficient matrix
M = output equation state coefficient matrix
N = output equation source coefficient matrix

For an nth-*order* system with n state variables and r sources we have

$$\dot{x}_1 = a_{11}x_1 + \cdots + a_{1n}x_n \cdots + b_{11}u_1 + \cdots + b_{1r}u_r \quad (1\text{A-}2\text{a})$$

$$\dot{x}_2 = a_{21}x_1 + \cdots + a_{2n}x_n \cdots + b_{21}u_1 + \cdots + b_{2r}u_r \quad (1\text{A-}2\text{b})$$

...

Using the general matrix notation according to Eqs. (1A-2) and (1A-3), we obtain

$$\begin{bmatrix} \dot{x}_1 \\ \dot{x}_2 \\ \cdot \\ \dot{x}_n \end{bmatrix} = \begin{bmatrix} a_{11} & a_{12} & \cdot & a_{1n} \\ a_{21} & a_{22} & \cdot & a_{2n} \\ \cdot & \cdot & \cdot & \cdot \\ a_{n1} & a_{n2} & \cdot & a_{nn} \end{bmatrix} \begin{bmatrix} x_1 \\ x_2 \\ \cdot \\ x_n \end{bmatrix} + \begin{bmatrix} b_{11} & b_{12} & \cdot & b_{1r} \\ b_{21} & b_{22} & \cdot & b_{2r} \\ \cdot & \cdot & \cdot & \cdot \\ b_{n1} & b_{n2} & \cdot & b_{nr} \end{bmatrix} \begin{bmatrix} u_1 \\ u_2 \\ \cdot \\ u_r \end{bmatrix}$$

Theory dictates that the generalized transfer function $T(s)$ of an nth-order *linear* passive system is

$$T(s) = [\mathbf{M}(s\mathbf{I} - \mathbf{A})^{-1}\mathbf{B} + \mathbf{N}] \quad (1\text{A-}4)$$

where $\mathbf{I} = \begin{pmatrix} 1 & 0 \\ 0 & 1 \end{pmatrix}$ corresponds to the unity matrix, which when multiplied by the Laplace operator s becomes $s\mathbf{I} = \begin{pmatrix} s & 0 \\ 0 & s \end{pmatrix}$.

Let us begin the analysis by drawing the *RLC* network featuring all state and output variables (see Fig. 1A-2).

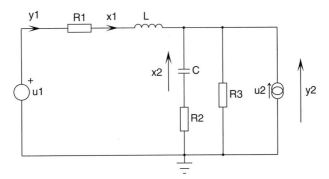

FIGURE 1A-2 The *RLC* network updated with pertinent variables.

State equations:

$$u_1 = R_1 x_1 + L\dot{x}_1 + x_2 + R_2 C \dot{x}_2 \tag{1A-5}$$

Rearranging gives

$$\dot{x}_1 = -\frac{R_1}{L} x_1 - \frac{1}{L} x_2 + \frac{1}{L} u_1 - \frac{R_2 C}{L} \dot{x}_2 \tag{1A-6}$$

The current flowing into C and R_3 is actually made of the sum of u_2 and x_1:

$$x_1 + u_2 = C\dot{x}_2 + \frac{x_2 + R_2 C \dot{x}_2}{R_3} \tag{1A-7}$$

Developing and factoring this equation give

$$(x_1 + u_2)R_3 = CR_3 \dot{x}_2 + x_2 + R_2 C \dot{x}_2 \tag{1A-8}$$

$$x_1 R_3 + u_2 R_3 - x_2 = C\dot{x}_2 (R_3 + R_2) \tag{1A-9}$$

$$\dot{x}_2 = \frac{R_3}{(R_3 + R_2)C} x_1 - \frac{1}{(R_2 + R_3)C} x_2 + \frac{R_3}{(R_2 + R_3)C} u_2 \tag{1A-10}$$

We can now plug Eq. (1A-10) into Eq. (1A-6) and factor all terms in a way to fit definition (1A-2):

$$\dot{x}_1 = -\frac{1}{L}\left(R_1 + \frac{R_2 R_3}{R_2 + R_3}\right) x_1 + \frac{1}{L}\left(\frac{R_2}{R_2 + R_3} - 1\right) x_2 + \frac{1}{L} u_1 - \frac{R_2 R_3}{(R_2 + R_3)L} u_2 \tag{1A-11}$$

Equations (1A-10) and (1A-11) now fit our second-order system arrangement described by Eqs. (1A-2a) and (1A-2b). Filling the **A** and **B** matrixes becomes an easy part:

$$\mathbf{A} = \begin{bmatrix} -\dfrac{R_1 R_2 + R_1 R_3 + R_2 R_3}{(R_2 + R_3)L} & \dfrac{-R_3}{(R_2 + R_3)L} \\ \dfrac{R_3}{(R_2 + R_3)L} & -\dfrac{1}{(R_2 + R_3)C} \end{bmatrix}$$

$$\mathbf{B} = \begin{bmatrix} \dfrac{1}{L} & \dfrac{-R_2 R_3}{(R_2 + R_3)L} \\ 0 & \dfrac{R_3}{(R_2 + R_3)C} \end{bmatrix}$$

Output equations:
Observing Fig. 1A-2 shows that

$$y_1 = x_1 \tag{1A-12}$$

$$y_2 = C\dot{x}_2 R_2 + x_2 \tag{1A-13}$$

$$y_2 = R_2\left(x_1 \frac{R_3}{R_2 + R_3} + u_2 \frac{R_3}{R_2 + R_3} - x_2 \frac{1}{R_2 + R_3}\right) + x_2 \tag{1A-14}$$

Rearranging Eq. (1A-14) under the wanted format leads to

$$y_2 = x_1 \frac{R_2 R_3}{R_2 + R_3} + x_2 \left(1 - \frac{R_2}{R_2 + R_3}\right) + u_2 \frac{R_3 R_2}{R_2 + R_3} \qquad (1A\text{-}15)$$

From the source Eqs. (1A-12) and (1A-15), we can fill up the **M** and **N** matrixes by identifying the right coefficients:

$$\mathbf{M} = \begin{bmatrix} 1 & 0 \\ \dfrac{R_3 R_2}{R_2 + R_3} & \dfrac{R_3}{R_2 + R_3} \end{bmatrix}$$

$$\mathbf{N} = \begin{bmatrix} 0 & 0 \\ 0 & \dfrac{R_3 R_2}{R_2 + R_3} \end{bmatrix}$$

We can now apply Eq. (1A-4) through a Mathcad spreadsheet to obtain *all* pertinent parameters in *one* shot:

- $T_{2,1}$—transfer function [Eq. (1A-16)]

$$T(s) = \frac{R_3}{R_1 + R_3} \cdot \frac{1 + sR_2 C}{s^2 LC \left(\dfrac{R_3 + R_2}{R_1 + R_3}\right) + s \dfrac{L + C(R_2 R_3 + R_1 R_3 + R_2 R_1)}{R_1 + R_3} + 1}$$

- $T_{2,2}$—output impedance [Eq. (1A-17)]

$$Z_{out}(s) = \frac{R_3}{R_1 + R_3} \cdot \frac{(sR_2 C + 1)(sL + R_1)}{s^2 LC \left(\dfrac{R_3 + R_2}{R_1 + R_3}\right) + s \dfrac{L + C(R_2 R_3 + R_1 R_3 + R_2 R_1)}{R_1 + R_3} + 1}$$

- $T_{1,1}$—input admittance [Eq. (1A-18)]

$$Y_{in}(s) = \frac{1}{R_1 + R_3} \cdot \frac{sC(R_2 + R_3) + 1}{s^2 LC \left(\dfrac{R_3 + R_2}{R_1 + R_3}\right) + s \dfrac{L + C(R_2 R_3 + R_1 \times R_3 + R_2 \times R_1)}{R_1 + R_3} + 1}$$

- $T_{1,2}$—output current susceptibility [Eq. (1A-19)]

$$T_{1,2}(s) = \frac{R}{R_1 + R_3} \cdot \frac{1 - sR_2 C}{s^2 LC \left(\dfrac{R_3 + R_2}{R_1 + R_3}\right) + s \dfrac{L + C(R_2 R_3 + R_1 R_3 + R_2 R_1)}{R_1 + R_3} + 1}$$

APPENDIX 1B THE CAPACITOR EQUIVALENT MODEL

Capacitors that we connect as filtering elements do not always behave as capacitors. Because of their manufacturing process, parasitic elements appear and alter the original capacitor characteristics: you expect a transient response based on a purely capacitive device, but you finally have something different. Figure 1B-1 depicts the well-known equivalent circuit of a capacitor.

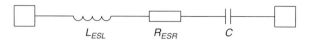

FIGURE 1B-1 The equivalent circuit of a capacitor shows the presence of parasitic elements.

If you sweep this arrangement by a current source, the impedance plot will reveal several portions, depending on the domination of a particular element (Fig. 1B-2):

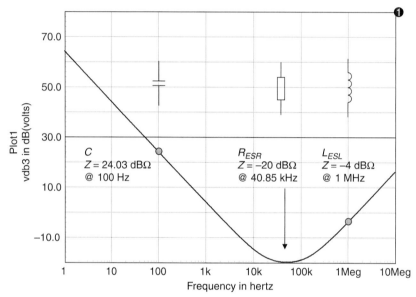

FIGURE 1B-2 An impedance sweep of the equivalent capacitor model shows the dominance of the various parasitic elements. In this example, $C = 100\ \mu F$, $R_{ESR} = 100\ m\Omega$, and $L_{ESL} = 100\ nH$.

- In the left area, the low-frequency domain, the capacitor value dominates and the impedance goes down as the frequency increases. The 100 Hz point confirms the 100 μF impedance value.
- In the middle of the explored spectrum, the resistive portion takes over and the curve flattens: we are at the resonant point of the tuned LC filter. You read the ESR at this point.
- Finally, as we keep increasing the modulation frequency, the inductive term dominates and the impedance goes up: the capacitive effect has long gone and the parasitic inductor remains.

The switching converter uses a capacitor placed on its output. When the converter is subjected to an output current step, the presence of the parasitic elements affects the response shape. Figure 1B-3 portrays a simplified representation of the converter when dynamically loaded. If an abrupt output step appears (Fig. 1B-4), then, before the converter loop reacts, the output voltage solely relies on the output capacitor. Hence, the total voltage drop can be expressed as follows:

$$\Delta V_{out} = I_c(t) R_{ESR} + L_{ESL} \frac{dI_c(t)}{dt} + \frac{1}{C} \int_0^{\Delta t} I_c(t) dt \qquad (1B\text{-}1)$$

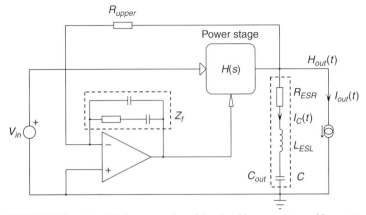

FIGURE 1B-3 A simplified representation of the closed-loop converter and its output capacitor.

Because the converter is bandwidth limited, it cannot compensate the voltage drop. Thus, during the load current rise time, the energy is entirely delivered by the output capacitor. In other words, $I_c(t) = I_{out}(t)$.

The load current evolves in time according to Eq. (1B-2):

$$I_{out}(t) = \Delta I \frac{t}{\Delta t} = S_I t \quad (1B\text{-}2)$$

Replacing $I_c(t)$ in Eq. (1B-1) gives us

$$\Delta V_{out} = R_{ESR} S_I t + L_{ESL} S_I + \frac{1}{C} \int_0^t S_I t \cdot dt \quad (1B\text{-}3)$$

which finally gives

$$\Delta V_{out} = R_{ESR} S_I t + L_{ESL} S_I + \frac{S_I t^2}{2C} \quad (1B\text{-}4)$$

Replacing t by its value Δt and factoring S_I, we obtain

$$\Delta V_{out} = S_I \left(R_{ESR} \Delta t + L_{ESL} + \frac{\Delta t^2}{2C} \right) \quad (1B\text{-}5)$$

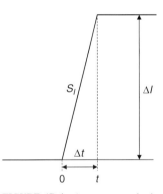

FIGURE 1B-4 A current step loads the output converter.

This equation describes the voltage drop you obtain from a fast-rising current step. Please note that we discuss a drop and not an undershoot. The term under/overshoot solely relates to a loop response whereas we deal in this case with parasitic elements only. Suppose you need to design a buck powering a central processor unit (CPU) core. In these applications, current rise times of 1 kA/μs are common. Facing this kind of event, the converter tries to push the current in the inductor which cannot increase at a pace faster than $\frac{V_L(t)}{L}$, with $V_L(t)$ being the instantaneous voltage across the inductor L. During this time, Eq. (1B-5) applies and helps you to select the right capacitor whose parasitic combination ensures that the spike stays within acceptable limits.

As Eq. (1B-5) shows, the output voltage spike is made of three components:

- The inductive kick brought by the ESL—the real spike, depending on the current slope
- The ohmic drop given by the ESR
- The capacitive integration

Observing the transient response of your converter with an oscilloscope reveals the dominant terms on your selected capacitor. Figure 1B-5 portrays voltage-mode simulation results ($f_c = 10$ kHz) where we have split the contribution of each of the parasitic terms.

Now, if we sum all terms through Eq. (1B-5), we have

$$\Delta V_{out} = 400k \left(100m \times 1u + 100n + \frac{1p}{2 \times 100u} \right) = 82\,mV \qquad (1B\text{-}6)$$

This is what we can finally read in Fig. 1B-5. You can note the small contribution brought by the capacitor, compared to that of the ESR and the ESL.

In applications where the ESL can be neglected, that is, when the rate of the output current rise is not significantly strong, an approximated formula exists to predict the output undershoot (thus feedback dependent). Assuming the converter time constant ($\frac{1}{f_c}$, with f_c the crossover frequency is much smaller than the $R_{ESR}C$ time constant, then the drop is mostly dictated by the crossover frequency and the output capacitor:

$$\Delta V_{out} = \frac{\Delta I_{out}}{2\pi f_c C} \qquad (1B\text{-}7)$$

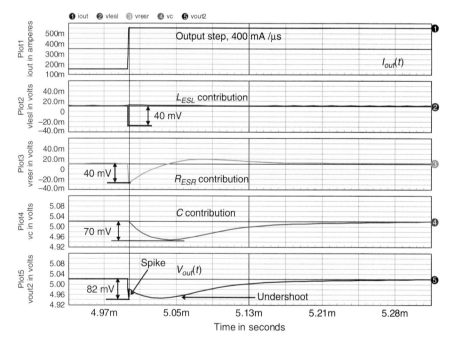

FIGURE 1B-5 Simulation results of a CCM voltage-mode buck loaded by a 400 mA step. $S_l = 400$ mA/μs. The element values are $C = 100$ μF, $R_{ESR} = 100$ mΩ, and $L_{ESL} = 100$ nH.

As we said, this equation holds only as long as the capacitive term dominates against the ESR. Otherwise stated, Eq. (1B-7) holds if

$$R_{ESR} \leq \frac{1}{2\pi f_c C} \qquad (1B\text{-}8)$$

With a 10 kHz crossover frequency and our 100 μF capacitor (100 m(ESR), Eq. (1B-8) is satisfied:

$$100 \text{ m}\Omega \leq 160 \text{ m}\Omega$$

The loop undershoot can thus be estimated via Eq. (1B-7):

$$\Delta V_{out} = \frac{0.4}{6.28 \times 10k \times 100u} = 64\, mV$$

Figure 1B-5 plot 4 gives 70 mV, very close to this number.

In conclusion, depending on the type of converter specification, you might start your capacitor selection using Eq. (1B-7). Then further check the effects of the parasitic terms once a capacitor has been chosen and find adequate combinations to reduce the ESR and ESL effects.

APPENDIX 1C POWER SUPPLY CLASSIFICATION BY TOPOLOGIES

Depending on the application and the power level in play, some topologies are better suited than others. The below array gathers the most popular topologies and proposes a selection based on the available output power.

Topology	Applications	Power	Benefits	Negative aspects	Cost
Flyback	Cell phone chargers (<10 W) Notebook adapters (<100 W) CRT power supplies to (<150 W)	<150 W	Ease of implementation Well-documented converter Operates on wide mains Large controller offer	High peak currents Leakage inductance difficult to manage Poor EMI signature Large output ripple, large ac content (conduction losses)	Low
Single-switch forward	ATX power supplies (<250 W) dc-dc converters for telecom	<300 W	Good cross-regulation with coupled inductors Good EMI signature Low ac content, low conduction losses	Stress on the power MOSFET Difficult to operate on wide mains Requires transformer reset Duty cycle clamped to 50%	Moderate

(Continued)

Topology	Applications	Power	Benefits	Negative aspects	Cost
Two-switch forward	ATX power supplies (<500 W) dc-dc converter for telecom Servers (<500 W)	100–500 W	Good cross-regulation with coupled inductors Good EMI signature MOSFET stress clamped to V_{in}	Difficult to operate on wide mains Requires transformer reset Duty cycle clamped to 50% Requires high-side drive	Moderate
Half-bridge	ATX power supplies (<500 W) dc-dc converter for telecom	100–500 W	Good cross-regulation with coupled inductors Good EMI signature MOSFET stress clamped to V_{in} Duty cycle <100%	Difficult to operate on wide mains Requires high-side drive Cannot easily work with current mode	Moderate
Half-bridge LLC	Medical power supplies LCD or plasma TVs	<500 W	Excellent EMI signature Can work in no load Smooth waveforms, zero voltage switching (ZVS) possible	High-side drive Large rms content Narrow mains operation Dangerous short-circuit	Moderate
Full bridge	Server and mainframe power supplies High-power dc-dc converters for telecom	>500 W	Good cross-regulation with coupled inductors Resonant operation via phase shift Good EMI signature MOSFET stress clamped to V_{in} Duty cycle <100%	Difficult to operate on wide mains Requires two high-side drive circuits Four MOSFETs to drive	High
Push-pull	Dc-dc converters	<200 W	MOSFETs control is ground-referenced Duty cycle <100%	Voltage stress of $2V_{in}$ Center-tapped primary	Moderate

CHAPTER 2
SMALL-SIGNAL MODELING

In Chap. 1, we strove to describe and simulate the basic structures (buck, boost, and buck-boost) in the time domain using switching elements. This is the *switched* model approach where we implement switching elements such as MOSFETs and diodes: the variable *t* rules the circuit operation. Simulation conditions are thus fixed via a *.TRAN* statement, implying a transient analysis. In the time domain, SPICE works like a sample- and-hold system featuring a variable step: it first fixes a step (or a time interval between two simulated points, the so-called time step) and tries to converge toward the right solution, reached when the result (i.e., the final error) fits the simulation tolerance options. This internal variable is labeled *Tstep*. If it fails to calculate the point, the engine reduces *Tstep* and tries to converge again. If it does, the point is accepted and stored; otherwise the point is rejected. As a result, on constant or regular slope signals, SPICE chooses a large time step because almost no changes occur. On the contrary, as soon as a slope transition appears, the engine reduces the time interval to capture enough data points. Figure 2-1 clearly shows this typical behavior.

When you are simulating small portions of working time, e.g., a few milliseconds, this technique delivers quick analysis results. Unfortunately, when you need to assess the transient response of a low-bandwidth system switching at 100 kHz, you might need to simulate tens—if not hundreds—of milliseconds before you obtain a usable result. With a variable time step system such as SPICE, it can easily take tens of minutes (sometimes hours) before you get a single waveform. Once the simulation has ended, it might even take a few minutes more before you can extract and display data buried into the 100 Mo file built on the disk by the simulator! A fixed time step simulator such as PSIM elegantly solves this problem and it represents an excellent solution when this phenomenon hampers your simulation time. A typical example is the power factor correction (PFC) circuit which switches at a high frequency but implements a 10 Hz bandwidth. We will come back to this later.

Another solution consists of deriving an *averaged* model. An averaged model implies the disappearance of any switching event (hence discontinuities) to the benefit of a smoothly varying, continuous signal. To clarify the concept, let us explain what the term *averaging* means: averaging a periodic time function $f(t)$ usually means that $f(t)$ is integrated over a cycle of duration T and divided by the duration of that cycle. This is the well-known definition

$$\langle f(t) \rangle_T = \frac{1}{T} \int_0^T f(t)\,dt \qquad (2\text{-}1)$$

where $\langle f(t) \rangle_T$ represents the average value of $f(t)$ over a period of T duration.

Applying Eq. (2-1) to a time-evolving signal would lead to a succession of separate discrete values, all representing the average value of $f(t)$ calculated at consecutive periods: 0–T, T–T_1, T_1–T_2, . . . If you link all these points with a *continuous* function that has the same values as the averaged function at the end of each period and is essentially smooth, then you get

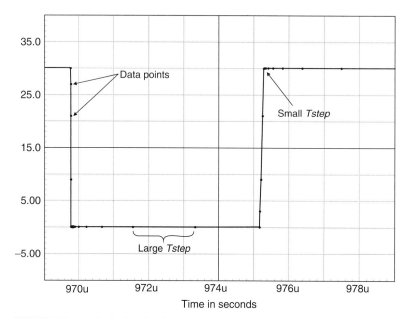

FIGURE 2-1 On slowly changing signals, SPICE takes large time intervals. As changes occur, SPICE reduces its time step.

an "averaged and continuous" representation of the original function. This is what your eye naturally does when you observe a program on TV: the moving image you can see is actually made of a succession of fixed frames, but they are displayed at a much quicker pace than your eye can sustain. You therefore observe a continuous flow of images.

Also, thanks to the averaging process, the ripple has disappeared. This is so because we assume that the modulating period is much larger than the switching period: your eye response time is larger than the pace at which images are displayed (20 ms in Europe). This concept is described by Fig. 2-2 where you see how a sinusoidal waveform modulates the duty cycle and imposes a current variation in the inductor. The duty cycle is modulated according to the following law:

$$d(t) = D_0 + D_{mod}\sin\omega_{mod}t \qquad (2\text{-}2)$$

Here D_0 represents the steady-state duty cycle corresponding to a given operating point, whereas D_{mod} describes the modulating peak value. Both D_0 and D_{mod} are considered constant. The modulation period $\frac{1}{f_{mod}}$ is also much larger than the converter switching period F_{sw}. The averaged and continuous function is similar to the filtered waveform, but is not exactly the same, since it is a mathematical abstraction rather than a real time-dependent physical variable: as Fig. 2-2 shows, the ripple has been neglected.

Thanks to the above technique, we will have a way to describe the behavior of a particular structure (a buck, for instance) via averaged equations. Once linearized across an operating point, these equations will lead to the so-called averaged small-signal model, useful to unveil the ac response of the structure under study or check the stability once the loop is closed and compensated.

As you can imagine, both models, switched and averaged, have their pros and cons. Here is a list:

SMALL-SIGNAL MODELING

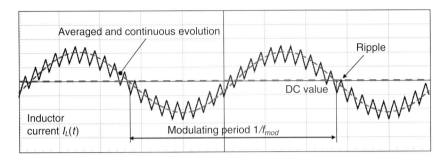

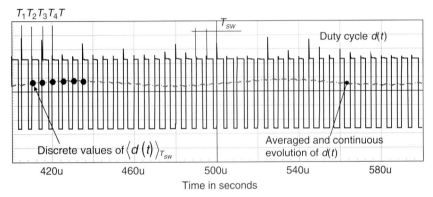

FIGURE 2-2 An ac modulated inductive current with its slowly varying low-frequency component. Note the difference between the modulating frequency and the switching frequency: $f_{mod} \ll F_{sw}$.

Averaged models

- Regarding small-signal response, draw Bode or Nyquist plots in a snapshot and assess the stability of your converter, get the input impedance variations, and check if the filter is compatible with your design without damping.
- There is no switching component. Simulation results are immediate! You can visualize long transient effects of several tens of milliseconds (e.g., in a low-bandwidth system such as a power factor corrector) and see if the output overshoot is within the specifications or if the distortion is close to the initial specification.
- It is difficult to see the effects of parasitic elements, such as inductor ohmic losses or diode forward drops, although some recent models now include them.
- You cannot evaluate the switching losses of semiconductors.

Transient models

- They can include parasitic elements. See the effect of the leakage inductance and quantify the main switch voltage stress or the poor resulting cross-regulation in multiple output converters.
- They reflect reality. If the right component models are used, you can build a virtual breadboard where ripple levels, conduction losses, and so on can easily be assessed. Switching losses require more precise models featuring parasitic elements, also available from SPICE editors.

- Because of the numerous switching events, the simulation time can be very long.
- It is difficult and tedious to draw a Bode plot from a transient simulation.
- Transient response is difficult and long to assess in low-bandwidth applications.

As you will discover in this chapter, several options exist to model or derive the small-signal equivalent circuit of a given topology: (1) either you write the *converter* state equations corresponding to the two switch positions and average them over a switching cycle, or (2) you average the *switch* waveforms only and create an equivalent model out of it. The common denominator of those methods remains the final small-signal linearization needed to extract the equivalent model. The first method is described through the state-space averaging (SSA) technique while the other option uses the averaged switch modeling technique. Let's begin with the first one.

2.1 STATE-SPACE AVERAGING

The SSA technique applied to power converters was first described by Dr. Ćuk [1] in the mid-1970s and then documented in another founding paper written with Dr. Middlebrook [2]. A lot of books discuss in detail this modeling technique [3, 4], and we will not reintroduce it in this chapter. However, to give you a small taste of the SSA technique, we have represented a simple buck converter and highlighted its *state* variables in Fig. 2-3.

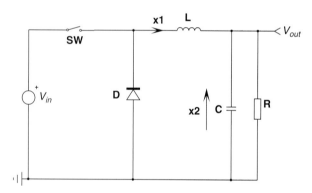

FIGURE 2-3 Highlighted state variables for a buck converter.

State variables are usually associated with storage elements such as capacitors and inductors. If we know the state of the circuit at a given time, e.g., at $t = t_0$, the so-called initial conditions, then we should be able to solve the system equations for other $t > t_0$. For a circuit including a capacitor and an inductor, the state variables will be, respectively, the capacitor voltage, denoted by x_2 and the inductor current, denoted by x_1.

The state equations of a system S can be put into compact form:

$\dot{x} = \mathbf{A}x(t) + \mathbf{B}u(t)$ general form of state equation (2-3)

$y(t) = \mathbf{M}x(t) + \mathbf{N}u(t)$ general form of output equation (2-4)

where $x(t)$ represents the state vector and contains all states variables of the circuit under study. In an *LC* circuit, it would contain variables x_1 and x_2 discussed above. The input vector $u(t)$ contains the system independent inputs, for instance, the input voltage V_{in}, noted u_1 in this example. It could also contain an excitation source, such as u_2 in App. 1A, which is a current source necessary to derive the output impedance. Even if we do not use u_2 here (equal to zero at the end), we keep it for the sake of the example.

For a second-order system as the buck circuit, we have

$$\dot{x}_1 = a_{11}x_1 + a_{12}x_2 + b_{11}u_1 + b_{12}u_2 \tag{2-3a}$$

$$\dot{x}_2 = a_{21}x_1 + a_{22}x_2 + b_{21}u_1 + b_{22}u_2 \tag{2-3b}$$

Using the general matrix notation according to Eqs. (2-3a) and (2-3b), we obtain

$$\begin{bmatrix} \dot{x}_1 \\ \dot{x}_2 \end{bmatrix} = \begin{bmatrix} a_{11} & a_{12} \\ a_{21} & a_{22} \end{bmatrix} \begin{bmatrix} x_1 \\ x_2 \end{bmatrix} + \begin{bmatrix} b_{11} & b_{12} \\ b_{21} & b_{22} \end{bmatrix} \begin{bmatrix} u_1 \\ u_2 \end{bmatrix} \tag{2-3c}$$

$y(t)$ is called the output vector, and it offers a way to place external waveforms that can be expressed as linear combinations of the state and input vectors. Any dependent signal can be put in $y(t)$; think of it as a probe, for instance, the output voltage of the *RLC* circuit in App. A1, which differs from the capacitor voltage since an ESR appears in series. We also have selected the input current y_1 because the input impedance is of interest.

$\dot{x} = \dfrac{dx}{dt}$ = time derivative of the state variable x

A = state coefficient matrix

u = source coefficient matrix

B = source coefficient matrix

M = output coefficient matrix

N = output source coefficient matrix

As Eqs. (2-3) and (2-4) describe the system S in the time domain t and provided that they are linear, we can derive the Laplace transfer function of the system. Thanks to the Laplace property of derivatives, we can write

$$sX(s) = \mathbf{A}X(s) + \mathbf{B}U(s) \tag{2-5}$$

$$Y(s) = \mathbf{M}X(s) + \mathbf{N}U(s) \tag{2-6}$$

Rearranging Eq. (2-5) gives

$$sX(s) - \mathbf{A}X(s) = \mathbf{B}U(s) \tag{2-7}$$

$$X(s)(s\mathbf{I} - \mathbf{A}) = \mathbf{B}U(s) \tag{2-8}$$

and finally

$$X(s) = (s\mathbf{I} - \mathbf{A})^{-1}\mathbf{B}U(s) \tag{2-9}$$

where **I** represents the identity matrix. Now plugging Eq. (2-9) into Eq. (2-6), we obtain

$$Y(s) = \mathbf{M}(s\mathbf{I} - \mathbf{A})^{-1}\mathbf{B}U(s) + \mathbf{N}U(s) \tag{2-10}$$

Factoring $U(s)$ gives

$$Y(s) = [\mathbf{M}(s\mathbf{I} - \mathbf{A})^{-1}\mathbf{B} + \mathbf{N}]U(s) = T(s)U(s) \tag{2-11}$$

Hence the generalized transfer function of an *N*th-order linear system is

$$T(s) = [\mathbf{M}(s\mathbf{I} - \mathbf{A})^{-1}\mathbf{B} + \mathbf{N}] \tag{2-12}$$

When further manipulating these equations, one can show that

- $\det(s\mathbf{I} - \mathbf{A})^{-1}$ represents the characteristic polynomial.
- $\det(s\mathbf{I} - \mathbf{A})^{-1} = 0$ is the characteristic equation whose roots are the transfer function poles.

- The Routh–Hurwitz criterion implies that stability is met if all coefficients of the characteristic polynomial are positive. This technique is shown to be extremely useful, for instance, when you try to ascertain whether a linear system is stable without deriving its complete transfer function: write the state equation matrix **A**, evaluate det $(s\mathbf{I} - \mathbf{A})^{-1}$, and look at the coefficients.

The first sequence of the SSA thus consists in identifying matrix coefficients corresponding to all the converter states. In CCM, we can identify two states (power switches on or off) during which all state variables exist within state durations. In DCM, however, a third state exists as both switches are now open. In this third state, the inductor state variable x_1 is zero since it has disappeared at the end of the diode conduction time. Each state will lead to a set of linear equations, formatted via the matrix approach: $\mathbf{A}_1$ and $\mathbf{B}_1$ for the on state, $\mathbf{A}_2$ and $\mathbf{B}_2$ for the off state. We will see how to link them later.

2.1.1 SSA at Work for the Buck Converter—First Step

To begin, we write all node and mesh equations pertinent to Fig. 2-3 in the two distinct switch positions, on and off. The converter operates in CCM, and V_{in} is replaced by u_1 to better stick to the SSA notation.

State 1. SW is closed and the diode is open (Fig. 2-4a). Solve the equation with the state variables x_1 and x_2 to find their respective derivatives:

$$u_1 = L\dot{x}_1 + x_2 \tag{2-13}$$

Once it is rearranged, we have

$$\dot{x}_1 = -\frac{1}{L}x_2 + \frac{1}{L}u_1 \tag{2-14}$$

$$C\dot{x}_2 = x_1 - \frac{1}{R}x_2 \tag{2-15}$$

After manipulation

$$\dot{x}_2 = \frac{1}{C}x_1 - \frac{1}{RC}x_2 \tag{2-16}$$

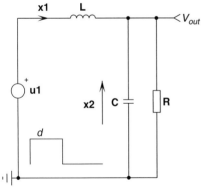 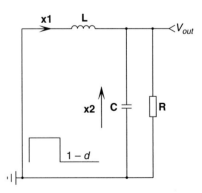

FIGURE 2-4a The switch *SW* is in the closed state.   **FIGURE 2-4b** The diode *D* is now conducting.

From Eq. (2-3c), we can write

$$\begin{bmatrix}\dot{x}_1\\\dot{x}_2\end{bmatrix}=\begin{bmatrix}0 & -\frac{1}{L}\\\frac{1}{C} & -\frac{1}{RC}\end{bmatrix}\begin{bmatrix}x_1\\x_2\end{bmatrix}+\begin{bmatrix}\frac{1}{L} & 0\\0 & 0\end{bmatrix}\begin{bmatrix}u_1\\u_2\end{bmatrix} \qquad (2\text{-}17)$$

Which leads to the following matrix identification:

$$\mathbf{A}_1=\begin{bmatrix}0 & -\frac{1}{L}\\\frac{1}{C} & -\frac{1}{RC}\end{bmatrix}\qquad \mathbf{B}_1=\begin{bmatrix}\frac{1}{L} & 0\\0 & 0\end{bmatrix}$$

State 2. SW is now open and the diode conducts (Fig. 2-4b). Let's solve the equation with the state variables x_1 and x_2 to find their respective derivatives:

$$0 = L\dot{x}_1 + x_2 \qquad (2\text{-}18)$$

$$\dot{x}_1 = -\frac{1}{L}x_2 \qquad (2\text{-}19)$$

$$x_2 = R(x_1 - \dot{x}_2 C) \qquad (2\text{-}20)$$

After a different arrangement we find

$$\dot{x}_2 = \frac{1}{C}x_1 - \frac{1}{RC}x_2 \qquad (2\text{-}21)$$

Again sticking to Eq. (2-3c) gives

$$\begin{bmatrix}\dot{x}_1\\\dot{x}_2\end{bmatrix}=\begin{bmatrix}0 & -\frac{1}{L}\\\frac{1}{C} & -\frac{1}{RC}\end{bmatrix}\begin{bmatrix}x_1\\x_2\end{bmatrix}+\begin{bmatrix}0 & 0\\0 & 0\end{bmatrix}\begin{bmatrix}u_1\\u_2\end{bmatrix} \qquad (2\text{-}22)$$

from which we can unveil matrices $\mathbf{A}_2$ and $\mathbf{B}_2$:

$$\mathbf{A}_2=\begin{bmatrix}0 & -\frac{1}{L}\\\frac{1}{C} & -\frac{1}{RC}\end{bmatrix}\qquad \mathbf{B}_2=\begin{bmatrix}0 & 0\\0 & 0\end{bmatrix}$$

At this point, we need to link both states: $\mathbf{A}_1$ and $\mathbf{A}_2$, $\mathbf{B}_1$ and $\mathbf{B}_2$. If you look at the buck schematic and the equations we wrote, you certainly remark that $\mathbf{A}_1$ and $\mathbf{B}_1$ apply for the first (on) interval, or during d of the switching time, while $\mathbf{A}_2$ and $\mathbf{B}_2$ exist during the $1-d$ (off) switching time interval. Thus, we can combine both matrices by using the following equations:

$$\mathbf{A} = \mathbf{A}_1 d + \mathbf{A}_2(1-d) \qquad (2\text{-}23)$$

$$\mathbf{B} = \mathbf{B}_1 d + \mathbf{B}_2(1-d) \qquad (2\text{-}24)$$

Thanks to the multiplication by d and $1-d$, Eqs. (2-23) and (2-24) smooth the discontinuity associated with the switching event and become continuous. Updating Eq. (2-3) leads to

$$\dot{x} = [\mathbf{A}_1 d + \mathbf{A}_2(1-d)]x(t) + [\mathbf{B}_1 d + \mathbf{B}_2(1-d)]u(t) \qquad (2\text{-}25)$$

Applying the previous equations to fit Eqs. (2-23) and (2-24) leads to

$$\mathbf{A} = \mathbf{A}_1 d + \mathbf{A}_2(1-d) = \begin{bmatrix} 0 & -\dfrac{d}{L} \\ \dfrac{d}{C} & -\dfrac{d}{RC} \end{bmatrix} + \begin{bmatrix} 0 & -\dfrac{1-d}{L} \\ \dfrac{1-d}{C} & -\dfrac{1-d}{RC} \end{bmatrix} = \begin{bmatrix} 0 & -\dfrac{1}{L} \\ \dfrac{1}{C} & -\dfrac{1}{RC} \end{bmatrix} \qquad (2\text{-}26)$$

$$\mathbf{B} = \mathbf{B}_1 d + \mathbf{B}_2(1-d) = \begin{bmatrix} \dfrac{d}{L} & 0 \\ 0 & 0 \end{bmatrix} + \begin{bmatrix} 0 & 0 \\ 0 & 0 \end{bmatrix} = \begin{bmatrix} \dfrac{d}{L} & 0 \\ 0 & 0 \end{bmatrix} \qquad (2\text{-}27)$$

We now have a set of *continuous* state equations which look like this:

$$\begin{bmatrix} \dot{x}_1 \\ \dot{x}_2 \end{bmatrix} = \begin{bmatrix} 0 & -\dfrac{1}{L} \\ \dfrac{1}{C} & -\dfrac{1}{RC} \end{bmatrix} \begin{bmatrix} x_1 \\ x_2 \end{bmatrix} + \begin{bmatrix} \dfrac{d}{L} & 0 \\ 0 & 0 \end{bmatrix} \begin{bmatrix} u_1 \\ u_2 \end{bmatrix} \qquad (2\text{-}28)$$

If we expand the equation set (2-28), we get

$$\dot{x}_1 = -\dfrac{1}{L} x_2 + \dfrac{d}{L} u_1 \qquad (2\text{-}29)$$

$$\dot{x}_2 = \dfrac{1}{C} x_1 - \dfrac{1}{RC} x_2 \qquad (2\text{-}30)$$

Comparing these two equations to Eqs. (2-14) and (2-16), we can see that the only difference lies in the presence of the d term affecting the input voltage u_1. Because d can potentially vary via the feedback loop as a combination of u_1, x_2, or x_1, Eqs. (2-29) and (2-30) are nonlinear. However, we can use them to build a nonlinear, also called large-signal, model as depicted by Fig. 2-4c, identifying u_1 as V_{in}:

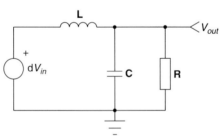

FIGURE 2-4c A continuous nonlinear model of the buck converter.

2.1.2 The DC Transformer

Nevertheless, from Fig. 2-4c, in reality dV_{in} is not the input source. Actually V_{in} is. To improve the visibility of Fig. 2-4c, we use a so-called dc transformer. This transformer does not have any physical meaning, as everyone knows that transformers only work in alternating current. But in our case, it will multiply V_{in} by the ratio d and will reflect the output current back to the

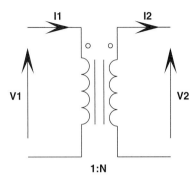

FIGURE 2-5 A dc transformer in which the magnetizing inductance is considered infinite.

source, as a transformer would normally do. Figure 2-5 shows its representation affected by a ratio N to start with. Please note the absence of magnetizing and leakage inductances, respectively, as we consider them infinite and null.

A transformer turns ratio N is usually defined as

$$N = \frac{N_s}{N_p} \tag{2-31a}$$

and it is often denoted on electrical schematics as

$$N_p:N_s \tag{2-31b}$$

where N_p and N_s are, respectively, the primary and secondary turns. To ease impedance reflection or current/voltage multiplications, it is often convenient to normalize all ratios to the primary N_p. Therefore, if we now divide all turns ratios by N_p, we obtain a different expression:

$$\frac{N_p}{N_p}:\frac{N_s}{N_p} = 1:N \tag{2-32}$$

This is what Fig. 2-5 already expresses. The (perfect) transformer equations can now be easily derived and are

$$P_{in} = P_{out} \tag{2-33}$$
$$V_1 I_1 = V_2 I_2 \tag{2-34}$$
$$I_1 = \frac{V_2}{V_1} I_2 = N I_2 \tag{2-35}$$
$$V_2 = \frac{I_1}{I_2} V_1 = N V_1 \tag{2-36}$$

In other words, the input current I_1 is the output current I_2 multiplied by N, whereas the output voltage V_2 corresponds to the input voltage V_1 also multiplied by N. This configuration can easily be arranged with a few SPICE elements, such as a current-controlled current source (the F SPICE primitive) and a voltage-controlled voltage source (the E SPICE primitive). Figure 2-6a shows how to wire them together [6]:

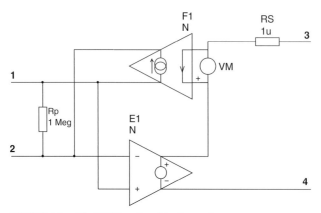

FIGURE 2-6a The SPICE version of the dc transformer.

R_s and R_p are added to avoid convergence problems when associating the transformer with other elements. Experience shows that this model is rather robust, however. The VM source measures the output current delivered by E_1 which is fed back to the primary via F_1, as dictated by Eq. (2-35). In SPICE, a current flowing within an element is "seen" positive when it enters via the element "+" and leaves it by its "−" hence VM's polarity. The primary lies between nodes 1 and 2 and the secondary between nodes 3 and 4.

Unfortunately, in this model, the ratio N is fixed and passed as a parameter. Since we want a ratio of d, with d being variable in nature, Fig. 2-6a needs a slight update via modern in-line equations; see Fig. 2-6b.

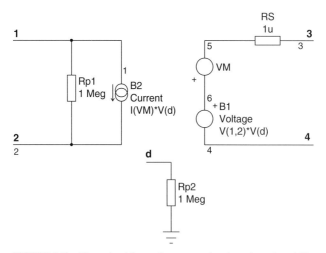

FIGURE 2-6b The updated dc transformer accepting d as a dynamic variable.

From this model, the B_1 and B_2 sources perform the same duty as E_1 and F_1, but their multiplier is now a node voltage, which is d. If we hook a source to d and sweep it, we directly sweep the transformer ratio.

2.1.3 Large-Signal Simulations

Thanks to this new device, we can update Fig. 2-4c. In Fig. 2-7 we have drawn the SPICE version of the buck converter in large-signal conditions.

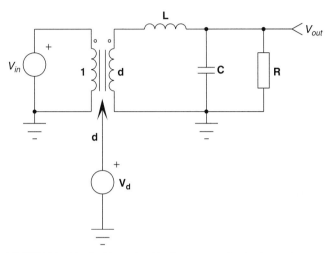

FIGURE 2-7 The final large-signal buck converter using the dc transformer.

Yes, we can simulate it. Suppose we want to deliver 5 V from a 12 V source; then d shall be fixed to $\frac{5}{12} = 416$ mV. If we consider a 100% duty cycle corresponding to 1 V, the dc transformer input will receive a 416 mV dc level representative of a 41.6% duty cycle. Running a simulation over Fig. 2-7 with true values reveals the bias point calculated by SPICE. See Fig. 2-8.

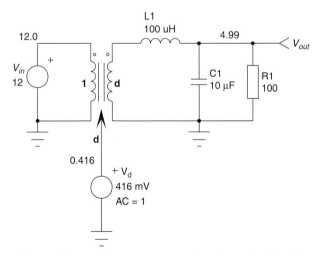

FIGURE 2-8 After simulation, calculated bias points can be reflected on the schematic.

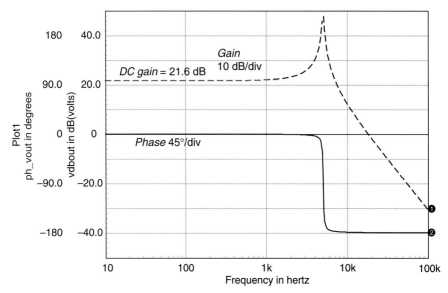

FIGURE 2-9 An ac plot is quickly obtained from the large-signal model as SPICE linearizes it for us.

Now, if we superimpose an ac signal on top of V_d, we can sweep the transfer function $\dfrac{V_{out}(s)}{d(s)}$. To do this, just add the SPICE keyword AC = 1 to the source V_d and you are done. A 1 V level might be an enormous level to start with, but SPICE automatically linearizes all the circuit to make it a small-signal version where the ac amplitude no longer has any influence. Remember, SPICE only solves linear equations ... Hence the 1 V value is just here to avoid further manipulations when we are looking at the selected transfer function. For instance, in this case, as we want to reveal $\dfrac{V_{out}(s)}{d(s)} = \dfrac{V_{out}(s)}{V_d(s)}$ since $V_d(s)$ is 1 V, just plotting V_{out} is enough! Figure 2-9 reveals the ac plot, showing the peaking typical of the LC configuration.

Since we deliver 5 V from a d input of 416 mV, the dc gain is simply

$$20 \log_{10}\left(\frac{5}{0.416}\right) = 21.6 \, dB \tag{2-37}$$

2.1.4 SSA at Work for the Buck Converter, the Linearization—Second Step

Equations (2-29) and (2-30) would be linear if d and $1-d$ were constant. However, in a normal application, this is not the case since some of the state variables (x_2 in the buck) are fed back to a control integrated circuit (IC). This chain continuously adjusts d in order to keep one of the state variables constant, i.e., the output voltage x_2. In brief, we have transformed a set of two distinct linear equations into a set of *nonlinear* but *continuous* equations.

To proceed with the process, we need to linearize the system across a given operating point. If a system exhibits highly nonlinear behavior over a large-signal excursion, it is usually possible to obtain linear behavior if the excitation amplitude is reduced across the same operating point. As an example, Fig. 2-10 portrays a voltage source forward biasing a diode.

If we dc bias the circuit with a 1 V level, then the diode starts conducting with a current limited by R_1. This current represents the diode operating point also called the bias point. This is a static, dc value. Now applying a sinusoidal excitation of amplitude A over the 1 V level will move the operating point up and down to a pace and amplitude imposed by V_{ac}. As you know, the *I–V* curve of a diode includes a knee located close to the forward voltage. This knee imposes a nonlinear relationship between *I* and *V*. If you reduce the modulating amplitude, you naturally reduce the effects of the knee or from another nonlinear region and the device characteristics appear to be linear: distortion no longer affects the resulting signal. The first operating mode refers to *large-signal* operation whereas the other one reflects *small-signal* operation. Figure 2-11 displays the results for large-signal modulation versus small-signal operation.

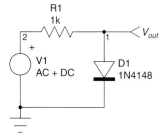

FIGURE 2-10 A diode is biased by a sinusoidal excitation straddling on a dc level.

In the SSA technique, we will split all variables with a dc term imposing the operating point, to which we add an ac modulating signal (sometimes described as a perturbation) sufficiently small in amplitude to keep the system linear. Figure 2-12a depicts the idea. This ac term is usually denoted with a small "hat" ^ (called the caret) but can also appear with a tilde ~. Hence, the steady-state dc values now become

$$d = D + \hat{d} \tag{2-38}$$

where *D* represents the dc term (V_{dc} from Fig. 2-12) and $\hat{d}$ the ac small-signal modulation (V_{ac} from Fig. 2-12).

$$x = x_0 + \hat{x} \tag{2-39}$$

$$u = u_0 + \hat{u} \tag{2-40}$$

If we now insert these definitions into the updated nonlinear Eq. (2-25), we have the following definition:

$$\dot{x}+\dot{\hat{x}} = \left[\mathbf{A}_1(\hat{d}+D)+\mathbf{A}_2(1-\hat{d}-D)\right](x_0+\hat{x})+\left[\mathbf{B}_1(\hat{d}+D)+\mathbf{B}_2(1-\hat{d}-D)\right](u_0+\hat{u}) \tag{2-41}$$

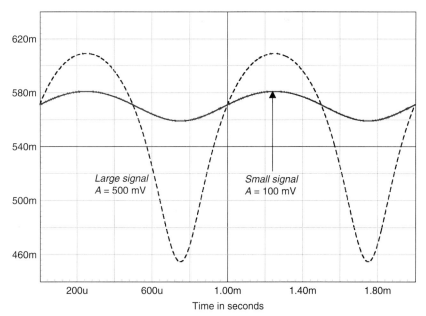

FIGURE 2-11 Large-signal operation shows nonlinearity whereas small-signal excitation reveals a linear relationship between the output and the input.

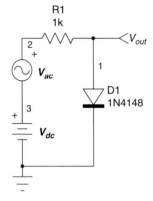

FIGURE 2-12a V_{dc} fixes the operating point, whereas V_{ac} represents the modulation signal.

If we expand all members, we end up with various terms:

- A dc term multiplied by an ac term becomes an ac term.
- Ac terms' cross-products can be neglected (small multiplied by small gives an even smaller result).
- We can collect all dc terms and ac terms to form two distinct equations.
- For the dc equation, dc meaning steady-state, all derivative terms are zero, implying $\dot{x} = 0$.

Based on these facts, the dc and ac equations are as follows:

Dc equation:

$$0 = [\mathbf{A}_1 D + \mathbf{A}_2(1 - D)]x_0 + [\mathbf{B}_1 D + \mathbf{B}_2(1 - D)]u_0 \quad (2\text{-}42)$$

Ac equation:

$$\dot{\hat{x}} = [\mathbf{A}_1 D + \mathbf{A}_2(1 - D)]\hat{x} + [\mathbf{B}_1 D + \mathbf{B}_2(1 - D)]\hat{u}$$
$$+ [(\mathbf{A}_1 - \mathbf{A}_2)x_0 + (\mathbf{B}_1 - \mathbf{B}_2)u_0]\hat{d} \quad (2\text{-}43)$$

Similarities exist between Eqs. (2-42) and (2-43), and we can rewrite them by defining

$$\mathbf{A}_0 = [\mathbf{A}_1 D + \mathbf{A}_2(1 - D)] \quad (2\text{-}44)$$
$$\mathbf{B}_0 = [\mathbf{B}_1 D + \mathbf{B}_2(1 - D)] \quad (2\text{-}45)$$
$$\mathbf{E} = (\mathbf{A}_1 - \mathbf{A}_2)x_0 + (\mathbf{B}_1 - \mathbf{B}_2)u_0 \quad (2\text{-}46)$$

Therefore, Eqs. (2-42) and (2-43) are updated by summing the ac and dc terms including Eqs. (2-44) through (2-46):

$$0 = \mathbf{A}_0 x_0 + \mathbf{B}_0 u_0 \quad (2\text{-}47)$$

$$\dot{\hat{x}} = \mathbf{A}_0 \hat{x} + \mathbf{B}_0 \hat{u} + \mathbf{E}\hat{d} \quad (2\text{-}48)$$

Back to the notation of Eqs. (2-38) through (2-39), we finally have

$$\dot{\hat{x}} = \mathbf{A}_0 x + \mathbf{B}_0 u + \mathbf{E}\hat{d} \quad (2\text{-}49)$$

This equation represents the final stage to let us unveil the linearized small-signal model of the buck converter.

2.1.5 SSA at Work for the Buck Converter, the Small-Signal Model—Final Step

By simply manipulating matrices $\mathbf{A}_1$, $\mathbf{B}_1$ and $\mathbf{A}_2$, $\mathbf{B}_2$ via Eqs. (2-44) through (2-46), linearized equations will naturally appear. Let us first evaluate $\mathbf{A}_0$ and $\mathbf{B}_0$, following Eqs. (2-44) and (2-45):

$$\mathbf{A}_0 = \begin{bmatrix} 0 & -\dfrac{1}{L} \\ \dfrac{1}{C} & -\dfrac{1}{RC} \end{bmatrix} D + \begin{bmatrix} 0 & -\dfrac{1}{L} \\ \dfrac{1}{C} & -\dfrac{1}{RC} \end{bmatrix}(1 - D) = \begin{bmatrix} 0 & -\dfrac{1}{L} \\ \dfrac{1}{C} & -\dfrac{1}{RC} \end{bmatrix} \quad (2\text{-}50)$$

$$\mathbf{B}_0 = \begin{bmatrix} \dfrac{1}{L} & 0 \\ 0 & 0 \end{bmatrix} D + \begin{bmatrix} 0 & 0 \\ 0 & 0 \end{bmatrix}(1 - D) = \begin{bmatrix} \dfrac{D}{L} & 0 \\ 0 & 0 \end{bmatrix} \quad (2\text{-}51)$$

$$\mathbf{E} = \left[\begin{bmatrix} 0 & -\frac{1}{L} \\ \frac{1}{C} & -\frac{1}{RC} \end{bmatrix} - \begin{bmatrix} 0 & -\frac{1}{L} \\ \frac{1}{C} & -\frac{1}{RC} \end{bmatrix} \right] \begin{bmatrix} x_{10} \\ x_{20} \end{bmatrix} + \begin{bmatrix} \frac{1}{L} & 0 \\ 0 & 0 \end{bmatrix} - \begin{bmatrix} 0 & 0 \\ 0 & 0 \end{bmatrix} \begin{bmatrix} u_{10} \\ u_{20} \end{bmatrix} = \begin{bmatrix} \frac{u_{10}}{L} \\ 0 \end{bmatrix} \quad (2\text{-}52)$$

Combining these matrices according to Eq. (2-49) gives

$$\begin{bmatrix} \hat{\dot{x}}_1 \\ \hat{\dot{x}}_2 \end{bmatrix} = \begin{bmatrix} 0 & -\frac{1}{L} \\ \frac{1}{C} & -\frac{1}{RC} \end{bmatrix} \begin{bmatrix} \hat{x}_1 \\ \hat{x}_2 \end{bmatrix} + \begin{bmatrix} \frac{D}{L} & 0 \\ 0 & 0 \end{bmatrix} \begin{bmatrix} \hat{u}_1 \\ \hat{u}_2 \end{bmatrix} + \begin{bmatrix} \frac{u_{10}}{L} \\ 0 \end{bmatrix} \hat{d} \quad (2\text{-}53)$$

Please note the absence of dc values as their summed combination gives 0, as seen in Eq. (2-47). Now, if we develop Eq. (2-53), we obtain a set of two linearized equations corresponding to the small-signal model description we are looking for:

$$\hat{\dot{x}}_1 = \frac{1}{L}\hat{x}_2 + \frac{D}{L}\hat{u}_1 + \frac{\hat{d}}{L}u_{10} \quad (2\text{-}54\text{a})$$

$$\hat{\dot{x}}_2 = \frac{1}{C}\hat{x}_1 - \frac{1}{RC}\hat{x}_2 \quad (2\text{-}54\text{b})$$

The difficult exercise now consists of building an equivalent schematic whose mesh and node equations lead to Eqs. (2-54a) and (2-54b), exactly as Fig. 2-4c implements Eqs. (2-29) and (2-30). First, we have a static dc term u_{10}. It is rather easy to derive it as it corresponds to the input voltage V_{in}. Comparing Eqs. (2-54b) and (2-30) shows that they are identical, implying the same RC architecture as in Fig. 2-4c. Now, Eq. (2-54a) also looks similar to Eq. (2-29) except that the left end of the inductor no longer receives DV_{in} alone, but a variable term $\hat{d}V_{in}$ comes in series with it. You have understood it, $\hat{d}$ will correspond to the duty cycle input node for the ac sweep. D represents the static duty cycle given by $D = \frac{V_{out}}{V_{in}}$ and will be passed to the circuit as a parameter, or via a fixed dc source. Figure 2-12b displays

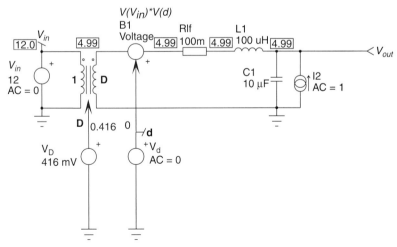

FIGURE 2-12b The final buck small-signal model including the inductor ohmic losses.

this final small-signal model for the buck converter operating in CCM where we purposely added some parasitic elements.

The V_D source sets the static duty cycle (D or D_0) whereas the V_d generator ac modulates the duty cycle. You could easily hook up an error amplifier and build a closed-loop buck converter. This is exactly what we have done, as Fig. 2-12c shows. A 60 dB gain error amplifier

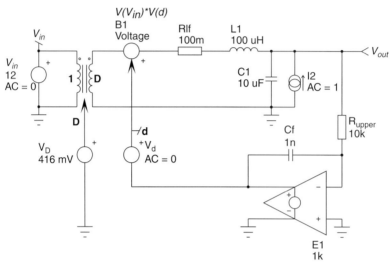

FIGURE 2-12c We can quickly add an error amplifier to close the feedback loop.

monitors the output with a simple feedback capacitor C_f, making it an integrating compensator together with R_{upper}. In open-loop, the output impedance, as expected, is the inductor series resistance of 100 mΩ or −20 dBΩ. Closing the loop with a total gain of 12,000 (1000 × 12) leads to a new closed-loop output impedance of Eq. (1-14)

$$R_{s,CL} = \frac{R_{s,OL}}{1 + T} = \frac{100m}{12,001} = -101.6 \, dB\Omega$$

which is confirmed by Fig. 2-13.

A few remarks are in order concerning this small-signal model:

- There is no lower resistor R_{lower} as in the previous examples. Why? Because, in this particular model arrangement, what fixes the dc operating point is D (416 mV). Hence the error amplifier is solely there for the ac feedback, not the dc one. thus the lack of R_{lower} (which does not play a role in the ac transfer function of the error amplifier by the way) and the reference voltage.
- Suppose an input filter is needed because of noise issues. In that case, we would need to rederive all state equations accounting for the presence of this filter! This is so because we are averaging the state variables of the *whole* converter. This is the pain of the SSA technique.

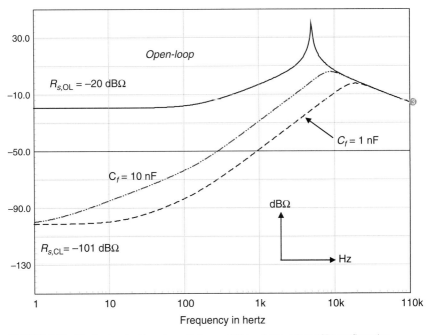

FIGURE 2-13 The buck output impedance Z_s in open (OL) and closed-loop (CL) configuration.

2.2 THE PWM SWITCH MODEL—THE VOLTAGE-MODE CASE

The SSA technique offers an interesting but complicated way for deriving small-signal models of power converters. The difficulty is mainly due to the fact that the SSA is carried over the entire converter, manipulating numerous state variables than can easily be omitted or lost during the derivation process. Fortunately, different techniques exist that really simplify the converter small-signal studies. Among these techniques, the PWM switch model plays an important role.

In 1986, Dr. Vatché Vorpérian, from Virginia Polytechnic Institute (VPEC, United States), developed the concept of the pulse width modulation (PWM) switch model [7]. At about the same time, Larry Meares from Intusoft (San Pedro, Calif.) also presented a paper in which the approach of the PWM switch was explored, although in a less comprehensive manner since CCM was the only case Meares covered [8]. As the diode and the power switch introduced the nonlinearity, these gentlemen considered modeling the switch network alone, to finally replace it with an equivalent small-signal three-terminal model (nodes A, P, and C) in exactly the same way as when you study the transfer function of a bipolar amplifier. The analysis was considerably simplified given that no average or linearization process of the whole converter was required: put the small-signal model in place and solve the equations to derive the parameters of your choice. This model is also invariant, meaning that once derived, it fits all topologies simply through different model rotations! With this method, Vorpérian demonstrated, among other results, that the flyback converter operating in DCM was still a second-order system, affected by a high-frequency right-half-plane zero. This result was not correctly predicted by SSA. Recent papers [9] have, however, revisited the SSA technique, and its reexamination led to slightly different results compared to those found by Vorpérian and Meares. Anyway, in the author's opinion, the PWM switch model offers the simplest way of deriving and understanding the dynamics of switch-mode converters. This section details how it was obtained and how to efficiently use it. Also, new autotoggling models (DCM/CCM) will be presented and tested against reality.

2.2.1 Back to the Good Old Bipolars

A similar small-signal problem also arose many years ago for the bipolar transistors: how to solve the transfer function of a simple amplifier built around bipolar transistors? J. J. Ebers and J. L. Moll derived a small-signal model whose electrical equivalent subcircuit was simply inserted in the original amplifier schematic, in place of the bipolar symbol. That way, since the circuit was made of resistors, capacitors together with a linear representation of the transistor, traditional analysis methods were applicable, such as the Laplace transform. Figure 2-14 illustrates this principle.

In ac, as the dc line voltage derivative $\dfrac{dV_g(t)}{dt}$ value becomes null (simply because it is held constant for the analysis), node 5 goes to ground and biasing resistors are in parallel, connected to the ground. The same applies for R_c, also folded to ground. Hence, if we write the mesh and nodes equations, we have

$$i_b = \frac{V_{in} - V_e}{h_{11}} \tag{2-55}$$

$$V_e = (\beta + 1)i_b R_e \| C_e = (\beta + 1)i_b \frac{R_e}{1 + R_e C_e s} \tag{2-56}$$

Inserting Eq. (2-56) into Eq. (2-55), we obtain the base current definition:

$$i_b = \frac{V_{in}}{h_{11} + \dfrac{R_e}{1 + R_e C_e s}(\beta + 1)} \tag{2-57}$$

The output voltage is simply

$$V_{out} = -\beta i_b R_c \tag{2-58}$$

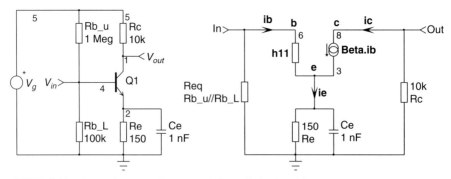

FIGURE 2-14 The transistor symbol is replaced by its small-signal model.

Substituting Eq. (2-57) into Eq. (2-58) gives

$$\frac{V_{out}(s)}{V_{in}(s)} = -\frac{\beta R_c}{h_{11} + \frac{R_e}{1 + R_e C_e s}(\beta + 1)} \approx -\frac{R_c}{R_e}(1 + R_e C_e s) \qquad (2\text{-}59)$$

Inserting new elements such as filters or attenuators, would not change the Ebers–Moll model but would simply require one to rewrite updated mesh and node equations. That is the advantage of linearizing the transistor alone, rather than the whole circuit. This is exactly what Dr. Vorpérian thought about: since the nonlinearity comes from the discontinuity imposed by the switches' operation, let us linearize this switching network alone: the PWM switch model was born.

2.2.2 An Invariant Internal Architecture

The term *invariant* means that the PWM switch structure is identical, in its electrical definition, whatever two-switch converters we talk about. The switch actually lumps into a single model the combined action of the main power switch (SW in the previous illustrations) and the diode D. If we go back to the basic structures, buck, buck-boost, and boost as shown in Figs. 2-15, 2-16 and 2-17, we can clearly see a switch arrangement where the following nodes appear:

- The *active* node, the switch terminal not connected to the diode
- The *passive* node, the diode terminal not connected to the switch
- The *common* node, the junction between the diode and the power switch terminals

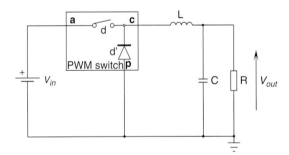

FIGURE 2-15 The PWM switch in a buck converter.

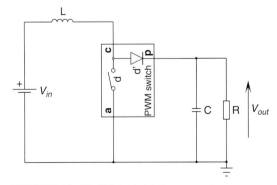

FIGURE 2-16 The PWM switch in boost converter.

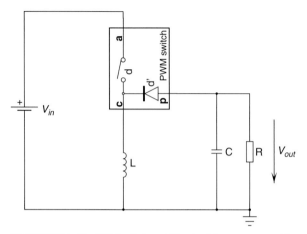

FIGURE 2-17 The PWM switch model in a buck-boost converter. The diode polarity has changed.

Please note that this arrangement does not depend on the operating mode (CCM or DCM). Also, it is sometimes necessary to "scratch" one's head in order to discover the PWM switch model position, for instance, for the flyback converter or the SEPIC as we will see later. The switch and diode configuration can also be better understood via Fig. 2-18. We can actually see a switch (SW) activated during a dT_{sw} period of time, whereas the diode D is activated during a $d'T_{sw}$ period of time. All their respective losses will also be weighted by the same amount of time.

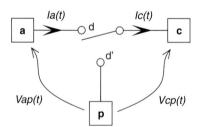

FIGURE 2-18 A single-pole double-throw configuration can help to better visualize the PWM switch.

Exactly as we did in the SSA example, we first need to identify the PWM switch variables, average them, and perturb them to extract a small-signal model. As you can see, these key-words always come back when you are talking about small-signal modeling.

2.2.3 Waveform Averaging

Understanding this term is key to proceeding with the PWM switch derivation. Waveform averaging actually consists of identifying the waveforms across the terminals of interest, e.g., their voltage and current, and averaging them over a switching period. We then toggle from instantaneous values to averaged values. If we take the waveform in Fig. 2-19a as an example,

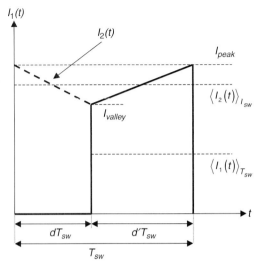

FIGURE 2-19a A typical instantaneous waveform.

we can derive its average value fairly quickly: the original waveform shows a signal $I_1(t)$ which is zero during the on time but jumps to $I_2(t)$ during the off time. Physically, this can be seen as a switch interrupting a current $I_2(t)$ to generate an output current $I_1(t)$ alternating between 0 and $I_2(t)$ (Fig. 2-19b). To simplify the equation, the current $I_2(t)$ can be represented by its average value $\langle I_2(t) \rangle_{T_{sw}}$, which is nothing but $\dfrac{I_{peak} + I_{valley}}{2}$. As a result, the output alternates between 0 and $\langle I_2(t) \rangle_{T_{sw}}$: 0 during dT_{sw} and $\langle I_2(t) \rangle_{T_{sw}}$ during $(1-d)T_{sw}$ or $d'T_{sw}$. We could also say that $I_1(t)$ represents the sampled version of $I_2(t)$. Mathematically, this description can be written as

$$\langle I_1(t) \rangle_{T_{sw}} = I_1 = \frac{1}{T_{sw}} \int_0^{T_{sw}} I_1(t)\, dt = d' \langle I_2(t) \rangle_{T_{sw}} = d' I_2 \qquad (2\text{-}60)$$

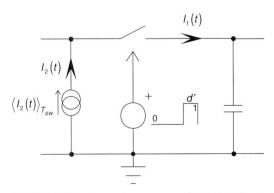

FIGURE 2-19b A sampling system routing $I_2(t)$ during d' only.

Please note the change from $I_2(t)$, an instantaneous value, to I_2, which is the averaged value $\langle I_2(t) \rangle_{T_{sw}}$. Here I_2 represents a dc level, but it can be ac modulated, as we explained in the beginning of this chapter. The next step will consist of identifying the PWM switch currents and voltages.

2.2.4 Terminal Currents

To describe the terminal currents, let us consider the Fig. 2-18 representation, now dropped into a buck converter as Fig. 2-20 shows.

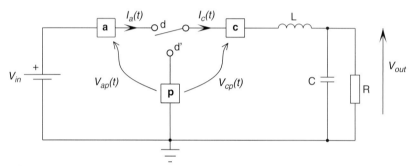

FIGURE 2-20 The PWM switch in a buck converter, showing its double-pole single-throw architecture.

By inspection, we can see that $I_a(t)$ equals $I_c(t)$ when the switch closes, i.e., during dT_{sw}. For the remaining portion of time, $I_a(t)$ becomes zero, whereas $I_c(t)$ keeps flowing through the p terminal. The graphical representation of these currents is shown in Fig. 2-21.

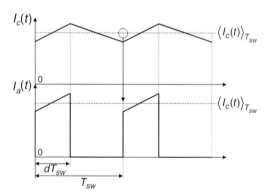

FIGURE 2-21 Idealized time variations of $I_a(t)$ and $I_c(t)$ currents.

Again, $I_c(t)$ can be represented as a sampled version of $I_a(t)$, described via the following equation:

$$\langle I_a(t) \rangle_{T_{sw}} = I_a = \frac{1}{T_{sw}} \int_0^{T_{sw}} I_a(t) dt = d \langle I_c(t) \rangle_{T_{sw}} = dI_c \tag{2-61}$$

Now, if you look at Figs. 2-16 and 2-17 and identify waveforms as we did above, you will find the same relationships: terminal currents are invariant.

2.2.5 Terminal Voltages

We can now run the same exercise with terminal voltages, again observing Fig. 2-20. Regardless of the position of the switch, $V_{ap}(t)$ is always a dc level: V_{in} for the buck, $V_{out} - V_{in}$ for the buck-boost and $-V_{out}$ for the boost. Figure 2-22 graphically represents these waveforms.

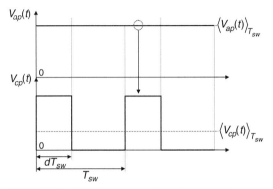

FIGURE 2-22 Graphical representation of terminal voltages.

The average derivation of V_{cp} is straightforward, as it is the average voltage of the square wave signal of $\langle V_{ap}(t) \rangle_{T_{sw}}$ peak amplitude:

$$\langle V_{cp}(t) \rangle_{T_{sw}} = V_{cp} = \frac{1}{T_{sw}} \int_0^{T_{sw}} V_{cp}(t)\, dt = d \langle V_{ap}(t) \rangle_{T_{sw}} = dV_{ap} \qquad (2\text{-}62)$$

2.2.6 A Transformer Representation

From the two main equations, (2-61) and (2-62), we can try to draw a simple schematic of current and voltage sources, illustrating their relationship. This is what Fig. 2-23 offers: a controlled current source on the input and a controlled voltage source on the output. Does it ring a bell? Of course, the Fig. 2-7 dc transformer of variable turns ratio d used in the buck converter small-signal model! Figure 2-24 reproduces this idea.

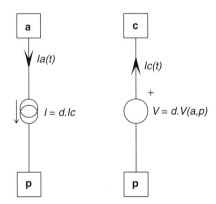

FIGURE 2-23 Electrical representations of Eqs. (2-61) and (2-62).

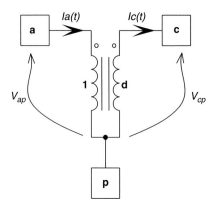

FIGURE 2-24 A simple $1:d$ transformer: the PWM switch!

This electrical representation requires a few comments, however:

- Based on Figs. 2-21 and 2-22, the transformer ratio ($1:d$) is only representative of CCM.
- We are in presence of a large-signal nonlinear model.

2.2.7 Large-Signal Simulations

Since we have a large-signal model, we can immediately use it in a simulation as SPICE will automatically reduce it to a linear small-signal approximation. Let us try a boost converter this time. The PWM switch is placed according to Fig. 2-25, and the dc transformer described in SSA is used here. We can very quickly place a dc bias to fix the duty cycle (1 V represents 100% so 0.4 is 40%) and superimpose an ac source. Plotting V_{out} immediately gives us the transfer function $\dfrac{V_{out}(s)}{d(s)}$ as portrayed by Fig. 2-26. Further to an ac analysis, we can also perform a dc sweep where SPICE will vary a given source from an initial point to a final value via a selected step. If we select V_{bias} as the swept source, then we can reveal the static transfer function of the boost converter (Fig. 2-27):

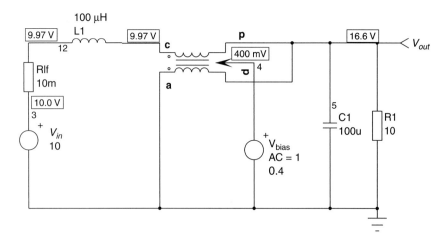

FIGURE 2-25 A boost converter using the PWM switch model in CCM.

SMALL-SIGNAL MODELING

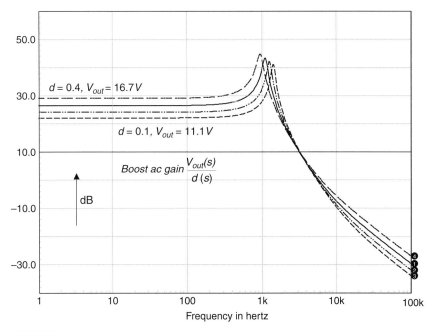

FIGURE 2-26 Ac sweeps showing the low-frequency gain and peaking variations.

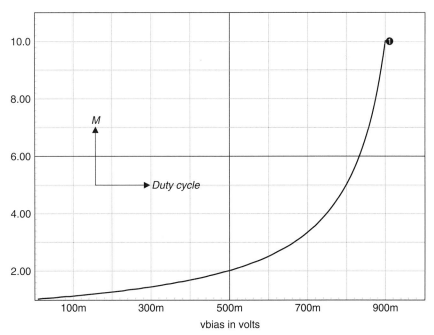

FIGURE 2-27 The boost dc transfer function depending on the duty cycle d. No ohmic losses.

```
.DC Vbias 0.01 0.9 0.01    ;dc sweep the source Vbias from 0.01 V to
                           ;0.9 V by 10 mV steps
```

We can also derive it by hand as we know that for a dc bias point analysis, SPICE short-circuits all inductors and opens all capacitors. The schematic now becomes a simple transformer delivering voltage to load R (Fig. 2-28). Let us write the pertinent mesh equations:

$$V_{in} - V_{cp} = V_{out} \tag{2-63}$$

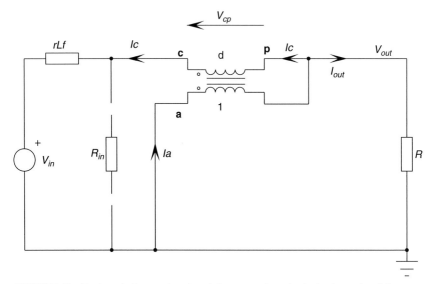

FIGURE 2-28   The boost in dc operation where inductors are short-circuited and capacitors left open.

Observing the transformer position, we clearly see that $-V_{out}$ appears on the primary of ratio 1, thus

$$V_{cp} = -V_{out}d \tag{2-64}$$

Replacing V_{cp} in Eq. (2-63) gives

$$V_{in} + V_{out}d = V_{out} \tag{2-65}$$

and factoring V_{out} finally gives

$$\frac{V_{out}}{V_{in}} = \frac{1}{1-d} = \frac{1}{d'} \tag{2-66}$$

which is the boost dc transfer function M.

In Fig. 2-28, a resistor R_{in} appears. It is supposed to represent the boost dc input resistance. We are going to derive it quickly in a few simple lines, demonstrating the power of the PWM switch model. The dc transformer being a perfect element, we can write

$$N_1 I_c = N_2 I_a \tag{2-67}$$

Recognizing that $N_1 = d$ and $N_2 = 1$, we can update the above equation as

$$dI_c = I_a \tag{2-68}$$

SMALL-SIGNAL MODELING

In the configuration, I_c represents the boost input current flowing through the input source V_{in}. If R_{in} plays the role of the equivalent input resistance we are looking for, then

$$I_c = -\frac{V_{in}}{R_{in}} \quad (2\text{-}69)$$

Applying Kirchhoff's law on the output currents, we have

$$I_c = I_a - I_{out} \quad (2\text{-}70)$$

Now combining Eqs. (2-68) to (2-70), we can solve the following equation:

$$-\frac{V_{in}}{R_{in}} = -\frac{dV_{in}}{R_{in}} - \frac{V_{out}}{R} \quad (2\text{-}71)$$

Factoring gives

$$-\frac{V_{in}}{R_{in}} = \frac{-dV_{in}R - V_{out}R_{in}}{R_{in}R} \quad (2\text{-}72)$$

Simplifying by R_{in} gives

$$-V_{in} = \frac{-dV_{in}R - V_{out}R_{in}}{R} = -dV_{in} - V_{out}\frac{R_{in}}{R} \quad (2\text{-}73)$$

or

$$\frac{V_{in}}{V_{out}}(1 - d) = \frac{R_{in}}{R} \quad (2\text{-}74)$$

Finally, calling on Eq. (2-66), we obtain the boost input dc resistance

$$R_{in} = R(1 - d)^2 = Rd'^2 \quad (2\text{-}75)$$

This result will be of great help in reflecting elements on the other side of the transformer.

2.2.8 A More Complex Representation

The boost converter, as any other inductor-based devices, can be upgraded with the inductor ohmic losses. This is done by simply inserting a resistor in series with L as in Fig. 2-29. Also,

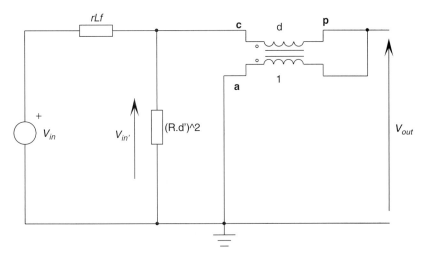

FIGURE 2-29 Updated boost converter in dc conditions, featuring the inductor ohmic losses.

to simplify the equations, why not reflect R on the other side of the transformer? It is easy, thanks to Eq. (2-75).

By looking at this circuit, we can derive the transfer function a snapshot since we have a resistive divider followed by a transformer obeying Eq. (2-66):

$$\frac{V_{out}}{V_{in}} = \frac{1}{d'} \frac{Rd'^2}{Rd'^2 + r_{Lf}} = \frac{1}{d'} \frac{1}{1 + \frac{r_{Lf}}{Rd'^2}} \quad (2\text{-}76)$$

In this equation, if r_{Lf} is small compared to R, then the boost transfer function is almost not disturbed. On the contrary, if this condition is not met, then the boost conversion ratio severely suffers. Figure 2-30 shows an update of the Fig. 2-27 simulation results with r_{Lf} swept from 1 Ω down to 100 mΩ.

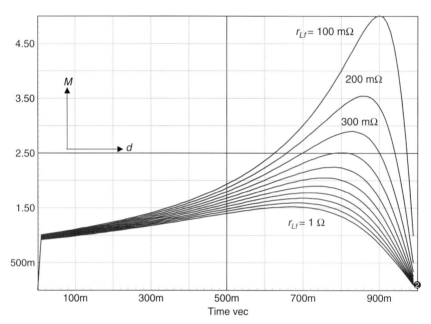

FIGURE 2-30 If inductor ohmic losses are too large, the boost ratio exhibits a maximum with respect to the duty cycle.

At a certain time, if the designer requires a higher conversion ratio, the boost converter simply collapses its output. We can evaluate the point at which the output folds back. We can derive Eq. (2-76) and look at which value of d' nulls it:

$$\frac{d}{dd'}\left(\frac{1}{d'} \frac{1}{1 + \frac{r_{Lf}}{Rd'^2}} \right) = 0 \quad (2\text{-}77)$$

$$\frac{Rr_{Lf} - R^2 d'^2}{\left(Rd'^2 + r_{Lf}\right)^2} = 0 \quad (2\text{-}78)$$

Solving for d' gives

$$d' = \frac{1}{R}\sqrt{Rr_{Lf}} = \sqrt{\frac{r_{Lf}}{R}} \qquad (2\text{-}79)$$

If we reinject Eq. (2-79) into Eq. (2-76), we obtain the maximum transfer ratio we can get from a certain load/inductor ohmic losses configuration:

$$M_{max} = \frac{R}{2\sqrt{Rr_{Lf}}} = \frac{1}{2}\sqrt{\frac{R}{r_{Lf}}} \qquad (2\text{-}80)$$

If, from Fig. 2-30, r_{Lf} equals 100 mΩ, then Eq. (2-79) gives a duty cycle d of 90% (900 mV) and Eq. (2-80) gives us a maximum gain M_{max} of 5. This is confirmed by the upper curve.

2.2.9 A Small-Signal Model

We will apply the same technique as for the SSA small-signal model, but this time directly on the model. We will then perturb Eqs. (2-61) and (2-62), applying the following definitions:

- $d = D + \hat{d}$ where D is steady-state duty cycle and $\hat{d}$ is ac small-signal modulation
- $I_a = I_{a0} + \hat{I}_a$ where I_{a0} is the steady-state direct current and $\hat{I}_a$ the ac perturbation
- $I_c = I_{c0} + \hat{I}_c$ where I_{c0} is the steady-state direct current and $\hat{I}_c$ the ac perturbation

It can be rather weird to talk about ac values for already averaged signals, such as I_a and I_c. But if we look back at Fig. 2-2, after all, the averaged value undergoes a modulation as the original variable does. We can then identify a dc portion and a modulating envelope.

For Eq. (2-61) we then have

$$I_{a0} + \hat{I}_a = (D + \hat{d})(I_{c0} + \hat{I}_c) = DI_{c0} + D\hat{I}_c + \hat{d}I_{c0} + \underbrace{\hat{d}\hat{I}_c}_{\approx 0} \qquad (2\text{-}81)$$

As we did before for the SSA, we can split Eq. (2-81) into two equations, ac and dc, neglecting ac cross-products (small multiplied by small becomes even smaller):

$$I_{a0} = DI_{c0} \qquad (2\text{-}82)$$

$$\hat{I}_a = D\hat{I}_c + \hat{d}I_{c0} \qquad (2\text{-}83)$$

The same applies for the voltage relationship, Eq. (2-62):

$$V_{cp0} + \hat{V}_{cp} = (D + \hat{d})(V_{ap0} + \hat{V}_{ap}) = DV_{ap0} + D\hat{V}_{ap} + \hat{d}V_{ap0} + \underbrace{\hat{d}\hat{V}_{ap}}_{\approx 0} \qquad (2\text{-}84)$$

Sorting out a dc and an ac equation yields

$$V_{cp0} = DV_{ap0} \qquad (2\text{-}85)$$

and $$\hat{V}_{cp} = D\hat{V}_{ap} + \hat{d}V_{ap0} \qquad (2\text{-}86)$$

Equations (2-83) and (2-86) describe the small-signal behavior of the PWM switch model operating in voltage mode and in continuous conduction mode. In light of these new definitions, we can update the original model. The dc values (subscript 0) will be either passed parameters or computed during dc point analysis whereas $\hat{d}$ will connect to the ac modulation for frequency sweep purposes. Figure 2-31 illustrates this upgraded model.

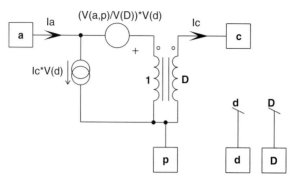

FIGURE 2-31 The small-signal model featuring an ac input (d) and a dc bias input (D).

To compute by hand the dc operating point, null all small-signal sources (open current sources, short voltage sources) and compute the mesh/loop equations in the converter under study. Then use them to compute the selected ac transfer function. Let's start by deriving the output-to-input voltage transfer function $\frac{V_{out}(s)}{V_{in}(s)}$ for the boost converter. We can reuse the small-signal model and set all ac sources to zero as d is fixed for the input perturbation analysis. The new schematic is shown in Fig. 2-32.

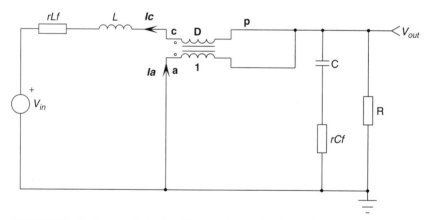

FIGURE 2-32 The boost small-signal model arranged to reveal its input rejection transfer function.

As it is, the sketch does not offer much insight. Thanks to Eqs. (2-66) and (2-75), we can rearrange all elements in a more convenient way.

From Fig. 2-33, we recognize an *RLC* filter, followed by a block (the transformer) featuring a "gain" of $\frac{1}{D'}$ [Eq. (2-66)]. We have already solved such a passive combination in Chap. 1,

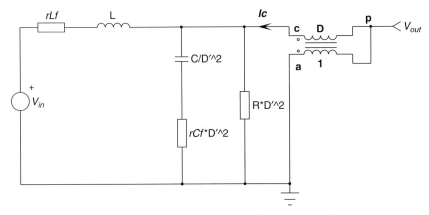

FIGURE 2-33 The boost rearranged further to the primary reflection of the loading elements.

App. 1A. The final result is

$$\frac{V_{out}(s)}{V_{in}(s)} = \frac{1}{D'} \frac{1}{\frac{R_1}{R_3} + 1} \frac{1 + sR_2C}{s^2 LC\left(\frac{R_3 + R_2}{R_1 + R_3}\right) + s\frac{L + C(R_2R_3 + R_1R_3 + R_2R_1)}{R_1 + R_3} + 1} \quad (2\text{-}87)$$

with $R_3 = RD'^2$, $R_2 = r_{Cf}D'^2$, $R_1 = r_{Lf}$, and $C = C/D'^2$. This equation represents a second-order system following the form

$$T(s) = M \frac{1 + \frac{s}{s_{z1}}}{\frac{s^2}{\omega_0^2} + \frac{s}{\omega_0 Q} + 1} \quad (2\text{-}88)$$

We can thus identify all pertinent terms by replacing R_1, R_2, and R_3 by their respective values:

$$M = \frac{1}{D'} \frac{1}{1 + \frac{r_{Lf}}{D'^2 R}} \quad (2\text{-}89)$$

$$s_{z1} = \frac{1}{r_{Cf} C_f} \quad (2\text{-}90)$$

$$\omega_0 = \frac{1}{\sqrt{LC}} \sqrt{\frac{R_1 + R_3}{R_2 + R_3}} = \frac{1}{\sqrt{L\frac{C}{D'^2}}} \sqrt{\frac{r_{Lf} + D'^2 R}{D'^2(r_{Cf} + R)}} = \frac{1}{\sqrt{LC}} \sqrt{\frac{r_{Lf} + D'^2 R}{r_{Cf} + R}} \quad (2\text{-}91)$$

$$Q = \frac{\omega_0}{\frac{r_{Lf}}{L} + \frac{1}{C(r_{Cf} + R)}} \quad (2\text{-}92)$$

Now, we can use the small-signal model and derive the complete ac transfer function $\frac{V_{out}(s)}{d(s)}$ for the boost converter. Please note that we now have added an ESR to the capacitor r_{Cf}.

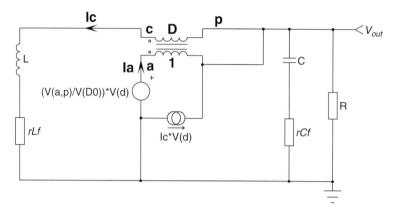

FIGURE 2-34 The small-signal boost converter arranged to derive $\dfrac{V_{out}(s)}{d(s)}$.

First, as in the bipolar example, set the input voltage to 0 since it won't move in ac. Redrawing the boost converter with the Fig. 2-31 model gives Fig. 2-34.

As it is now, solving equations directly would be quite tedious given the transformer position. Capitalizing on the previous findings, we can rotate the elements and build a compatible circuit, giving us greater insight into the operation:

- The source $(V(a, p)/V(D0))*V(d)$ can slide along the lower leg and come in series with r_{Lf}. D_0 represents the static duty cycle, the 0 being added as SPICE does not distinguish between nodes labeled D and d. As the source goes on the other side of the transformer, it must be divided by D [or $V(D0)$ in SPICE].
- The node a now goes to ground, hence $V(a,p)$ becomes $V(0,p)$.
- The equivalent small-signal current source can also go on the other side. After Eq. (2-75), we simply need to divide it by D' or $1-V(D0)$ in SPICE.
- The same applies for the loading elements r_{Cf}, R and C which are, respectively, multiplied by D'^2 (resistive elements) and divided by D'^2.

The updated sketch in Fig. 2-35 now looks simpler to understand (are we sure?).

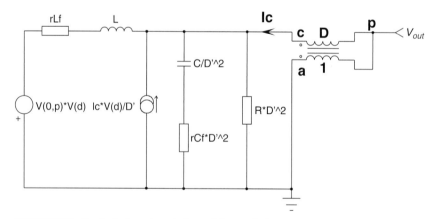

FIGURE 2-35 The final schematic ready for mesh/loop analysis.

Several techniques are available to obtain the final transfer function. For instance, as we are dealing with linear elements, superposition holds. However, given the numerous elements and the two independent sources, it would inevitably lead to complex equations. Complex equations are not a problem in themselves, actually. Programs such as Mathcad can easily numerically extract phase and gain values to quickly generate Bode plots. What poses the problem is to rearrange these equations in order to reveal the position of zeros or poles and then see what elements affect their position and how the stability could be impacted through their respective variations.

Dr. Vorpérian, in his original paper, derived the position of the poles and zeros using fast analytical techniques that he recently thoroughly documented [10]. He applied the fact that regardless of the means you use to excite a multi-input plant (the power supply in our case), the denominator $D(s)$ is always the same in every transfer function. For instance, when one is comparing $\dfrac{V_{out}(s)}{V_{in}(s)}$ and $\dfrac{V_{out}(s)}{d(s)}$, both functions can be put over a common denominator $D(s)$. The roots of this $D(s)$ are nothing else than what we already derived when dealing with Eq. (2-88). The exercise solely consists of finding the position of the zeros induced by the Fig. 2-35 circuitry. If poles can be seen as roots of the denominator that brings the transfer function to infinity, then zeros, on the contrary, null the numerator, hence the transfer function. The exercise usually lies in "inspecting" the network architecture and locating the elements' associations that could bring the numerator to zero. In Fig. 2-35, one can see that if the network made of the series connection of $\dfrac{C}{D'^2}$ and $r_{Cf}D'^2$ imposes a zero impedance, all the transfer function cancels. Therefore, we can write

$$\frac{1}{sCD'^2} + r_{Cf}D'^2 = 0 \tag{2-93}$$

whose solution is

$$s_{z1} = -\frac{1}{Cr_{Cf}} \tag{2-94}$$

The other zero, s_{z2}, is located in the right half portion of the s plane. This is often referred to in the literature as an RHPZ (right half-plane zero). The s plane describes the positions of zeros and poles, based on their respective imaginary (y axis) and real (x axis) portions. Positioning zeros and poles in this s plane gives stability indications for the system under study. For instance, we know that if all poles of the characteristic equation of a system ($D(s) = 0$) lie in the left half-plane, the system is said to be stable. The RHPZ, on the other hand, differs from a traditional zero (which resides in the left half, an LHPZ) in the sense that instead of boosting the phase as an LHPZ would, it lags it further down, often destroying the phase margin and engendering oscillations. Appendix 2B gives you a quick overview of poles and zeros and why their positions are important.

In a boost converter operating in CCM (you won't have a noticeable RHPZ effect in DCM), a sudden increase in the duty cycle (due to a load change, for instance) will first decrease the output voltage, thus sending it in the wrong direction. Why is that, since the feedback instructs it to increase? The answer is simply that the boost belongs to the indirect energy transfer converters family (like the buck-boost and the flyback): You first store energy during DT_{sw}, then dump it into the output capacitor during $(1 - D)T_{sw}$. The diode output average current depends on the $(1 - D)$ term whereas the transistor peak current depends on the D term. If you increase D in response to a load step, you reduce $(1 - D)$, hence the average output current and the voltage go low. It is impossible to fight this RHPZ which often moves in relation to the duty cycle. Therefore, you must fold back the cutoff frequency before its phase lag becomes noticeable. Chapter 3, dedicated to feedback, will review this topic.

Finally, based on Ref. 7, the second zero s_{z2} is defined as follows:

$$s_{z2} = \frac{D'^2}{L}\left(R - r_{Cf}\|R\right) - \frac{r_{Lf}}{L} \tag{2-95}$$

The low-frequency "gain" depends on static parameters such as D, the dc duty cycle or D_0 in our previous notations. We can obtain it by computing $\frac{dV_{out}}{dD}$, where V_{out} equals $V_{in}M$:

$$\frac{dV_{out}}{dD} = V_{in}\frac{dM}{dD} \approx \frac{V_{in}}{D'^2} \tag{2-96}$$

Based on these equations, the complete equation looks like this:

$$\frac{V_{out}(s)}{d(s)} = \frac{V_{in}}{D'^2}\frac{\left(1 + \frac{s}{s_{z1}}\right)\left(1 - \frac{s}{s_{z2}}\right)}{\frac{s^2}{\omega_0^2} + \frac{s}{\omega_0 Q} + 1} \tag{2-97}$$

2.2.10 Helping with Simulation

Fortunately, SPICE shields us from any hand derivation. As an example, this subcircuit can now be directly dropped in place of the large-signal model to demonstrate its working abilities. An independent source generates the necessary dc bias for the D_0 variable, but we can upgrade the circuit by adding the simplest error amplifier we can find: an E_1 source. Figure 2-36 shows the corresponding architecture.

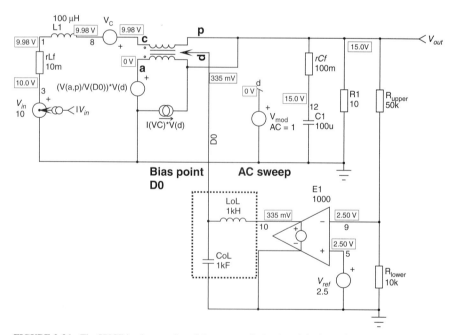

FIGURE 2-36 The SPICE implementation of the new small-signal model using a simple voltage-controlled voltage source as an error amplifier (E_1).

On this small-signal boost, the E_1 source amplifies by 60 dB the difference between a portion of V_{out} and the 2.5 V reference. However, if we directly connected its output to the D_0 input, we would run the circuit in a closed-loop configuration, which we do not want. For that reason, we can install an LC filter featuring an extremely low cutoff frequency: during bias point analysis, LoL is short-circuited, CoL is opened, and the right duty cycle is computed, according to the output voltage set point (given by R_{upper}, R_{lower}, and V_{ref}). Here, we can read 419 mV or 41.9%. The nice thing is that when you move V_{in} or R_1, the duty cycle will automatically adjust at the beginning of the ac simulation (during the bias point calculation) to deliver 15 V (of course, given that the input/output conditions are compatible with the converter operation). Now, when the ac analysis starts, LoL and CoL form a filter which will stop any ac excitation: it is open-loop. We will come back to this technique later. Here, we can plot the output voltage in decibels which directly corresponds to $20\log_{10}\left(\dfrac{V_{out}}{V_d}\right)$ since the stimulus level is 1 V. Figure 2-37 displays the results in a Bode plot for various values of the inductor series resistance r_{Lf}:

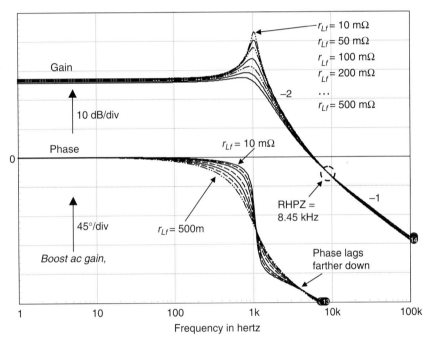

FIGURE 2-37 Transfer function variations depending on the damping introduced by r_{Lf}.

2.2.11 Discontinuous Mode Model

So far, we described the PWM switch in a continuous current environment. In his second paper, Dr. Vorpérian also tackled the DCM analysis, but using a different switch arrangement. If you look at Fig. 2-20, the PWM switch is configured in a "common passive" configuration (a bit like a common emitter for a bipolar stage). For various reasons, Vorpérian derived its discontinuous model in a different configuration, the "common common" configuration. As you can easily understand, merging both models into a single one to cover all possible modes was a difficult exercise. Microsim (the *PSpice* editor) did it, but using two toggling switches which hampered convergence (SWI_RAV.LIB, Sept. 1993). As we will see later, reconfiguring the DCM PWM switch in its original common passive configuration leads to an extremely simple model, naturally toggling between DCM and CCM.

Figure 2-38a represents the buck converter operating in DCM. As we explained, the discontinuous mode gives rise to a third state, where the inductor current has gone to zero. For this study, we will purposely redefine the duty cycle instants as follows:

- $d_1 T_{sw}$: the on time during which SW closes and current increases in the inductor—the magnetization time
- $d_2 T_{sw}$: the off time, during which SW is open and the diode D freewheels—the demagnetization time
- $d_3 T_{sw}$: the dead time (DT) when both switches are opened

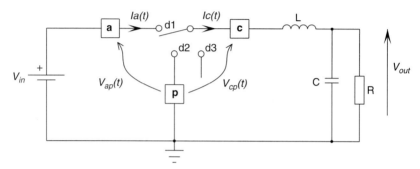

FIGURE 2-38a The buck converter operating in the discontinuous mode of operation.

The relationship that links all events is

$$1 = d_1 + d_2 + d_3 \tag{2-98}$$

The exercise remains the same: identify the PWM switch waveforms and average them over a switching cycle. We will thus draw all the pertinent waveforms and display them in Fig. 2-38b. From these waveforms, we can write a few basic equations. Dealing with triangles makes averaging an easy exercise:

$$I_a = \frac{I_{peak} d_1}{2} \tag{2-99}$$

Now, the average value of I_c is found by summing the half-triangle areas, since the DT area ($d_3 T_{sw}$) is null:

$$I_c = \frac{I_{peak} d_1}{2} + \frac{I_{peak} d_2}{2} = \frac{I_{peak}(d_1 + d_2)}{2} \tag{2-100}$$

From Eq. (2-100), we can derive I_{peak}

$$I_{peak} = \frac{2 I_a}{d_1} \tag{2-101}$$

and substitute it into Eq. (2-100). We obtain the relationship between I_a and I_c:

$$I_c = \frac{2 I_a}{d_1} \frac{d_1 + d_2}{2} = I_a \frac{d_1 + d_2}{d_1} \tag{2-102}$$

From Eq. (2-98), the difference between CCM and DCM lies in the presence of $d_3 T_{sw}$. When this term vanishes to zero, the PWM switch enters CCM. In Eq. (2-102), if $d_2 = 1 - d_1$, or $d_3 T_{sw} = 0$, it simplifies to $I_a = I_c d_1$ which is Eq. (2-61).

SMALL-SIGNAL MODELING

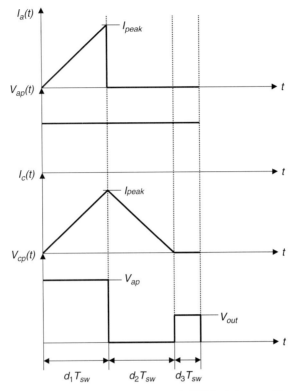

FIGURE 2-38b Waveforms pertinent to the PWM switch operated in DCM.

We can now average the V_{cp} waveform, seeing that the $d_3 T_{sw}$ period offers a high impedance state, letting V_{out} freely appear across the p and c terminals:

$$d_3 = 1 - d_1 - d_2 \tag{2-103}$$

$$V_{cp} = V_{ap} d_1 + V_{cp}(1 - d_1 - d_2) \tag{2-104}$$

$$V_{cp} - V_{cp}(1 - d_1 - d_2) = V_{ap} d_1 \tag{2-105}$$

$$V_{cp} = V_{ap} \frac{d_1}{d_1 + d_2} \tag{2-106}$$

Again, this equation simplifies to $V_{cp} = V_{ap} d_1$ when $d_2 = 1 - d_1$. The equation becomes the CCM one [see Eq. (2-62)].

If we look back at Eqs. (2-102) and (2-106), we again have a simple transformer whose turns ratio now depends upon $\frac{d_1}{d_1 + d_2}$. Figure 2-39 shows the new configuration for the DCM switch model.

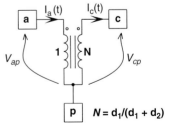

FIGURE 2-39 A simple dc transformer affected by a $\frac{d_1}{d_1 + d_2}$ ratio.

2.2.12 Deriving the d_2 Variable

If $d_1 T_{sw}$ is imposed by the controller, d_2 needs to be computed as its value depends on the demagnetization time, hence current and inductance value. So d_2 can be derived by looking at Fig. 2-38a, using the buck configuration. By observing that $V_a = V_{in}$ and $V_c = V_{out}$ on average, a second equation for the peak current definition can be derived as the switch closes during the on time:

$$V_{ac} = L \frac{I_{peak}}{d_1 T_{sw}} \qquad (2\text{-}107)$$

From Eq. (2-100), we can see that

$$I_{peak} = \frac{2 I_c}{d_1 + d_2} \qquad (2\text{-}108)$$

Extracting I_{peak} from Eq. (2-107) and substituting it into Eq. (2-108) lead to

$$\frac{2 I_c}{d_1 + d_2} = \frac{V_{ac} d_1 T_{sw}}{L} \qquad (2\text{-}109)$$

Solving for d_2 gives us the final equation we are looking for:

$$d_2 = \frac{2 I_c L - V_{ac} d_1^2 T_{sw}}{V_{ac} d_1 T_{sw}} = \frac{2 L F_{sw}}{d_1} \frac{I_c}{V_{ac}} - d_1 \qquad (2\text{-}110)$$

We are all set! We know that both founding Eqs. (2-102) and (2-106) naturally toggle from DCM to CCM when the Eq. (2-110) result hits $1 - d_1$. By simply clamping the d_2 generator [Eq. (2-110)] between 0 and $1 - d_1$, the model transitions between CCM and DCM automatically. When d_2 computed by Eq. (2-110) hits $1 - d_1$, it simply means that the switch is at the limit between DCM and CCM.

We could also add a dependent source to tell us in which mode the model stays. In CCM, the off duty cycle d_2 equals $1 - d_1$. However, in DCM, the off duty cycle d_2 accounts for the presence of the dead time d_3 where the inductor current has collapsed to zero. We can therefore write two simple equations, implying these variables:

$$d_{2,CCM} = 1 - d_1 \qquad (2\text{-}111)$$

$$d_{2,DCM} = 1 - d_1 - d_3 \qquad (2\text{-}112)$$

From Eq. (2-112), we can see that $d_{2,DCM}$ is smaller than $d_{2,CCM}$, since d_3 is further subtracted. Therefore, if we check what the simulator computes for d_2 and compare it to Eq. (2-111), we can deduce the operating mode. A simple in-line equation can do the job for us, where $V(d_2)$ comes from Eq. (2-110) and $V(d)$ represents the control node:

```
Bmode mode 0 V= (2*{L}*{Fs}*I(VM)/(V(dc)*V(a,c)+1u)) - V(dc)<
  1-v(dc) ? 0 : 1
IF (2*{L}*{Fs}*I(VM)/(V(dc)*V(a,cx)+1u)) - V(dc) is smaller
   than 1-V(dc)
THEN V(mode) equals 0 (we are in DCM)
ELSE V(mode) equals 1 (we are in CCM)
```

We could include it in the next model as an output variable giving the operating mode.

2.2.13 Clamping Sources

We have seen from the above that a need exists to clamp the various generators. First, the model duty cycle input node d (same as d_1 for DCM) must evolve between 0 and 100%. If we select 1 V to be 100%, then the control circuit will not deliver more than this value. Otherwise either nonconvergence issues can arise or wrong operating points can be calculated. To clamp a source in SPICE, several options exist. The first one implements a single in-line equation. Suppose we want to limit the duty cycle excursion between 10 mV (1%) and 999 mV (99%). The equation could look like this:

```
Bd dc 0 V = V(d) < 10m ? 10m : V(d) > 999m ? 999m : V(d) ; IsSpice
    ;syntax
IF the "d" node is smaller than 10 mV, THEN the Bd source
    delivers 10 mV
ELSE
IF the "d" node is greater than 999 mV, THEN the Bd source
    delivers 999 mV
ELSE the Bd source delivers the "d" node value
```

In *PSpice*, you have the choice between two equations:

```
Ed dc 0 Value ={IF (V(d) < 10m, 10m, IF (V(d) > 999m, 999m,
+ V(d) ))} ; PSpice
```

or, using the keyword *table*,

```
Ed dc 0 TABLE {V(d)} ((10m,10m) (999m,999m)) ; PSpice
```

However, experience shows that this sort of equation can sometimes bring convergence issues. For this reason, we recommend using an active clamp built with a few passive elements that make the model more robust at the end. We recommend this because the diodes start clamping smoothly compared to an in-line equation, which is extremely sharp in essence. Figure 2-40 shows this circuitry.

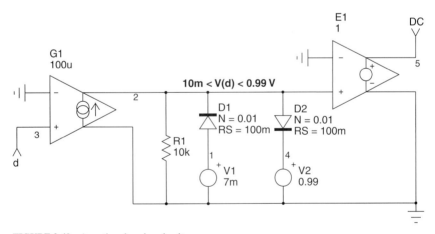

FIGURE 2-40 An active clamping circuit.

The *N* parameter on diodes allows one to reduce the forward drop to nearly zero. This is the so-called emission coefficient. The output E_1 amplifier buffers the clamped output, which can be freely used elsewhere in the model.

The second clamp, for d_2, works in almost the same manner except that the positive clamp depends on $1 - d_1$ now. See Fig. 2-41. The input d2NC gets a voltage computed by Eq. (2-110), and its excursion is limited by the B_1 source to $1 - d$. Please note the 6.687 mV term that perfectly compensates for the diode D_2 forward drop.

134 CHAPTER TWO

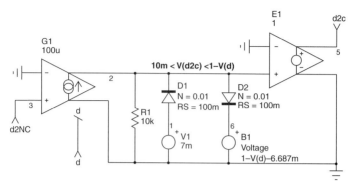

FIGURE 2-41 The clamp circuit now limits the d_2 generator.

2.2.14 Encapsulating the Model

Now that the equations are properly clamped, we can put together the pieces and write the final voltage-mode, autotoggling, PWM switch model. This is what the following lines describe in both *IsSpice* and *PSpice*:

IsSpice netlist:

```
.SUBCKT PWMVM a c p d {L=75u Fs=100k}
*
*This subckt is a voltage-mode DCM-CCM model
*
.subckt limit d dc params: clampH=0.99 clampL=16m
Gd 0 dcx d 0 100u
Rdc dcx 0 10k
V1 clpn 0 {clampL}
V2 clpp 0 {clampH}
D1 clpn dcx dclamp
D2 dcx clpp dclamp
Bdc dc 0 V=V(dcx)
.model dclamp d n=0.01 rs=100m
.ENDS
*
.subckt limit2 d2nc d d2c
Gd 0 d2cx d2nc 0 100u
Rdc d2cx 0 10k
V1 clpn 0 7m
BV2 clpp 0 V=1-V(d)-6.687m
D1 clpn d2cx dclamp
D2 d2cx clpp dclamp
B2c d2c 0 V=V(d2cx)
.model dclamp d n=0.01 rs=100m
.ENDS
*
Xd d dc limit params: clampH=0.99 clampL=16m
BVcp 6 p V=(V(dc)/(V(dc)+V(d2)))*V(a,p)
BIap a p I=(V(dc)/(V(dc)+V(d2)))*I(VM)
Bd2 d2X 0 V=(2*I(VM)*{L}-v(a,c)*V(dc)^2*{1/Fs}) / ( v(a,c)*V(dc)
+ *{1/Fs}+1u)
```

SMALL-SIGNAL MODELING

```
Xd2 d2X dc d2 limit2
VM 6 c
.ENDS
```

PSpice netlist:

```
.SUBCKT PWMVM a c p d params: L=75u Fs=100k
*
*auto toggling between DCM and CCM, voltage-mode
*
Xd d dc limit params: clampH=0.99 clampL=16m
EVcp 6 p Value = {(V(dc)/(V(dc)+V(d2)))*V(a,p)}
GIap a p Value = {(V(dc)/(V(dc)+V(d2)))*I(VM)}
Ed2 d2X 0 value = {(2*{L}*{Fs}*I(VM)/(V(dc)*V(a,c)+1u)) -
+ V(dc)}
Xd2 d2X dc d2 limit2
VM 6 c
*
.ENDS
****** subckts *****
.subckt limit d dc params: clampH=0.99 clampL=16m
Gd 0 dcx VALUE = {V(d)*100u}
Rdc dcx 0 10k
V1 clpn 0 {clampL}
V2 clpp 0 {clampH}
D1 clpn dcx dclamp
D2 dcx clpp dclamp
Edc dc 0 value={V(dcx)}
.model dclamp d n=0.01 rs=100m
.ENDS
********
.subckt limit2 d2nc d d2c
*
Gd 0 d2cx d2nc 0 100u
Rdc d2cx 0 10k
V1 clpn 0 7m
E2 clpp 0 Value = {1-V(d)- 6.687m}
D1 clpn d2cx dclamp
D2 d2cx clpp dclamp
Edc d2c 0 value={V(d2cx)}
.model dclamp d n=0.01 rs=100m
.ENDS
```

The basic circuit to operate the model is shown in Fig. 2-42. The model is assembled for a buck configuration. The V_{bias} source fixes the dc duty cycle but also modulates the input via the AC keyword. The inductance parameter ($L =$) represents the external inductance value which, together with the switching frequency ($Fs =$), helps determine the operating mode. As you can see, the calculated operating points are back reflected to the schematic. Always display them or look for them in output file (extension .OUT) to see whether SPICE succeeded in finding the right dc point. For example, we are looking for a 5 V buck converter, and it seems that the output voltage is good.

As you can see, the input source V_{bias} fixes the operating point. However, if you change the load or the input voltage, V_{bias} might require a tweak to readjust V_{out} to its original value. There is a little trick introduced a while ago, which helps to automatically adjust the duty cycle to the right point. Figure 2-43 shows how it works. It was originally presented in Fig. 2-36 and appears in Fig. 2-43.

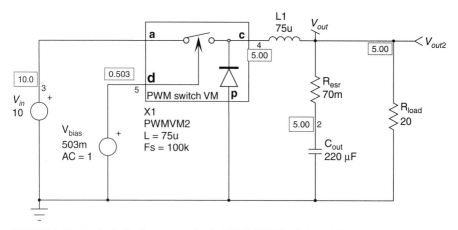

FIGURE 2-42 The basic circuit to operate the CCM-DCM PWM switch model in a voltage-mode type of converter.

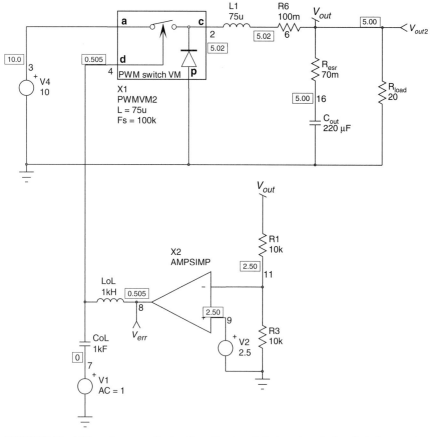

FIGURE 2-43 Adding a large capacitor together with a large inductor automatically adjusts the operating point for you.

We have inserted a resistor in series with L_1, so the dc feedback has automatically adjusted the duty cycle to 50.5% to keep V_{out} to 5 V despite increased ohmic losses. Before any simulation, SPICE must compute a bias point. This is its starting point at $t = 0$. To do so, the simulator short-circuits all inductors and opens all capacitors. In the circuit, this implies that node 8 (the op amp output) is directly connected to the duty cycle input. Therefore, the feedback loop plays its role and increases the duty cycle until the output target is reached. If it cannot regulate, the op amp is pushed to its upper stops (or lower stops depending on the way V_{out} moves), and you see that it failed when observing the output file *.OUT (or looking at the reflected dc points if your simulator allows it). You could also put an E_1 source in place of the op amp, if you wish. Then once the bias point is found, the simulator starts to sweep in ac. But as we have installed the $LoL - CoL$ (extremely) low-pass filter, no ac perturbation can go through: we are in open-loop. If you move R_{load} or V_{in}, the op amp will readjust the duty cycle to reach the output voltage set point. Having the stimulus source placed here offers the ability to ac probe any node with respect to $d(s)$ since the excitation level is 1 V: probe $V(V_{out})$ and you immediately have $\frac{V_{out}(s)}{d(s)}$. Sometimes, some kind of noise appears during the ac sweep, and a small resistor of 100 mΩ can be inserted in series with LoL.

Another solution consists of placing the ac stimulus in series with the operational amplifier, or anywhere else where the following conditions are met:

- The − of the ac source must connect to a low output impedance point.
- The + of the ac source must go to a high input impedance node.

Figure 2-44 shows the connection.

The op amp output is a low-impedance node and so is the duty cycle input. We could also have placed the source in series with R_1. With the series ac source, you must purposely instruct the analysis tool (*IntuScope*, *Probe*, etc.) to display the ratio of interest: VdB (V_{out}/V_{in}) for instance.

How can a source inserted in series in a closed-loop system help display its open-loop characteristics? It can because the stimulus voltage being kept constant during the sweep, that is, $V(V_{in}) - V(V_{err}) = $ constant level, naturally forces the system to stick to this constant characteristic. That is,

- At low frequencies, the open-loop gain is large, thus $V(V_{in})$ is small (the input needs a small excitation to obtain some output) and $V(V_{err})$ is large.
- Close to the 0 dB point, $V(V_{in})$ equals $V(V_{err})$: the gain is unity.
- At higher frequency, as the gain goes low, $V(V_{in})$ is large (the input requires greater signal to observe a signal on the output) but $V(V_{err})$ is smaller.

Figure 2-45 shows how the source was wired, mainly in series with the output voltage, a low-impedance point. We used a voltage controlled oscillator (VCO) model whose output undergoes an attenuation to avoid saturating the circuit. A sinusoidal waveform confirms the good linear behavior as Fig. 2-46 testifies. The resonance can clearly be seen around 3 kHz. Please note that in these transient sweeps, the supply under study must be stable; otherwise the error amplifier will saturate and no useful signal can be observed.

In Fig. 2-45, the *parameter* text section automatically calculates the stabilizing elements given a particular bandwidth wish. It uses the *k* factor, a powerful tool we describe later in the feedback loop section.

The right portion of Fig. 2-45 represents the usual test feature, implemented in the laboratory when you need to measure power supply bandwidth. The SPICE source B_1 is actually replaced by a transformer whose output comes across a 20 Ω resistor. This resistor, installed on the board, offers a useful way to connect the ac source via the transformer and closes the dc loop in the absence of measurement. The ac amplitude must stay reasonable and not bring any

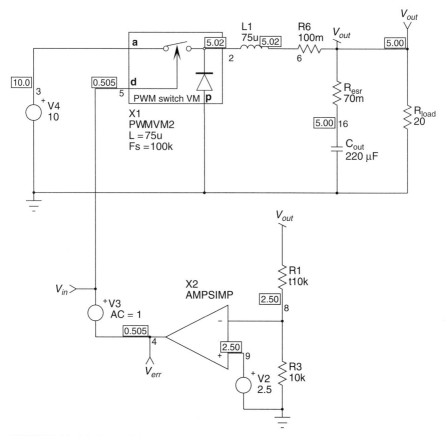

FIGURE 2-44 It is also possible to place the stimulus source in series.

element into saturation: this is a small-signal analysis. An oscilloscope probe can be hooked to the SMPS output or to the error amplifier output to avoid any doubt about the linearity. The V_a and V_b signals go to the network analyzer that performs $20 \log_{10}\left(\dfrac{V_b}{V_a}\right)$ to reveal the open-loop Bode plot. As we have previously stated, this option requires the power supply to be stable. If it is not, a different technique can be used, as we will see.

2.2.15 The PWM Modulator Gain

A voltage-mode power supply regulates its output by generating a pulse width variable pattern. The right width is obtained by comparing the dc error signal (the op amp output) to a fixed-amplitude sawtooth. In Chap. 1, Fig. 1-16a described this method where the comparator output represents the final switching signal. Figure 2-47a and b reproduces it again, updated with critical signals.

The sawtooth is a linear voltage increase defined by

$$V_{saw} = V_{peak}\frac{t}{T_{sw}} \tag{2-113}$$

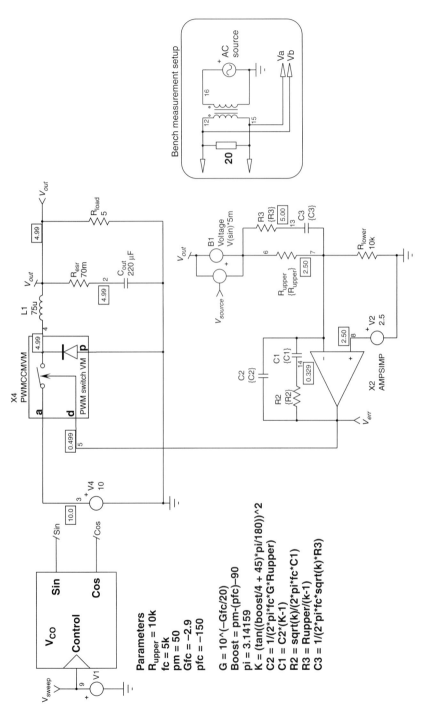

FIGURE 2-45 The power supply test fixture for transient sinusoidal sweep.

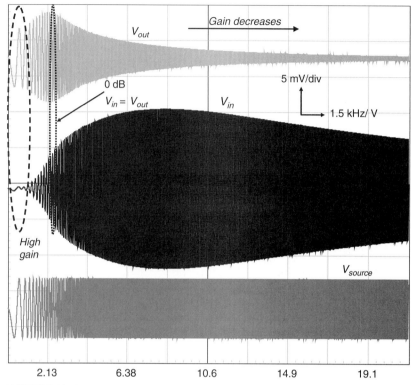

FIGURE 2-46 Transient sweep results obtained from a CCM buck converter.

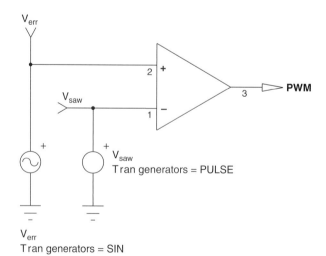

FIGURE 2-47a The PWM modulator, made of a fast comparator and a sawtooth generator.

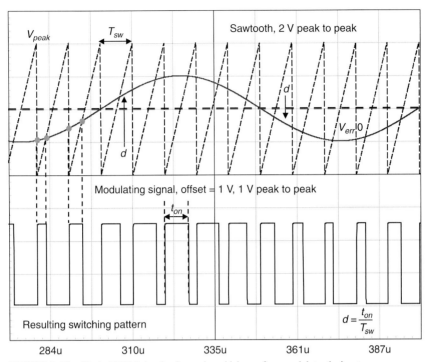

FIGURE 2-47b Typical PWM signals when a sinusoidal waveform modulates the input.

Observing Fig. 2-47b shows that transition points occur when the error signal crosses the sawtooth. At some extremes, if no transition can be observed, the comparator either stays permanently high (output is 1, 100% duty cycle) or permanently low (output is 0, 0% duty cycle):

- If $V_{err} > V_{saw}$, $d = 1$.
- If $V_{err} < 0$, $d = 0$.

From Eq. (2-113), the toggling point, or the on time ending point, occurs when $V_{err}(t_1) = V_{saw}(t_1)$. This can be written as

$$V_{err}(t) = V_{peak}\frac{t_{on}}{T_{sw}} \qquad (2\text{-}114)$$

Recognizing that the right-hand term is the duty cycle $d(t)$, we can update Eq. (2-114) by

$$V_{err}(t) = V_{peak}d(t) \qquad (2\text{-}115)$$

or, otherwise stated,

$$d(t) = \frac{V_{err}(t)}{V_{peak}} \qquad (2\text{-}116)$$

The small-signal model can still be obtained by perturbing the error voltage $V_{err}(t)$ and the duty cycle $d(t)$, as done numerous times before:

$$d(t) = D_0 + \hat{d}(t) \qquad (2\text{-}117)$$

$$V_{err}(t) = V_{err0} + \hat{V}_{err}(t) \qquad (2\text{-}118)$$

We can separate two equations, one dc and one ac:

$$D_0 = \frac{V_{err0}}{V_{peak}} \quad (2\text{-}119)$$

$$\hat{d}(t) = \frac{\hat{V}_{err}(t)}{V_{peak}} = K_{PWM}\hat{V}_{err}(t) \quad (2\text{-}120)$$

where $K_{PWM} = \frac{1}{V_{peak}}$ and K_{PWM} is called the PWM modulator gain.

Applying Eq. (2-119) to Fig. 2-47b shows a quiescent duty cycle of 50%. From Eq. (2-120), the small-signal gain of the modulator is $\frac{1}{2}$ = 0.5 or –6 dB. Figure 2-48 shows how to insert it in the average model implementation through the addition of a gain block.

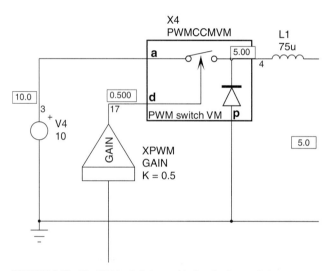

FIGURE 2-48 The PWM gain is inserted before the duty cycle input.

2.2.16 Testing the Model

When a model has been derived, it is necessary to test its results against reality. One good way of doing it uses a cycle-by-cycle model, supposed to replicate real hardware behavior. Experience shows that SPICE cycle-by-cycle simulations are extremely close to the real-world reality, provided parasitic elements are taken into account. In our case, since the average model does not yet include these parasitic elements, the transient buck model will be rather simple. Figure 2-49 shows its implementation.

The averaged buck model follows Fig. 2-50 and includes the PWM modulator. All passive elements feature the same values, and the step load circuit is rigorously identical to that of the cycle-by-cycle circuit. The compensating element values are those of Fig. 2-45, and they offer a 5 kHz cutoff frequency:

$$C_1 = 20.6 \text{ nF}$$
$$C_2 = 2.3 \text{ nF}$$
$$C_3 = 9.1 \text{ nF}$$

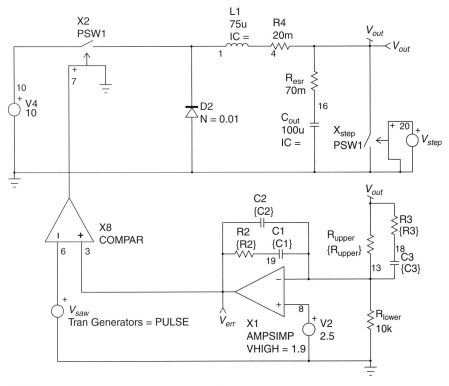

FIGURE 2-49 A simple cycle-by-cycle buck converter.

$$R_{upper} = 10 \text{ k}\Omega$$
$$R_3 = 1.1 \text{ k}\Omega$$
$$R_2 = 4.9 \text{ k}\Omega$$

The output undergoes a quick 100/10 Ω load variation, and one observes both the error and the output voltages. A stabilization period is necessary to let the switching converter reach steady state before the load step. The step occurs for $t = 10$ ms. Figure 2-51 unveils the test results and the difference between the two models is almost invisible: they perfectly superimpose on each other! The model works.

2.2.17 Mode Transition

One of the important mode features lies in the automatic transition between CCM and DCM. We can use the buck converter and sweep the output current to check if the toggling point corresponds to the right load condition. A probe connected to the generator implementing Eqs. (2-111) and (2-112) will reveal the transition event.

Back to Eq. (1-74b), we can calculate the critical load:

$$\mathbf{R}_{critical} = 2F_{sw}L\frac{V_{in}}{V_{in} - V_{out}} = 2 \times 100k \times 75u \times \frac{10}{5} = 30 \, \Omega$$

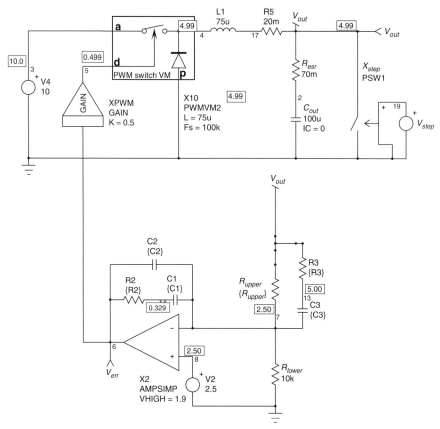

FIGURE 2-50 The complete averaged model test fixture.

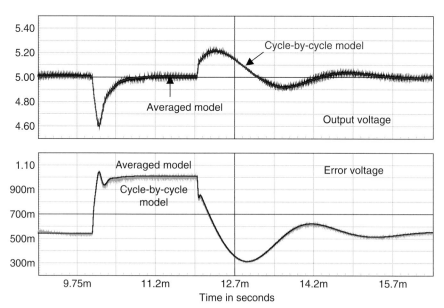

FIGURE 2-51 Final results, cycle-by-cycle model versus the averaged model.

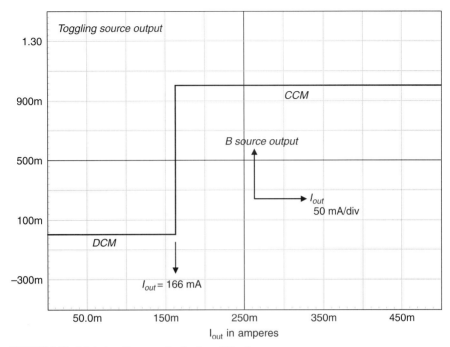

FIGURE 2-52 Mode transition occurring for $I_{out} = 166$ mA.

Hence, as long as the output current stays below $5/30 = 166$ mA, we are in DCM; and above this value, we go in CCM. Figure 2-52 displays the evolution of the toggling source. The toggling point confirms our calculation, which also implies the good behavior of the self-transitioning model.

2.3 THE PWM SWITCH MODEL— THE CURRENT-MODE CASE

Current-mode control represents one of the most popular control methods. Dr. Vorpérian described how he derived the CC-PWM switch model for the current control (CC) case in CCM in Ref. 11, but he never published his work on the DCM case. In this section, we will offer a simple derivation of the DCM model, leading to a new autotoggling CCM-DCM model.

Figure 1-17a and b described a practical implementation of current-mode control (CMC) converters. Actually, what differs between current mode and voltage mode is the way the duty cycle is elaborated. In voltage mode (VM), the error voltage is compared to a sawtooth to create a switching pattern, independently of the inductor current. In modern controllers, however, the inductor current is monitored for safety, but the circuitry stays transparent as long as the current stays below a limit value. In current mode (CM), first a switching clock closes the power switch. Then the error voltage V_{err} fixes the peak value at which the inductor current must increase before opening the switch. A comparator detects the current level via a sensing resistor R_i and resets the latch at the set point fixed by the error voltage. Figure 2-53 portrays a CM version of the PWM switch architecture.

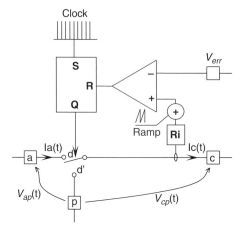

FIGURE 2-53 The CM version of the PWM switch.

Figure 2-53 shows a ramp added to the current sense information brought by R_i. This ramp participates in the stabilization of the converter in CCM situations where the duty cycle approaches 50%. Let us see why this instability takes place and how to cure it.

2.3.1 Current-Mode Instabilities

Current-mode instabilities, also called subharmonic oscillations, have been the subject of many studies. In this paragraph, we will quickly show how the current-mode converter is inherently unstable if the duty cycle exceeds 50% in CCM (in DCM, there are no subharmonic instabilities). Figure 2-54 shows the current inside the inductor (regardless of the topology), starting from an initial point and ending up at the same point value since it is in steady state: $I_L(0) = I_L(T_{sw})$. From this picture, we can derive a few equations:

$$I_L(T_{sw}) = I_L(0) + S_1 dT_{sw} - S_2 d'T_{sw} \qquad (2\text{-}121)$$

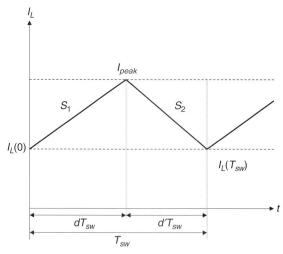

FIGURE 2-54 Inductor current at steady state.

SMALL-SIGNAL MODELING

and since $I_L(0) = I_L(T_{sw})$ at equilibrium,

$$S_2 d'T_{sw} = S_1 dT_{sw} \qquad (2\text{-}122)$$

Rearranging Eq. (2-122) gives

$$\frac{S_2}{S_1} = \frac{d}{d'} \qquad (2\text{-}123)$$

Now, suppose the system reacts to a small perturbation, meaning that the cycle starts no longer from $I_L(0)$ but from $I_L(0) + \Delta I_L$, assuming that $\Delta I_L \ll I_L(0)$. In that case, the controller will strive to keep the power switch on until the current reaches the peak value I_{peak}. Then, further to the internal latch reset, the switch opening will make the current fall with an S_2 slope until a new cycle begins. As you can imagine, the current level at which the new cycle reactivates the switch differs from the previous $I_L(T_{sw})$. Figure 2-55 describes this operation, purposely exaggerating the perturbation amplitude for the sake of readability. Again, a few simple equations are necessary to describe the behavior of Fig. 2-55.

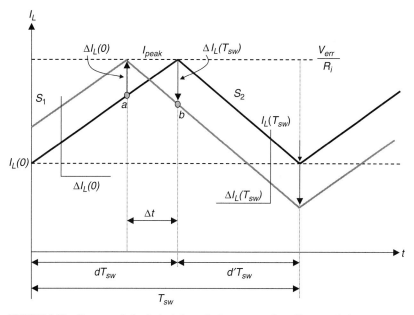

FIGURE 2-55 Current evolution in the inductor in the presence of a sudden perturbation.

If we call Δt the time difference separating the points at which both curves touch the peak set point I_c, actually I_{peak}, then

$$I_{peak} = a + S_1 \Delta t \qquad (2\text{-}124)$$

$$b = I_{peak} - S_2 \Delta t \qquad (2\text{-}125)$$

Solving for Δt in both equations gives

$$\frac{I_{peak} - a}{S_1} = \frac{I_{peak} - b}{S_2} \qquad (2\text{-}126)$$

Replacing with the right values, we get

$$\frac{\Delta I_L(0)}{S_1} = \frac{\Delta I_L(T_{sw})}{S_2} \qquad (2\text{-}127)$$

We can also see that the perturbation changes its polarity when carried over the switching cycle. Hence

$$\Delta I_L(T_{sw}) = \Delta I_L(0)\left(-\frac{S_2}{S_1}\right) \qquad (2\text{-}128)$$

If we agree that $\Delta I_L \ll I_L(0)$, then Eq. (2-123) still holds. Hence

$$\Delta I_L(T_{sw}) = \Delta I_L(0)\left(-\frac{d}{d'}\right) \qquad (2\text{-}129)$$

Equation (2-129) describes the perturbation amplitude after one switching cycle. What about the amplitude in the second cycle?

$$\Delta I_L(2T_{sw}) = \Delta I_L(T_{sw})\frac{d}{d'} = \Delta I_L(0)\left(-\frac{d}{d'}\right)^2 \qquad (2\text{-}130)$$

After n switching cycles, we obtain the following general expression:

$$\Delta I_L(nT_{sw}) = \Delta I_L(0)\left(-\frac{d}{d'}\right)^n \qquad (2\text{-}131)$$

In the above equation, what matters is the $\frac{d}{d'}$ ratio. If this ratio stays below 1, then when it is elevated to the power n, it naturally tends to zero: the perturbation dies out after a few cycles. This is what Fig. 2-56 shows.

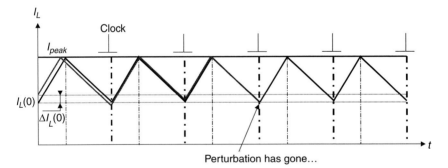

FIGURE 2-56 The perturbation disappears if the duty cycle stays below 50%.

On the contrary, if this quotient exceeds 1 or above, that is, when the duty cycle goes beyond 50%, then the perturbation no longer dies out and a steady-state oscillation occurs (Fig. 2-57a).

If we were to extract the PWM modulation envelope of a converter subject to subharmonic oscillations, we would observe a periodic modulation, pulsing at one-half the switching frequency. This is what Fig. 2-57b suggests.

SMALL-SIGNAL MODELING   **149**

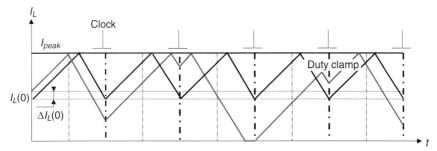

FIGURE 2-57a An initial perturbation becoming permanent in fixed-frequency, CCM current-mode operation.

To further stress the effect of the duty cycle, we can use a spreadsheet to plot various modulation results depending on it. Figure 2-58a, b, c, and d shows the waveform you would obtain if you could demodulate the PWM pattern at different duty cycles. For the case of Fig. 2-58d, the oscillation amplitude clearly diverges as Fig. 2-57a predicted.

In all these cases below where d stays at 50%, the oscillation naturally decays as a damped LC filter would do—an LC filter affected by a quality coefficient Q whose value depends on the duty cycle! It was a path first explored in the 1990s to explain this phenomenon [12]. Then came a more mathematical explanation using sampled data analysis. We will return to this topic later on.

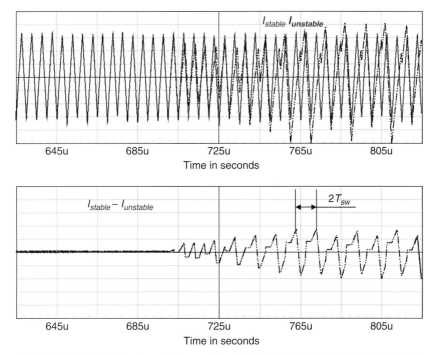

FIGURE 2-57b Subtracting the unstable current from the steady-state current shows a periodic signal pulsing at one-half of the switching frequency.

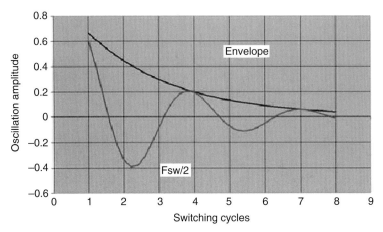

FIGURE 2-58a Subharmonic decrease for $d = 0.4$.

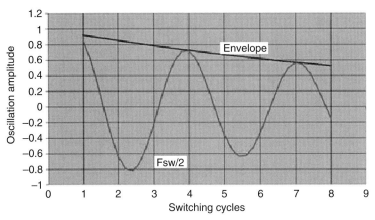

FIGURE 2-58b Subharmonic decrease for $d = 0.48$.

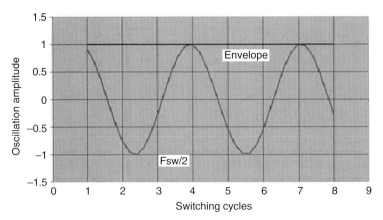

FIGURE 2-58c Subharmonic steady state for $d = 0.5$.

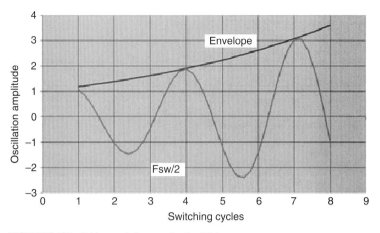

FIGURE 2-58d Subharmonic increase for $d = 0.54$.

Please note that this type of instability would also occur with a voltage-mode controller featuring a peak current limit: at start-up or in fault conditions, the inductor current decides the on time termination, and provided the duty cycle exceeds 50%, subharmonic oscillations can occur. Oscillations actually occur because the on-time duration affects the off-time length in situations where the duty cycle is above 50%. It is exactly like the case of a right half-plane zero. In topologies where the on time or the off time is fixed by design, subharmonic oscillations do not appear. The feedback loop thus regulates the power transfer by modulating, respectively, the off-time or the on-time, and you do not need to worry about the current-loop oscillations.

2.3.2 Preventing Instabilities

Subharmonic oscillations are not lethal to the converter. They unfortunately induce output ripple and sometimes audible noise in the transformer: they must be prevented by some means. Injection of what is called *ramp compensation* also called *slope compensation* is a method known to damp these oscillations, ensuring stability over a large duty cycle range. Ramp compensation can be either summed to the current sense information (usually delivered by a sense resistor or a current transformer) or subtracted from the feedback signal V_{err}. Our study will consider this solution, and it appears in Fig. 2-59. The drawing is a little bit more complicated compared to the construction of Fig. 2-55, but equations can easily be written.

Starting from the origin, we can define the point b value:

$$b = I_L(0) + \Delta I_L(0) + S_1(dT_{sw} - \Delta t) \qquad (2\text{-}132)$$

From the end of the switching period, we can still define point b:

$$b = I_L(T_{sw}) - \Delta I_L(T_{sw}) + S_2(d'T_{sw} + \Delta t) \qquad (2\text{-}133)$$

Since Eqs. (2-132) and (2-133) are equal, we can extract $I_L(0)$:

$$I_L(0) = -\Delta I_L(0) - S_1 dT_{sw} + S_1 \Delta t + I_L(T_{sw}) - \Delta I_L(T_{sw}) + S_2 \Delta t + S_2 d'T_{sw} \qquad (2\text{-}134)$$

Let us do the same for point c:

$$c = I_L(0) + S_1 dT_{sw} \qquad (2\text{-}135)$$

From the end of the switching period we have

$$c = I_L(T_{sw}) + S_2 d' T_{sw} \tag{2-136}$$

As we did above, we can equate Eqs. (2-135) and (2-136) to extract $I_L(0)$:

$$\Delta I_L(0) = -S_1 dT_{sw} + I_L(T_{sw}) + S_2 d' T_{sw} \tag{2-137}$$

As we now have two definitions for $I_L(0)$, Eqs. (2-134) and (2-137) are equal. Rearranging the final result gives

$$\Delta t (S_1 + S_2) - \Delta I_L(0) = \Delta I_L(T_{sw}) \tag{2-138}$$

The equation becomes simpler to manipulate, but we are missing the Δt definition. If we observe Fig. 2-59, we can quickly fix this.

$$\Delta t = \frac{b - c}{S_a} \tag{2-139}$$

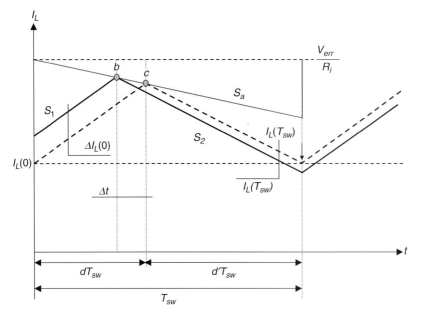

FIGURE 2-59 Ramp compensation S_a is subtracted from the current set point imposed by the feedback voltage.

Respectively replacing b and c by Eqs. (2-132) and (2-135) gives

$$\Delta I_L(0) + S_1 dT_{sw} - S_1 \Delta t - S_1 dT_{sw} = S_a \Delta t \tag{2-140}$$

Solving for Δt leads to

$$\Delta t = \frac{\Delta I_L(0)}{S_1 + S_a} \tag{2-141}$$

SMALL-SIGNAL MODELING

Now substituting Eq. (2-141) into Eq. (2-138) delivers the final information we are looking for:

$$\Delta I_L(0)\left[\frac{S_1 + S_2}{S_1 + S_a}\right] - \Delta I_L(0) = \Delta I_L(T_{sw}) \quad (2\text{-}142)$$

After rearranging Eq. (2-142) and again observing that the perturbation $\Delta I_L(0)$ generates a signal $\Delta I_L(T_{sw})$ of reversed polarity, we find

$$\Delta I_L(T_{sw}) = \Delta I_L(0)\left[-\frac{S_2 - S_a}{S_1 + S_a}\right] \quad (2\text{-}143)$$

Further to what we obtained from Eq. (2-129), we can generalize Eq. (2-143):

$$\Delta I_L(nT_{sw}) = \Delta I_L(0)\left[-\frac{S_2 - S_a}{S_1 + S_a}\right]^n \quad (2\text{-}144)$$

To find the stability condition, we can factor the above quotient term and replace $\frac{S_1}{S_2}$ by the Eq. (2-123) definition:

$$\Delta I_L(nT_{sw}) = \Delta I_L(0)\left[-\frac{1 - \frac{S_a}{S_2}}{\frac{S_1}{S_2} + \frac{S_a}{S_2}}\right]^n = \Delta I_L(0)\left[-\frac{1 - \frac{S_a}{S_2}}{\frac{d'}{d} + \frac{S_a}{S_2}}\right]^n = \Delta I_L(0)(-a)^n \quad (2\text{-}145)$$

Equation (2-145) can be unstable if the coefficient a goes above 1. To make sure this does not happen for all duty cycles up to 100% (meaning $d' = 0$), let us see what ramp compensation value guarantees it:

$$\left|\frac{1 - \frac{S_a}{S_2}}{0 + \frac{S_a}{S_2}}\right| < 1 \quad (2\text{-}146)$$

$$\frac{S_2}{S_2} - \frac{S_a}{S_2} < \frac{S_a}{S_2} \quad (2\text{-}147)$$

Finally,

$$S_a > \frac{S_2}{2} \quad \text{or} \quad S_a > 50\% S_2 \quad (2\text{-}148)$$

This is the minimum value that guarantees the stability for all operating duty cycles. In the literature, other choices are often proposed, such as compensation levels up to 75%. One should keep in mind that over compensating a converter seriously hampers its dynamic behavior but also reduces its maximum peak current capability and hence the available power (look at points b and c, compared to the set point imposed by the feedback loop level V_{err}).

2.3.3 The Current-Mode Model in CCM

We now know that current-mode converters operating in CCM together with a duty cycle above 50% will need a stabilization ramp. This is typical of fixed-frequency PWM converters. However, from an operating point of view, when you are observing static waveforms on an oscilloscope, nothing can tell you what control method the converter uses. For this reason, the invariant properties already derived for the voltage-mode PWM switch still apply for current-mode:

$$I_a = dI_c \qquad (2\text{-}149)$$

$$V_{cp} = dV_{ap} \qquad (2\text{-}150)$$

Later we will see a good illustration of this statement when using the voltage-mode PWM switch model to which is added a current-mode duty cycle factory.

To derive the model, the methodology remains the same: identify all pertinent waveforms and average them over a switching cycle. This will lead us to a nonlinear average model that can be directly simulated by SPICE. Let us start by inserting the Fig. 2-53 representation in a buck converter. We can plot the inductor current and define its slopes as Figs. 2-61 and 2-62 suggest.

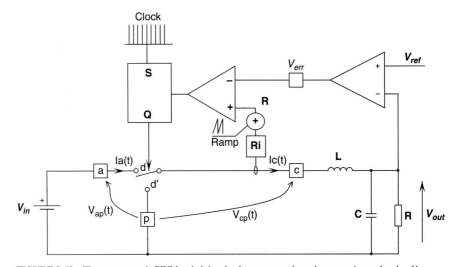

FIGURE 2-60 The current-mode PWM switch in a buck converter where the output is regulated to V_{ref}.

The up and down slopes can be derived by observing Fig. 2-60

$$S_1 = \frac{V_{ac}}{L} \qquad (2\text{-}151)$$

$$S_2 = \frac{V_{cp}}{L} \qquad (2\text{-}152)$$

and plotted in Fig. 2-61.

Now that all inductor slopes are known, we can redraw the picture by adding the compensation ramp S_a, described earlier. Figure 2-62 shows the updated graph. Please note that the whole figure represents currents, hence the scaling factor brought to the error voltage V_{err}, but also to the compensation ramp S_a which is, by nature, a voltage ramp.

From this representation, we need to extract the average value of the current flowing through the c terminal I_c. As usual, we can proceed step by step and find our way from point a to point b.

$$b = a - \frac{S_a}{R_i} dT_{sw} \qquad (2\text{-}153)$$

SMALL-SIGNAL MODELING 155

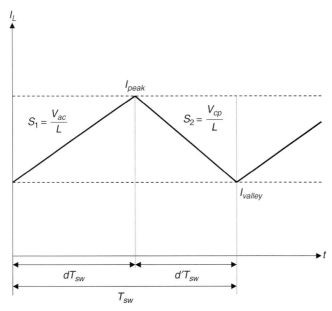

FIGURE 2-61 Typical invariant operating slopes of the PWM switch in CCM. This graph applies to current-mode but also to voltage-mode control.

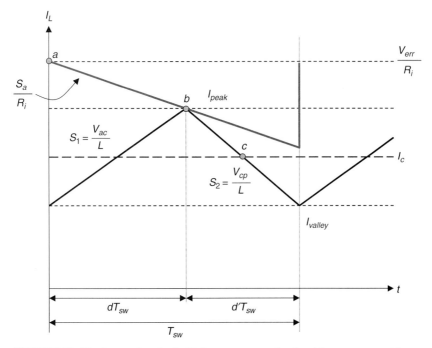

FIGURE 2-62 The slope graph updated with the ramp compensation S_a and the average current I_c.

If we want to reach the point c (actually the average current I_c) from b, then it is simply

$$c = b - \frac{S_2 d' T_{sw}}{2} \tag{2-154}$$

Replacing a, b, and c by their respective values and considering time-dependent waveforms, we have

$$I_c(t) = \frac{V_{err}(t)}{R_i} - \frac{S_a}{R_i} d(t) T_{sw} - \frac{S_2 d'(t) T_{sw}}{2} \tag{2-155}$$

However, Eq. (2-155) is like a camera snapshot, representing the instantaneous current $I_c(t)$ during a given switching period. To form an average model, we need to manipulate averaged values. Hence, Eq. (2-155) can simply be rewritten by considering the average of the current smoothed over the period of observation, which is much larger than the switching period:

$$I_c = \frac{V_{err}}{R_i} - \frac{S_a}{R_i} dT_{sw} - \frac{S_2 d' T_{sw}}{2} \tag{2-156}$$

If we make use of Eq. (2-152), we can update Eq. (2-156):

$$I_c = \frac{V_{err}}{R_i} - \frac{S_a}{R_i} dT_{sw} - V_{cp}(1-d)\frac{T_{sw}}{2L} \tag{2-157}$$

We can see a current generator $\frac{V_{err}}{R_i}$ from which is subtracted a compensating current (via the S_a ramp) and a third term. Based on this arrangement, we can think of a simple SPICE model made of current sources only! Figure 2-63 shows what it could look like.

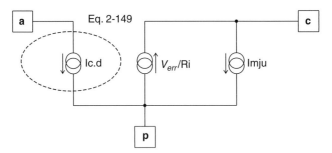

FIGURE 2-63 The skeleton of the CC-PWM switch model for the large-signal representation.

In this figure, for compactness purposes, the I_μ source (Imju on the sketch) can be rewritten as

$$I_\mu = \frac{S_a}{R_i} dT_{sw} + V_{cp}(1-d)\frac{T_{sw}}{2L} \tag{2-158}$$

This represents the large-signal model of the PWM switch operated in current-mode control. As it is now, it can be simulated under SPICE. A true small-signal version of this model can

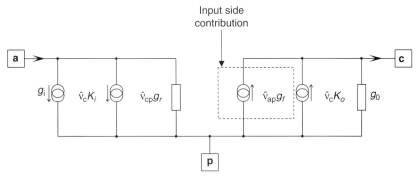

FIGURE 2-64 The small-signal version of the CC-PWM switch.

be found by perturbing the founding Eqs. (2-157) and (2-149) and neglecting all ac cross-products. Once everything is rearranged, Fig. 2-64 shows the final small-signal model of the CC-PWM switch model as described in Ref. 11. The source definitions are the following, where the 0-indiced values are static, dc parameters:

$$k_i = \frac{D_0}{R_i} \tag{2-159}$$

$$g_i = D_0\left(g_f - \frac{I_{c0}}{V_{ap0}}\right) \tag{2-160}$$

$$g_r = \frac{I_{c0}}{V_{ap0}} - g_0 D_0 \tag{2-161}$$

$$k_0 = \frac{1}{R_i} \tag{2-162}$$

$$g_0 = \frac{T_{sw}}{L}\left(D'_0\frac{S_a}{S_1} + \frac{1}{2} - D_0\right) \tag{2-163a}$$

$$g_f = D_0 g_0 - \frac{D_0 D'_0 T_{sw}}{2L} \tag{2-163b}$$

Given these definitions, but also by glancing at the small-signal model, we can make a few statements:

- Equation (2-162) describes the current-loop gain. A current-mode converter actually embeds two loops: an outer loop, the voltage loop which keeps V_{out} constant, and an inner loop, the current loop.
- Equation (2-163b) describes the g_f coefficient. This coefficient illustrates the input side transmission (via the V_{ap} term). It can be shown that a 50% ramp compensation injected in a buck converter operated in current-mode control nullifies the g_f term, bringing an infinite input rejection.
- Equation (2-161) represents the reverse current gain of the CC-PWM switch model and describes the action of the c-p port voltage, seen on the primary side.

But, in the above equations, nothing indicates an instability as discovered at the beginning of this section. When carefully observing Fig. 2-59, one can see that despite a perturbation in the control signal V_{err}, the falling slope S_2 will not instantaneously change. Therefore, the perturbation in the next cycle will depend on the previous S_2 slope. If we recall that S_2 depends on the V_{cp} port voltage, it suggests that an element placed on the terminal mimics a kind of "memory" effect, imposing the same slope on the following switching event. What can store, or "hold," a voltage from one cycle to the other? Yes, a simple capacitor appearing between terminals c and p! Also, if we look back at Fig. 2-58, as the converter already includes an inductor, the resonating behavior implies the presence of a simple capacitor which resonates together with the inductor at some point.

2.3.4 Upgrading the Model

How can we determine the value of this resonating capacitor? If we keep all control voltages and input/output voltages constant, then their small-signal values fade away in Eqs. (2-159) to (2-163). Hence, we can update Fig. 2-60 by shorting the a and p terminals (V_{in} is constant so its ac value is null, same for V_{out}) and removing all ac-dependent sources from Fig. 2-64. Only g_0 remains in parallel with the added resonating capacitor C_s and the converter inductor L: this is an LC network. Figure 2-65 illustrates how these elements are arranged together in a high-frequency ac model.

If we now derive the impedance of this parallel RLC network, we obtain the following equations:

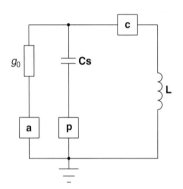

FIGURE 2-65 Considering all voltages constant implies shorting the a and p terminals while canceling all internal small-signal ac sources. An LC network appears.

$$Y_{RLC} = g_0 + C_s s + \frac{1}{Ls} \qquad (2\text{-}164)$$

$$Y_{RLC} = \frac{g_0 Ls + LC_s s^2 + 1}{Ls} \qquad (2\text{-}165)$$

Or, by taking the reverse of the admittance, we obtain the impedance

$$Z_{RLC} = \frac{Ls}{g_0 Ls + LC_s s^2 + 1} \qquad (2\text{-}166)$$

Dr. Vorpérian showed in his paper that the denominator $D(s)$ of Eq. (2-166) appears in all transfer functions pertinent to the CC-PWM switch model. This is an invariant equation:

$$D(s) = g_0 Ls + LC_s s^2 + 1 = 1 + \frac{s}{\omega_n Q} + \frac{s^2}{\omega_n^2} \qquad (2\text{-}167)$$

We know that the subharmonic instability equals one-half of the switching pulsation ω_s. From Eq. (2-167), we can identify the resonant pulsation ω_n and calculate C_s in such a way that $\omega_n = \frac{\omega_s}{2}$:

$$\frac{\omega_s}{2} = \frac{1}{\sqrt{LC_s}} \qquad (2\text{-}168)$$

Extracting C_s leads to

$$C_s = \frac{4}{L\omega_s^2} \qquad (2\text{-}169)$$

SMALL-SIGNAL MODELING

The roots of this characteristic equation are the poles affecting the stability. Looking at Fig. 2-65, we can see that the only element damping the *LC* network is the g_0 term (remember the EMI *LC* filter...). Otherwise stated, if this transconductance term cancels or becomes negative, we have oscillation conditions. Let us consider the definition of g_0 and see what conditions that nullify it. First, we suppress any compensation ramp, meaning $S_a = 0$:

$$g_0 = \frac{T_{sw}}{L}\left(\frac{1}{2} - D_0\right) \qquad (2\text{-}170)$$

We can immediately see that if D_0 is equal to or larger than 0.5, then g_0 respectively becomes null or turns negative: oscillations appear. This justifies the subharmonic oscillations. To evaluate the amount of ramp stabilization, we can check what S_a level prevents g_0 from being null.

From Eq. (2-163b),

$$D'_0 \frac{S_a}{S_1} + \frac{1}{2} - D_0 = 0 \qquad (2\text{-}171)$$

$$D'_0 \frac{S_a}{S_1} = D_0 - \frac{1}{2} \qquad (2\text{-}172)$$

Thanks to Eq. (2-123), we can replace the D'_0 value:

$$\frac{S_1}{S_2} \frac{S_a}{S_1} = D_0 - \frac{1}{2} \qquad (2\text{-}173)$$

Simplifying and rearranging give

$$S_a = S_2(D_0 - 0.5) \qquad (2\text{-}174)$$

For a duty cycle up to 100%, it becomes

$$S_a \geq \frac{1}{2} S_2 \qquad (2\text{-}175)$$

This is exactly what we found in Eq. (2-148).

Now, if we reveal the quality coefficient of Eq. (2-167), the term g_0 naturally appears. When it goes to zero, Q turns to infinity: the circuit oscillates.

From Eq. (2-167):

$$sLg_0 = \frac{s}{\omega_n Q} \qquad (2\text{-}176)$$

we know that $\omega_n = \frac{\omega_s}{2}$. Thus

$$Lg_0 = \frac{1}{\frac{\omega_s}{2} Q} \qquad (2\text{-}177)$$

Solving for Q gives

$$Q = \frac{1}{Lg_0 \frac{2\pi}{2T_{sw}}} = \frac{1}{\pi\left(D'_0 \frac{S_a}{S_1} + \frac{1}{2} - D_0\right)} \qquad (2\text{-}178)$$

where S_a represents the compensation ramp, S_1 the inductor on slope, and D_0, the dc (static) duty cycle.

Given the definition of Q, it becomes wiser to damp the resonating work at a given operating point, rather than for all duty cycles up to 100% as we did with Eq. (2-148). Otherwise, the current loop could become overdamped, leading to mediocre performance. Why? Because injecting more ramp pushes the current-mode converter to behave as a voltage-mode type. Without manipulating equations, Fig. 2-66 shows the arrangement around the current-sense comparator when ramp compensation is needed. The comparator toggles when the sum of the weighted ramp compensation and the actual sensed current equals the error voltage. If at a certain point the artificial ramp becomes much larger than the real current, that is, if R_r diminishes with respect to R_s, then you come closer and closer to Fig. 1-16a in Chap. 1. The converter operates in voltage mode.

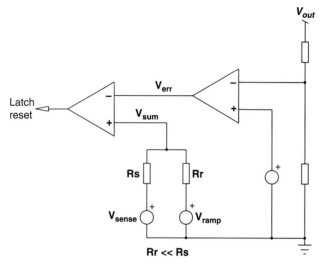

FIGURE 2-66 Injecting more ramp turns the CM converter into a VM converter, for instance, if V_{sum} becomes mostly dictated by V_{ramp}.

To reasonably damp the equivalent LC network, let us put its Q to 1 or less. Hence, from Eq. (2-178), it implies

$$\pi\left(D'_0 \frac{S_a}{S_1} + \frac{1}{2} - D_0\right) = 1 \qquad (2\text{-}179)$$

$$D'_0 \frac{S_a}{S_1} = \frac{1}{\pi} - \frac{1}{2} + D_0 \qquad (2\text{-}180)$$

Solving for S_a leads to

$$S_a \geq \left(\frac{1}{\pi} - \frac{1}{2} + D_0\right)\frac{S_1}{D'_0} \qquad (2\text{-}181)$$

This equation will help us define the optimum ramp compensation level for the converter under study. If performance is not so critical, but stability is, then 50% from the off slope will do the job.

SMALL-SIGNAL MODELING

In 1990, Dr. Raymond Ridley worked on a current-mode model based on the PWM switch model [13]. However, he based his study on sampled data analysis using z transform equations. Results found by Dr. Vorpérian fully agreed with those found by Dr. Ridley, but the Q expression arranged by Ridley slightly differed from Eq. (2-178). However, they are perfectly similar:

$$Q = \frac{1}{\pi(m_c D'_0 - 0.5)} \qquad (2\text{-}182)$$

where

$$m_c = 1 + \frac{S_a}{S_1} \qquad (2\text{-}183)$$

A 50% compensation ramp is obtained when $M_c = 1.5$. Hence, $Q = 1$ if

$$m_c = \frac{\frac{1}{\pi} + 0.5}{D'_0} \qquad (2\text{-}184)$$

In all the above equations, please keep in mind the following notations:

D_0 (or D'_0) represents the steady-state or dc duty cycle.

S_a is the compensation ramp, also often named S_e in the literature (e for external).

S_1 and S_2 are, respectively, the *on* and *off* slopes, denoted by S_n and S_f elsewhere and in the literature.

The fact that we have a quality coefficient implies that we have a peaking (at one-half of the switching frequency). If this peaking crosses the 0 dB (e.g., if Q is large) axis when the loop is closed (please see later in the chapter dedicated to feedback), oscillations might occur. Adding ramp compensation will damp the quality coefficient, naturally reducing risks of oscillations.

2.3.5 The Current-Mode Model in DCM

Our goal, once again, is to derive a model able to automatically toggle between the two modes CCM and DCM. Dr. Vatché Vorpérian did derive the DCM model, but he never published it. By examining the methodology he used to obtain the CCM model, a DCM model was successfully derived. As a good surprise, the obtained model smoothly transitions from one mode to the other! Here we go . . .

As we did in Fig. 2-62, we would like to obtain an expression for the average c terminal current when the converter operates in DCM. Figure 2-67 portrays the signal when this mode is entered. Please note the presence of the ramp as some applications still require a compensation despite the DCM mode.

We need to find a relationship which offers an analytical description of I_c. If we start from $t = 0$, we have V_{err}, trying to impose a peak current set point of $\frac{V_{err}}{R_i}$. However, the external ramp compensation S_a (in volts per second) diminishes this set point by subtracting from V_{err}. Therefore, the real peak current I_{peak} is

$$I_{peak} = \frac{V_{err} - d_1 T_{sw} S_a}{R_i} \qquad (2\text{-}185)$$

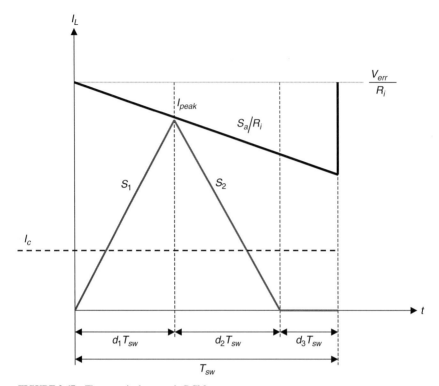

FIGURE 2-67 The c terminal current in DCM.

From this point I_{peak}, we can reach I_c via the off slope S_2, expressed in amperes per second:

$$I_c = \frac{V_{err} - d_1 T_{sw} S_a}{R_i} - \alpha d_2 T_{sw} S_2 \qquad (2\text{-}186)$$

The next step is to determine the value of α, which actually links I_{peak} and I_c via the equation (Fig. 2-68)

$$\alpha I_{peak} = I_{peak} - I_c \qquad (2\text{-}187)$$

I_c can be found by summing up the Fig. 2-67 half-triangle areas, with the $d_3 T_{sw}$ portion being null. Therefore,

$$I_c = \frac{I_{peak} d_1}{2} + \frac{I_{peak} d_2}{2} = \frac{I_{peak}(d_1 + d_2)}{2} \qquad (2\text{-}188)$$

If we now substitute I_c into Eq. (2-187) and rearrange it, we obtain Eq. (2-189):

$$\alpha I_{peak} = I_{peak} - \frac{I_{peak}(d_1 + d_2)}{2} \qquad (2\text{-}189)$$

SMALL-SIGNAL MODELING

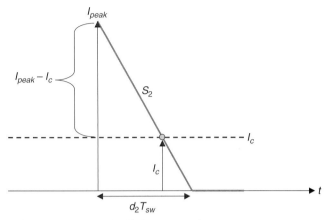

FIGURE 2-68 The relationship between I_c and I_{peak}.

Dividing through by I_{peak} gives

$$\alpha = 1 - \frac{d_1 + d_2}{2} \tag{2-190}$$

From that point, we can update Eq. (2-186):

$$I_c = \frac{V_{err}}{R_i} - \frac{d_1 T_{sw} S_a}{R_i} - d_2 T_{sw} S_2 \left(1 - \frac{d_1 + d_2}{2}\right) \tag{2-191}$$

Knowing that the off slope S_2 is $\frac{V_{cp}}{L}$, we obtain the final I_c definition:

$$I_c = \frac{V_{err}}{R_i} - \frac{d_1 T_{sw} S_a}{R_i} - d_2 T_{sw} \frac{V_{cp}}{L} \left(1 - \frac{d_1 + d_2}{2}\right) \tag{2-192}$$

To better stick to the CCM current-mode PWM switch model original definition:

$$I_c = \frac{V_{err}}{R_i} - I_\mu \tag{2-193}$$

where I_μ is simply

$$I_\mu = \frac{d_1 T_{sw} S_a}{R_i} + d_2 T_{sw} \frac{V_{cp}}{L} \left(1 - \frac{d_1 + d_2}{2}\right) \tag{2-194}$$

Please note that this equation simplifies to the CCM Eq. (2-158) when $d_2 = 1 - d_1$. This is a natural CCM-DCM toggling effect!

2.3.6 Deriving the Duty Cycles d_1 and d_2

Thanks to the Fig. 2-69 waveforms, we can work out the rest of the needed equations. From the DCM voltage-mode model we know

$$V_{cp} = V_{ap} \frac{d_1}{d_1 + d_2} \tag{2-195}$$

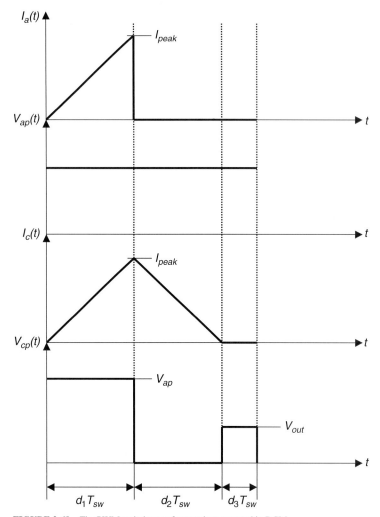

FIGURE 2-69 The PWM switch waveforms when operated in DCM.

or

$$\frac{V_{cp}}{V_{ap}}(d_1 + d_2) = d_1 \tag{2-196}$$

Again, this equation simplifies to $\frac{V_{cp}}{V_{ap}} = d_1$ when $d_2 = 1 - d_1$. This is the CCM equation.

Solving Eq. (2-196) for d_1 gives

$$d_1 = \frac{d_2 V_{cp}}{V_{ap} - V_{cp}} \tag{2-197}$$

From Fig. 2-69, we can write a few basic equations:

$$I_a = \frac{I_{peak} d_1}{2} \qquad (2\text{-}198)$$

From Eq. (2-198), we can get I_{peak} which equals $\frac{2I_a}{d_1}$ and plug it into Eq. (2-188). We obtain the relationship between I_a and I_c:

$$I_c = \frac{2I_a}{d_1}\frac{d_1 + d_2}{2} = I_a \frac{d_1 + d_2}{d_1} \qquad (2\text{-}199)$$

$$I_a = I_c \frac{d_1}{d_1 + d_2} \qquad (2\text{-}200)$$

Again, this equation simplifies to $I_a = I_c d_1$ when $d_2 = 1 - d_1$. This is the original CCM equation.

2.3.7 Building the DCM Model

We can now assemble the DCM model according to Fig. 2-70, where current sources are arranged as in the original CCM model.

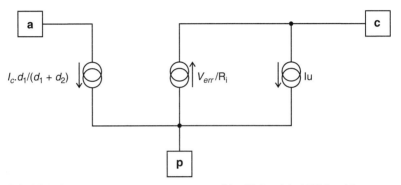

FIGURE 2-70 A current source arrangement compatible with the original CCM model.

We are almost finished. Unfortunately, we still need to derive the d_2 value that we are lacking. d_2 can be derived by observing Fig. 2-69, still using the buck configuration. This helps us derive a second equation for the peak current definition during the on time:

$$V_{ac} = L \frac{I_{peak}}{d_1 T_{sw}} \qquad (2\text{-}201)$$

which on average makes sense since the inductance voltage is null and V_{out} appears across the c terminal.

From Eq. (2-188), we can see that

$$I_{peak} = \frac{2I_c}{d_1 + d_2} \qquad (2\text{-}202)$$

Extracting I_{peak} from Eq. (2-201) and setting it equal to Eq. (2-202) leads to

$$\frac{2I_c}{d_1 + d_2} = \frac{V_{ac} d_1 T_{sw}}{L} \qquad (2\text{-}203)$$

Solving for d_2 gives us the final equation:

$$d_2 = \frac{2I_c L - V_{ac} d_1^2 T_{sw}}{V_{ac} d_1 T_{sw}} = \frac{2LF_{sw}}{d_1} \frac{I_c}{V_{ac}} - d_1 \qquad (2\text{-}204)$$

We are done with the DCM model derivation. We just need to write the netlist code which includes the two clamping circuits (see Figs. 2-40 and 2-41) and the resonating capacitor C_s. As highlighted several times in the above text, not only does the arrangement of the sources look very similar to that of the CC-PWM switch model, but also the equation factors reduce to d_1 when d_2 hits its upper limit, which is $1 - d_1$: it is in CCM. Therefore, thanks to this natural equation arrangement, the model automatically toggles between CCM and DCM.

In the DCM model, an equation will compute the duty cycle d_2 as Eq. (2-204) suggests. When this equation hits $1 - d_1$, the model toggles in CCM. Therefore, a single in-line equation detects in which mode the model stands:

```
Bmode mode 0 V = (2*{L}*{Fs}*I(VM)/(V(dc)*V(a,cx)+1u)) - V(dc) <
+ 1-v(dc) ? 1 : 0
IF (2*{L}*{Fs}*I(VM)/(V(dc)*V(a,cx)+1u)) - V(dc) is smaller than
1-V(dc)
THEN V(mode) equals 1 (we are in DCM)
ELSE V(mode) equals 0 (we are in CCM)
```

Why do we need to detect the modes since the model automatically transitions between them? Because of the C_s capacitor which will be present in CCM but absent in DCM. An equation will give it the right value for the CCM mode and reduce it to almost zero in DCM:

```
C1 c p C=V(mode) > 0.1 ? {4/((L)*(6.28*Fs)^2)} : 1p
Bmode mode 0 V= (2*{L}*{Fs}*I(VM)/(V(dc)*V(a,cx)+1u)) < 1 ? 0 : 1
; IsSpice syntax
```

In *PSpice*, the exercise becomes a bit more complicated as you cannot pass logical equations to the capacitor. We therefore have been obliged to create a voltage-controlled capacitor via a dedicated subcircuit. We will see later on how we derived it (Chap. 4).

Also, in both implementations, the dc output node delivers the duty cycle computed by the model.

IsSpice netlist:

```
.SUBCKT PWMCM a c p vc dc {L=35u Fs=200k Ri=1 Se=100m}
*
* This subckt is a current-mode DCM-CCM model
*
.subckt limit d dc params: clampH=0.99 clampL=16m
*
Gd 0 dcx d 0 100u
Rdc dcx 0 10k
V1 clpn 0 {clampL}
V2 clpp 0 {clampH}
D1 clpn dcx dclamp
D2 dcx clpp dclamp
Bdc dc 0 V=V(dcx)
.model dclamp d n=0.01 rs=100m
.ENDS
*
.subckt limit2 d2nc d d2c
```

SMALL-SIGNAL MODELING

```
Gd 0 d2cx d2nc 0 100u
Rdc d2cx 0 10k
V1 clpn 0 7m
BV2 clpp 0 V=1-V(d)
D1 clpn d2cx dclamp
D2 d2cx clpp dclamp
B2c d2c 0 V=V(d2cx)
.model dclamp d n=0.01 rs=100m
.ENDS
*
Bdc dcx 0 V=v(d2)*v(cx,p)/(v(a,p)-v(cx,p)+1u)
Xdc dcx dc limit params: clampH=0.99 clampL=7m
Bd2 d2X 0 V=(2*I(VM)*{L}-v(a,c)*V(dc)^2*{1/Fs}) / ( v(a,c)* V(dc)*
  {1/Fs}+1u )
Xd2 d2X dc d2 limit2
BIap a p I=(V(dc)/(V(dc)+V(d2)+1u))*I(VM)
BIpc p cx I=V(vc)/{Ri}
BImju cx p I= {Se}*V(dc)/({Ri}*{Fs}) + (v(cx,p)/{L})*V(d2)* {1/Fs}*
  (1-(V(dc)+V(d2))/2)
Rdum1 dc 0 1Meg
Rdum2 vc 0 1Meg
VM cx c
C1 cx p C=V(mode) > 0.1 ? {4/((L)*(6.28*Fs)^2)} : 1p
Bmode mode 0 V= (2*{L}*{Fs}*I(VM)/(V(dc)*V(a,cx)+1u)) < 1 ? 0 : 1
  ; connect or disconnects the resonating capacitor
*
.ENDS
```

PSpice netlist:

```
.SUBCKT PWMCM a c p vc dc params: L=35u Fs=200k Ri=1 Se=100m
*
* auto toggling between DCM and CCM, current-mode
*
Edc dcx 0 value = { v(d2)*v(cx,p)/(v(a,p)-v(cx,p)+1u) }
Xdc dcx dc limit params: clampH=0.99 clampL=7m
Ed2 d2X 0 value = { (2*{L}*{Fs}*I(VM)/(V(dc)*V(a,cx)+1u)) - V(dc) }
Xd2 d2X dc d2 limit2
GIap a p value = { (V(dc)/(V(dc)+V(d2)+1u))*I(VM) }
GIpc p cx value = { V(vc)/{Ri} }
GImju cx p value = {{Se}*V(dc)/({Ri}*{Fs}) + (v(cx,p)/{L})*V
  (d2)*{1/Fs}*(1-(V(dc)+V(d2))/2) }
Rdum1 dc 0 1Meg
Rdum2 vc 0 1Meg
VM cx c
XC1 cx p mode varicap      ; voltage-controlled capacitor
Emode mode 0 Value = { IF ((2*{L}*{Fs}*I(VM)/(V(dc)*V(a,cx)+1u))
  + < 1, 1p , {4/((L)*(6.28*Fs)^2)}) } ; connect or disconnects the
  resonating capacitor
*
.ENDS
********
.subckt limit d dc params: clampH=0.99 clampL=16m
Gd 0 dcx VALUE = { V(d)*100u }
Rdc dcx 0 10k
```

```
V1 clpn 0 {clampL}
V2 clpp 0 {clampH}
D1 clpn dcx dclamp
D2 dcx clpp dclamp
Edc dc 0 value={ V(dcx) }
.model dclamp d n=0.01 rs=100m
.ENDS
********
.subckt limit2 d2nc d d2c
*
Gd 0 d2cx d2nc 0 100u
Rdc d2cx 0 10k
V1 clpn 0 7m
E2 clpp 0 Value = { 1-V(d) }
D1 clpn d2cx dclamp
D2 d2cx clpp dclamp
Edc d2c 0 value={ V(d2cx) }
.model dclamp d n=0.01 rs=100m
.ENDS
********
.SUBCKT VARICAP 1 2 CTRL
R1 1 3 1u
VC 3 4
EBC 4 2 Value = { (1/v(ctrl))*v(int) }
GINT 0 INT Value = { I(VC) }
CINT INT 0 1
Rdum INT 0 10E10
.ENDS
*******
```

You need to feed the model subcircuit with several parameters, such as the inductor value (L, in henrys), the switching frequency (F_s, in hertz), the sense resistor (R_i, in ohms), and finally the external ramp compensation (S_e in volts per second). Please note that we use S_a or S_e indiscriminately (as in the available literature), as they relate to the same compensating ramp.

2.3.8 Testing the Model

We have several ways to test this current-mode model. First, we can check its frequency response against known models such as Ridley's, in both modes. Then we can run a cycle-by-cycle simulation and check the load step response. Let's start with an ac analysis.

Figure 2-71 shows the current-mode PWM switch model wired in a buck configuration. The loop is closed in dc via the error amplifier, but the *LoL* and *CoL* network opens the loop in ac.

The same buck can now be simulated by using Ridley models, however, only in ac. Figure 2-72 represents the DCM buck model using the sampled data version. Despite its inability to compute any dc value, we added inside the subcircuit a source manually computing the duty cycle. This is the only dc point calculated by the model which agrees with Fig. 2-71 quite well.

Figure 2-73 portrays the ac response delivered by both models. The gain plots agree with each other, and the phase traces show a little difference in the higher portion of the spectrum.

The next step consists of checking the transient load response of a DCM cycle-by-cycle buck converter working in current mode. Figure 2-74 shows the compensated averaged model whose output will undergo a load step. Figure 2-75 portrays the switched model, made of comparators and logic gates, whose output will experience the same transient load change. Please

SMALL-SIGNAL MODELING

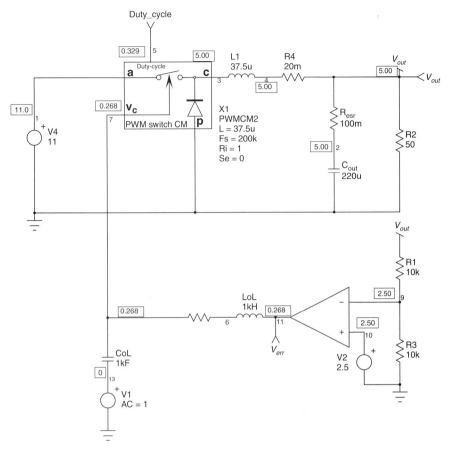

FIGURE 2-71 The implementation of the CC-PWM switch model in a DCM buck.

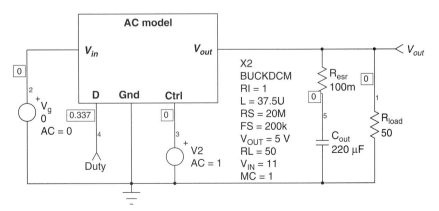

FIGURE 2-72 Same DCM buck using Ridley models.

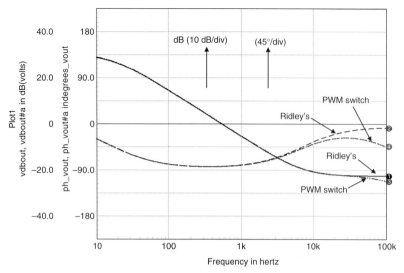

FIGURE 2-73 Comparison of the ac responses between Ridley models and the CC-PWM switch operated in DCM.

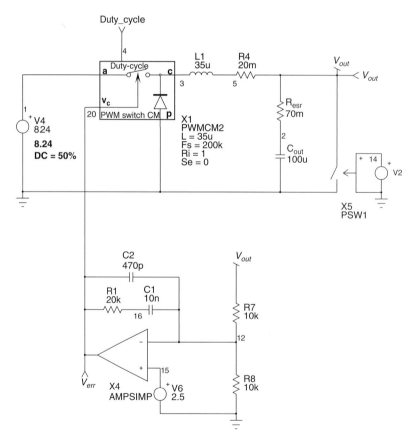

FIGURE 2-74 The averaged compensated DCM buck converter experiencing an output step load.

SMALL-SIGNAL MODELING   171

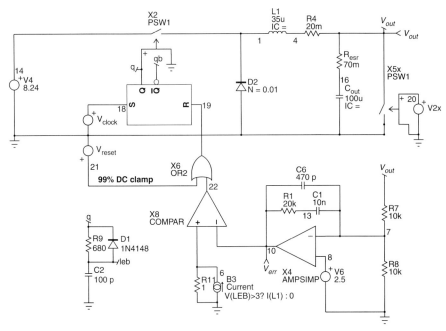

FIGURE 2-75 The cycle-by-cycle version used to test the averaged model results.

note the presence of a leading edge blanking (LEB) circuit whose purpose is to clean up the current-sense information to deliver noiseless information, naturally reducing false triggers.

Figure 2-76 reveals the results delivered by both models: they are almost identical! This is a good demonstration of the model validity.

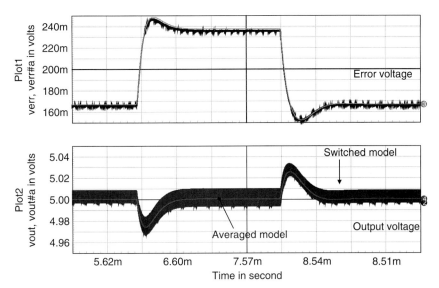

FIGURE 2-76 Both models deliver similar results.

When you are comparing results of models, it becomes extremely important to reduce all elements in the cycle-by-cycle model that could affect its ability to reproduce the perfect buck function. We say perfect because the averaged model, so far, does not include any parasitic element effects such as the $R_{DS(on)}$ or the freewheel diode forward drop V_f.

2.3.9 Buck DCM, Instability in DC

In DCM, the current-mode versions of the boost and buck-boost converters are stable without the addition of the compensation ramp. It is, however, less known that the current-mode buck operated in DCM shows an instability when its conversion ratio M exceeds 2/3. The converter small-signal model includes a pole expression whose denominator becomes negative when M exceeds 2/3. References 14 and 15 document this interesting result and conclude that a little ramp can stabilize the converter for all duty cycles up to 100%. One must inject a ramp S_a of amplitude greater than $0.086 S_2$. The DCM model can also predict this strange behavior. Figure 2-77 portrays the open-loop gain of the DCM buck used in current-mode control for different input voltages, hence different transfer ratios M. It can clearly be seen that a sudden phase reversal occurs as M approaches and goes beyond 2/3. From $M = 0.66$ to $M = 0.71$, the gain difference is nearly 20 dB. If we now inject a little bit of ramp compensation, the problem fades away, as Fig. 2-78 shows.

2.3.10 Checking the Model in CCM

The circuit in Fig. 2-73 was rearranged to fit CCM conditions, as Fig. 2-79 shows. Here, the frequency was arbitrarily reduced to 50 kHz. The critical load for a buck is given by Eq. (1-74b):

$$R_{critical} = 2F_{sw} L \frac{V_{in}}{V_{in} - V_{out}} = \frac{2 \times 50k \times 37.5u \times 11}{11 - 5} = 6.87 \, \Omega$$

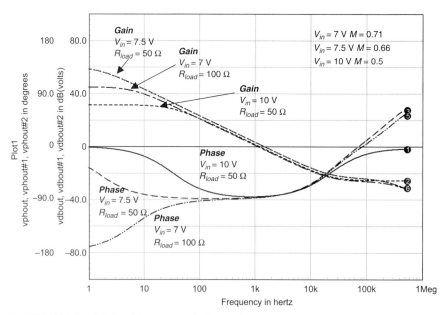

FIGURE 2-77 The DCM buck in current mode shows an instability as M approaches and exceeds 2/3.

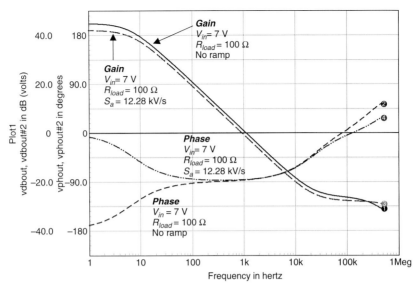

FIGURE 2-78 Injecting ramp compensation cures the buck instability for M ratios above 2/3.

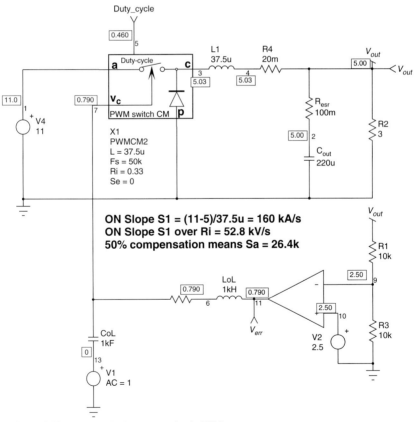

FIGURE 2-79 The same buck now operating in CCM.

In this particular example, it implies the following mode changes:

$$R_{load} < 6.87\,\Omega \quad \text{then CCM}$$

$$R_{load} > 6.87\,\Omega \quad \text{then DCM}$$

Given the working conditions of $V_{in} = 11$ V, $V_{out} = 5$ V, and $L = 37.5$ μH, let us calculate the ramp compensation S_a via the on slope S_1. However, S_1 must be reflected over the sense resistor R_i:

$$S_1 = \frac{V_{in} - V_{out}}{L} R_i = \frac{11 - 5}{37.5u} \times 0.33 = 52.8\,\text{kV/μs} \tag{2-205}$$

If we select 50% of this value [Eq. (2-183)], then the ramp compensation amplitude S_a equals 26.4 kV/s. We will thus pass $26.4k$ as a ramp parameter to the model.

Figure 2-80 now shows the same buck implemented with the Ridley CCM model. At first, there is no ramp compensation on any of these converters: $S_a = 0$ and $M_c = 1$ [M_c is defined by Eq. (2-183)]. Figure 2-81 shows the ac response of both the Ridley and the CC-PWM switch DCM/CCM models. They are identical!

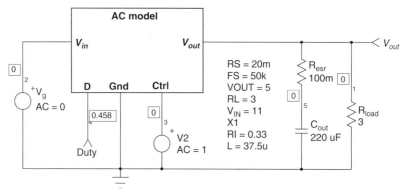

FIGURE 2-80 The Ridley model computes a similar duty cycle value in CCM.

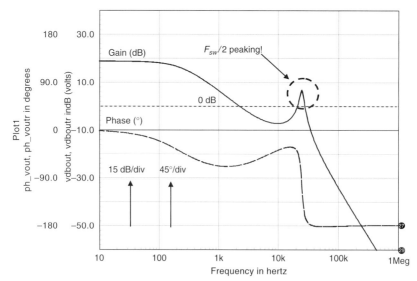

FIGURE 2-81 Both models predict the same ac response, no compensation ramp.

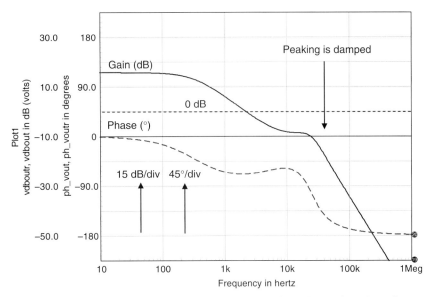

FIGURE 2-82 The CC-PWM switch model when compensated by a 50% on slope shows the same response as the Ridley model.

Now, if we inject the calculated ramp compensation S_a = 26.4k (50% compensation) and M_c = 1.5, Fig. 2-82 reveals the new results. Again, there is very good agreement.

Also, it was tested that the B_{mode} source, inside the model, properly toggles when R_{load} transitions between 6.5 and 7 Ω.

2.4 THE PWM SWITCH MODEL—PARASITIC ELEMENTS EFFECTS

When deriving the PWM switch model, in current mode or in voltage mode, we considered the V_{ap} voltage as a flat dc voltage. In reality, depending on the model configuration, this statement is not true. Very often the dc input voltage passes through an *LC* filter whose output capacitor includes an equivalent series resistor (ESR). This is the case for a buck configuration, for instance. In a boost architecture, V_{ap} directly represents the output voltage, whose loading elements are the load plus the output capacitor, again affected by an ESR. This creates a pulsating voltage inducing a ripple superimposed on V_{ap}.

In both cases, V_{ap} is affected by a voltage drop, and the previous equations need to be slightly upgraded. To understand how the phenomenon takes place, let us take a look at the three basic structures together with their associated waveforms. This is what Fig. 2-83a, b, and c represents where the ripple amplitude depends on a resistor value R_e whose value differs as the topology under study changes.

For the boost configuration, please note the presence of the negative signs on all variables. This is due to the particular switch arrangement inherent to the PWM switch model for which the same notation must be kept: this is true for I_c which leaves the original model (it enters it in the boost) but also for I_a which normally enters the model (it leaves it in the boost).

To determine the value of the resistor r_e, imagine the model in high frequency. This is where the ripple becomes dominant. Therefore, short-circuit all capacitors, open all inductors, and the resistor will reveal itself. Here are the values you will need to feed the updated model:

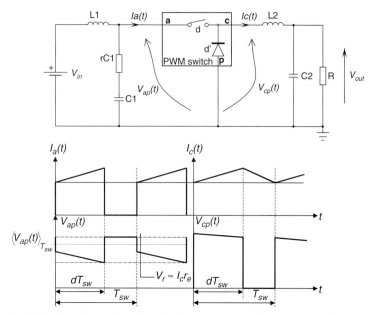

FIGURE 2-83a The buck configuration showing the ripple voltage induced on V_{ap}.

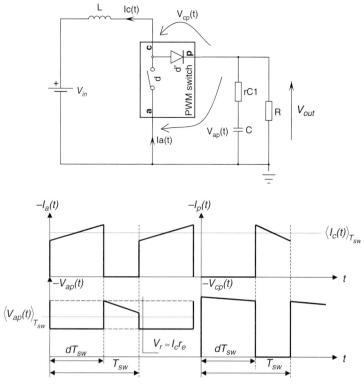

FIGURE 2-83b The PWM switch model in the boost where the ripple appears on the output port.

SMALL-SIGNAL MODELING

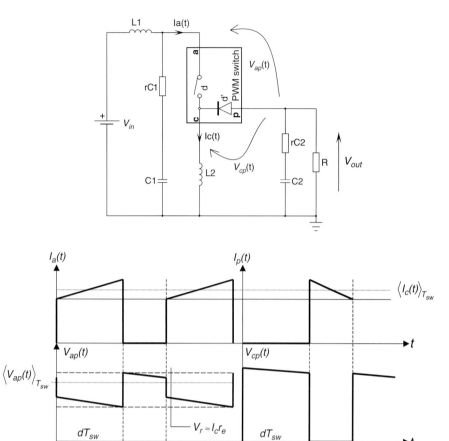

FIGURE 2-83c The buck-boost converter featuring a pulsating voltage on both input and output.

- Buck: $r_e = r_{c1}$
- Boost: $r_e = r_{C1} || R$
- Buck-boost: $r_e = r_{C1} + r_{C2} || R$

If we neglect the little droop present on $V_{ap}(t)$ (nonhorizontal plateaus), redrawing the waveform will help us find the ripple value (Fig. 2-84). What we need is to derive V_{ap} during the on time. First, in steady state, we can observe that the signal is centered on V_{ap}. Therefore, areas above and below V_{ap} are equal (remember Fig. 1-14). If we call α_1 the first drop amplitude and α_2 the second voltage increase, we can write the following relationships:

$$\alpha_1 + \alpha_2 = V_r \tag{2-206}$$

According to the equal-surfaces relationship,

$$dT_{sw}\alpha_1 = d'T_{sw}\alpha_2 \tag{2-207}$$

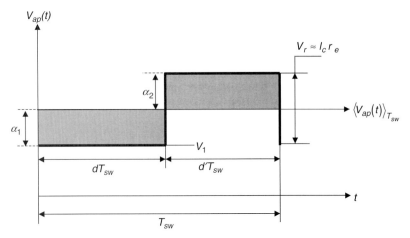

FIGURE 2-84 Neglecting the voltage drop helps to derive the voltage V_1 value.

The voltage we are looking for is $V_{ap}(t)$ during the on time or dT_{sw}. Glancing at Fig. 2-84, we see this is

$$V_1 = V_{ap} - \alpha_1 \quad (2\text{-}208)$$

From Eq. (2-206), we can extract α_2 and insert it into Eq. (2-207):

$$dT_{sw}\alpha_1 = d'T_{sw}(V_r - \alpha_1) \quad (2\text{-}209)$$

$$\alpha_1 = d'V_r \quad (2\text{-}210)$$

Hence, from Eq. (2-208)

$$V_1 = V_{ap} - d'V_r = V_{ap} - d'I_c r_e \quad (2\text{-}211)$$

If we now recall the invariant relationship that links V_{ap} and V_{cp} [Fig. 2-22, Eq. (2-62)], we can replace V_{ap} in Eq. (2-211):

$$V_{cp} = dV_1 = d(V_{ap} - d'I_c r_e) = dV_{ap} - dd'I_c r_e \quad (2\text{-}212)$$

The link between I_a and I_c does not vary and is still $I_a = dI_c$. Therefore, the parasitic term appearing in Eq. (2-212) (actually a voltage drop) can simply be modeled by a resistor placed in series with the c terminal of the large-signal PWM switch. Figure 2-85 portrays the updated

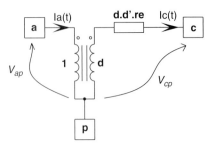

FIGURE 2-85 The updated large-signal model featuring the ESR-related parasitic effect.

model featuring this resistor. The value of r_e must be determined depending on the topology as highlighted above. This parasitic effect can often be considered of second order when you are using the large-signal model to reveal an open-loop Bode plot. Therefore, this effect will only be inserted in the CCM PWM switch and not in the autotoggling subcircuit, for the sake of simplicity.

2.4.1 A Variable Resistor

As Fig. 2-85 shows, we now have inserted a resistor whose value depends on a fixed parameter, but also from d and d'. We thus need to create a subcircuit emulating a resistor whose value is permanently adjusted by the dd' product. This can be done by considering Fig. 2-86. In this picture, we can see that the equivalent circuit of a passive resistor is nothing other than a current source I of value equal to the resistive drop divided by the resistor value:

$$I = \frac{V(1,2)}{R} \tag{2-213}$$

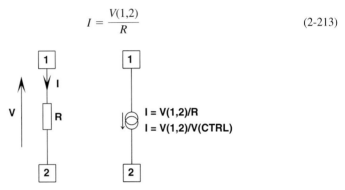

FIGURE 2-86 A controlled resistor is made of a current source.

Passing a voltage level V via the control node will emulate a resistor of $V\,\Omega$. The netlist is quite simple as exemplified below:

IsSpice
```
.subckt VARIRES 1 2 CTRL
R1 1 2 1E10
B1 1 2 I=V(1,2)/(V(CTRL)+1u)
.ENDS
```

PSpice
```
.subckt VARIRES 1 2 CTRL
R1 1 2 1E10
G1 1 2 Value = {V(1,2)/(V(CTRL)+1u)}
.ENDS
```

In the current source expression, the 1 μ quantity avoids divide-by-zero overflows in the case of extremely low control values. If V(CTRL) equals 100 kV, then the equivalent resistor will be 100 kΩ. If we now go back to the voltage-mode CCM netlist, we need to update the model with a few more lines:

```
Bdum 100 0 V= V(dc)*(1-V(dc))*{re}
*
.subckt Re 1 2 3
```

```
R1 1 2 1E10
B1 1 2 I=V(1,2)/(V(3)+1u)
.ENDS
*
```

where the B_{dum} source is inserted in series with the terminal c. Please note that $d' = 1 - d$ is only valid for the CCM mode.

2.4.2 Ohmic Losses, Voltage Drops: The VM Case

In the previous voltage-mode models, we considered null voltage drops across the conducting switches, SW and the diode D. In reality, the MOSFET $R_{DS(on)}$ or the bipolar $V_{ce(sat)}$ plays a role in the final efficiency, as well as the diode voltage drop V_f. Including these effects in the model is straightforward. Looking back at Fig. 2-21, we see the voltage drop over the power switch SW occurs during the on time when I_a crosses the switch. The average current flowing through SW is thus defined by

$$I_{sw} = \frac{\langle I_a \rangle_{T_{sw}}}{d} \qquad (2\text{-}214)$$

But since $\langle I_a \rangle_{T_{sw}} = d \langle I_c \rangle_{T_{sw}}$, then

$$I_{sw} = I_c \qquad (2\text{-}215)$$

We can therefore imagine a current source biasing a resistor (or directly the selected MOSFET) to develop the right SW drop voltage V_{SW} (Fig. 2-87).

If we write the equations for the diode, we end up with the same result as portrayed by Fig. 2-88 where the drop is called V_{DIO}. We can now update Fig. 2-22 including these two voltage drops (Fig. 2-89).

As a result, Eq. (2-62) also needs an update since the final average value changes:

$$V_{cp} = d(V_{ap} - V_{SW}) - d'V_{DIO} \qquad (2\text{-}216)$$

Rearranging the equation leads to

$$V_{cp} = dV_{ap} - (dV_{SW} + d'V_{DIO}) \qquad (2\text{-}217)$$

The second term of this equation can be expressed as an independent voltage source permanently subtracted from the original c terminal. Hence, as the relationship between the output and input current remains the same, we can now add Eq. (2-217) in the Fig. 2-24 original model.

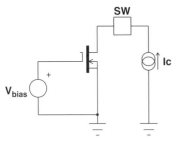

FIGURE 2-87 Generating the MOSFET voltage drop.

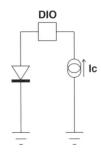

FIGURE 2-88 Same configuration for the diode.

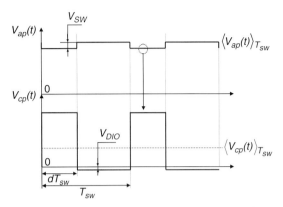

FIGURE 2-89 Voltage drops are seen on the input voltage via the SW element and during the off time through the diode forward drop.

Fig. 2-90 represents a CCM large-signal model including the various voltage drops. To make it toggle in DCM, we need to replace d' by the computed $d2$ value, actually clamped to $1 - d$ in CCM.

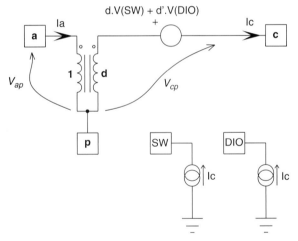

FIGURE 2-90 The series voltage source illustrates the drops incurred by the power switch SW and the freewheel diode D.

To test it, we will insert the lossy PWM switch model into a buck configuration featuring an off-the-shelf MOSFET and a Schottky diode. This is what Fig. 2-91 portrays. From the dc study of this configuration, we are able to reveal the conversion ratio accounting for the presence of the lossy devices:

$$\frac{V_{out}}{V_{in}} = D \left[\frac{1}{1 + \frac{R_{ds(on)}}{R}D + \frac{r_{Lf}}{R} - \frac{V_f}{V_{out}}D'} \right] \qquad (2\text{-}218)$$

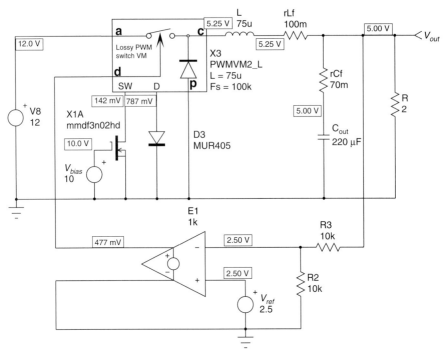

FIGURE 2-91 Voltage drop elements such as the power MOSFET and a diode are now inserted in the model.

One can see that each element is weighted according to its presence time inside the circuit: the diode during the off time and the MOSFET during the on time. The resistive inductor loss appears on both events as it is connected to the common terminal c. If we extract the duty cycle definition from Eq. (2-218), we obtain the following value:

$$D = \frac{\frac{R + r_{Lf}}{R} V_{out} - V_f}{V_{in} - V_f - \frac{R_{ds(on)} V_{out}}{R}} \tag{2-219}$$

From Fig. 2-91, we measure an $R_{DS(on)}$ of 56 mΩ and a voltage drop of –787 mV. If we insert these values into Eq. (2-219), we obtain $D = 0.477$, exactly the predicted value if you observe the error amplifier output.

2.4.3 Ohmic Losses, Voltage Drops: The CM Case

In current mode, the situation does not really differ. We still have drops incurred in the presence of the MOSFET and the freewheel diode. The parasitic source of Eq. (2-217) remains the same and will still be inserted in series with the c terminal. Figure 2-92 portrays the updated model. Please note that in this case, the duty cycle computation looks for the original c terminal voltage (c' in the drawing) and not the one connected to the inductor. Otherwise the source drop would not induce any change at all in the model.

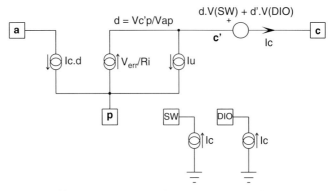

FIGURE 2-92 The current-mode model updated with conduction losses.

Please note, and the remark holds for the voltage-mode case as well, that the d' factor becomes the computed d_2 value, given by Eqs. (2-204) (CM) and (2-110) (VM).

2.4.4 Testing the Lossy Model in Current Mode

To test the validity of the approach, we have captured a buck converter operated in current mode in both CCM and DCM conditions. To change a little from a step load test, we have simulated a start-up sequence where the input source is suddenly applied to the converter. The nonlinear PWM switch model lends itself pretty well to large excursion transient simulations. However, as the numerical solver is heavily solicited during simulations, nonconvergence issues can arise. In case these errors occur, one can call on the ITL1 and ITL4 options. The first one, ITL1, can be increased to 1000 if the simulator fails to find its dc operating point. The other option, ITL4, can also be increased to allow more points rejection. The increase is up to 1000 with *IsSpice*, around 100 for *PSpice*.

Figure 2-93 represents the buck converter we have selected. Since we operate in CCM in the first case, we need ramp compensation. To simplify, let us apply 50% of the downslope S_2 on the current-sense information:

$$S_2 = \frac{V_{out}}{L} = \frac{5}{75} = 66 \text{ mA/}\mu\text{s} \qquad (2\text{-}220)$$

$$50\% \, S_2 = 33 \text{ mA/}\mu\text{s} \quad \text{or} \quad 33 \text{ mV/}\mu\text{s as } R_{sense} = 1 \, \Omega \qquad (2\text{-}221)$$

The compensation ramp and the current-sense information are combined through an adder subcircuit affected by two coefficients K_1 and K_2. As K_2 routes the inductor current, its value equals 1. K_2 must inject 33 mV/μs if we want to satisfy Eq. (2-221). The V_{ramp} source in Fig. 2-93 delivers a peak voltage of 2 V over a 10 μs period of time. Its slope is therefore 200 mV/μs. To satisfy Eq. (2-221), K_1 must equal 33/200 = 0.165.

The average lossy model appears in Fig. 2-94. All passed parameters comply with the Fig. 2-93 operating parameters: 100 kHz switching frequency, 75 μH inductor, 1 Ω sense resistor, and a 33 kV/s compensation ramp. The MOSFET on resistance has been replaced by a fixed 200 mΩ component. The simulation results appear in Fig. 2-95, and the agreement between the curves is excellent! Measuring the duty cycle on the Fig. 2-93 circuit gives 81% whereas the lossy average model delivers 82.4%. If we replace it with the nonlossy model, the error between the computed duty cycle and the real one increases beyond 10%.

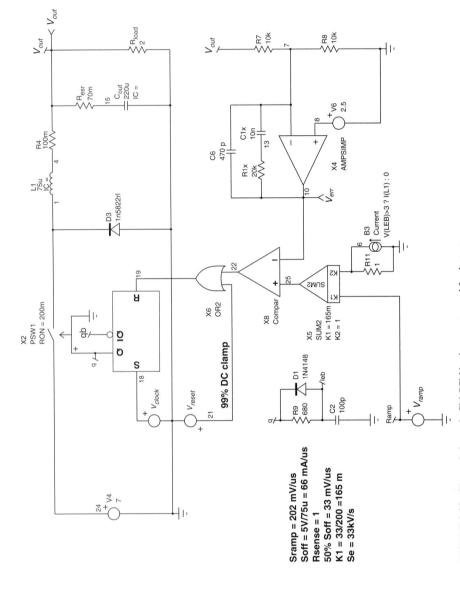

FIGURE 2-93 The cycle-by-cycle CM CCM buck converter used for the test.

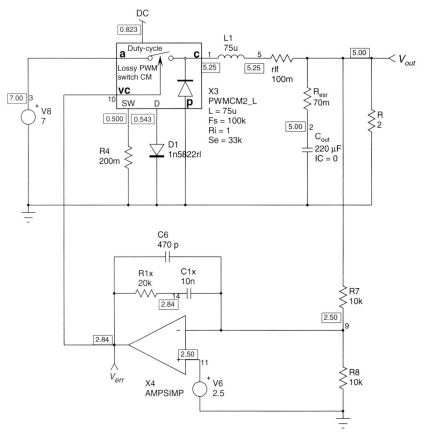

FIGURE 2-94 The lossy CM PWM switch model.

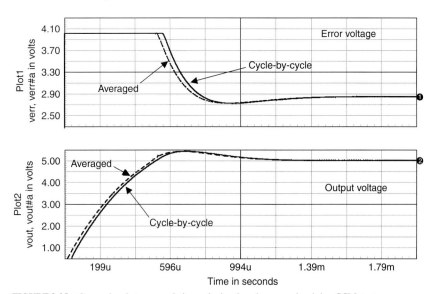

FIGURE 2-95 Comparison between cycle-by-cycle signals and average signals in a CCM start-up sequence.

185

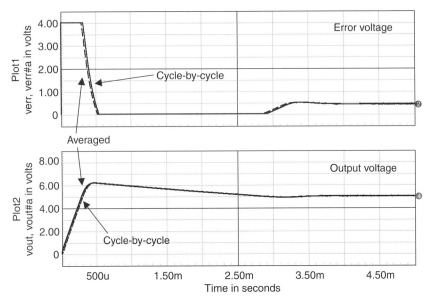

FIGURE 2-96 Comparison between cycle-by-cycle signals and average signals in a DCM start-up sequence.

The DCM operation is then simulated, by increasing the load resistor to 50 Ω. Again, Fig. 2-96 unveils the results and confirms the validity of the model. The current set point (the error voltage divided by the sense resistor) is exactly the same for both approaches.

Are these lossy models of optimum accuracy? Certainly not, simply because the watts dissipated in the transistor, resistive conduction losses in this case, depend on the rms current and not the average current. However, as the diode dissipation depends on its forward drop and the average current (neglecting the dynamic resistance), the results stay within acceptable limits. Hence, assuming a low-ripple situation, otherwise stated as a weak ac contribution to Eq. (1-26), then the difference between the average and rms levels becomes less important.

2.4.5 Convergence Issues with the CM Model

The CM model, lossy or normal, includes a resonating capacitor, active in CCM operation only. This capacitor disappears in DCM since subharmonic oscillations are absent in this mode. The connection or disconnection of this capacitor relies on a dedicated in-line equation, dynamically testing the mode of operation. This is B_{mode} in the *IsSpice* model and E_{mode} in the *PSpice* model. When you are running ac simulations, the presence of this capacitor does not create any trouble, simply because the operating point remains constant throughout the run. The decision to select CCM or DCM is made during the dc bias point calculation and no longer varies. In this ac configuration, the model has proved to be robust. When you need to use the averaged autotoggling CM model in transient situations, where a mode transition occurs during the simulation (e.g., a step load from light to heavy conditions), then nonconvergence issues are likely to arise—simply because the connection or disconnection of the resonating capacitor creates a discontinuity, lethal to the simulation engine. A solution consists of inserting a little "*" character in front of the CS (*IsSpice*) or CX1 (*PSpice*) to remove the problem. If the transient simulation does not imply a mode change, then the resonating capacitor can stay in place. Under *OrCAD*, the model with the resonating capacitor disconnected is PWMCMX. You can use this model when you need to run transient simulations in current mode.

2.5 PWM SWITCH MODEL IN BORDERLINE CONDUCTION

Borderline conduction mode (BCM), also called *critical conduction mode*, offers numerous advantages. By detecting the moment at which the inductor is reset (i.e., its current reaches zero) prior to reactivating the power switch, we are able to implement a converter offering several particular advantages:

- The circuit never enters the CCM mode; therefore no reverse recovery effects bother the designer, even in short-circuit or during the start-up sequence.
- As the circuit stays DCM, it remains a first-order system, easing the design task to close its feedback loop.
- If a small delay appears before activation of the power switch, the transistor can use the natural drain-source oscillating sine wave to bring its voltage close to zero and make it operate in zero voltage switching (ZVS) conditions. This technique is often implemented in flyback converters and in Power Factor correction (PFC) circuits.

However, one drawback lies in the variable frequency which depends on line and load conditions. Given the popularity of this structure, we will derive a special version of the PWM switch in BCM.

2.5.1 Borderline Conduction—The Voltage-Mode Case

Borderline conduction mode converters operate without an internal clock. The switching frequency is naturally dependent upon the external input/output conditions to force the converter to operate at the boundary between CCM and DCM. The switching events are the following:

1. An internal signal cranks up the controller: the power switch closes.
2. The current ramps up until the PWM modulator instructs the switch to open (voltage mode). In current mode, the switch opening occurs when the peak current set point is reached.
3. The inductor current keeps circulating (or reaches the secondary for a flyback) and decreases with a constant slope.
4. When it reaches zero, the inductor core is told to be "reset." The controller detects this state and reactivates the power switch. Go to item 1 for the next cycle.

Figure 2-97 represents the internal arrangement for a voltage-mode BCM controller. The error voltage V_{err}, driven by the feedback loop, is permanently compared to a sawtooth, itself reset when the drive is *off*. Therefore, the feedback voltage amplitude dictates the on-time duration: as the driver goes high, the capacitor is freed and can linearly charge up. When its level intersects with that of V_{err}, the latch receives a reset signal and the drive goes low. Now the coil current starts to ramp down via a freewheel action. Thanks to Lenz' law, when this current reaches zero ($d\varphi = 0$), the voltage developed on the right terminal of resistor R_{limit} collapses. As soon as this voltage passes below a 65 mV threshold (it could be simply zero as well, this is just an industry standard), the reset detector goes high and restarts a new cycle.

Converters using this operating principle are often used in so-called quasi-resonant (QR) topologies, but the proper term remains *quasi square wave* topologies. We can find them in flybacks but also in PFC boost configurations where they benefit from large popularity. As a natural ringing takes place at the switch opening (due to the presence of an inductor and parasitic capacitors which form a ringing *LC* network), the designer often inserts a small delay prior to reactivating the power switch. As such, if the delay is set to the right value, the switch closes right in the minimum of the drain-source signal. This is a so-called valley switching operation. Figure 2-98 portrays a typical operating signal obtained on a boost converter. You can clearly

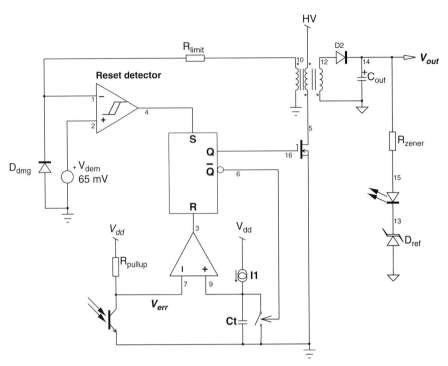

FIGURE 2-97 The internal structure of these specific controllers when wired in a flyback application.

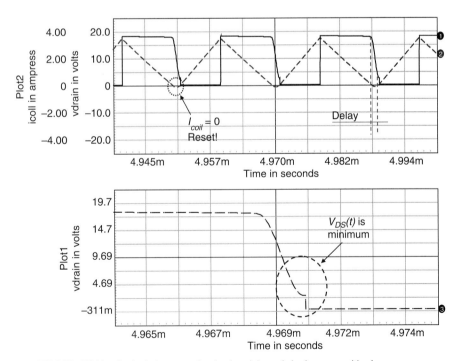

FIGURE 2-98 Waiting for the drain-source signal to be minimum helps lower capacitive losses.

see the small delay inserted in front of the reset detector which ensures a nice valley switching operation.

Please note that the model we are going to derive will not account for this dead time which, once properly adjusted, can ensure zero voltage switching. The derivation exercise, as usual, consists of writing average equations for the BCM PWM switch model's pertinent waveforms. Figure 2-99 displays them. A quick glance shows that these waveforms are the same as those

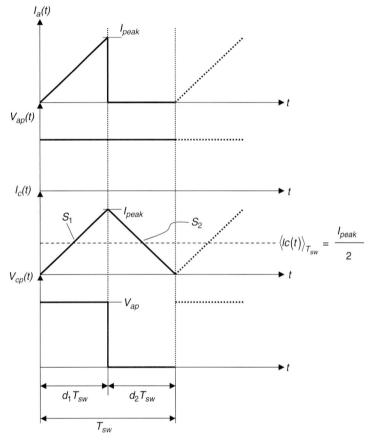

FIGURE 2-99 The BCM waveforms are the same as those of the CCM voltage mode.

from the voltage-mode PWM switch operating in CCM when the ac amplitude (the current ripple) becomes maximum. We can therefore reuse the same model as depicted by Fig. 2-24. The difference lies in the input signal. Rather than driving a duty cycle, we will enter an on time t_{on}, further transformed into a duty cycle by the BCM model. The equations are the following:

$$I_c = \frac{I_{peak}}{2} \qquad (2\text{-}222)$$

$$d_2 T_{sw} = t_{off} = \frac{L}{V_{cp}} I_{peak} \qquad (2\text{-}223)$$

Extracting I_{peak} from Eq. (2-222) and plugging it into Eq. (2-223), we obtain the t_{off} generator equation:

$$t_{off} = \frac{2LI_c}{V_{cp}} \qquad (2\text{-}224)$$

If we consider the model input as being an on time t_{on}, then we can easily construct the duty cycle, knowing that it is in BCM.

$$t_{on} + t_{off} = T_{sw} \qquad (2\text{-}225)$$

Then the duty cycle comes easily:

$$d_1 = \frac{t_{on}}{T_{sw}} = \frac{t_{on}}{t_{on} + t_{off}} \qquad (2\text{-}226)$$

The model thus includes an off-time generator depending on the control input, the on time. However, we do not want to manipulate microvolt-type levels as control voltages (10 μs would then be modeled as 10 μV...). For that reason, a scaling factor of 1 Meg will be internally used to obtain 1 V = 1 μs. Once the off-time generator has undergone a clamp, we are all set. That is all!

The model includes several key outputs:

- The peak current I_p, computed via the formula $I_{peak} = \frac{V_{ac}t_{on}}{L}$ where t_{on} is the control voltage.
- The duty cycle $dc = \frac{t_{on}}{t_{on} + t_{off}}$. Both t_{on} and t_{off} are known through Eq. (2-224).
- And finally, the switching frequency F_{sw}, by the inversed sum of $t_{on} + t_{off}$.

The model accepts only one parameter, the inductor value. Its implementation in *IsSpice* and *PSpice* is described below.

IsSpice netlist:

```
.SUBCKT PWMBCMVM a c p vc dc fsw ip {L=1.2m}
*
* This subckt is a voltage-mode BCM model
*   -> 1V for vc = 1us for ton
*
.subckt limit d dc params: clampH=0.99 clampL=16m
Gd 0 dcx d 0 100u
Rdc dcx 0 10k
V1 clpn 0 {clampL}
V2 clpp 0 {clampH}
D1 clpn dcx dclamp
D2 dcx clpp dclamp
Bdc dc 0 V=V(dcx)
.model dclamp d n=0.01 rs=100m
.ENDS
*
Bdc dcx 0 V = V(vc)*1u/(V(vc)*1u + V(toff))
Xd dcx dc limit params: clampH=0.99 clampL=16m
BVcp 6 p V=V(dc)*V(a,p)
BIap a p I=V(dc)*I(VM)
Btoff toff 0 V = 2*I(VM)*{L}/V(c,p) < 0 ? 0 :
+ 2*I(VM)*{L}/V(c,p)
Bfsw fsw 0 V =  (1/(V(vc)*1u + V(toff)))/1k
Bip ip 0 V=abs(V(a,c)*V(vc)*1u)/{L}
VM 6 c
*
.ENDS
```

PSpice netlist:

```
.SUBCKT PWMBCMVM a c p vc dc fsw ip params: L=1.2m
* Borderline Conduction Mode, voltage-mode
* -> 1V for vc = 1us for ton
EBdc dcx 0 Value = { V(vc)*1u/(V(vc)*1u + V(toff)) }
Xd dcx dc limit params: clampH=0.99 clampL=16m
EBVcp 6 p Value = { V(dc)*V(a,p) }
GBIap a p Value = { V(dc)*I(VM) }
EBtoff toff 0 Value = { IF ( 2*I(VM)*{L}/V(c,p) ) < 0, 0,
+ 2*I(VM)*{L}/V(c,p) ) }
EBfsw fsw 0 Value = { (1/(V(vc)*1u + V(toff)))/1k }
EBip ip 0 Value = { abs(V(a,c)*V(vc)*1u)/{L} }
VM 6 c
*
.ENDS
.subckt limit d dc params: clampH=0.99 clampL=16m
Gd 0 dcx VALUE = { V(d)*100u }
Rdc dcx 0 10k
V1 clpn 0 {clampL}
V2 clpp 0 {clampH}
D1 clpn dcx dclamp
D2 dcx clpp dclamp
Edc dc 0 value={ V(dcx) }
.model dclamp d n=0.01 rs=100m
.ENDS
```

2.5.2 Testing the Voltage-Mode BCM Model

To test the model, we will first capture a boost converter operating in voltage-mode. To ease the control circuit, we will use the generic free-run QR controller, meant to work in current mode. Nothing prevents us from using a current-mode controller as a voltage-mode controller: just connect a ramping voltage on its current-sense input, and you have a voltage-mode operation! Figure 2-100 portrays the control circuit we have used here. The sawtooth generator is made by pulling a capacitor (C_1) by current source (I_1). The ruling equation for the t_{on} generation is pretty simple:

$$t_{on} = \frac{V_{err} C_t}{I_1} \tag{2-227}$$

The model takes the feedback voltage (pin 2), internally divides it by 3, and internally clamps it to $V_{err} = 1$ V maximum. Therefore, with $C_1 = 4.7$ nF and $I_1 = 100$ μA, the corresponding maximum on time will be

$$\frac{1 \times 4.7n}{100u} = 47 \; \mu s \tag{2-228}$$

After running the Fig. 2-100 circuit, we have found the following static values for a 10 V input voltage and a 17.2 V output level:

- Switching frequency: 15.6 kHz
- Peak current: 12.8 A
- Duty cycle: 45%

To check the model's validity, Fig. 2-101 displays the adopted configuration. It perfectly reproduces the Fig. 2-100 arrangement, however, without including the switch and diode losses. This explains the slight difference in the operating points:

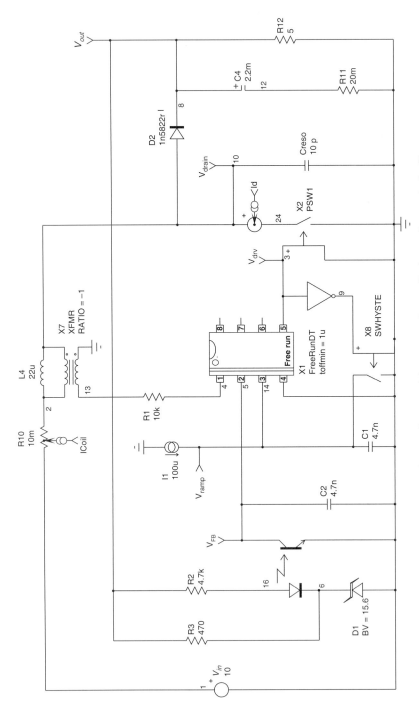

FIGURE 2-100 By adding a ramp to the current-sense input, we have turned the current-mode controller into a voltage-mode circuit!

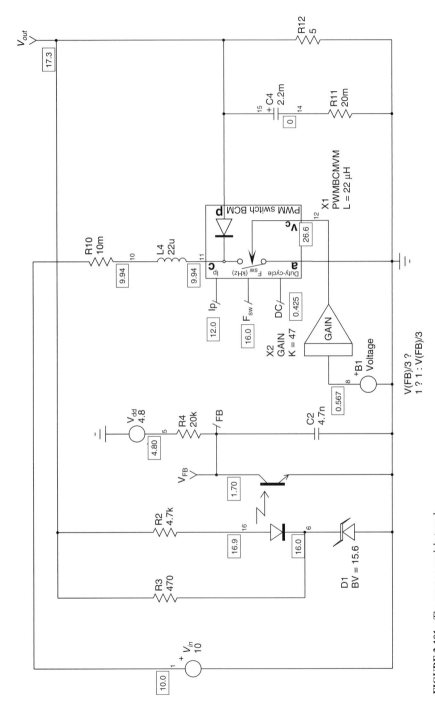

FIGURE 2-101 The average model at work.

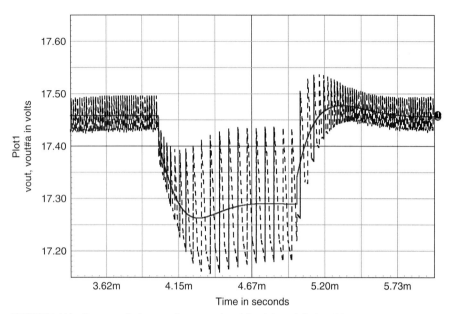

FIGURE 2-102 Output results between the averaged model and the switched model.

- Switching frequency: 16 kHz
- Peak current: 12 A
- Duty cycle: 42.5%

If one includes additional losses as described in a previous section, static numbers come very close to one another. It is now time to step load the output and compare both answers, cycle-by-cycle and averaged. This test can quickly reveal any model mistakes or incorrect principles. Figure 2-102 unveils the result and shows excellent agreement between both approaches, confirming the validity of the technique. The ac study of QR converters will come later, in the section dedicated to using a different modeling technique.

2.5.3 Borderline Conduction—The Current-Mode Case

The current-mode model will follow the exact same steps that we used for the voltage-mode derivation. Figure 2-103 shows the internal arrangement of the BCM current-mode controller. What matters now is the inductor peak current imposed by the control voltage. A resistor R_{sense} delivers a voltage image of the current flowing through the inductor. The maximum excursion of this current depends on the maximum error voltage present on the inverting input of the current-sense comparator. In most controllers, this voltage equals 1 V.

Regarding the average model, the waveforms in Fig. 2-99 still hold because nothing tells us the type of control mode (VM or CM) by solely looking at the static waveforms. One noticeable point, however, is that the inductor average current in BCM always equals the peak divided by 2:

$$\langle I_c(t) \rangle_{T_{sw}} = I_c = \frac{I_{peak}}{2} \qquad (2\text{-}229)$$

SMALL-SIGNAL MODELING

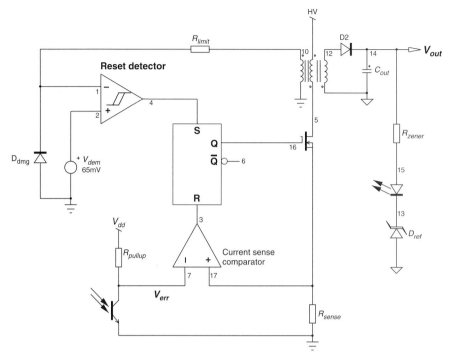

FIGURE 2-103 The internal circuitry of a BCM current-mode controller.

Since we now deal with the peak current, and no longer t_{on}, as an input set point, we can write the classical current-mode equation

$$I_{peak} = \frac{V_{err}}{R_i} \tag{2-230}$$

where R_i represents the sense resistor and V_{err} the control voltage brought by the feedback loop.

Following the original CCM methodology, I_c crosses the off slope, just in its middle [see Eq. (2-229)]. Therefore, the average value of I_c can be found from

$$I_c = \frac{V_{err}}{R_i} - d_2 T_{sw} \frac{V_{cp}}{2L} = \frac{V_{err}}{R_i} - V_{cp}(1 - d_1)\frac{T_{sw}}{2L} \tag{2-231}$$

From Fig. 2-99, we can find the relationship that links I_a and I_c:

$$I_a = \frac{I_{peak} d_1}{2} \tag{2-232}$$

$$I_c = \frac{I_{peak}(d_1 + d_2)}{2} \tag{2-233}$$

In true BCM (without any delay), as in CCM, $d_1 + d_2 = 1$; therefore, if we plug I_{peak} given by Eq. (2-232) into Eq. (2-233) and replace d_2 with $1 - d_1$, we obtain the same CCM relationship between I_a and I_c:

$$I_a = d_1 I_c \tag{2-234}$$

Since the control voltage V_{err} imposes a peak current as Eq. (2-230) defines, we can reformulate the duty cycle d_1 definition:

$$I_{peak} = \frac{V_{err}}{R_i} = \frac{V_{ac}}{L} d_1 T_{sw} \qquad (2\text{-}235)$$

Solving for d_1 gives

$$d_1 = \frac{V_{err}}{R_i} \frac{L}{V_{ac} T_{sw}} \qquad (2\text{-}236)$$

We now need to express the switching period T_{sw} by summing up the t_{on} and t_{off} events:

$$T_{sw} = t_{on} + t_{off} \qquad (2\text{-}237)$$

Based on the respective *on* and *off* slope definitions, we have

$$t_{on} = \frac{V_{err}}{R_i} \frac{L}{V_{ac}} \qquad (2\text{-}238)$$

$$t_{off} = \frac{V_{err}}{R_i} \frac{L}{V_{cp}} \qquad (2\text{-}239)$$

$$T_{sw} = \frac{V_{err}}{R_i} \frac{L}{V_{ac}} + \frac{V_{err}}{R_i} \frac{L}{V_{cp}} \qquad (2\text{-}240)$$

Factoring $\frac{V_{err}}{R_i}$ gives

$$T_{sw} = \frac{V_{err} L}{R_i} \left(\frac{1}{V_{ac}} + \frac{1}{V_{cp}} \right) \qquad (2\text{-}241)$$

Based on these equations, we have everything we need to create the large-signal averaged model. Figure 2-104 shows the final structure where, based on Eq. (2-231), I_μ (Imju) equals

$$I_\mu = V_{cp} (1 - d_1) \frac{T_{sw}}{2L} \qquad (2\text{-}242)$$

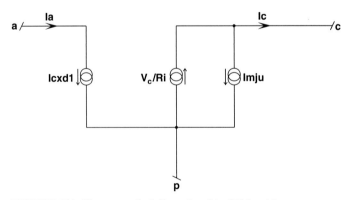

FIGURE 2-104 The structure is similar to that of the CCM model.

SMALL-SIGNAL MODELING

However, since SPICE will calculate the numerical values (T_{sw}, d_1, etc.) via sources, we can easily have switching periods around a few microseconds that will be coded into voltage sources delivering a few microvolts. To avoid large-voltage dynamics within the model, all these sources manipulating microvolts will be multiplied by a 1E6 factor (1 Meg), further divided by the same value during mathematical manipulations. Below is the resulting SPICE coding:

```
Btsw tsw 0 V = ((V(vc)*{L}/{Ri})*((1/(v(a,cx)+1u))+(1/(v(cx,p)
+1u))))*1Meg
```

The above code calculates the switching period based on Eq. (2-241).

```
Bdc dcx 0 V=V(vc)*{L}/({Ri}*V(a,cx)*(V(tsw)/1Meg)+1u)
```

The above code evaluates the duty cycle d_1 based on Eq. (2-236).

```
BImju cx p I= v(cx,p)*(1-V(dc))*v(tsw)/(2*{L}*1Meg)
```

The above codes Eq. (2-242).

```
Bton ton 0 V = V(dc)*v(tsw)/1Meg
```

The above codes Eq. (2-238).

```
Bfsw fsw 0 V = (1/(V(tsw)/1Meg))/1k
```

The above code outputs key variables from the model such as the on time and the switching frequency, expressed in kilohertz.

The complete current-mode BCM model in both *IsSpice* and *PSpice* appears below; the input parameters are the sense resistor R_i and the inductor L.

IsSpice netlist:

```
.SUBCKT PWMBCMCM a c p vc ton fsw {L = 1.2m Ri=0.5}
*
* This subckt is a current-mode BCM model, version 1
*
.subckt limit d dc params: clampH=0.99 clampL=16m
*
Gd 0 dcx d 0 100u
Rdc dcx 0 10k
V1 clpn 0 {clampL}
V2 clpp 0 {clampH}
D1 clpn dcx dclamp
D2 dcx clpp dclamp
Bdc dc 0 V=V(dcx)
.model dclamp d n=0.01 rs=100m
.ENDS
*
Btsw tsw 0 V= ((V(vc)*{L}/{Ri})*(1/v(a,cx) + 1/v(cx,p))) *1Meg
Bdc dcx 0 V=V(vc)*{L}/({Ri}*V(a,cx)*(V(tsw)/1Meg))
Xdc dcx dc limit params: clampH=0.99 clampL=7m
BIap a p I=V(dc)*I(VM)
BIpc p cx I=V(vc)/{Ri}
BImju cx p I= v(cx,p)*(1-V(dc))*v(tsw)/(2*{L}*1Meg)
Bton ton 0 V = V(dc)*v(tsw)
Bfsw fsw 0 V = (1/(V(tsw)/1Meg))/1k
Rdum1 vc 0 1Meg
```

```
VM cx c
*
.ENDS
```

PSpice netlist:

```
.SUBCKT PWMBCMCM a c p vc dc fsw ip params: L=1.2m Ri=0.5
*
* This subckt is a current-mode BCM model, version 1
*
EBtsw tsw 0 Value = {(((V(vc)*{L}/{Ri}) * (1/v(a,cx))
+ 1/v(cx,p) ))*1Meg)}
EBdc dcx 0 Value = { V(vc)*{L}/({Ri}*V(a,cx)*(V(tsw)/1Meg)) }
Xdc dcx dc limit params: clampH=0.99 clampL=7m
GBIap a p Value = { V(dc)*I(VM) }
GBIpc p cx Value = { V(vc)/{Ri} }
GBImju cx p Value = { v(cx,p)*(1-V(dc))*v(tsw)/(2*{L}*1Meg) }
EBton ton 0 Value = { V(dc)*v(tsw) }
EBfsw fsw 0 Value = { (1/(V(tsw)/1Meg))/1k }
EBip ip 0 Value = { abs(V(a,c)*V(ton)*1u)/{L} }
Rdum1 vc 0 1Meg
VM cx c
*
.ENDS
.subckt limit d dc params: clampH=0.99 clampL=16m
Gd 0 dcx VALUE = { V(d)*100u }
Rdc dcx 0 10k
V1 clpn 0 {clampL}
V2 clpp 0 {clampH}
D1 clpn dcx dclamp
D2 dcx clpp dclamp
Edc dc 0 value={ V(dcx) }
.model dclamp d n=0.01 rs=100m
.ENDS
********
```

2.5.4 Testing the Current-Mode BCM Model

Using almost the same template as for the voltage-mode version, we can remove all the ramp generation-related portion to now connect the current-sense signal to its dedicated pin, pin 3. This is what Fig. 2-105 displays.

The averaged model template implementing the BCM current-mode technique appears in Fig. 2-106, showing exactly the same configuration. Internally, the feedback voltage divides by 3, and a kind of zener diode limits its maximum excursion to 1 V. Hence, the maximum peak current cannot exceed $1/R_{sense}$ or 15.2 A in both examples. The static operating point does not differ from what we have already noted for the voltage-mode version. If we now compare both output load responses, Fig. 2-107 confirms the good agreement between the cycle-by-cycle model and the averaged model.

There is something worth mentioning concerning Fig. 2-106: the sense resistor value is negative! Why is this? It is so simply because the I_c current polarity differs in a boost structure from that shown in Fig. 2-104. In a boost, the current must enter the model. This is true for the fixed-frequency model but also for the BCM type. One can rotate the model to match the right polarity, but passing a negative value for R_i gives the same result without altering the original

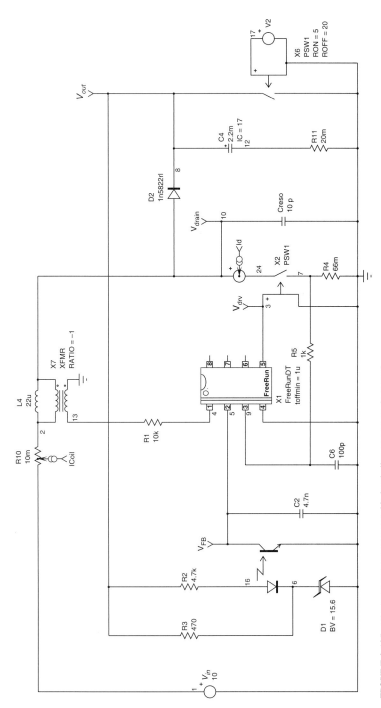

FIGURE 2-105 The boost converter operated in borderline current-mode control.

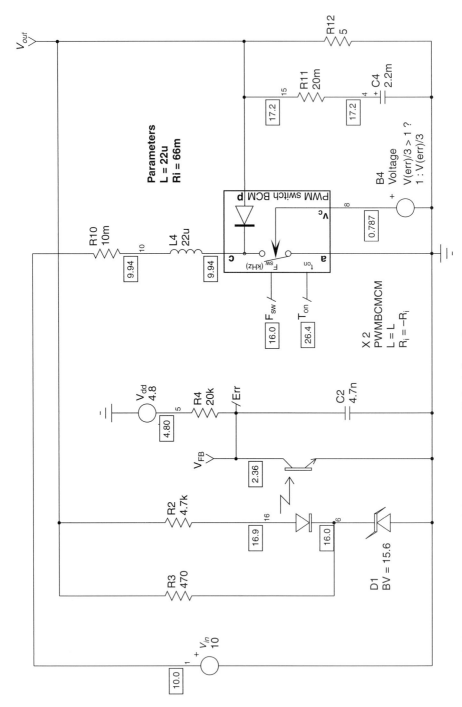

FIGURE 2-106 The averaged current-mode boost converter operated in borderline mode.

SMALL-SIGNAL MODELING

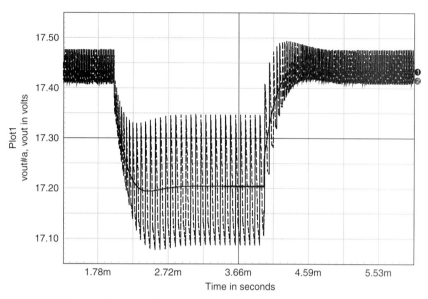

FIGURE 2-107 A good agreement between cycle-by-cycle and averaged model responses to an output step load.

configuration. This is what all current-mode boost PWMCM-based configurations will feature.

One other test consists of stepping the input voltage from 10 to 14 V and observing the output voltage variation. This is what Fig. 2-108 represents. On this graph, both VM and CM responses appear. The CM response does not present any overshoot, whereas the VM does slightly.

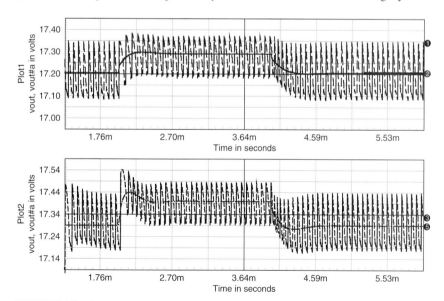

FIGURE 2-108 Output response to an input step, from 10 to 14 V. The upper curve depicts the current-mode stage answer, the lower curve the voltage-mode one.

Please note that the vertical scales are similar. The output excursion in voltage mode reaches 110 mV between plateaus (not inclusive of overshoots) and stays below 80 mV for the current mode. This confirms the inherent better input voltage rejection of the current-mode technique.

PWMBCM also suffers from nonconvergence issues as the fixed-frequency current mode does. To be able to use the model in transient PFC circuits or multioutput flybacks, we have rederived another model, PWMBCM2, actually based on the voltage-mode version to which we added a front-end block. This block merely computes the on time based on the set point and input conditions. This added equation looks like this:

PSpice:

```
EBton ton 0 Value = {((V(vc)*{L}/({Ri}*abs(v(a,c))))*1Meg) + 2}
; min Ton = 2us
```

IsSpice:

```
Bton ton 0 V = ((V(vc)*{L}/({Ri}*abs(v(a,c))))*1Meg + 2)
```

We have compared results delivered by both PWMBCM and PWMBCM2 models, and they looked identical. The second one no longer needs to see a negative sense resistor.

2.6 THE PWM SWITCH MODEL—A COLLECTION OF CIRCUITS

Now that we have derived a complete set of models based on the PWM switch, it is time to show how to implement them in various applications. Since terminal connections do not differ between the various modes VM, CM, and BCM, we will sometimes use a voltage-mode model for the sake of simplicity; or, if some tricks are necessary, a current-mode model will be exemplified (for the forward or the boost, for instance). Let us first review the available models and their related parameters. These names are common and do not differ between *IsSpice* and *PSpice*:

PWMCCMVM	The original PWM switch model for the CCM voltage-mode control. The passed parameter is only the element R_e described in the parasitic element section.
PWMDCMVM	The newly derived model for the discontinuous mode of operation, voltage mode only. Parameters to pass are the inductor value L and the switching frequency F_s.
PWMVM	The autotoggling voltage-mode model, transitioning from DCM to CCM. As above, you will need to pass the inductor value L and the switching frequency F_s.
PWMDCMCM	The newly derived current-mode model operating in DCM. As in any discontinuous model, the parameters to pass are the inductor value L, the switching frequency F_s, the sense resistor R_i, and finally the compensation ramp S_e in volts per second.
PWMCM	The autotoggling current-mode model, transitioning from DCM to CCM. The needed parameters are the same as above.
PWMCMX	This is the same as the above except that the resonating capacitor has been removed to avoid convergence issues in transient mode. Use PWMCM for ac analysis and PWMCMX for transient studies.
PWMBCMVM	The borderline (or critical mode) model operated in voltage mode. The only passed parameter remains the inductor value L.

PWMBCMCM   The borderline (or critical mode) model operated in current mode. Passed parameters are the inductor value L and the sense resistor R_i.

PWMCM_L   The autotoggling current-mode lossy model. Passed parameters are similar to those of the regular PWMCM model.

PWMVM_L   The autotoggling voltage-mode lossy model. Passed parameters are similar to those of the regular PWMVM model.

Most of the circuits appearing in this compilation will be the object of a more comprehensive description in the topology design and simulation dedicated chapters (5, 6, 7 and 8). Hence the lack of details at this time, except the dc transfer functions for the tapped versions of buck and boost which are not very common!

2.6.1 The Buck

This is the simplest circuit and appears in Fig. 2-109a. We have nothing to add as this circuit has been described numerous times previously. Its SPICE implementation follows the recommendations in Fig. 2-109b.

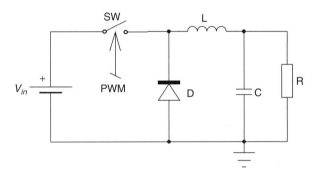

FIGURE 2-109a A classical buck converter.

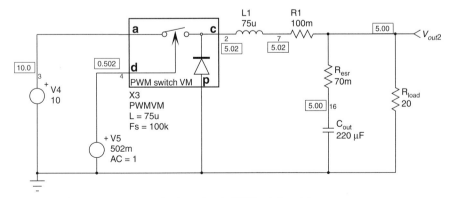

FIGURE 2-109b A buck converter implementing the PWM switch model.

2.6.2 The Tapped Buck

The tapped buck does not benefit from great popularity. Actually a buck using an inductor featuring an intermediate terminal (hence the "tapped" descriptor), this converter can play a role in conversion chains where the input voltage exceeds the output voltage by a factor of 50 to 100. With a classical buck structure, the duty cycle would stabilize to a ridiculously low value, say, 1.6% to deliver 5 V from a 300 V rail. Low duty cycle can lead to instabilities, particularly in current-mode architectures where the propagation delay associated with the leading edge blanking (LEB) severely limits the low excursions. The tapped buck can be declined in two different ways, as portrayed in Fig. 2-110a.

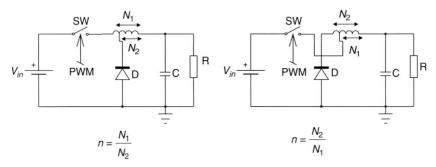

FIGURE 2-110a A tapped buck can be wired in two different manners.

We can see two different approaches, one where the diode is tapped (left side of Fig. 2-110a, $N_1 > N_2$) or one where the switch directly taps the inductor ($N_2 > N_1$). This brings two distinct operating modes as discussed in Ref. 10. Fortunately, the PWM switch implementation does not differ between the two options: the turns ratio n just needs to be chosen to be the right value. Figure 2-110b represents the tapped buck usage with the PWM switch model. Note the relationship between the duty cycle and the output voltage, leading to a duty cycle of 7.5% this time (7.8 calculated by the model):

$$M = \frac{D}{D + \dfrac{D'}{n}} \tag{2-243}$$

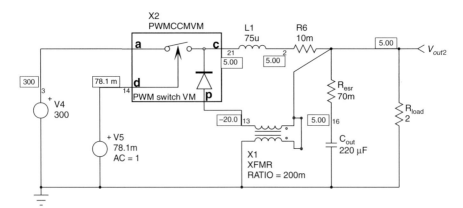

FIGURE 2-110b A tapped buck converter using the CCM version of the PWM switch model. $n = 200\text{m}$ in this example.

2.6.3 The Forward

The forward can work in off-line applications but also in more conventional dc-dc converters. Figure 2-111a represents a simplified version of this popular topology. There are several possibilities to model a forward converter as shown in Figs. 2-111b and 2-111c. The simplest one consists of inserting the transformer ratio between the input and the buck model. But the sense resistor must also be affected by the transformer since it now appears after it. Hence the multiplication by N.

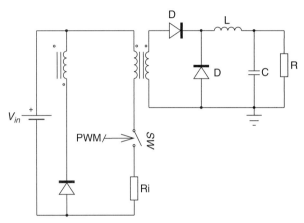

FIGURE 2-111a A simplified forward representation.

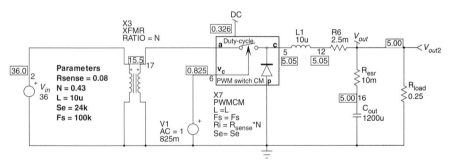

FIGURE 2-111b A forward model where the transformer appears on the left side.

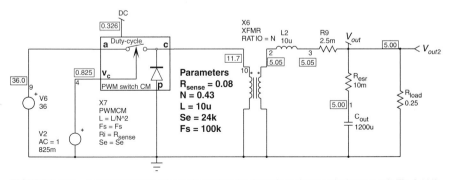

FIGURE 2-111c A different approach for the forward model. Operating points remain the same as in Fig. 2-111b.

In multioutputs forwards, this configuration cannot be easily used. We will therefore stick to a different arrangement where the transformer stays at its original place. R_i can keep its original value, but the inductor must now be reflected to the primary side.

This particular implementation will come back in the forward dedicated section in a multioutput example.

2.6.4 The Buck-Boost

Buck-boosts are used to deliver a negative voltage, still referenced to the input ground. Figure 2-112a portrays its practical implementation whereas Fig. 2-112b shows the PWM switch in action.

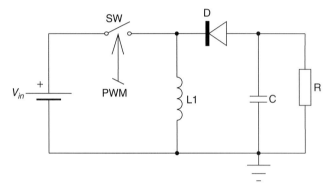

FIGURE 2-112a The buck-boost converter delivering a negative output voltage.

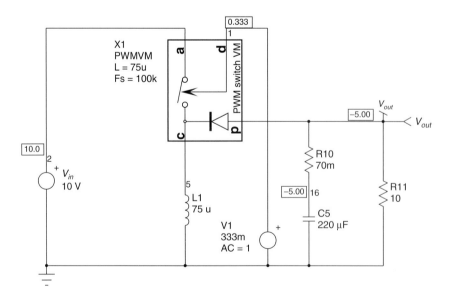

FIGURE 2-112b The buck-boost converter delivering a negative voltage.

2.6.5 The Flyback

The flyback finds its root in the buck-boost converter. A transformer brings the necessary isolation together with a possibility to increase or decrease the output voltage. Its output polarity can be either positive or negative. Figure 2-113a depicts a flyback converter, Fig. 2-113b its implementation with the PWM switch.

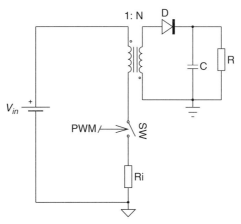

FIGURE 2-113a A flyback converter. Note the reverse dot polarity on the coupling coefficient.

Thanks to this implementation, stacking up transformers on node 3 allows the creation of simple multioutput converters. The turns ratio is normalized to the primary and, as such, directly appears as a parameter in Fig. 2-113b. Please note its negative polarity to deliver a positive output voltage.

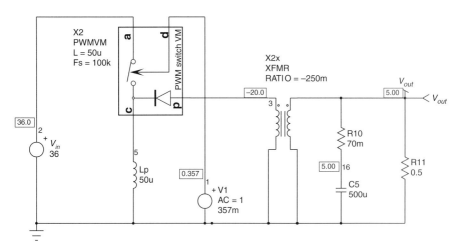

FIGURE 2-113b A single-output flyback converter.

2.6.6 The Boost

As previously noted, the boost in current mode requires a little attention since the sense resistor value must be negative. This is to reflect the right polarity for the output current I_c. Once this is understood, the model works fine. Figure 2-114a represents a classical boost converter and Fig. 2-114b represents a boost in current mode.

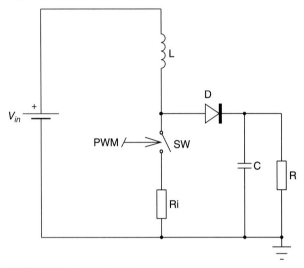

FIGURE 2-114a A classical boost converter.

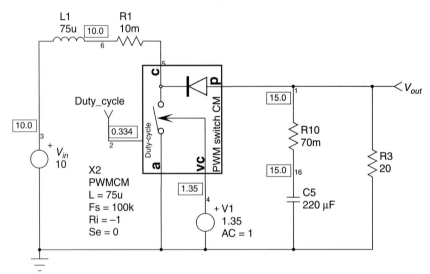

FIGURE 2-114b A boost in current mode featuring a negative resistor value to reflect the original I_c polarity.

2.6.7 The Tapped Boost

Sometimes employed in power factor correction circuits, the tapped boost uses an inductor featuring a tap, exactly as for the tapped buck version. Figure 2-115a depicts a tapped boost circuit, again available in two configurations featuring different turns ratios.

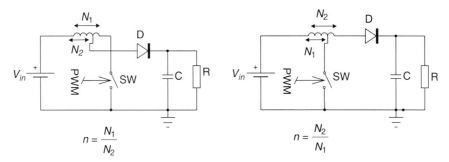

FIGURE 2-115a The tapped boost converter implements an inductor with a tap.

Its implementation with the PWM switch appears in Fig. 2-115b. Short-circuiting the inductor and opening the capacitor help us to obtain the dc transfer function

$$M = 1 + \frac{nD}{D'} \tag{2-244}$$

Note that this transfer function simplifies to that of the boost when n equals 1.

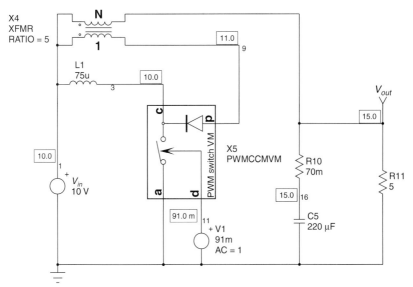

FIGURE 2-115b The PWM switch in a tapped boost converter.

2.6.8 The Nonisolated SEPIC

The single-ended primary inductance converter (SEPIC) finds applications where the output voltage is above or below the input voltage (Fig. 2-116a). As we will see later, its ac response creates difficulties because of the various resonances. Figure 2-116b represents the PWM switch in a nonisolated SEPIC. To minimize the input ripple, both inductors L_1 and L_2 can be coupled.

One difficulty lies with this implementation: the PWM switch model floats. As in any floating configuration, it might necessitate some tricks to help it converge. This is the reason for the presence of R_6 (a dc leak to the ground) and R_7. Always check the validity of the bias point

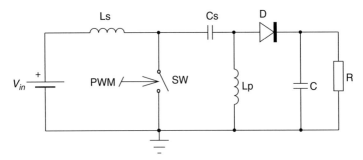

FIGURE 2-116a The SEPIC is often used in portable applications.

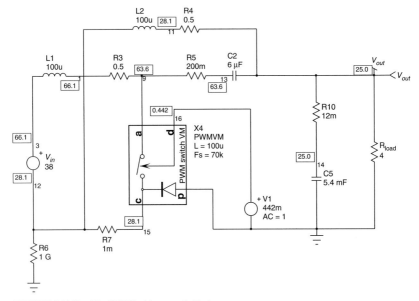

FIGURE 2-116b The SEPIC with uncoupled inductors.

before considering the ac response! Keeping the inductor in the same configuration, one can add a coupling factor k and see the resulting effect.

2.6.9 The Isolated SEPIC

The isolated SEPIC includes a transformer which provides isolation between the secondary and primary grounds. The implementation in Fig. 2-116b still holds, except that a few reflections are necessary. Take the element values in Fig. 2-117, and apply the following transformations before updating any Fig. 2-116b references.

$V_{in} = NV_{in} \rightarrow$ take V_{in} from Fig. 2-117 and multiply it by N; enter result in Fig. 2-116b

$L_1 = L_1 N^2$

$C_1 = \dfrac{C_1}{N^2}$

$L_2 = L_s$

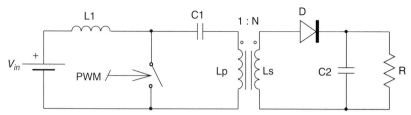

FIGURE 2-117 The isolated SEPIC electrical sketch.

2.6.10 The Nonisolated Ćuk Converter

The Ćuk converter, also sometimes called a boost-buck converter, theoretically offers input/output ripple cancellation if its inductors Ls_1 and Ls_2 are properly coupled (Fig. 2-118a). The PWM switch does lend itself very well to a Ćuk converter simulation. Figure 2-118b shows how to wire this element, here with uncoupled inductors L_1 and L_2. The output voltage is negative.

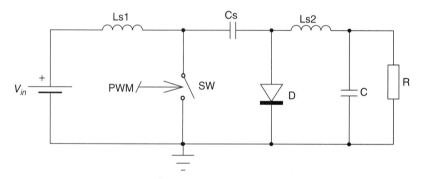

FIGURE 2-118a The nonisolated Ćuk converter delivers a negative voltage.

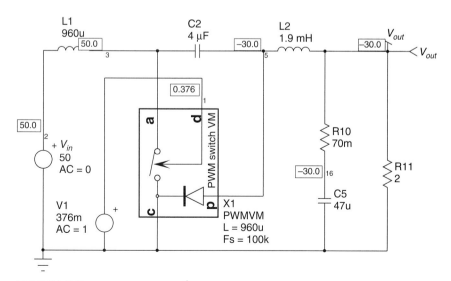

FIGURE 2-118b The PWM switch in a Ćuk converter.

2.6.11 The Isolated Ćuk Converter

This converter can also be declined in an isolated version, implementing a transformer, as Fig. 2-119a shows.

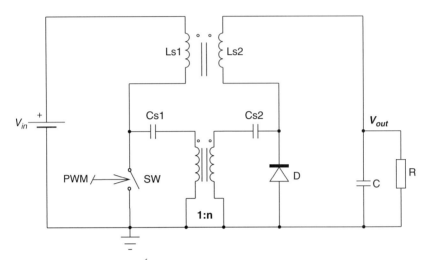

FIGURE 2-119a The isolated Ćuk converter can deliver a positive voltage this time.

The updated version using the PWM switch is displayed in Fig. 2-119b. Note the coupling between the input and output inductances brought by k_1, the SPICE coupling coefficient. However, because of the PWM switch configuration being reduced to a nonisolated version, the converter delivers a negative output. When one is observing the output phase, a simple $-180°$ translation will be necessary.

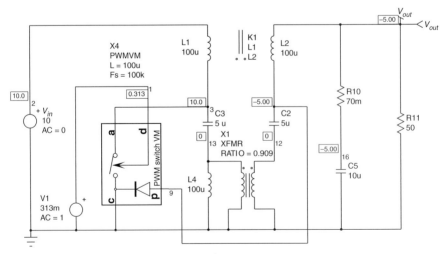

FIGURE 2-119b The PWM switch in an isolated Ćuk converter.

2.7 OTHER AVERAGED MODELS

2.7.1 Ridley Models

In 1990, Dr. Raymond Ridley of the Virginia Tech University (VPEC) showed through a sampling model that a current-mode control (CMC) power stage was best modeled by a third-order polynomial form [13]: one low-frequency pole ω_p and a double pole ω_n, located at one-half the switching frequency. The low-frequency pole ω_p moves in relation to the duty cycle and the external compensation ramp, when present. For a CCM buck operated in peak current-mode control, Eq. (2-245) defines its position:

$$\omega_p = \frac{1}{CR} + \frac{T_{sw}}{LC}(m_c D' - 0.5) \tag{2-245}$$

The sampling action on the inductor current can be modeled as an equivalent second-order filter affected by a quality coefficient Q and a double pole located at ω_n. This is what Eq. (2-246) describes:

$$F_h(s) = \frac{1}{1 + \frac{s}{\omega_n Q_p} + \frac{s^2}{\omega_n^2}} \tag{2-246}$$

The value of this quality coefficient depends on the compensating ramp and the duty cycle. Ridley derived Eq. (2-182) via his sampling model approach, and we showed how Vorpérian later found the same result through his PWM switch theory. Ridley models actually use the PWM switch operating in voltage mode to which a loop elaborating the duty cycle in current mode has been added: the only difference between a current-mode and a voltage-mode converter lies in the way the duty cycle is elaborated, with the final power stage remaining the same as Ridley demonstrated.

The presence of the subharmonic double pole ω_n in the $\dfrac{V_{out}(s)}{V_{err}(s)}$ transfer function can be explained by the sampling process of the inductance current. Actually, this process creates a pair of right half-plane zeros in the current loop which are responsible for the boost in gain at $\dfrac{F_{sw}}{2}$ but also stress the phase lag at this point. Once included in the voltage loop, these zeros turn into poles. If the gain margin vanishes at this frequency, because of the undamped peaking, any perturbation in the current will make the system unstable since both voltage and current loops are embedded. This is shown in Figs. 2-56 to 2-58. You can fight the problem by providing the converter with an external compensation ramp, as we have seen. This will oppose the duty cycle action by lowering the current-loop dc gain. As more external ramp is added, the low-frequency pole ω_p moves to higher frequencies while the double poles present at ω_n will split into two distinct poles. One goes to high frequency, whereas the other one moves toward lower frequencies until it joins and combines with the first low-frequency pole. If we talk about a buck converter, both poles ω_p and the low-frequency side of ω_n unite at the LC network resonant frequency. At this point, the converter behaves as if it were operating in voltage mode. Thus, if compensating the CCM current-mode converter obviously represents a necessity to avoid subharmonic oscillations, overcompensating it brings it closer to a traditional voltage-mode operation, losing current-mode benefits.

2.7.2 Small-Signal Current-Mode Models

If Ridley models can predict subharmonic oscillations, they were originally written in SPICE2 and were not easy to implement: you need to evaluate the values in Eqs. (2-159) to (2-163b) for each configuration, with the right polarity. Thanks to the recent parameter passing features and definition keywords, new SPICE3 models proposed in this book are really easy to use. There is one drawback, however, they are ac models only and cannot toggle between CCM and DCM. You select the model according to your operating mode, and you can only visualize ac results, no dc bias. Fortunately, in-line equations automatically calculate all the dc values, e.g., the operating duty cycle for the desired output voltage at a given input voltage. You just need to pass the constant numbers such as inductor value, switching frequency, etc. Another good point is that Ridley models are universal: if the sense resistor R_i offers a finite nonnull value, it is in current mode. In *IsSpice,* if you decrease R_i to a small value (1 µΩ, for instance), you operate in voltage mode. In *PSpice*, a star must be placed in front of a specific line (see below) to turn the model in voltage mode. All the available subcircuits are built on the original PWM-CCM and PWMDCM models [15] to which calculated parameters are passed after a topology adjustment. Below is an example for the boost model operated in CCM. You can see that all parameter calculations are done in the *.PARAM* keywords. This is a *PSpice* listing:

```
.SUBCKT BOOSTCCM Vin Vout Gnd Control D PARAMS: RI=0.4 L=140U
+ RS=190M FS=100K VOUT=100V RL=50 VIN=48V MC=1.5 VR=2V
* To toggle into Voltage-Mode, put RI=0 and VP becomes VR (The
* PWM sawtooth amplitude)
.PARAM D={(VOUT-VIN)/VOUT}   ; DC duty cycle for Continuous mode
.PARAM VAP={-VOUT}
.PARAM VAC={-VIN}
.PARAM VCP={-VOUT+VIN}
.PARAM IA={-((VOUT^2)/RL/VIN)*D}
.PARAM IP={-VOUT/RL}
.PARAM IC={-VOUT/RL/(1-D)}
.PARAM VP={-VAC*(1/FS)*RI*(MC-1)/L}
* .PARAM VP=VR ; Put RI=0 and remove this start (while putting it
*at the above line) turns into VM
EBD D 0 VALUE = {D}
RL Vin LL {RS}
L LL C {L}
X1 Gnd Vout C Vin Control PWMCCM PARAMS: RI={-RI} L={L} FS={FS}
+ D={D} VAP={VAP} VAC={VAC} IC={IC} VP={VP}
.ENDS
```

2.7.3 Ridley Models at Work

Ridley models require a few parameters to operate. Here are the details you need to know before working with them:

RI	The sense resistor, expressed in ohms.
L	The inductor in play within the circuit. For a flyback, pass the secondary inductor value.
RS	The inductor ohmic losses.
FS	The switching frequency.
VOUT	The output voltage. Remember, this is an ac model, so static parameters must be supplied to evaluate the small-signal sources.
RL	The load resistor.
VIN	The input voltage.
MC = 1.5	The ramp compensation. Please refer to Eq. (2-183) for more details. $M_c = 1.5$ implies a 50% compensation.
VR = 2V	In case the sense resistor comes close to zero (pass 10 µ, for instance), the model toggles to VM with a modulator ramp amplitude equal to *VR*.

In the library (RIDLEY.LIB for *PSpice* and *IsSpice*), the following models are available:

PWMCCMr	The original Ridley CCM invariant current-mode model called from models below
PWMDCMr	The original Ridley DCM invariant current-mode model called from models below
BOOSTDCM	Boost in DCM
BOOSTCCM	Boost in CCM
BUCKDCM	Buck in DCM
BUCKCCM	Buck in CCM
FLYBACKCCM	Flyback in CCM, you pass the secondary inductance, the turns ratio is 1:N
FLYBACKDCM	Flyback in DCM, same as above
FWDDCM	Forward in DCM, implements Fig. 2-111b configuration
FWDCCM	Forward in CCM, same as above

We are going to plot the ac response of a buck converter in current mode using OrCAD's *PSpice* in this example. The circuit is shown in Fig. 2-120. You can see the lack of dc value for

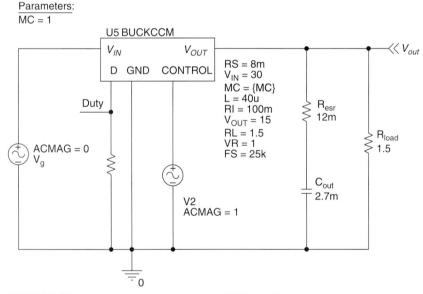

FIGURE 2-120 A current-mode buck converter using Ridley model.

the input voltage V_g, for instance. This is, again, due to the ac nature of the model. If you display the voltage bias point, the duty node will actually tell you the computed duty cycle, but the rest of the nodes might only display 0. If we now sweep the ramp compensation parameter M_c from 1 (no ramp) to $M_c = 10$, for instance, the plot in Fig. 2-121 appears. With a lack of ramp compensation, the double poles get so lightly damped that a strong peaking occurs at one-half the switching frequency (12.5 kHz) and jeopardizes the stability. By adding some compensation (50%) we immediately damp the double poles and the gain margin goes back to a safer value.

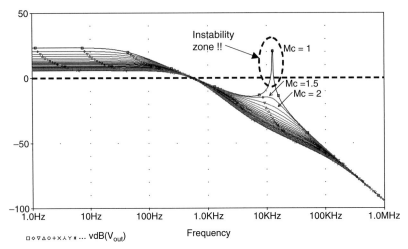

FIGURE 2-121 Sweeping the ramp compensation level gives a good indication of the overall stability in different operating conditions.

2.7.4 CoPEC Models

Dr. Dragan Maksimovic and Dr. Robert Erickson from the Colorado Power Electronic Center (CoPEC) have developed a set of averaged models which are thoroughly described in several publications [3, 16]. The idea remains the same—averaging the pertinent waveforms of the power switch and the diode. However, in this approach, the authors have strived to keep the switch and the diode separated, which sometimes eases their insertion into the converter under study. Following their derivations, the CCM averaged model of the switch network appears in Fig. 2-122.

In DCM, the model uses the concept of a lossless two-port network. It implies that the instantaneous input power $P_{in}(t)$ equals the output power $P_{out}(t)$. This concept is illustrated in Fig. 2-123

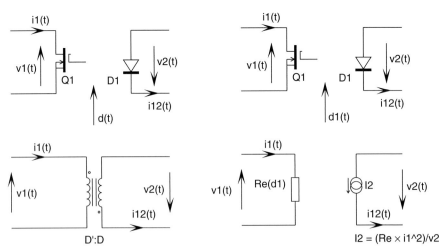

FIGURE 2-122 The switch network equivalent model in CCM as derived by the CoPEC authors.

FIGURE 2-123 The DCM model uses the concept of the loss-free resistor: 100% of the absorbed input power is transmitted to the output.

where we can see a resistor dissipating (absorbing) $P_{in}(t)$ but 100% of the input power is transmitted to the output port. The input resistor derives from the DCM equations and naturally depends on $d_1(t)$, the on duty cycle in DCM. The output power source is made via a controlled current source delivering a value proportional to the input power. The authors then show how to combine both CCM and DCM models into one single subcircuit, making the final result an attractive circuit. The netlist below shows the final encapsulation in *PSpice* of the CCM-DCM1 model. This model does not include a transformer ratio but CCM-DCM2 does. The needed parameters are the switching frequency (F_S) and the inductance value (L), necessary to find the toggling point between CCM and DCM.

```
.SUBCKT CCM-DCM1 1   2   3   4 CTRL params: FS=100k L=75u
Xd CTRL 5 limit params: clampH=0.95 clampL=16m
Vd 1 1x
Et 1x 2 Value = {(1-V(u))*V(3,4)/(V(u)+1u)}
Gd 4 3 Value = {(1-V(u))*I(Vd)/(V(u)+1u)}
Ga 0 a Value = {IF (I(Vd) > 0, I(Vd), 0)}
Va a b
Rdum1 b 0 10
Eu1 100 0 Value = {V(5)}
Eu2 20X 0 Value = {V(5)*V(5)/((V(5)*V(5)+2*{L}*{FS}*I(Vd)/ V(3,4))+
+ 1u)}
Xd2 20X 5 200 limit2
D1 200 u DN
D2 100 u DN
.model DN D N=0.01 RS=100m
.ENDS
********
.subckt limit d dc params: clampH=0.99 clampL=16m
Gd 0 dc VALUE = {V(d)*100u}
Rdc dc 0 10k
V1 clpn 0 {clampL}
V2 clpp 0 {clampH}
D1 clpn dc dclamp
D2 dc clpp dclamp
.model dclamp d n=0.01 rs=100m
.ENDS
********
.subckt limit2 d2nc d d2c
*
Gd 0 d2c d2nc 0 100u
Rdc d2c 0 10k
V1 clpn 0 7m
E2 clpp 0 Value = {1-V(d)}
D1 clpn d2c dclamp
D2 d2c clpp dclamp
.model dclamp d n=0.01 rs=100m
.ENDS
********
```

The *limit/limit2* subcircuit smoothly clamps the duty cycle excursions and is very similar to those developed for the PWM switch model. CoPEC models are available in the COPEC.LIB library file in both *PSpice* and *IsSpice* versions. The current-mode version is also included in the library file (the CPM block), but we found the implementation slightly more complicated than the PWM switch. Also, the model cannot predict the subharmonic oscillations. Let us now see how to wire the model through several examples.

2.7.5 CoPEC Models at Work

The first example depicts a voltage-mode flyback converter shown in Fig. 2-124a. It uses the CCM-DCM2 model to which the transformer ratio has been passed. After running the simulation, PROBE unveils the complete Bode plot of the circuit, as Fig. 2-124b shows. The model converges quite well and has been extensively tested by the authors. One can see that the averaged model arrangement follows the real flyback architecture, as the averaged switch and diodes are not tied together by a common point. This really eases the implementation in a complicated structure.

The next example depicts a voltage-mode nonisolated SEPIC. Its schematic appears in Fig. 2-125.

Again, there are no particular difficulties in inserting the model into the simulation fixture. Just follow the real schematic, and replace the switch and the diode by their respective averaged connections. Figure 2-126 displays the gain and phase evolutions.

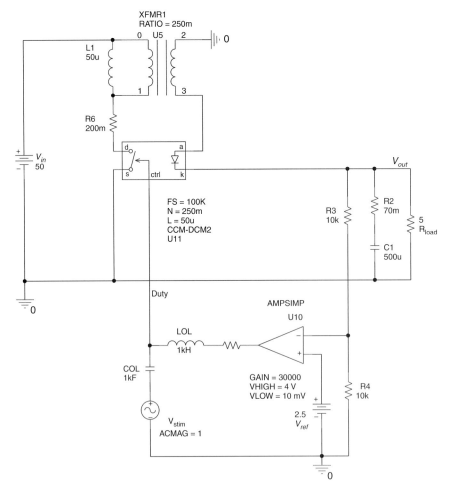

FIGURE 2-124a A flyback converter using the CoPEC model.

SMALL-SIGNAL MODELING

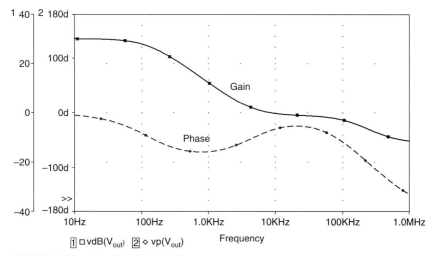

FIGURE 2-124b Bode plot of the flyback converter.

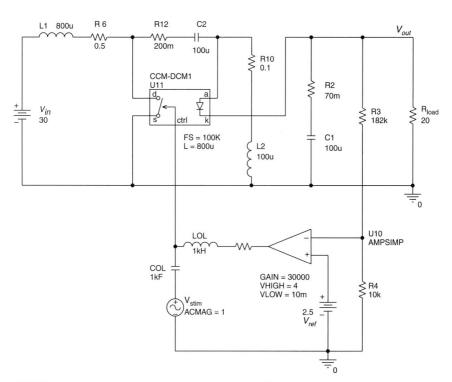

FIGURE 2-125 A nonisolated SEPIC implementing the CoPEC model.

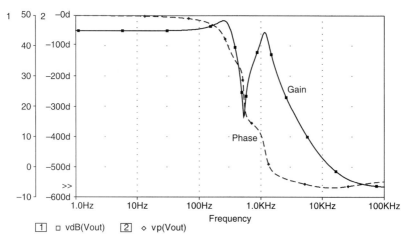

FIGURE 2-126 SEPIC Bode plot.

2.7.6 Ben-Yaakov Models

In the 1990s, Sam Ben-Yaakov from the Ben-Gurion University of the Negev (Israel) introduced the concept of the switched inductor model (SIM) [17]. His approach was close to the PWM switch concept except that Ben-Yaakov included the inductor as part of the model. The idea was to represent the switching cell as a single-pole double-throw (SPDT) device connected to an inductor. Figure 2-127 shows the concept.

Figure 2-128 represents the SIM implemented into a buck-boost converter. Based on this configuration, we can quickly derive a few equations. For instance, if we consider the instantaneous inductor current to be

$$\frac{dI_{L_f}}{dt} = \frac{V_{L_f}}{L_f} \qquad (2\text{-}247)$$

then its average current can be expressed by

$$\frac{d\langle I_{L_f}\rangle_{T_{sw}}}{dt} = \frac{\langle V_{L_f}\rangle_{T_{sw}}}{L_f} \qquad (2\text{-}248)$$

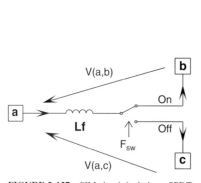

FIGURE 2-127 SIM circuit includes a SPDT switch connected to an inductor.

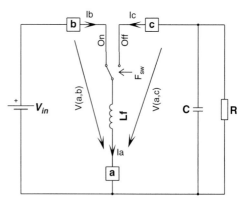

FIGURE 2-128 SIM inside a buck-boost.

Based on this expression, we can see that the total average voltage across the inductor is the sum of both the *on* and *off* inductor average voltages:

$$\langle V_{L_f} \rangle_{T_{sw}} = \frac{\langle V(a,b) \rangle_{T_{sw}} t_{on} + \langle V(a,c) \rangle_{T_{sw}} t_{off}}{T_{sw}} = \langle V(a,b) \rangle_{T_{sw}} D + \langle V(a,c) \rangle_{T_{sw}} D' \quad (2\text{-}249)$$

This equation holds for both CCM and DCM since in the latter no current flows in the inductor during the third interval, hence no voltage appears across the inductor. The inductor thus acts as a set of three independent current generators: the average inductor current G_a, the on-time current G_b, and the off-time current G_c. Figure 2-129 gathers them into the generic switch inductor model (GSIM), which represents the foundations of the eponymous average models.

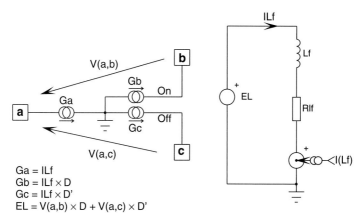

Ga = ILf
Gb = ILf × D
Gc = ILf × D'
EL = V(a,b) × D + V(a,c) × D'

FIGURE 2-129 The foundation of the GSIM.

As for any model, the duty cycle generators are coded into voltage sources D_{on} and D_{off}, respectively, for D and D'. Based on these remarks, it is possible to write some equations describing both operating modes, as already derived in Ref. 17:

$$G_a = I(L_f) \quad (2\text{-}250)$$

$$G_b = \frac{VD_{on}}{VD_{on} + VD_{off}} I(L_f) \quad (2\text{-}251)$$

$$G_c = \frac{VD_{off}}{VD_{on} + VD_{off}} I(L_f) \quad (2\text{-}252)$$

$$EL = V(a,b)D_{on} + V(a,c)D_{off} \quad (2\text{-}253)$$

$$VD_{off} = \frac{2I(L_f)L_f F_{sw}}{V(a,b)VD_{on}} - VD_{on} \quad (2\text{-}254)$$

If we classically clamp Eq. (2-254) between $10m$ and $1 - VD_{on}$, we have a model naturally toggling between CCM and DCM. The following lines give the voltage-mode netlist for the GSIM in *IsSpice* and in *PSpice*:

IsSpice netlist:

```
.SUBCKT GSIM_VM a b c DON {FS=100k L=50u RS=100m}
BGB 0 br I=I(VIL)*V(DON)/(V(DON)+V(DOFF)+1u)
BGC 0 cr I=I(VIL)*V(DOFF)/(V(DON)+V(DOFF)+1u)
BGA ar 0 I=I(VIL)
RDUMA ar a 1u
RDUMB br b 1u
RDUMC cr c 1u
BEL el 0 V=V(a,b)*V(DON)+V(a,c)*V(DOFF)
L1 el erl {L}
RL erl vil {RS}
vil vil 0 0
BEDOFF doffc 0 V=(2*I(VIL)*{L}*{FS}/((V(a,b)*V(DON))+1n))-V(DON)
Rdoffc doffc 0 10E10
Rlmt doffc doff 1
VCLP vc 0 9m
Dlmtz vc doff DBREAK
Dlmtc doff doffm DBREAK
Bclamp doffm 0 V=1-V(DON)-9m
Rdoff doff 0 10E10
.Model Dclamp D N=0.01
.ENDS
```

PSpice netlist:

```
.SUBCKT GSIM_VM a b c DON params: FS=100k L=50u RS=100m
GB 0 br value = {I(VIL)*V(DON)/(V(DON)+V(DOFF)+1u)}
GC 0 cr value = {I(VIL)*V(DOFF)/(V(DON)+V(DOFF)+1u)}
GA ar 0 value = {I(VIL)}
RDUMA ar a 1u
RDUMB br b 1u
RDUMC cr c 1u
EL el 0 Value = {V(a,b)*V(DON)+V(a,c)*V(DOFF)}
L1 el erl {L}
RL erl vil {RS}
vil vil 0 0
EDOFF doffc 0 Value = {(2*I(VIL)*{L}*{FS}/((V(a,b)*V(DON))+1n))-
+ V(DON)}
Rdoffc doffc 0 10E10
Rlmt doffc doff 1
VCLP vc 0 9m
Dlmtz vc doff DBREAK
Dlmtc doff doffm DBREAK
Eclamp doffm 0 Value = {1-V(DON)-9m}
Rdoff doff 0 10E10
.Model Dclamp D N=0.01
.ENDS
```

These models have recently been upgraded to account for conduction losses [18]. In the current-mode version, a duty cycle factory is added in front of the voltage-mode model to generate the *D* variable. It leads to a more complicated model, which is unable to predict the subharmonic instabilities. Also, if the GSIM as described in Fig. 2-129 lends itself very well to simulating simple structures such as boost, buck, flyback, and so on, it cannot be used as it is to implement SEPIC or Ćuk converters, for instance. Below in Fig. 2-130 lies an application of the GSIM in a boost converter featuring losses. Figure 2-131 displays its complete Bode plot.

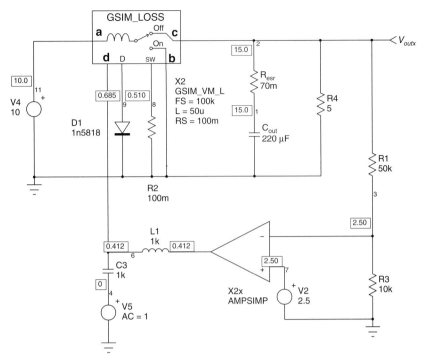

FIGURE 2-130 A lossy boost converter using the GSIM approach.

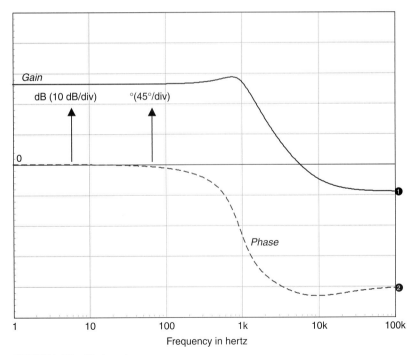

FIGURE 2-131 The lossy boost ac response.

WHAT I SHOULD RETAIN FROM CHAP. 2

1. State-space averaging (SSA) offers a way to derive the transfer function of converters, but it is a long and complicated process. As the analysis is carried over the entire converter, the addition of new elements (such as filtering devices) is required to restart the derivation from scratch.
2. Analyzing the converter reveals that both *on* and *off* states of the power switches imply linear networks. Unfortunately, the brutal transition between these states creates a discontinuity that needs to be smoothed. The process of waveform *averaging* helps to remove this discontinuity and leads to a large-signal *averaged* model implementing nonlinear equations. A linearization process across a dc point makes a small-signal model out of it. From this small-signal model, we can extract the frequency response to further compensate the converter.
3. As the switching transistor and the diode are "guilty" of this nonlinearity, why not simply make a small-signal model of this association? The PWM switch is born! Identify the PWM switch in your converter under study, insert the small-signal PWM model as you would with a bipolar transistor model, and directly solve *linear* equations.
4. Different control methods exist such as voltage mode or current mode. Voltage mode does not need to sense the inductor current and can be seen as a simpler implementation compared to current mode. However, it suffers from several drawbacks such as poor input voltage rejection. It is also more difficult to stabilize a converter operated in CCM compared to the same converter driven in current mode. Current mode permanently senses the inductor current and adjusts its value depending on the output power demand. In CCM, the current-mode architecture can suffer from subharmonic oscillations at duty cycle levels above 50%: the converter requires a ramp compensation signal to gain stability. Both methods require a particular PWM switch model able to transition between CCM and DCM. These models are described in detail.
5. It is necessary to compensate the current-mode converter by an external ramp, but overcompensating it (too much ramp) brings it closer to voltage-mode operation.
6. Thanks to the averaged models, one can easily extract the Bode plot response of the converter under study and check the effects of parasitic element impacts such as output capacitor ESRs.

REFERENCES

1. S. Ćuk, "Modeling, Analysis and Design of Switching Converters," Ph.D. thesis, Caltech, November 1976.
2. D. Middlebrook and S.Ćuk, "A General Unified Approach to Modeling Switching-Converter Power Stages," *International Journal of Electronics*, vol. 42, no. 6, pp. 521–550, June 1977.
3. R. Erickson and D. Maksimovic, *Fundamentals of Power Electronics*, Kluwers Academic Press, 2001.
4. D. Mitchell, *Dc-dc Switching Regulator Analysis,* McGraw-Hill, New York, 1988. Available as a reprint from http://www.ejbloom.com/.
5. D. Mitchell, "Switching Regulators Design and Analysis Methods," Modern Power Conversion Design Techniques, ej Bloom course, Portsmouth UK, 1996.
6. L. G. Meares and C. E. Hymovitz, "Improved Spice Model Simulates Transformer's Physical Processes," *EDN*, August 19, 1996.
7. V. Vorpérian, "Simplified Analysis of PWM Converters Using the Model of the PWM Switch, Parts I (CCM) and II (DCM), " *Transactions on Aerospace and Electronics Systems*, vol. 26, no. 3, May 1990.
8. L. G. Meares, "New Simulation Techniques Using Spice," *Proceedings of the IEEE 1986 Applied Power Electronics Conference*, pp. 198–205, New Orleans 1986.

9. J. Sun, Dan Mitchell, M. Greuel, P. T. Krain, and R.M. Bass, "Average Modeling of PWM Converters in Discontinuous Conduction Mode: A Reexamination," *IEEE Power Electronics Specialists Conference*, Fukoka Japan, 1998, pp. 615–622.
10. V. Vorpérian, "*Fast Analytical Techniques for Electrical and Electronic Circuits,*" Cambridge University Press, 2002.
11. V. Vorpérian, "Analysis of Current-Controlled PWM Converters Using the Model of the Current Controlled PWM Switch," *Power Conversion and Intelligent Motion Conference,* 1990, pp. 183–195.
12. B. Holland, "Modelling, Analysis and Compensation of the Current-Mode Converter," *Powercon* 11, 1984.
13. R. B. Ridley, "A New Continuous-Time Model for Current-Mode Control," *IEEE Transactions of Power Electronics*, vol. 6, April 1991, pp. 271–280.
14. R. Erickson and D. Maksimovic, "Advances in Averaged Switch Modeling and Simulation," *Professional seminars, Power Electronics Specialists Conference*, Charleston USA, 1999.
15. R. B. Ridley, "A New Small-Signal Model for Current-Mode Control," Ph.D. dissertation, Virginia Polytechnic Institute and State University, 1990.
16. R. Erickson and D. Maksimovic, "Advances in Averaged Switch Modeling and Simulation," *Power Electronics Specialist Conference, 1999*. Download at: http://ece.colorado.edu/~pwrelect/publications.html.
17. Sam Ben-Yaakov, "Average Simulation of PWM Converters by Direct Implementation of Behavioral Relationships," *IEEE Applied Power Electronics Conference (APEC'93)*, pp. 510–516.
18. Sam Ben-Yaakov, "Generalized Switched Inductor Model (GSIM): Accounting for Conduction Losses," *Aerospace and Electronic Systems, IEEE Transactions,* vol. 38, no. 2, April 2002, pp. 681–687.

APPENDIX 2A   BASIC TRANSFER FUNCTIONS FOR CONVERTERS

Further to the analysis of converters using the PWM switch, we have gathered in this appendix the transfer equations of the three basic converters operated in fixed switching frequency, DCM or CCM, voltage-mode or current-mode control. You will sometimes see two forms of equations. The first one corresponds to the simplified version of the second one, whose expression does not lend itself easily to an immediate implementation. Thanks to the SPICE implementation via small-signal models (or linearized large-signal models), you can avoid manipulating them. However, keep in mind that the key to stabilizing a power supply lies in the knowledge of the pole-zero locations and how they move in relation to stray elements or input/output parameters.

In all the following equations, we will have

V_{peak}	Sawtooth amplitude for the voltage-mode PWM modulator
r_{Cf}	Output capacitor ESR
r_{Lf}	Inductor ESR
R	Load resistor
C	Output capacitor
L	Inductor
M	Conversion ratio $\frac{V_{out}}{V_{in}}$
D	*On* duty cycle
D'	*Off* duty cycle, also denoted by $1 - D$, depending on the mood
T_{sw}	Switching period

F_{sw} Switching frequency
R_i Sense resistor in current mode
S_1 or S_n Inductor on slope, e.g., $\dfrac{V_{in}}{L}$ for a boost converter
S_2 or S_f Inductor off slope, e.g., $\dfrac{V_{out} - V_{in}}{L}$ for a boost converter
S_a or S_e External ramp compensation slope
m_c Compensation ramp according to Ridley notation [2]: $m_c = 1 + \dfrac{S_e}{S_n}$

2A.1 Buck

Voltage-mode, CCM:

Reference 1 equations:

$$\frac{V_{out}(s)}{V_{err}(s)} = \frac{V_{in}}{V_{peak}} K_c \frac{1 + \dfrac{s}{\omega_{z1}}}{1 + \dfrac{s}{Q\omega_0} + \left(\dfrac{s}{\omega_0}\right)^2} \tag{2A-1}$$

$$\frac{V_{out}(s)}{V_{in}(s)} = D \frac{1 + \dfrac{s}{\omega_{z1}}}{1 + \dfrac{s}{Q\omega_0} + \left(\dfrac{s}{\omega_0}\right)^2} \tag{2A-2}$$

$\omega_{z1} = \dfrac{1}{r_{Cf}C}$

$\omega_{z2} = \infty$ no RHPZ for the CCM buck

$K_c = \dfrac{R}{r_{Lf} + R}$ if $r_{Lf} = r_{Cf} \approx 0$ $K_c = 1$

$\omega_0 = \dfrac{1}{\sqrt{LC \dfrac{R + r_{Cf}}{R + r_{Lf}}}}$ if $r_{Lf} = r_{Cf} \approx 0$ $\omega_0 = \dfrac{1}{\sqrt{LC}}$

$Q = \dfrac{1}{\dfrac{Z_o}{r_{Lf} + R} + \dfrac{r_{Cf} + r_{Lf} \| R}{Z_o}}$ if $r_{Lf} = r_{Cf} \approx 0$ $Q = R\sqrt{\dfrac{C}{L}}$

with $Z_o = \sqrt{\dfrac{L}{C}}$, the LC network characteristic equation.

Voltage-mode, DCM

Reference 1 equations:

$$\frac{V_{out}(s)}{V_{err}(s)} = \frac{V_{in}}{V_{peak}} \frac{K_1\left(1 + \dfrac{s}{\omega_{z1}}\right)}{\left(1 + \dfrac{s}{\omega_{p1}}\right)} \tag{2A-3}$$

$$\frac{V_{out}(s)}{V_{in}(s)} = \frac{V_{out}}{V_{in}} \frac{\left(1 + \frac{s}{\omega_{z1}}\right)}{\left(1 + \frac{s}{\omega_{p1}}\right)} \tag{2A-4}$$

$$K_1 = \frac{2(1-M)}{2-M}\sqrt{\frac{1-M}{K}} \quad \text{with } K = \frac{2L}{RT_{sw}}$$

$$\omega_{z1} = \frac{1}{r_{Cf}C}$$

$$\omega_{p1} = \frac{2-M}{1-M}\frac{1}{RC}$$

Current-mode, CCM:

Reference 1 equations:

$$\frac{V_{out}(s)}{V_{err}(s)} \approx \frac{R}{R_i} \frac{\left(1 + \frac{s}{\omega_{z1}}\right)}{\left(1 + \frac{s}{\omega_{p1}}\right)} \frac{1}{1 + \frac{s\left[\left(1 + \frac{S_a}{S_1}\right)D' - 0.5\right]}{F_{sw}} + \frac{s^2}{(\pi F_{sw})^2}} \tag{2A-5}$$

$$\frac{V_{out}(s)}{V_{in}(s)} = 0 \tag{2A-6}$$

$$\omega_{z1} = \frac{1}{r_{Cf}C}$$

$$\omega_{p1} = \frac{1}{RC}$$

for $S_a = 50\% \, S_2$.

Reference 2 equations:

$$\frac{V_{out}(s)}{V_{err}(s)} \approx \frac{R}{R_i} \frac{1}{1 + \frac{RT_{sw}}{L}\left[m_c D' - 0.5\right]} F_p(s) F_h(s) \tag{2A-7}$$

$$\frac{V_{out}(s)}{V_{in}(s)} \approx \frac{D\left[m_c D' - \left(1 - \frac{D}{2}\right)\right]}{\frac{L}{RT_{sw}} + \left[m_c D' - 0.5\right]} F_p(s) F_h(s) \tag{2A-8}$$

$$F_p(s) = \frac{1 + \frac{s}{\omega_{z1}}}{1 + \frac{s}{\omega_{p1}}}$$

$$\omega_{z1} = \frac{1}{r_{Cf}C}$$

$$\omega_{p1} = \frac{1}{RC} + \frac{T_{sw}}{LC}\left[m_c D' - 0.5\right]$$

$$F_h(s) = \frac{1}{1 + \dfrac{s}{\omega_n Q_p} + \dfrac{s^2}{\omega_n^2}}$$

$$\omega_n = \frac{\pi}{T_{sw}}$$

$$Q_p = \frac{1}{\pi(m_c D' - 0.5)}$$

$$m_c = 1 + \frac{S_e}{S_n}$$

Current-mode, DCM:

Reference 2 equations:

$$\frac{V_{out}(s)}{V_{err}(s)} = F_m H_c \frac{1 + \dfrac{s}{\omega_{z1}}}{\left(1 + \dfrac{s}{\omega_{p1}'}\right)\left(1 + \dfrac{s}{\omega_{p2}'}\right)} \quad (2A\text{-}9)$$

$$\omega_{z1} = \frac{1}{r_{Cf}C}$$

$$\omega_{p1}' = \frac{1}{RC}\frac{2m_c - (2 + m_c)M}{m_c(1 - M)}$$

$$\omega_{p2}' = 2F_{sw}\left(\frac{M}{D}\right)^2$$

$$H_c = \frac{2M_c V_{out}}{D}\frac{1 - M}{2m_c - (2 + m_c)M}$$

$$F_m = \frac{1}{S_n m_c T_{sw}}$$

$$M = \frac{2}{1 + \sqrt{1 + \dfrac{8\tau_L}{D^2}}} \quad \text{with } \tau_L = \frac{L}{RT_{sw}}$$

For a forward topology in voltage mode, all transfer functions and component definitions remain the same, except V_{in} which becomes NV_{in}, with $N = \dfrac{N_s}{N_p}$. In current mode, adopting the scaling in V_{in}, the sense resistor R_i must also account for the transformer presence and must be changed to $R_i' = R_i N$. Also, some half-bridge topologies require the input voltage to be divided by 2 given the transformer connection (e.g., via a capacitive bridge).

2A.2 Boost

Voltage-mode, CCM

Reference 1 equations:

$$\frac{V_{out}(s)}{V_{err}(s)} = K_d \frac{\left(1 + \frac{s}{\omega_{z1}}\right)\left(1 - \frac{s}{\omega_{z2}}\right)}{1 + \frac{s}{Q\omega_0} + \left(\frac{s}{\omega_0}\right)^2} \qquad (2A\text{-}10)$$

$$\frac{V_{out}(s)}{V_{in}(s)} = M \frac{1 + \frac{s}{\omega_{z1}}}{1 + \frac{s}{Q\omega_0} + \left(\frac{s}{\omega_0}\right)^2} \qquad (2A\text{-}11)$$

$$\omega_{z1} = \frac{1}{r_{Cf}C}$$

$$\omega_{z2} = \frac{D'^2}{L}\left(R - r_{Cf}\|R\right) - \frac{r_{Lf}}{L} \quad \text{if } r_{Lf} = r_{Cf} \approx 0 \quad \omega_{z2} = \frac{RD'^2}{L} \quad \text{(RHPZ)}$$

$$K_d = \frac{V_{in}}{V_{peak}D'^2} = \frac{V_{out}^2}{V_{in}V_{peak}}$$

$$M = \frac{1}{D'}\frac{1}{1 + \frac{r_{Lf}}{D'^2R} + \frac{(r_{Cf}\|R)D}{RD'}} \quad \text{if } r_{Lf} = r_{Cf} \approx 0 \quad M = \frac{1}{D'}$$

$$\omega_0 = \frac{1}{\sqrt{LC}}\sqrt{\frac{r_{Lf} + D'^2R}{r_{Cf} + R}} \quad \text{if } r_{Lf} = r_{Cf} \approx 0 \quad \omega_0 = \frac{1-D}{\sqrt{LC}}$$

$$Q = \frac{\omega_0}{\frac{r_{Lf}}{L} + \frac{1}{C(r_{Cf} + R)}} \quad \text{if } r_{Lf} = r_{Cf} \approx 0 \quad Q = R(1-D)\sqrt{\frac{C}{L}}$$

Voltage-mode, DCM

Reference 1 equations:

$$\frac{V_{out}(s)}{V_{err}(s)} = \frac{V_{in}}{V_{peak}} \frac{K_1\left(1 + \frac{s}{\omega_{z1}}\right)}{\left(1 + \frac{s}{\omega_{p1}}\right)} \qquad (2A\text{-}12)$$

$$\frac{V_{out}(s)}{V_{in}(s)} = \frac{V_{out}}{V_{in}} \frac{\left(1 + \frac{s}{\omega_{z1}}\right)}{\left(1 + \frac{s}{\omega_{p1}}\right)} \qquad (2A\text{-}13)$$

$$K_1 = \frac{2}{2M-1}\sqrt{\frac{M(1-M)}{K}} \quad \text{with } K = \frac{2L}{RT_{sw}}$$

$$\omega_{z1} = \frac{1}{r_{Cf}C}$$

$$\omega_{p1} = \frac{2M-1}{M-1}\frac{1}{RC}$$

Current-mode, CCM:

Reference 1 equations:

$$\frac{V_{out}(s)}{V_{err}(s)} = \frac{R}{2R_i}\frac{V_{in}}{V_{out}}\frac{\left(1+\frac{s}{\omega_{z1}}\right)}{\left(1+\frac{s}{\omega_{p1}}\right)} \frac{\left(1-\frac{s}{\omega_{z2}}\right)}{1+\frac{s\left[\left(1+\frac{S_a}{S_1}\right)D'-0.5\right]}{F_{sw}}+\frac{s^2}{(\pi F_{sw})^2}} \quad (2A\text{-}14)$$

$$\frac{V_{out}(s)}{V_{in}(s)} = \frac{V_{out}}{2V_{in}}\frac{\left(1+\frac{s}{\omega_{z1}}\right)}{\left(1+\frac{s}{\omega_{p1}}\right)} \quad (2A\text{-}15)$$

$$\omega_{z1} = \frac{1}{r_{Cf}C}$$

$$\omega_{z2} = \frac{RD'^2}{L} \quad \text{(RHPZ)}$$

$$\omega_{p1} = \frac{2}{RC}$$

Reference 2, 3, 4, and 5 equations:

$$\frac{V_{out}(s)}{V_{err}(s)} = \frac{R}{R_i}\frac{1}{2M+\frac{RT_{sw}}{LM^2}\left(\frac{1}{2}+\frac{S_e}{S_n}\right)}\frac{\left(1+\frac{s}{\omega_{z1}}\right)\left(1-\frac{s}{\omega_{z2}}\right)}{\left(1+\frac{s}{\omega_{p1}}\right)\left(1+\frac{s}{\omega_n Q_p}+\frac{s^2}{\omega_n^2}\right)} \quad (2A\text{-}16)$$

$$\omega_n = \frac{\pi}{T_{sw}}$$

$$Q_p = \frac{1}{\pi(m_c D' - 0.5)}$$

$$m_c = 1 + \frac{S_e}{S_n}$$

$$\omega_{z1} = \frac{1}{r_{Cf}C}$$

$$\omega_{z2} = \frac{RD'^2}{L} \quad \text{(RHPZ)}$$

$$\omega_{p1} = \frac{\frac{2}{R} + \frac{T_{sw}}{LM^3}\left(1 + \frac{S_e}{S_n}\right)}{C}$$

Current-mode, DCM:

Reference 2, 3, 4, and 5 equations:

$$\frac{V_{out}(s)}{V_{err}(s)} = F_m H_d \frac{\left(1 + \frac{s}{\omega_{z1}}\right)\left(1 - \frac{s}{\omega_{z2}}\right)}{\left(1 + \frac{s}{\omega_{p1}}\right)\left(1 + \frac{s}{\omega_{p2}}\right)} \tag{2A-17}$$

$$\omega_{z1} = \frac{1}{r_{Cf}C}$$

$$\omega_{z2} = \frac{R}{M^2 L} \quad \text{(high-frequency RHPZ)} \qquad \text{In relation to } \omega_{p2}: \quad \omega_{z2} = \frac{\omega_{p2}}{1 - \frac{1}{M}} > 2F_{sw}$$

$$\omega_{p1} = \frac{1}{RC}\frac{2M - 1}{M - 1}$$

$$\omega_{p2} = 2F_{sw}\left(\frac{1 - \frac{1}{M}}{D}\right)^2 \geq 2F_{sw}$$

$$H_d = \frac{2V_{out}}{D}\frac{M - 1}{2M - 1}$$

$$F_m = \frac{1}{S_n m_c T_{sw}}$$

$$M = \frac{1 + \sqrt{1 + \frac{2D^2}{\tau_L}}}{2} \quad \text{with } \tau_L = \frac{L}{RT_{sw}}$$

2A.3 Buck-Boost

Voltage-mode, CCM

Reference 1 equations:

$$\frac{V_{out}(s)}{V_{err}(s)} = -\frac{V_{in}}{(1-D)^2 V_{peak}} \frac{\left(1 + \frac{s}{\omega_{z1}}\right)\left(1 - \frac{s}{\omega_{z2}}\right)}{1 + \frac{s}{Q\omega_0} + \left(\frac{s}{\omega_0}\right)^2} \tag{2A-18}$$

$$\frac{V_{out}(s)}{V_{in}(s)} = -\frac{V_{out}}{V_{in}} \frac{\left(1 + \frac{s}{\omega_{z1}}\right)}{1 + \frac{s}{Q\omega_0} + \left(\frac{s}{\omega_0}\right)^2} \quad (2A-19)$$

$\omega_{z1} = \dfrac{1}{r_{Cf}C}$

$\omega_{z2} = \dfrac{D'^2 R}{DL}$ (RHPZ)

$\omega_0 = \dfrac{D'}{\sqrt{LC}}$

$Q = D'R\sqrt{\dfrac{C}{L}}$

Voltage-mode, DCM

Reference 1 equations:

$$\frac{V_{out}(s)}{V_{err}(s)} = \frac{V_{in}}{V_{peak}} \frac{K_1\left(1 + \frac{s}{\omega_{z1}}\right)}{\left(1 + \frac{s}{\omega_{p1}}\right)} \quad (2A-20)$$

$$\frac{V_{out}(s)}{V_{in}(s)} = \frac{V_{out}}{V_{in}} \frac{\left(1 + \frac{s}{\omega_{z1}}\right)}{\left(1 + \frac{s}{\omega_{p1}}\right)} \quad (2A-21)$$

$K_1 = -\dfrac{1}{\sqrt{K}}$ with $K = \dfrac{2L}{RT_{sw}}$

$\omega_{z1} = \dfrac{1}{r_{Cf}C}$

$\omega_{p1} = \dfrac{2}{RC}$

Current-mode, CCM:

Reference 1 equations:

$$\frac{V_{out}(s)}{V_{err}(s)} = \frac{R}{R_i} \frac{V_{in}}{(V_{in} - 2V_{out})} \frac{\left(1 + \frac{s}{\omega_{z1}}\right)\left(1 - \frac{s}{\omega_{z2}}\right)}{\left(1 + \frac{s}{\omega_{p1}}\right)\left\{1 + \frac{s\left[\left(1 + \frac{S_a}{S_1}\right)D' - 0.5\right]}{F_{sw}} + \frac{s^2}{(\pi F_{sw})^2}\right\}} \quad (2A-22)$$

$$\frac{V_{out}(s)}{V_{in}(s)} = \frac{V_{out}^2}{V_{in}^2 - 2V_{in}V_{out}} \frac{\left(1 + \frac{s}{\omega_{z1}}\right)}{\left(1 + \frac{s}{\omega_{p1}}\right)} \quad (2A-23)$$

SMALL-SIGNAL MODELING

$$\omega_{z1} = \frac{1}{r_{Cf} C}$$

$$\omega_{z2} = \frac{D'^2 R}{DL} \quad \text{(RHPZ)}$$

$$\omega_{p1} = \frac{V_{in} - 2V_{out}}{V_{in} - V_{out}} \frac{1}{RC}$$

Reference 2, 3, 4, and 5 equations:

$$\frac{V_{out}(s)}{V_{err}(s)} = \frac{\left(1 + \frac{s}{\omega_{z1}}\right)\left(1 - \frac{s}{\omega_{z2}}\right)\left(1 + \frac{s}{\omega_{z3}}\right)}{\left(1 + \frac{s}{\omega_{p1}}\right)} A_c F_h(s) \quad \text{(2A-24)}$$

$$\frac{V_{out}(s)}{V_{in}(s)} = \frac{\left(1 + \frac{s}{\omega_{z1}}\right)}{\left(1 + \frac{s}{\omega_{p1}}\right)} A_l F_a(s) F_h(s) \quad \text{(2A-25)}$$

$$\omega_{z1} = \frac{1}{r_{Cf} C}$$

$$\omega_{z2} = \frac{D'^2 R}{DL} \quad \text{(RHPZ)}$$

$$\omega_{z3} = \frac{1}{RC_s D'} > \frac{F_{sw}}{2} \quad C_s \text{ is the resonant capacitor calculated on CCM CC-PWM switch model}$$

$$\omega_{p1} = \frac{D' \frac{K_c}{K}\left(1 + 2\frac{S_e}{S_n}\right) + 1 + D}{RC} \quad \text{for deep CCM:} \quad \omega_{p1} \approx \frac{1 + D}{RC}$$

$$F_h(s) = \frac{1}{1 + \frac{s}{\omega_n Q_p} + \frac{s^2}{\omega_n^2}}$$

$$F_a(s) = \left(1 + \frac{s}{\omega_a Q_a} + \frac{s^2}{\omega_a^2}\right)$$

$$A_l = M \frac{\frac{K_c}{K}\left(M - 2\frac{S_e}{S_n}\right) - M}{\frac{K_c}{K}\left(1 + 2\frac{S_e}{S_n}\right) + 2M + 1}$$

$$A_c = -\frac{R}{R_i} \frac{1}{\frac{K_c}{K}\left(1 + 2\frac{S_e}{S_n}\right) + 2M + 1}$$

$$\omega_n = \frac{\pi}{T_{sw}}$$

$$Q_p = \frac{1}{\pi(m_c D' - 0.5)}$$

$$\omega_a = \pi F_{sw} \sqrt{\frac{\frac{2K_c}{K}\left(\frac{1}{2} - \frac{S_e}{MS_n}\right) - 1}{\frac{2K_c}{K}\left(\frac{S_e}{S_n} - \frac{M}{2}\right) - 1}}$$

$$Q_a = \frac{1}{\pi D}\left(M - \frac{2K_c}{K}\left(\frac{M}{2} - \frac{S_e}{S_n}\right)\right)$$

$$K_c = D'^2$$

$$K = \frac{2LF_{sw}}{R}$$

Current-mode, DCM

Reference 2, 3, 4, and 5 equations:

$$\frac{V_{out}(s)}{V_{err}(s)} = F_m H_d \frac{\left(1 + \frac{s}{\omega_{z1}}\right)\left(1 - \frac{s}{\omega_{z2}}\right)}{\left(1 + \frac{s}{\omega_{p1}}\right)\left(1 + \frac{s}{\omega_{p2}}\right)} \qquad (2A\text{-}26)$$

$$\omega_{z1} = \frac{1}{r_{Cf}C}$$

$$\omega_{z2} = \frac{R}{M(1+M)L} \quad \text{(high-frequency RHPZ)} \quad \text{In relation to } \omega_{p2}:$$

$$\omega_{z2} = \omega_{p2}(1 + 1/M) > 2F_{sw}$$

$$\omega_{p1} = \frac{2}{RC}$$

$$\omega_{p2} = 2F_{sw}\left(\frac{1/D}{1 + 1/M}\right)^2 \geq 2F_{sw}$$

$$H_d = \frac{V_{in}}{\sqrt{K}}$$

$$F_m = \frac{1}{S_n m_c T_{sw}}$$

$$M = -D\sqrt{\frac{1}{2\tau_L}} \quad \text{with } \tau_L = \frac{L}{RT_{sw}}$$

$$K = \frac{2LF_{sw}}{R}$$

For a flyback topology, we can use the buck-boost equations. However, a few manipulations are necessary as the primary inductance L_p and the sense resistor R_i are located on the primary side whereas the load and the output capacitor reside on the secondary side.

1. Keep L_p (the L parameter in the equations) and R_i (for current mode) on the primary side, and reflect C and R to the primary side via the following equations:

   $R' = R/N^2$

   $C' = CN^2$

   with $N = \dfrac{N_s}{N_p}$

2. Calculate the secondary-side inductor value $L_s = L_p N^2$ and use the result as the parameter L in the equations. For current-mode control, reflect the primary-side sense resistor R_i to the secondary side via $R' = R/N^2$. Thus C and R can be kept at their original values.

References

1. A. S. Kislovski, R. Redl, and N. O. Sokal, *Dynamic Analysis of Switching-Mode DC/DC Converters*, Van Nostrand Reinhold, New York, 1991.
2. R. B. Ridley, "A New Continuous-Time Model for Current-Mode Control," *IEEE Transactions of Power Electronics*, vol. 6, April 1991, pp. 271–280.
3. V. Vorpérian, "Analytical Methods in Power Electronics," In-house Power Electronics Class, Toulouse, France, 2004.
4. V. Vorpérian, "Simplified Analysis of PWM Converters Using the Model of the PWM Switch, Parts I (*CCM*) and II (*DCM*)," *Transactions on Aerospace and Electronics Systems*, vol. 26, no. 3, *May* 1990.
5. V. Vorpérian, *Fast Analytical Techniques for Electrical and Electronic Circuits*, Cambridge University Press, 2002.

APPENDIX 2B POLES, ZEROS, AND COMPLEX PLANE—A SIMPLE INTRODUCTION

When one is discussing stability, it is important to know why the positions of the poles and zeros in the complex plane are important. The general expression of the loop gain is usually found in the form

$$T(s) = \frac{b_0 + b_1 s + \cdots + b_m s^m}{a_0 + a_1 s + \cdots + a_n s^n} = \frac{N(s)}{D(s)} \qquad (2\text{B-}1)$$

where $N(s)$ and $D(s)$, respectively, represent the numerator and denominator polynomials. One condition for the system to be stable is that the degree of the numerator always be smaller than the degree of the denominator: $m < n$. This condition, called *properness*, implies that $\lim_{s \to \infty} T(s) = 0$. A transfer function not adhering to this rule is said to be *improper*, for instance, if $\lim_{s \to \infty} T(s) = \infty$.

For certain values of s, the numerator or the denominator will go to zero. Finding the numerator and denominator roots help you to locate, respectively, the zeros and poles affecting the transfer function $T(s)$. To find the positions of the zeros, you need to find the roots verifying $N(s) = 0$. Zeros are frequency values that actually null the transfer function. For instance, suppose $N(s) = (s + 5k)(s + 30k)$. Then one zero occurs at 795 Hz ($5k/2\pi$) and another zero at 4.77 kHz ($30k/2\pi$). Note that the roots are negative since $s_{z1} = -5k$ and $s_{z2} = -30k$.

On the other hand, poles are found by solving the equation $D(s) = 0$. When the denominator cancels, the gain $T(s)$ goes to infinity. Suppose we have $D(s) = s^2(s - 22k)$. Then calculation of its roots leads us to find a double pole at $s = 0$; this is a double pole placed at the origin.

Then another pole is located at 3.5 kHz ($22k/2\pi$). In this particular case, the root is positive as $D(s) = 0$ when $s_{p1} = +22$ k.

The need for negative roots

To obtain the time-domain response of a given transfer function $T(s)$ to a particular input, we take its inverse Laplace transform. If we multiply $T(s)$ by $\frac{1}{s}$ and look for its inverse Laplace transform, we will have an idea of its time-domain response to a step input. Then it can be shown that its time-domain response will be made up of a sum of exponential terms whose exponents are the roots of the equation $D(s) = 0$. For instance, suppose we have the following transfer function:

$$T(s) = \frac{2}{(s+1)(s+2)} \qquad (2\text{B-}2)$$

We multiply it by $\frac{1}{s}$ to seek its answer to a step input:

$$T(s) = \frac{1}{s}\frac{2}{(s+1)(s+2)} \qquad (2\text{B-}3)$$

Rearranging this equation gives

$$T(s) = \frac{k_1}{s} + \frac{k_2}{s+1} + \frac{k_3}{s+2} \qquad (2\text{B-}4)$$

Which, after all coefficients are identified, leads to

$$T(s) = \frac{1}{s} - \frac{2}{s+1} + \frac{1}{s+2} \qquad (2\text{B-}5)$$

In this example, finding k_1 to k_3 is fairly simple and requires a set of three equations to get the three unknowns. In complicated equations with multiple zeros, specific methods exist to quickly find all the coefficients.

Now, to find the time-domain response of Eq. (2B-5), we take the inverse Laplace transform of the individually selected quotients and sum them (the Laplace transform is a linear operator). This gives the following result:

$$T(t) = 1 - 2e^{-t} + e^{-2t} \qquad (2\text{B-}6)$$

From the above equation, we can see the roots of the characteristic equations appearing as the exponents of the exponentials: first root $s_{p1} = 0$, second root $s_{p2} = -1$, and third root $s_{p3} = -2$. A necessary and sufficient condition for this system to be stable is that all roots have a negative real part, which is the case here. We know that a function such as $f(t) = Ae^{-\frac{t}{\tau}}$ naturally decays in time in response to an input step because $\lim_{t \to \infty} f(t) = 0$. If the exponential coefficient were positive, then the function would diverge as t increased.

In the above example, all roots are real. When the solution of the characteristic equation $D(s) = 0$ leads to declare s^2 to be a negative value, for instance, the associated roots become complex, featuring an imaginary and a real part. For instance, if we have

$$T(s) = \frac{5}{(s+0.8)[(s+2.5)^2+4]} \qquad (2\text{B-}7)$$

the solution to $D(s) = 0$ gives the following roots:

$s_{p1} = -2.5 - 2j$

$s_{p2} = -2.5 + 2j$

$s_{p3} = -0.8$

These are two conjugate roots plus one real root. Again, if we need the step response, then we multiply Eq. (2B-7) by $\frac{1}{s}$ and either extract by hand the time-domain response or use a dedicated software such as Mathcad. This is what Eq. (2B-8) shows us:

$$T(t) = 0.6 - 0.907e^{-0.8t} + 0.297e^{-2.5t}\cos(2t) + 0.0088e^{-2.5t}\sin(2t) \qquad (2B\text{-}8)$$

Graphing this equation can be done via Mathcad® or through *PSpice* which can directly handle Laplace equations. The following netlist shows how to generate the necessary code:

```
.tran 10m 10s
.probe
V1 1 0 pwl 0 0 1u 1
;stepinput
E1 2 0 LAPLACE {V(1, 0)} = {5/(((s+2.5)^2+4)*(s+0.8))}
;Laplace generator
.end
```

Figure 2B-1a and b, respectively, compare the results delivered by Probe and Mathcad: they are identical for this particular case.

To comment on the stability of a transfer function, it is often convenient to place its numerator and denominator roots (zeros and poles) on the *s* plane. This is what Fig. 2B-2 represents for Eq. (2B-7) where crosses illustrate pole positions and circles illustrate zeros, if any. By inspecting the position of the transfer function poles, we can immediately see if an instability may exist. Here, as they are all located in the left half-plane (LHP), no oscillation will exist upon a transient step input. These poles being in the LHP implies that all exponents have a

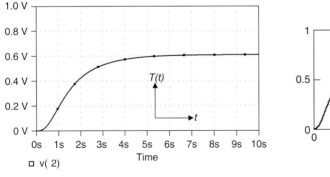
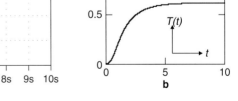

FIGURE 2B-1a, b Simulation and calculation results from Probe and Mathcad.

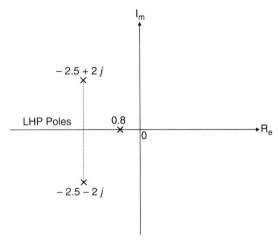

FIGURE 2B-2 Roots positioned on the s-plane. Crosses represent poles and circles zeros, if any.

negative sign, therefore leading to a decay in time. This is not something obvious when you see the presence of sine and cosine in Eq. (2B-8)!

Moving poles, root locus plots

Figure 2B-3 shows a classic compensated plant—typically the converter power stage—featuring a unity gain feedback path and an error gain of k. From this sketch, it is easy to derive the closed-loop transfer function:

$$\frac{V_{out}(s)}{V_{ref}(s)} = \frac{kG(s)H(s)}{1 + kG(s)H(s)} \qquad (2\text{B-9})$$

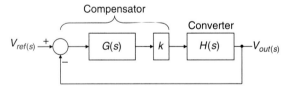

FIGURE 2B-3 A compensated plant featuring a unity gain feedback path.

If we now consider separate expressions for $G(s)$ and $H(s)$ as follows, we can rearrange Eq. (2B-9):

$$G(s) = \frac{N_G(s)}{D_G(s)}$$

$$H(s) = \frac{N_H(s)}{D_H(s)}$$

SMALL-SIGNAL MODELING

$$\frac{V_{out}(s)}{V_{ref}(s)} = \frac{k\dfrac{N_G(s)}{D_G(s)}\dfrac{N_H(s)}{D_H(s)}}{1 + k\dfrac{N_G(s)}{D_G(s)}\dfrac{N_H(s)}{D_H(s)}} = \frac{kN_G(s)N_H(s)}{D_G(s)D_H(s) + kN_G(s)N_H(s)} \quad (2\text{B-}10)$$

Equation (2B-10) reveals that the poles and zeros you introduced during the open-loop gain analysis (to shape the ac response for adequate bandwidth and phase margin) now, respectively, appear via $D_G(s)$ and $N_G(s)$. They appear in the denominator of the closed-loop transfer function and affect the roots of the closed-loop characteristic equation. Hence, if you push the compensation zeros in the lowest portion of the spectrum, e.g., to boost the phase at the crossover point, it might slow down the response as these will now appear as poles in the closed-loop characteristic equation:

$$\chi(s) = D_G(s)D_H(s) + kN_G(s)N_H(s) = 0 \quad (2\text{B-}11)$$

Hence, depending on the value of k, several cases can exist:

- k is small and $\chi(s) = D_G(s)D_H(s) = 0$. The closed-loop poles are those already present in the open-loop expression.
- If k now increases, since the poles are a continuous function of k, they move away from their open-loop definition to fully satisfy Eq. (2B-11).
- If k increases further, Eq. (2B-11) becomes $\chi(s) = kN_G(s)N_H(s) = 0$ and the closed-loop poles tend to reach the open-loop zeros!

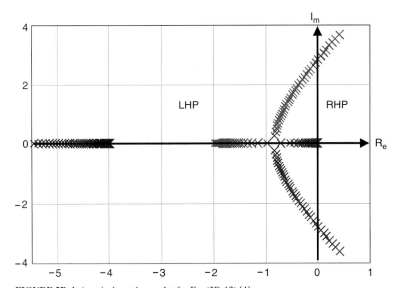

FIGURE 2B-4 A typical root locus plot for Eq. (2B-12) [4].

Examination of the moving poles position in the s plane is called *root-locus* analysis. It can be done automatically by dedicated software or graphically by hand. Depending on the equation, it can be a long and difficult process. A typical root locus plot is shown in Fig. 2B-4.

It graphs the evolution of the poles versus the values of the gain k in the following characteristic equation:

$$\chi(s) = s^3 + 6s^2 + 8s + k \qquad (2\text{B-}12)$$

There is one real pole and two complex poles. By studying the way poles evolve on the graph, several bits of information can be obtained. For instance, the k value at which the two trajectories cross the imaginary axis gives the gain margin in closed-loop.

More information can be found in the references below. A few websites contain useful data, particularly Ref. 2, which includes a lot of applets.

References

1. H. Özbay, *Introduction to Feedback Control Theory,* CRC Press, 1999.
2. http://www.facstaff.bucknell.edu/mastascu/eControlHTML/RootLocus/RLocus1A.html#Introduction.
3. http://virtual.cvut.cz/dynlabmodules/syscontrol/node1.html.
4. A. Stubberud, I. Williams, and J. DiStefano, *Schaum's Outline of Feedback and Control Systems,* McGraw-Hill, New York, 1994.
5. V. Vorpérian, *Fast Analytical Techniques for Electrical and Electronic Circuits,* Cambridge University Press, 2002.

CHAPTER 3
FEEDBACK AND CONTROL LOOPS

Feedback theory has been the object of numerous textbooks, and this chapter does not have the arrogance to compete with any of them. On the contrary, we will focus on known and simple results to introduce compensation techniques that help to quickly stabilize power converters. We invite the reader interested by more theoretical analyses to consult the references given at the end of this section.

For nearly 100% of the applications, a switch-mode converter delivers a parameter—a voltage or a current—whose value must remain constant, independent of various operating conditions, such as the input voltage, the output loading, the ambient temperature. To perform such a task, a portion of the circuit must be insensitive to any of the above variations. This portion is called the *reference*, usually a voltage source, V_{ref}, which is precise and well stable over temperature. A fraction (α) of the converter output variable (for instance, the output voltage V_{out}) is permanently compared to this reference. Thanks to a loop that feeds the information back to it, hence the term *feedback loop*, the controller strives to maintain the theoretical equality between these two levels:

$$V_{out} = \frac{V_{ref}}{\alpha} \qquad (3\text{-}1)$$

If you go through some power electronics books, you often see the converter modeled using the classical feedback representation. This approach can sometimes confuse the reader as the network divider featuring the α ratio appears in the chain. Unfortunately, if the ratio surely plays a role in setting the dc output, because of the op amp and its virtual ground, its action disappears in the ac analysis. This is discussed in greater detail in App. 3D.

Figure 3-1a portrays the simplified switch-mode converter as it appears on your bench. As we said, thanks to the error amplifier, the whole loop strives to satisfy Eq. (3-1).

Based on Fig. 3-1a, we can draw Fig. 3-1b as a simplified static representation where we purposely reduce the whole chain in a unity-return configuration. From this new drawing, we can write a few equations:

$$\left[\frac{V_{ref}}{\alpha} - V_{out}\right] G(0)H(0) = V_{out} \qquad (3\text{-}2a)$$

$$\frac{V_{ref}}{\alpha} G(0)H(0) = V_{out}\left[1 + G(0)H(0)\right] \qquad (3\text{-}2b)$$

$$\frac{V_{out}}{V_{ref}} = \frac{1}{\alpha}\frac{G(0)H(0)}{1 + G(0)H(0)} = \frac{1}{\alpha}\frac{T(0)}{1 + T(0)} \qquad (3\text{-}2c)$$

$$\frac{V_{out}}{V_{ref}} \approx \frac{1}{\alpha} \quad \text{for } \|T(0)\| \gg 1 \qquad (3\text{-}2d)$$

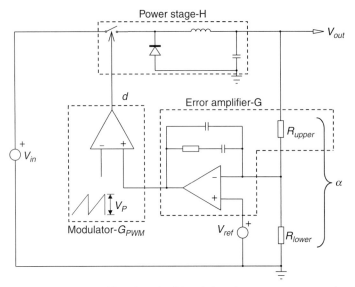

FIGURE 3-1a A simplified schematic of the switch-mode power converter operating in closed-loop mode.

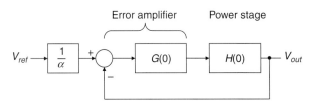

FIGURE 3-1b A simplified static representation of the switch-mode converter.

Here $T(0)$ represents the static loop gain linking V_{out} and V_{ref}. The term $\dfrac{T(0)}{1 + T(0)}$ illustrates the *static error* between the theoretical output value you want $\left(\dfrac{V_{ref}}{\alpha}\right)$ and the final measurement you read with a voltmeter once the loop is closed. This is something already seen in Chap. 1. Thus, using a large open-loop gain op amp is key to reducing the static error, but it also helps to provide enough low-frequency gain to fight the rectification ripple in off-line supplies.

Figure 3-2 illustrates the small-signal representation of the power supply. As we explained, the op amp keeps its noninverting pin to zero in small-signal conditions. Therefore, R_{lower} naturally fades away, and the loop gain is solely fixed by R_{upper} and Z_f. Changing the divider network ratio (α) has no effect on the loop gain, as demonstrated in App. 3D. This is what Fig. 3-2 suggests, around a familiar buck converter. The output signal feeds the inverting input of the operational amplifier (op amp) whose frequency response is affected by a compensation network made around Z_f and R_{upper}. The purpose of this compensation network is to tailor the converter frequency response to make it stable once operated in closed-loop conditions. We will find more complex arrangements later, however. The output of this op amp, $V_{err}(s)$, flows through the PWM gain block to finally generate the *control variable*, the duty cycle of the power stage. The power stage is affected by a transfer function $H(s)$. In this configuration, the loop gain is simply

$$T(s) = H(s)G(s)G_{PWM} \qquad (3\text{-}2e)$$

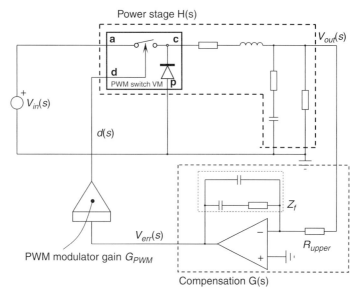

FIGURE 3-2 A small-signal representation of the switch-mode converter.

The power converter output, unfortunately, does not solely depend on the control variable d. Some external perturbations contribute to its deviation from the imposed target: they are the input voltage V_{in} and the output current I_{out}. We have seen in Chap. 1 how negative feedback reduces these effects. Loop analysis will consist of studying the open-loop gain/phase response of the total chain (the transfer function), most of the time through a Bode plot, and shaping it via the compensation network to stabilize the power supply over the various input/output conditions the converter will encounter in its lifetime.

3.1 OBSERVATION POINTS

Various methods exist to reveal the transfer function of a converter. The simplest one consists of opening the loop—we say breaking the loop—to inject an input signal and observe what is obtained on the other side of the opened path. Capitalizing on the previous figure, we can propose a simple way that the next analysis sketches will adopt (Fig. 3-3).

In this example, a dc source fixes the operating point, and the ac modulation comes in via a coupling capacitor. A network analyzer monitors V_{stim} and V_{err} and computes the gain by $20 \log_{10}\left(\dfrac{V_{err}}{V_{stim}}\right)$. If this method works great in a SPICE environment, it suffers from a major problem linked to bias point runaway in the presence of high open-loop gains. We therefore do not recommend it for practical experiments.

Observing the voltage on the power stage output will reveal a dc gain and a phase starting from $0°$, then becoming negative down to $-180°$ in the case of a second order system (e.g. our voltage-mode CCM buck). If we now look at the voltage delivered by the op amp output, given its inverting configuration, we will add another $-180°$ phase shift to the inverted output stage signal. In classical technical literature, the phase representation of an open-loop system is plotted between $0°$ and $-360°$. When it comes to phase margin study, very often, authors purposely omit the $-180°$ inversion brought by the op amp and display the response between $0°$ and $-180°$. However, unlike the classical representations, SPICE often bounds its phase display between $-180°$ and $+180°$. It considers a complete phase rotation when the phase hits $0°$. This modulo 2π representation

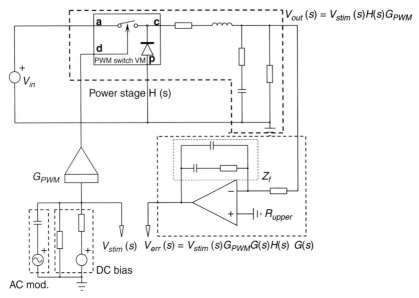

FIGURE 3-3 Opening the loop before the modulating point using an external bias source.

explains why, later on in the text, the phase margin is read as the distance between the open-loop phase trace and the 0° axis. After all, if you observe two waveforms W_1 and W_2 on an oscilloscope, stating that W_2 lags W_1 by $-270°$ is similar to say that W_2 leads W_1 by $+90°$!

Figure 3-4 shows a typical Bode plot for the CCM buck operated in voltage mode. The upper section depicts the power stage only, $H(s)$. The lower trace represents the total loop

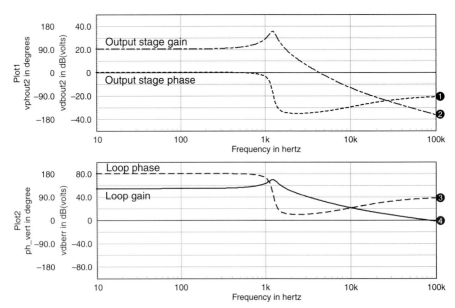

FIGURE 3-4 A typical Bode plot of a CCM buck power stage, followed by the op amp circuit response (no compensation, no origin pole).

gain $T(s)$ after compensation. As indicated in the above paragraph, the observed loop phase starts from 180° in the low frequencies and heads towards 0° if no proper compensation is provided.

Exploring the ac behavior of Fig. 3-3 can be performed by SPICE via the familiar network LoL and CoL, as already studied before. Figure 3-5 shows it again for reference. Sometimes,

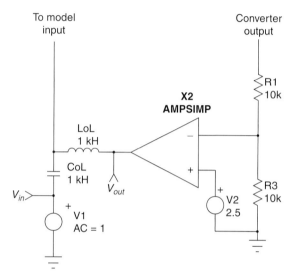

FIGURE 3-5 The inductor closes the loop in dc (during bias point calculation) and opens it in ac, blocking all modulation signals coming from the error chain.

a kind of noise appears in ac sweeps using this technique. Adding a series resistor with either CoL or LoL will cure the problem (typically 100 mΩ).

The important improvement brought by Fig. 3-5 is the closed-loop dc point you will not have in Fig. 3-3. Any change in the load or the input voltage will automatically adjust the duty cycle to keep the output constant (within the converter capabilities, of course).

As we explained in the previous lines, if you take a power supply and follow Fig. 3-3 opening recommendations, in other words, you physically open the loop and fix a bias point via an external dc source (that you tweak to obtain the right output voltage), then you might encounter difficulties in maintaining the right operating point in the presence of a high dc gain (feedback with an op amp, for instance). A shift of a few millivolts on the external supply due to temperature variations, and the error amplifier hits one of its stops (if it is the upper stop, a loud noise is usually heard!). The transformer method, already tackled in Figs. 2-45 and 2-46, is actually the best and recommended as best practical measurement practice.

In SPICE, this solution consists of inserting the ac source in series with either the error signal or the output signal. Figure 3-6a and b portrays these possibilities; please note the ac source polarity which appears in series with the preceding signal. The source connections must have satisfied some impedance requirements, however. The − shall connect to a low-impedance point, and the + must go to a high-impedance point. Based on the author's experience, using the LoL/CoL method offers the easiest way with SPICE to probe the transfer function at any point. This is so because the excitation source refers to the ground and does not float. To obtain the loop gain in Fig. 3-5 using *IntuScope* (*IsSpice* graphical tool) or *Probe* (*PSpice* graphical tool), you simply need to type the following commands:

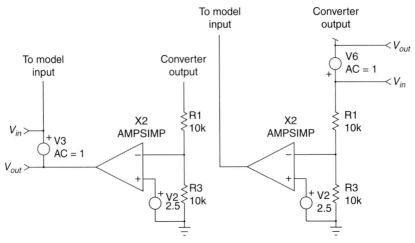

FIGURE 3-6a and b Inserting the ac source in series represents another viable solution.

Gain:

IsSpice: click on dB V(vout)
Probe: type dB(V(vout))

Phase:

IsSpice: click on phase V(vout)
Probe: type Vp(vout)

No further signal manipulations are required. The same applies if you want to probe another node, for instance, the output stage signal, before the op amp divider.

On the contrary, if you use the approaches of Fig. 3-6a and b, since the source floats, you need to apply imaginary signal algebra:

Gain:

IsSpice: click on dB V(vout), it gives waveform 1 (W1); click on dB V(vin), it gives waveform 2 (W2). Now plotting W1-W2, you obtain the transfer function.
Probe: type dB(V(vout)/V(vin))

Phase:

IsSpice: click on phase V(vout), it gives waveform 1 (W1), click on phase V(vin), it gives waveform 2 (W2). Now plot W1-W2, you obtain the phase transfer function.
Probe: type Vp(vout) − Vp(vin)

Note: under *IntuScope*, pressing the keyboard letter b directly plots the Bode diagram on the screen.

As you can see, the floating source requires some manipulations compared to the *LoL/CoL* method. However, to run a transient analysis, you must absolutely reduce *LoL/CoL* to 1 pH and 1 pF in order to not disturb the loop. Hence, toggling between ac and transient analysis can quickly become a tedious task. Something you really do not care about with the floating source method as a transient analysis automatically puts this ac source to 0. It thus shields you from toggling between ac and transient schematics. For the sake of simplicity, we will use the *LoL/CoL* method in this book, but the floating source can be applied the same way.

3.2 STABILITY CRITERIA

Among stability tools (Nyquist, Nichols, . . .), Bode's approach is probably the most popular owing to its simplicity. When other methods require manipulating data in the complex plan, the Bode diagram offers an immediate insight as the transfer function amplitude appears in the frequency domain.

We know that a feedback system takes a portion of the output variable and compares it to a stable reference. It then further "amplifies" the error between these signals, via the loop gain, to generate a corrective action. In other words, if the output voltage deviates from its target—let us assume it increases—the error signal must reduce to instruct the converter to diminish its output. On the contrary, if the output voltage stays below the target, the error voltage will increase to let the converter know that there is a demand for more output voltage. The control action consists of opposing the variation observed on the regulated output, hence the term *negative feedback*. As the frequency increases, the converter output stage $H(s)$ introduces further delay (we say it "lags") and its gain drops. Combined with the correction loop $H(s)$, a case might quickly appear where the total phase difference between the control signal and the output signal vanishes to 0°. Theory thus shows that if, for any reason, both output and error signals arrive in phase while the gain loop reaches unity (or 0 dB in a log scale), we have built a *positive feedback* oscillator, delivering a sinusoidal signal at a frequency fixed by the 0 dB crossover point.

When we compensate a power supply, the idea is not to build an oscillator! The design work will thus consist of shaping the correction circuit $G(s)$ to make sure that (1) when the loop gain crosses the 0 dB axis, there exists sufficient phase difference between the error and the output signal and (2) $G(s)$ offers a high gain value in the dc portion to reduce the static error and the output impedance and to improve the input line rejection. This phase difference is called the *phase margin* (PM). How much phase margin must be selected? Usually, 45° represents the absolute minimum, but rock-solid designs aim for around 70° to 80° phase margin, offering good stability and a fast nonringing transient response.

Figure 3-7 represents the loop gain of a compensated CCM buck voltage-mode converter and highlights the phase margin. We can read a PM greater than 50° and a 0 dB crossover frequency (or a bandwidth) of 4.2 kHz. Please note that PM on this drawing is read as the distance between the phase curve and the 0° line. Sometimes, in textbooks, PM is assessed as the distance between the phase curve and the −180° line. It leads to the same interpretation either way.

Note that a null phase margin at above or below the 0 dB point offers what is called *conditional stability*. That is, if the gain moves up or down (the phase shape remaining the same), the unity gain crossover point can coincide with a 0 phase margin, engendering oscillations. What matters is the distance between the 0 dB axis and the point at which danger can occur. This situation can be seen on the right side of Fig. 3-7 where both error and output signals are in phase (0°). If the gain increased by 20 dB, we would be in trouble. . . The gain increase (or decrease in some cases) necessary to reach the 0 dB axis is called the *gain margin* (GM). Good designs ensure at least a 10 to 15 dB margin to cope with any gain variations, due to loading conditions, component dispersions, ambient temperature, and so on.

Figure 3-8 represents the same CCM buck but now featuring a reduced phase margin of 25° at the crossover frequency. This is too low. Furthermore, the phase almost hits 0° around 2 kHz. If the gain reduces by 20 dB and crosses 0 dB at that particular point, oscillations will occur, this is the conditional stability described above.

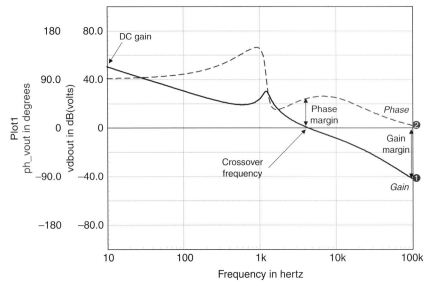

FIGURE 3-7 The compensated loop gain of a compensated CCM voltage-mode buck converter.

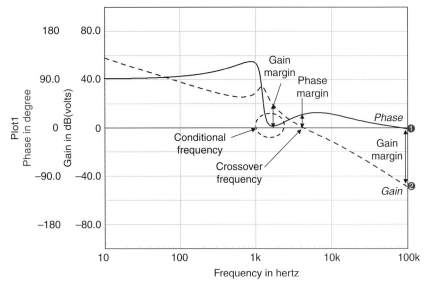

FIGURE 3-8 In this configuration, the bandwidth did not change, but the phase margin did.

3.3 PHASE MARGIN AND TRANSIENT RESPONSE

A relationship exists between the phase margin of a second-order closed-loop system and the quality coefficient Q of its transfer function [1]. If the phase margin is too small, the peaking induces high output ringing, exactly as in an RLC circuit. On the contrary, if the phase margin becomes too large, it slows down the system: the overshoot goes away but to the detriment of response and recovery speed. An equivalent quality coefficient of 0.5 brings a theoretical phase margin of 76° as highlighted in [1]. It leads to a critically-damped converter, combining response speed and lack of overshoot. Based on this statement, the converter phase margin target must be set to 70°, with a worst case of 45°.

The CCM buck featuring a 4.3 kHz bandwidth was simulated in a transient load step together with various phase margins imposed by the compensation network. Figure 3-9 has collected all transient responses. We can see that a weak phase margin gives birth to oscillations and large overshoots: the system becomes undamped. This is obviously not an acceptable design. As the phase margin increases, the response time slows down a little, but the overshoot fades away. For a 76° phase margin, the overshoot keeps within a 0.5% window.

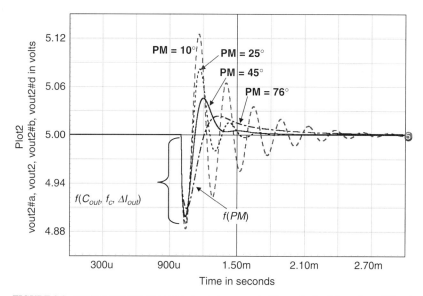

FIGURE 3-9 Various step load responses versus phase margin. The undershoot depends on different parameters, including the output capacitor, but the recovery time links to the phase margin.

3.4 CHOOSING THE CROSSOVER FREQUENCY

The crossover frequency is chosen depending on various design factors and constraints. In a power converter, it is possible to approximate its closed-loop output impedance by the output capacitor impedance at the crossover frequency f_c. Therefore, the output voltage undershoot level V_p occurring during an output transient step ΔI_{out} can be approximated by the following formula [2]

$$V_p \approx \frac{\Delta I_{out}}{C_{out} 2\pi f_c} \tag{3-3}$$

where C_{out} is the output capacitor and f_c the crossover frequency. Note that this equation holds as long as the output capacitor ESR is less than the reactance of C_{out} at the crossover frequency, implying that the capacitor is solely held responsible for the undershoot. This condition can be expressed by

$$\text{ESR}_{C_{out}} \leq \frac{1}{2\pi f_c C_{out}} \tag{3-4}$$

As Fig. 3-9 shows, the undershoot depends on Eq. (3-3), but the recovery time mostly depends on the phase margin at the crossover frequency.

Equation (3-3) can help you make the decision on the crossover value once the output capacitor has been selected based on the needed ripple performance and its rms current capability, for instance. However, there are other limiting factors that you need to consider. For instance, if your converter features a RHP zero, like in CCM boost, buck-boost or flyback converters, then the crossover frequency f_c cannot be higher than 30% of its worse- case lowest position. It quickly closes the debate! In voltage-mode operated converters, the peaking of the LC network (L or L_e) also bounds the crossover frequency: trying to fix f_c too close to the resonant frequency f_0 of the LC network will bring obvious stability troubles given the phase lag at resonance. Make sure to select f_c at least equal to three times f_0 in worse-case conditions.

In absence of RHP zero, however, one-tenth to one-fifth of the switching frequency (10% to 20% of F_{sw}) looks like a possible target. Extending the crossover frequency can bring additional problems such as noise pick-up: a theoretical design might show adequate PM and GM at the chosen cutoff frequency, but the reduction to practice can show instability because of noise susceptibility brought by the wide loop bandwidth. Do not push the cutoff frequency beyond what you really need to avoid this problem: there is no need for a 15 kHz crossover if a 1 kHz one can do the job transient wise!

3.5 SHAPING THE COMPENSATION LOOP

The stability exercise requires shaping the compensation circuit $G(s)$ in order to provide adequate phase margin at the selected crossover point, together with a high gain in dc. To do so, several compensation circuits can be used, assembling poles and zeros. What we usually need is a *phase boost* at the crossover frequency to provide the right phase margin. This is done by forcing the loop to cross over with a -1 slope, or -20 dB/decade in the vicinity of the crossover frequency. However, the needed boost is sometimes so large that you cannot reach the crossover frequency you have in mind. You must revise your goal and adopt a more humble target. Let us review the basics about passive filters first, quickly followed by operational amplifier-based circuits.

3.5.1 The Passive Pole

Figure 3-10a represents an RC circuit producing a so-called passive single-pole response. Also known as a low-pass filter, it introduces phase lag (or delay) as the frequency increases. Its Laplace transfer function has the following form:

$$\frac{V_{out}(s)}{V_{in}(s)} = \frac{1}{1 + sRC} = \frac{1}{1 + \frac{s}{\omega_0}} \tag{3-5}$$

The cutoff pulsation of this passive filter, that is, when the dc "gain" reduces by -3 dB, is given by the classical formula

$$\omega_0 = \frac{1}{RC} \tag{3-6}$$

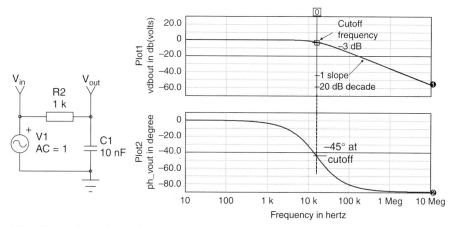

FIGURE 3-10a, b A single-pole *RC* network and its frequency response.

Figure 3-10a represents the electrical construction of such a low-pass filter whereas Fig. 3-10b shows its Bode plot.

A single pole is often inserted in compensation circuits to roll off the gain at a certain point. The rate at which the amplitude goes down is –20 dB by decade. That is, after the cutoff frequency, the amplitude difference at frequencies f_1 and f_2, where $f_2 = 10f_1$, will be -20 dB. On the Bode plot, this is shown as a "-1" slope, whereas a -40 dB by decade circuit (typically a second-order network) would be designated as a -2 slope.

A pole corresponds to a root in the transfer function denominator $D(s)$. Solving for the roots gives an indication of the system stability (please see App. 2B for more details).

3.5.2 The Passive Zero

If the transfer function contains a zero, it appears in the numerator $N(s)$. At the zero frequency, the numerator cancels and nulls the transfer function. Equation (3-7) describes the generalized form of the zero:

$$G(s) = 1 + \frac{s}{\omega_0} \tag{3-7}$$

Such an expression describes a 0 dB "gain" in dc, followed by a $+20$ dB/decade slope (a $+1$ slope) occurring at the zero location. The phase is now positive, as seen in Fig. 3-11a. This is the property of a zero that actually "boosts" the phase, compared to a pole that "lags" the phase. Zeros are thus introduced in $G(s)$ to compensate for excessive phase lag occurring in the power stage response.

Back to the passive circuits, Fig. 3-11b represents a high-pass filter. The transfer function of such a simple RC circuit also contains one pole and one zero, but placed at the origin. It looks like

$$\frac{V_{out}(s)}{V_{in}(s)} = \frac{sRC}{1 + sRC} = \frac{\dfrac{s}{\omega_0}}{1 + \dfrac{s}{\omega_0}} \tag{3-8}$$

where the cutoff pulsation is the same as Eq. (3-6).

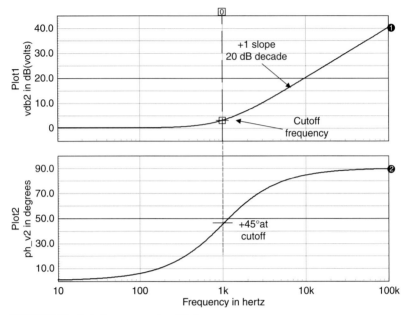

FIGURE 3-11a A single zero network and its frequency response.

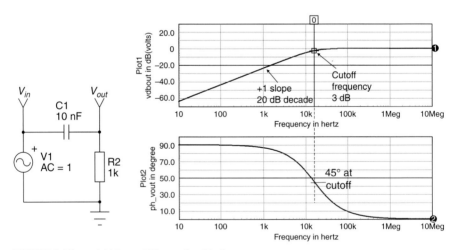

FIGURE 3-11b, c A high-pass *RC* network and its frequency response.

This filter features a low-frequency asymptote of a +20 dB/decade slope (a +1 slope) and high-frequency gain of 1 or 0 dB. Here we have a zero located at the origin nulling the transfer function in dc (s = 0) as Fig. 3-11c portrays.

In all the above equations, a negative numerator root sign signifies a *left half-plane zero* (LHPZ) position. In some converters, a right half-plane zero (RHPZ) can exist and stability is jeopardized.

3.5.3 Right Half-Plane Zero

The RHPZ is not part of the loop shaping toolbox. You actually undergo a RHPZ rather than create it for stability purposes! Its general form looks pretty much like Eq. (3-7) except that a negative sign appears:

$$G(s) = 1 - \frac{s}{\omega_0} \tag{3-9a}$$

A RHPZ can be formed by using the circuit in Fig. 3-12a where we can see an active high-pass filter whose inverting output (the negative sign) is summed with the input. The transfer function is easy to derive

$$V_{out}(s) = V_{in}(s) - V_{in}(s)\frac{R_1}{\frac{1}{sC_1}} = V_{in}(s)\left(1 - \frac{s}{\omega_0}\right) \tag{3-9b}$$

with Eq. (3-6) again ruling the cutoff frequency.

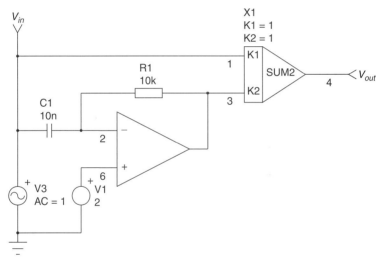

FIGURE 3-12a A RHPZ artificially created via an active high-pass filter and an adder.

As Fig. 3-12b shows, the gain output looks like a traditional zero: a +1 slope of 20 dB/decade with a cutoff frequency imposed by R_1 and C_1. The difference lies in the phase diagram. Instead of a phase boost, the perfidious RHPZ gives you a phase lag and further degrades the phase margin you strived to save.

Right half-plane zeros usually exist in indirect energy transfer converters where energy is first stored (*on* time) prior to being dumped in the output capacitor (*off* time). If we take the example of the boost converter, the average diode current equals the load dc current. This diode current I_d actually equals the inductor current I_L during the *off* time, or $d'T_{sw}$. Its average value can thus be written

$$\langle I_d \rangle_{T_{sw}} = I_{out} = \langle I_L \rangle_{T_{sw}} d' \tag{3-10}$$

Suppose we have a 40% duty cycle in a CCM-operated boost converter. A sudden load step occurs which, via the feedback loop, pushes the duty cycle to 50%. The current in the inductor

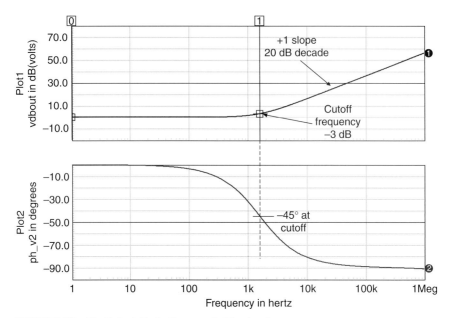

FIGURE 3-12b The Bode plot looks like a zero, but the phase lags.

increases accordingly, but what about the average output current in the diode? It drops because as $d' = 1 - d$, if d increases, d' shrinks and the output capacitor first discharges instead of increasing! The change goes in the wrong direction until the current builds up in the inductor, bringing the average diode current to its right value and finally lifting up the output voltage. Since it is a closed-loop system, the converter becomes unstable. There is nothing you can do about it, except to severely roll off the bandwidth in order to not undergo the RHPZ additional phase lag.

The RHPZ frequency position unfortunately changes with the duty cycle. The typical rule of thumb recommends that you select a crossover frequency to be around one-third of the lowest RHPZ position. If you try to increase the bandwidth closer to the RHPZ location, you might encounter a problem as the phase lag becomes too large. The RHPZ occurs in CCM operated converters such as buck-boost, boost, or flybacks. The RHPZ disappears in DCM although some academic studies state the presence of one in DCM, but relegated to higher frequencies.

Figure 3-13 plots a converter featuring a RHPZ. In the presence of a load step, the duty cycle suddenly increases. As a result, the inductor current increases as the switch stays closed for a longer time. But as d' has diminished, the diode average current now goes down. This situation translates into a decreasing output voltage, the opposite of what the loop is asking for. Then the average current eventually catches up with the inductor current, and the output voltage rises.

The above lines showed how passive poles and zeros help to shape the loop gain. Unfortunately, used on their own, they do not provide any dc gain, which is badly needed for a low static error, good input rejection, and so on. Associated with operational amplifiers (op amps), these so-called active filters provide the necessary transfer function together with the required amplification. Three different types have been identified.

FEEDBACK AND CONTROL LOOPS

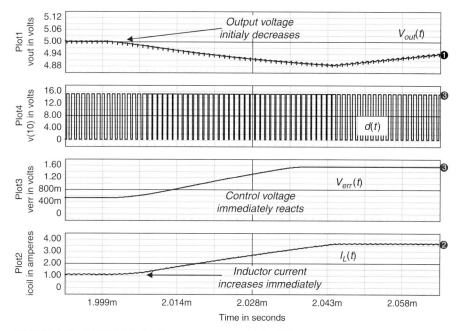

FIGURE 3-13   A RHPZ effect in a boost converter operating in CCM.

3.5.4   Type 1 Amplifier—Active Integrator

Implementing the largest dc gain naturally pushes the usage of an op amp as part of the corrective loop. Rather than cascading the passive networks followed by the high-gain op amp, designers often combined them to form active filters. This is the case for the pure integrator shown in Fig. 3-14a. Note that all configurations are inverting the input signal, but we omitted the minus sign for the sake of clarity.

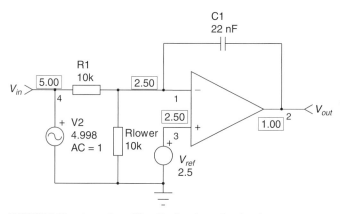

FIGURE 3-14a   A type 1 amplifier. No phase boost, just dc gain.

The transfer function of this pure integral compensator is easy to derive:

$$G(s) = \frac{1}{sR_1C_1} \quad (3\text{-}11)$$

It features an origin pole, given by R_1 and C_1. In dc mode, when the capacitor is open, the op amp open-loop gain fixes the gain. We purposely put it to 60 dB in all the op amp models appearing in the following examples. Then, as the frequency rises, the capacitor impedance drops and reduces the gain with a -1 slope, or -20 dB/decade. The phase curve stays flat and does not provide any boost in phase: cumulating the $-180°$ phase reversal due to the inverting op amp configuration plus the $-90°$ brought by the origin pole, the type 1 compensator permanently rotates the input phase by $-270°$ or $+90°$ when displayed modulo 2π in SPICE Figure 3-14b shows the resulting frequency sweep brought by such a configuration. Note the action of the op amp *origin pole* on the dashed curves (30 Hz, 60 dB open-loop gain). It further degrades the phase in higher frequencies and must be accounted for in the final design. Fortunately, SPICE does it naturally as we can pick the op amp model we have selected for the loop design.

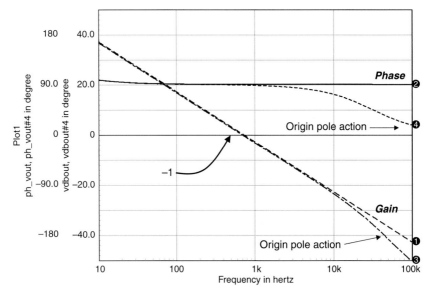

FIGURE 3-14b The resulting Bode plot of a type 1 amplifier. Note the action of the op amp origin pole, here active around 30 Hz.

Note that R_{lower} does not play a role in the ac response as long as the op amp ensures a virtual ground. Why? Simply because the op amp maintains 0 V on the inverting pin, thus making R_{lower} useless for the ac analysis. However, R_{lower} helps to select the needed dc output voltage together with R_1. See App. 3D for more details.

3.5.5 Type 2 Amplifier—Zero-Pole Pair

The previous amplifier type did not provide us with any phase boost, which we badly need if the phase margin is too low at the desired crossover frequency. Figure 3-15a depicts such

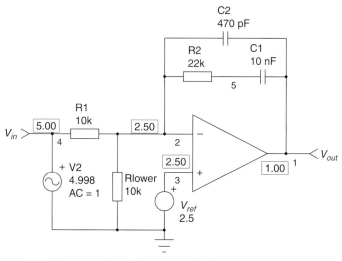

FIGURE 3-15a A type 2 amplifier can boost the phase.

a compensator, referenced as a type 2 amplifier. It produces an integrator together with a zero-pole pair.

Its transfer function can be obtained via a few lines of Laplace equations:

$$G(s) = \frac{1 + sR_2C_1}{sR_1(C_1 + C_2)\left(1 + sR_2\frac{C_1C_2}{C_1 + C_2}\right)} \quad (3\text{-}12)$$

We immediately can see a zero

$$\omega_z = \frac{1}{R_2C_1} \quad (3\text{-}13)$$

a pole at the origin (the integrator)

$$\omega_{p1} = \frac{1}{R_1(C_1 + C_2)} \quad (3\text{-}14)$$

and a high-frequency pole

$$\omega_{p2} = \frac{1}{R_2\left(\frac{C_1C_2}{C_1 + C_2}\right)} \quad \text{but if } C_2 \ll C_1 \text{ then} \quad \omega_{p2} = \frac{1}{R_2C_2} \quad (3\text{-}15)$$

Figure 3-15b shows how the phase and gain evolve with frequency. We can clearly see the phase increasing between the pole and zero locations. The phase boost depends on the distance between these two points, as we will discover in a few moments. The phase boost peaks right in the geometric mean of ω_z and ω_{p2} which occurs at a pulsation equal to $\sqrt{\omega_z\omega_{p2}}$.

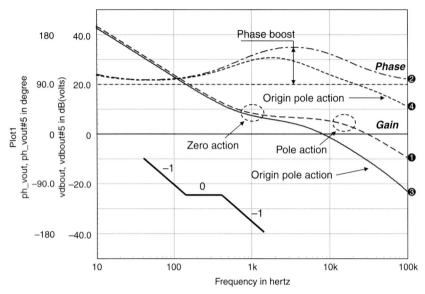

FIGURE 3-15b Type 2 amplifier response. The dashed line illustrates the op amp origin pole.

3.5.6 Type 2a—Origin Pole Plus a Zero

By suppressing capacitor C_2, it is possible to get rid of the high-frequency pole and change the frequency response of the compensation network. Figure 3-16a shows how the type 2 amplifier transforms.

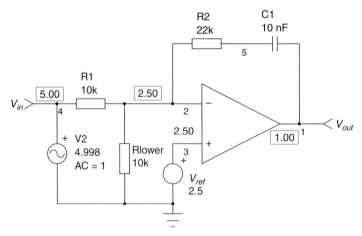

FIGURE 3-16a Suppressing C_2 gives a different compensation network and a different Bode plot shape.

The transfer function now becomes

$$G(s) = \frac{1 + sR_2C_1}{sR_1C_1} \qquad (3\text{-}16)$$

with a zero described by Eq. (3-17) and a pole at the origin induced by R_1 and C_1:

$$\omega_z = \frac{1}{R_2C_1} \qquad (3\text{-}17)$$

$$\omega_p = \frac{1}{R_1C_1} \qquad (3\text{-}18)$$

As the frequency increases, the equation reduces to a gain imposed by the two resistors:

$$\lim_{s\to\infty} G(s) = \frac{R_2}{R_1} \qquad (3\text{-}19)$$

Figure 3-16b plots the resulting frequency sweep, again showing the op amp origin pole effect:

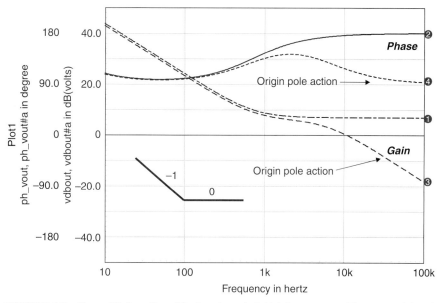

FIGURE 3-16b The modified type 2 amplifier featuring a single high-frequency zero. The op amp origin pole effect clearly appears on the graph.

3.5.7 Type 2b—Proportional Plus a Pole

Another variation of the type 2 amplifier consists of adding a resistor to make a proportional amplifier and removing the integral term present with the two previous configurations. Figure 3-17a depicts such an arrangement where a capacitor C_1 placed in parallel with the resistor R_1 introduces a high-frequency gain, necessary to roll off the gain. The transient response imposed by this type of amplifier resembles that of Fig. 1-9b, bringing less overshoot in steep load steps. This type of amplifier offers a flat gain imposed by R_2 and R_1, until

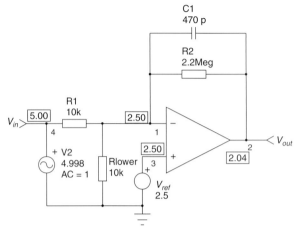

FIGURE 3-17a A type 2b amplifier where proportional control is necessary.

the pole imposed by C_1 starts to act. The transfer function is

$$G(s) = \frac{R_2}{R_1} \frac{1}{1 + sR_2C_1} \qquad (3\text{-}20)$$

The pole obeys the classical formula

$$\omega_p = \frac{1}{R_2C_1} \qquad (3\text{-}21)$$

Figure 3-17b portrays the ac response brought by such a configuration.

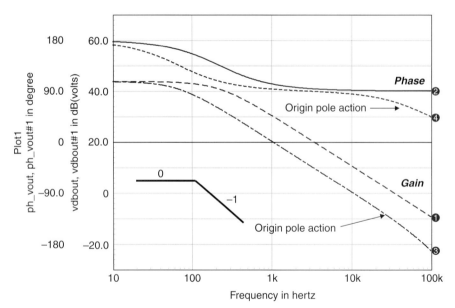

FIGURE 3-17b The dc gain is flat until the high-frequency pole starts to act and imposes a −1 slope decay.

3.5.8 Type 3—Origin Pole Plus Two Coincident Zero-Pole Pairs

The type 3 amplifier is used where a large phase boost is necessary, for instance, in CCM voltage-mode operation of converters which feature a second-order response. Its transfer function can be quickly derived, by calculating the impedance made of $Z_f = \frac{1}{sC_2} \| \left(R_2 + \frac{1}{sC_1} \right)$, divided by the input series impedance $Z_i = R_1 \| \left(R_3 + \frac{1}{sC_3} \right)$. See Fig. 3-18a.

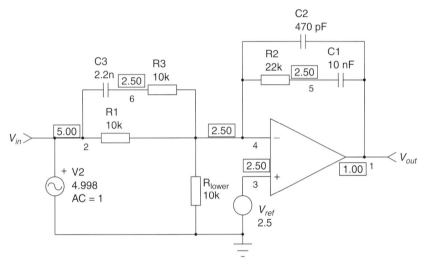

FIGURE 3-18a The type 3 amplifier circuitry. Two coincident pole-zero pairs associated with an integrator.

We obtain the following expression, highlighting the pole and zero definitions:

$$G(s) = \frac{Z_f}{Z_i} = \frac{sR_2C_1 + 1}{sR_1(C_1 + C_2)\left(1 + sR_2\frac{C_1C_2}{C_1 + C_2}\right)} \cdot \frac{sC_3(R_1 + R_3) + 1}{(sR_3C_3 + 1)} \quad (3\text{-}22)$$

Assuming $C_2 \ll C_1$ and $R_3 \ll R_1$

$$\omega_{z1} = \frac{1}{R_2C_1} \quad (3\text{-}23)$$

$$\omega_{z2} = \frac{1}{R_1C_3} \quad (3\text{-}24)$$

$$\omega_{po} = \frac{1}{R_1C_1} \quad (3\text{-}25a)$$

$$\omega_{p1} = \frac{1}{R_3C_3} \quad (3\text{-}25b)$$

$$\omega_{p2} = \frac{1}{R_2C_2} \quad (3\text{-}26)$$

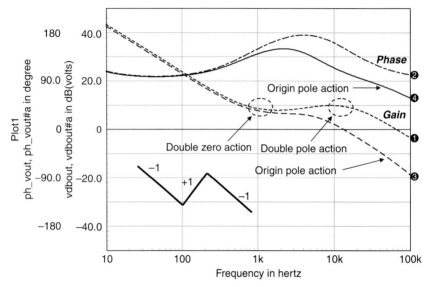

FIGURE 3-18b The type 3 amplifier introduces an integrator, a double zero, and a double pole.

Figure 3-18b plots the frequency response of the Fig. 3-18a amplifier and shows the slope evolution.

3.5.9 Selecting the Right Amplifier Type

Both the converter type and the transient response you need for your design will guide you through the selection of one particular compensation type.

- **Type 1:** As it does not offer any phase boost, the type 1 amplifier can be used in converters where the power stage phase shift is small, e.g., in an application where you would like to roll off the gain far away from the resonant frequency of a second-order filter. As in any integral type compensation, it brings the largest overshoot in the presence of a sudden load change. This type is widely used in power factor correction (PFC) applications, for instance, via a transconductance amplifier.
- **Type 2:** This amplifier is the most widely used and works fine for power stages lagging down to −90° and where the boost brought by the output capacitor ESR must be canceled (to reduce the gain in high frequency). This is the case for current-mode CCM and voltage-mode (direct duty cycle control) converters operated in DCM.
- **Type 2a:** The application field looks the same as for type 2, but when the output capacitor ESR effect can be neglected, e.g., the zero is relegated to the high-frequency domain, then you can use a type 2a.
- **Type 2b:** By adding the proportional term, it can help reduce the under- or overshoots in severe design conditions. We have seen that it prevents the output impedance from being too inductive, therefore offering superior transient response. Nevertheless, you pay for it by a reduction of the dc gain, hence a larger static error.
- **Type 3:** You use this configuration where the phase shift brought by the power stage can reach −180°. This is the case for CCM voltage-mode buck or boost-derived types of converters.

3.6 AN EASY STABILIZATION TOOL—THE K FACTOR

How can we easily position the pole(s) and zero(s) to cross over at a selected frequency with a specified phase margin? In the 1980s a gentleman named Dean Venable introduced the concept of the k factor [3]. It consists of deriving a number k based on observation of the open-loop Bode plot of the converter we want to stabilize. This k factor indicates the necessary separation (the distance) between the frequency position of the pole(s) and zero(s) brought by the compensation network (see Fig. 3-19a, b, and c). Then, by selecting the desired crossover frequency f_c and the amount of phase margin you need at f_c, the k factor automatically places the poles and zeros to make f_c the geometric mean between their respective locations: this is the place where the highest phase boost occurs. Depending on the value of k, different phase boosts are thus brought at the crossover frequency.

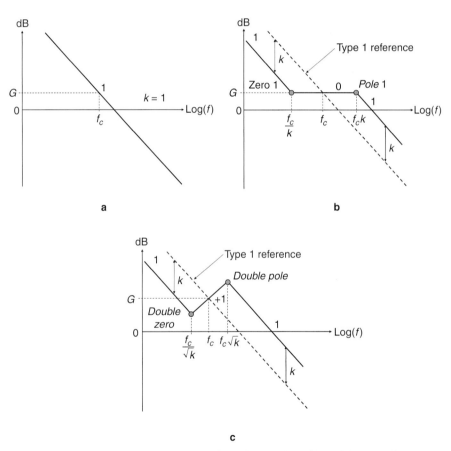

FIGURE 3-19a, b, and c The k factor adjusts the distance between the positions of the poles and zeros to get the desired phase boost at the crossover frequency.

Once k is calculated together with different data values coming from the open-loop Bode plot of the converter to stabilize, the derivation of all compensation elements for types 1, 2, and 3 is straightforward. Follow the steps.

3.6.1 Type 1 Derivation

The type 1 amplifier, with its pole at the origin (pure integrator), always features a k factor of 1 and introduces a permanent phase delay of $-270°$. It implies that both zeros and poles occupy the same position on the frequency axis. Here G is an indication of the gain needed to cross the 0 dB axis at the select crossover frequency. It must compensate (in both ways, either by amplifying or by attenuating) the gain developed by the power stage at the considered crossover frequency. Suppose you need a crossover frequency f_c of 1 kHz. Looking at the open-loop Bode plot, you read at 1 kHz a gain Gf_c of -18 dB. Therefore, the capacitor C must be calculated so that the type 1 amplifier delivers a gain G of $+18$ dB at 1 kHz:

$$G = 10^{\frac{Gf_c}{20}} \tag{3-27a}$$

$$C = \frac{1}{2\pi f_c G R_1} \tag{3-27b}$$

3.6.2 Type 2 Derivation

A complex number $a + jb$ features an argument equal to

$$\arg(a + jb) = \arctan\left(\frac{b}{a}\right) \tag{3-28}$$

If we consider a transfer function featuring one pole and one zero, we can also calculate the phase boost that the function introduces:

$$\arg(T(f)) = boost = \arg\left(\frac{1 + \frac{f}{f_{z0}}}{1 + \frac{f}{f_{p0}}}\right) = \arctan\left(\frac{f}{f_{z0}}\right) - \arctan\left(\frac{f}{f_{p0}}\right) \tag{3-29}$$

Yes, if both the zero and the pole are coincident, there is no phase boost at all since $\arg(T(f)) = 0$. But let us assume we place a zero at frequency $\frac{f}{k}$ and a pole at a frequency kf. Equation (3-29) can then be updated:

$$Boost = \arctan(k) - \arctan\left(\frac{1}{k}\right) \tag{3-30}$$

We know from the past trigonometric classes that the following identity is true:

$$\arctan(x) + \arctan\left(\frac{1}{x}\right) = 90° \tag{3-31}$$

Now, if we extract $\arctan\left(\frac{1}{x}\right)$ from Eq. (3-31) and inject it into Eq. (3-30), we obtain

$$Boost = \arctan(k) - 90 + \arctan(k) = 2\arctan(k) - 90 \tag{3-32}$$

Rearranging leads to

$$\arctan(k) = \frac{boost}{2} + 45 \tag{3-33}$$

Solving for k gives

$$k = \tan\left(\frac{boost}{2} + 45\right) \tag{3-34}$$

Equation 3-34 links the value of k to the amount of needed phase boost occurring at the crossover frequency. The needed phase boost is obtained from reading the phase shift information (*PS*) on the open-loop Bode plot of the converter you have to stabilize and the phase margin (*PM*) you finally want. The first phase delay is given by the phase shift brought by the converter power stage. Then, you add the $-90°$ phase shift brought by the integrator (origin pole). After the addition, you have to calculate how much positive phase you need to add (the *boost*) to obtain the desired phase margin *PM* which keeps you away from the $-180°$ limit.

$$PS - 90 + boost = -180 + PM \qquad (3\text{-}35a)$$

solving for *boost* gives:

$$Boost = PM - PS - 90 \qquad (3\text{-}35b)$$

where: *PM* is the phase margin you want at f_c and *PS* is the negative phase shift brought by the converter, also read at f_c.

Now, based on these numbers [boost, G—via Eq. (3-27a)—f_c and k], Dean Venable linked the pole and zero locations via the following formulas, where component labels correspond to Fig. 3-15a:

$$C_2 = \frac{1}{2\pi f_c G k R_1} \qquad (3\text{-}36)$$

$$C_1 = C_2(k^2 - 1) \qquad (3\text{-}37)$$

$$R_2 = \frac{k}{2\pi f_c C_1} \qquad (3\text{-}38)$$

To show the effect of varying k, that is, changing the distance between the pole and zero, we can update Fig. 3-15b with a new plot (Fig. 3-20a), where k moves from 1 to 10. For $k = 1$, the

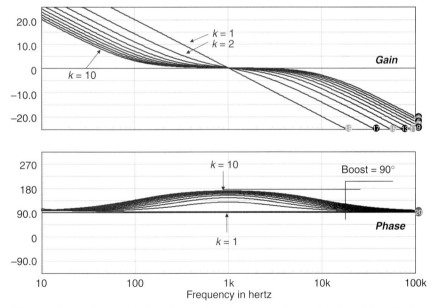

FIGURE 3-20a Adjusting the value of k allows the designer to modulate the phase boost. Here, G has been selected to be 1 (0 dB) and f_c to be 1 kHz. Note the decrease of dc gain as k increases.

pole and zero are coincident, and the phase boost is null [see Eq. (3-29)]. You can recognize the type 1 response. As k increases, so does the distance between the pole and zero locations, offering a greater boost at f_c. Unfortunately, increasing the phase boost comes at the price of reducing the dc gain. The k factor can thus be considered as a gain loss penalty that you pay for a greater phase boost. This statement also holds for the type 3 amplifier.

We will see soon the successive steps to successfully stabilize a converter using a type 2 amplifier.

3.6.3 Type 3 Derivation

Dean Venable also considered the type 3 amplifier and derived pole-zero equations based on his k factor. As with type 2, the k factor adjusts the distance between the pole-zero pairs and defines the phase boost it brings. Venable defined a zero at frequency $\dfrac{f}{\sqrt{k}}$ and a pole at frequency $\sqrt{k}f$. Capitalizing on results obtained by Eq. (3-30) gives

$$Boost = \arctan(\sqrt{k}) - \arctan\left(\dfrac{1}{\sqrt{k}}\right) \qquad (3\text{-}39)$$

However, we are going to place a double zero and a double pole. Therefore, the phase boost given by Eq. (3-39) must be multiplied by 2 if the double zeros and poles are, respectively, coincident. Therefore, the boost given by a type 3 amplifier is

$$Boost = 2\left[\arctan(\sqrt{k}) - \arctan\left(\dfrac{1}{\sqrt{k}}\right)\right] \qquad (3\text{-}40)$$

Using Eq. (3-31), we can write

$$Boost = 2\left[\arctan(\sqrt{k}) + \arctan\sqrt{(k)} - 90\right] = 4\arctan\sqrt{(k)} - 180 \qquad (3\text{-}41)$$

Solving for k gives

$$k = \left[\tan\left(\dfrac{boost}{4} + 45\right)\right]^2 \qquad (3\text{-}42)$$

Based on the definition of the phase boost in Eq. (3-35b), we can define all compensation values, where labels are coming from Fig. 3-18a:

$$C_2 = \dfrac{1}{2\pi f_c G R_1} \qquad (3\text{-}43)$$

$$C_1 = C_2(k-1) \qquad (3\text{-}44)$$

$$R_2 = \dfrac{\sqrt{k}}{2\pi f_c C_1} \qquad (3\text{-}45)$$

$$R_3 = \dfrac{R_1}{k-1} \qquad (3\text{-}46)$$

$$C_3 = \dfrac{1}{2\pi f_c \sqrt{k} R_3} \qquad (3\text{-}47)$$

Here f_c is the crossover frequency and G the needed gain at f_c.

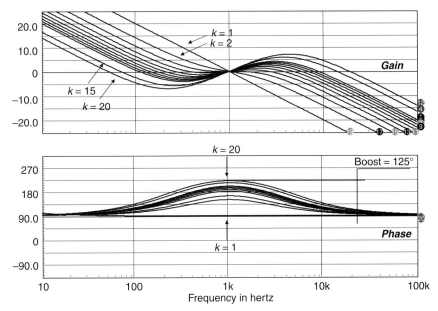

FIGURE 3-20b Adjusting the value of k allows the designer to modulate the phase boost. The phase boost, as in type 2, depends on the distance between the pole-zero pair. Note the decrease of dc gain as k increases.

If we enter these definitions into the simulator and sweep k, we generate the Fig. 3-20b plot, showing the effect of varying the k factor. A phase boost up to 180° can theoretically be attained.

3.6.4 Stabilizing a Voltage-Mode Buck Converter with the k Factor

As you will see, stabilizing a converter can be quite easy if you use the k factor technique. Let us follow the steps:

1. **Generate an open-loop Bode plot**. This open-loop plot can come from a laboratory sweep obtained via a network analyzer, or you can make it with an averaged SPICE model. If we take the example of a buck converter, Fig. 3-21 shows a typical simulation sketch. This is a 100 kHz CCM buck converter in voltage mode. The PWM block uses a 2 V peak-to-peak sawtooth V_{peak}, hence the −6 dB attenuation brought by the X_{PWM} subcircuit $\left(\dfrac{1}{V_{peak}}\right)$. The input voltage varies between 10 V and 20 V. The maximum output current is 2 A, and the minimum is 100 mA, implying a load variation between 2.5 Ω and 50 Ω. We will plot the Bode plot for a low line input, maximum current. This is what Fig. 3-22 represents.

2. **Select a crossover frequency and a phase margin**. As we operate to 100 kHz, experience shows that we could select a crossover frequency of one-fourth of the switching frequency, or 25 kHz. Let us be humble and select 5 kHz for the first step. The phase margin we want for this example is 45°.

3. **Read the Bode plot at the crossover frequency**. From Fig. 3-22, we can see a power stage phase shift of −146° and attenuation Gf_c of −9.2 dB, both measured at 5 kHz.

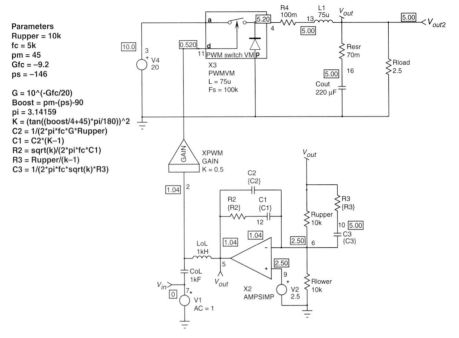

FIGURE 3-21 A buck converter where the calculation of the compensation network has been automated.

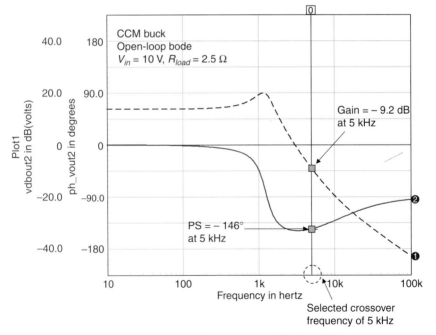

FIGURE 3-22 The open-loop Bode plot of the CCM buck converter. We select a 5 kHz crossover frequency.

4. **Select the amplifier type**. From the previous lines, we can see a power stage phase shift down to $-180°$, implying a type 3 amplifier. The *LC* network resonates at 1.2 kHz.
5. **Apply formulas**. Use Eqs. (3-35) and (3-42) to (3-47) to calculate the compensation elements:

    Phase boost = $101°$

    $k = 7.76$

    $G = 2.88$

    $C_1 = 7.5$ nF

    $C_2 = 1.1$ nF

    $C_3 = 7.72$ nF

    $R_2 = 11.9$ kΩ

    $R_3 = 1.5$ kΩ

    From calculations, the k factor places a double zero at 1.8 kHz and a double pole at 14 kHz.
6. **Sweep the open-loop gain with the above values**. The result appears in Fig. 3-23a and b as $R_{load} = 2.5$ Ω and $R_{load} = 50$ Ω at the two input voltage extremes.

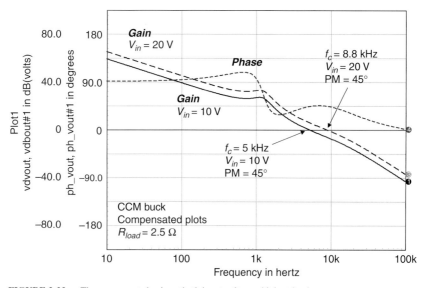

FIGURE 3-23a   The compensated gain, at both input voltages, highest load.

7. Check that phase margin *and* gain margin are within safe limits in all cases.
8. **Vary the output capacitor(s) ESRs**. The ESR varies with the capacitor age but also its internal temperature. Make sure the associated zero variations do not jeopardize point 7.
9. **Step load the output**. Replace the fixed load by a signal-controlled switch and step load it between two operating points at both input voltage extremes to check the stability. This is done in Fig. 3-24. A controlled current source can also perform the task. Following Eq. (3-3), we should have the following approximate undershoots:

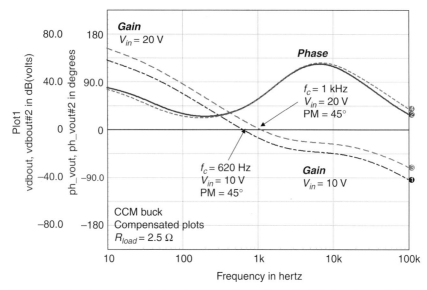

FIGURE 3-23b The compensated gain, at both input voltages, lightest load. The buck has toggled in the DCM mode thanks to the automatic mode transition of the model.

$$V_p = \frac{1.9}{6.28 \times 5k \times 220u} = 275 \text{ mV} \quad \text{hence a drop to } 5 - 0.275 = 4.72 \text{ V } @ f_c = 5 \text{ kHz}$$

$$V_p = \frac{1.9}{6.28 \times 8k \times 220u} = 171 \text{ mV} \quad \text{or a drop down to } 5 - 0.171 = 4.82 \text{ V } @ f_c = 8 \text{ kHz}$$

Figure 3-24 confirms these numbers.

As shown in various screen shots (Fig. 3-21, for example), the schematic capture *SNET* from Intusoft offers the ability of automating the *k* factor compensation elements. This is made through a text window on which the keyword *parameters* appears. The equations following the keyword then describe how the element values must be evaluated before the simulation is begun. You can thus quickly change the crossover frequency or the phase margin and see how it affects the transient response. *OrCAD* can also do that, and examples on the CD-ROM make use of it. For those who still would like to separately calculate the element values, a simple Excel spreadsheet lets you do so. It also includes the formulas described below for the manual pole and zero placements. The spreadsheet is described in App. 3A.

3.6.5 Conditional Stability

It is difficult to beat the *k* factor technique in terms of simplicity. However, the method does not represent a panacea for many people. Why is that? Because the technique is actually blind to the Bode plot shape, the open-loop diagram that you read before compensation. The *k* factor method ignores whatever resonance peaks or strange gain behaviors there are, since you only pick up the values at a given point, the desired crossover frequency. Therefore, compensation elements are computed to boost the phase right at the crossover frequency without actually knowing what is happening before, or after. Based on these comments, engineering judgment is necessary to

FEEDBACK AND CONTROL LOOPS

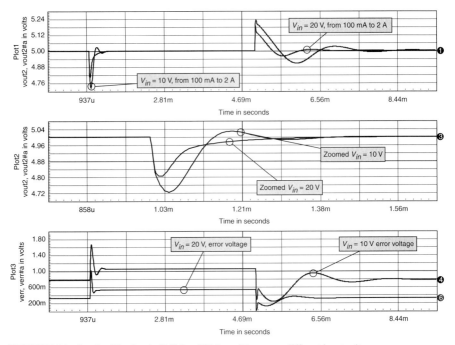

FIGURE 3-24 Step load forcing the DCM-to-CCM transition at two different input voltages.

check whether the final results fulfill all stability requirements. In some cases, and particularly in CCM voltage mode, conditional stability can occur. Let us push the Fig. 3-21 buck crossover frequency to 10 kHz. From the open-loop Bode plot (Fig. 3-22), we can see a power stage phase shift of $-132°$ and an attenuation Gf_c of -19.6 dB, both measured at 10 kHz. Applying type 3 formulas, we get the following values:

Phase boost = 87°

$k = 5.42$

$G = 9.55$

$C_1 = 736$ pF

$C_2 = 167$ pF

$C_3 = 3$ nF

$R_2 = 50.3$ kΩ

$R_3 = 2.3$ kΩ

From calculations, the k factor places a double zero at 4.3 kHz and a double pole at 23 kHz.

Sweeping the compensated open-loop gain reveals the picture in Fig. 3-25. We can clearly see an area where the phase margin collapses to zero while we still have gain. Is it dangerous? Well, if for any reason the gain curve crosses the 0 dB axis in this region, there is a chance (or bad luck!) that it will make the converter oscillate. However, we need to consider the corresponding gain margins (GMs), that is, the necessary variations that will push the gain curve up or down. In the above figure, the smallest GM is around 18 dB, which is quite a large number.

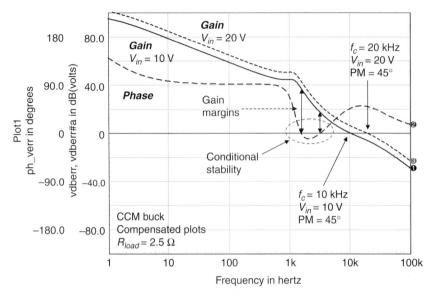

FIGURE 3-25 In CCM, the *k* factor has difficulty avoiding conditional stability.

In a buck converter, the control over output gain depends on the input voltage. If within the whole variation the Bode plot still exhibits a GM of 18 dB or more, then there is no problem. Conditional stability was a problem a long time ago when the amplifier's gain could change with temperature (warm-up time for tube-based designs), for instance. Now thanks to stable designs and existing ways (simulation, for example) to exactly bound the conditional stability zone, it is a less critical problem. Why do we have a conditional zone, by the way, in Fig. 3-25? Because the double zero occurs too far away from the resonance of the *LC* filter. Hence, despite the boost precisely brought at the crossover frequency, it cannot fight the phase shift engendered by the resonating filter before the appearance of the double zero. For some designers, it might represent a problem if customer design constraints specify an unconditionally stable design. Let us study an alternative compensation method.

3.6.6 Independent Pole-Zero Placement

In voltage-mode CCM designs, hence for buck-derived topologies, for instance, we know that a peak occurs at the resonant frequency. Then starts a -2 slope, often broken by the capacitor ESR action which introduces a zero. Sometimes this zero can be used to stabilize the converter (ESR of moderate value); at other times you cannot account for it because the ESR is too small, so it naturally relegates the zero to a higher-frequency domain. Thus, its occurrence has no phase boost effect in the bandwidth of interest. Figure 3-26 represents the asymptotic breakdown of the CCM voltage-mode gain curve and the proposed compensation shape. Let us study the overall curve.

- The power stage gain is flat at the beginning, $s = 0$. By inserting an origin pole to obtain high dc gain, the slope becomes -1 starting from nearly 0 Hz.
- The resonance now appears. The power stage imposes a -2 slope. We insert a double zero $(f_{z1} - f_{z2})$ imposing a $+2$ slope, now turning the total slope in: $-1 - 2 + 2 = -1$.

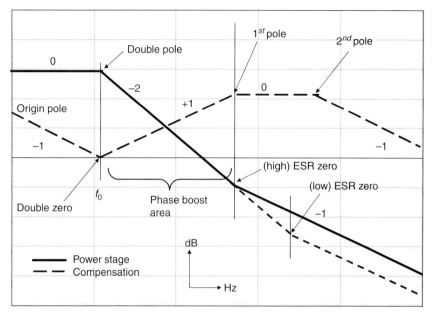

FIGURE 3-26 The asymptotic representation of a buck-derived CCM voltage-mode converter and the proposed compensation shape.

- The zero given by the capacitor ESR enters the action. It breaks the −2 power stage slope into a −1 slope. The occurrence of this slope breaking point can now move and depends on the ESR value, as shown in Fig. 3-26. Care must be taken to ensure enough PM in all possible ESR configurations. If the ESR action is within the band of interest (let us assume close to the crossover frequency), we must compensate it to force the gain to still decrease; otherwise it will simply flatten with a null slope. We place the first pole fp_1 here, right at the ESR frequency. On the contrary, if the ESR zero is out of the band of interest, e.g., away in high frequencies, then we place the first pole together with the second pole fp_2 at one-half of the switching frequency.

Please note that the compensation shape (in dashed lines) looks very similar to that of Fig. 3-19c where the poles are coincident. Figure 3-27 depicts the final shape where the gain curve crosses the 0 dB axis with a −1 slope, giving the guarantee of a good phase margin in the area.

With this method, you have the ability to place poles and zeros the way you wish, thus tailoring the loop gain as needed. How to select the component values to cross over at the selected frequency?

3.6.7 Crossing Over Right at the Selected Frequency

A type 3 amplifier follows a transfer function described by Eq. (3-22). By reworking it and rearranging the terms, it is possible to obtain a simpler expression [2]:

$$G(s) \approx \frac{R_2}{R_3} \frac{\left(\dfrac{1}{sR_1C_3} + 1\right)\left(\dfrac{1}{sR_2C_1} + 1\right)}{(1 + sR_2C_2)\left(\dfrac{1}{sR_3C_3} + 1\right)} \qquad (3\text{-}48)$$

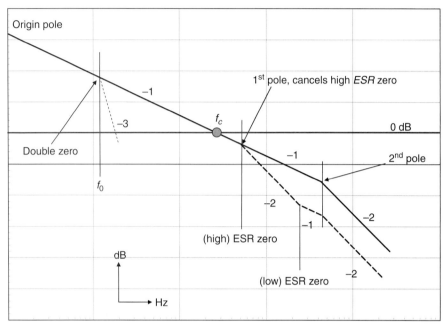

FIGURE 3-27 You shape the loop gain to force a crossover with a −1 slope, offering the best phase margin at this particular point.

If we rewrite this equation using pole and zero definitions, we obtain the following simplified module definition in the frequency domain:

$$G(f) \approx \left| \frac{R_2}{R_3} \frac{\left(1 + \frac{s_{z1}}{s}\right)\left(1 + \frac{s_{z2}}{s}\right)}{\left(1 + \frac{s}{s_{p1}}\right)\left(1 + \frac{s_{p2}}{s}\right)} \right| \tag{3-49a}$$

with

$$f_{z1} = \frac{1}{2\pi R_1 C_3} \tag{3-49b}$$

$$f_{z2} = \frac{1}{2\pi R_2 C_1} \tag{3-49c}$$

$$f_{p1} = \frac{1}{2\pi R_2 C_2} \tag{3-49d}$$

$$f_{p2} = \frac{1}{2\pi R_3 C_3} \tag{3-49e}$$

The exercise now consists of extracting the value of R_2 to get the right compensation level at the crossover frequency. For instance, what is the value for R_2 if we want $G(f) = 19.6$ dB @

10 kHz, compensating Fig. 3-22 for $f_c = 10$ kHz? Using Eq. (3-49a), we can calculate R_2 after reworking the Eq. (3-49a) module expression:

$$R_2 = \sqrt{\frac{(f_{p1}^2 + f_c^2)(f_{p2}^2 + f_c^2)}{(f_{z1}^2 + f_c^2)(f_{z2}^2 + f_c^2)}} \frac{Gf_c R_3}{f_{p1}} \quad (3\text{-}50)$$

If we decide to have coincident poles and zeros (respectively, f_p and f_z), then Eq. (3-50) simplifies to

$$R_2 = \frac{f_p^2 + f_c^2}{f_z^2 + f_c^2} \frac{Gf_c R_3}{f_p} \quad (3\text{-}51)$$

Let us now detail the calculating steps to stabilize Fig. 3-22, using the above formulas. We will place the double zero at 1.2 kHz (the resonant frequency), one pole at 14 kHz (the ESR zero), and the second one at one-half of the switching frequency

1. Calculate C_3 given f_{z1}. Use Eq. (3-49b).
2. Calculate R_3 given C_3 and f_{p2}. Use Eq.(3-49e).
3. As R_3 is known, as well as the desired pole and zero locations, extract R_2 via Eq. (3-50).
4. Calculate C_1 given R_2 and f_{z2}. Use Eq. (3-49c).
5. Calculate C_2 given R_2 and f_{p1}. Use Eq. (3-49d).

We obtain the following values:

$G = 9.55$

$C_1 = 9.4$ nF

$C_2 = 803$ pF

$C_3 = 13.3$ nF

$R_2 = 14.2$ kΩ

$R_3 = 240$ Ω

Figure 3-28 depicts the schematic capture using an automated calculation routine via the *parameters* keyword. Element values are evaluated by the formulas then finally assigned before simulation. The dc sources on the bottom offer an easy way to unveil the calculated results. It was already implemented in Fig. 3-21. Thanks to this feature, you can easily and quickly explore effects induced by changing pole and zero locations. Let us now sweep the control to output response, using these new values. Figure 3-29 displays the result. The conditional stability no longer exists, and the phase margin is greater than 80° at both input levels. This is a rock-solid design! A good thing is now to compare step load and input load responses between a *k* factor compensation and the above method.

3.6.8 The *k* Factor Versus Manual Pole-Zero Placement

To illustrate both behaviors, the output is stepped from 1 to 2 A in a 1 A/μs slew rate. The 10 V input voltage implies a similar crossover frequency of 10 kHz for the two circuits, and the phase margin on both designs is also the same. We boosted the phase margin to 80° on the *k* factor circuit to compare apples to apples: the *k* factor design versus manual poles/zeros placement. Figure 3-30a portrays both updated ac plots whereas Fig. 3-30b shows the resulting transient waveforms.

CHAPTER THREE

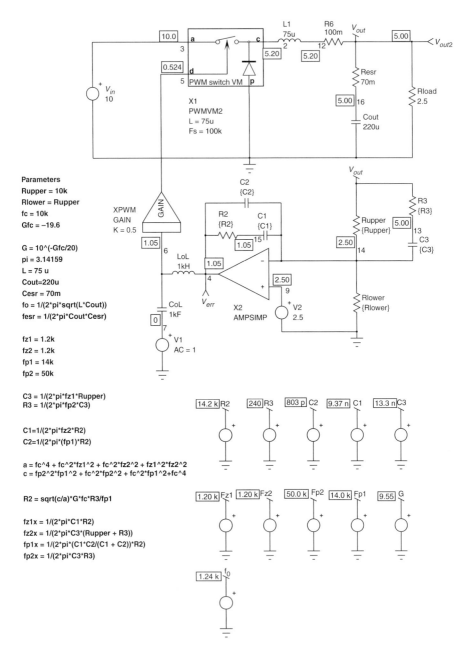

FIGURE 3-28 A fully automated calculation via the *parameters* keyword on Intusoft's *SPICENET*.

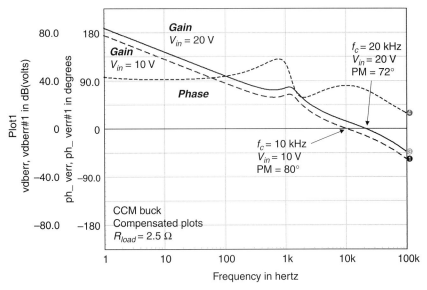

FIGURE 3-29 Using a different pole and zero placement, we reach a design where the conditional stability area has gone.

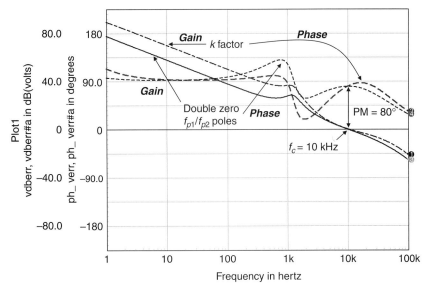

FIGURE 3-30a The k factor design has been changed to increase the phase margin to 80° as the manually compensated circuit provides.

Figure 3-30b certainly shows that the type of transient response you really want dictates the location of poles and zeros. Having zeros in the lower portion of the frequency spectrum (then close to the resonant frequency) inherently slows down the system despite similar crossover frequencies (10 kHz in this example). Why? In the closed-loop equation featuring

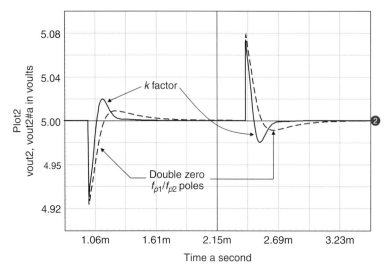

FIGURE 3-30b In output transient steps, the *k* factor design gives a little overshoot but recovers more quickly than the other design.

the compensation network $G(s)$, if we replace $G(s)$ by $G(s) = \dfrac{N(s)}{D(s)}$, we obtain the following expression:

$$\frac{V_{out}(s)}{V_{ref}(s)} = \frac{\dfrac{N(s)}{D(s)} H(s) G_{PWM}}{1 + H(s) G_{PWM} \dfrac{N(s)}{D(s)}} = \frac{N(s) H(s) G_{PWM}}{D(s) + N(s) H(s) G_{PWM}} \qquad (3\text{-}52)$$

In this new expression, the denominator, which now includes the closed-loop poles of the system, shows the presence of $N(s)$. It is the numerator of $G(s)$, the error amplifier compensation network, where we placed the zeros to shape the loop. Hence, putting these zeros in low-frequency domain naturally slows down the system as they now appear as poles of the closed-loop gain equation! Appendix 2B gives more details about this important fact. This is why the *k* factor gives the fastest response (we do not say the best), when its double zero is located at 2.6 kHz versus 1.2 kHz for the other network. The penalty is a small overshoot at both events, the load step reaction and the recovery at the load release. However, if the conditional stability did disappear, the low phase margin around 2 kHz (see Fig. 3-30a) could be a problem in the presence of large gain variations. In that particular case, a compensation using the separate pole-zero placement can be justified for an unconditionally stable design. As expected, the *k* factor gives the highest dc gain. We can therefore expect an improved input ripple rejection [Eq. (1-19)].

If we ac sweep the input voltage for a constant loading, the audio susceptibility can be obtained. As Eq. (1-19) has shown, the closed-loop audio susceptibility depends on the loop gain $T(s)$. At dc, the large gain implies an excellent line (input) rejection. As the gain decreases with frequency [the power stage but also the compensation network $G(s)$ you put in place], the input rejection behavior tends to be the same as in open-loop conditions. This is exactly what Fig. 3-31a displays where, at the end of the spectrum, all curves superimpose.

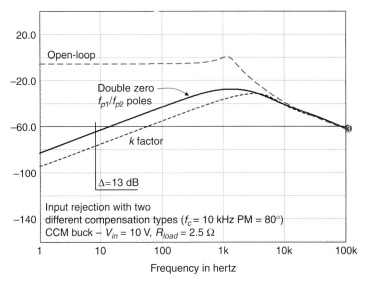

FIGURE 3-31a By bringing a larger dc gain, the *k* factor offers superior input rejection in this particular case.

Again, as the *k* factor type of compensation gives a better dc gain, the rejection it brings looks much better than what the other method gives. To test the transient response to the step input, let us vary the input source from 10 V to 20 V in 10 μs. Figure 3-31b displays the variations.

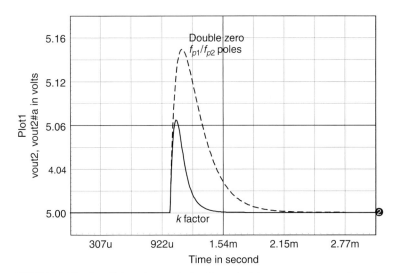

FIGURE 3-31b By offering a better rejection figure, the *k* factor response is superior to the separate pole-zero placement in this line rejection test.

3.6.9 Stabilizing a Current-Mode Buck Converter with the *k* Factor

The above example tackled a voltage-mode power supply exhibiting second-order behavior, justifying the implementation of a type 3 amplifier. Figure 3-32a shows the same buck converter but now operated in current mode. The PWM switch model we used is PWMCM. It features a duty cycle output that lets you know the static operating point. After simulation, we can observe the bias points on the schematic. A duty cycle of 52% is likely to induce subharmonic

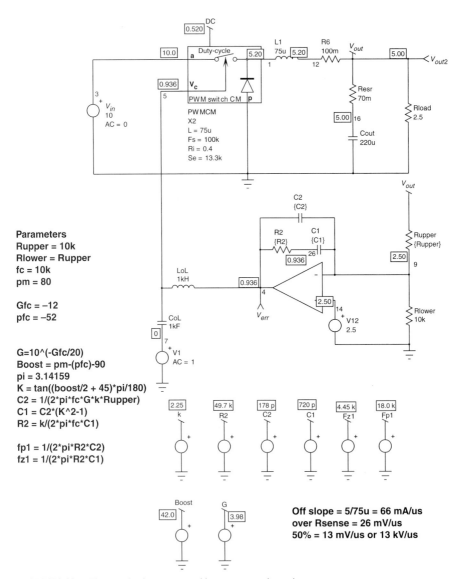

FIGURE 3-32a The same buck now operated in a current-mode version.

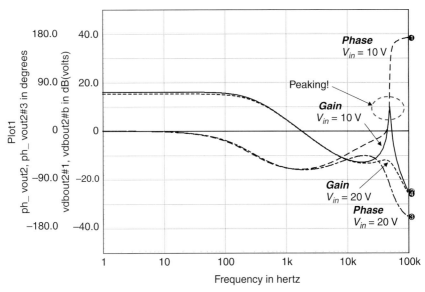

FIGURE 3-32b Subharmonic oscillations appear as the duty cycle exceeds 50% on this current-mode CCM buck (10 V input voltage). But they fade away in higher-input-voltage conditions (20 V).

oscillations. This is what Fig. 3-32b tells us. However, as soon as the input voltage reaches 20 V, the duty cycle goes back below 50% and the peaking (and its associated problems) fades away. In the 10 V input voltage configuration, the peaking crosses the 0 dB axis and must absolutely be damped by ramp compensation. As we discussed in Chap. 2, we can decide on the level of ramp compensation to apply. In this case, we will select one-half of the inductor downslope.

A few steps give us the right value that we are going to feed the model with:

- Inductor downslope: $S_{off} = \dfrac{V_{out}}{L} = 66 \text{ kA/s}$ or $66.6 \text{ mA/}\mu\text{s}$.
- Over the 0.4 Ω sense resistor R_i, it will develop $V_{sense} = S_{off}R_i = 66.6m \times 0.4 = 26.6 \text{ mV/}\mu\text{s}$.
- If we inject 50% of this slope, then the final compensating level is $26.6 \times 0.5 = 13.3 \text{ mV/}\mu\text{s}$ or 13.33 kV/s. We enter $13.3k$ to the model in the S_e line.

A new simulation can be run with the ramp compensation in place, and results are available in Fig. 3-32c.

From this picture, let us apply the k factor methodology:

1. **Select a crossover frequency and a phase margin.** Let us take 10 kHz to compare with the previous voltage-mode design and see how the current-mode buck performs. The required phase margin specification is 80°.
2. **Read the ramp compensated Bode plot at the crossover frequency.** From Fig. 3-32c, we can see a power stage phase shift of $-52°$ and an attenuation Gf_c of -12 dB, both measured at 10 kHz.
3. **Select the amplifier type.** From the Bode plot, we can see a power stage phase shift down to $-90°$ in the low-frequency portion of the spectrum, implying a type 2 amplifier.

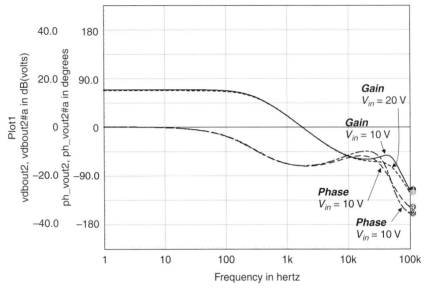

FIGURE 3-32c Injecting ramp compensation nicely damps the subharmonic poles: the peaking has left.

4. **Apply formulas.** Use Eqs. (3-36) to (3-38) to calculate the compensation elements.

   Phase boost = $42°$

   $k = 2.25$

   $G = 3.98$

   $C_1 = 720$ pF

   $C_2 = 178$ pF

   $R_2 = 49.7$ kΩ

   From calculations, the k factor places a single zero at 4.5 kHz and a single pole at 18 kHz.

5. **Run the simulations with the above values.** This is what has been done in Fig. 3-33a and b, respectively, with $R_{load} = 2.5$ Ω and $R_{load} = 50$ Ω at the two input voltage extremes.
6. Check that phase margin and gain margin are within safe limits in all cases.
7. **Vary the output capacitor(s) ESRs.** The ESR varies with the capacitor age but also its internal temperature. Make sure the associated zero variations do not jeopardize point 6.
8. **Step load the output.** Replace the fixed load by a signal-controlled switch, and step load it between two operating points at both input voltage extremes to check the stability. A controlled current source can also perform the task.

As Fig. 3-33c shows, it is difficult to distinguish between the responses obtained at different input voltage cases. This is the natural strength of current-mode designs versus voltage-mode ones. This can be easily shown by plotting the audio susceptibility of the current-mode CCM buck and comparing it to that of Fig. 3-31a (voltage mode). As we previously discussed, applying a ramp compensation made of 50% from the inductor downslope theoretically nullifies the input sensitivity, implying an infinite rejection. This is exactly what Fig. 3-34a shows with extremely high rejection level (-150 dB!). Of course, in reality, smaller ratios are reached

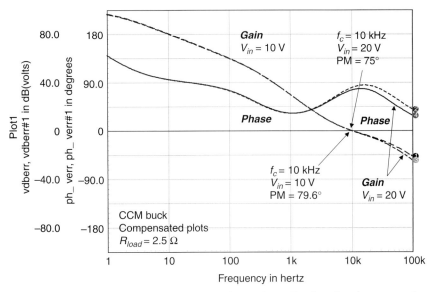

FIGURE 3-33a Compensated CCM buck at the two input voltages. Note how the gain curve remains unchanged.

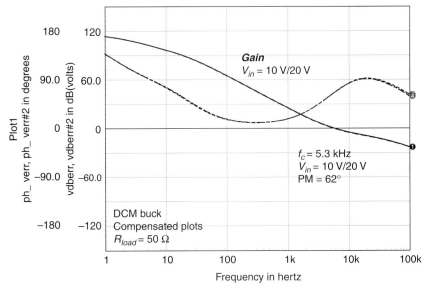

FIGURE 3-33b Compensated DCM buck at the two input voltages. Despite a crossover change, there is almost no difference between phase and gain curves at both operating voltages.

but still make the current-mode buck design superior to its voltage-mode counterpart when we are talking about input voltage rejection capability.

Finally, we can now compare the voltage mode and current mode, both compensated with the k factor, in a step load and step line test. In both designs, the crossover frequency is 10 kHz, the input voltage is 10 V, and the phase margin is 80°. Figure 3-34b shows both transient

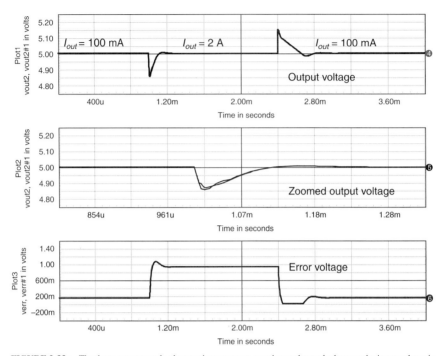

FIGURE 3-33c Thanks to current mode, the transient response remains unchanged whatever the input voltage is.

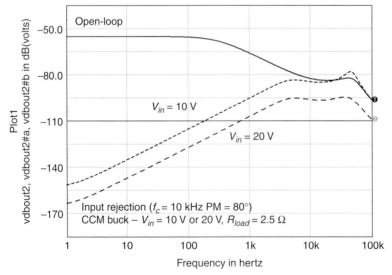

FIGURE 3-34a The input rejection in current cannot compare with that of voltage mode. With a 50% ramp compensation, the audio susceptibility is theoretically nulled.

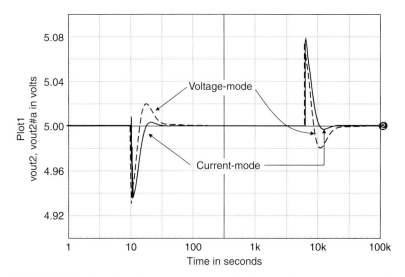

FIGURE 3-34b Comparison between current mode and voltage mode in a step load response. The current mode remains a first-order system in the low-frequency range and shows less overshoot.

responses to an output step. The undershoots are almost equivalent, but the current-mode rings less than the voltage-mode circuit. Well, a voltage mode in CCM remains a second-order system, where the current mode behaves as a first-order system in the lower frequency portion of the spectrum, naturally generating less ringing. Regarding input rejection, the current-mode design is simply excellent (Fig. 3-34c): on a similar scale, you cannot notice a perturbation

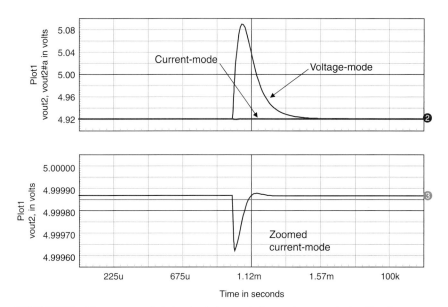

FIGURE 3-34c The current-mode design simply ignores the input voltage perturbation. Note the response polarity typical of the current-mode design.

compared to the voltage-mode design that overshoots quite a bit. If we now zoom in on the output voltage for the current-mode design, we can see a negative answer: this is typical of current mode. If we injected more compensation ramp, the current-mode undershoot would increase until it became similar to that of the voltage mode. It would be an overcompensated design.

We can see through this example the good behavior of the k factor with current-mode supplies, mainly first-order systems in the low-frequency portion (before subharmonic poles). In some cases, designers might prefer to manually place the poles and zeros after the crossover frequency has been chosen. This is possible with the type 2 amplifier, and the derivation has been documented in App. 3C. You mainly select your pole and zero locations and calculate the resistor R_2 to generate the right gain at f_c.

3.6.10 The Current-Mode Model and Transient Steps

The current-mode model, in its CCM, DCM, or autotoggling versions, puts a rather heavy burden on the SPICE numerical solver. As a result, simulations can sometimes fail to properly converge when the step occurs in transient runs. The ac analysis does not usually show any problem since the operating point is calculated before the simulation starts. But in transient, some particular modes can lead to trouble. Here is a quick guide to get rid of the convergence issues in transient:

- This is one drawback of the current-mode autotoggling model: if during the transient you transition between the two modes, CCM or DCM, then the resonating capacitor connection or disconnection can lead to convergence issues. The best thing is to then put a star at the beginning of the SPICE code that describes the capacitor expression in the PWMCM subcircuit, as suggested below:

    ```
    * C1  c p C=V(mode) > 0.1 ? {4/((L)*(6.28*Fs)^2)} : 1p ; IsSpice
    * XC1 c p mode varicap                             ; PSpice
    ```

- Increase the transient iteration limit which is the number of trials before a data point is rejected. By default, *ITL4* = 100. Increasing to 300 to 500 helps to solve "Time step too small" errors.
- Relax the relative tolerance *RELTOL* to 0.01.
- If it still fails, relax the absolute current and voltage error tolerance, respectively, *ABSTOL* and *VNTOL*. Values such as 1 μ for *ABSTOL* and 1 m for *VNTOL* usually help a lot.
- Increase *GMIN* to 1 nS or 10 nS. *GMIN* is the minimum conductance in each branch. It helps convergence in deep nonlinear circuits by linearly routing some current out of the nonlinear element.

3.7 FEEDBACK WITH THE TL431

Showing compensation circuits around an op amp is an interesting thing, but the industrial design world differs in reality. The TL431 presence in feedback systems is overwhelming, and few designs still use a true operational amplifier. Why? Because the TL431 already includes a stable and precise reference voltage with an error amplifier. Even if its open-loop gain cannot compete against a true op amp, it is good enough for the vast majority of product definitions. What exactly is a TL431? Figure 3-35a shows the internal arrangement of the device. You can observe a reference voltage of 2.5 V biasing an operational amplifier inverting input.

FEEDBACK AND CONTROL LOOPS

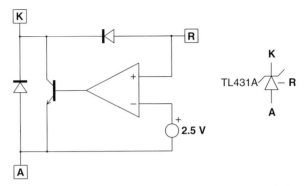

FIGURE 3-35a The internal schematic of the TL431 featuring a 2.5 V reference voltage.

The output drives a bipolar transistor, actually making the TL431 a shunt regulator: when the voltage on the reference pin (R) is below 2.5 V, the transistor remains open, and the TL431 is transparent to the circuit. As soon as the voltage exceeds the reference, the transistor starts to conduct and a current circulates inside the device. If an optocoupler LED appears in series with the cathode, it becomes possible to build an opto-isolated feedback system. Figure 3-35b shows how most of today's power supplies implement a TL431: here, in a typical flyback converter.

The TL431 also exists in different precision versions, depending on what you are looking for. In some cases where you need output voltages below 2.5 V, the TLV431 might be a good choice. The latter also features a smaller minimum biasing current compared to the TL431. It can be a good advantage in low-standby-power designs. The following array compares all versions.

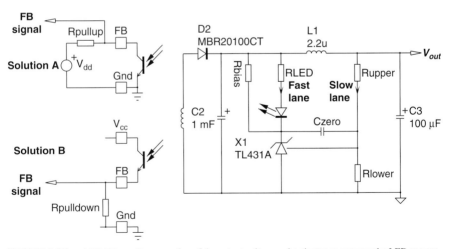

FIGURE 3-35b A TL431 monitors a portion of the output voltage and activates an optocoupler LED to transmit the feedback information to the nonisolated primary side.

Reference	V_{ref}	$I_{bias,min}$	Precision	Max. voltage	Max. current
TL431I	2.495 V	1 mA	± 2% @ 25 °C	36 V	100 mA
TL431A	2.495 V	1 mA	± 1% @ 25 °C	36 V	100 mA
TL431B	2.495 V	1 mA	± 0.4% @25 °C	36 V	100 mA
TLV431A	1.24 V	100 μA	± 1% @25 °C	18 V	20 mA
TLV431B	1.24 V	100 μA	± 0.5% @25 °C	18 V	20 mA

Appendix 3B describes the SPICE models of a behavioral TL431, which we extensively used in all the book examples. This model can work on *IsSpice* or *PSpice* and has proved to properly reflect reality.

In Fig. 3-35b, the resistive network $R_{upper}-R_{lower}$ senses the output voltage and biases the TL431 reference pin. When the output is above the reference, the TL431 reduces its cathode voltage and increases the LED current. This, in turn, reduces the feedback set point, and the converter delivers less power. On the contrary, when the output is below the target, the TL431 almost leaves the cathode open and stops pumping current into the LED. As a result, the primary feedback allows for more output power, pushing the converter to increase the output voltage until the TL431 detects the target is reached. The converter can accept two different optocoupler configurations, described as solutions A and B:

- *Solution A*: This is a common emitter configuration as found on popular controllers such as ON Semiconductor's NCP1200 series. Bringing the FB pin down reduces the peak current in this current-mode controller. This solution also exists on UC384X-based designs where the collector can directly drive the output of the internal op amp.
- *Solution B*: In this common collector configuration, the emitter pulls high the FB pin to reduce the duty cycle or the peak current set point. This option usually requires an inverting amplifier inside the controller.

As you can see in Fig. 3-35b, the LED branch is called the "fast lane" whereas the divider network is tagged "slow lane." The slow lane uses the internal op amp to drive the TL431 output transistor and fixes the dc operating point via the resistive network divider $R_{upper} - R_{lower}$. Thanks to the presence of the capacitor C_{zero}, it is possible to introduce an origin pole and thus roll off the gain as a standard type 1 amplifier would do. Alas! Above a certain frequency range, because C_{zero} has completely rolled off the gain, the shunt regulator no longer behaves as a controlled zener diode. The internal op amp still fixes the dc bias point but no longer ac controls the shunt regulator as its gain has gone to a low value via the impedance of C_{zero}. The sketch thus simplifies to Fig. 3-36 (with solution A, for instance) where the TL431 becomes a simple zener diode. For the small-signal study, we can replace this diode by a fixed voltage in series with its internal impedance, the LED undergoing the same translation. However, as the sum of these dynamic resistors remains small compared to R_{LED}, we can easily neglect them in the final calculation.

From Kirchhoff's law,

$$V_{FB}(s) = -I_1 R_{pullup} \text{CTR} \qquad (3\text{-}53)$$

where CTR represents the optocoupler current transfer ratio, a gain linking the quantity of photons collected by the transistor base and the collector current they engender: $I_c = I_1 \text{CTR}$. Considering constant the LED voltage drop and the zener voltage, their derivative terms are zero in the small-signal analysis, therefore:

$$I_1 = \frac{V_{out}(s)}{R_{LED}} \qquad (3\text{-}54)$$

Substituting Eq. (3-54) into Eq. (3-53) gives

$$\frac{V_{FB}(s)}{V_{out}(s)} = -\frac{R_{pullup}}{R_{LED}} \text{CTR} \qquad (3\text{-}55)$$

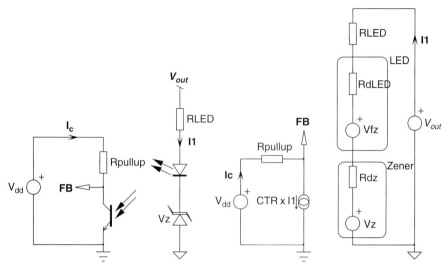

FIGURE 3-36 The small-signal model includes various dynamic resistors, but they are of low value compared to the series resistor R_{LED}.

This equation describes the "fast lane" gain that you simply cannot roll off. It also sheds light on the selection of R_{LED}: it obviously cannot be made solely on bias current considerations as it affects the loop gain. Bias current will rely on the resistor R_{bias} whose current does not cross the optocoupler LED and thus does not play a role in the gain definition. For Fig. 3-35b solution B, the result is almost similar to the Eq. (3-55) result, except that there is no phase reversal as with the common collector structure.

Once the operating principle is understood, the final TL431 representation makes more sense, as Fig. 3-37 shows. You can see the standard op amp having a capacitor C_{zero} but followed by an adder network representative of the fast lane. Note in Fig. 3-35b that the LED

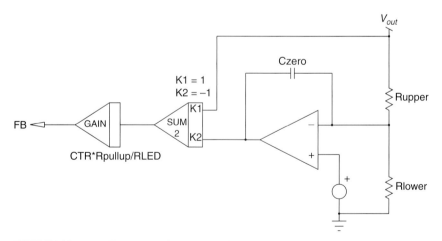

FIGURE 3-37 From this sketch, the fast lane can easily be identified.

connects *before* the secondary *LC* filter. This is done to avoid offering gain at high frequency when the *LC* network starts to resonate. This configuration is typical of flyback converters featuring high-frequency noise reduction via an *LC* filter. Make sure the resonant frequency of this filter is placed at least 10 times above the selected crossover frequency to avoid any interaction.

As it is placed, C_{zero} capacitor looks as if it introduces an origin pole. However, if we derive the transfer function of this complete feedback chain, we get the following results:

$$V_{FB}(s) = \left(V_{out}(s)\frac{1}{sR_{upper}C_{zero}}\right) + V_{out}(s)\frac{R_{pullup}}{R_{LED}}\text{CTR} \tag{3-56}$$

Rearranging the equation gives

$$\frac{V_{FB}(s)}{V_{out}(s)} = -\left(\frac{1}{sR_{upper}C_{zero}} + 1\right)\frac{R_{pullup}}{R_{LED}}\text{CTR} = -\left(\frac{sR_{upper}C_{zero} + 1}{sR_{upper}C_{zero}}\right)\frac{R_{pullup}}{R_{LED}}\text{CTR} \tag{3-57}$$

Equation (3-57) reveals the presence of an origin pole, f_{po} (as suspected . . .) plus a zero, f_z, introduced by the fast lane configuration. Unfortunately, if we want a type 2 amplifier, which is the most common type, we need a pole f_p somewhere. How can we obtain it? Simply by placing a capacitor from the output node to ground (Fig. 3-38a and b). If we update Eq. (3-57) with the presence of this pole, we obtain the final equation ruling the TL431 network as presented by Fig. 3-35b:

$$G(s) = \frac{V_{FB}(s)}{V_{out}(s)} = -\left(\frac{sR_{upper}C_{zero} + 1}{sR_{upper}C_{zero}}\right)\left(\frac{1}{1 + sR_{pullup}C_{pole}}\right)\frac{R_{pullup}}{R_{LED}}\text{CTR} \tag{3-58}$$

As usual, we can evaluate poles and the zero positions through the following definitions:

$$f_{po} = \frac{1}{2\pi R_{upper}C_{zero}} \tag{3-59}$$

$$f_z = \frac{1}{2\pi R_{upper}C_{zero}} \tag{3-60a}$$

$$f_p = \frac{1}{2\pi R_{pullup}C_{pole}} \tag{3-60b}$$

$$G = \frac{R_{pullup}}{R_{LED}}\text{CTR} \tag{3-61}$$

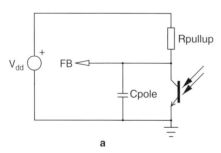
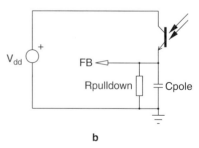

FIGURE 3-38a A simple capacitor placed between the feedback pin and ground introduces a pole.

FIGURE 3-38b The same applies to a common collector configuration.

As shown by Eqs. (3-59) and (3-60a), the origin pole f_{po} and the zero f_z are coincident. This implies a slope change right on the 0 dB axis, as confirmed by Fig. 3-38c. Thanks to this arrangement naturally brought by the TL431, the needed midband gain G at the crossover frequency is simply governed by Eq. (3-61), independent of the positions of f_z and f_p.

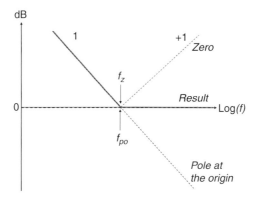

FIGURE 3-38c When the zero is coincident with the origin pole, the slope change occurs on the 0 dB axis.

The next step is to find a way to organize the pole-zero placement for a good compensation.

3.7.1 A Type 2 Amplifier Design Example with the TL431

The goal is to apply the k factor methodology to the TL431 network, as it is a simple and straightforward way to stabilize power supplies. However, if you do not like the k factor technique, you are free to assign the pole and zero locations as you wish, using Eqs. (3-59) to (3-61). Looking back at Fig. 3-19b, we see that the k factor places the pole and zero at the following locations:

$$f_z = \frac{f_c}{k}$$

$$f_p = kf_c$$

The pull-up resistor depends on the controller and can sometimes be integrated inside. If it is externally placed, the designer can select it to increase the optocoupler bias current in order to reach a higher bandwidth. If we use the ON Semiconductor NCP1200 controller, this resistor is internally fixed at 20 kΩ. Let us assume that R_{upper} equals 10 kΩ for the sake of the example, where we need to create a type 2 amplifier offering the following parameters:

- Crossover frequency = 1 kHz
- Needed phase margin = 100°
- Gain attenuation read at crossover frequency $Gf_c = -20$ dB
- Phase observed at crossover = $-55°$
- The k factor computes: $k = 4.5$, $f_z = 222$ Hz, and $f_p = 4.5$ kHz
- $G = 10^{\frac{-Gf_c}{20}} = 10$

- CTR = 1
- $R_{pullup} = 20$ kΩ
- $R_{upper} = 10$ kΩ

From Eqs. (3-59) through (3-61),

$$C_{zero} = \frac{1}{2\pi R_{upper} f_z} = 71.8 \text{ nF} \tag{3-62}$$

$$C_{pole} = \frac{1}{2\pi R_{pullup} f_p} = 1.76 \text{ nF} \tag{3-63}$$

$$R_{LED} = \frac{CTR \cdot R_{pullup}}{G} = 2 \text{ k}\Omega \tag{3-64}$$

where G is the midband gain (or attenuation) wanted at the crossover frequency [Eqs. (3-27a) and (3-61)].

Figure 3-39 gathers all solutions to compare their ac responses. At the top of the picture, you can recognize the traditional op amp–based type 2 amplifier, tailored to deliver +20 dB at 1 kHz. This is the reference. On the right side, this is the equivalent internal arrangement of the TL431, featuring the various electrical paths. Finally, on the left side, appears the complete chain, implementing a simplified optocoupler network (a current-controlled current source F_2) without a pole. When the simulator runs, it delivers Bode plots, all assembled in Fig. 3-40, which shows perfect agreement between all graphical representations. This terminates the type 2 configuration built around a TL431. It is now time to explore the type 3 version.

3.7.2 A Type 3 Amplifier with the TL431

The type 3 version is a bit more complicated, again because of the fast lane presence. We could get rid of this input by inserting a zener diode or a bipolar device to avoid any interference with the output voltage. This is shown in Figs. 3-41 and 3-42 [4]. However, it requires external components and it slightly complicates the design. How do we place a zero in the TL431 chain? In parallel with R_{upper} as on the traditional op amp–based solution? No, because of the fast lane, the solution does not work. The only solution is to place an RC network in parallel with R_{LED}. This is what Fig. 3-43 portrays. Fortunately, the transfer function of this new arrangement does not differ too much from Eq. (3-58). The only difference lies in the R_{LED} expression since an RC network now comes in parallel. The equivalent arrangement features the following impedance:

$$Z_{eq} = \frac{R_{LED}(sR_{pz}C_{pz} + 1)}{sC_{pz}(R_{LED} + R_{pz}) + 1} \tag{3-65}$$

Once Eq. (3-65) is inserted into Eq. (3-58) in place of R_{LED}, the complete transfer function becomes

$$G(s) = \frac{V_{FB}(s)}{V_{out}(s)} = -\left(\frac{sR_{upper}C_{zero1} + 1}{sR_{upper}C_{zero1}}\right)\left(\frac{1}{1 + sR_{pullup}C_{pole2}}\right)\frac{[sC_{pz}(R_{LED} + R_{pz}) + 1]}{(sR_{pz}C_{pz} + 1)}\frac{R_{pullup}}{R_{LED}} CTR \tag{3-66}$$

As usual, we can calculate the poles and zeros from the following definitions:

$$f_{z1} = \frac{1}{2\pi R_{upper} C_{zero1}} \tag{3-67}$$

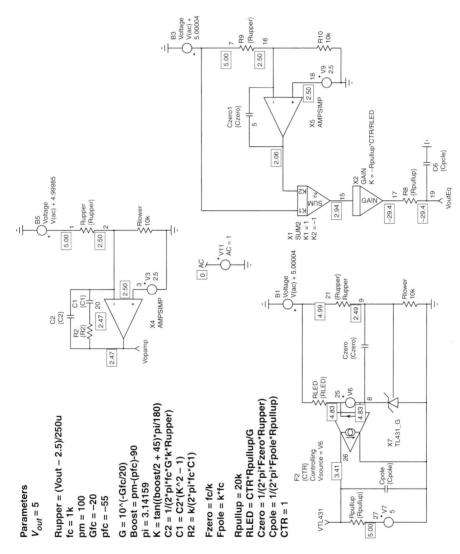

FIGURE 3.39 The comparison between the original op amp-based type 2 amplifier and the TL431 implementation.

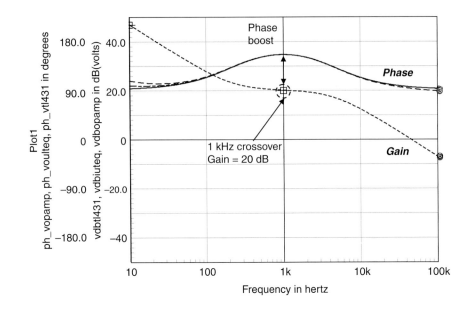

FIGURE 3-40 All Bode plots give identical results.

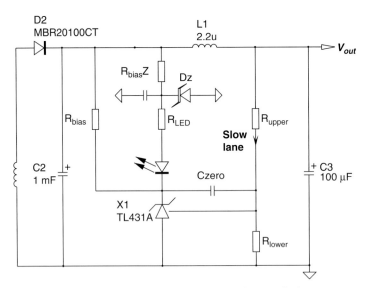

FIGURE 3-41 Adding a zener helps to get rid of the fast lane contribution.

FEEDBACK AND CONTROL LOOPS

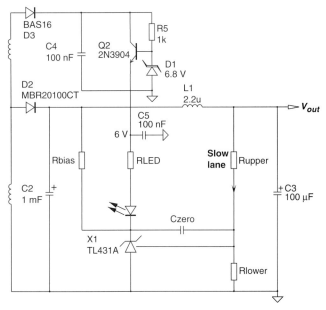

FIGURE 3-42 A bipolar can also do the job.

$$f_{p1} = \frac{1}{2\pi R_{pz}C_{pz}} \tag{3-68a}$$

$$f_{po} = \frac{1}{2\pi R_{upper}C_{zero1}} \tag{3-68b}$$

$$f_{z2} = \frac{1}{2\pi(R_{LED} + R_{pz})C_{pz}} \tag{3-69}$$

$$f_{p2} = \frac{1}{2\pi R_{pullup}C_{pole2}} \tag{3-70}$$

$$G = \frac{R_{pullup}}{R_{LED}}\text{CTR} \tag{3-71}$$

The first exercise consists of finding the value of R_{LED} to get the right gain (or attenuation) at the selected crossover frequency f_c. Equation (3-66) can be rewritten as follows, highlighting the positions of the poles and zeros:

$$G(s) \approx \text{CTR}\frac{R_{pullup}}{R_{LED}}\frac{\left(1+\dfrac{s}{s_{z1}}\right)\left(1+\dfrac{s}{s_{z2}}\right)}{\dfrac{s}{s_{z1}}\left(1+\dfrac{s}{s_{p1}}\right)\left(1+\dfrac{s}{s_{p2}}\right)} \tag{3-72}$$

Parameters
$V_{out} = 5$
$R_{upper} = (V_{out} - 2.5)/250u$
$fc = 1k$
$pm = 100$
$Gfc = -20$
$pfc = -55$

$G = 10^{\wedge}(-Gfc/20)$
Boost = pm−(pfc)−90
pi = 3.14159
$K = (tan((boost/4 + 45)*pi/180))^{\wedge}2$
$C2 = 1/(2*pi*fc*G*Rupper)$
$C1 = C2*(K-1)$
$R2 = sqrt(k)/(2*pi*fc*C1)$
$R3 = Rupper/(k-1)$
$C3 = 1/(2*pi*fc*sqrt(k)*R3)$

Fzero = fc/sqrt(k)
Fpole = sqrt(k)*fc

Rpullup = 20k

$a = (fpole^{\wedge}2 + fc^{\wedge}2)*(fc^{\wedge}2 + fzero^{\wedge}2)*(fpole^{\wedge}2 + fc^{\wedge}2)*(fc^{\wedge}2 + fzero^{\wedge}2)$
$b = fpole^{\wedge}2*fpole^{\wedge}2 + fpole^{\wedge}2*fc^{\wedge}2 + fc^{\wedge}2*fpole^{\wedge}2 + fc^{\wedge}4$
$Rled = (sqrt(a)/b)*Rpullup*fpole*fpole/(fzero*fc*G)$

$Czero1 = 1/(2*pi*Fzero*Rupper)$
$Cpole2 = 1/(2*pi*Fpole*Rpullup)$
$Cpz = (fpole-fzero)/(2*fzero*fpole*Rled*pi)$
$Rpz = 1/(2*pi*Fpole*Cpz)$
CTR = 1

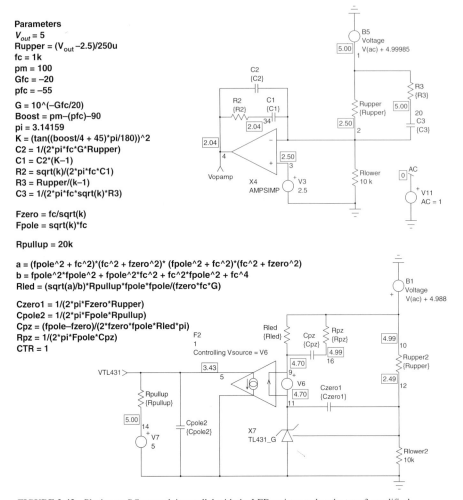

FIGURE 3-43 Placing an *RC* network in parallel with the LED resistor makes the type 3 amplifier!

Extracting the value of R_{LED} from the module expression leads to the following (complicated) result:

$$R_{LED} = \frac{\sqrt{(f_{p1}^2 + f_c^2)(f_{z1}^2 + f_c^2)(f_{p2}^2 + f_c^2)(f_{z2}^2 + f_c^2)}}{f_{p1}^2 f_{p2}^2 + f_{p2}^2 f_c^2 + f_{p1}^2 f_c^2 + f_c^4} \frac{CTR \cdot f_{p1} f_{p2} R_{pullup}}{f_{z2} f_c G} \quad (3\text{-}73)$$

Fortunately, if poles and zeros are coincident (respectively f_p and f_z), the formula simplifies to

$$R_{LED} = \frac{(f_z^2 + f_c^2)}{(f_p^2 + f_c^2)} \frac{f_p^2 R_{pullup} CTR}{f_c f_z G} \quad (3\text{-}74)$$

Thanks to the above equations, we can derive the value of R_{LED}, given the gain needed at the crossover frequency. Now, manipulating Eqs. (3-68a) and (3-69), we can compute a value for C_{pz}, as this capacitor plays a role for the zero (but also for the pole position). From Eq. (3-68a), we can extract R_{pz}:

$$R_{pz} = \frac{1}{2\pi f_{p1} C_{pz}} \quad (3\text{-}75)$$

Inserting this result into Eq. (3-69) gives

$$f_{z2} = \frac{1}{2\pi \left(R_{LED} + \frac{1}{2\pi f_{p1} C_{pz}}\right) C_{pz}} \quad (3\text{-}76)$$

Solving for C_{pz} leads us to the final result:

$$C_{pz} = \frac{f_{p1} - f_{z1}}{2\pi f_{z1} f_{p1} R_{LED}} \quad (3\text{-}77)$$

As we did for the type 2 amplifier, let us imagine that R_{upper} equals 10 kΩ. Then, for the sake of the example, we will create a type 3 amplifier offering the following parameters:

- Crossover frequency = 1 kHz
- Needed phase margin = 100°
- Gain needed at crossover = +20 dB (hence $G = 10$)
- Phase observed at crossover = –55°
- $k = 3.32$ given by the k factor tool

First, the k factor for the type 3 amplifier computes the following coincident pole and zero locations [Eq. (3-42)]. Of course, nothing prevents you from placing individual poles and zeros as discussed above. In that case, use Eq. (3-73) instead.

$$f_z = \frac{f_c}{\sqrt{k}} = 549 \text{ Hz}$$

$$f_p = f_c \sqrt{k} = 1.8 \text{ kHz}$$

Via Eq. (3-74), we obtain the value for R_{LED}: 3.6 kΩ with R_{pullup} = 20 kΩ and CTR = 1.
From Eq. (3-77), C_{pz} = 55.6 nF. The rest of the elements are easily calculated via Eqs. (3-67), (3-68a), and (3-70):

$$C_{zero1} = \frac{1}{2\pi R_{upper} f_{z1}} = 29 \text{ nF} \quad (3\text{-}78)$$

$$R_{pz} = \frac{1}{2\pi f_{p1} C_{pz}} = 1.57 \text{ k}\Omega \quad (3\text{-}79)$$

$$C_{pole2} = \frac{1}{2\pi R_{pullup} f_{p2}} = 4.37 \text{ nF} \quad (3\text{-}80)$$

Figure 3-43 portrays the test circuitry where all elements are automatically computed. This also works in *OrCAD*. Once the simulation has finished, Fig. 3-44 shows the results: the TL431 plots perfectly match those coming from the type 3 op amp–based circuit!
   As we will see in some of the examples, the TL431 does not lend itself well to the type 3 implementation. This is so because the LED resistor acts in the gain definition and the pole-zero position. Depending on the pull-up resistor, an R_{LED} value ensuring the right gain at the crossover frequency and enough bias in light-load conditions can sometimes lead to an impossible solution. In that case, the solution might require the use of an operational amplifier. The implementation of one of the solution depicted by Fig. 3-41 or Fig. 3-42 is also possible.

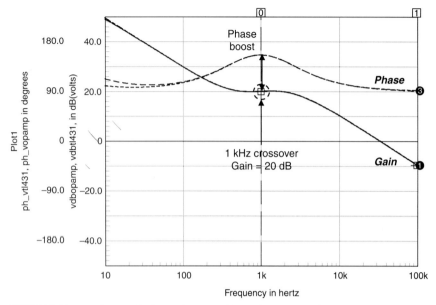

FIGURE 3-44 Waveforms perfectly superimpose, evidence of the good architecture of the TL431 type 3 version.

In that case, R_{LED} plays the role of the collector resistor for the TL431 and affects the dc gain only. It no longers participate to the pole-zero positions and a traditional type 3 op amp-based configuration can be used.

3.7.3 Biasing the TL431

The TL431 requires a minimum of biasing current to make it fulfill its data sheet parameters. This current must be larger than 1 mA. If you fail to properly bias the device, it may happen in serious open-loop gain degradation. Figure 3-45 shows a real power supply behavior with and without

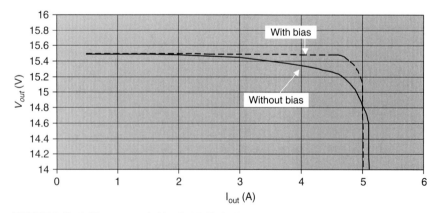

FIGURE 3-45 Failing to properly bias the TL431 leads to open-loop gain degradation, increasing the output impedance.

FEEDBACK AND CONTROL LOOPS   299

bias. Without bias, you can clearly see the voltage-current graph bending, evidence of poor output impedance. As we inject the right level of bias current, the curve takes on a better shape.

Very often, designers believe that R_{LED} sets the bias current in the TL431. This is incorrect. The bias current is only set by the feedback current on the primary side and the optocoupler CTR. For instance, if your dc point imposes a 600 µA primary feedback current together with a 150% CTR, then the current flowing via R_{LED} on the secondary side will be 400 µA, well below the minimum biasing current. To increase the bias current independent of the regulation loop, Fig. 3-46a and b shows two possible configurations:

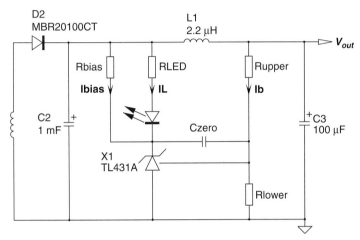

FIGURE 3-46a A complete feedback circuit showing the external biasing circuit.

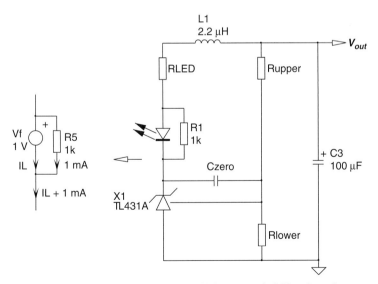

FIGURE 3-46b A resistor placed in parallel with the optocoupler LED makes a cheap constant-current reference.

1. Figure 3-46a: you force an external bias current from a resistor connected to the output voltage, and it biases the TL431 as soon as the output voltage starts to rise.
2. Figure 3-46b: you use the LED forward drop ($\approx$ 1 V) to implement a constant-current generator biasing the TL431. Designers usually put a 1 kΩ resistor there. As the bias only occurs when the TL431 starts to activate (when V_{out} reaches its target), it can induce a slight overshoot. The major advantage of this solution lies in its simplicity of design.

Focusing on the Fig. 3-46a option, you need to wire a resistor R_{bias} from the output and impose a current inside the TL431 independent of the LED path. Therefore, the feedback loop is not affected by the presence of this resistor. Another solution would be to force a higher feedback current on the primary side, but the LED current would still depend on the CTR variations (see below). Figure 3-46a shows the component arrangement. To better understand how to calculate R_{bias}, Fig. 3-47 offers a closer look at the elements in play. On the primary side, we can see a pull-up resistor hooked to an internal V_{dd} rail (usually 5 V for integrated circuits) and a

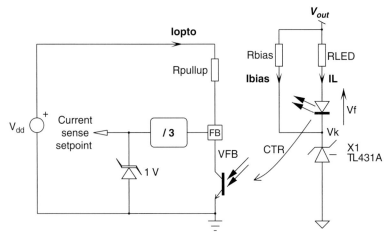

FIGURE 3-47 The optocoupler collector current together with the CTR fixes the current in the LED in the absence of external bias.

feedback pin going through a divider (typically a divide-by-3 as in the UC384X series). Thanks to an active clamp (the zener symbol), the output of this divider cannot swing above 1 V: this is the maximum current set point imposed by the controller in the absence of feedback closure (during short-circuit, start-up sequence, etc.). Hence, if the current pin connects to a 1 Ω resistor, the maximum allowed peak current is 1 A. As such, the useful dynamic on the feedback pin (node V_{FB}) is constrained between $V_{ce(sat)}$ and 3 V. Above 3 V, the feedback pin cannot impose more current as the 1 V zener clamp gets activated. The collector current varies between a minimum and a maximum:

$$I_{opto, max} = \frac{V_{dd} - V_{ce(sat)}}{R_{pullup}} \qquad (3\text{-}81)$$

$$I_{opto, min} = \frac{V_{dd} - 3}{R_{pullup}} \qquad (3\text{-}82)$$

The optocoupler is also affected by a current transfer ratio varying a lot between manufacturers' batches, temperature, collector current, age, and so on. Below are some typical CTR variations of popular off line optocouplers for 10, 1 (between parenthesis) and 5 mA bias currents:

Reference	Manufacturer	CTR$_{min}$ (%)	CTR$_{typ}$ (%)	CTR$_{max}$ (%)
SFH615A-1	Vishay	40 (13)	(30)	80
SFH615A-2	Vishay	63 (22)	(45)	125
PC817A	Sharp	80 (5 mA)		160 (5 mA)
PC817B	Sharp	130 (5 mA)		260 (5 mA)

The variations, as you can observe, are extremely large! You should watch out for these when exploring the effects on the feedback loop. Thus, depending on the CTR and the feedback level, a certain current I_L will flow in the LED. The worst case corresponds to Eq. (3-82), that is, maximum power delivered on the output. If your controller features a different arrangement (common collector configuration as in Fig. 3-38b, or a shunt regulator), the principle remains the same: identify the lowest primary feedback current condition to see the bias variations on the LED. Sticking to the common-emitter configuration, the current seen by the LED depends on the optocoupler CTR:

$$I_{L,\,min} = \frac{I_{opto,\,min}}{CTR_{max}} = \frac{V_{dd} - 3}{R_{pullup} CTR_{max}} \quad (3\text{-}83)$$

Of course, this is the maximum CTR that imposes the lowest LED current. Without external bias conditions, the current circulating in the TL431 could be as low as Eq. (3-83) describes. To force a larger current, let us calculate the necessary R_{bias} value:

$$R_{bias} = \frac{V_{out} - V_k}{I_{bias}} = \frac{R_{LED} I_{L,\,min} + V_f}{I_{bias}} \quad (3\text{-}84)$$

or, via Eq. (3-83)

$$R_{bias} = \frac{R_{LED} \dfrac{V_{dd} - 3}{R_{pullup} CTR_{max}} + V_f}{I_{bias}} \quad (3\text{-}85a)$$

This additive current will force the circulation of $I_{bias} + I_L$ inside the TL431. The forward voltage of the optocoupler LED is around 1 V. We now have everything to calculate the resistor via a practical example:

$V_{out} = 12$ V

$R_{LED} = 2.2$ kΩ

$R_{pullup} = 20$ kΩ

$V_{dd} = 4.8$ V

SFH-615A-1, CTR$_{typ} = 30\%$ (bias current of 1 mA)

Optocoupler $V_{ce(sat)} = 200$ mV

$V_f = 1$ V

Forced bias current = 2 mA

$$R_{bias} = \frac{2.2k \dfrac{4.8 - 3}{20k \cdot 0.3} + 1}{2m} = 830\ \Omega \quad (3\text{-}85b)$$

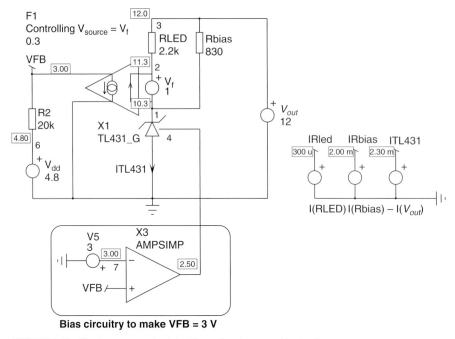

FIGURE 3-48 The dc sources on the right side confirm the correct bias levels.

Figure 3-48 represents a simple simulation circuit which helps verify the calculations: The current circulating in the TL431 corresponds to what we wanted, 2 mA minimum. The maximum current the device will see occurs in output overshoot conditions, that is, when the cathode voltage drops to 2.5 V to force the converter to heavily brake. If for any reason the output voltage overshoots to 15 V, the current seen by the TL431 will be

$$I_{TL431,\,max} = I_{L,\,max} + I_{bias,\,max} = \frac{V_{out,\,max} - 2.5 - V_f}{R_{LED}} + \frac{V_{out,\,max} - 2.5}{R_{bias}}$$

$$= \frac{15 - 2.5 - 1}{2.2k} + \frac{15 - 2.5}{503} = 30 \text{ mA} \qquad (3\text{-}85c)$$

This is below the maximum current accepted by the TL431 ($I_{max} = 100$ mA), and furthermore, the largest portion of this current (24.8 mA) does not cross the LED but R_{bias}.

The final equation to see how R_{LED} fits the circuitry is to check that enough current can circulate in the LED when the converter needs to brake. In Fig. 3-47, the maximum current needed to bring the feedback pin low is defined by Eq. (3-81). The worst case occurs in the presence of the lowest CTR, imposing the largest current in the LED. Hence, once R_{LED} has been calculated according to Eq. (3-64) or (3-73), the designer must check the following equality:

$$\frac{V_{out} - 2.5 - V_f}{R_{LED}} \geq \frac{V_{dd} - V_{ce,\,sat}}{R_{pullup} CTR_{min}} \qquad (3\text{-}86)$$

According to the optocoupler parameter sheet, the minimum CTR being 13% (1 mA bias current), Eq. (3-86) gives us the following numerical results: 3.86 mA > 1.76 mA which is okay. If you could not reach the right current level, you will need to select an optocoupler offering a better CTR range. Note that large pull-up resistors on the controller side naturally bring large R_{LED} values, especially if you do not need a lot of gain. Rather than select a better CTR optocoupler, you can reduce the controller pull-up resistor by connecting a resistor from the FB pin to V_{cc} (common-emitter configuration) or from the FB pin to ground in case of a common collector configuration.

Equation (3-86) represents the major limiting factor when you are designing compensation circuits around the TL431, especially for the type 3 configuration. This is so because the midband gain depends on the LED resistor, and as a function of the pull-up resistor, it can force the adoption of large LED resistor values, leading to an impossible operation. A few iterations are thus necessary to either change the crossover frequency (thus a different midband gain) or alter the internal pull-up resistor by connecting an external resistor in parallel with it (e.g., another pull-up from the FB pin to a reference voltage or V_{cc}).

3.7.4 The Resistive Divider

The feedback loop works by monitoring the output voltage via a resistive divider. The op amp then strives to maintain equality between the reference voltage (2.5 V for a TL431, 1.25 V for a TLV431) and the bridge node. Figure 3-49 shows what this configuration looks like.

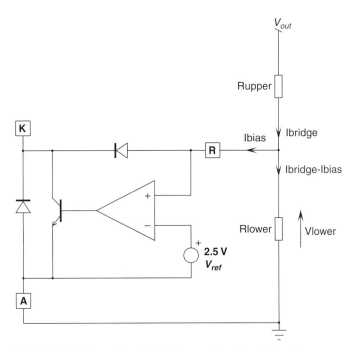

FIGURE 3-49 The resistive divider brings a portion of the output voltage, permanently compared to an internal stable reference voltage.

To calculate the value of R_{upper} and R_{lower}, let us first consider the bias current entering the TL431 (6 μA over the temperature range). This current appears in Fig. 3-49 as I_{bias}. Writing Kirchhoff's law, we have

$$V_{lower} = (I_{bridge} - I_{bias})R_{lower} \qquad (3\text{-}87)$$

where V_{lower} is the voltage appearing across R_{lower}. Of course, this is so because we assume a steady-state closed-loop configuration where V_{lower} equals the reference voltage V_{ref}. The second equation links the upper resistor and the bridge current:

$$I_{bridge} = \frac{V_{out} - V_{lower}}{R_{upper}} \tag{3-88}$$

Now, using this equation within Eq. (3-87) leads to

$$V_{lower} = \left(\frac{V_{out} - V_{lower}}{R_{upper}} - I_{bias}\right) R_{lower} = \left(V_{out} - V_{lower} - I_{bias} R_{upper}\right) \frac{R_{lower}}{R_{upper}} \tag{3-89}$$

Rearranging this equation gives us the final expression linking all elements:

$$V_{out} = V_{ref}\left(\frac{R_{upper}}{R_{lower}} + 1\right) + R_{upper} I_{bias} \tag{3-90}$$

Equation (3-90) reveals the role played by the bias current. It is thus the designer's duty to select a total bridge current greater than this bias current to make its presence a negligible term. Also, reducing the bridge impedance not only reduces the potential bias error, but also increases the noise immunity by diminishing the drive impedance at node *R*. Typical currents ranging from 250 µA up to a few milliamperes can be used, depending also on the acceptable bridge power dissipation. If you chase every milliwatt for a low-standby power converter, you cannot afford to waste 100 mW in the feedback bridge, for instance.

As a quick design example, suppose we want to stabilize the output of the converter to 12 V with a TL431. The steps are as follows:

1. Select a bridge current. Here, we chose 1 mA, we can thus neglect I_{bias}.
2. Calculate R_{lower}:

$$R_{lower} = \frac{V_{ref}}{I_{bridge}} = \frac{2.5}{1m} = 2.5 \text{ k}\Omega$$

3. R_{upper} immediately becomes

$$R_{upper} = \frac{V_{out} - V_{ref}}{I_{bridge}} = \frac{12 - 2.5}{1m} = 9.5 \text{ k}\Omega$$

4. The bridge dissipation is

$$\frac{V_{out}^2}{R_{lower} + R_{upper}} = 12 \text{ mW}$$

3.8 THE OPTOCOUPLER

An optocoupler provides an optical link between a primary side, usually connected to the mains, and a secondary side isolated from the primary for obvious safety reasons. The link is made via the light emitted by an LED (photons), pointing to a bipolar transistor's base that will collect the photons. This gives rise to a collector current whose intensity depends on the injected current in the LED and thus the luminous flux intensity it emits. The amount of current circulating in the collector links to the current flowing in the LED via the current transfer ratio (CTR). This CTR, as previous arrays showed, depends on the LED current, the optocoupler age, and junction temperature. This is the reason why the optocoupler vendor gives you two limits within which the converter must be stable. Since the collector-base junction in a phototransistor must work as a photodetector, the manufacturer makes it large, to collect the largest quantity of photons. This unfortunately brings a large capacitance between the collector and the base, dramatically reducing the available bandwidth. Of course, the bandwidth depends on the collector pull-up resistor and the associated bias current. Reference 4 gives good information about optocouplers.

3.8.1 A Simplified Model

Figure 3-50 shows a simplified representation [5] of the optocoupler where a capacitor is simply added to the output of a current-controlled current source. This capacitor is evidence of the presence of the pole and can be evaluated given the fall time of the optocoupler measured in the typical switching circuit as most of data sheets describe. The data sheet gives a fall time

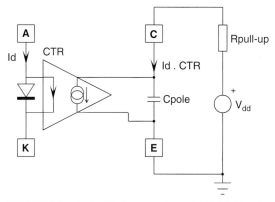

FIGURE 3-50 A simplified optocoupler model highlighting the capacitor position.

measured at a given collector resistor. For instance, on the SFH615A-X, SHARP specifies a fall-time of 15 μs obtained with a 1 kΩ pull-up resistor. Hence, the capacitor value can be obtained via the simplified formula

$$C_{pole} = \frac{T_{fall}}{2.2 \cdot R_{pullup}} = 6.8 \text{ nF} \tag{3-91a}$$

To refine the model, a diode featuring a null V_f (N = 0.01) and a 30 V breakdown voltage will avoid negative collector values when too much current is injected in the LED. Also, a simple voltage source in series represents the transistor saturation voltage. Figure 3-51 shows the updated model.

A second option to determine C_{pole} consists of extracting the optocoupler pole when it is biased the same way as the controller will. Once the pole is known, we can easily calculate the equivalent C_{pole} value, using the controller pull-up resistor value:

$$C_{pole} = \frac{1}{2\pi R_{pullup} f_{pole}} \tag{3-91b}$$

Looking back at the various TL431-based circuits (Fig. 3-38a and b), we can see that the compensation capacitor we add to the collector or the emitter of the optocoupler is in parallel with the optocoupler equivalent capacitor C_{pole}. You understood that if this equivalent capacitor

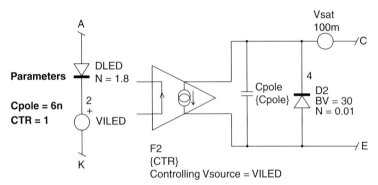

FIGURE 3-51 Adding a diode helps to avoid convergence issues when too much current flows in the LED.

is already larger than the one you plan to put in parallel, then the final pole introduced depends on the optocoupler and no longer on the capacitor you have placed. For this reason, when you design wide-bandwidth power supplies, it is necessary to adopt small pull-up (1 to 4.7 kΩ) resistors and thus higher bias currents.

3.8.2 Extracting the Pole

To extract the optocoupler pole, you need to know the pull-up resistor you will hook it to as well as the corresponding operating bias current (20 kΩ for the NCP1216, for instance, and a 5 V dc source). Then wire the optocoupler on the bench as proposed by Fig. 3-52a. As you can see, a dc supply biases the LED to position the V_{FB} node to a value corresponding to the nominal converter power: around 3 V in this case, bringing the real bias current. Adjust V_{bias} (or R_{bias}) to get there. Then inject a sinusoidal voltage via a dc block capacitor (C_{dc}) that you will observe on V_{FB}. Make sure the modulation amplitude of V_{rms} is low enough to (1) keep V_{FB}

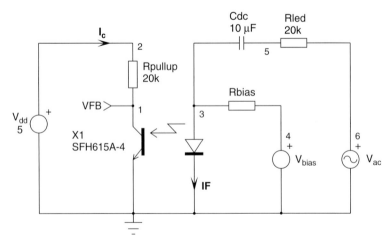

FIGURE 3-52a Biasing the LED to get the right point and ac sweeping the input reveal the optocoupler bandwidth.

undistorted and (2) stay in small-signal analysis. Observe the V_{FB} envelope and increase the modulating frequency. As soon as the envelope goes down by a ratio of 0.707 (−3 dB), read the frequency generator: this is your pole position.

On SPICE, if you pick up the optocoupler model from the SPICE vendor's available list, you can wire it as Fig. 3-52b shows. Selecting the proper bias on V_{bias} leads to the right operating current (V_{FB} = 3 V). Then ac sweep from 10 Hz to 100 kHz, and plot V_{FB}. When the gain drops by −3 dB, this is your pole position (Fig. 3-53). Then, reading the current through R_{pullup} (I_c) and R_{LED} (I_F) gives you the CTR value of the optocoupler under study.

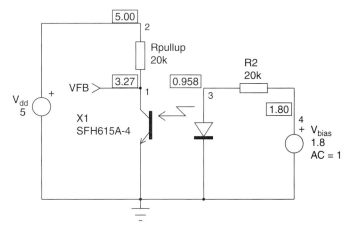

FIGURE 3-52b SPICE simplifies the architecture necessary to obtain the optocoupler pole position.

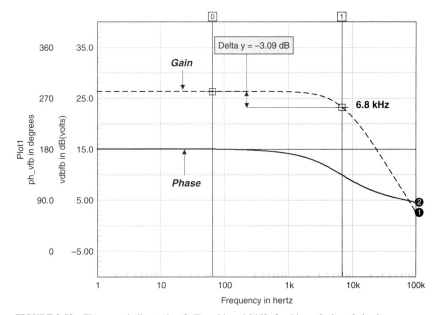

FIGURE 3-53 The cursor indicates the −3 dB position: 6.8 kHz for this particular subcircuit.

It now becomes obvious that despite all the work you have put into designing the type 2 or 3 amplifier, inserting another pole in the system can ruin all your past efforts to get the right bandwidth and phase margin. And this an important point: the optocoupler plays the role of the traitor here! It is absolutely imperative that you account for its pole position and its natural CTR variations when compensating the supply. If you have the choice, make sure that enough current flows in the optocoupler to relegate this pole toward higher frequencies. For instance, we took a SFH615A-1 and varied the pull-up resistor, while measuring the pole position:

$$R_{pullup} = 20\text{ k}\Omega \quad f_{pole} = 5.4\text{ kHz}$$
$$R_{pullup} = 10\text{ k}\Omega \quad f_{pole} = 8.7\text{ kHz}$$
$$R_{pullup} = 1.5\text{ k}\Omega \quad f_{pole} = 48\text{ kHz}$$

In some situations, decreasing the pull-up resistor could engender a higher consumption in standby (higher bias current derived from V_{cc}), and some tradeoffs then have to be found. On the other hand, if you want a large bandwidth, you need to have a low collector (or emitter) resistor.

3.8.3 Accounting for the Pole

To simplify the compensation technique, an idea consists of including the pole in the open-loop gain during the first ac sweep. That is, rather than checking, at the end, the effect of the pole, let us put it in the open-loop gain transfer function and then apply the compensation technique over it. Figure 3-54 represents a flyback converter where we placed the above optocoupler pole in series with the sweep path (subcircuit X_5). This is a simple Laplace equation, but an equivalent RC filter would do the same job.

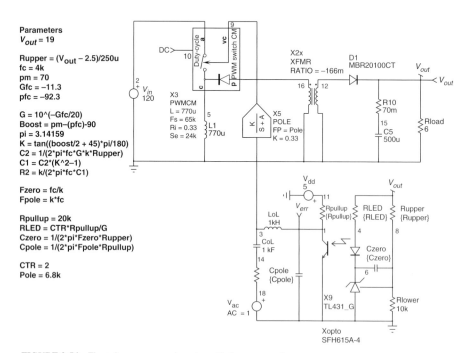

FIGURE 3-54 The pole now appears in series with the sweep path.

The k coefficient accounts for the internal division by 3 that takes place in popular current-mode controllers, such as the UC384X or the NCP1200 series. This application portrays a classical notebook adapter, delivering 19 V at 3 A. After sweep, the Bode plot including the optocoupler pole appears in Fig. 3-55.

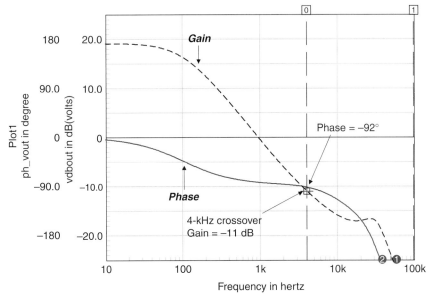

FIGURE 3-55 The complete ac response, inclusive of the optocoupler pole.

From Fig. 3-52a, observing the current in R_{LED} and R_{pullup} via the voltage bias points gives us the model's CTR at this particular bias point: $\text{CTR} = \dfrac{I_{R_{pullup}}}{I_{R_{LED}}} = 2$. Of course, knowing the real CTR from the manufacturer data sheet, you would use this number instead for the compensation calculations. Given the first-order response, we will select a type 2 amplifier and apply the k factor technique to Fig. 3-55. As we operate the converter in CCM, the flyback converter exhibits a RHPZ. The worst case of this RHPZ, meaning how deep it folds back in the low-frequency portion, occurs at the minimum input voltage, the maximum duty cycle, and the highest output current. Depending on this RHPZ position, we will be forced to adopt a crossover point placed at 30% of its lowest location. The RHPZ location for a buck-boost or a flyback converter is the following (see App. 2A):

$$F_{z2} = \frac{(1-D)^2 R_{load}}{2\pi D L_{sec}} \tag{3-92}$$

For a flyback, L_{sec} represents the secondary inductance together with the load R_{load}. It could also be the primary inductance, but R_{load} then needs to be reflected on the primary. Let us use the secondary inductor calculation where L_p is the primary (magnetizing) inductance:

$$L_{sec} = \frac{L_p}{\left(\dfrac{N_p}{N_s}\right)^2} = \frac{770u}{\left(\dfrac{1}{166m}\right)^2} = 21\ \mu\text{H} \tag{3-93a}$$

From Eq. (3-92), we can compute the various RHPZ locations. The simulation gives us the following duty cycle variations for an arbitrary 50 V input span:

$$V_{in} = 100 \text{ V}, D = 0.54$$

$$V_{in} = 150 \text{ V}, D = 0.44$$

With a 6 Ω maximum load, the RHPZ moves between

$$F_{z2,\text{min}} = \frac{(1 - 0.54)^2 \cdot 6}{2\pi \cdot 0.54 \cdot 21u} = 17.8 \text{ kHz} \qquad (3\text{-}93\text{b})$$

and

$$F_{z2,\text{max}} = \frac{(1 - 0.44)^2 \cdot 6}{2\pi \cdot 0.44 \cdot 21u} = 32.4 \text{ kHz} \qquad (3\text{-}93\text{c})$$

From the minimum value, we can see that 30% of it gives a theoretical maximum usable bandwidth of 5 kHz. Going beyond this value could possibly engender oscillations. We will aim for 4 kHz, including some safety margin. Using Eqs. (3-62) to (3-64) helps to calculate all the compensation elements for a 70° phase margin and a compensation at 0 dB of 11.3 dB.

The compensation calculation places a pole and a zero at the following frequencies:

$$f_z = \frac{f_c}{k} = 269 \text{ Hz} \qquad (3\text{-}93\text{d})$$

$$f_p = f_c k = 59 \text{ kHz} \qquad (3\text{-}93\text{e})$$

$$R_{upper} = \frac{19 - 2.5}{250u} = 66 \text{ k}\Omega \qquad (3\text{-}93\text{f})$$

$$C_{zero} = \frac{1}{2\pi R_{upper} f_z} = 3.9 \text{ nF} \qquad (3\text{-}93\text{g})$$

$$C_{pole} = \frac{1}{2\pi R_{pullup} f_p} = 310 \text{ pF} \qquad (3\text{-}93\text{h})$$

$$R_{LED} = \frac{CTR \cdot R_{pullup}}{G} = 11.3 \text{ k}\Omega \qquad (3\text{-}93\text{i})$$

Once these values are passed to the components to which they correspond, a final ac sweep can be performed on V_{err} (the optocoupler collector) to see the obtained bandwidth after compensation. In this case, make sure the optocoupler pole is removed from the path. Figure 3-56 portrays the result. We can observe a 3.8 kHz crossover point together with a 71° phase margin. The gain margin is around 10 dB and could be improved by slightly increasing the ramp compensation level (50% of the off slope in this example, 24 kV/s) or reducing the bandwidth. Since R_{LED} has a rather high value, it is important to check all the bias conditions highlighted in previous lines that enough current circulates in the circuitry. A transient step will finally be performed to verify the converter's stability. Thanks to the automated compensation calculation, it is easy to alter the phase margin and see how the transient response moves. Figure 3-57 portrays the application schematic where the 3 V zener clamps the maximum current excursion, exactly as a UC384X or NCP1200 would do. The output is stepped from 1 to 4 A in 1 μs. Different load responses with three different phase margins appear in Fig. 3-58.

Now we address the practical aspects of this technique:

1. The LED resistor plays an important role here as it sets the midband gain. Few people actually pay attention to it, but if you want to be serious about loop compensation, then you simply cannot ignore its role. The above example states a value for R_{LED} of 11.3 kΩ. Over 19 V, it lets the LED current increase up to $(19 - 1 - 2.5)/11.3k = 1.4$ mA. You now need to check

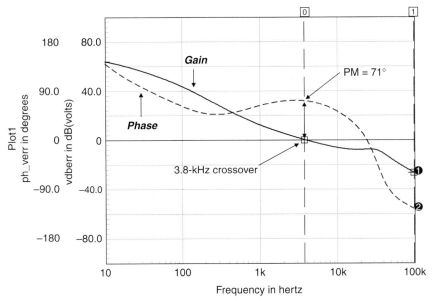

FIGURE 3-56 The final ac sweep is performed after the artificial pole removal.

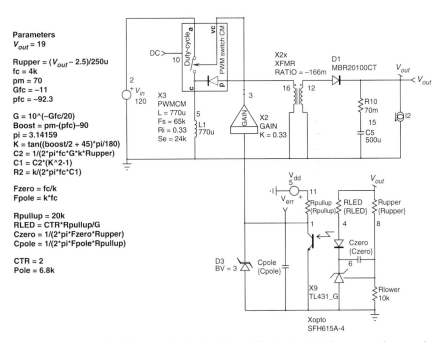

FIGURE 3-57 A step load gives a good indication of the stability obtained with the compensation network.

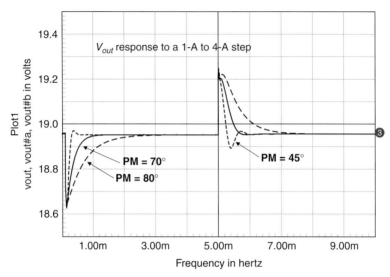

FIGURE 3-58 Altering the phase margin induces different recovery times.

if this current, once multiplied by the optocoupler CTR, allows a complete feedback level swing. Otherwise, you might need to select a higher-CTR device, such as the SFH615A–3 or −4 (100/200% or 160/320%). External bias to force above 1 mA in the TL431 is obviously mandatory here. A possible option lies in externally reducing the pull-up resistor by adding another resistor from the feedback pin either to a reference level, if available, or to the V_{cc} pin. In the above example, reducing R_{pullup} to 8 kΩ (e.g., using an NCP1200) gives an R_{LED} of 4.51 kΩ, with the rest of the components also being updated of course.

2. It is important to discuss the position of the pole imposed by the k factor or any other technique. As the capacitor placing this pole comes in parallel with the internal equivalent optocoupler capacitor, it can sometimes have little effect, except for noise filtering. For instance, the above values recommend a C_{pole} value of 310 pF wired in parallel with the 6.8 nF equivalent pole. Ac sweeping the loop gain with or without C_{pole} shows no difference, and this is normal. However, it the copper link from the feedback pin to the optocoupler is long, a small capacitor of 330 pF wired very close to the integrated circuit between the pin and the controller ground won't do any harm to the noise immunity.

3. Subharmonic oscillations can plague the gain margin if the double poles at $F_{sw}/2$ look too peaky. You can either inject more ramp, with the risk of transforming the current mode into voltage mode, or roll off the gain earlier.

3.9 SHUNT REGULATORS

The vast majority of controllers implement a voltage-type feedback input. That is, the voltage level present on this pin sets the duty cycle level (or the peak current set point). However, you can find on the market some controllers/switchers that expect a feedback in current. This is, for instance, the case for the *TOPSwitch* series from Power Integration, where you need to inject current in the feedback pin. However, at the end, an internal current-to-voltage converter transforms the current information into a voltage, classically compared to a sawtooth. This is what Fig. 3-59 shows.

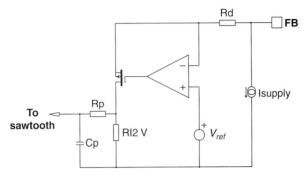

FIGURE 3-59 Some switchers only accept a feedback in current.

The circuit actually implements a zener diode affected by a breakdown voltage of V_{ref} and featuring an internal resistance of R_d. When the voltage on pin FB reaches V_{ref}, the MOSFET starts to conduct and maintains a low-impedance V_{ref} level over the feedback pin. As the feedback loop injects more current, the voltage starts to grow over $RI2V$, instructing for a duty cycle reduction since the sawtooth appears on the inverting pin of the PWM comparator. For compensation purposes, the manufacturer has installed a low-pass filter offering a corner frequency of 7 kHz. Together with an external low frequency zero, it will create the type 2 circuit, somehow limited due to the fixed position of the internal pole.

3.9.1 SPICE Model of the Shunt Regulator

A model of the shunt regulator appears in Fig. 3-60. It now includes a duty cycle limitation between a few percentage points and 67% as stated in the data sheet of the *TOPSwitch* series.

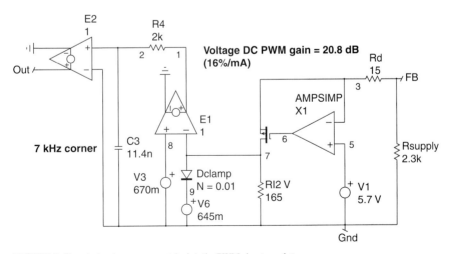

FIGURE 3-60 A simple arrangement depicts the PWM shunt regulator.

If we dc sweep the input of the modulator and observe the output voltage, we can plot the duty cycle level (between 0 V and 1 V—0% to 100%) versus the input current, as Fig. 3-61 shows.

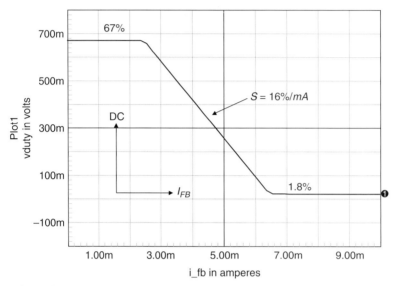

FIGURE 3-61 The duty cycle variation versus the injected current. Note the plateau presence linked to the product current consumption.

3.9.2 Quickly Stabilizing a Converter Using the Shunt Regulator

A DCM flyback converter using the shunt regulator appears in Fig. 3-62. In this circuit, several conditions are necessary to make the converter work properly:

1. The V_{cc} capacitor that creates a pole together with R_d must be large enough to supply the circuit during the start-up sequence. A value around 47 µF is recommended, but it introduces a fixed low-frequency pole of $f_p = \dfrac{1}{2\pi R_d C_{Vcc}} = 225$ Hz.

2. The LED resistor will authorize enough current to power the controller, but it should allow the duty cycle to reduce by injecting at least up to 7 mA in the LED (or more, depending on the CTR). This is design constraint which implies a careful selection of the LED resistor.

From Fig. 3-62, we can run an ac sweep. After the bias point calculation, the duty cycle sets to 23.5%. The open-loop gain together with the recommended value appears in Fig. 3-63. It reveals a crossover frequency around 1.3 kHz, together with a poor phase margin: the circuit is unstable.

To boost the phase around 1 kHz, we have the ability to insert a resistor R_{zero} in series with the V_{cc} capacitor. The addition of this resistor slightly shifts the pole and creates a second zero:

$$f_p = \dfrac{1}{2\pi(R_d + R_{zero})C_{Vcc}} \quad (3\text{-}94)$$

$$f_z = \dfrac{1}{2\pi R_{zero} C_{Vcc}} \quad (3\text{-}95)$$

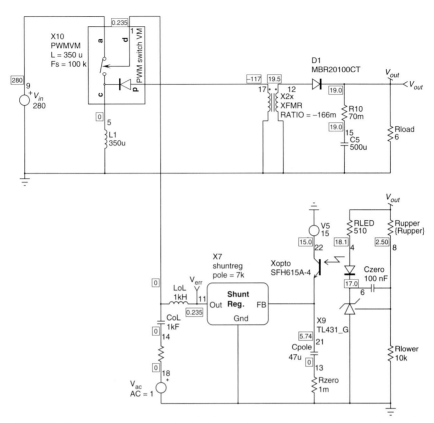

FIGURE 3-62 A complete converter using the shunt regulator together with the PWM switch model.

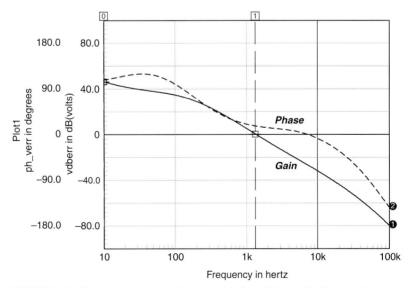

FIGURE 3-63 The circuit shows a poor phase margin and would be unstable if powered.

By inserting a 6 Ω resistor in series with the V_{cc} capacitor (R_{zero} in Fig. 3-62), we locally boost the phase and give the converter a better margin. This is what Fig. 3-64 shows. With shunt-regulated types of circuits, since a lot of variables are fixed, it is more difficult to precisely

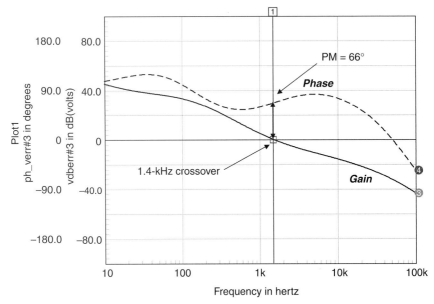

FIGURE 3-64 After the insertion of a zero, the phase margin looks much better.

chose a given bandwidth as we did before. The problem is complicated in CCM mode where you would need to apply the techniques described for the type 3 amplifier using the TL431. The equations have been derived for the type 3 using a shunt regulator but will not be described here.

3.10 SMALL-SIGNAL RESPONSES WITH PSIM AND SIMPLIS

All the above ac responses that we have obtained so far are based on the use of an averaged model dedicated to a particular power topology. Sometimes the topology under study does not have an averaged model, and obtaining its ac response becomes a tedious exercise. An idea could be to use SPICE with a cycle by cycle model operated in transient analysis. It has been recently described in an Intusoft newsletter [6], and despite good results, it requires a long simulation time and a large memory volume. This is so because SPICE permanently adjusts its time step to cope with fast transitions. Hence, simulating a 10 Hz stimulated power supply switching at a 100 kHz pace demands a large collection of stored data points and requires hours of computer power.

Introduced on the market in 1993, Powersim Technology's PSIM [7] uses a different operating mode. On the contrary of SPICE, PSIM keeps a fixed time step during the simulation and considers all the elements perfect with no associated conduction or switching losses. For instance, the voltage generated by an inductance is commonly calculated by $V_L = L\dfrac{dI_L}{dt}$. After a discretization procedure, PSIM finally calculates the voltage by using $V_L = I_L R_{eq}$, where R_{eq} corresponds to an equivalent resistor evaluated by the software at a given time. When the simulator encounters a nonlinear zone, it uses a piecewise linear algorithm which cuts the nonlinear behavior into linear slices. As you can imagine, the simulation time is flashing!

PSIM offers for a few years now, a possibility to ac sweep a cycle by cycle model. Thanks to its execution speed, we can directly build transient models and sweep them using a variable-frequency source. This is what Fig. 3-65a proposes where you can see a classical buck structure operated in voltage mode. The duty cycle is set by the 1.5 V dc source and regulates the output to 15 V (50% dc). The simulator starts to inject a sinusoidal signal of 100 Hz and slowly increments its frequency while analyzing the output voltage. The ac sweep window sets the number of points, and you also have the ability to instruct the sweep processor to slow down in the vicinity of the resonant frequency: it helps to collect a larger number of points and improve the peak plot. This is what Fig. 3-65b shows.

After a few seconds (14 s on a 3 GHz machine), PSIM gives you the ac solution as it appears in Fig. 3-66. Once the open-loop plot is obtained, we can apply the compensation technique and insert an op amp. Figure 3-67 shows how to do that. An ac source is now inserted in series with the sensing bridge. By observing the signals on both sides of the ac source, we can obtain the small-signal response of the buck (see again Figs. 3-6a and 3-6b). The simulator gave the final answer (Fig. 3-68) in 40 s on a 3 GHz computer, which is difficult to beat with SPICE. This is a fully compensated response, showing a crossover frequency of 5 kHz. The k factor compensation uses a type 3 amplifier, and, again, a conditional stability appears around 1 kHz. Manual pole-zero placement would easily get rid of this conditional stability area.

Once the converter is compensated, PSIM can also give you its transient response to an input or output step. The application schematic appears in Fig. 3-69 and shows load generators connected in series with the input voltage or loading the output. The design gives a 5 kHz

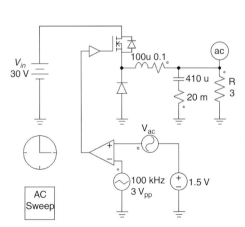

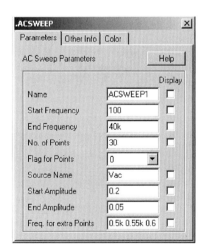

FIGURE 3-65a The buck in open-loop test.   **FIGURE 3-65b** The ac control window.

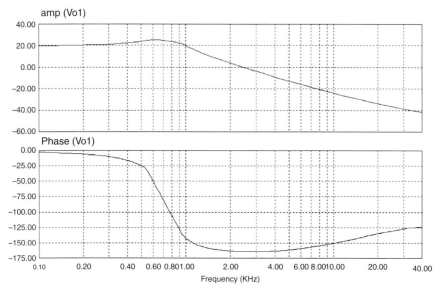

FIGURE 3-66 PSIM gives the small-signal ac solution in a snapshot!

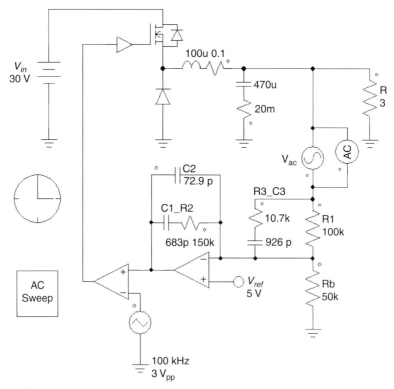

FIGURE 3-67 PSIM can simulate a compensated power supply in a very easy manner.

FEEDBACK AND CONTROL LOOPS 319

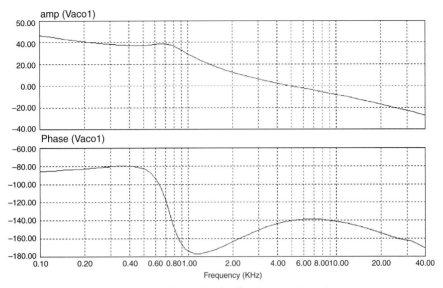

FIGURE 3-68 The buck ac response delivered by PSIM in a few tens of seconds.

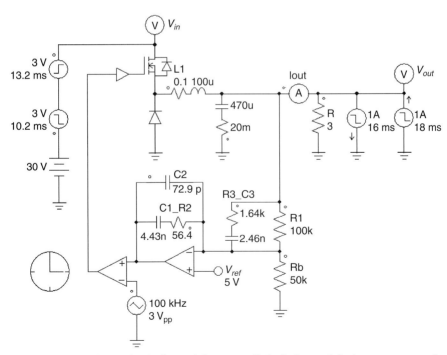

FIGURE 3-69 Adding transient loading can help you to verify the final system behavior once compensated.

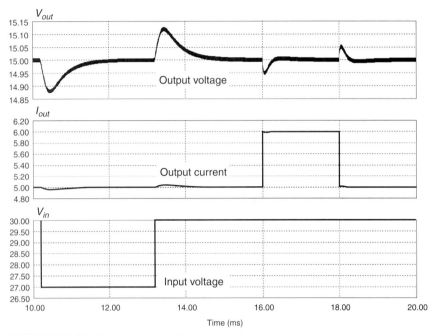

FIGURE 3-70 The final step response of the compensated converter.

crossover frequency with an 80° phase margin. After 4 s of simulation time, PSIM delivers the input and output transient responses. This is what Fig. 3-70 shows, with delayed events. The first step stimulates the converter input, whereas the second one loads the output. In both cases, the converter exhibits good stability.

The above examples are extracted from a tutorial course given by Dr. Richard Redl [8] on behalf of PowerSys, the Powersim Technology European representative [9]. A complimentary PSIM 6.0 demonstration version is included with this book, and some simple examples will run on this limited version. The rest will need the full version. The author wishes to express his thanks to Dr. Redl and PowerSys for their kind agreement to authorize the distribution of the examples and models included in this book.

Transim's SIMetrix/SIMPLIS also lets the user run an ac analysis from a cycle-by-cycle simulation. This is extremely useful in the presence of complex architecture for which you do not necessarily have an average model. This is the case of some resonant structures such as the popular LLC converter. Particularly well suited for power electronics, SIMetrix/SIMPLIS offers all the design environment including a powerful schematic capture tool. The demonstration version included in the book CD is supplied with numerous symbols and will give you a taste of what this state-of-the-art simulation suite can do for you [10]. Figure 3-71 depicts a ringing-choke converter, a classical structure found in cell phone battery chargers.

The software runs a cycle-by-cycle simulation, and thanks to the stimulus source V_3, it can also deliver the ac small-signal response of the converter. This is what Fig. 3-72 shows you where a phase margin of 50° will guarantee a stable operation.

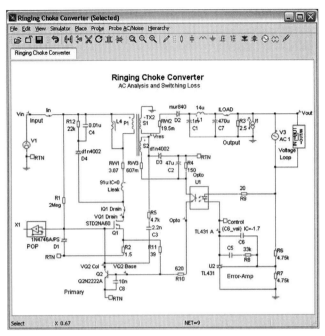

FIGURE 3-71 A ringing-choke converter drawn with SIMetrix schematic capture.

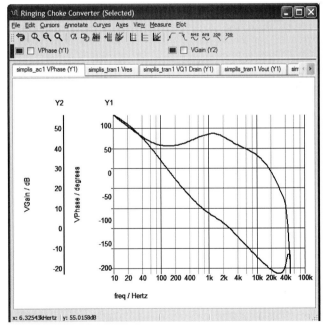

FIGURE 3-72 Because of its powerful engine, the software can deliver ac results in a minute. The V_3 source injects a sinusoidal modulation in series with the feedback loop, and probes are placed on both of its terminals.

WHAT I SHOULD RETAIN FROM CHAP. 3

1. The resistive network adjusting the dc output, R_{lower} and R_{upper}, does not modify the loop gain. In ac, R_{lower} disappears and R_{upper} alone participates in the loop gain. To stabilize a power converter, the keywords are high dc gain, for good output precision, low output impedance, and better input rejection.
   - Optimum crossover frequency for improved transient response with reduced undershoot. Picking up a high crossover frequency can induce instabilities due to noise picking.
   - A phase margin around 70° for a nonovershooting response. 45° is an absolute minimum below which the converter might deliver an unacceptably high ringing response.
   - Gain margin above 10 to 15 dB to cope with unavoidable gain variations (input voltage, load, etc.).
2. A simple method exists to quickly place poles and zeros: the k factor. This method does not represent a panacea since it focuses solely on the crossover frequency, without knowing what is before or after it. Engineering judgment is mandatory to assess the recommend pole and zero placements. If the compensation using type 2 amplifiers in DCM voltage-mode or DCM/CCM current-mode designs usually leads to acceptable results, type 3 compensation for voltage-mode CCM operations can sometimes deliver compensated plots being conditionally stable. In that particular case, manually placing individual poles and zeros gives better results.
3. Compensating the power supply without keeping in mind the optocoupler pole position is useless. Always measure or compute the pole position and include it in the loop design process to avoid last- minute bad surprises.
4. Once it is compensated, *always* measure the loop bandwidth in the laboratory using a network analyzer. Some old equipment such as the HP4195 may do the job perfectly, but more modern devices are also available (see www.ridleyengineering.com, for example). Very often we have seen boards quickly tweaked by using the step-load technique but that later failed in production because ESR dispersions were not accounted for at the early design stage.

REFERENCES

1. R. Erickson and D. Maksimovic, *Fundamentals of Power Electronics*, Kluwers Academic Press, 2001.
2. V. Vorpérian, "Analytical Methods in Power Electronics," Power Electronics Class, Toulouse, France, 2004.
3. D. Venable, "The k-Factor: A New Mathematical Tool for Stability Analysis and Synthesis," *Proceedings of Powercon 10*, 1983, pp. 1–12. Not mentioned in IEEE papers.
4. B. Mamano, "Isolating the Control Loop," SEM 700 seminar series, http://focus.ti.com/lit/ml/slup090/slup090.pdf
5. R. Kollman and J. Betten, "Closing the Loop with a Popular Shunt Regulator," *Power Electronics Technology*, September 2003, pp. 30–36.
6. Intusoft newsletter issue 76, April 2005, http://www.intusoft.com/nlhtm/nl76.htm
7. http://www.powersimtech.com/
8. R. Redl, "Using PSIM for Analyzing the Dynamic Behavior and Optimizing the Feedback Loop Design of Switch Mode Power Supplies," a training course on behalf of POWERSYS.
9. POWERSYS, Les Grandes Terres, 13650 Meyrargues, France; www.powersys.fr
10. http://www.transim.com/

APPENDIX 3A AUTOMATED POLE-ZERO PLACEMENT

Even if some software allows one to automate the calculation of the compensating elements for type 2 and 3 error amplifiers, some designers prefer to compute these values in a different way. An Excel spreadsheet represents an alternative solution. We have compiled all possible combinations of type 1, 2, and 3 amplifiers, using either classical op amps with k factor pole-zero placement or manual placement, leaving pole-zero locations to your choice. We did not forget the TL431 for type 2 and 3, this being offered later in manual placement as well. An OTA configuration has even been added for PFC compensations.

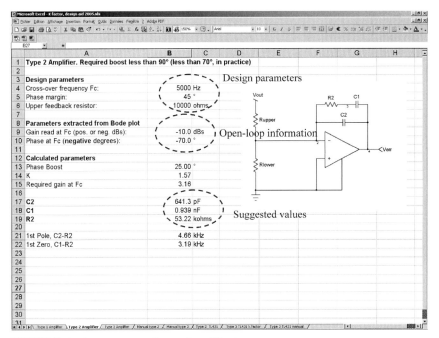

FIGURE 3A-1 A screen capture of a type 2 amplifier sheet.

Figure 3A-1 shows a typical screenshot of the type 2 amplifier calculation sheet. The design methodology used on this design follows the guidelines given below:

1. Extract the converter open-loop Bode plot at the operating point of interest. It is usually in the worst-case conditions, e.g., minimum input voltage and maximum output current. This open-loop plot can be obtained through simulation (SPICE averaged models, PSIM) or by using a network analyzer on the bench.
2. Select a crossover frequency f_c and measure the needed gain at the chosen crossover frequency. In the example, it was a 5 kHz cutoff frequency, and the attenuation was -10 dB at this point. Then extract the phase rotation at f_c (here $-70°$) and feed the sheet with this value.
3. The phase margin needed depends on the system response you want. A 70° phase margin will give you rock-solid stability, but lower values can also be experimented, with 45° being the absolute minimum.

324 CHAPTER THREE

4. Then capacitor and resistor values appear, with suggested values for pole-zero locations. You can paste these values into the various *IsSpice/OrCAD* templates and check for the right compensated Bode plot shape.
5. Once compensation is done, always check for conditional stability zones with SPICE as we explored in the text. Sweep all stray elements, input voltages, and loads between their minimum and maximum values to check for the right phase/gain margin in all cases.

Let's illustrate the method with a simple example.

Figure 3A-2 shows a voltage-mode boost converter operated in CCM. This 15 V converter powers a 5 Ω load from a 10 V input source featuring an operating duty cycle of 33% (obtained

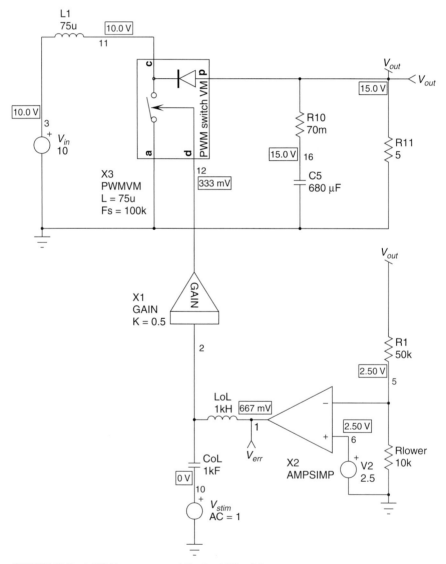

FIGURE 3A-2 A CCM boost converter delivering 15 V at 3 A.

further to a bias point analysis). From this electrical diagram, we are going to launch an ac analysis to reveal the open-loop Bode plot. This is what Fig. 3A-3 shows. The crossover frequency we can reach depends on the RHPZ location and the resonant frequency of the LC filter. From Chap. 2, we know where the RHPZ is hidden.

$$f_{z2} \approx \frac{RD'^2}{2\pi L} = \frac{5 \times 0.67^2}{2 \times 3.14 \times 75u} = 4.7 \text{ kHz} \quad (3\text{A-1})$$

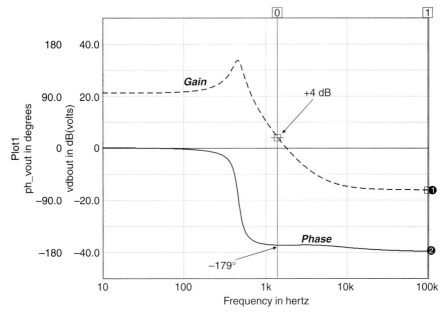

FIGURE 3A-3 The open-loop Bode plot shows us a gain excess of +4 dB at 1.4 kHz.

As we explained, given the phase stress brought by the LC filter, it is important to assess the worst case resonant frequency to place the crossover point far enough from the peak. Otherwise, the phase stress brought by the filter might be difficult to compensate at the crossover frequency. Usually, selecting a crossover point above three times the resonant frequency gives sufficient margin. The resonant frequency of the CCM boost depends on the duty cycle value:

$$f_0 \approx \frac{D'}{2\pi\sqrt{LC}} = \frac{0.67}{2 \times 3.14 \times \sqrt{75u \times 680u}} = 472 \text{ Hz} \quad (3\text{A-2})$$

The crossover frequency will be selected below 30% of the RHPZ and above three times the resonant frequency. Therefore, a 1.4 kHz seems to be a reasonable value. Reading Fig. 3A-3 shows an excess of gain of +4 dB at 1.4 kHz. The capacitor ESR contributes a zero located at

$$f_{zero} = \frac{1}{2\pi r_{Cf} R} = \frac{1}{6.28 \times 70m \times 680u} = 3.3 \text{ kHz} \quad (3\text{A-3})$$

We are going to place a double zero slightly below the resonant frequency (350 Hz), a pole at 3.3 kHz (to compensate the ESR zero), and finally a high-frequency pole at 50 kHz ($F_{sw}/2$).

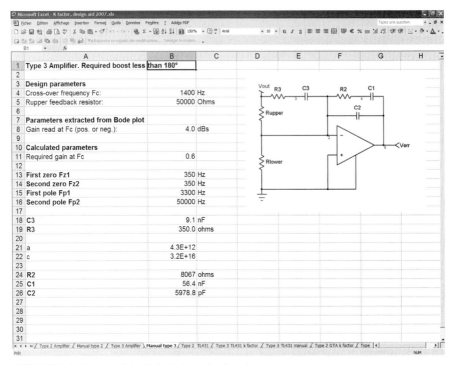

FIGURE 3A-4 The sheet fed with the design and collected parameters.

Let's use the manual type 3 Excel spreadsheet where we enter all the design parameters, as shown in Fig. 3A-4. We then use the suggested values and paste them in the compensation network placed over the op amp model X_2. After sweeping the compensated converter, we unveil Fig. 3A-5, which is evidence of a 1.4 kHz crossover frequency and a phase margin of 55°. Different pole-zero positions can quickly be explored to see their effects on the resulting phase margin, for instance.

APPENDIX 3B   A TL431 SPICE MODEL

This appendix details how to create a simple SPICE model for the TL431. Based on simple elements, it includes enough information to satisfy first-order simulations.

3B.1   A Behavioral TL431 Spice Model

The TL431 data sheet frontpage displays the internal circuitry of the shunt regulator. Based on this simplified circuit (it does not include the V_{be} curvature correction, for instance), it would be possible to create a model properly predicting the TL431 behavior, including bias current information. However, using numerous bipolar transistors could hamper the convergence of

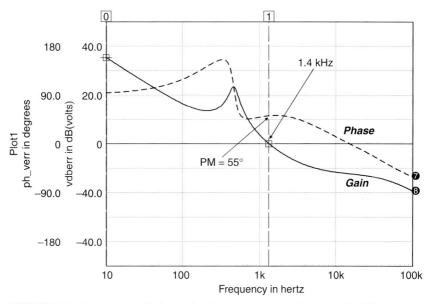

FIGURE 3A-5 The compensated Bode plot showing the right crossover frequency of 1.4 kHz.

the circuit under simulation. That is the reason why we selected a simpler approach. Using Fig. 3-35a, a behavioral model of the TL431 can be derived. This is what Fig. 3B-1 shows.

This figure shows a transconductance amplifier G_1 biasing a Darlington output. An external zener diode fixes the breakdown voltage of the device, whereas C_1 and C_2 were tweaked to get the right frequency response. The Q_2 area number (number of bipolars stacked in parallel in an integrated circuit) has been adjusted to obtain the right output impedance, as defined by the data sheet. D_2 mimics the reverse biasing of the TL431 and its breakdown voltage. It has

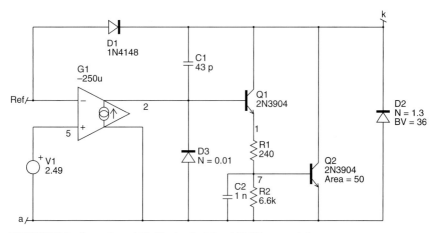

FIGURE 3B-1 A generic model built using the internal TL431 representation.

been added for the sake of respecting the data sheet only, as the designer must ensure that this breakdown voltage is never exceeded in reality.

After encapsulation in the adequate graphical symbol, it is time to test the model versus the data sheet typical characterization curves. This is an important step that will enable you to decide whether the derived model precision suits your needs.

3B.2 Cathode Current Versus Cathode Voltage

This test wires the TL431 as if it were a classical 2.5 V zener diode with the reference pin connected to the cathode. The bias takes place via an external current-limiting resistor. As Fig. 3B-2a shows, the positive current brutally increases as the input voltage reaches the reference level. On the contrary, in negative bias, the zener starts to conduct in the opposite direction, but its dynamic resistance is rather poor. Figure 3B-2b shows what the model delivers. There is good

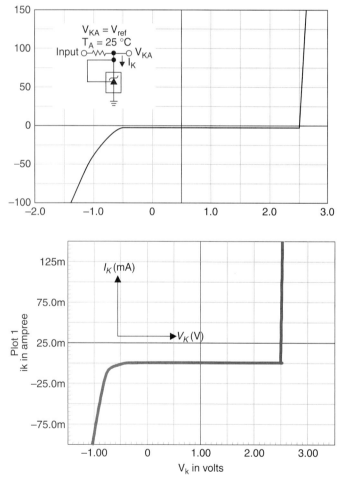

FIGURE 3B-2a, b Testing the controlled zener behavior of the generic model.

agreement, despite a slight variation in the negative portion. Again, this is not a problem as normal design never uses the TL431 in this region:

3B.3 Output Impedance

The data sheet specifies a static output impedance, 0.22 Ω typically, but nothing indicates the bias current at which it has been measured. This statement also holds for the ac output impedance, as shown by Fig. 3B-3a. Nonetheless, a similar plot was obtained when sweeping the model between 1 kHz and 10 MHz (Fig. 3B-3b).

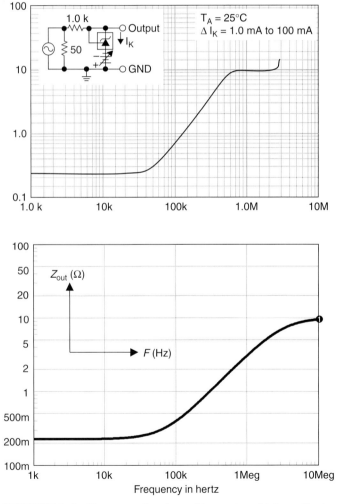

FIGURE 3B-3a, b   The ac output impedance of the generic model. $I_{bias} = 40$ mA.

3B.4 Open-Loop Gain

The test circuit uses the TL431 wired as a controlled zener. An ac signal couples to the reference pin via a dc block capacitor. The simulated bandwidth matches the 0 dB crossover but is unable to predict the flat response at the beginning of the sweep (Fig. 3B-4a and b). However, it does not impact the transient response as the next plots illustrate.

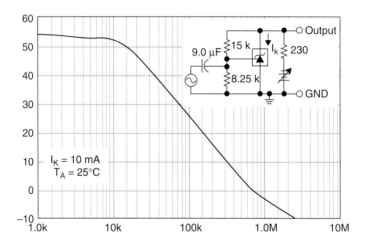

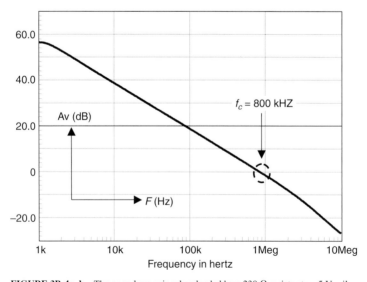

FIGURE 3B-4a, b The open-loop gain when loaded by a 230 Ω resistor to a 5 V rail.

3B.5 Transient Test

An interesting test consists of testing the TL431'a ability to maintain a fixed voltage on its cathode, despite input current pulses. The results highlight a little overshoot, but they look in line with the characterization curves (Fig. 3B-5a and b).

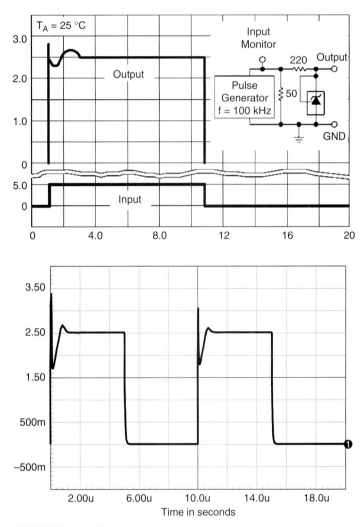

FIGURE 3B-5a, b The transient response to external current pulses.

3B.6 Model Netlist

The netlist appears below and can be used equally well with *PSpice*, *IsSpice*, or any other simulator as all elements are SPICE2-compatible:

```
.SUBCKT TL431 a k ref
C1 7 a 1n
Q1 k 2 1 QN3904
R1 1 7 240
R2 7 a 6.6k
Q2 k 7 a QN3904 50
D1 ref k D4mod
D2 a k D2mod
V1 5 a DC=2.49
D3 a 2 D3mod
C4 k 2 43p
G2 a 2 5 ref -250u
*
.MODEL QN3904 NPN AF=1.0 BF=300 BR=7.5 CJC=3.5PF CJE=4.5PF
+ IKF=.025 IS=1.4E-14 ISE=3E-13 KF=9E-16 NE=1.5 RC=2.4
+ TF=4E-10 TR=21E-9 VAF=100 XTB=1.5
.MODEL D3mod D N=0.01
.MODEL D2mod D BV=36 CJO=4PF IS=7E-09 M=.45 N=1.3 RS=40
+ TT=6E-09 VJ=.6V
.MODEL D4mod D BV=100V CJO=4PF IS=7E-09 M=.45 N=2 RS=.8
+ TT=6E-09 VJ=.6V
*
.ENDS
*********
```

APPENDIX 3C   TYPE 2 MANUAL POLE-ZERO PLACEMENT

This appendix details how to manually place poles and zeros on a type 2 amplifier. Figure 3C-1 shows the classical amplifier configuration, used when the power stage maximum phase shift reaches $-90°$.

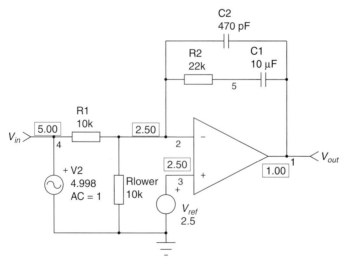

FIGURE 3C-1   A classical type 2 amplifier featuring one pole and one zero.

Its transfer function can be easily obtained by calculating the equivalent impedance Z_1 made of C_2, R_2, and C_1, further divided by the simple resistor R_1 (remember, R_{lower} does not play a role in the *closed-loop* ac configuration):

$$Z_1 = \frac{1}{sC_2} \| \left(R_2 + \frac{1}{sC_1}\right) = \frac{\frac{1}{sC_2}\left(R_2 + \frac{1}{sC_1}\right)}{\frac{1}{sC_2} + \left(R_2 + \frac{1}{sC_1}\right)} = \frac{R_2 + \frac{1}{sC_1}}{1 + sR_2C_2 + \frac{C_2}{C_1}} \quad (3C\text{-}1)$$

If we now divide Z_1 by R_1 and consider that $C_2 \ll C_1$, we have the simplified gain expression

$$G(s) \approx \frac{R_2}{R_1} \frac{1 + \frac{1}{sR_2C_1}}{1 + sR_2C_2} = \frac{R_2}{R_1} \frac{1 + \frac{\omega_z}{s}}{1 + \frac{s}{\omega_p}} \quad (3C\text{-}2)$$

where

$$\omega_z = \frac{1}{R_2C_1} \quad (3C\text{-}3)$$

and

$$\omega_p = \frac{1}{R_2C_2} \quad (3C\text{-}4)$$

The ratio $\frac{R_2}{R_1}$ brings the midband gain we actually look for at the crossover frequency f_c. Equation (3C-2) can be reformulated in the frequency domain via its module definition:

$$G(f) \approx \left| \frac{R_2}{R_1} \frac{1 + \frac{s_z}{s}}{1 + \frac{s}{s_p}} \right| \quad (3C\text{-}5)$$

A mathematical spreadsheet such as Mathcad can help us to quickly plot the transfer function described by this equation (Fig. 3C-2).

Using Eq. (3C-5) and its module, we now extract the value of R_2 that will bring us the desired gain (or attenuation) at the selected crossover frequency:

$$R_2 = \frac{\sqrt{(f_c^2 + f_z^2)(f_c^2 + f_p^2)}}{f_c^2 + f_z^2} \frac{R_1 G f_c}{f_p} \quad (3C\text{-}6)$$

As a simple example, Fig. 3C-3 shows the automated SPICE example of manual pole-zero placement. We purposely largely split them to form a flat midband gain region. One pole is selected at 50 kHz where the zero occurs at 150 Hz. The needed midband gain is +15 dB in this example. Figure 3C-4 displays the resulting Bode plot, showing a midband gain of 15 dB at 2 kHz, as we wanted.

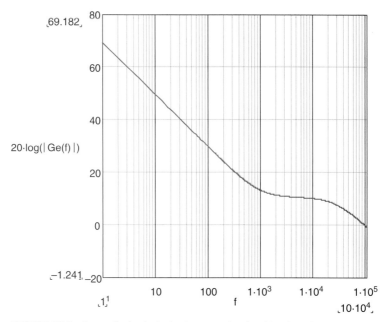

FIGURE 3C-2 An amplitude plot in the frequency domain with pole and zero, respectively, placed at 27 kHz and 880 Hz, with a midband gain of 10 dB.

Parameters

Rupper = 10k
fc = 2k
Gfc = –15

G = 10^(–Gfc/20)
pi = 3.14159

fp = 50k
fz = 150

a = sqrt((fc^2 + fp^2)*(fc^2 + fz^2))
c = (fc^2 + fz^2)

R2 = (a/c)*Rupper*fc*G/fp

C2 = 1/(2*pi*fp*R2)
C1 = 1/(2*pi*fz*R2)

With this type 2 amplifier, you can manually place the pole and zero at locations of your choice and the software will evaluate the value of R2 to get the right gain at Fc.

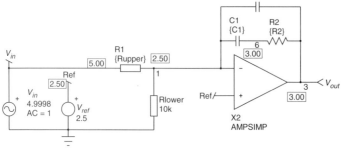

FIGURE 3C-3 A simulation example with pole and zero manually selected.

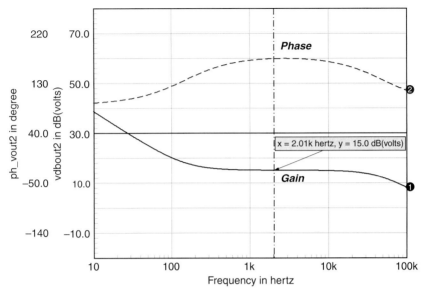

FIGURE 3C-4 Simulation results showing the right gain in the flat region.

APPENDIX 3D UNDERSTANDING THE VIRTUAL GROUND IN CLOSED-LOOP SYSTEMS

In designing the feedback loop of the converter, you often use a divider network to obtain the right dc operating point. Figure 3D-1 shows a typical proportional error amplifier. When you derive the equations, you actually discover that the resistor R_{lower} plays a role solely to adjust the dc bias point but goes off in ac.

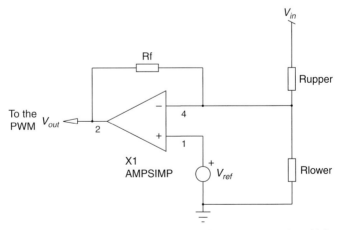

FIGURE 3D-1 A typical op amp used in the feedback loop together with its divider network.

To derive the loop gain brought by this structure, we need to make a few assumptions:
- The op amp open-loop gain A_{OL} is large.
- A direct implication of the above fact relates to the voltage difference between pin $-$ and $+$. Called ε and equal to $\varepsilon = V(+) - V(-) = \dfrac{V_{out}}{A_{OL}}$, its value is negligible: both inputs thus exhibit the same potential when the op amp operates in closed-loop conditions. Should you remove R_f, this is no longer true.

Applying the superposition theorem gives us the transfer function in a snapshot:

1. V_{in} is short-circuited to ground:

$$V_{out} = V_{ref}\left(\frac{R_f}{R_{upper}||R_{lower}} + 1\right) \tag{3D-1}$$

2. V_{ref} is short-circuited to ground, $V(-)$ is null, and R_{lower} goes off the picture. The output voltage is that of a classical inverting op amp:

$$V_{out} = -\frac{R_f}{R_{upper}}V_{in} \tag{3D-2}$$

Adding Eqs. (3D-1) and (3D-2), we obtain the output voltage

$$V_{out} = V_{ref}\left(\frac{R_f}{R_{upper}||R_{lower}} + 1\right) - \frac{R_f}{R_{upper}}V_{in} \tag{3D-3}$$

This is a dc equation. To obtain the ac equation, simply derive it with respect to V_{in}. As V_{ref} is constant, it becomes zero in ac. We then have

$$\frac{V_{out}(s)}{V_{in}(s)} = -\frac{R_f}{R_{upper}} \tag{3D-4}$$

As Eq. (3D-4) demonstrates, the bottom resistor R_{lower} does not play a role in the transfer function. However, it fixes the dc point as Eq. (3D-3) shows.

3D.1 Numerical Example

Figure 3D-2 shows the same configuration as above, together with an op amp lacking external compensation. Remember, if there is no compensation on the op amp (dc or ac), the virtual ground is lost and the divider ratio plays a role in this particular case.

In the Fig. 3D-2, the dc gain solely depends on R_1 and R_3 a) the open-loop gain does not play a role b) R_2 fixes the dc bias, that is all. We can thus read a dc gain ($s = 0$) of 20 dB and an output voltage defined by Eq. (3D-3) which gives 2.49 V. In the bottom case, the feedback resistor has gone and so has the virtual ground. The output voltage is now fixed by the divider network and equals

$$V_{out} = \left(2.5 - V_{in}\frac{R_5}{R_5 + R_6}\right)A_{OL} \tag{3D-5}$$

FEEDBACK AND CONTROL LOOPS 337

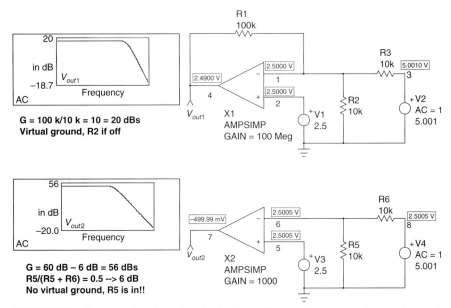

FIGURE 3D-2 Op amp configurations where the virtual ground is present in one case and gone in the second.

Using numerical values, we have $V_{out} = \left[2.5 - \left(5.001 \times \dfrac{10}{20}\right)\right] \times 1000 = -500$ mV. The dc gain is simply

$$\dfrac{V_{out(0)}}{V_{in(0)}} = 20 \log_{10}\left(\dfrac{R_5}{R_5 + R_6}\right) + 20 \log_{10} A_{OL} \tag{3D-6}$$

It is 56 dB, as shown on the left side of the figure.

3D.2 Loop Gain Is Unchanged

It is little known that changing the divider ratio does not affect the ac loop gain or the crossover frequency: we can thank the op amp virtual ground for this nice performance. Figure 3D-3a and b represents two classical buck stages followed by a type 3 compensator. The aim is to deliver 5 V, and two different dc configurations are adopted:

1. R_{upper} and R_{lower} are equal to 10 kΩ and $V_{ref} = 2.5$ V.
2. $R_{upper} = 10$ kΩ, $R_{lower} = 50$ kΩ, and $V_{ref} = 4.166$ V.

In the first configuration, the divider ration α equals 0.5. In the second, $\alpha = 0.83$. All dc points are equal, except, of course, those around the reference nodes as displayed on the schematic. Then an ac sweep is performed, and its result appears in Fig. 3D-4.

In this figure, it is not possible to distinguish both ac sweep results. The ac loop gain stays unchanged despite the modifications in the divider ratio. The op amp open-loop gain is around 90 dB in this example and operates in its linear region.

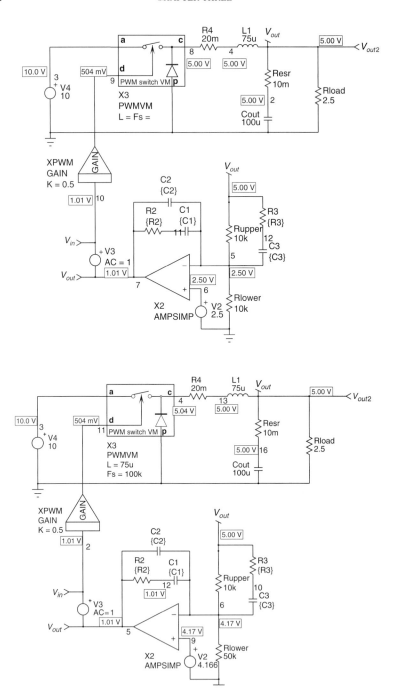

FIGURE 3D-3a and b The same buck stage but featuring two different dc configurations on the divider network and the reference voltage. Both offer a similar output level and an identical duty cycle.

FEEDBACK AND CONTROL LOOPS

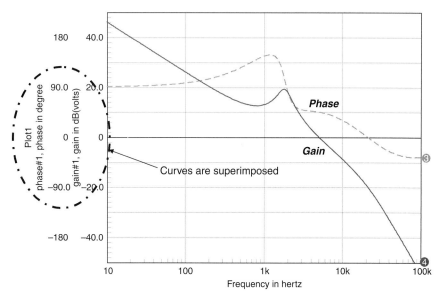

FIGURE 3D-4 Both curves perfectly superimpose, so the ac gain is unchanged by the modification in the divider network!

Again, it is important to note that the gain curve remains unaffected by the divider network as long as we have a virtual ground. This is no longer the case for a loop closed with a transconductance amplifier (OTA) as in PFC stages, for instance (MC33262).

CHAPTER 4
BASIC BLOCKS AND GENERIC SWITCHED MODELS

4.1 GENERIC MODELS FOR FASTER SIMULATIONS

Often you want to simulate a circuit idea quickly or test the results given by component values. But before starting to capture the electrical diagram, you must ask yourself, "Do I need full-order models or are simple models good enough for what I need to study?" We have listed in bullets the implications of using full-order models versus simpler models.

- **Execution speed**: Complexity implies a lot of internal nodes, SPICE primitives, or *.model* that load the simulation engine and lengthen the computational time. Why bother with an op amp consumption current if you just need the primitive function of an error amplifier affected by an origin pole and an open-loop gain? A simplified op amp requires 10 elements, and you can tailor its open-loop gain or origin pole to see the impact on the overall operation.
- **Availability**: The controller or the topology you have in mind does not necessarily have a dedicated model because this is a recent circuit or the semiconductor vendor did not create any subcircuit for it. Build your own generic circuit library and see how changing some key default values affects the circuit operation.
- **Convergence**: A model featuring a lot of internal subcircuits or primitives is inherently prone to convergence problems. Simplicity is the key for quick and efficient SPICE simulations!
- **Flexibility**: You need a voltage-controlled oscillator, but an LM566 does not correspond to what you are looking for (by the way, who remembers the LM566?). Build your own VCO and add to or tailor its behavior as you wish.

This chapter describes some key elements needed when you want to simulate power supplies. Of course, once the circuit works properly, you can easily replace a generic op amp with the final model you are going to pick. The debugging starts with simple models, but it is your choice to make it more complicated, closer to the real world. We first build the toolbox, made of generic models and subcircuits, and then see how to link them to build generic PWM controllers. Let us start with in-line equations, the heart of analog behavioral modeling (ABM).

4.1.1 In-Line Equations

The nonlinear controlled source, or B element, is part of the Berkeley SPICE3 engine, which was released to the public domain in 1986. Depending on the compatibility of the SPICE3 simulator, the corresponding syntax may vary significantly. B elements can be linear or nonlinear current or voltage sources. Some vendors have expanded the B element syntax to include BOOLEAN and IF-THEN-ELSE functions. In *IsSpice*, the writing of I or V math equations using B elements is the same because both are SPICE3-compatible. For example, current/voltage generators whose current depends on various nodes can be expressed as

```
B1 1 0 V = V(2,3) * 4 ; multiply voltage of (2,3) by four and
; delivers voltage
B1 1 0 I = I(V1) * 5   ; the current flowing through V1 is multi
; plied by 5
```

PSpice has departed from the Berkeley standard and uses a different syntax. *PSpice* modifies the dependent voltage-controlled sources (E and G elements) to achieve the effects of the SPICE3 B elements. The equivalent *PSpice* examples are as follows:

```
E1 1 0 Value = { V(2,3) * 4 }
G1 1 0 Value = { I(V1) * 5 }
```

SPICE introduces the notion of conditional expression. That is, if a condition is fulfilled, then an action is undertaken. Or an additional condition is tested to end up with a final action. These expressions are called *if-then-else* expressions:

```
IsSpice   B1 1 0 V = V(3) > 5 ? 10 : 100m
PSpice    E1 1 0 Value = { IF ( V(3) > 5, 10, 100m ) }
```

You read this as follows: IF V(3) is greater than 5 V, THEN V(1,0) = 10 V ELSE V(1,0) = 100 mV. It is not recommended to pass the units in these ABM expressions. Expressions can now be nested and arranged as you want. Here is a limiter device clamping the input voltage between 5 V and 100 mV:

```
IsSpice   B1 1 0 V = V(3) > 5 ? 5 : V(3) < 100m ? 100m : V(3)
PSpice    E1 1 0 Value = { IF ( V(3) > 5, 5, IF ( V(3) < 100m, 100m,
+ V(3) ) ) }
```

You read this as follows: IF V(3) is greater than 5 V, THEN V(1,0) = 5 V ELSE IF V(3) is less than 100 mV, THEN V(1,0) = 100 mV ELSE V(1,0) = V(3).

Let us go to more complicated expressions where you will see that *IsSpice* handles the equations in a real easy way.

IsSpice

```
B1 69 14 V = V(27,14) > V(18,14)/2 ? V(18,14) : V(26,14) > 0.44 ?
+ V(18,14) : (V(13,14)+V(26,14)+V(12,14)) > V(31,14) ? V(18,14)
+ : 0
```

PSpice

```
E1 69 14 VALUE = { IF ( V(27,14) > V(18,14)/2, V(18,14), IF
+ ( V(26,14) > 0.44, + V(18,14), IF ( (V(13,14)+V(26,14)+V(12,14))
+ > V(31,14), V(18,14), 0 ) ) ) }
```

Note the number of parentheses for *PSpice* which complicates the writing of nested expressions compared to *IsSpice*. But this is just my opinion.

If *IsSpice* accepts parameter passing for SPICE primitives such as E (voltage-controlled voltage source) or G (voltage-controlled current source), *PSpice* does not and you need to adapt them with a VALUE keyword:

```
IsSpice   E1 1 2 3 4 { gain }
          G1 1 2 3 4 { gm }
```

```
PSpice    E1 1 2 VALUE = { V(3,4)*gain }
          G1 1 2 VALUE = { V(3,4)*gm }
```

A B element source is essentially a zero time span. In other words, the transition from one state to the other is immediate. This characteristic may create convergence problems in transitions associated with these perfect sources. We recommend that you tailor the output switching times in a more realistic manner via a simple *RC* network, for instance (10 Ω/100 pF is good).

4.2 OPERATIONAL AMPLIFIERS

Unless you really need to simulate the true operational amplifier (op amp) used in the circuit (including offset voltages, bias currents, slew rates, and so on), simple generic models are, most of the time, sufficient to highlight first-order effects. They simulate fast, and by feeding them with parameter values, you tailor them to fit the actual op amp specs. Figure 4-1a and b describe how to build very simple devices.

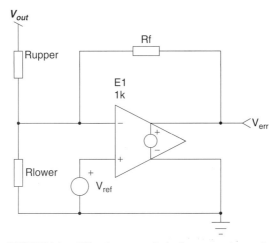

FIGURE 4-1a This voltage-controlled voltage source is a possible simple solution.

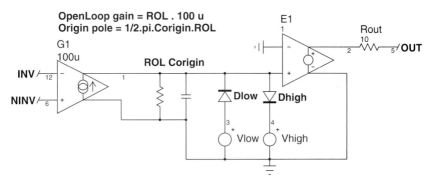

FIGURE 4-1b A better option uses a transconductance amplifier and two clamping diodes.

Figure 4-1a is attractive because of the few elements it implies. However, we would recommend its implementation where ac sweeps only are being used (e.g., with Ridley models). Why? Simply because of the lack of output clamping levels that could cause convergence problems when the op amp is pushed into its upper or lower limits:

$$V_{out} = \left[V_{(+)} - V_{(-)}\right] A_{OL} = \left[V_{(+)} - V_{(-)}\right] \cdot 1000 \qquad (4\text{-}1)$$

where A_{OL} represents the open-loop gain, 60 dB for this particular example. If the difference between both inputs reaches 1 V for any reason, the source develops 1 kV. Trying to clamp the output by an in-line equation will bring convergence problems when the op amp starts to clamp.

If clamping represents a real concern, then the example of Fig. 4-1b offers a better structure by combining a transconductance amplifier and two clamping diodes. If you alter the diode emission coefficient N and set it to 0.01, then you obtain a perfect diode with a null V_f. By tweaking the V_{clamp} source in series with these perfect diodes, you select the lowest and highest output levels of the generic op amp. Because you only clamp a few hundred microamperes through these elements, it does not bother the simulator. Could you do it any simpler? No.

The open-loop gain is set by adjusting the gm of the voltage-controlled current source and the R_{OL} resistor (R_1 in the netlist). With an arbitrary 100 µS transconductance value, then a 10 MΩ resistor gives a 60 dB open-loop gain. Finally, tailor the C_{origin} capacitor (C_1 in the netlist) to position the origin pole. If the upper and lower clamping levels are not very precise, they surely are good enough for the vast majority of cases. Below is the corresponding netlist, compatible with both *PSpice* and *IsSpice*:

```
.SUBCKT AMPSIMP 1 5 7    params:  POLE=30  GAIN=30000  VHIGH=4V
+ VLOW=100mV
*               + -  OUT
G1 0 4 1 5 100u
R1 4 0 {GAIN/100u}
C1 4 0 {1/(6.28*(GAIN/100u)*POLE)}
E1 2 0 4 0 1
Rout 2 7 10
Vlow 3 0 DC={VLOW}
Vhigh 8 0 DC={VHIGH}
Dlow 3 4 DCLP
Dhigh 4 8 DCLP
.MODEL DCLP D N=0.01
.ENDS
```

The working parameters are therefore

POLE	In hertz, it places the origin pole by calculating the right value for C_1.
GAIN	It represents the open-loop gain. 30000 corresponds to $20 \cdot \log_{10} 30000 = 89.5$ dB.
VHIGH and VLOW	Respectively, they limit the highest and lowest output levels.

4.2.1 A More Realistic Model

This model accounts for the output current capability of the operational amplifier. In some PWM integrated circuits (ICs) hosting an error amplifier, the internal op amp is an open-collector type: it can sink current, but its limitation in current source allows the user to easily bypass it via an optocoupler connection. That is the case for the UC384X controller family, for instance.

Figure 4-1c describes the model we adopted in all the following generic subcircuits. The parameters that you enter via the schematic capture automatically adjust the internal component value to shape the op amp performance as needed. The netlist appears below and works for

BASIC BLOCKS AND GENERIC SWITCHED MODELS

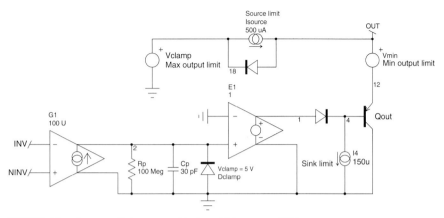

FIGURE 4-1c A more realistic op amp with given sink/source capabilities.

both *IsSpice* and *PSpice*. Some components have been added to ease convergence such as R_{in} which fixes the input impedance.

```
.SUBCKT ERRAMP 20 8 3 21 params: ISINK= 15M ISOURCE=500U VHIGH=2.8
+ VLOW=100M POLE=30 GAIN=30000
*   +  -  OUT GND
RIN 20 8 8MEG
CP 11 21 {1/(6.28*(GAIN/100U)*POLE)}
E1 5 21 11 21 1
R9 5 2 5
D14 2 13 DMOD
IS 13 21 {ISINK/100}; mA
Q1 21 13 16 QPMOD
ISRC 7 3 {ISOURCE}; uA
D12 3 7 DMOD
D15 21 11 DCLAMP
G1 21 11 20 8 100U
V1 7 21 {VHIGH-0.6V}
V4 3 16 {VLOW-38MV}
RP 11 21 {GAIN/100U}
.MODEL QPMOD PNP
.MODEL DCLAMP D (RS=10 BV=10 IBV=0.01)
.MODEL DMOD D (TT=1N CJO=10P)
.ENDS
```

Its use is similar to that of the previous type of op amp except that

- *ISINK* tailors the current that the op amp can sink to ground before falling out of regulation.
- *ISOURCE* represents the maximum current the device can deliver before dropping its output voltage.

A quick simulation confirms the sink/source capability of our generic op amp (Fig. 4-1d).

4.2.2 A UC384X Error Amplifier

The above circuit lends itself to simulations where a single pole serves the designer's needs. In some popular controllers, such as the UC384X series, there is a second pole, as seen in Fig. 4-1e.

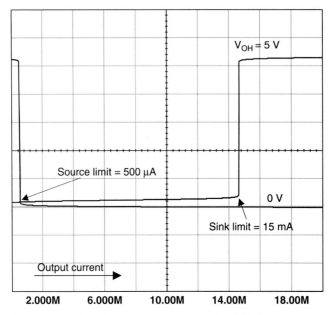

FIGURE 4-1d Simulations results showing the limited sink/source capability.

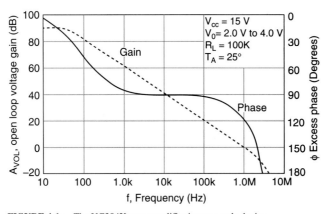

FIGURE 4-1e The UC384X error amplifier is a two-pole device.

The open-loop gain reaches 90 dB with a unity gain bandwidth of 1 MHz, and the phase keeps decreasing after 1 MHz. We need to slightly update the original circuit by buffering the first pole and tweaking the sink and source capabilities. This is what Fig. 4-1f shows. Once it is swept in open-loop conditions, Fig. 4-1g appears, confirming the 1 MHz gain bandwidth as detailed in the data sheet. The dc sources were also tweaked to offer adequate sink and source capabilities. Finally, a 2.5 V reference voltage represents the UC384X internal architecture. The netlist appears below, fully compatible with *IsSpice* and *PSpice*:

```
.SUBCKT OP384X In out gnd
Vmin out 2 DC=0.6
```

BASIC BLOCKS AND GENERIC SWITCHED MODELS 347

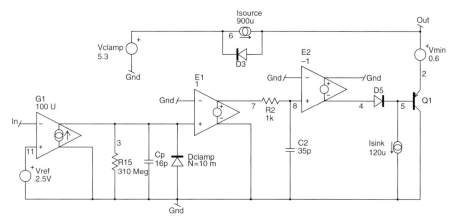

FIGURE 4-1f The original circuit is updated to place a second pole around 5 MHz.

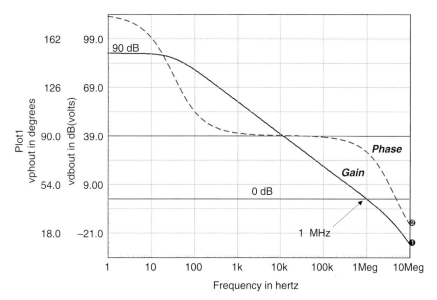

FIGURE 4-1g Open-loop Bode plot of the UC384X error amplifier.

```
R15 3 gnd 310Meg
Cp 3 gnd 16p
Dclamp gnd 3 DCLP3
D5 4 5 DCLP1
G1 gnd 3 11 In 100U
Isink 5 gnd DC=120u
Isource 6 out DC=900u
E1 7 gnd 3 gnd 1
Vclamp 6 gnd DC=5.3
Q1 gnd 5 2 QP
R2 7 8 1k
```

```
C2 8 gnd 35p
E2 gnd 4 8 gnd -1
D3 out 6 DCLP2
Vref 11 gnd DC=2.5V
*
.MODEL DCLP3 D BV=5 N=10m
.MODEL DCLP2 D BV=100V CJO=4PF IS=7E-09 M=.45 N=2 RS=.8
+ TT=6E-09 VJ=.6V
.MODEL QP PNP
.MODEL DCLP1 D BV=5
.ENDS
```

4.3 SOURCES WITH A GIVEN FAN-OUT

Some applications, such as internal voltage references, cannot always be modeled as a simple source in series with a static resistor. Another important parameter is the fan-out, or the maximum output current that the source can supply before it is out of regulation. The simplest way is to describe this behavior is to use a single in-line equation. If the reference level is made dependent on the input node (imagine the reference block is supplied from a V_{cc} line), we naturally describe its dc audio susceptibility (also called line ripple rejection).

Suppose that we create a 5 V source affected by a static output impedance R_{out} of 2.5 Ω and a line ripple rejection of −60 dB. If the output current is less than a given value (1.5 mA is this example), the output level is not affected, except by the normal loss incurred by the presence of R_{out}. The in-line equation or the analog behavioral modeling (ABM) description could look like the following in *IsSpice* format:

```
B1 INT GND V=I(Vdum) < 1.5m ? ({Vref} + {Vref} * 1m)
```

IF the output current is less than 1.5 mA, THEN the source delivers $V_{ref} + V_{ref} \times 1m$ in series with a 2.5 Ω resistor. The term $V_{ref} \times 1m$ represents the −60 dB input susceptibility.

Now we need to define how the output voltage decreases, or what resistive slope R_{slope} affects the output in overcurrent condition. If we select a 10 kΩ slope, the final equation looks like this:

```
B1 INT GND  0 V = I(Vdum)>1.5m ? ({Vref} + V(in) * 1m) - 10k *
+ I(Vdum) + 15 : {Vref}+V(in)*1m
```

To avoid the sudden discontinuity in V_{ref} when the current increases beyond 1.5 mA, a term is added in series within the equation. This terms makes the drop equal to 0 when the current reaches the limit (1.5 mA in the example) and forces the voltage to smoothly decrease with a slope of 1 V/100 μA, imposed by the 10 kΩ resistor. The term is simply equal to $1.5m \times 10k = 15$ and avoids the discontinuity at the transition point. A subcircuit can now be constructed, featuring the following lines:

IsSpice

```
.SUBCKT REFVAR  In  Out  Gnd params: Vref=5  Zo=2.5  Slope=10k
+ Imax=1.5m Ripple=1m
Rout 5 6 {Zo}
Vdum 6 Out
Bout 5 Gnd V=I(Vdum) < {Imax} ? ({Vref}-V(in)*{Ripple}) :
+ ((({Vref}-V(in)*{Ripple}) - {Slope}*I(Vdum))+({Slope}*{Imax}))
.ENDS
```

PSpice

```
.SUBCKT REFVAR    In  Out  Gnd  params:  Vref=5  Zo=2.5  Slope=10k
+ Imax=1.5m Ripple=1m
Rout 5 6 {Zo}
Vdum 6 Out
Eout  5  Gnd  Value  =  { IF ( I(Vdum) < {Imax}, ({Vref}-V(in)*
+ {Ripple}),   +   ((({Vref}-V(in)*{Ripple})   -   {Slope}*I(Vdum))+
+ ({Slope}*{Imax})) ) }
.ENDS
```

The operating parameters are the following:

Vref	the nominal output voltage
Zo	the static output impedance
Slope	the output resistance beyond the current limit
Imax	the maximum current limit
Ripple	input rejection, positive or negative, 1m equals −60 dB, for instance

The application represents the limited current capability voltage source hooked up to a voltage-controlled resistor as shown in Fig. 4-2a. We will see below how to construct such a device.

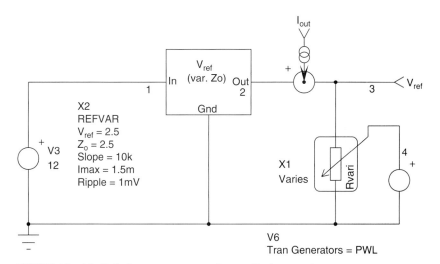

FIGURE 4-2a   The limited output current source in an application example.

The current is then swept between a few hundred microamperes up to a few milliamperes. The result appears in Fig. 4-2b and correctly models the 1.5 mA break point.

4.4 VOLTAGE-ADJUSTABLE PASSIVE ELEMENTS

Very often, when one is simulating an electric circuit with SPICE, a need arises for a variable passive element such as a resistor, a capacitor, or an inductor. If an electric source could externally control the value of the above devices, it would naturally open the door to ABM expressions

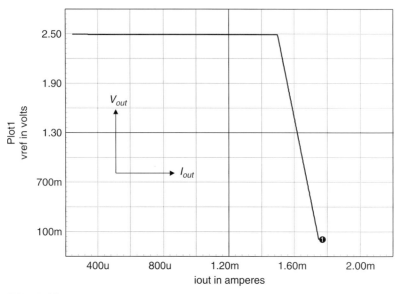

FIGURE 4-2b The X–Y plot delivered by the loaded reference voltage.

for capacitors and inductors: nonlinear behaviors, inductance variations with current, etc. Unfortunately, few SPICE-based simulators accept in-line equations for passive elements. To work around this restriction, this section describes passive elements whose values can be adjusted via an external voltage source.

4.4.1 The Resistor

A resistor R when multiplied by a current I develops a voltage V. Yes, everyone knows Ohm's law, derived by Georg Simon Ohm (1789–1854), a German physicist. The same resistor R can be represented by a current source I

$$I = \frac{V(1, 2)}{R} \qquad (4\text{-}2)$$

where 1 and 2 represent the resistor terminals. Figure 4-3a shows this representation.

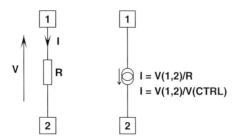

FIGURE 4-3a A resistor can be portrayed as a controlled current source.

BASIC BLOCKS AND GENERIC SWITCHED MODELS

By using this simple equation, it becomes possible to construct a variable resistor subcircuit, in Intusoft's *IsSpice* or CADENCE's *PSpice*, where R in Eq. (4-2) will be directly driven by a control source, via the CTRL node:

IsSpice

```
.subckt VARIRES 1 2 CTRL
R1 1 2 1E10
B1 1 2 I=V(1,2)/(V(CTRL)+1u)
.ENDS
```

PSpice

```
.subckt VARIRES 1 2 CTRL
R1 1 2 1E10
G1 1 2 Value = { V(1,2)/
+ (V (CTRL)+1u) }
.ENDS
```

In the current source expression, the 1 µ quantity avoids divide-by-zero overflows in case of extremely low control values. If V(CTRL) equals 100 kV, then the equivalent resistor will be 100 kΩ. Figure 4-3b shows the results of a simple resistive divider built around the subcircuit imposing a 1 Ω resistor. Now, a complex voltage source can be built for V_3, and nonlinear relationships can easily be envisioned.

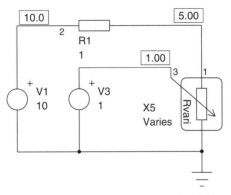

FIGURE 4-3b Dropped in a simple resistive divider, it generates a 1 Ω resistor.

4.4.2 The Capacitor

As we did before with the resistor, a capacitor can be portrayed by a voltage source obeying the following law:

$$V_C(t) = \frac{1}{C} \int I_C(t)\,dt \qquad (4\text{-}3)$$

That is, if we integrate the current flowing into the equivalent subcircuit capacitor and multiply it by the inverse of a control voltage V, we obtain a capacitor value C equal to the control voltage V. Unfortunately, there is no integral primitive under SPICE since it involves the variable t, continuously varying. Therefore, why not capitalize on Eq. (4-3) and force the subcircuit current into a 1 F capacitor? By observing the resulting voltage over this 1 F capacitor, we have integrated $I_C(t)$! Figure 4-4a shows how we can build the subcircuit.

The dummy source V routes the current into the 1 F capacitor which develops the integrated voltage on the "int" node (Fig. 4-4b). Then once it is multiplied by the inverse of the CTRL node voltage, it mimics the variable capacitor. Figure 4-4c displays voltages and currents obtained from both the real capacitor and the variable one. There is no difference between plots.

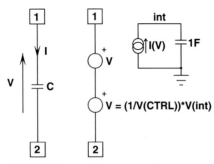

FIGURE 4-4a Building an equivalent capacitor involves an integration built on top of a 1 F capacitor.

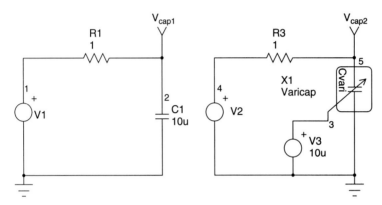

FIGURE 4-4b Test circuit pulsing a 10 μF capacitor with a square wave source.

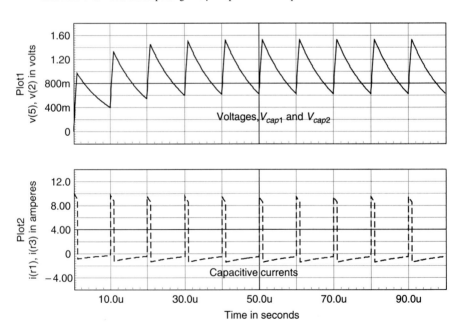

FIGURE 4-4c The variable and the classical capacitor models produce similar waveforms.

Below are the models in both *IsSpice* and *PSpice*:

IsSpice

```
.SUBCKT VARICAP 1 2 CTRL
R1 1 3 1u
VC 3 4
BC 4 2 V=(1/v(ctrl))*v(int)
BINT 0 INT I=I(VC)
CINT INT 0 1
.ENDS
```

PSpice

```
.SUBCKT VARICAP 1 2 CTRL
R1 1 3 1u
VC 3 4
EC  4  2  Value  =  {  (1/v(ctrl))
+ *v(int) }
GINT 0 INT Value = { I(VC) }
CINT INT 0 1
.ENDS
```

Simulations were also performed in ac analysis where the model confirms its ability to work well in the frequency domain too.

4.4.3 The Inductor

When an inductor is energized, it struggles to keep the ampere-turns constant, acting as a real current source. That is the way we are going to model the variable inductor. If we apply Lenz' law, formulated by Heinrich Friedrich Emil Lenz, a Russian physicist (1804–1865), we can state that

$$V_L(t) = L\frac{di_L(t)}{dt} \tag{4-4}$$

Rearranging it leads to

$$V_L(t)dt = Ldi_L(t) \tag{4-5}$$

If we now integrate both parts of this equation, we obtain

$$\int V_L(t)dt = \int Ldi_L(t) \tag{4-6}$$

Since L is constant, we have

$$\int V_L(t)dt = Li_L(t) \tag{4-7}$$

or

$$I_L(t) = \frac{1}{L}\int V_L(t)dt \tag{4-8}$$

Equation (4-8) simply means that we need to integrate the voltage present across the equivalent inductor and divide it by the control voltage aimed to simulate L. Figure 4-5a shows the equivalent subcircuit.

The voltage integration is made by transforming terminal voltages into a current and, again, injecting this equivalent current into a 1 F capacitor. The subcircuit netlist can be found below:

IsSpice

```
.SUBCKT VARICOIL 1 2 CTRL
BC 1 2 I=V(INT)/V(CTRL)
BINT 0 INT I=V(1,2)
CINT INT 0 1
.ENDS
```

PSpice

```
.SUBCKT VARICOIL 1 2 CTRL
GC 1 2 Value = {V(INT)/V(CTRL)}
BGINT 0 INT Value = {V(1,2)}
CINT INT 0 1
.ENDS
```

Of course, adjusting *LC* filters can easily be constructed for complex ac analysis. If we now simulate Fig. 4-5b, we obtain Fig. 4-5c waveforms which are the dual of those in Fig. 4-4c.

354 CHAPTER FOUR

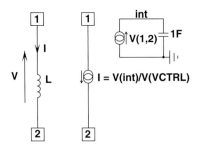

FIGURE 4-5a The equivalent L subcircuit.

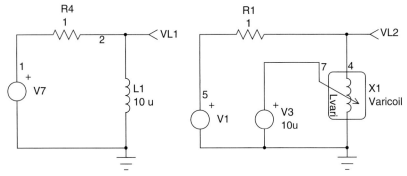

FIGURE 4-5b The test circuit with the equivalent inductor.

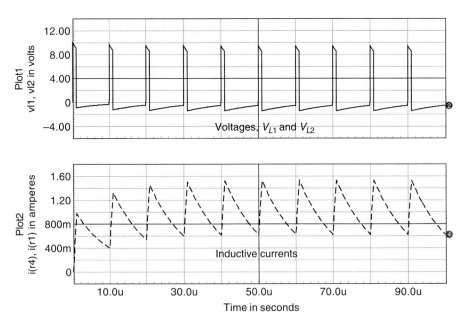

FIGURE 4-5c Equivalent L simulations, revealing dual waveforms with capacitive results.

4.5 A HYSTERESIS SWITCH

IsSpice naturally offers the SPICE primitive *S*, a switch affected by hysteresis. That is, the switch toggles from an off state to an on state, having a resistor immediately moving from the value R_{off} to the value R_{on} when the switch control voltage reaches a level equal to $V_T + V_H$. Here V_T represents the threshold level and V_H the hysteresis width. When the control voltage falls below $V_T - V_H$, the switch goes back to the open state and features a resistor equal to R_{off}. The transition time is 0, and care must be taken to avoid convergence issues (such as adding small *RC* time constants). Figure 4-6a depicts how to wire this component in a simple comparator architecture whereas Fig. 4-6b shows the resulting V_{out} versus V_{in} curve.

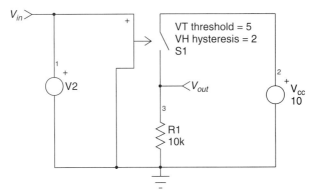

FIGURE 4-6a This circuit toggles at 7 V and V resets at 3 V.

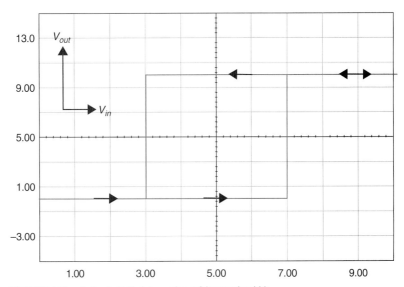

FIGURE 4-6b A simple X-Y plot reveals a ±2 hysteresis width.

On the contrary, *PSpice* only features a *.model VSWITCH* which also offers *on* and *off* resistors, but the parameter voltages V_{ON} and V_{OFF} just set the boundaries within which the switch resistor will vary. If $V_{ON} = 5$ V and $V_{OFF} = 3$ V, then the resistor offered by the switch will smoothly vary between these two levels: from 5 V to 3 V, the resistor changes from R_{ON} to R_{OFF}, and from 3 V to 5 V, the resistor will move from R_{OFF} to R_{ON}. Internal resistor expressions ensure a smooth resistor transition between these states (good convergence); however, no hysteresis is provided. *OrCAD* has actually put a hysteresis switch in recent versions, but it is more fun to make one yourself!

A simple circuit arrangement can be used to offer a hysteresis switch in *PSpice*. It will actually combine a smoothly transitioning switch and some programmable hysteresis. We need to add a bunch of Behavioral ABM sources to help tailor the moving switching events. Figure 4-6c portrays the way in which we have derived the new device with adjustable hysteresis.

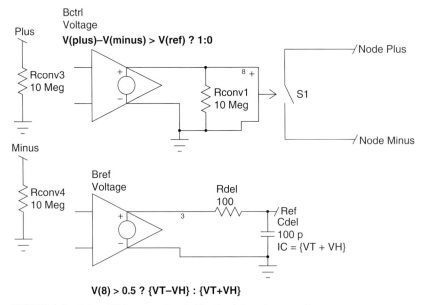

FIGURE 4-6c A few ABM sources, the primitive *S* switch, and you are all set.

The "plus" and "minus" inputs route the switch control signal to the B_{ctrl} behavioral element (a B element with *IsSpice* becomes a "value" *E* source under *PSpice*). When the switch is open (B_{ctrl} delivers 0), the reference (node "ref") is set to the highest toggling point (7 V in Fig. 4-6a). When the node V_{in} increases, it crosses the ref node level and B_{ctrl} goes high. The switch S_1 closes and authorizes the current to flow in its terminals via the R_{ON} parameter. At this time, the B_{ref} source has detected that S_1 is closed and now modifies its reference node to the second level, 3 V in the example. This action creates hysteresis in the model. When V_{in} diminishes, it crosses 7 V again, but no action takes place since node ref has changed to 3 V. As V_{in} finally crosses 3 V, S_1 opens and applies R_{OFF} between its terminals. The $R_{del} - C_{del}$ network smoothes the transitions of node ref whereas the *.IC* statement of C_{del} properly positions the switch at start-up. If convergence problems should arise, it could be necessary to increase ITL4 to 20 through the following statement: .OPTION ITL4 = 20.

The complete *PSpice* netlist is given below:

```
.subckt  SWhyste  NodeMinus  NodePlus  Plus  Minus  PARAMS:  RON=1
+ ROFF=1MEG VT=5 VH=2
S5 NodePlus NodeMinus 8 0 smoothSW
```

```
EBcrt1 8 0 Value = { IF ( V(plus)-V(minus) > V(ref), 1, 0 ) }
EBref ref1 0 Value = { IF ( V(8) > 0.5, {VT-VH}, {VT+VH} ) }
Rdel ref1 ref 100
Cdel ref 0 100p   IC={VT+VH}
Rconv1 8 0 10Meg
Rconv2 plus 0 10Meg
Rconv3 minus 0 10Meg
.model smoothSW VSWITCH (RON={RON} ROFF={ROFF} VON=1 VOFF=0)
.ends SWhyste
```

The parameter passing is similar to that of the SPICE primitive S switch as implemented by *IsSpice*: $V_{ON} = V_T + V_T$ and $V_{OFF} = V_T - V_H$. Both R_{ON} and R_{OFF} remain the same.

Figure 4-6d shows how a simple test RC oscillator can be built thanks to the hysteresis we added. Once the simulation is finished, Fig. 4-6e reveals the results.

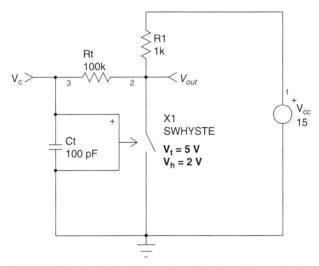

FIGURE 4-6d Combining the hysteresis switch with an RC network makes a cheap oscillator.

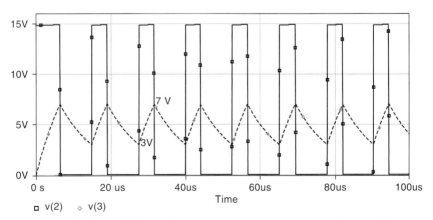

FIGURE 4-6e *PSpice* simulation results confirming the 4 V hysteresis as passed to the subcircuit.

4.6 AN UNDERVOLTAGE LOCKOUT BLOCK

The hysteresis switch can be used in a lot of ways. One is to use it as comparator featuring hysteresis. There is no need to manipulate references or in-line equations. Just use the SWhyste switch model described before and an *undervoltage lockout* (UVLO) circuit can be quickly made. UVLO circuits are used in PWM controllers or other circuits to ensure that a sufficient supply voltage exists to guarantee reliable operation of the converter. This is done because all internal references (such as bandgaps) need enough voltage headroom to operate, or you must bias the MOSFET with a sufficient level of gate voltage. The UVLO circuit permanently monitors the V_{cc} and shuts down the controller below a certain threshold. It reactivates the circuit when V_{cc} goes up again and crosses a second threshold. Typical levels could be 12 V to start pulsing and around 8 V when the controller decides to stop driving the MOSFET. These are common levels found on off-line controllers, but there are many other possible combinations.

Figure 4-7a depicts the hysteresis switch in a simple configuration where the control node receives V_{cc}. The other control node goes to the node gnd that can be the 0 V point or another

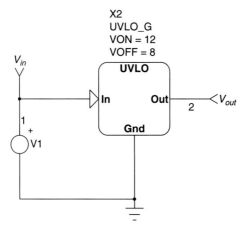

FIGURE 4-7a A simple UVLO implementation monitors the V_{in} source.

potential if you have a floating circuit. Based on the required *on* and *off* voltages, the switch parameters V_T and V_H can quickly be derived:

$$V_{ON} = V_T + V_H \tag{4-9}$$

$$V_{OFF} = V_T - V_H \tag{4-10}$$

By simple manipulations we obtain

$$V_H = \frac{V_{ON} - V_{OFF}}{2} \tag{4-11}$$

$$V_T = V_{ON} - V_H \tag{4-12}$$

A pull-up resistor to a 5 V internal source will deliver the internal logic level to indicate a good or bad V_{cc}. The internal diagram actually looks similar to that in Fig. 4-6a and corresponds to the following netlist:

PSpice

```
.SUBCKT UVLO_G  1   2   30 params: VON=12 VOFF=10
*     VIN OUT Gnd
X1 1 3 1 30 SWhyste RON=1 ROFF=1E6 VT={((VON-VOFF)/2) + VOFF}
+ VH={(VON-VOFF)/2}
RUV 3 30 100K
B1 4 0 V=V(3,30) > 5V ? 5V : 0
RD 4 2 100
CD 2 0 100P
.ENDS UVLO_G
```

The *IsSpice* netlist uses the internal *S* primitive which already features hysteresis. A simulation featuring a triangular waveform feeding the input model delivered Fig. 4-7b results, highlighting the 4 V hysteresis zone.

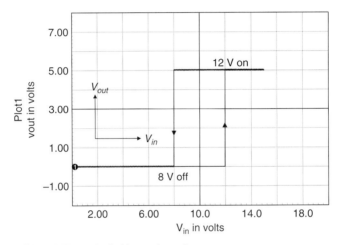

FIGURE 4-7b A classical hysteresis graph.

4.7 LEADING EDGE BLANKING

A switch-mode power supply is a noisy converter in essence. A variety of events can generate spikes such as capacitor charging or parasitic short-circuit moments, for instance, when a t_{rr}-affected diode is suddenly blocked. In a flyback converter, every time the main switch closes and discharges primary parasitic capacitors or brutally stops the conduction of the secondary diode in CCM, a current spike is generated in the primary branch. This spike is seen by the current-sensing elements and is routed to the controller. Depending on the conditions, this spike can interfere with the current limit section and can reset the comparator in current-mode structures. An adequate *RC* network can cure this default, but many of today's circuits implement *leading edge blanking* (LEB) circuitry. It consists of transmitting the sensed current pulse only a few hundred of nanoseconds after it has started. That way, if a big spike occurs at the switch closing, the LEB naturally blinds the system for the given period and fully transmits the information afterward. Figure 4-8a shows how to simply connect a delay line together with a comparator to build an efficient LEB. Figure 4-8b details the final behavior.

By feeding the delay line subcircuit with a given transfer time, you tailor the LEB at your convenience. The delay line can sometimes significantly increase the simulation time. In that case, one can easily replace this subcircuit by a simple *RC* network introducing a similar delay.

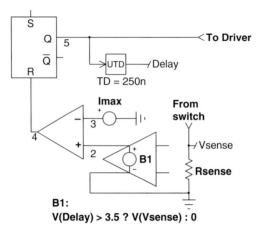

FIGURE 4-8a B_1 blanks the sense information until *Delay* goes high.

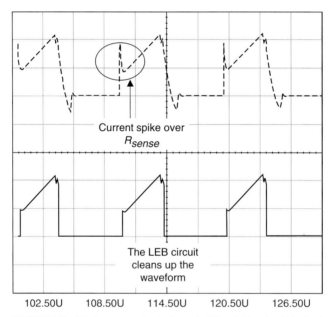

FIGURE 4-8b The lower trace clearly shows the 250 ns LEB action. This signal appears on B_1 output.

To improve safety in the presence of lethal spikes (e.g., winding short-circuits), some high-end controllers include a second fast comparator which will reset or even latch off the controller when the current-sense information exceeds the maximum authorized current by a certain amount ($\times$ 2, for example) before the LEB period is over.

4.8 COMPARATOR WITH HYSTERESIS

Analog designers rarely include a fast comparator without introducing a bit of hysteresis. This feature improves the noise immunity and ensures clean, sharp transitions, especially when the open-loop gain of the comparator is high. If adding resistors across the comparator allows a quick hysteresis implementation, the method requires a bit of calculation to determine the true hysteresis value you finally use.

Figure 4-9a represents a simple arrangement made of two in-line equations. The first one builds a dc voltage—the hysteresis—put in series with the noninverting pin. This hysteresis

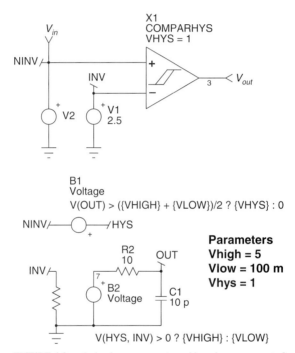

FIGURE 4-9a A simple arrangement provides a known amount of hysteresis.

actually depends on the output level. The output delivers two different voltages that you can pass as parameters to tweak the comparator levels to your needs. A small *RC* network gives the output transition a reasonable duration and reduces convergence problems. Figure 4-9b shows the results with a 1 V hysteresis passed as the parameter. The netlist is given below:

IsSpice

```
.SUBCKT COMPARHYS NINV INV OUT params: VHIGH=5 VLOW=100m VHYS=50m
B2 HYS NINV V = V(OUT) > {(VHIGH+VLOW)/2} ? {VHYS} : 0
B1 4 0 V = V(HYS,INV) > 0 ?   {VHIGH}   :   {VLOW}
R0 4 OUT 10
C0 OUT 0 100PF
.ENDS
```

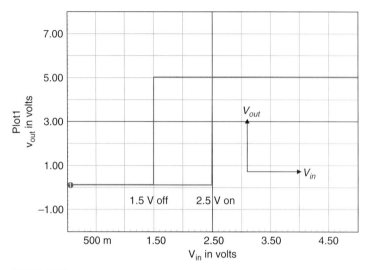

FIGURE 4-9b In this example, the system introduces 1 V of hysteresis.

PSpice

```
.SUBCKT COMPARHYS NINV INV OUT params: VHIGH=5 VLOW=100m VHYS=50m
E2 HYS NINV Value = {IF(V(OUT) > {(VHIGH+VLOW)/2},{VHYS},0 )}
E1 4 0 Value = {IF(V(HYS,INV) > 0, {VHIGH}, {VLOW} )}
RO 4 OUT 10
CO OUT 0 100PF
.ENDS
```

4.9 LOGIC GATES

Without redefining a complete logic library, it is often desirable to define its own simple logic gates. A single in-line equation can do the job and simulates much more quickly than a complex, full-featured gate, for instance. Also, when building a model, you will include not the simulator library gates but rather individual logic functions that you have defined yourself. Below are a few examples of dual input basic gates that can be expanded to multi-input devices quite easily. Note that the input thresholds are $\frac{V_{cc}}{2}$ by default and can be altered if the 5 V output value needs to be changed.

IsSpice AND

```
.SUBCKT AND2 1 2 3
B1 4 0 V= (V(1)>2.5) & (V(2)>2.5) ? 5 : 100m
*   (V(1)>2.5) & (V(2)>2.5) ? 100m : 5  is a NAND
RD 4 3 100
CD 3 0 10P
.ENDS AND2
```

PSpice AND

```
.SUBCKT AND2 1 2 3
E1 4 0 Value = { IF (  (V(1)>2.5) & (V(2)>2.5), 5, 100m ) }
```

```
*   { IF (   (V(1)>2.5) &  (V(2)>2.5), 100m, 5 ) } is a NAND
RD 4 3 100
CD 3 0 10P
.ENDS AND2
```

IsSpice OR2

```
.SUBCKT OR2 1 2 3
B1 4 0 V= (V(1)>2.5) | (V(2)>2.5) ? 5 : 100m
*  (V(1)>2.5) | (V(2)>2.5) ? 100m : 5   is a NOR
RD 4 3 100
CD 3 0 10P
.ENDS OR2
```

PSpice OR2

```
.SUBCKT OR2 1 2 3
E1 4 0 Value = { IF (   (V(1)>2.5) | (V(2)>2.5), 5, 100m ) }
*   { IF (   (V(1)>2.5) | (V(2)>2.5),100m, 5 ) } is a NOR
RD 4 3 100
CD 3 0 10P
.ENDS OR2
```

Combining these gates leads to the creation of logic circuits such as latches, flip-flops, and so on. Here is an RS latch:

IsSpice latch

```
.SUBCKT LATCH 6 8 2 1
*       S R Q Qb
BQB   10 0 V=(V(8)<2.5) & (V(2)>2.5) ? 100m : 5
BQ    20 0 V=(V(6)<2.5) & (V(1)>2.5) ? 100m : 5
RD1   10 1 100
CD1   1 0 10p IC=5
RD2   20 2 100
CD2   2 0 10p IC=100m
.ENDS LATCH
```

PSpice latch

```
.SUBCKT LATCH 6 8 2 1
*       S R Q Qb
EQB   10 0 Value = {IF (  (V(8)<2.5) & (V(2)>2.5) ? 100m : 5 ) }
EQ    20 0 Value = {IF (  (V(6)<2.5) & (V(1)>2.5) ? 100m : 5 ) }
RD1   10 1 100
CD1   1 0 10p IC=5
RD2   20 2 100
CD2   2 0 10p IC=100m
.ENDS LATCH
```

IsSpice accepts so-called nested subcircuits. It means that you can include a subcircuit within a subcircuit model and it stays confined within the netlist. *PSpice* does not implement this method and subcircuits called within a main subcircuit model have to be placed after the main .ENDS statement. A problem arises when *PSpice* processes the libraries and consider all subcircuits as global definitions. If the simulator finds two subcircuits sharing the name AND2 but featuring different definitions, it will take the first one and ignore the other. If the pin count differs between the subcircuits, an error is generated and you can quickly correct the mistake.

However, if the pin count is the same and models are inherently different but sharing a similar name, completely unexpected results can occur. Good modeling practice therefore includes the addition of a suffix to all subcircuits you build within a model netlist. For instance, a latch implemented in a UC3843 subcircuit would be named LATCH_UC3843 to avoid any confusion with other existing models of latch.

4.10 TRANSFORMERS

To model a simple dual-winding transformer, we can use the SPICE primitive k which describes the coupling ratio between a primary and a secondary. Figure 4-10a shows how to place the inductors around this coupling element. However, as simple as this can be, you do not gain much insight into the transformer elements: you have to derive the leakage inductor value from k and the turns ratio too. It can sometimes be unclear and painful when iterations are needed (e.g., you may adjust the leakage to assess the effect of a clamping network). In this figure, Lss means L_s is short-circuited and Lso implies L_s is left open unconnected: glancing at a figure where the coupling coefficient is used does not reveal the complete picture!

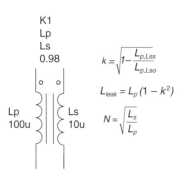
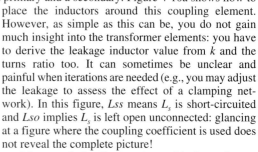

FIGURE 4-10a Using the SPICE primitive k does not simplify the handling of transformers.

A better solution lies in using an ideal transformer as we depicted in Chap. 2 (remember, the d-dependent ratio) [1]. Figure 4-10b represents the principle of modeling where the simple founding equations appear below.

$$V_2 = V_1 \frac{N_s}{N_p} \quad (4\text{-}13)$$

$$I_1 = I_2 \frac{N_s}{N_p} \quad (4\text{-}14)$$

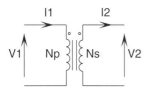

FIGURE 4-10b The current and voltage relationships of an ideal transformer.

If we use voltage-controlled voltage sources and current-controlled current sources whose parameters relate to the windings ratios, we can make an ideal transformer model sticking to Eqs. (4-13) and (4-14). This is what Fig. 4-10c represents. In this transformer, the magnetizing current is null (infinite magnetizing inductance). It thus offers the ability to model the magnetizing and the leakage inductance separately. These two elements can be found by solving a few equations after specific measurements have been carried out on the transformer. Once they are known, the ideal transformer can be assembled as Fig. 4-10d suggests.

In the vast majority of the design examples in the remaining chapters, this equivalent model is used to simulate the transformer operation for two reasons. One is that experience shows that it converges in a better way than a coupling coefficient. The other reason is that the turns ratio and leakage inductance immediately appear. Please note that in this subcircuit, the turns ratio is normalized to the primary (for example, $N_p = 20$ and $N_s = 3$ would mean a 1:0.15 ratio, thus a value of 0.15 passed to the model).

Below are the two necessary netlists to implement the Fig. 4-10c model in *PSpice* and *IsSpice* syntax.

BASIC BLOCKS AND GENERIC SWITCHED MODELS

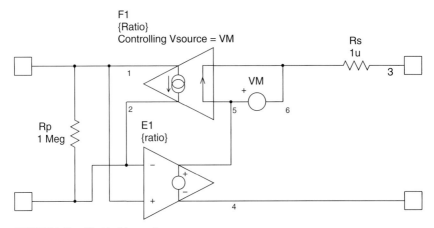

FIGURE 4-10c The ideal dc transformer.

IsSpice

```
.SUBCKT XFMR 1 2  3   4  params: RATIO=1
RP 1 2 1MEG
E 5 4 1 2 {RATIO}
F 1 2 VM {RATIO}
RS 6 3 1U
VM 5 6
.ENDS
```

PSpice

```
.SUBCKT XFMR1 1 2 3 4 PARAMS: RATIO=1
RP 1 2 1MEG
E 5 4 VALUE = { V(1,2)*RATIO }
G 1 2 VALUE = { I(VM)*RATIO }
RS 6 3 1U
VM 5 6
.ENDS
```

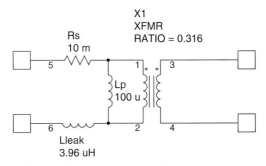

FIGURE 4-10d Once known, external elements such as magnetizing and leakage inductance can be added to the model. The turns ratio is the only parameter passed to the subcircuit.

Different transformers will be used in the examples such as multioutputs or center-tap for push-pull applications. They are available on the OrCAD example libraries and as a standard model under Intusoft schematic capture, *SpiceNET*.

4.10.1 A Simple Saturable Core Model

Since the primary inductance appears separately, it is possible to replace it by a saturable type, actually accounting for the material used in the manufacturing process. A lot of models are available for simulating these kind of inductors, featuring saturation effects and hysteresis. *PSpice* uses the Jiles-Atherton model [2], together with a coupling coefficient k to which is associated a ferrite type, B51, 3C8, and so on. *PSpice* reference manual includes a lot of modeling information related to their saturable core model, its reading is extremely valuable. Needless to say the inclusion of saturation effects quickly hampers simulation time and can bring serious convergence issues. We will rarely use this feature in the design examples, however, a simple circuit is available in the OrCAD files supplied with this book.

IsSpice has purposely departed from the Jiles and Atherton model and created a complete simple behavioral model made of a subcircuit. This subcircuit was also part of the Ref. 1 publication. Its ability to quickly and easily converge is interesting when the user wants a simple model for modeling saturation effects. We have seen, in the modeling section of passive elements, how to artificially create an inductor of variable value. We force the circulation of a current between terminals 1 and 2 of the subcircuit, this current being proportional to the integral of the voltage across these terminals. This is nothing other than applying the formula already given:

$$I_L(t) = \frac{1}{L}\int V_L(dt) \qquad (4\text{-}15)$$

If L is given by a voltage V, we actually create an inductor of value V. Now, if the result of Eq. (4-15) circulates in a resistor R [or if the voltage delivered by a source implementing Eq. (4-15) is applied over a resistor R], then the emulated inductor becomes the value of R. This is what Fig. 4-11a shows you.

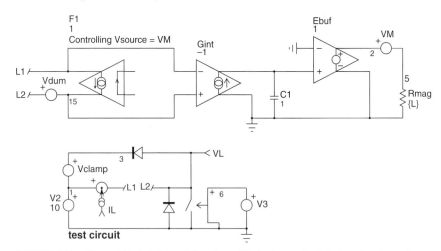

FIGURE 4-11a The variable inductor model can be modified to impose the inductor value via a resistor.

In this figure, the integrated voltage across C_1 is buffered by E_{buf} and applied across R_{mag}. The circulating current in R_{mag} is then forced to flow between terminals L_1 and L_2 to emulate the inductor action. This is what Fig. 4-11b shows you where a simple switching test fixture is wired. The current flowing through L is observed via the dummy I_L source.

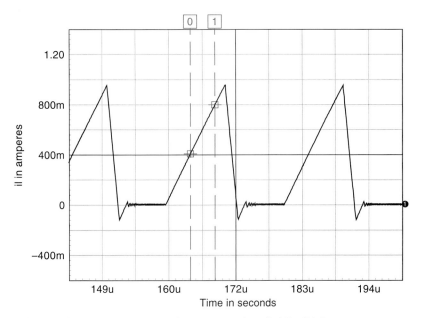

FIGURE 4-11b The emulated inductor forces a current slope of a 100 μH inductor.

If this model offers the necessary flexibility for a standard inductor, it does not give any saturation limit. During saturation, the material relative permeability μ_r drops to 1, causing the inductance to fall to a value fixed by the number of turns and the bobbin geometry: it almost becomes an air-wound inductor. What would be needed is to actually detect a certain voltage over C_1 and activate a resistor coming in parallel with R_{mag}: L would be dramatically reduced, artificially emulating the saturation effect. To create it, we can connect another resistor R_{sat} in series with a voltage source V_{sat} and a diode. If the voltage developed by C_1 (actually the flux) does not reach V_{sat} (whose value represents the saturation flux), the exhibited inductor is set by R_{mag}. On the contrary, if the diode conducts, R_{mag} is affected by R_{sat} and the inductor value goes down: the component is saturated. Figure 4-11c shows this intermediate step. Core losses

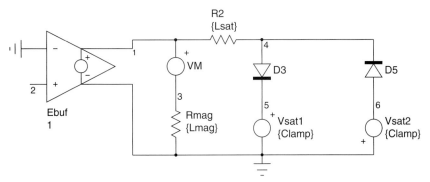

FIGURE 4-11c Activating a resistor when the volt-seconds are too big (the voltage across C_1) mimics a saturated inductor.

brought by eddy currents (circulating currents creating losses because of the material resistivity) are usually simulated by adding a resistor across the inductor. However, this simple option does not emulate frequency-dependent losses. Another way would consist of adding a capacitor in parallel with R_{mag}. If this capacitor varied in a nonlinear relationship with the voltage across it, it could emulate amplitude and frequency-dependent flux losses. Manipulating nonlinear equations is always a problem for SPICE convergence. The idea has been finally to use the diode junction capacitance C_{jo}. As this parameter nonlinearly varies with the diode bias voltage, it naturally fulfils the function we are looking for. Let us review the parameters we will pass to the model:

- **VSEC**: This is the flux capacity in volt-seconds that the inductor can accept before saturating. This is a founding equation when you design inductors:

$$VSEC = NA_e B_{sat} > L_{mag} I_{peak} \qquad (4\text{-}16)$$

where A_e represents the material core area (m^2), N the turns number, B_{sat} the saturating flux density (in teslas), and I_{peak} the maximum peak current allowed to circulate in the inductor before saturation. Suppose the inductor is made of 10 turns on a core featuring an area of 0.000067 m^2 and accepts up to 300 mT of maximum flux density. In that case, VSEC equals 201u.

- **ISEC**: This is a residual flux level you can pass to the model if you would like the model to start with a residual value. As is VSEC, it is expressed in volt-seconds.

- L_{mag}: The magnetizing inductance of the an iron-core inductor with an air gap may be expressed via

$$L_{mag} = \frac{N^2 A_e \mu_e}{l_m} \qquad (4\text{-}17a)$$

with

$$\mu_e = \frac{\mu_r \mu_0}{1 + \mu_r \left(\dfrac{l_g}{l_m}\right)} \qquad (4\text{-}17b)$$

where μ_e is the effective permeability, l_g is the gap length (meters), l_m is the total magnetic path (meters), μ_0 is the air permeability ($4\pi 10^{-7}$ henrys per meter), and μ_r is the core relative permeability. The model can actually be written to accept directly either the magnetizing inductance value or the core physical parameters. We prefer to consider this last option. Please note that when the ratio $\mu_r \left(\dfrac{l_g}{l_m}\right)$ is much bigger than 1, then μ_r and l_m disappear from the equation and the gap dimension l_g dominates the inductor properties.

- L_{sat}: This is the inductor value when the material is saturated. Therefore, the material relative permeability μ_r drops to 1, and the inductor expression becomes

$$L_{sat} = \frac{N^2 A_e \mu_0}{l_m + l_g} \qquad (4\text{-}18)$$

In saturation, the inductance drops to a low value as if the inductor became an air-wound type.

- **Reddy**: The loss parameter depends on a graph usually given for magnetic materials. It describes the permeability variation versus the excitation frequency. You identify a point on the graph where the permeability drops by 3 dB. The frequency is then noted and passed to the model. See Fig. 4-11d.

- **Flux density and field probes**: It is interesting to monitor two important parameters: the flux density, denoted by B (expressed in teslas) and the field H, expressed in ampere-turns per meter. If we want to plot the B-H curves, we need independent sources delivering computed values for B and H. They can be obtained simply through the following in-line equations:

```
VB B 0 Value = V(phi) /{N}*{Ae} ; tesla
VH H 0 Value = {(N/lm)}*I(VDUM) ; amp turns per meter - A-T/m
```

BASIC BLOCKS AND GENERIC SWITCHED MODELS 369

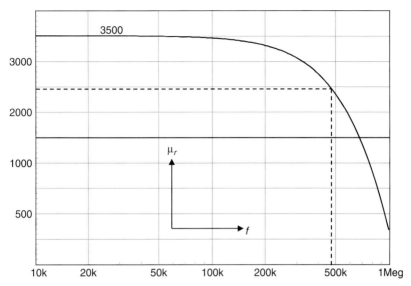

FIGURE 4-11d An example of the permeability drop with frequency. Here, the −3 dB is around 480 kHz.

Here *V(phi)* is the voltage image of the flux developed across C_1, and *I(VDUM)* is the current circulating in the inductor. Finally, capitalizing on all the above definitions, the model appears in Fig. 4-11e.

Note the presence of two current sources across the inductance. They are here to model the hysteresis at low frequencies and represent an improvement brought by Ref. 5. You need to know the coercive field value H_c to make use of this improvement. The complete netlist appears below and is compatible with *IsSpice*. A working example is also available under OrCAD.

```
.subckt coresat L1 L2 110 100 params: Feddy=25k IVSEC=0 Ae=
+ 0.000067 lm=0.037
+lg=0 Bsat=350m ur=6000 N=15 Hc=50
*
.param VSEC=N*Ae*Bsat
.param u0=1.25u
.param VSEC={N*Ae*Bsat}
.param u={u0*ur/(1+ur*(lg/lm))}
.param Lmag={u*N^2*Ae/lm}
.param Lsat={u0*N^2*Ae/(lm+lg)}
.param IHyst={Hc*lm/N}
.param Cjo={3*VSEC/(6.28*Feddy*clamp*Lmag)}
.param clamp=250
*
F1 L1 12 VM 1
Gint 0 phi 12 L1 -1
C1 phi 0 {VSEC/Clamp} IC={IVSEC/VSEC*clamp}
Ebuf 5 0 phi 0 1
Rmag 8 0 {Lmag*clamp/VSEC}
VM 5 8
D3 2 9 D2mod
```

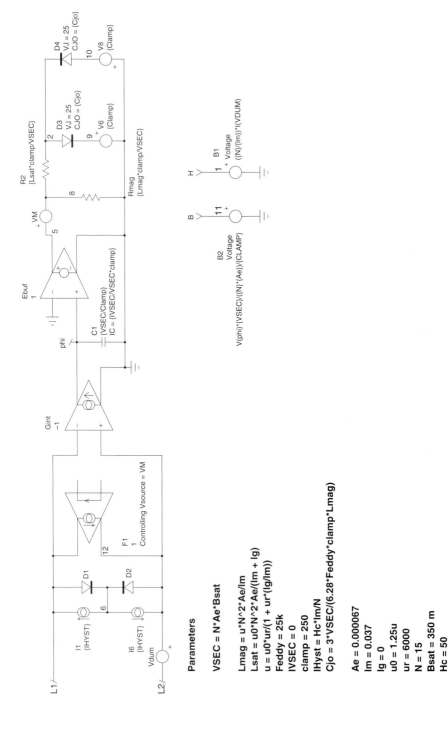

FIGURE 4-11e The final saturating inductor model featuring hysteresis losses.

```
V6 9 0 DC={Clamp}
R2 2 8 {Lsat*clamp/VSEC}
V8 0 10 DC={Clamp}
Vdum 12 L2
D4 10 2 D2mod
I1 6 L1 DC={IHYST}
B1 100 0 V=({N}/{lm})*I(VDUM)
B2 110 0 V=V(phi)*{VSEC}/({N}*{Ae})/
+ {CLAMP}
I6 6 12 DC={IHYST}
D1 L1 6 Dmod
D2 12 6 Dmod
.MODEL Dmod D N=1
.MODEL D2mod D CJO={Cjo} VJ=25
.ENDS
```

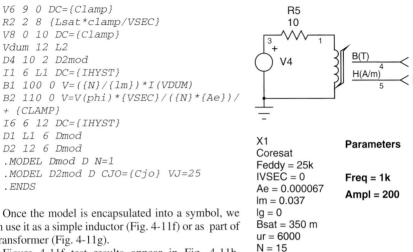

Once the model is encapsulated into a symbol, we can use it as a simple inductor (Fig. 4-11f) or as part of a transformer (Fig. 4-11g).

Figure 4-11f test results appear in Fig. 4-11h, showing hysteresis widening (actually losses) as the frequency increases. The source used to generate the plots is of sinusoidal type.

FIGURE 4-11f The saturable core can be tested alone.

This saturable model does offer an easy means to quickly model an inductor based on data book parameters. It can be used to first test a circuit featuring a saturable inductor, later replaced by a more comprehensive model once everything converges well. Both OrCAD and Intusoft propose more realistic models based on curve-fitting derivations, but they sometimes hamper the simulation time and make the circuit more prone to convergence issues.

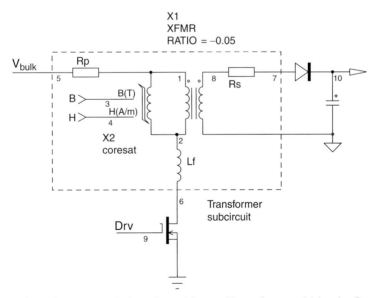

FIGURE 4-11g It can also be used to model a saturable transformer model, here in a flyback converter.

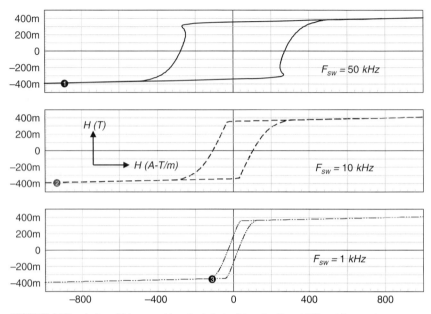

FIGURE 4-11h A sinusoidal source drives the inductor into saturation at different frequencies.

Literature on the magnetic components is plentiful, and references at the end of this chapter should help you to strengthen your knowledge in this area. However, a quick introduction to this world is given in App. 4A. Once you understand these concepts, you should find your way easily in more comprehensive books or articles.

4.10.2 Multioutput Transformers

A multioutput transformer model requires the association of the above transformer model, as Fig. 4-12 shows. The total leakage inductance appears on the primary side, and secondary side leakage inductors represent the coupling quality between the secondary windings. Appendix 4B describes how to measure a transformer and obtain the values with which to feed the model in Fig. 4-12. Unfortunately, extracting parameter models for the multioutput transformer subcircuit can be an extremely tedious job. Reference 4 discusses how to run the experiment for the so-called cantilever model, supposed to properly predict cross-regulation problems in multioutput structures. The model is also described in the Chap. 7 multioutput design example.

4.11 ASTABLE GENERATOR

An astable generator is often needed wherever switching cycles exist. Many methods exist to create a relaxation mechanism; Fig. 4-6d is a good example of one of them. Figure 4-13a proposes an alternative solution implementing a constant current source and the hysteresis comparator already introduced. The capacitor is charged to a level imposed by the reference voltage V_1,

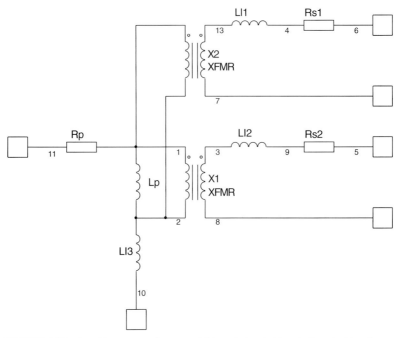

FIGURE 4-12 A multioutput transformer model is made by stacking single output transformer models, sharing the same primary inductance L_p.

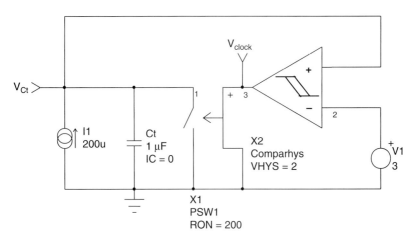

FIGURE 4-13a A simple ramp/square wave generator built around the hysteresis comparator.

and via a discharge orchestrated by X2, a new cycle takes place. This system could easily be used to generate a PWM ramp after adequate buffering. Figure 4-13b displays the simulation results. Please note the rather large R_{ON} value. It improves convergence by reducing the brutality with which C_t could be discharged with a lower value.

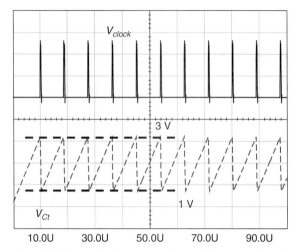

FIGURE 4-13b A 200 μA source together with a 2 V capacitor dynamic generate a 100 kHz pattern.

4.11.1 A Voltage-Controlled Oscillator

In resonant power supplies, the output power depends on the switching frequency. A *voltage-controlled oscillator* (VCO) sees its control input driven by the error amplifier which permanently monitors the output variable. The VCO is affected by two main parameters: a minimum and a maximum switching frequency. These parameters enable limiting the resonant converter to safe levels when the feedback is lost or when a default is detected.

Figure 4-14a represents the adopted architecture. Given the lack of control voltage, the frequency is set by the current source B_1. As soon as a voltage above 100 mV appears on the err node, the current source B_4 starts to activate on top of B_1: the frequency goes up and increases as the err node takes off. When its voltage reaches 5 V, the maximum frequency is attained. Of course, these two parameters can be altered to set a voltage excursion similar to that of the circuit. As an example, ON Semiconductor NCP1395 requires a full 5 V span to jump from the minimum to maximum switching frequency. The timing capacitor discharge is ensured by B_2 which sets the duty cycle to 50%. This is a classical I/2I oscillator. Finally, as we intend to use it for power electronic applications, we have added a power output made by the ABM source B_5. Two high-voltage diodes ensure current circulation in both directions when driving an inductive network.

Figure 4-14b shows the VCO once encapsulated driving an LLC resonant converter in open-loop conditions. Thanks to the simplicity of implementation, the simulation is extremely fast. The netlist is given below in *PSpice* syntax:

PSpice

```
.SUBCKT POWERVCO err out params: Fmin=50k Fmax=400k Vout=350
*
G4 0 1 value={ 2*100p*{Fmin} }
R1 err 0 100k
G1 0 1 Value = { 2*100p*({Fmax}-{Fmin})*V(CTRL)/5 }
G2 1 0 Value = { IF ( V(osc) > 2.5, (4*100p*{Fmin})+2*2*100p*
+ ({Fmax}-{Fmin})*V(CTRL)/5, 0 ) }
C1 1 0 100p IC=0
```

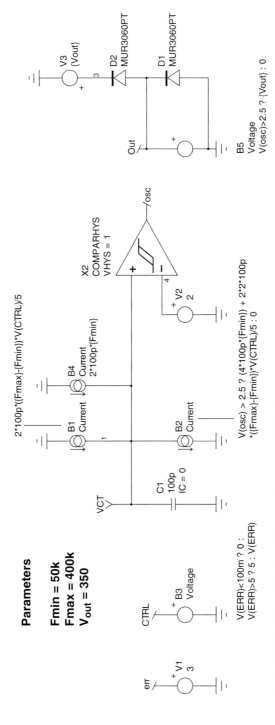

FIGURE 4-14a A VCO featuring minimum and maximum switching frequencies.

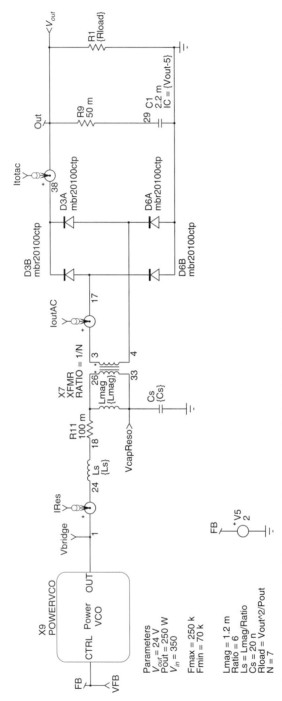

FIGURE 4-14b The power VCO drives an LLC resonant network which delivers around 24 V.

```
E3 CTRL 0 Value = {IF ( V(ERR)<100m, 0, IF ( V(ERR)>5, 5, V(ERR)
+ ) ) }
X2 1 4 osc COMPAR params: VHIGH=5 VHYS=1
V2 4 0 DC=2
E5 out 0 Value = { IF ( V(osc)>2.5, {Vout}, 0 ) }
V3 3 0 DC={Vout}
D1 0 out MUR3060
D2 out 3 MUR3060
*
.MODEL MUR3060 D BV=600 CJO=517P IBV=10U IS=235U M=.333
+ N=3.68 RS=35M TT=86.4N VJ=.75
*
.ENDS
****
.SUBCKT COMPAR NINV INV OUT params: VHIGH=12 VLOW=100m VHYS=50m
E2 HYS NINV Value = { IF ( V(OUT) > {(VHIGH+VLOW)/2}, {VHYS}, 0
+ ) }
E1 4 0 Value  = {IF (  V(HYS,INV) > 0, {VHIGH}, {VLOW} ) }
RO 4 OUT 10
CO OUT 0 100PF
.ENDS
****
```

4.11.2 A Voltage-Controlled Oscillator Featuring Dead Time Control

Resonant topologies make use of half- or full-bridge configurations. We can thus extend the power VCO by adding a dead time and complementary outputs. We will therefore be able to use it straightaway, without the need for a complicated controller model. The internal circuitry shown in Fig. 4-15a differs slightly from that of Fig. 4-14a, but the spirit is similar.

Dead time (DT) is a small amount of time during which both switches are open. This is needed for several reasons. First, if one switch is activated while the second one is still heading toward the blocking state, so-called shoot through can occur and the associated consumption dramatically increases. Second, in some resonant topologies, the dead time helps to achieve zero voltage switching (ZVS) and delays the activation of the transistor until its body diode fully conducts.

A soft start has also been added to avoid lethal overcurrent situations at power-on. It is programmable via a simple parameter change (SS). Figure 4-15b shows a frequency variation between 1 MHz and 100 kHz with a 100 μs soft start.

Output a and b delivers TTL-compatible logic levels and can drive a simple bipolar driver or directly a transformer, as we will see in a real application later.

4.12 GENERIC CONTROLLERS

Using a complete PWM model can sometimes lead to prohibitive simulation times simply because of inherent model complexity. In some situations you really need this level of detail. However, for a first-order approach, e.g., to quickly test a configuration or an idea, it is simply overkill. The following lines describe the book generic models and show how fast they simulate. These models will cover voltage-mode, current-mode, push-pull, half-bridge, two-switch versions, and so on. We will discover how to model the basic current-mode and voltage-mode subcircuits, leaving the description of the others to the topology-by-topology design recipes.

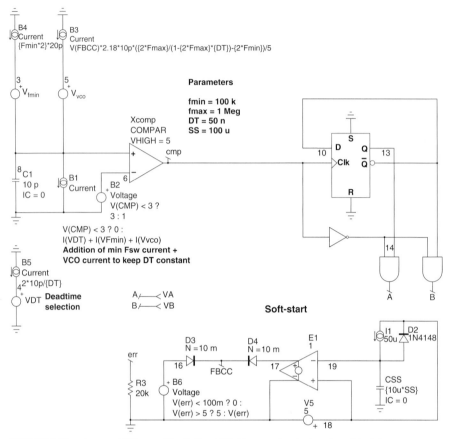

FIGURE 4-15a A power VCO controlling half or full bridges requires the insertion of dead time.

4.12.1 Current-Mode Controllers

A current-mode controller requires a few key elements to operate:

- A clock generator fixes the frequency at which the power supply is operated. Its output drives the set of the latch circuit. The clock generator is often built around a ramp signal, later on used for ramp compensation purposes. Finally, one of its outputs resets the latch in absence of proper signal coming from the current comparator: this is the duty cycle limit.
- The latch represents another important section of the controller. It transforms every single pulse coming from the clock into a steady-state signal, usually equal to V_{cc}. This signal will stay high until another clock cycle makes it toggle via the duty cycle limit or the current-sense comparator detects that the current has reached its set point.
- An error amplifier is sometimes present. In modern controllers, this element has gone since all the intelligence (error amplifier and reference) is kept on the secondary side, e.g., with a TL431. This is the case for the ON Semiconductor NCP120X series. However, in the UC384X series, the op amp is onboard and will therefore be kept in this model.
- A fast current comparator is important to reduce propagation delays and overshoots. The model does not include leading edge blanking, and a small *RC* network will be necessary to cleanse the voltage applied on the current-sense pin.

BASIC BLOCKS AND GENERIC SWITCHED MODELS 379

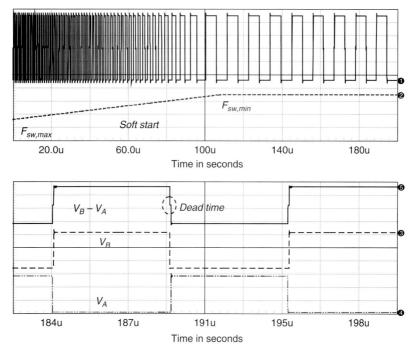

FIGURE 4-15b A start-up sequence showing frequency variations between 1 MHz and 100 kHz.

Based on these comments, the internal diagram of the controller appears in Fig. 4-16. Its simple symbol resides in the lower right corner.

For the sake of simplicity, most of the characteristics inherent to controllers have been omitted such as UVLO circuits, current consumption, driver stage, and so on. Should you need more

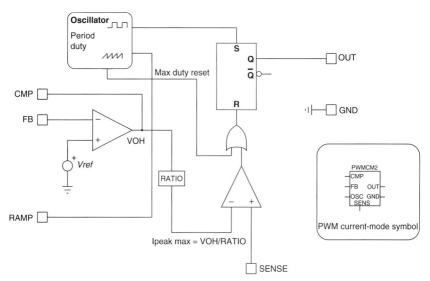

FIGURE 4-16 The internal circuitry of a generic single-output current-mode PWM controller.

comprehensive devices, some editors sell power electronic libraries in which real models such as the UC3842/43/44/45 are included and mimic most of the internal behaviors [6]. Below is a list of the alterable parameters that the model accepts. It is PWMCM2 under *IsSpice* and *PSpice*:

REF	This is the internal reference voltage, usually 2.5 or 1.25 V
PERIOD	Switching period at which the controller operates, 10 μs for 100 kHz, for instance
DUTYMAX	Maximum duty cycle in fault condition, start-up, and so on
RAMP	Pass the amplitude of the ramp you want to see on pin "RAMP"
VOUTHI	Driver output voltage in the high state
VOUTLO	Driver output voltage in the low state
ROUT	Driver output resistor
VHIGH	Op amp maximum output voltage
VLOW	Op amp minimum output voltage
ISINK	Op amp sink capability
ISOURCE	Op amp source capability
POLE	First pole position in hertz (origin pole)
GAIN	DC open-loop gain (default = 90 dB)
RATIO	Fixes the maximum peak voltage when the op amp delivers *VHIGH*. For instance, if *VHIGH* = 3 V and *RATIO* = 0.33, then the maximum sensed voltage is 1 V. If you put a 1 Ω resistor to monitor the current, the maximum peak current is 1 A.

4.12.2 Current-Mode Model with a Buck

A quick test can now be implemented using a classical buck converter. Figure 4-17a depicts the adopted structure. Since the power switch floats, a ground-referenced source B_1 routes the voltage image on the current-sense pin. The system delivers 5 V at 10 A.

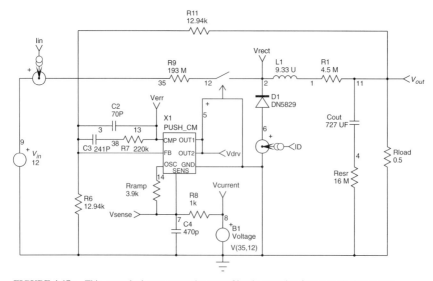

FIGURE 4-17a This example demonstrates the ease of implementation for a generic CM model.

The ramp compensation is accomplished by summing a fraction of the oscillator sawtooth and the current-sense information. It has several beneficial effects that we discuss later. A 500 µs simulation has been executed in 10 s using a 2 GHz computer. With this simulation speed, output response to load or input step can be accomplished rather quickly. Figure 4-17b shows the start-up conditions before the output voltage has reached its final value. A 200 kHz switching frequency was selected.

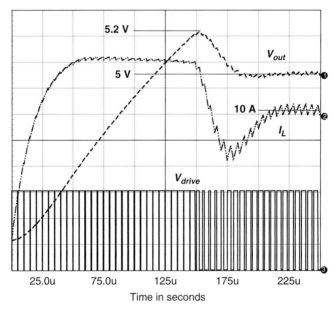

FIGURE 4-17b Simulation results of the current-mode buck converter, here a power-on sequence.

4.12.3 Current-Mode Instabilities

In Chap. 2, we discussed the instabilities introduced by the sampling action in the current loop. To highlight this phenomenon, R_{ramp} is elevated to 1 Meg to suppress any ramp compensation. If we abruptly change the input voltage from 18 V to 12.5 V, the $F_{sw}/2$ component appears (100 kHz) and stays there. Running an FFT over the coil current clearly reveals the subharmonic oscillation, as Fig. 4-17c portrays.

The input perturbation makes the entire system oscillate at $F_{sw}/2$, even if the loop gain shows a good phase margin at the 0 dB crossover frequency. The so-called gain peaking is attributed to the action of the high-Q double pole, which pushes the gain above the 0 dB line at $F_{sw}/2$ and produces an abrupt drop in phase at this point. If the duty cycle is smaller than 0.5, oscillations will naturally die out after a few switching cycles (see Chap. 2). On the contrary, if the duty cycle is greater, the oscillation will remain, as Fig. 4-17c demonstrates with the FFT of the inductor current. In conclusion, providing an external ramp is a wise solution, even if the SMPS duty cycle will be limited to 0.5: the $F_{sw}/2$ double pole will be damped, thereby preventing these subharmonic oscillations.

The audio susceptibility is also affected by slope compensation. Early studies showed that an external ramp whose slope is equal to 50% of the inductor downslope could nullify the audio susceptibility in a buck configuration. As previously stated, excessive ramp compensation makes the converter behave as if it were in voltage mode, with a resulting poor audio

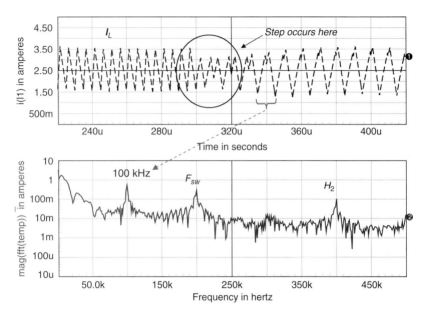

FIGURE 4-17c The inductor current shows a transition in its ripple frequency because of subharmonic oscillations.

susceptibility. Also, if minimal compensation or no ramp is provided, good input voltage rejection is achieved, and the phase of the resulting audio susceptibility is negative: an increase in input voltage will cause the output voltage to decrease. Figure 4-17d illustrates these behaviors when the input of the buck converter is stressed by a 6 V variation. The upper curve depicts the output voltage for critical ramp compensation. The voltage difference in the output envelope is only 10 mV for a 6 V input step, which leads to a (theoretical) $\Delta V_{out}/\Delta V_{in}$ of -55 dB. The middle curve shows how the response starts to degrade as additional ramp is added. The lower curve represents the error amplifier response when a slight ramp compensation is added. The decrease in the output voltage is clearly revealed by the rise in the error voltage.

4.12.4 The Voltage-Mode Model

The voltage-mode generic controller follows the description given in Fig. 4-18a.

The architecture allows the inclusion of a current limit circuit to reduce the on time of the external power switch when its peak current exceeds a user-defined limit. This option is strongly recommended to make rugged and reliable SMPS designs that can safely handle input or output overloads. By simply connecting the IMAX input to ground, you disable this option.

4.12.5 The Duty Cycle Generation

In this model, the duty cycle is no longer controlled by the current information (except in limitation mode). It is controlled by the pulse width modulator, which compares the error voltage with the reference sawtooth. The error amplifier output swing will then define the duty cycle limits. Since this output swing is user-dependent, the model will calculate the peak and valley voltages of the reference sawtooth so that the chosen duty cycle boundaries are not violated. Figure 4-18b depicts the well-known naturally sampled pulse width modulator.

BASIC BLOCKS AND GENERIC SWITCHED MODELS **383**

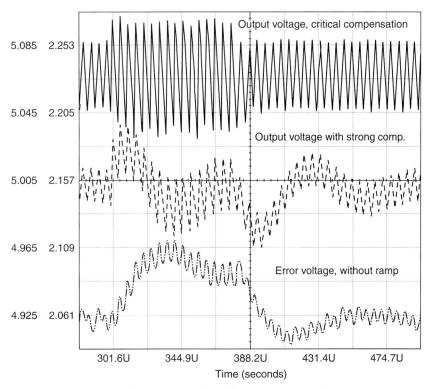

FIGURE 4-17d   The input ripple rejection is clearly affected by the compensation ramp.

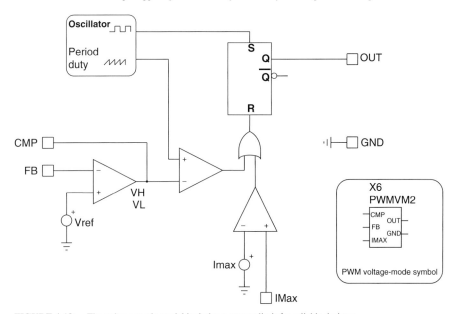

FIGURE 4-18a   The voltage-mode model includes a current limit for reliable designs.

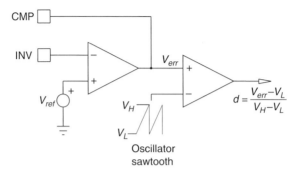

FIGURE 4-18b The model parameter will automatically adjust to satisfy the imposed duty cycle limits.

Since you will provide the main subcircuit with duty cycle limits and the error amplifier output swing, it is possible to calculate the corresponding sawtooth peak values V_{low} and V_{high}. In OrCAD's *PSpice* or Intusoft's *IsSpice*, it is easy to define some particular variables with a .*PARAM* statement. The reading of the remaining lines in the netlist is then considerably simplified:

```
.PARAM VP = { (VLOW*DUTYMAX-VHIGH*DUTYMIN+VHIGH-VLOW)/(DUTYMAX-
+ DUTYMIN) }
.PARAM VV = { (VLOW-DUTYMIN*VP)/(1-DUTYMIN) }
```

The sawtooth source then becomes

```
VRAMP 1 0   PULSE {VV} {VP} 0 {PERIOD-2N} 1N 1N {PERIOD}
```

In Fig. 4-18a, you can see a latch driving the output. This latch ensures so-called double pulse suppression, which is very useful in the presence of switching noise. Figure 4-18c explains how this latch avoids the appearance of a double pulse when a quick spike distorts the ramp level.

This technique is commonly used in modern voltage-mode controllers.

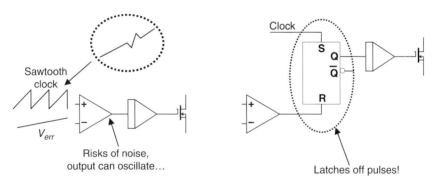

FIGURE 4-18c A latch protects the modulator against a double pulse occurring in a noisy environment.

4.12.6 A Quick Example with a Forward Converter

Since the remaining elements have already been defined (comparators, error amplifier, etc.), we are all set. The test circuit of Fig. 4-19a represents a forward converter which delivers 28 V at 4 A from a 160 V dc input source.

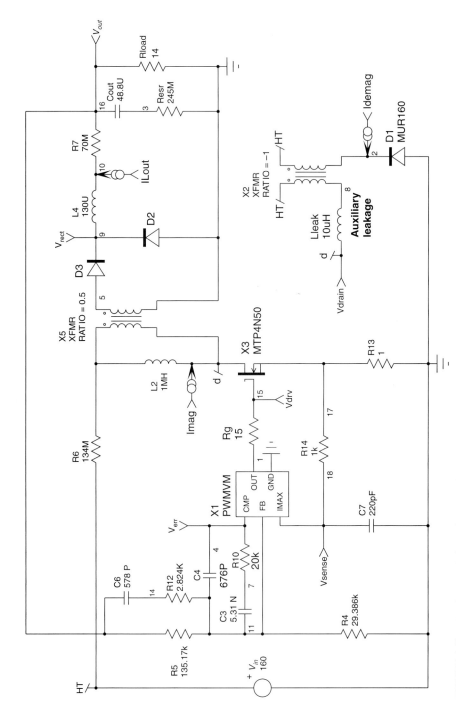

FIGURE 4-19a A voltage-mode forward converter built around the generic model.

The switching frequency is set at 200 kHz, with a maximum duty cycle of 0.45 because of the forward structure and the 1:1 power/demag turns ratio (more details will be given in the forward dedicated section). The power switch is modeled with a smooth transition element, as provided by OrCAD and Intusoft. Figure 4-19b depicts the curves which are obtained at steady state. The drain voltage swings to twice the input voltage and stays there until core reset occurs. You can see the ringing as evidence of the presence of a leakage inductor on the reset circuit.

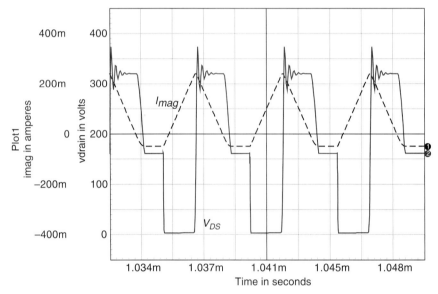

FIGURE 4-19b Simulation results of the voltage-mode forward converter.

The voltage-mode model appears under the names of PWMVM2 under *IsSpice* and *PSpice*. Their operating parameters are the following:

REF	This is the internal reference voltage, usually 2.5 or 1.25 V
PERIOD	Switching period at which the controller operates, 10 μs for 100 kHz, for instance
DUTYMAX	Maximum duty cycle in fault condition, start-up, and so on
DUTYMIN	Minimum duty cycle the controller can go down to (0.1 = 10%)
IMAX	Maximum authorized peak current (voltage across the sense resistor)
VOUTHI	Driver output voltage in the high state
VOUTLO	Driver output voltage in the low state
ROUT	Driver output resistor
VHIGH	Op amp maximum output voltage
VLOW	Op amp minimum output voltage
ISINK	Op amp sink capability
ISOURCE	Op amp source capability
POLE	First pole position in hertz (origin pole)
GAIN	DC open-loop gain (default = 90 dB)

4.13 DEAD TIME GENERATION

Bridge or half-bridge designs using MOSFETs or IGBTs need some dead time between the commutations to avoid any cross-conduction current spikes. Unfortunately, classical *PULSE* or *PWL* commands are often impractical, especially when either frequency or pulse width needs to change during the simulation run. For this reason, a simple block transforming a single clock input into two complemented signals is an interesting subcircuit. Associated with the above generic models, it will be a good companion to also implement synchronous rectification, active clamp, and so on.

Figure 4-20a offers a simple way to build a dead time generator. An *RC* network slows down on the inputs and creates the dead time we are looking for. If the AND gate threshold voltage corresponds to one-half of the high level, then it is easy to show that

$$t_{delay} = 0.693 \times RC \tag{4-19}$$

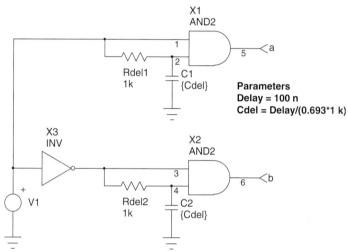

FIGURE 4-20a Three logical gates, *RC* networks, and you are all set to build a dead time generator.

The *PARAMETERS* statement in SNET calculates the capacitor value (the resistor being fixed to 1 kΩ) to reach the desired delay. Figure 4-20b shows the results, confirming the presence of a dead time between both outputs.

Typical applications include a half-bridge driver but also synchronous rectification. Figure 4-21a portrays a half-bridge driver featuring a gate-drive transformer together with a simple bipolar buffer. A third PNP transistor makes sure the MOSFET is properly blocked (and stays locked in this state) when the other one is activated. Figure 4-21b shows resulting waveforms, which indicate the presence of dead time between both driving signals.

4.14 LIST OF GENERIC MODELS

Several generic models have been built, and their use will be detailed in specific application examples. Here is the list of what you will find to work with under *IsSpice* or *PSpice*:

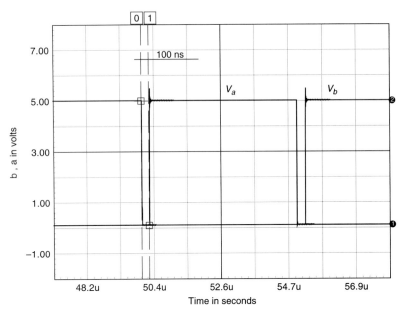

FIGURE 4-20b Both signals are free of any overlapping area.

- PWMCM2 or PWMCM: current-mode controller, single-ended
- PWMVM2 or PWMVM: voltage-mode controller, single-ended
- PWMCMS: current-mode controller featuring a secondary output for synchronous rectification or active clamp techniques
- 2SWITCHCM: current-mode controller, dual outputs
- PUSH_CM: current-mode controller for push-pull applications
- PUSH_VM: voltage-mode controller for push-pull applications
- HALF_CM: current-mode controller, half-bridge applications
- HALF_VM: voltage-mode controller, half-bridge applications
- FULL_CM: current-mode controller, full bridge applications
- FULL_VM: voltage-mode controller, full bridge applications
- FREERUN: a controller for QR applications. One includes a dead time (DT suffix), the other does not

4.15 CONVERGENCE OPTIONS

As can happen with any mathematical tool when one is looking for a solution, SPICE can sometimes fail to converge. Without detailing why this can happen (Refs. 7 and 8 do it thoroughly), we give below some options that you can use to tweak SPICE parameters and make the simulator finally converge.

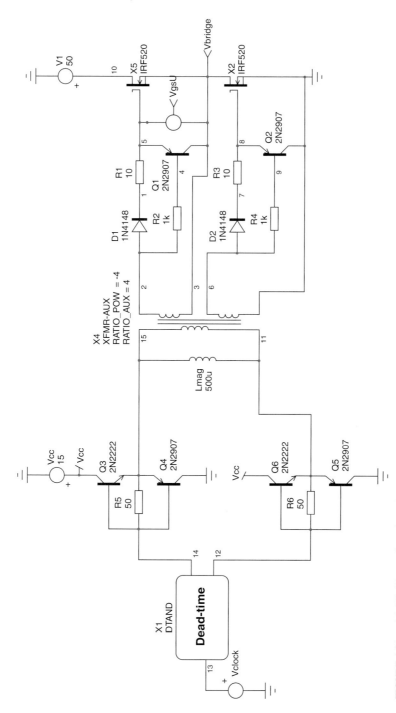

FIGURE 4-21a A half-bridge configuration using the dead time generator and a bipolar driver.

389

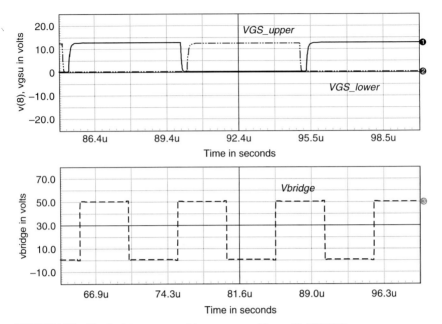

FIGURE 4-21b   The dead time generator drives an upper and lower side MOSFET.

TRANSIENT Simulations

.options RELTOL = 0.01 (default = 0.001)	RELTOL sets the relative error tolerance for convergence. It also obviously affects the simulation speed. You can relax it down to 1% (0.01) for power simulation where less precise results are needed.
.options ABSTOL = 1 μA (default =1 pA) .options VNTOL = 1 mV (default 1 μV)	ABSTOL and VNTOL, respectively, define the absolute voltage and current error tolerances. Once RELTOL has been defined, you need to evaluate the smallest current and voltages present in the circuit. Then apply the following relationship to set ABSTOL and VNTOL: ABSTOL = RELTOL * Smallest_current VNTOL = RELTOL * Smallest_voltage
.options GMIN = 100p (default = 1p)	GMIN ensures that a voltage applied across an active component always forces a minimum current to flow through it. If a device presented an infinite conductance ($di/dt = 0$), the iteration algorithm would simply fail. Typical relaxing values are $1n$ or $10n$.
.options ITL1 = 1k (default = 100)	ILT1 is used to increase the dc iteration limit. This the number of iterations SPICE will

BASIC BLOCKS AND GENERIC SWITCHED MODELS   391

perform before giving up during the bias point calculation. If you have a messagesuch as "No Convergence in dc analysis," then you should consider raising ITL1 to 1*k* or 1000 with *PSpice*.

.options RSHUNT = 10 Meg (default = 0)

Implemented by *IsSpice*, this option instructs SPICE to wire a resistor of RSHUNT ohms from every node to ground, thus always offering a dc path to ground.

AC Simulations

Ac simulations are less prone to nonconvergence problems. However, the simulator can sometimes fail during the bias point calculation, at the very beginning of the simulation. If you use B elements or other behavioral sources, be sure that no division by zero is possible. You can easily avoid that by adding some offset to any denominator: $V = V(3)/V(8) \rightarrow V = V(3)/(V(8) + 1u)$. *.options ITL1 = 1k* can also help in finding the correct dc bias point. It is not recommended to pass initial conditions to an average model. Nevertheless, it is possible to "advise" the simulator with an initial guess value. If you know that node 4 should be around 4 V, then you can add a statement such as *.nodeset V(4) = 4*. It will sometimes help you to debug nagging convergence issues.

WHAT I SHOULD RETAIN FROM CHAP. 4

This chapter helps you to build your own simulation toolbox. Once it is built according to your needs, you will discover how easy it becomes to assemble the blocks together and form working models. However, there are a few particular points you must focus on:

1. In-line equations (B elements with *IsSpice* and G/E values with *PSpice*) are analog behavioral model sources: they describe perfect behaviors via equations. Perfection is not part of the world, but is part of SPICE: when you have an equation $V=V(1) > 3 ? 5 : 0$, it implies a zero time span switching output as soon as the condition is fulfilled. Should you drive a capacitor with such a source, it would be likely to see a convergence issue because of the induced capacitive current. For that reason, always include parasitic elements (such as *RC* networks) to produce realistic subcircuit signals.

2. Divisions by zero can lead to numerical overflow. Always make sure the denominator will not reach zero with the inclusion of a small fixed value, such as 1u: $V(4)/V(5)$ becomes $V(4)/(V(5)+1u)$.

3. Do not use full-featured models unless you need to assess effects of slew rates, bias offsets, current, etc. First-order models are usually good enough to debug the circuit and see how the power supply behaves.

4. Magnetic core models exist and vary from simple ones to more complex ones (the generic core model and the Jiles and Atherton description, for instance). It is good practice to start with a simple form and gradually increase the model complexity once convergence is solid.

5. Use generic models for PWM simulations. As they do not implement UVLO circuits, discrete drive circuits, and so on, they can deliver results in a much faster way than the complete model would do.

REFERENCES

1. L. G. Meares, "New Simulation Techniques Using Spice," *Proceedings of the IEEE 1986 Applied Power Electronics Conference*, pp. 198–205.
2. D. C. Jiles and D. L. Atherton, "Theory of Ferromagnetic Hysteresis," *Journal of Magnetism and Magnetic Materials*, vol. 61, no. 48, 1986.
3. S. M. Sandler, *SMPS Simulations with SPICE3*, McGraw-Hill, 2005, New York.
4. D. Maksimovic and R. Erickson, "Modeling of Cross Regulation in Multiple-Output Flyback Converters," *IEEE Applied Power Electronics Conference*, Dallas, 1999.
5. L. Dixon, *Unitrode Magnetics Design Handbook*, MAG100A, www.ti.com (search for MAG100A).
6. http://www.aeng.com/PSpice.asp
7. R. Kielkowski, *Inside SPICE*, McGraw-Hill, New York, 1994.
8. A. Vladimirescu, *The SPICE Book*, Wiley, New York, 1993.

APPENDIX 4A AN INCOMPLETE REVIEW OF THE TERMINOLOGY USED IN MAGNETIC DESIGNS

This appendix offers a quick review (thus incomplete!) of the commonly used terms in magnetic designs. A lot of literature is available on the market which comprehensively covers the subject. We therefore strongly encourage readers to acquire some of the books cited in the References to help strengthen their knowledge of this particular domain.

4A.1 Introduction

Figure 4A-1 portrays an inductor L made of N turns wound in the air and driven by a dc source of amplitude V. If we were to manufacture this inductor and place some iron filings on top of a paper sheet above it, we would see the filings arranged along lines called magnetic force lines. These lines are going around the inductor and come back to the core through the air. In this drawing, the air is considered as the magnetic path, offering a way to "conduct" the magnetic force lines around the core. Depending on the literature, these lines are called field lines, flux lines, or force lines. Flux lines are always closed loops. In an arbitrary volume, the number of flux lines entering the volume must equal the number of lines leaving that volume. This is called flux conservation.

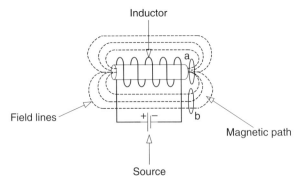

FIGURE 4A-1 A simple core wound in the air generates magnetic field lines around it.

The induction flux through a surface (a magnetic core section, for instance) represents the totality of line forces that cross this surface. The flux will be maximal when the surface stands normal to the line forces and will be null when this surface becomes parallel to the same line forces. The flux (Greek symbol φ), whose unit is the weber (abbreviation: Wb), only expresses the amount of line forces crossing the surface. Again, the flux can be designated in the literature as flux, magnetic flux, or magnetic induction flux.

The quantity of flux lines crossing a given area defines the flux density B, expressed in teslas (abbreviation: T). One tesla equals one weber per square meter. In presence of a uniform flux density B in a surface A_e, we have the following relationship linking these two variables:

$$\varphi = BA_e \quad (4\text{A-1})$$

In Fig. 4A-1 for instance, we can observe that surface a collects more force lines than surface b. Hence the flux density in surface a is greater than in b.

4A.2 Field Definition

Back to the illustration, a current I circulates in the inductor, proportional to the source voltage and the inductor resistance. The circulation of this current creates a so-called magnetizing force H, also called the magnetic field. This field can generate a force on other moving electrical charges. This is the case, for instance, in a cathode ray tube where the emitted electrons are deviated during vertical and horizontal sweeps.

Figure 4A-2 shows two points separated by a distance l and immersed into a magnetic field H. The force between these two points is called the magnetomotive force (*mmf*) and is obtained by the integral of the magnetic field along the distance l. When the magnetic field is of uniform strength along l, the force expression becomes

$$F = mmf = Hl \quad (4\text{A-2})$$

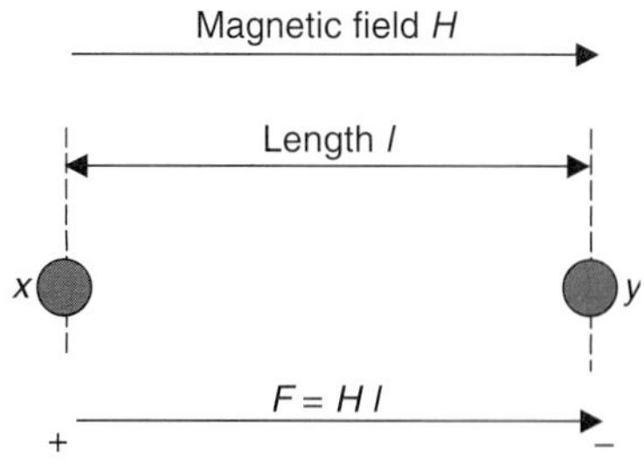

FIGURE 4A-2 The magnetic field generates a force, the magnetomotive force, between x and y.

The magnetomotive force *mmf* should not be confused with the magnetizing force H. The magnetomotive force is the cause, whereas the magnetizing force is the resulting effect. The *mmf* depends on the current amplitude I and the number of turns N. Current and *mmf* are linked via

$$mmf = NI \quad (4\text{A-3})$$

The *mmf* unit is the ampere-turn (A-T).

If the *mmf* represents a force, H is a magnetizing force per unit length. The relationship between both variables is described by the Ampere's law:

$$H = \frac{NI}{l} \quad (4\text{A-4})$$

The magnetizing force unit is the ampere per meter (A/m), l represents the magnetic path length (also denoted by l_m in the literature), sometimes called the mean magnetic path length (MPL).

4A.3 Permeability

In Fig. 4A-1, force field lines circulate through air. If we now put a block of glass in the force lines, we will not notice a modification in their trajectories. On the contrary, if we replace glass by a

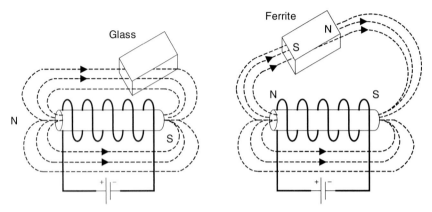

FIGURE 4A-3 A magnetic material has the ability to deviate line forces.

material having magnetic properties, e.g., ferrite, force lines are now deviated and cross the material. This is shown in Fig. 4A-3 where we can see the appearance of poles on the inserted material.

Materials characterized by magnetic properties have the ability to let magnetic field lines go through and confine them with a certain ease. As we have seen, air can be considered as a material or an environment through which line forces can move. Air is thus characterized by a certain *permeability* to line forces. This permeability is denoted by the Greek letter μ_0. Its unit of measure is henries per meter:

$$\mu_0 = 4\pi \cdot 10^{-7} = 1.257 \cdot 10^{-6} = 1.257u \tag{4A-5}$$

Magnetic materials are characterized by a permeability denoted by μ_r. The r stands for a *relative* permeability. Its value can be very high depending on the material type and its geometry (6000 is a common value, for instance). The material permeability depends on the applied magnetizing force. When the material is nonsaturated, its total permeability μ is that of free space multiplied by its relative permeability:

$$\mu = \mu_0 \mu_r \tag{4A-6}$$

Permeability expresses the material's ability to accept and confine flux lines when subjected to a magnetic field. Flux density, magnetic field, and permeability are thus linked via the following formula:

$$B = \mu H = \mu_0 \mu_r H \tag{4A-7}$$

The relative permeability value varies depending on the applied magnetic field. Figure 4A-4 portrays how permeability moves when the magnetizing force H is increased. At low values, Eq. (4A-7) is almost linear and the total permeability remains high: the steeper the slope, the higher the total permeability. When H continues to increase, the total permeability starts to decline and the slope tends to bend: μ_r quickly decreases and reaches 1, and the material is saturated with $\mu = \mu_0$. The inductor wound on the saturated magnetic material is almost the same as if it were wound on air. Saturation levels for ferrite material typically range from 0.25 to 0.5 T. It can go up to 1 T for powdered irons and molypermalloy materials. When several cycles are performed (H is positively increased, decreased to zero, and increased in the other direction), a hysteresis cycle is observed (Fig. 4A-5). Looking at this picture, we can define several parameters often encountered in magnetic materials data sheets:

- μ_i: the initial permeability describes the slope of the magnetization curve at the origin.
- μ_r: this is the permeability of a material relative to that of free space.

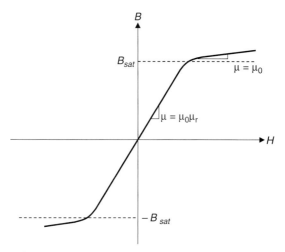

FIGURE 4A-4 The relative permeability depends on the applied magnetic field.

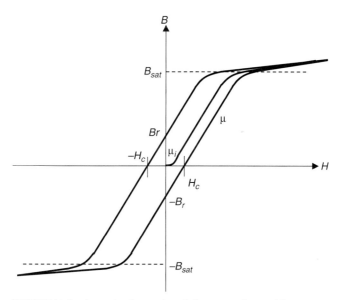

FIGURE 4A-5 A complete hysteresis cycle for a magnetic material.

- B_{sat}: this is the induction flux density at which μ_r drops to 1. This value depends on the temperature (negative coefficient).
- B_r: This is the remanent induction flux level when the magnetizing field is zero. For a flyback converter, the introduction of an air gap will make this value drop to almost zero, increasing the storage capability (LI^2) of the "transformer" (actually coupled inductors).
- H_c: this is the coercive field which brings the flux density back to zero.

4A.4 Founding Laws

James Clerck Maxwell introduced the uniform concept of electromagnetism in 1864 via a set of differential equations known as *Maxwell's equations*. Among these equations are found Faraday's and Ampere's laws, expressed here in their simplified forms:

Faraday's law:

$$V_L(t) = N \frac{d\varphi(t)}{dt} \qquad (4A\text{-}8)$$

This equation describes the voltage amplitude obtained from a winding of N turns subject to a flux variation $d\varphi$.

Ampere's law [Eq. (4A-4)]:

$$H = \frac{NI}{l}$$

This equation describes the resulting magnetizing force obtained by circulating a current I in a winding of N turns and of length l.

From Faraday's law, Lenz' law can be derived:

Lenz' law:

$$V_L(t) = -L \frac{dI_L(t)}{dt} \qquad (4A\text{-}9)$$

If you want to reduce the magnetic flux in an inductor, then an induced voltage across this inductor (electromagnetic force, *emf*) is produced in such a way that it opposes the decrease in flux. This is the well-known kickback resulting from the sudden interruption of current in a biased inductor. Without a freewheel diode, which ensures current continuity, a high voltage appears across the inductor.

4A.5 Inductance

With this set of equations on hand, it becomes possible to define the inductance of an inductor (unit of measure is the henry, abbreviated H) via a few lines:
From Eq. (4A-8) we know that

$$V_L(t) = N \frac{d\varphi(t)}{dt}$$

From Eq. (4A-1) we have

$$\varphi = BA_e$$

If we differentiate Eq. (4A-1) we obtain

$$\frac{d\varphi}{dt} = \frac{dB}{dt} A_e \qquad (4A\text{-}10)$$

From Eq. (4A-7), we know that

$$B = \mu H$$

Hence, inserting Eq. (4A-7) into (4A-10) gives

$$\frac{d\varphi}{dt} = \frac{dH}{dt} \mu A_e \qquad (4A\text{-}11)$$

Using Ampere's law, we obtain

$$\frac{d\varphi}{dt} = \frac{dI_L(t)}{dt}\frac{N\mu A_e}{l} \quad (4A\text{-}12)$$

If we reinject this definition into Eq. (4A-8), we have

$$V_L(t) = \frac{dI_L(t)}{dt}\frac{N^2\mu A_e}{l} \quad (4A\text{-}13)$$

Finally, using Lenz' law, we obtain the inductance formulation:

$$L\frac{dI_L(t)}{dt} = \frac{dI_L(t)}{dt}\frac{N^2\mu A_e}{l} \quad (4A\text{-}14)$$

After simplification, this becomes

$$L = \frac{N^2\mu A_e}{l} \quad (4A\text{-}15)$$

We have seen in this chapter that the addition of a small air gap in the core changes the permeability definition as suggested by Eq. (4-17b).

4A.6 Avoiding Saturation

The saturation of a magnetic element occurs when the flux density inside the core exceeds its saturation limit B_{sat}. For a true transformer, that is, when currents simultaneously circulate in both the primary and secondary (in a forward converter, for instance), it is important to watch the applied volt-second products across the primary. Volt-seconds, or the area of the primary voltage wave shape during the on time, directly relates to flux:
From Lenz' law [Eq. (4A-9)], we know that

$$V_L(t) = L\frac{dI_L(t)}{dt}$$

If we integrate both parts, we obtain

$$\int V_L(t) \cdot dt = \int L\frac{dI_L(t)}{dt} \cdot dt \quad (4A\text{-}16)$$

which after simplification gives

$$\int V_L(t) \cdot dt = LI_L(t) \quad (4A\text{-}17)$$

Now, by manipulating 4A-1, 4A-7 and 4A-15, we can write that:

$$N\varphi(t) = LI_L(t) \quad (4A\text{-}18)$$

Hence, Eqs. (4A-17) and (4A-18) show that integrating the voltage across an inductor (or a transformer primary) gives an indication of the flux level inside the core. Based on these definitions, a simple and powerful design equation can be quickly derived to check that the inductor (either on its own or the magnetizing inductance of a transformer) will stay out of saturation in its normal operating mode.
From Eq. (4A-1) we have

$$\varphi_{sat} = B_{sat}A_e$$

Thanks to Eq. (4A-18), we can write

$$N\varphi_{sat} = LI_{L,max} \quad (4A\text{-}19)$$

Inserting Eq. (4A-1) into (4A-19) gives us a simple design equation to avoid saturation when designing an inductor or flyback-type transformers:

$$NB_{sat}A_e > LI_{L,max} \quad (4A\text{-}20)$$

To avoid saturation of the design, you can either increase the number of turns or select a material featuring a larger core effective area. Adding an air gap will not improve the situation.

For a transformer design, e.g., in a forward design, the equation looks very similar. The magnetizing current ramps from 0 to $I_{mag,peak}$ and goes back to zero. It must remain DCM over a switching cycle; otherwise flux runaway may occur, leading to a lethal saturation. In a forward converter, the maximum magnetizing peak current is defined by

$$I_{mag,max} = \frac{V_{in}}{L} t_{on,max} \quad (4A\text{-}21)$$

If we substitute this definition into Eq. (4A-20), we obtain one of the fundamental design equations for a forward transformer:

$$NB_{sat}A_e > V_{in,max} t_{on,max} \quad (4A\text{-}22)$$

References

1. L. Dixon, *Unitrode Magnetics Design Handbook*, MAG100A, www.ti.com (search for MAG100A).
2. C. McLyman, *Transformer and Inductor Design Handbook*, Marcel Dekker, New York, 2004.
3. E. C. Snelling, *Soft Ferrites: Properties and Applications*, Butterworth-Heineman, 1988.
4. R. Erickson and D. Maksimovic, *Fundamentals of Power Electronic*, Kluwers Academic Press, 2001.
5. C. Mullet, "Magnetics in Switch-Mode Power Supplies," ON Semiconductor technical training in Asia, 2005.

APPENDIX 4B FEEDING TRANSFORMER MODELS WITH PHYSICAL VALUES

4B.1 Understanding the Equivalent Inductor Model

Inductance measurements are often made at a single frequency (1 kHz) by the manufacturer via an *LRC* meter. Unfortunately, given the equivalent model of an inductor (or transformer), the single frequency measurement can lead to significant errors. Figure 4B-1 portrays a simple equivalent inductor model valid at high frequency. Values are arbitrary for the sake of the example.

As you can imagine, this is going to resonate when sweeping the frequency. This is what Fig. 4B-2 confirms. In this figure, you can observe three distinct regions:

- *Ohmic region*: This is where none of the *LC* elements are changing the impedance. What is measured in this region is ohmic losses only.
- *Inductive region*: This is where *L* dominates *C*, the impedance increases with frequency. You should measure the inductance in this region.
- *Capacitive region*: *L* has been taken over by *C*, and impedance goes down as the frequency goes up.

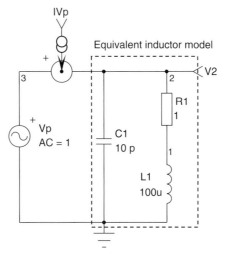

FIGURE 4B-1 An inductor equivalent model.

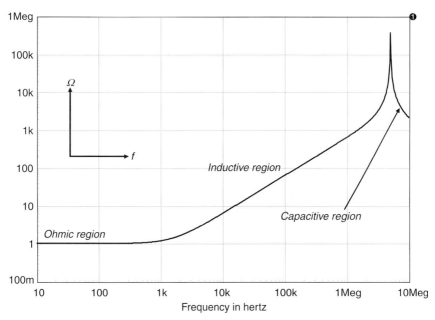

FIGURE 4B-2 Several zones can be highlighted when sweeping an inductor via a network analyzer.

If you adjust the *RLC* meter anywhere within the inductive region, you will obtain the right value. However, if you sweep within one of the other two regions, the measurement will be wrong. Therefore, it is best to use a network analyzer to unveil its impedance versus frequency plot. From this chart, the chances of making a measurement error will be greatly diminished. Otherwise, if there is no network analyzer, set the *RLC* meter to 1 to 10 kHz for magnetizing inductance (10 to 100 kHz for leakage) and verify that slightly changing the test frequency does not engender a large inductance variation. If this is the case, you are likely to be somewhere other than in the inductive region.

4B.2 Determining the Physical Values of the Two-Winding T Model

You often read, "short the secondary and measure the leakage inductance on the primary." This statement depends on the model you have adopted for simulation. For the "T" model depicted by Fig. 4B-3, if you follow the above recommendation, you can see that you actually measure the reflected secondary leakage L_{l2} in parallel with the primary inductance L_m, all in series with the primary leakage inductance L_{l1}. To obtain adequate results to feed the model with, follow the measurement procedure related to the two-winding the T-model.

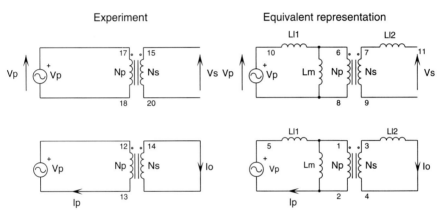

FIGURE 4B-3 The T model used in the transformer SPICE model.

1. Inject a sinusoidal voltage V_p on the primary and measure the open-circuit voltage on the secondary V_s. This what the upper portion of Fig. 4B-3 portrays. Compute the turns ratio by

$$N = \frac{N_p}{N_s} = \frac{V_p}{V_s} \tag{4B-1}$$

Note that this measurement neglects L_{l1} compared to L_m.

2. Now measure the primary inductance with a frequency around 1 kHz. This gives you L_{psopen}.
3. Repeat step 2, but the secondary is now short-circuited. Set the frequency to 10 kHz or above. Check that varying the frequency does not change the reading too much. You have $L_{psshort}$.
4. Compute the coupling coefficient k from

$$k = \sqrt{1 - \frac{L_{psshort}}{L_{psopen}}} \tag{4B-2}$$

5. Compute L_{l1} with:

$$L_{l1} = (1 - k)L_{psopen} \tag{4B-3}$$

6. Compute L_{l2} with

$$L_{l2} = (1 - k)L_{psopen}\frac{1}{N^2} \tag{4B-4}$$

7. Finally, compute the magnetizing inductance by

$$L_m = kL_{psopen} \tag{4B-5}$$

With an ohmmeter measure the primary and secondary dc resistances, respectively, R_p and R_s and enter these values into the transformer subcircuit as shown in Fig. 4B-4.

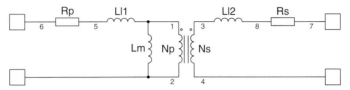

FIGURE 4B-4 The complete two-winding transformer SPICE model.

4B.3 The Three-Winding T Model

The final three-winding model appears in the Fig. 4B-5 drawing where three leakage elements take place in series with each winding. All secondary windings are normalized to

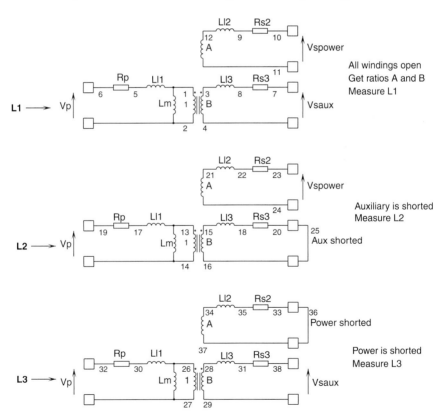

FIGURE 4B-5 The three-winding equivalent model showing all leakage elements and the successive measurement steps.

the primary winding by ratios A and B (1:A, 1:B). It can be shown [1] that the primary leakage element depends upon the primary leakage air path P_1 but also on the permeance P_{23} between both secondary windings. In other words, if you improve the coupling between both secondaries (e.g., by twisting the wires), you *increase* the primary leakage inductance. Reference 1 also demonstrated how the leakage elements stay practically independent of the air gap length: the coupling coefficient diminishes as the air gap increases (the magnetizing inductance becomes smaller) but leakage elements stay constant.

As you will discover through the following steps, the results express how you combine the various leakage elements from the measurement data. We built a transformer prototype, hence the numerical values given in the text.

1. Inject a sinusoidal voltage V_p on the primary, and measure the open-circuit voltages on the secondaries V_{spower} and V_{saux}. Compute

$$A = \frac{V_{spower}}{V_p} \tag{4B-6}$$

$$B = \frac{V_{saux}}{V_p} \tag{4B-7}$$

Measurements give $A = 0.0817$ and $B = 0.156$.

2. Measure the inductance L_1 seen from the primary, all secondaries open:

$$L_1 = L_{l1} + L_m = 3.62 \text{ mH} \tag{4B-8}$$

3. Measure the inductance L_2 seen from the primary with the power winding open and the auxiliary short-circuited:

$$L_2 = L_{l1} + \frac{L_m \frac{L_{l3}}{B^2}}{L_m + \frac{L_{l3}}{B^2}} = 199 \text{ μH} \tag{4B-9}$$

4. Measure the inductance L_3 seen from the primary with the power winding short-circuited and the auxiliary output open:

$$L_3 = L_{l1} + \frac{L_m \frac{L_{l2}}{A^2}}{L_m + \frac{L_{l2}}{A^2}} = 127 \text{ μH} \tag{4B-10}$$

5. Measure the inductance L_4 seen from the power winding, the auxiliary short-circuited and the primary open:

$$L_4 = L_{l2} + A^2 \left[\frac{L_m \frac{L_{l3}}{B^2}}{L_m + \frac{L_{l3}}{B^2}} \right] = 1.405 \text{ μH} \tag{4B-11}$$

We now have a system of four equations with four unknowns. Feeding a math processor with these equations gives the solutions in a snapshot:

$$L_{l1} = L_1 - \sqrt{L_3 L_2 - L_3 L_1 - L_1 L_2 + L_1^2 + \frac{L_4 L_1 - L_4 L_3}{A}} = 58.5 \text{ μH} \tag{4B-12}$$

BASIC BLOCKS AND GENERIC SWITCHED MODELS

$$L_{l2} = \frac{A^2(L_{l1} - L_1)(L_3 - L_{l1})}{L_3 - L_1} = 466 \text{ nH} \quad (4\text{B-}13)$$

$$L_{l3} = \frac{B^2(L_{l1} - L_1)(L_2 - L_{l1})}{L_2 - L_1} = 3.558 \text{ μH} \quad (4\text{B-}14)$$

$$L_m = L_1 - L_{l1} = 3.56 \text{ mH} \quad (4\text{B-}15)$$

Series resistances are measured with a 4-wire ohmmeter and included in the SPICE model that appears in Fig. 4B-6. The netlist of the perfect multioutput transformer is given below in both *IsSpice* and *PSpice* syntax. You will need to add external elements to account for the above calculated values.

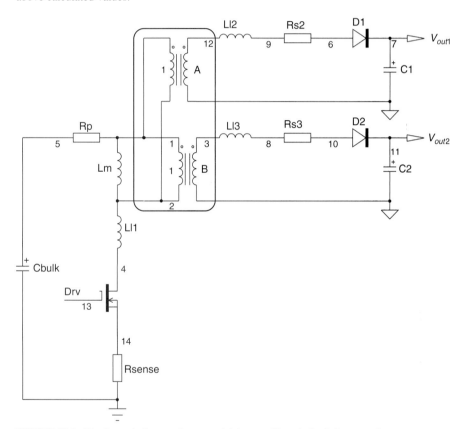

FIGURE 4B-6 The three-winding transformer model that we will use in the design examples.

IsSpice

```
.SUBCKT XFMR-AUX 1 2 3 4 10 11 params: RATIO_POW=1 RATIO_AUX=1
*Connections +Pri -Pri +SecP -SecP +SecA -SecA
*
* RATIO_POW = 1:A
* RATIO_AUX = 1:B
*
```

```
RP 1 2 1MEG
E1 5 4 1 2 {RATIO_POW}
F1 1 2 VM1 {RATIO_POW}
RS1 6 3 1U
VM1 5 6
E2 20 11 2 1 {RATIO_AUX}
F2 2 1 VM2 {RATIO_AUX}
RS2 21 10 1U
VM2 20 21
.ENDS
```

PSpice

```
.SUBCKT XFMR2 1 2 3 4 10 11 PARAMS: RATIO1=1 RATIO2=1
*
* RATIO1 = 1:A
* RATIO2 = 1:B
*
RP 1 2 1MEG
E1 5 4 VALUE = { V(1,2)*RATIO1 }
G1 1 2 VALUE = { I(VM1)*RATIO1 }
RS1 6 3 1U
VM1 5 6
E2 20 11 VALUE = { V(2,1)*RATIO2 }
G2 2 1 VALUE = { I(VM2)*RATIO2 }
RS2 21 10 1U
VM2 20 21
.ENDS
```

In some cases, it is interested to know quickly the amount of the available leakage term, as with a common-mode (CM) choke used for differential filtering purposes. Figure 4B-7 depicts the T model of such a CM choke where a 1:1 turns ratio links both windings. If we short-circuit terminals 1 and 3, the voltage across L_m should equal that on the secondary side since $n = 1$. As a result, since terminals 1 and 3 are linked, then nodes x and y share a similar level: seen from

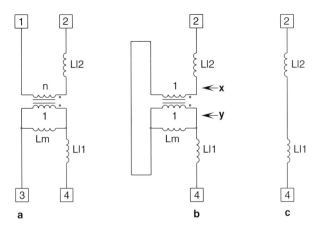

FIGURE 4B-7 When the common-mode choke is short-circuited on both of its left terminals, its right terminals appear to be connected together and the leakage terms naturally cumulate.

terminals 2 and 4, both leakage inductances are in series. This method is known to deliver quickly the leakage term of a common-mode inductor when you need help to filter differential currents. You obtain the total leakage term $L_{l2} + L_{l1}$ in one attempt, rather than L_{l2} in series with $L_m//L_{l1}$ if you decide to classically short-circuit terminals 3 and 4.

References

1. S-P. Hsu, R. D. Middlebrook, and S. Ćuk, "Transformer Modeling and Design for Leakage Control," *Advances in Switched-Mode Power Conversion*, vols. 1 and 2, TESLAco, 1983.
2. C. Basso, "AN1679/D How to Deal with Leakage Elements in Flyback Converters," www.onsemi.com.

CHAPTER 5
SIMULATIONS AND PRACTICAL DESIGNS OF NONISOLATED CONVERTERS

In this chapter, we will see how to design and simulate nonisolated switch-mode power converters via practical cases and examples. SPICE can be used as a real assistant when a design is made to check if the simulated circuit variables are within the ballpark of what you expect. Nevertheless, as often said, the simulation software represents another guiding tool just like a spreadsheet or an automated calculation program. Always question the delivered results via engineering judgment! This is the reason why an equation-based description is offered throughout the previous chapters. Please read it thoroughly. Otherwise, if you fail to understand the fundamental definitions, how can you check if results are correct? Let us start the series of examples via the simplest converter, the buck converter.

5.1 THE BUCK CONVERTER

The buck converter can be used in a variety of applications where the need to decrease a voltage exists: off-line power supply for white goods products, battery-powered circuits (e.g., cell phones), local regulators [so-called point-of-load (POL) regulators] and so on. When one looks for an increase in efficiency, a synchronous buck can be built or even a multiphase buck used if output currents are too large for a single stage. We can now study the first example with a simple voltage-mode circuit.

5.1.1 A 12 V, 4 A Voltage-Mode Buck from a 28 V Source

The specification states that the output must be 12 V when loaded by a 4 A current and powered by a source varying from 20 V to 30 V. The maximum peak-to-peak output voltage ripple must always stay below 125 mV, and the switching frequency is 100 kHz. Please note an ac input ripple current specification of 15 mA peak to peak:

$V_{in,min} = 20$ V

$V_{in,max} = 30$ V

$V_{out} = 12$ V

$V_{ripple} = \Delta V = 125$ mV

V_{out} drop $= 250$ mV maximum from $I_{out} = 200$ mA to 3 A in 1 μs

$I_{out,max} = 4$ A
$F_{sw} = 100$ kHz
$I_{ripple,peak} = 15$ mA, input current maximum ripple

First, we can search for the duty cycle extremes:

$$D_{min} = \frac{V_{out}}{V_{in,max}} = \frac{12}{30} = 0.4 \quad (5\text{-}1)$$

$$D_{max} = \frac{V_{out}}{V_{in,min}} = \frac{12}{20} = 0.6 \quad (5\text{-}2)$$

Rearranging Eq. (1-91), we can place the corner frequency of the LC filter depending on a ripple amplitude of 125 mV:

$$f_0 = F_{sw}\frac{1}{\pi}\sqrt{\frac{2\Delta V}{(1 - D_{min})V_{out}}} \quad (5\text{-}3a)$$

This gives us

$$f_0 = \frac{100k}{3.14}\sqrt{\frac{0.25}{(1 - 0.4) \cdot 12}} = 5.93 \text{ kHz} \quad (5\text{-}3b)$$

Equation (5-3a) defines the corner frequency placement but does not define L or C separately. We need another equation to obtain either C or L. Figure 5-1 represents the ripple current in the inductor.

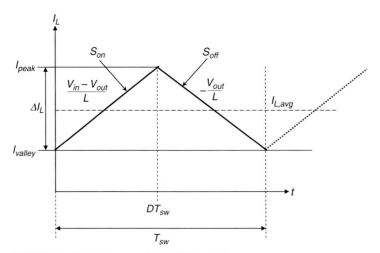

FIGURE 5-1 The ripple current in the CCM buck inductor.

Observing this figure, we see that the valley current can be expressed by

$$I_{valley} = I_{peak} - S_{off}t_{off} = I_{peak} - \frac{V_{out}(1 - D)}{LF_{sw}} \quad (5\text{-}4)$$

The total ripple current ΔI is defined by $I_{peak} - I_{valley}$, thus

$$\Delta I_L = I_{peak} - I_{valley} = I_{peak} - I_{peak} - \frac{V_{out}(1-D)}{LF_{sw}} = \frac{V_{out}(1-D)}{LF_{sw}} \qquad (5\text{-}5)$$

What matters is the ripple amplitude compared to the average current (actually the output dc current). Therefore, we can reformulate Eq. (5-5) by dividing both sides of it by I_{out}:

$$\frac{\Delta I_L}{I_{out}} = \delta I_r = \frac{V_{out}(1-D)}{LF_{sw}I_{out}} \qquad (5\text{-}6)$$

From this equation, the normalized ripple increases as D goes down in value. Therefore, we can extract the inductor value that we look for to minimize the ripple current at high line:

$$L = \frac{V_{out}(1 - D_{min})}{\delta I_r F_{sw} I_{out}} \qquad (5\text{-}7a)$$

If we select a maximum ripple amplitude δI_r of 10%, then we find the inductance value must be

$$L = \frac{12(1 - 0.4)}{0.1 \times 100k \times 4} = 180\,\mu H \qquad (5\text{-}7b)$$

At maximum load current, the peak current will increase to

$$I_{peak} = I_{out} + \frac{V_{out}(1 - D_{min})}{2LF_{sw}} = 4 + \frac{12(1 - 0.4)}{2 \times 180u \times 100k} = 4.2\,A \qquad (5\text{-}8)$$

Combining Eq. (1-87) (the resonant frequency definition) and Eqs. (5-3), we can now calculate the capacitor value:

$$C_{out} = \frac{1}{[2\pi f_0]^2 L} = \frac{1}{39.5 \times 5.93k^2 \times 180u} = 4\,\mu F \qquad (5\text{-}9)$$

What matters for the final choice of a capacitor is the rms current that crosses it (for obvious thermal reasons). For a CCM buck, this current equals the ac portion of the rms inductor current. We can show that its value is

$$I_{C_{out},rms} = I_{out}\frac{1 - D_{min}}{\sqrt{12\tau_L}} \qquad (5\text{-}10)$$

where $\tau_L = \frac{L}{R_{load}T_{sw}}$ as shown in Chap. 1. Applying numerical values, we found an rms current of 115 mA. Looking at the data sheet specification of some capacitor vendors, we pick a ZL series from Rubycon [1] which features the following characteristics:

$C = 33\,\mu F$
$I_{C,rms} = 210\,mA\ @\ T_A = 105\,°C$
$R_{ESR,low} = 0.45\,\Omega\ @\ T_A = 20\,°C, 100\,kHz$
$R_{ESR,high} = 1.4\,\Omega\ @\ T_A = -10\,°C, 100\,kHz$
ZL series, 16 V

5.1.2 Ac Analysis

These capacitors are well suited for switch-mode power supplies as they feature a low impedance and can work up to a high ambient temperature. As we can see, the ESR varies quite a bit depending on this operating temperature. We know by definition that its value affects the associated zero position. The first simulation is therefore the small-signal response of the buck together with the calculated *LC* elements. This is what Fig. 5-2a shows, with an operational amplifier in the feedback path.

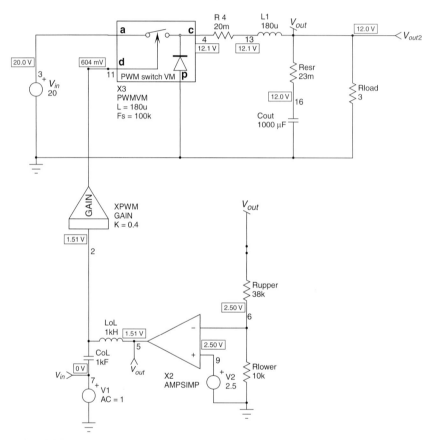

FIGURE 5-2a The voltage-mode buck converter using the calculated *LC* elements.

In this picture, you now recognize the op amp arrangement whose ac output is blocked via *LoL* but not its dc level for the bias calculation (remember, *LoL* is a short-circuit in dc). Also, the PWM gain of 0.4 indicates that the selected controller features a sawtooth of 2.5 V amplitude. The small-signal Bode plot appears in Fig. 5-2b with two different input values.

Looking at the specification, we can see an output drop requirement of 250 mV for a ΔI_{out} of 2.8 A, swept in 1 μs. Applying Eq. (3-3), we can estimate the needed bandwidth to satisfy the voltage drop specification:

$$f_c = \frac{\Delta I_{out}}{2\pi C_{out} \Delta V_{out}} \qquad (5\text{-}11)$$

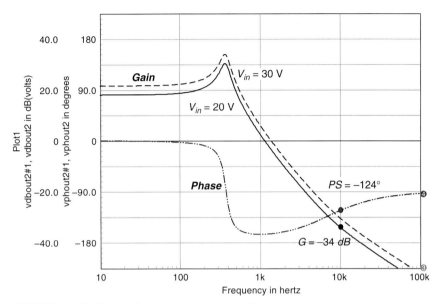

FIGURE 5-2b The buck small-signal response for the two input voltages.

Numerical application gives us a cutoff frequency of 54 kHz, a fairly large value prone to opening the door to external noises. As a preliminary conclusion, asking for a 250 mV drop from a 33 µF capacitor undergoing a 2.8 A peak load is not realistic. Therefore, we need to pick up a capacitor value that will lead to a more reasonable crossover frequency, around 10 kHz, for instance. This time, we can directly apply Eq. (3-3) and obtain the required capacitor value:

$$C_{out} = \frac{\Delta I_{out}}{2\pi f_c \Delta V_{out}} = \frac{2.8}{6.28 \times 10k \times 250m} = 178 \,\mu F \tag{5-12a}$$

It is important to note that this approximate definition only holds when the ESR effects are small compared to those of the capacitor. This situation was noted in Eq. (3-4).

This new value represents quite a change compared to what we previously estimated. Still from Rubycon, a 330 µF capacitor is now chosen. This capacitor features the following characteristics:

$C = 330 \,\mu F$
$I_{C,rms} = 760 \text{ mA} @ T_A = 105\,°C$
$R_{ESR,low} = 72 \text{ m}\Omega @ T_A = 20\,°C$
$R_{ESR,high} = 220 \text{ m}\Omega @ T_A = -10\,°C$
ZL series, 16 V

Unfortunately, the low-temperature ESR (220 mΩ) of this capacitor does not satisfy Eq. (3-4) as shown below:

$$\text{Is } R_{ESR} \leq \frac{1}{2\pi f_c C_{out}} ?? \tag{5-12b}$$

$$220 \text{ m}\Omega \leq Z_{Cout} @ 10 \text{ kHz } (48 \text{ m}\Omega) \rightarrow \text{No!}$$

We must therefore select a capacitor whose worst-case ESR stays below the capacitor impedance at the crossover frequency in order to limit its contribution to the transient output drop. A 1000 μF ZL-series capacitor seems to do the job:

$C = 1000$ μF
$I_{C,rms} = 1820$ mA @ $T_A = 105\,°C$
$R_{ESR,low} = 23$ mΩ @ $T_A = 20\,°C$
$R_{ESR,high} = 69$ mΩ @ $T_A = -10\,°C$
ZL series, 16 V

Still, the ESRs seem a little high, compared to the 1000 μF capacitor reactance value at 10 kHz (16 mΩ), but we will check in the transient response plots how it influences the waveform. What actually matters here is the current step amplitude times the complex impedance made up of C_{out} and its ESR. Diminishing this ESR naturally reduces its contribution to the output undershoot.

From the ESR values, we have a zero whose position varies between two values:

$$f_{z,low} = \frac{1}{2\pi R_{ESR,high} C_{out}} = \frac{1}{6.28 \times 69m \times 1m} = 2.3\,\text{kHz} \tag{5-13}$$

$$f_{z,high} = \frac{1}{2\pi R_{ESR,low} C_{out}} = \frac{1}{6.28 \times 23m \times 1m} = 6.9\,\text{kHz} \tag{5-14}$$

The new LC circuit resonates at a frequency given by

$$f_0 = \frac{1}{2\pi \sqrt{LC}} = \frac{1}{6.28 \times \sqrt{1m \times 180u}} = 375\,\text{Hz} \tag{5-15}$$

A CCM voltage-mode converter operated in CCM is usually stabilized in the following way:

- Place a pole at the origin to get the highest dc gain.
- Place a pair of zeros located at or slightly before the lowest resonant frequency.
- Place a first pole compensating the ESR zero. If this zero occurs at too high a frequency, look for a RHPZ, if any, and place the pole at its location. Otherwise, place the pole at one-half the switching frequency.
- In this buck application, the second pole finds its place at one-half the switching frequency, usually for noise filtering purposes. In converters where a RHPZ exists (boosts and flyback designs etc.), the second pole can take place at this RHPZ location in case the previous pole was placed elsewhere.

To stabilize the circuit via a type 3 circuit, we are going to use the manual pole and zero method described in Chap. 3:

1. For a 12 V output from a 2.5 V reference voltage, let us adopt a 250 μA bridge current. The lower and upper resistor values are therefore

$$R_{lower} = \frac{2.5}{250u} = 10\,\text{k}\Omega$$

$$R_{upper} = \frac{12 - 2.5}{250u} = 38\,\text{k}\Omega$$

2. Open-loop sweep the voltage-mode buck at the two input levels (Fig. 5-2b results).
3. From the Bode plot, we can see that the required gain at 10 kHz is 34 dB worst case.

4. To cancel the *LC* filter peaking, place a double zero at the resonant frequency, 375 Hz.
5. Place a first pole at the highest zero position to force the gain to roll off, 7 kHz.
6. Place a second pole at one-half of the switching frequency to avoid noise pickup, 50 kHz.
7. Using the manual placement method described in Chap. 3, evaluate all the compensator elements.

$R_2 = 127$ kΩ

$R_3 = 285$ Ω

$C_1 = 3.3$ nF

$C_2 = 180$ pF

$C_3 = 12$ nF

These results are delivered by the automated schematic capture (see Fig. 5-2c *parameters* table), but the CD-ROM spreadsheet would give similar results. We can now plot the ac response featuring both ESRs at the two different input voltages (high and low). The application circuit is portrayed by Fig. 5-2c where ac responses are gathered in Fig. 5-2d. In the worst case, the cutoff frequency is 10 kHz, as expected, and the phase margin never drops below 60°. The best test to check whether the design is stable and offers the right dropout characteristic is to step its output. Connecting a load to the output via a switch gives the needed current sweep from 200 mA to 4 A in 1 µs. Results appear in Fig. 5-2e with different combinations of input voltages and ESRs. All cases are perfectly stable and the voltage drop is within specifications, even with the ESR values of the ZL capacitor type. Let us now take a look at the operating waveforms.

5.1.3 Transient Analysis

Transient analysis can be carried out using either a real buck controller circuit, e.g., a monolithic switcher such as the LM257X from ON Semiconductor, or the generic voltage-mode controller we have depicted in a preceding chapter. This is what is shown in Fig. 5-3a. You can recognize the upper side switch and the freewheel diode. For the sake of simplicity, we picked up a perfect switch whose on-state resistor can be modified. Of course, you could place a P-channel MOSFET (easy to control) or a cheaper N-channel but driven through either a charge pump or a bootstrap circuit. The inductor current is difficult to sense in a floating configuration as with a buck. Again, as many different options exist, we decided to use an in-line equation, B_1, to route the floating resistor current through a ground-referenced source. This output enters the generic model via its *IMAX* input. By the way, this model features the following parameter changes:

$PERIOD = 10u$	Switching period of 10 µs (100 kHz)
$DUTYMAX = 0.9$	Maximum duty cycle up to 90%
$DUTYMIN = 0.01$	Minimum duty cycle down to 1%
$REF = 2.5$	Internal reference voltage of 2.5 V
$IMAX = 5$	Reset occurs when the voltage on the *IMAX* pin exceeds 5 V
$VOL = 100m$	Error amplifier output when low
$VOH = 2.5$	Error amplifier output when high

The rest of the parameters are defaulted. The internal ramp amplitude computation depends on the duty cycle limits and the op amp excursion (*VOL* and *VOH*). For a maximum duty cycle

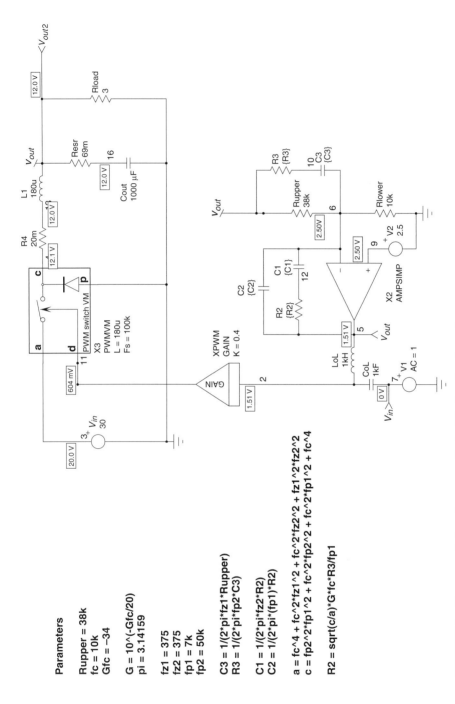

FIGURE 5-2c The automated schematic capture really simplifies the pole and zero placement.

SIMULATIONS AND PRACTICAL DESIGNS OF NONISOLATED CONVERTERS

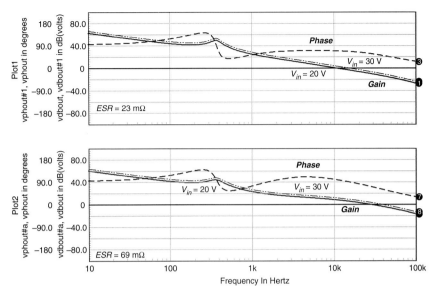

FIGURE 5-2d The Bode plot reveals a good phase margin at a 10 kHz minimum cutoff frequency. Phase curves are superimposed on each other.

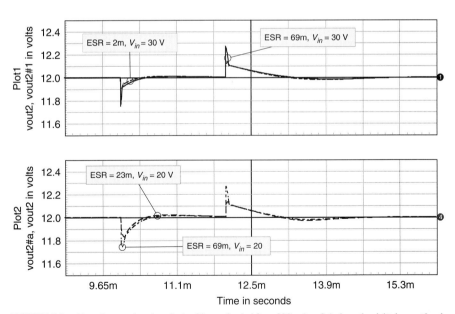

FIGURE 5-2e Transient results when the load is step loaded from 200 mA to 3 A show the right dropout level.

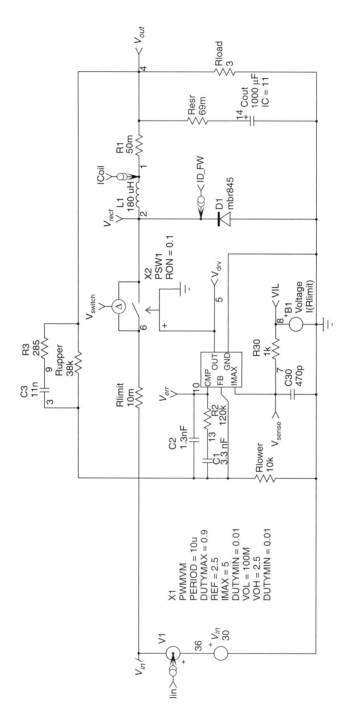

FIGURE 5-3a The buck transient simulation using the generic model.

close to 100% (as here) and an op amp output of 2.5 V, the ramp amplitude is 2.5 V. If *VOH* were 3 V, then the ramp would have increased to 3 V to satisfy 100% duty cycle. Conversely, if we have a duty cycle limit to 50% and we select a *VOH* of 3 V, the ramp amplitude becomes 6 V.

A 3 ms simulation (notice the initial condition on the output capacitor) takes 17 s to complete (1.8 GHz computer). From this simulation, we can obtain lot of information. Let us explore it.

5.1.4 The Power Switch

Figure 5-3b portrays the switch waveforms, mainly the current and voltage across it. Of course, with a lack of parasitic elements, they are quite idealized! Nonetheless, we can extract useful information that will guide us in the final MOSFET selection. First, we can see the contribution

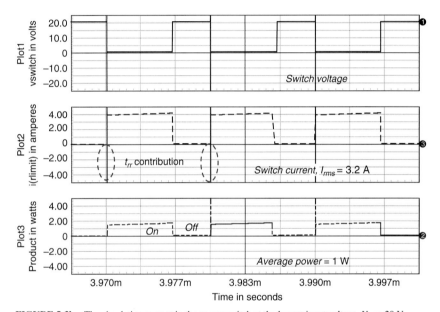

FIGURE 5-3b The circulating current in the power switch at the lowest input voltage, $V_{in} = 20$ V.

of the freewheel diode recovery time t_{rr}: when a diode conducts current and you brutally block it by a reverse bias, the diode becomes a short-circuit until it recovers its blocking capability. In the buck, the current imputed to the recovery time is seen as a narrow spike in the MOSFET current. As the switching frequency increases, it contributes to elevate losses in the switch and the diode. Alright, but here we have a Schottky and this type of diode does not exhibit any recovery characteristics. True, but a Schottky features a large parasitic capacitance that also contributes to increase the switch burden. Also, sometimes under heavy biases, its guard rings are activated and need to recover as well. Exploring these loss mechanisms goes beyond the scope of this book. Reference 2, page 96. gives a good description of the phenomenon in play.

To measure the switch conduction losses, first isolate one switching period and extract the rms value. Here we read 3.2 Arms. As the switch $R_{DS(on)}$ is arbitrarily fixed to 10 mΩ (always take the $R_{DS(on)}$ at the highest junction temperature, e.g., 110 °C), then the conduction loss is found to be

$$P_{cond,MOSFET} = R_{DS(on),max} I_{D,rms}^2 = 102 \text{ mW} \tag{5-16}$$

The diode loss contribution can be significant in a buck converter. To get an idea of the average power dissipated by the switch, just multiply the voltage across the switch by its current: you obtain the instantaneous power. Now average the result around 1 switching period, and you have the average power. Here we read 1 W. Of course, this number is not very precise as capacitive parasitic elements (in particular the nonlinear junction capacitors) and recovery time are not optimally modeled in SPICE. Hence, bench measurements and oscilloscope observations are essential to obtain an accurate power loss value. However, in this particular case, the major losses on the power switch are those attributed to switching. Thus, careful selection of the diode is extremely important to reduce the dissipated power on the switch.

5.1.5 The Diode

The diode "sees" current during the off time only. As a result, in CCM, its dissipation increases as the duty cycle goes lower (off time goes up), at the highest input voltage. Figure 5-3c displays

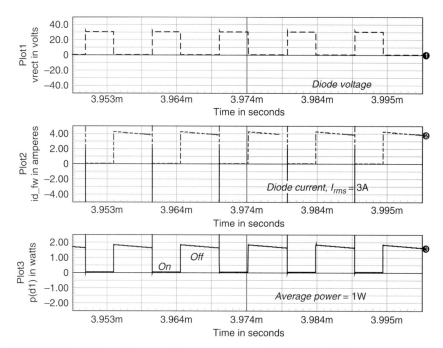

FIGURE 5-3c Diode waveforms, revealing an average power of 1 W for $V_{in} = 30$ V.

the diode's waveforms, current, voltage, and instantaneous power obtained as mentioned before. The total conduction losses for a diode are given by

$$P_{cond} = V_{T0}I_{d,avg} + R_d I_{d,rms}^2 \approx V_f I_{d,avg} \quad (5\text{-}17)$$

where $I_{d,avg}$ = average current in the diode
V_f = forward voltage at the considered diode current

V_{T0} = forward voltage at which the diode starts to conduct ($\approx$ 0.4 V for a Schottky, $\approx$ 0.6 V for a PN diode)

R_d = dynamic resistance of junction at considered averaged current (look at V–I curve in diode data sheet)

$I_{d,rms}$ = rms current flowing through diode

We can extract all these data from the diode data sheet, but a quick integration of the $P_d(t)$ waveform gives the total average dissipated power, including switching losses: 1 W. Again, this value is indicative as recovery or capacitive effects on simple Schottky diodes are not well modeled. The Intusoft newsletter NL32 discusses a new model approach brought by Lauritzen in 1993 [3]. It shows improvement over the traditional SPICE primitive, but as it uses a complex subcircuit, it can sometimes be the source of convergence issues.

It is important to make a remark, here, about the diode selection. Usually, I/V characteristics of diodes are very similar. If you look at the specifications of a 20 A diode and a 3 A diode, forward drops are very close at a 1 A average current:

MBR20H100CT V_f = 0.3 V for a 1 A current and a junction at 100 °C.
MBRA340T3 V_f = 0.25 V for a 1 A current and a junction at 100 °C.

However, the difference between the two devices lies in the diode die size and therefore its ability to evacuate heat. The information to look for, then, is the thermal performance of the diode via its various R_θ characteristics:

MBR20H100CT $R_{\theta JA}$ = 60 °C/W in free air
MBRA340T3 $R_{\theta JA}$ = 81 °C/W when soldered on minimum pad area

As you can see, the 3 A diode (MBRA340T3) features a decent thermal resistance provided that you reserve a copper area of 1 in square area. Depending on the board's physical dimensions that you have to deal with, it might become a difficult layout exercise to find the right place for the device. On the contrary, the MBR20H100CT can dissipate more power in free air, without any heat sink, naturally easing the layout phase. The 3 A diode might be the correct choice as indicated by analytical calculations, but, in the end, the thermal performance dictates the final selection.

5.1.6 Output Ripple and Transient Response

One of the first design criteria is the ripple amplitude. The LC filter time constant was actually evaluated given this important parameter. However, because of undershoot considerations, the output capacitor has been increased, given the authorized bandwidth. Figure 5-3d displays the output ripple at 4 A, with two different input levels. We read 30 mV peak to peak, well within the original specifications. As the figure shows, worst case occurs at V_{in} = 30 V [see Eq. (5-3)].

Thanks to the compensation network we have put in place, we can also test the transient response to an output load test. It can be done by a standard current source using a PWL statement or a simple voltage-controlled switch. The test is carried out at both minimum and maximum input voltages and with different ESR combinations. Figure 5-3e confirms the good stability of the compensated converter for a 69 mΩ ESR. As we can see, the selected capacitor offers the right voltage drop (around 200 mV), giving a 50 mV margin compared to the starting specification of 250 mV.

The efficiency can be measured by displaying the product of $I(V_{in})$ and $V(V_{in})$. This gives you the instantaneous input power. Average it over one switching cycle, and you obtain the average input power P_{in}. Display the output voltage, get the average value, square it, and divide it by the load: this is the output power P_{out}. The efficiency is simply $\frac{P_{out}}{P_{in}}$. We read 93.5% at low line and 92.7% at high line.

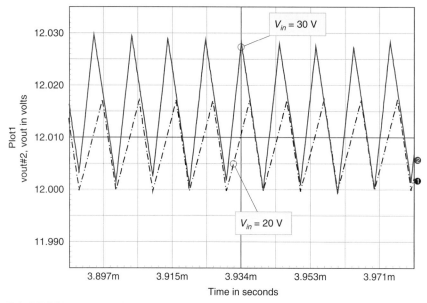

FIGURE 5-3d The output ripple comfortably fits the specifications.

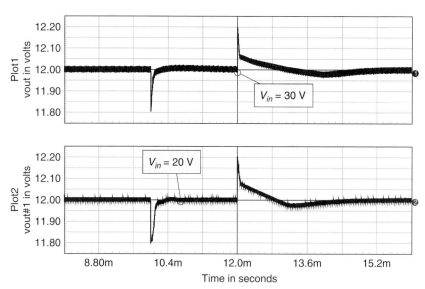

FIGURE 5-3e The buck output voltage when suddenly loaded from 200 mA to 3 A confirms its good stability performance.

5.1.7 Input Ripple

For conducted EMI reasons, the input current ac content must be kept below 15 mA peak. Therefore, an input filter has to be installed. The first operation consists of evaluating the buck input current signature. Thanks to SPICE, it is rather easy. Figure 5-3f displays the resulting waveform and its associated FFT. From the picture, we can read a peak fundamental value of 2.45 A. Let us now break down the filter design into successive steps, as we did in Chap. 1:

1. The specification limits the peak input ripple to 15 mA. Therefore, the required attenuation needs to be

$$A_{filter} < \frac{15m}{2.45} < 6m \quad \text{or better than a 44 dB attenuation}$$

2. Position the cutoff frequency of the LC filter by using Eq. (1-198):

$$f_0 < \sqrt{0.006 \times F_{sw}} < 7.7 \text{ kHz}$$

Let us select $f_0 = 7$ kHz.

From the Fig. 5-3f measurements, we can evaluate the rms current circulating in the filtering capacitor via the simulator measurement tools. The result represents one of the key selection criteria for filtering capacitors:

$$I_{ac} = \sqrt{I^2_{rms} - I^2_{dc}} = \sqrt{2.7^2 - 1.7^2} = 2.1 \text{ Arms,}$$

a rather large value in this case.

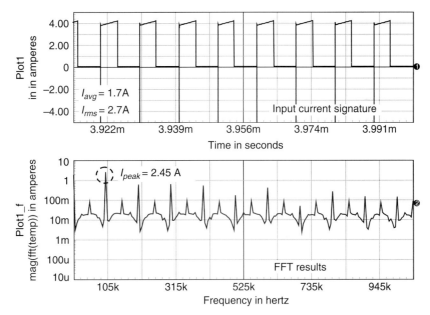

FIGURE 5-3f The buck input current signature.

From the step 2 result, we have two variables to play with: L and C. Most of the time, what matters is the total combined volume brought by the two elements. Once completed, a check must be made to see whether the important variables are compatible with the component maximum ratings: rms current for the capacitor, peak current for the inductor, and so on. Let us choose a 100 μH inductor, and we can quickly determine the capacitor value via

$$C = \frac{1}{4\pi^2 f_0^2 L} = 5.2 \,\mu F$$

or 5.6 μF for the normalized value. We checked the data sheet of a metallized polypropylene type and found that a Vishay 731P [4] could do the job. The specification sheet states a ripple current of 5.1 Arms at 85 °C, which corresponds to what we are looking for.

3. From the vendor catalog, the data sheet unveils the parasitic components of L and C: $r_{Cf} = 14 \,m\Omega$ (R_2) and $r_{Lf} = 10 \,m\Omega$ (R_1).
4. We now plug these parasitic resistors into Eq. (1-196) to check if the final attenuation remains within the limit, e.g., below 10m, despite the addition of the parasitic elements:

$$L = 100 \,\mu H \qquad R_1 = 10 \,m\Omega \qquad C = 5.6 \,\mu F \qquad R_2 = 14 \,m\Omega$$

$$\left\|\frac{I_{in}}{I_{out}}\right\| = \sqrt{\frac{0.014^2 + \frac{1}{(3.51)^2}}{(24m)^2 + \frac{1}{(3.51)^2} - \frac{200u}{5.6u} + (62.8)^2}} = 4.56\,m$$

which is below the 6m specification (see item 1).

5. Applying Eq. 1-185 gives the maximum filter output impedance:

$$\left\|Z_{outFILTER}\right\|_{max} = \sqrt{\frac{(5.6u \times 14m^2 + 100u)(5.6u \times 10m^2 + 100u)}{5.6u^2(14m + 10m)^2}} = 744 \,\Omega \text{ or } 57 \,dB\Omega$$

6. The converter static input resistance can be found by using Eq. (1-181b) which gives 8.3 Ω or 18.4 dBΩ at low line. Looking at the above result, where the output impedance peaks to 744 Ω, we will find overlapping areas between the filter output impedance and the power supply input impedance: a damping action seems obvious in this case to ensure stability. We can unveil instability in two different ways. The first way is by displaying on the same graph both the SMPS ac input impedance plot and the filter output impedance. This is what Fig. 5-3g offers where the overlapping area immediately pops up. The second option consists of ac sweeping the converter open-loop gain with and without the EMI front-end filter. Figure 5-3h has retained this solution. One can clearly see the filter resonant action which engenders the stiff phase dip: the power supply could not be compensated.

7. The damping element, a series resistor, needs to be placed across the output capacitor to damp the peak. To avoid permanent heat dissipation, a capacitor appears in series with the resistor to block the dc component. To evaluate the series resistor value, you can either use Eq. (1-193) or place an arbitrary 1 Ω resistor in series with a capacitor of 4 times the filter capacitor. Then sweep the resistor value until the peak dampens. This what we did in Fig. 5-3i. A combination made up of a 4 Ω resistor and 22 μF capacitor gives adequate results in this case.

8. Once the filter is properly damped, you can run the Fig. 5-2d circuit to which the filter is added. This is what we did, and we found the transient response was not affected.

9. Final check: Install the *RLC* filter in Fig. 5-3a, run a simulation again, and see whether the source input current amplitude $I(V_{in})$ appears to be within specification. The result appears in Fig. 5-3j; we read $I_{in,peak} = 11.4$ mA, so we are within the initial requirements. A transient load step can be undertaken as a final check if needed.

This ends the design example for the buck in voltage mode.

SIMULATIONS AND PRACTICAL DESIGNS OF NONISOLATED CONVERTERS   **423**

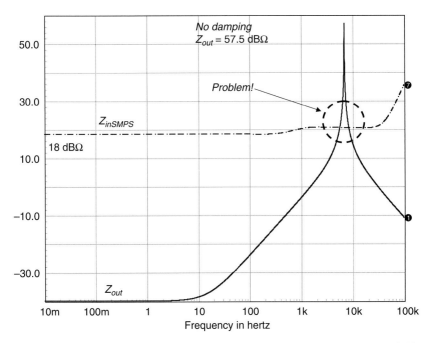

FIGURE 5-3g Both ac converter input impedance and filter output impedance plots are shown in this graph.

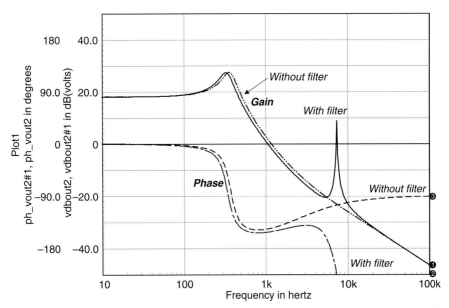

FIGURE 5-3h The open-loop gain, once the filter is added, reveals a clear phase degradation associated with the *RLC* filter peaking.

424 CHAPTER FIVE

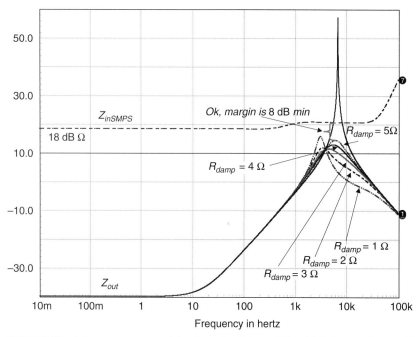

FIGURE 5-3i A few ac sweeps suggest damping values between 3 Ω and 5 Ω. Lower values would modify the resonant frequency.

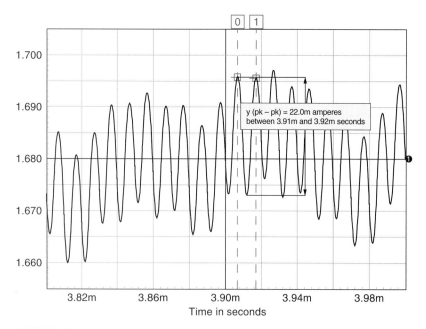

FIGURE 5-3j After installation of the input filter, the peak input current stays below 15 mA, as the initial design requirements request.

5.1.8 A 5 V, 10 A Current-Mode Buck from a Car Battery

As the electronic load share increases in the automotive environment, a need exists to safely power logic blocks from a stable 5 V or 3.3 V source. We assume the design requirements are as follows:

$V_{in,min} = 10$ V
$V_{in,max} = 15$ V
$V_{out} = 5$ V
$V_{ripple} = \Delta V = 25$ mV peak to peak
V_{out} drop $= 200$ mV maximum from $I_{out} = 1$ A to 10 A in 1 μs
$I_{out,max} = 10$ A
$F_{sw} = 250$ kHz
$I_{ripple,peak} =$ no input current requirement

Despite the current-mode architecture, the passive element selections do not differ too much compared to the voltage-mode technique. We look first for the converter's duty cycle variations:

$$D_{min} = \frac{V_{out}}{V_{in,max}} = \frac{5}{15} = 0.33 \tag{5-18}$$

$$D_{max} = \frac{V_{out}}{V_{in,min}} = \frac{5}{10} = 0.5 \tag{5-19}$$

In light of Eq. (5-19), we can see a maximum operating duty cycle of 50%. Since we selected a current-mode structure, we will surely need ramp compensation. Now that the duty cycle excursion is known, we see what kind of cutoff frequency satisfies the output ripple requirements [Eq. (5-3a)]:

$$f_0 = F_{sw}\frac{1}{\pi}\sqrt{\frac{2\Delta V}{(1-D_{min})V_{out}}} = \frac{250k}{3.14}\sqrt{\frac{0.05}{(1-0.33)5}} = 9.7 \text{ kHz} \tag{5-20}$$

We have already determined that the filter inductor value must satisfy Eq. (5-7a). Again, if we take a 10% ripple, we obtain the following inductor value:

$$L = \frac{V_{out}(1-D_{min})}{\delta I_r F_{sw} I_{out}} = \frac{5(1-0.33)}{0.1 \times 250k \times 10} = 13.4 \text{ μH} \tag{5-21}$$

We will select the closest normalized value which is 10 μH, at the expense of a slightly higher ripple current. The current seen by the power switch peaks to [Eq. (5-8)]

$$I_{peak} = I_{out} + \frac{V_{out}(1-D_{min})}{2LF_{sw}} = 10 + \frac{5(1-0.33)}{2 \times 10u \times 250k} = 10.7 \text{ A} \tag{5-22}$$

In a current-mode control architecture, the controller measures the instantaneous current flowing through the inductor. The simplest technique uses a resistive drop (a shunt) across which the voltage is measured. The controller thus defines the maximum shunt voltage the feedback authorizes when the maximum power is requested. For instance, assume the maximum sense voltage is 1 V; then, across a 1 Ω shunt, the maximum authorized current will be 1 A. In the example, a direct current of 10 A flows through the inductor. We therefore are interested in selecting a controller with a low sense voltage, e.g., in the vicinity of 100 mV maximum. Otherwise permanent losses on the shunt will hamper efficiency. Therefore, using a 12 A peak current and assuming a maximum sense voltage of 100 mV, we find the shunt value via

$$R_{sense} = \frac{100m}{12} = 8.3 \, \text{m}\Omega \tag{5-23}$$

The 12 A value includes the necessary margin to cope with shunt and maximum sensed voltage dispersions. These are often overlooked design parameters which lead to a poor design, unable to fulfill the specifications because of natural parameter spreads. Fortunately, the simulator lets you assign tolerances to key components, and a Monte Carlo analysis helps you to run simulations representative of dispersions. This is a lengthy simulation, but worthwhile if you want to get a production-proof design.

The capacitor, as we have seen, depends on the *LC* cutoff frequency but also on the final voltage drop. If we apply Eq. (5-9), the recommended capacitor value for the ripple only specification is

$$C_{out} = \frac{1}{4\pi^2 f_o^2 L} = \frac{1}{39.5 \times 9.7k^2 \times 10u} = 27 \, \mu\text{F} \tag{5-24}$$

We operate the converter at 250 kHz. A 15 kHz crossover frequency does not look too ambitious. Of course, you always have the possibility of increasing the crossover frequency, but since an automobile engine compartment is a noisy area, noise pickup is likely to occur if this number is too large. Plugging 15 kHz into Eq. (5-12a), we obtain the capacitor value leading to the required drop of 200 mV:

$$C_{out} = \frac{\Delta I_{out}}{2\pi f_c \Delta V_{out}} = \frac{9}{6.28 \times 15k \times 0.2} \approx 480 \, \mu\text{F} \tag{5-25}$$

Before we can choose the right capacitor, its ripple current has to be known. From Eq. (5-10) we find

$$I_{C_{out},rms} = I_{out} \frac{1 - D_{min}}{\sqrt{12\tau_L}} = 10 \frac{1 - 0.33}{\sqrt{12 \times 5}} = 387 \, \text{mA} \tag{5-26}$$

where $\tau_L = \dfrac{L}{R_{load} T_{sw}}$

In this application, since the output current variation is large (9 A), we absolutely need to minimize the ESR of the selected capacitor, despite the wide temperature range required by the automotive market for this design. One of the capacitor types known for low parasitic resistance is the OS-CON series from Sanyo [5]. Another good characteristic of these devices is an almost constant ESR when the temperature changes. We select an SEQP/F13 type featuring the following characteristics:

$C = 560 \, \mu\text{F}$
$I_{C,rms} = 1.6 \, \text{A} \, @ \, T_A = 105 \text{ to } 125 \, °\text{C}$
$R_{ESR,low} = 13 \, \text{m}\Omega \, @ \, T_A = 20 \, °\text{C}, 100 \text{ to } 300 \text{ kHz}$
$R_{ESR,high} = 16 \, \text{m}\Omega \, @ \, T_A = -55 \, °\text{C}, 100 \text{ to } 300 \text{ kHz}$
10SEQP560M, 10 V

5.1.9 Ac Analysis

The buck current-mode converter acts as a third-order system, as detailed in Chaps. 2 and 3: a low-frequency pole, a zero linked to the ESR of the output capacitor and a double pole located at half the switching frequency. The zero, given by the combination of the output capacitor ESR and the capacitor itself, can help to gain some phase boost at the crossover frequency. However, as the ESR varies with temperature and age, it is important to assess what the associated zero displacements will be:

$$f_{z,low} = \frac{1}{2\pi R_{ESR,high} C_{out}} = \frac{1}{6.28 \times 16m \times 560u} = 17.8 \, \text{kHz} \quad (5\text{-}27)$$

$$f_{z,high} = \frac{1}{2\pi R_{ESR,low} C_{out}} = \frac{1}{6.28 \times 13m \times 560u} = 21.9 \, \text{kHz} \quad (5\text{-}28)$$

We can see that as these zeros are rather highly positioned, they will not be of any help to offer a natural phase boost at 15 kHz.

The LC circuit resonates at a frequency given by

$$f_0 = \frac{1}{2\pi \sqrt{LC}} = \frac{1}{6.28 \times \sqrt{10u \times 560u}} = 2.1 \, \text{kHz} \quad (5\text{-}29)$$

The ac analysis uses the current-mode PWM switch model, as illustrated in Fig. 5-4a. We can see the compensation circuit where all elements are automatically computed by the *parameters* macro. This is very useful to explore the effects of changing the bandwidth or reducing the phase margin at crossover. The pole-zero placement uses the k factor technique, but individual pole-zero placement also works, of course. The k factor technique can be explored for a type 2 compensation (DCM or current-mode CCM), but we do not recommend its use for type 3 compensation as conditional stability can occur (see Chap. 3).

Here the type 2 output goes through a divider. Why? Simply to ensure enough dynamics in the op amp output, as the shunt resistor is around 8 mΩ. We do not want the op amp voltage to move between 0 and 100 mV to respectively program 0 A or 12.5 A. The divider by 30 thus provides adequate swing on the op amp whose output now needs to deliver 3 V to impose the 100 mV drop on the shunt. The presence of this divider is seen on the total power stage gain. By observing the node V_{out2}, we plot the open-loop Bode plot, shown in Fig. 5-4b. The schematic reveals an operating duty cycle of 51% which is likely to induce subharmonic oscillations.

This is what we can immediately observe, the gain peaking at one-half of the switching frequency. This sudden gain increase must be damped via ramp compensation; otherwise instability will occur. We will inject 50% of the current inductor downslope (the slope during t_{off}), a classical value for a buck converter. Ramp calculation requires only a few simple analytical steps:

1. Evaluate the inductor downslope.

$$S_{off} = \frac{V_{out}}{L} = \frac{5}{10u} = 500 \, \text{mA/}\mu\text{s} \quad (5\text{-}30)$$

2. Reflect it as a voltage on the sense resistor.

$$S_{off}' = S_{off} R_{sense} = 0.5 \times 8m = 4 \, \text{mV/}\mu\text{s} \quad (5\text{-}31)$$

3. Apply a compensation ratio of $\frac{S_{off}'}{2}$, expressed as $S_e = 2$ mV/μs or 2 kV/s.

Enter this compensation value into the model and rerun the ac analysis. The result appears at the bottom of Fig. 5-4b.

As we can see, once compensated, the open-loop gain looks more friendly. Reading the needed gain at the 15 kHz crossover frequency point gives a value of -20 dB and a phase

428 CHAPTER FIVE

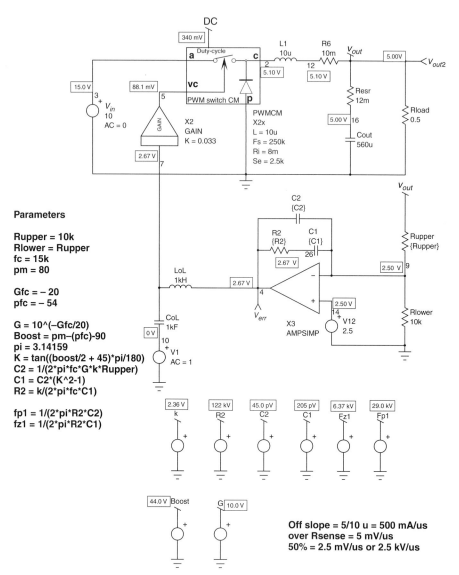

FIGURE 5-4a This buck in current mode requires a type 2 amplifier for compensation purposes.

rotation of $-54°$. These values are entered into the schematic capture macro (also made available in OrCAD), and a new ac plot can be generated to check the phase and gain margins at both input levels with different output capacitor ESRs. This is what Fig. 5-4c displays. We have not individually labeled the curves as you almost cannot distinguish them from one another. As expected, despite input voltage variations, the crossover keeps constant. This was not the case when the buck was operated using a voltage-mode control regulation scheme (Fig. 5-2c).

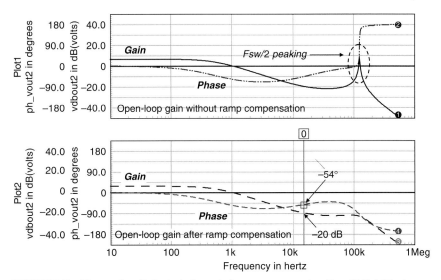

FIGURE 5-4b The open-loop Bode plot before and after ramp compensation, V_{in} = 10 V, full load.

5.1.10 Transient Analysis

For the transient analysis, the electric circuit does differ from that of the voltage-mode buck (Fig. 5-4d). The controller has changed, and we now use PWMCM2. The diode is now a 20 A type given the output current.

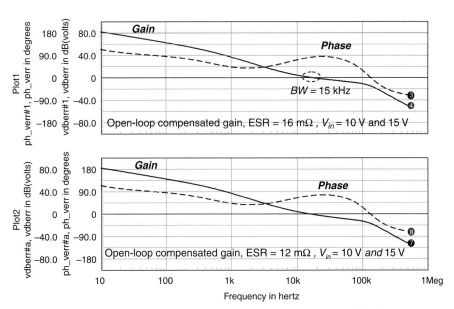

FIGURE 5-4c The compensated loop shows an excellent phase margin together with a 15 kHz crossover frequency. ESR variations induce no significant effects.

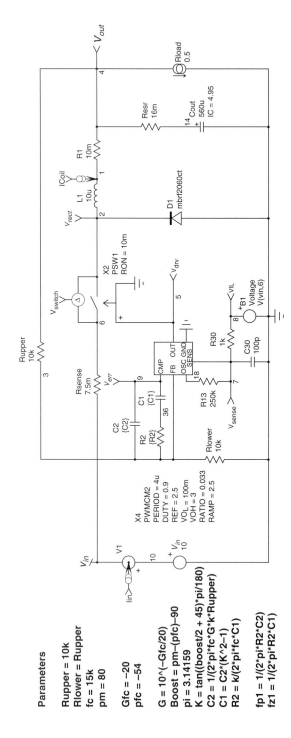

FIGURE 5-4d The transient cycle-by-cycle current-mode buck converter.

The controller parameters are the following:

PERIOD = 4u   Switching period of 4 μs (250 kHz)
DUTYMAX = 0.9   Maximum duty cycle up to 90%
DUTYMIN = 0.01   Minimum duty cycle down to 1%
REF = 2.5   Internal reference voltage of 2.5 V
RAMP = 2.5   Internal oscillator sawtooth amplitude, 0 to 2.5 V
VOL = 100m   Error amplifier output when low
VOH = 3   Error amplifier output when high
RATIO = 0.033   Feedback signal is internally divided by 30 (3/30 = 100 mV)

We have seen that the injected ramp compensation level must be around 2 mV/μs. Given that the controller delivers a 2.5 V peak ramp amplitude, how can we calculate the ramp resistor? Figure 5-4e represents the actual electric circuit leading to the current-sense node.

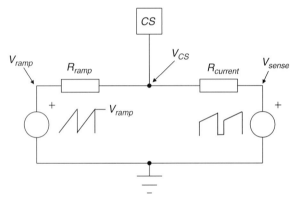

FIGURE 5-4e The current-sense pin receives information from two different signal sources.

As you can see, the current-sense signal combines a portion of the actual circuit current (its voltage image via the sense resistor) and a fraction of the oscillator ramp. The final voltage on the controller current-sense pin (CS) can be calculated by using the superposition theorem:

$V_{ramp} = 0$:

$$V_{CS} = V_{sense} \frac{R_{ramp}}{R_{ramp} + R_{current}} \qquad (5\text{-}32)$$

$V_{sense} = 0$:

$$V_{CS} = V_{ramp} \frac{R_{current}}{R_{ramp} + R_{current}} \qquad (5\text{-}33)$$

Combining Eqs. (5-32) and (5-33), we obtain the final voltage level.

$$V_{CS} = V_{sense} \frac{R_{ramp}}{R_{ramp} + R_{current}} + V_{ramp} \frac{R_{current}}{R_{ramp} + R_{current}} \qquad (5\text{-}34)$$

We can rearrange Eq. (5-34) in a more convenient way:

$$V_{CS} = \frac{R_{ramp}}{R_{ramp} + R_{current}}\left(V_{sense} + V_{ramp}\frac{R_{current}}{R_{ramp}}\right) \quad (5\text{-}35)$$

Equation (5-35) deals with voltages, but if we divide every term by t_{on}, we can express slopes instead.

$$S_{CS} = \frac{R_{ramp}}{R_{ramp} + R_{current}}\left(S_{sense} + S_{ramp}\frac{R_{current}}{R_{ramp}}\right) \quad (5\text{-}36)$$

By factoring the first term as k, Eq. (5-36) can be now be rewritten as

$$S_{CS} = k(S_{sense} + M_r S'_{off}) \quad (5\text{-}37)$$

where $k = \dfrac{R_{ramp}}{R_{ramp} + R_{current}}$ and M_r represents the percentage of the off slope we want to inject. Identifying the right-side terms gives the final equation:

$$S_{ramp}\frac{R_{current}}{R_{ramp}} = M_r S'_{off} \quad (5\text{-}38)$$

Solving for R_{ramp} leads to

$$R_{ramp} = \frac{S_{ramp}}{M_r S'_{off}} R_{current} \quad (5\text{-}39)$$

where M corresponds to the ramp coefficient we need, here 50%.
If we apply the numerical values, we have:

$S'_{off} = 4\,\text{mV}/\mu\text{s}$
$S_{ramp} = 625\,\text{mV}/\mu\text{s}$ (a 2.5 V amplitude over a 4 μs period)
$R_{current} = 1\,\text{k}\Omega$ (arbitrarily fixed)
$M_r = 50\%$ as discussed above

Equation (5-39) recommends $R_{ramp} = 312\,\text{k}\Omega$ (R_{13} on the simulation diagram).

To test the circuit, we applied an output step from 1 to 10 A in 1 μs. Figure 5-4f depicts the response obtained at two different input voltages, verifying the good transient behavior. The voltage drops to 4.8 V which is right at the specification level. If necessary, the output capacitor can be slightly raised to the upper value, or the bandwidth pushed a little bit more. The lower portion of the graph includes the transient response delivered by the averaged model. Curves compare favorably except for the drop where the absence of losses (switch and diode in the averaged model) makes the model behave in a more optimistic way than the switched version. If bench experiments confirm marginal step load results, a slight increase in the output capacitor value (820 μF, for instance) will cure the problem. In that case, the designer needs to go back to the stability study and check the phase margin.

After measurement, the output ripple stays below 30 mV peak to peak for $V_{in} = 15$ V. All power dissipations and remarks formulated for the buck in voltage mode are perfectly valid for this design as well. In particular, should you need an input filter, you can follow the same steps: at the end, do not forget to test the final compensated open-loop gain once the filter is added. This problem is often overlooked and leads to oscillations on the final board. Finally, the low line efficiency was measured around 86.8% and decreases down to 83.6% at an input voltage of 15 V. This is normal; most nonisolated converters deliver their best efficiency figure (we

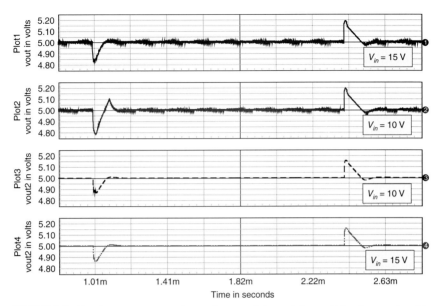

FIGURE 5-4f Transient responses at two input voltages, cycle-by-cycle (upper side) and averaged models (lower side).

can talk about the buck efficiency *acme*!) when the difference between the input and output voltages is the smallest.

5.1.11 A Synchronous Buck Converter

The above current-mode buck converter efficiency can be improved by using a synchronous MOSFET in place of the commutation diode. In the previous application circuit, the diode ensured the freewheel portion during the off time, where the common point switch-diode drops to almost zero. Actually, it is not exactly zero as various diode elements are active. This illustrated in Fig. 5-5a. Here the diode's forward drop V_f and its dynamic resistance R_d have been added to the circuit.

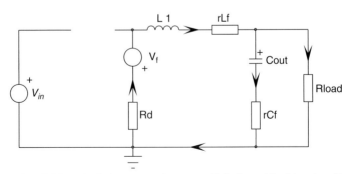

FIGURE 5-5a The diode acts as a voltage source V_f (its forward drop) in series with its dynamic resistance R_d.

Because of these elements, the efficiency of the converter clearly suffers, especially if the diode forward drop reaches a significant value compared to the load voltage V_{out}. For instance, a 0.6 forward drop with a 5 V output has greater efficiency impact than a 0.7 V drop with an 18 V output. To solve this problem, a simple solution consists of replacing the diode with a low-resistance power switch, a driven MOSFET, as Fig. 5-5b portrays.

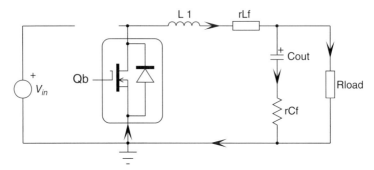

FIGURE 5-5b A MOSFET short-circuits the diode during the off time, reducing the forward drop.

When the $R_{DS(on)}$ of the selected MOSFET goes sufficiently low, the drop clearly approaches a few tens of millivolts, giving the efficiency a boost at low input voltages. The Q_b label implies that the synchronous MOSFET receives a driving signal out of phase compared to the main power switch. However, care must be taken to avoid turning on these two switches at the same moment, or else shoot-through will occur. For this reason, a dead time always exists to let the synchronous MOSFET body diode first conduct, then to later activate the MOSFET at almost zero V_{DS}. To help simulate this structure, a generic controller has been built (PWMCMS under *IsSpice* and *PSpice*) and appears in the current-mode battery-operated converter (Fig. 5-5c).

The new model parameter is simply

$$DT = 100 \text{ ns} \quad \text{inserted dead time between activation of two switches}$$

In this simulation circuit, we now observe the inductor current and not the switch current as we did before. This helps to eliminate all spurious pulses linked to shoot-through or reverser recovery effects. The op amp is a fast type featuring a fixed gain of 3. The compensation is similar to that of the previous buck.

After the simulation has been performed, Fig. 5-5d unveils all the typical waveforms. We can clearly see the presence of the dead time between the actions of the switches. In the middle curve, at the main switch opening, the current first circulates either in synchronous MOSFET body diode or in the diode you have kept in place (here, this is D_1). Some designers always keep a Schottky in place, thus diverting current from the body diode, considered too lazy for a proper job. The diode conduction can be seen as a small negative voltage step, verifying its forward drop. Then the synchronous switch closes, and the voltage approaches zero given the $R_{DS(on)}$ we adopted here (10 mΩ). Compared to the previous circuit, the efficiency has raised to 93.8% at low line, an increase of 7%. This is a bit optimistic as new losses (such as gate charge losses that must also be added) make a practical increase closer to 4 to 5% in real circuits. The transient response stays similar to that of the nonsynchronous version.

5.1.12 A Low-Cost Floating Buck Converter

Some applications require simple nonisolated buck directly operated from the rectified mains. This is the case for white goods apparatus (the washing machine is part of the white goods family) where

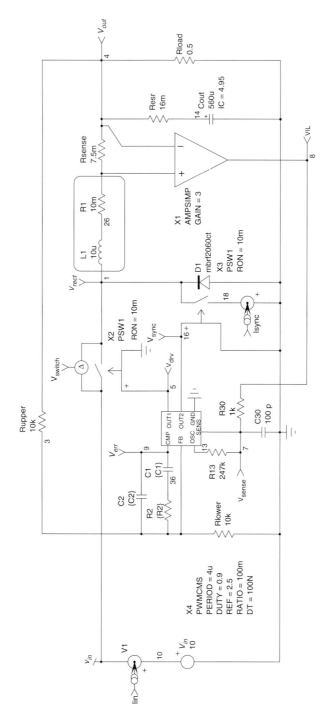

FIGURE 5-5c A current-mode buck operated with a synchronous switch.

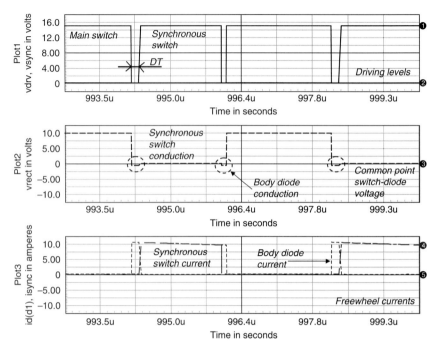

FIGURE 5-5d Typical waveforms of the synchronous buck converter.

a microprocessor gives life to the front panel. In these designs, the current-mode topology does not represent the panacea as small duty cycles are in play. The initial specifications are as follows:

$V_{in,min}$ = 85 Vrms, or a rectified 120 Vdc rail, neglecting the ripple
$V_{in,max}$ = 265 Vrms, or a rectified 374 Vdc rail
V_{out} = 5 V ± 10%
$I_{out,max}$ = 300 mA

Suppose we want to deliver a 5 V output from a universal input. The duty cycle variations can then be expressed as the equations below show (expressed in percentage to highlight the small resulting ratio):

$$D_{min} = \frac{V_{out}}{V_{in,max}} = \frac{5}{374} = 1.33\% \qquad (5\text{-}40)$$

$$D_{max} = \frac{V_{out}}{V_{in,min}} = \frac{5}{120} = 4.16\% \qquad (5\text{-}41)$$

An inherent advantage of the current mode, however, lies in its ability to cope with large input voltage variations, such as the low line rectified ripple, for instance. In universal applications, nothing competes to it in that respect. Also, recent controllers such as the ON Semiconductor NCP1200 incorporate a so-called skip-cycle capability which slices the switching pattern in bunches at low output power demands. Hence, rather than reduce the duty cycle to a limit value (thus bringing instabilities), the controller prefers to interrupt switching pulses to regulate the power flow. This is done by observing the feedback voltage. When the potential drops too low, meaning the output voltage has reached its target, a comparator stops the pulses. As V_{out} now starts to drop (because the switching is interrupted), the feedback goes up again and reactivates the pulses. This hysteretic technique also brings excellent standby power results, as will be discussed in the flyback dedicated section.

The nonisolated buck is shown in Fig. 5-6a. This time, we no longer use a generic model but a real switcher model, an NCP1013 from ON Semiconductor operating at a 65 kHz frequency [7]. This circuit includes a current-mode controller like the NCP1200 but associated with a high-voltage MOSFET, hence the term *switcher*. The transistor features a BV_{DSS} of 700 V and, as such, perfectly suits off-line applications.

The design methodology remains similar to what we have seen. The difference lies in the fixed peak current, inside the NCP1013, of 350 mA. Hence, the maximum output current the converter will be able to deliver is (Fig. 5-1)

$$I_{out,max} = I_{peak} - \frac{\Delta I_L}{2} \tag{5-42}$$

If the ripple increases (the inductor gets smaller, for instance), the output current capability will go down. If we combine the above equation with Eq. (5-5), a simple inductor definition shows up:

$$(I_{peak} - I_{out})2 = \frac{V_{out}(1 - D_{min})}{LF_{sw}} \tag{5-43}$$

$$L \geq \frac{V_{out}(1 - D_{min})}{2(I_{peak} - I_{out})F_{sw}} \tag{5-44}$$

The above equations imply a CCM operation. Equation 1-75a/b must therefore be used to check the critical inductor limit, then implement Eq. 5-44 to obtain the final value.

Suppose we need to deliver 300 mA from our 5 V source. Equation 1-75a recommends an inductor greater than 136 μH to operate in CCM in high line. Equation 5-44 states that the selected inductor must be larger than 760 μH, naturally satisfying equation 1-75a. We can select a 1 mH inductor value.

We could calculate the amount of needed capacitor given some required ripple, as we did with Eqs. (5-3a) and (5-9). However, there is something different in this design. Equation (5-40) states

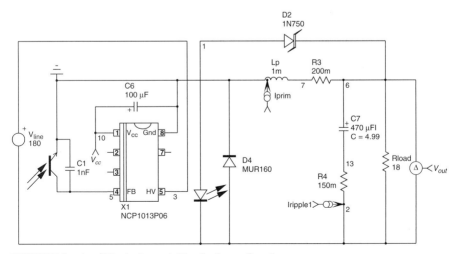

FIGURE 5-6a An off-line buck operated in a floating configuration.

a minimum duty cycle of 1.33% for $V_{in} = 374$ Vdc. Over a 15 µs switching period, it corresponds to an on time of 200 ns. Unfortunately, most current-mode controllers suffer from a minimum on time limit, and the NCP1013 is no exception to that rule. This is linked to the various propagation delays inside the logic and the LEB circuit which make the controller react more slowly than it should. Typical propagation delay values are around a few hundred nanoseconds (360 ns for the NCP1013). Therefore, you can clearly see that the controller will have extreme difficulties to cope with the minimum duty cycle at high line: the inductor current increases with a fast slope, but the controller cannot react faster than 360 ns when it has detected the overcurrent. The only way to stop the controller is via the skip-cycle feature, which, at the end, ensures a proper regulation in large input levels. However, this is to the detriment of the output ripple (this is typical of hysteretic regulators). This situation naturally fades away at low line where the inductor slope clearly diminishes. The corresponding on time of around 700 ns ($D = 4.2\%$) is closer to the circuit capability: the skip cycle disappears and a more traditional switching pattern takes place. Therefore, the traditional way of calculating the output capacitor does not really fit this structure, especially at high line.

However, to give us an idea of the needs, we can try to apply Eq. (5-12a) to low line conditions. We can therefore roughly estimate a capacitor value, considering a measured bandwidth around 1 kHz and a maximum drop of 100 mV:

$$C_{out} \geq \frac{\Delta I_{out}}{2\pi f_c \Delta V_{out}} = \frac{0.3}{6.28 \times 1k \times 0.1} = 477 \, \mu F \quad (5\text{-}45)$$

Then experiments on the bench or via simulation are carried out to check if the output ripple fits the specifications. This is what Fig. 5-6b and c tells us.

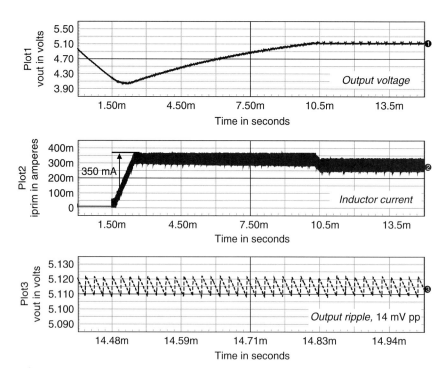

FIGURE 5-6b The start-up sequence at low line. The current is well limited by the controller.

SIMULATIONS AND PRACTICAL DESIGNS OF NONISOLATED CONVERTERS 439

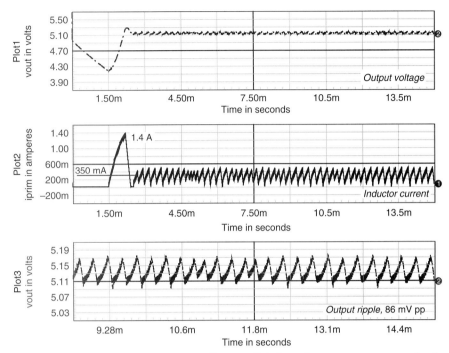

FIGURE 5-6c In high line conditions, the controller cannot limit the maximum current upon start-up, but the feedback takes over in a hysteretic mode.

At low line, the inductor slope and the duty cycle let the controller properly control the peak current. The limit is kept to 350 mA, and the output voltage slowly increases. In a steady state, the output ripple reaches about 14 mV peak- to- peak. On the contrary, at high line, the inductor slope is so high that the controller tries to reduce its on time to the minimum, but cannot fight the current increase every switching cycle. The inductor current increases (here to 1.4 A) to quickly reduce as the output voltage approaches the target. At this time, the controller enters the so-called skip cycle and hysteretic regulation occurs. The output ripple is slightly degraded but keeps within 2% of V_{out}. A final test via a load step confirms the stability of the converter at both input levels (Fig. 5-6d). Given the absence of a real compensating element, we can see the proportional type of response, compared to the integral type as already observed in Fig. 5-2e. The static error exists once loaded, but there is almost no undershoot.

5.1.13 Component Constraints for the Buck Converter

To end the buck designs, we have gathered the constraints seen by the key elements used in the buck. These data should help you select adequate breakdown voltages of the diode and the power switch. All formulas relate to the CCM operation.

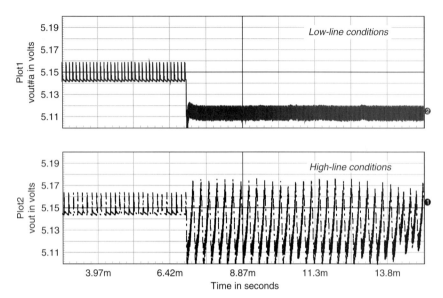

FIGURE 5-6d A load step at both input levels does not show any problem. The undershoot is well controlled.

MOSFET
$BV_{DSS} > V_{in,max}$ Breakdown voltage
$I_{D,max} > I_{out} + \dfrac{\Delta I_L}{2}$ Maximum peak current
$I_{D,rms} = I_{out}\sqrt{\left[D_{max}\left(1 + \dfrac{1}{12}\left(\dfrac{1-D_{max}}{\tau_L}\right)^2\right)\right]}$
Diode
$V_{RRM} > V_{in,max}$ Peak repetitive reverse voltage
$I_{F,avg} = I_{out}(1 - D_{min})$ Continuous current
$I_{F,rms} = I_{out}\sqrt{\left[1 - D_{min}\left(1 + \dfrac{1}{12}\left(\dfrac{1-D_{min}}{\tau_L}\right)^2\right)\right]}$ with. $\tau_L = \dfrac{L}{R_{load}T_{sw}}$.
Capacitor
$I_{C_{out},rms} = I_{out}\dfrac{1 - D_{min}}{\sqrt{12\tau_L}}$ same τ_L as above.

A comprehensive list of the buck equations in both DCM and CCM appears in Ref. 6. They are extremely useful to understand how a given converter variable (e.g., the duty cycle) can affect a parameter. However, SPICE naturally delivers all characteristics such as root mean square values, average or peak results: run the simulation worst case, and you can have a pretty good image of the currents and voltages in play. However, always consider the parasitic effects (capacitors or stray inductors) perhaps not included in your simulation fixture. These elements can add lethal spikes during switching events and need to be damped. We will come back to these data in other examples.

5.2 THE BOOST CONVERTER

Unlike the buck, the boost finds its place in systems where users need a higher voltage than the input can deliver. It can be the case in battery-powered systems (e.g., a 12 V battery supplying an audio amplifier), in a cellular phone (to deliver enough bias to the RF amplifier), or locally to provide adequate high-side bias to a dedicated circuitry. Let us start the examples with an automotive application.

5.2.1 A Voltage-Mode 48 V, 2 A Boost from a Car Battery

High-power audio applications require a fairly large input voltage to deliver a certain quantity of watts, necessary to amplify Lemmy's bass performance on stage. In this example, a boost converter is designed to deliver around 100 W from a 12 V car battery. As the load widely changes depending on the audio level, the converter will naturally cross the DCM and CCM mode boundary. However, to reduce the rms current in the capacitor, we will make sure the converter enters DCM in light-load conditions only. The specifications are as follows:

$V_{in,min} = 10$ V
$V_{in,max} = 15$ V
$V_{out} = 48$ V
$V_{ripple} = \Delta V = 250$ mV
V_{out} drop $= 0.5$ V maximum from $I_{out} = 1$ to 2 A in 10 μs
$I_{out,max} = 2$ A
$F_{sw} = 300$ kHz

As in previous examples, we first define the duty cycle excursion during the boost operation. Since it is in a continuous conduction mode by definition, we can apply Eq. (1-98):

$$D_{min} = \frac{V_{out} - V_{in,max}}{V_{out}} = \frac{48 - 15}{48} = 0.69 \quad (5\text{-}46)$$

$$D_{max} = \frac{V_{out} - V_{in,min}}{V_{out}} = \frac{48 - 10}{48} = 0.79 \quad (5\text{-}47)$$

To help define the needed inductor values, let us observe Fig. 5-7a which depicts its current variations. The peak current is defined by the excursion from the average inductor current (actually the input current of the boost converter) to one-half of the excursion ΔI_L. Mathematically, this current can be defined as

$$I_{peak} = I_{in,avg} + \frac{\Delta I_L}{2} \quad (5\text{-}48)$$

where ΔI_L represents the inductor current excursion from the valley to the peak. It is accomplished during the on time of the converter power switch. Hence

$$\Delta I_L = \frac{V_{in} D T_{sw}}{L} \quad (5\text{-}49a)$$

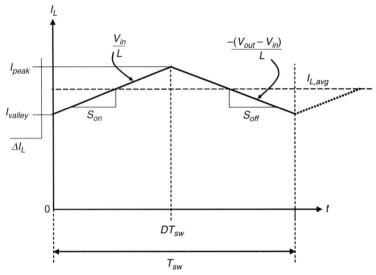

FIGURE 5-7a The boost inductor current waveform.

From this equation, it becomes possible to evaluate the point at which the ripple is maximum or minimum, to be used as a guideline in this design. Equation (5-49a) can be rewritten by replacing D with its above definition and then taking the derivative with respect to the input voltage.

$$\frac{d\Delta I_L}{dV_{in}} = \frac{d}{dV_{in}} \frac{V_{in} T_{sw}}{L} \left(1 - \frac{V_{in}}{V_{out}}\right) = \frac{T_{sw}}{L}\left(1 - \frac{2V_{in}}{V_{out}}\right) \quad (5\text{-}49\text{b})$$

Now solving for V_{in} when the derivative value is zero gives us the condition for maximum ripple:

$$V_{in} = \frac{V_{out}}{2} \quad (5\text{-}49\text{c})$$

We can also plot Eq. (5-49a) as a function of the input voltage value and check the ripple evolution. This is what Fig. 5-7b shows, confirming 24 V as being the worst case when $V_{out} = 48$ V. In this particular example, the worst case occurs for the maximum input voltage and thus the minimum duty cycle. We will see in the next example that it is not always the case depending on which side of the curve you operate. If we now link the inductor peak-to-peak current to the average input current by a ripple value called δI_r, we can reformulate Eq. (5-49a):

$$\frac{\Delta I_L}{I_{in,avg}} = dI_r = \frac{V_{in}}{I_{in,avg}} \frac{D}{F_{sw} L} \quad (5\text{-}50)$$

We know by definition that

$$\frac{P_{out}}{\eta} = V_{in} I_{in,avg} \quad (5\text{-}51)$$

Extracting $I_{in,avg}$ and substituting it into Eq. (5-51) yield

$$\delta I_r = \frac{\eta V_{in}^2}{P_{out}} \frac{D}{F_{sw} L} \quad (5\text{-}52)$$

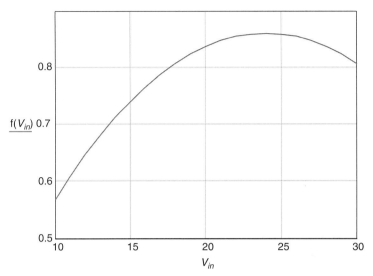

FIGURE 5-7b The ripple evolution versus the input voltage when $V_{out} = 48$ V, $L = 46$ μH, and $F_{sw} = 300$ kHz.

Solving for L leads to the needed inductor value, given a selected ripple amplitude. As shown above, V_{in} is selected at its highest level:

$$L = \frac{\eta V_{in}^2 D}{\delta I_r F_{sw} P_{out}} \quad (5\text{-}53)$$

If we select a 10% ripple value, assuming a 90% efficiency, maximum input voltage conditions, then Eq. (5-53) suggests an inductor value of

$$L = \frac{0.9 \times 15^2 \times 0.69}{0.1 \times 300k \times 100} = \frac{140}{3Meg} = 46.6 \text{ μH} \quad (5\text{-}54)$$

Now combining Eqs. (5-48), (5-49a), and (5-51), let us calculate the maximum peak current at the lowest input voltage:

$$I_{peak} = \frac{P_{out}}{V_{in,min}\eta} + \frac{V_{in,min}D_{max}}{2LF_{sw}} = \frac{96}{0.85 \times 10} + \frac{10 \times 0.79}{2 \times 46.6u \times 300k} = 11.6 \text{ A} \quad (5\text{-}55a)$$

With this value, we can now check at which output level the converter enters the discontinuous mode of operation. We derived the critical inductor equation in Eq. (1-119a):

$$R_{critical} = \frac{2F_{sw}L}{D_{min}(1 - D_{min})^2} = \frac{600k \times 46.6u}{0.69 \times (1 - 0.69)^2} = 421 \text{ Ω} \quad (5\text{-}55b)$$

This resistor value corresponds to an output current of $48/421 = 114$ mA or 5.7% of the nominal current. In this configuration, thanks to Eq. (1-114), the peak current will reach 221 mA for $V_{in} = 10$ V.

The output capacitor can now be evaluated by using Eq. (1-128b) and the 250 mV output ripple specification:

$$C_{out} \geq \frac{D_{min}V_{out}}{F_{sw}R_{load}\Delta V} \geq \frac{0.69 \times 48}{300k \times 24 \times 0.25} \geq 18 \text{ μF} \quad (5\text{-}56)$$

This element gives us enough capacitance to fulfil the output ripple requirement without ESR. However, what matters here, for the final selection, is the rms current flowing in this output capacitor. Equation (5-57) analytically expresses this rms value as

$$I_{C_{out},rms} = I_{out}\sqrt{\frac{D}{D'} + \frac{D}{12}\left(\frac{D'}{\tau_L}\right)^2} = 2\sqrt{\frac{0.79}{0.21} + \frac{0.79}{12}\left(\frac{0.21}{582m}\right)^2} = 3.88\,A \quad (5\text{-}57)$$

with $\tau_L = \dfrac{L}{R_{load}T_{sw}} = 582m$

The maximum occurs at the minimum input voltage or the maximum duty cycle. To handle such current, we need a capacitor able to operate above 48 V, sustaining the necessary ripple level. We select then a YXG type of capacitor from Rubycon with the following characteristics:

$C = 1500\,\mu F$
$I_{C,rms} = 2330\,mA$ @ $T_A = 105\,°C$
$R_{ESR,low} = 0.036\,\Omega$ @ $T_A = 20\,°C$, 100 kHz
$R_{ESR,high} = 0.13\,\Omega$ @ $T_A = -10\,°C$, 100 kHz
YXG series, 63 V

Given the available ripple capability, two of these parts will be put in parallel to satisfy the converter requirement, thus offering adequate margin.

5.2.2 Ac Analysis

The ac analysis should reveal the open-loop Bode plot of the boost when operating in its extreme areas. Also, the ESR must be swept to different values after or before the compensation, to check that its effects are well accounted for. Talking about ESRs, let us compute the zero positions they bring ($C_{out} = 2 \times 1500\,\mu F$, the equivalent ESR is one-half of each part's ESR):

$$f_{z,low} = \frac{1}{2\pi R_{ESR,high} C_{out}} = \frac{1}{6.28 \times 65m \times 3m} = 816\,\text{Hz} \quad (5\text{-}58)$$

$$f_{z,high} = \frac{1}{2\pi R_{ESR,low} C_{out}} = \frac{1}{6.28 \times 18m \times 3m} = 2.95\,\text{kHz} \quad (5\text{-}59)$$

Therefore these zeros will move depending on the temperature and the capacitor dispersions. Another important point concerns the RHPZ position. We have already explained its effects as it severely limits the available bandwidth to 20 to 30% of its lowest frequency position. It can be defined by

$$f_{z2} = \frac{R_{load}D'^2}{2\pi L} \quad (5\text{-}60)$$

Substitution of numerical values gives

$$f_{z2} = \frac{24 \times (1 - 0.79)^2}{6.28 \times 46.6u} = 3.61\,\text{kHz} \quad (5\text{-}61)$$

In other words, the crossover frequency must be selected around 1 kHz.

The boost converter filter also features a resonant point, just as the buck design did, thanks to the presence of the LC filter. However, the inductance here does not have a value of L but an equivalent inductance L_e calculated by

$$L_e = \frac{L}{(1-D)^2} \quad (5\text{-}62)$$

Hence, the resonant frequency is

$$f_0 = \frac{L_e}{2\pi\sqrt{LC}} = \frac{1-D}{2\pi\sqrt{LC}} \quad (5\text{-}63)$$

In a CCM boost design, D is theoretically independent of the output current value. Therefore, the resonant frequency will follow the input voltage range:

$$f_{0,low} = \frac{1-D_{max}}{2\pi\sqrt{LC}} = \frac{1-0.79}{6.28\times\sqrt{46.6u\times 3m}} = 89\,\text{Hz} \quad (5\text{-}64)$$

$$f_{0,high} = \frac{1-D_{min}}{2\pi\sqrt{LC}} = \frac{1-0.69}{6.28\times\sqrt{46.6u\times 3m}} = 132\,\text{Hz} \quad (5\text{-}65)$$

As in the buck example, a double zero will be placed right at the resonant peak to compensate the severe phase lag occurring at this point. Some designers recommend placing the double zero location slightly below: thanks to the automated simulation sheet, it is easy to test the results. Now, given the allocated bandwidth (1 kHz) and the output capacitor value, we can assess the drop in the presence of the output current variation (1 A in 10 μs):

$$\Delta V_{out} \approx \frac{\Delta I_{out}}{2\pi f_c C_{out}} \approx 53\,\text{mV} \quad (5\text{-}66)$$

Is 18 mΩ ≤ Z_{Cout} @ 1 kHz (53 mΩ)?? Yes

It is time to observe the small-signal response with all the elements installed around the PWM switch. The revised model is shown in Fig. 5-8a, whereas Fig. 5-8b and c portrays the ac sweep results. In this particular case, the modulator sawtooth waveform features a 2 V amplitude, hence the 0.5 coefficient present in the ac analysis ($G_{PWM} = 0.5$) We can see the influence of the ESR as it increases the gain when the zero occurs lower in the frequency spectrum. Compensating Fig. 5-8c thus represents the solution, however slightly detrimental to the bandwidth in case the ESR will decrease.

Dealing with a second-order system forces us to implement a type 3 compensation, as shown in Fig. 5-8a. To stabilize the circuit via a type 3 circuit, we are going to use the manual pole and zero method described in Chap. 3:

1. For a 48 V output from a 2.5 V reference voltage, let us adopt a 250 μA bridge current. The lower and upper resistor values are therefore

$$R_{lower} = \frac{2.5}{250u} = 10\,\text{k}\Omega$$

$$R_{upper} = \frac{48-2.5}{250u} = 182\,\text{k}\Omega$$

2. Open-loop sweep the voltage-mode buck at the two input levels (Fig. 5-8b/c results).
3. From the Bode plot, we can see that the required gain at 1 kHz is −3 dB worst case.
4. To cancel the *LC* filter peaking, place a double zero at a resonant frequency of 89 Hz. We could also place this double zero at 140 Hz to compensate for the variation of the resonant peak, especially when it comes closer to f_c. However, we have margin given the selected 1 kHz crossover frequency (seven times f_0).
5. Place a first pole at the highest zero position to force the gain to roll off, 3 kHz.
6. Place a second pole at the RHPZ zero, to force the gain to further roll off, 4 kHz.
7. Using the manual placement method described in Chap. 3, evaluate all the compensator elements.

446 CHAPTER FIVE

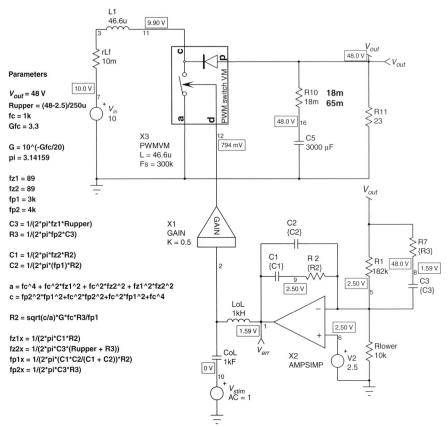

FIGURE 5-8a The PWM switch model in the boost operated in voltage mode. The type 3 calculation is automated, as usual.

$R_2 = 12 \text{ k}\Omega$
$R_3 = 4 \text{ k}\Omega$
$C_1 = 150 \text{ nF}$
$C_2 = 4.4 \text{ nF}$
$C_3 = 9.8 \text{ nF}$

Following compensation, we can explore the Bode plot, generated at the two extremes of the input voltage and the ESR variations. The plot is shown in Fig. 5-8d. The minimum bandwidth lies around 600 Hz and still satisfies Eq. (5-66). In all cases, we have a comfortable phase margin, the guarantee of a stable transient response. The gain margin can be found to be a little low when the ESR reaches 65 mΩ as we have 8 dB only. This margin can be improved by reducing the bandwidth to 20% of the RHPZ position.

It is now time to step load the output from 1 to 2 A, as in the original specification. Figure 5-8e shows the final results with all input voltage/ESR combinations. We can measure a maximum voltage deviation of 116 mV in the worst case, due to the ESR presence.

If we now reduce the minimum current to 50 mA instead of 1 A, the converter must react while in DCM. Unfortunately, in DCM, the double pole goes away and a single pole remains. One zero plays its role, and the other, becoming a pole of the closed-loop system,

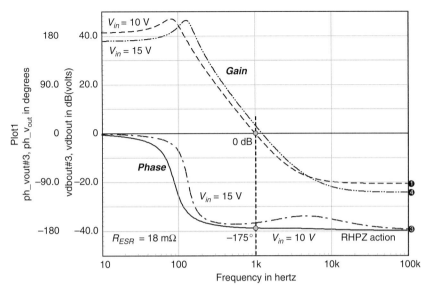

FIGURE 5-8b Ac sweep results of the CCM boost operated in voltage mode with an 18 mΩ ESR.

now hampers the response time, as Fig. 5-8f demonstrates. It would therefore be more desirable to keep operating in CCM during load steps. It is nevertheless interesting to note that the undershoot stays below 2% of V_{out}, which remains an acceptable performance for this audio converter.

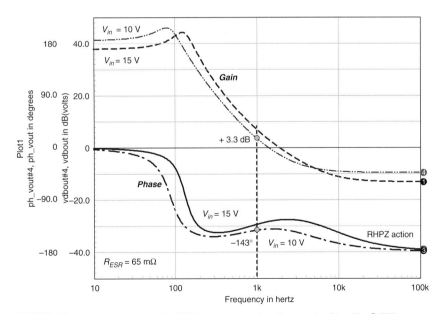

FIGURE 5-8c Ac sweep results of the CCM boost operated in voltage mode with a 65 mΩ ESR.

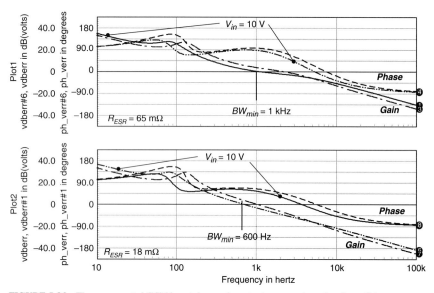

FIGURE 5-8d The compensated CCM boost shows adequate phase margin under all conditions.

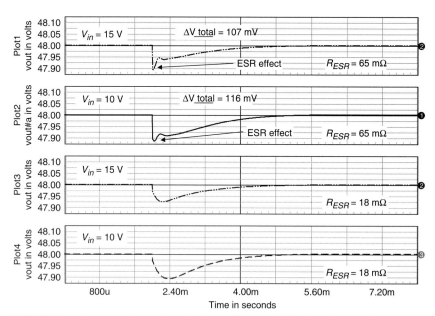

FIGURE 5-8e The step load results show that the specification is fulfilled in all cases, with an excellent stability.

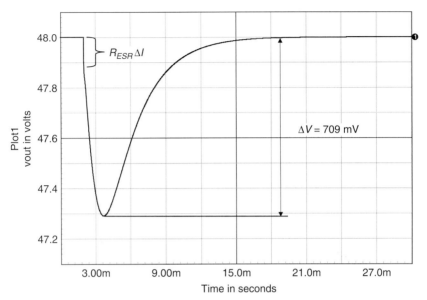

FIGURE 5-8f The DCM step shows a particular slow response due to the lack of available control bandwidth.

5.2.3 Transient Analysis

As we did for the buck example, we will use the generic controller to simulate the boost converter in a transient analysis. Thanks to the final results, we should be able to verify the initial assumptions as well as some other parameters. Figure 5-9a depicts the boost converter circuit for this exercise.

Equation (5-55a) indicates a peak current of 11.6 A. To give some margin, we will set the maximum authorized peak to 15 A. This is accomplished by selecting an *IMAX* parameter of 150 mV and a sense resistor R_{sense} of 10 mΩ. A simple *RC* network (R_4C_1) filters out any spurious high-frequency signals that could be generated, for instance, by the diode recovery current spikes. The compensation elements are directly copied from the ac analysis via the *parameters* macro. Hence, given an unsatisfactory response, the designer can quickly test new values in the ac analysis and immediately bring them back to the transient analysis. Starting from zero, the simulation time can be rather long given the large output capacitor value here. Some initial conditions are therefore necessary to reduce the computation time to a few minutes. This is the obvious case for the output capacitor C_{out} but also for C_5, directly placed in the feedback path. The inductor current also receives an initial current condition of 15 A. After simulation for V_{in} = 10 V, key signals appear in Fig. 5-9b.

As we can see immediately, the output ripple reaches the design constraint. This is so because the ESR value together with an 11 A peak current simply dictates the total ripple, the capacitive contribution to ripple voltage with a 3000 μF capacitor being extremely small. This result will obviously worsen with the 65 mΩ simulation. A solution lies in a different capacitor arrangement to still provide the right capacitive value (for the step load response) but featuring a much smaller total ESR to cope with the worst-case output ripple.

From the results, we can extract the total losses on the MOSFET and diode, including conduction losses plus switching losses. Of course, they are indicative numbers only since (1) the MOSFET $R_{DS(on)}$ does not depend on the junction temperature (therefore the model $R_{DS(on)}$ is at 25 °C, whereas calculations should include $R_{DS(on)}$ at T_j = 125 °C) and (2) switching losses on both

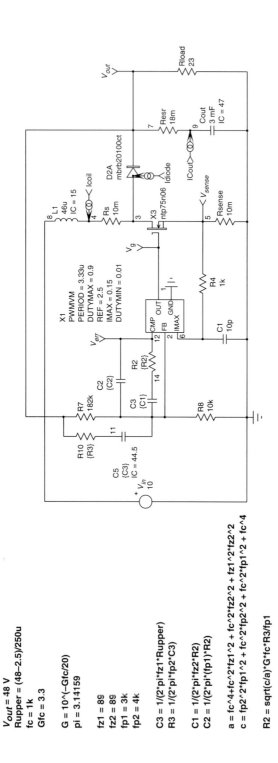

Parameters

V_{out} = 48 V
Rupper = (48–2.5)/250u
fc = 1k
Gfc = 3.3

G = 10^(–Gfc/20)
pi = 3.14159

fz1 = 89
fz2 = 89
fp1 = 3k
fp2 = 4k

C3 = 1/(2*pi*fz1*Rupper)
R3 = 1/(2*pi*fp2*C3)

C1 = 1/(2*pi*fz2*R2)
C2 = 1/(2*pi*(fp1)*R2)

a = fc^4+fc^2*fz1^2 + fc^2*fz2^2 + fz1^2*fz2^2
c = fp2^2*fp1^2 + fc^2*fp2^2 + fc^2*fp1^2 + fc^4

R2 = sqrt(c/a)*G*fc*R3/fp1

fz1x = 1/(2*pi*C1*R2)
fz2x = 1/(2*pi*C3*(Rupper + R3))
fp1x = 1/(2*pi*(C1*C2/(C1 + C2))*R2)
fp2x = 1/(2*pi*C3*R3)

FIGURE 5-9a The boost converter operated around the generic model PWMVM.

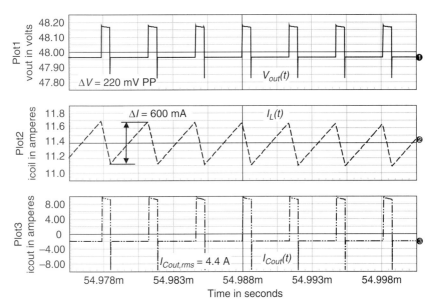

FIGURE 5-9b Transient signals of the CCM boost converter operated at $V_{in} = 10$ V, $R_{ESR} = 18$ mΩ.

semiconductors must include parasitic elements and recovery effects, particularly meaningful for a CCM boost converter. Figure 5-9c represents the MOSFET losses whereas Fig. 5-9d shows the diode losses. These losses should mainly be of capacitive origins as the recovery phenomenon does not exist in a Schottky diode. The first case reveals a total loss of 10 W including about 1.6 W for the diode. In total, the efficiency for this boost converter reaches 87.5% at the lowest input voltage and goes up to 95% at a 15 V input.

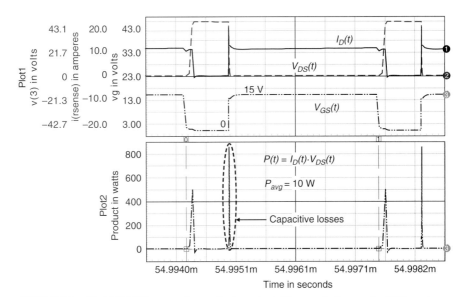

FIGURE 5-9c The MOSFET losses totaling conduction and switching contributions reach a 10 W worst case.

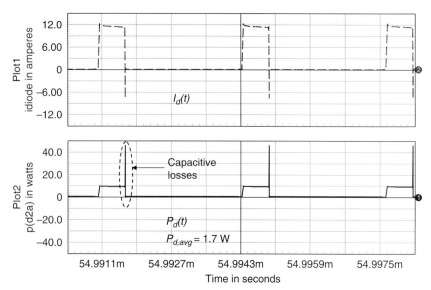

FIGURE 5-9d Diode losses are much lower and contribute only 1.6 W to the total loss budget.

5.2.4 A Current-Mode 5 V, 1 A Boost from a Li-Ion Battery

In battery-powered applications, it is often necessary to deliver energy to logic circuits operated from a 5 V rail. In this particular application, the battery also shares its energy with a RF emitter, forcing the boost converter to limit its ripple current pollution on the dc line. The specifications are as follows:

$V_{in,min} = 2.7$ V
$V_{in,max} = 4.2$ V
$V_{out} = 5$ V
$V_{ripple} = \Delta V = 50$ mV
V_{out} drop $= 0.1$ V maximum from $I_{out} = 100$ mA to 1 A in 1 μs
$I_{out,max} = 1$ A
$F_{sw} = 1$ MHz
$I_{ripple,peak} = 1$ mA, input current maximum ripple

Despite the current-mode structure, the passive element selections do not differ much from to the voltage-mode technique. We first look for the duty cycle extremes.

$$D_{min} = \frac{V_{out} - V_{in,max}}{V_{out}} = \frac{5 - 4.2}{5} = 0.16 \tag{5-67}$$

$$D_{max} = \frac{V_{out} - V_{in,min}}{V_{out}} = \frac{5 - 2.7}{5} = 0.46 \tag{5-68}$$

In accord with Eq. (5-49c), the maximum ripple will occur at one-half the output voltage, 2.5 V in this particular example. Note that this is the lowest input voltage when compared to the previous design example. Hence, if we select an arbitrary inductor ripple current of 12%, the component value can be determined by using Eq. (5-53), this time featuring the lowest input level:

$$L = \frac{0.9 \times 2.7^2 \times 0.46}{0.12 \times 1Meg \times 5} = \frac{3.02}{600k} = 5\,\mu H \tag{5-69}$$

Thanks to Eq. (5-48), the inductor value will induce a worst-case operating peak current of ($V_{in} = V_{in,min}$)

$$I_{peak} = \frac{P_{out}}{\eta V_{in}} + \frac{V_{in}D}{F_{sw}2L} = \frac{5}{0.9 \times 2.7} + \frac{2.7 \times 0.46}{2Meg \times 5u} = 2.17\,A \tag{5-70}$$

To keep a good transient response in CCM, we can assess the load point at which the boost enters in DCM operation:

$$R_{critical} = \frac{2F_{sw}L}{D_{min}(1-D_{min})^2} = \frac{2Meg \times 5u}{0.16 \times (1-0.16)^2} = 88.6\,\Omega \tag{5-71}$$

This load corresponds to a 56.4 mA load current, slightly below the design specification of 100 mA. As we stay in CCM, we should expect a good transient response over the whole current range.

According to the allowable ripple, the output capacitor can now be evaluated as we did with Eq. (5-56):

$$C_{out} \geq \frac{D_{min}V_{out}}{F_{sw}R_{load}\Delta V} \geq \frac{0.16 \times 5}{1Meg \times 5 \times 0.05} \geq 3.2\,\mu F \tag{5-72}$$

The rms current value represents the final capacitor criterion for its selection. The formula given before applies without any problem for a current-mode converter.

$$I_{C_{out},rms} = I_{out}\sqrt{\frac{D}{D'} + \frac{D}{12}\left(\frac{D'}{\tau_L}\right)^2} = 1\sqrt{\frac{0.46}{0.54} + \frac{0.46}{12}\left(\frac{0.54}{1}\right)^2} = 929\,mA \tag{5-73}$$

with

$$\tau_L = \frac{L}{R_{load}T_{sw}} = 1$$

For a compact boost converter, we cannot stack electrolytic capacitors as in an open-frame adapter, for instance, where size may not be an issue. In this example, the boost can go into a portable phone, and its physical dimensions must respect a few limits, in particular the board height. We have selected a multilayer capacitor from TDK [8], a C series part with the following parameters:

$C = 10\,\mu F$
$I_{C,rms} = 3\,A\ @\ T_A = 105\,°C, F_{sw} = 1\,MHz$
$R_{ESR} = 3\,m\Omega\ @\ T_A = 20\,°C, 1\,MHz$
C series, 10 V − C3216XR51A106M

As we can see, its extremely low ESR allows for a high ripple current despite its small size (L × W = 3.2 mm × 1.6 mm). Its associated high-frequency zero will not provide any help to boost the phase at the crossover frequency. Let us now see what bandwidth we do need to fulfill the step load requirement using this 10 μF capacitor:

$$f_c \approx \frac{\Delta I_{out}}{2\pi \Delta V_{out}C_{out}} = \frac{0.9}{6.28 \times 0.1 \times 10u} = 143\,kHz \tag{5-74}$$

Can we afford such a high bandwidth? Certainly not, especially in the CCM boost design where the RHPZ severely limits the selection of the crossover frequency. For this design, the RHPZ lies at

$$f_{z2} = \frac{R_{load}D'^2}{2\pi L} = \frac{5 \times (1 - 0.46)^2}{6.28 \times 5u} = 46.4 \text{ kHz} \quad (5\text{-}75)$$

Unfortunately, because of the various losses inherent in the semiconductors, the duty cycle is likely to increase, naturally lowering the RHPZ position. Allowing for some margin, we will select a crossover frequency of 8 kHz. Given this new number, let us recompute the amount of needed output capacitance:

$$C_{out} \approx \frac{\Delta I_{out}}{2\pi \Delta V_{out} f_c} = \frac{1}{6.28 \times 0.1 \times 8k} \approx 200 \text{ μF} \quad (5\text{-}76)$$

This shows that the original capacitor choice was not correct. Searching for a new capacitor gives us the following reference from TDK. Three of these capacitors will be put in parallel to satisfy Eq. (5-76):

$C = 68$ μF
$I_{C,rms} = 3$ A @ $T_A = 105$ °C, $F_{sw} = 1$ MHz
$R_{ESR} = 2.5$ mΩ @ $T_A = 20$ °C, 1 MHz
1A series, 10 V − C5750X5R1A686M

As the ESR of this capacitor is an extremely low value, we cannot rely on it for its phase boost action at the crossover frequency.

5.2.5 Ac Analysis

As the boost operates with low input and output voltages, a model including on losses and forward drops would lead to better dc results. This is what Fig. 5-10a shows, implementing the autotoggling current-mode model including the MOSFET and diode forward drops. Note the presence of the negative sense resistor value as the boost configuration features reversed current polarities compared to the original buck model (see Chap. 2 for more details). Choices for the active components are the following devices from ON Semiconductor [9]:

MOSFET	NTS3157N	$BV_{DSS} = 20$ V, $R_{DS(on)}$ @ [$V_{GS} = 2.5$ V, $T_j = 125$ °C] = 140 mΩ
Diode	MBRA210ET3	$V_{RRM} = 10$ V, V_f @ [$I_f = 1$ A, $T_j = 25$ °C] = 0.48 V

We selected an N-channel MOSFET whose on resistance is guaranteed at a low V_{GS} of 2.5 V, and the specification even states a 1.8 V value: low input voltage conditions will not be a problem then. The diode represents the latest generation in terms of breakdown voltage (10 V) and leakage current (500 μA at high temperature and 5 V). Again, it is possible to implement synchronous rectification if efficiency is part of the design parameters. Since no SPICE model existed for the transistor, we replaced it with a fixed resistor featuring its high-temperature $R_{DS(on)}$ value. The diode model, available from ON Semiconductor website, was inserted in the boost application circuit.

Equation (5-70) defines the necessary peak current. If we assume a maximum sensed voltage of 100 mV and a maximum allowable peak of 2.5 A, then the sense resistor value is found to be

$$R_{sense} = \frac{100m}{2.5} = 40 \text{ mΩ} \quad (5\text{-}77)$$

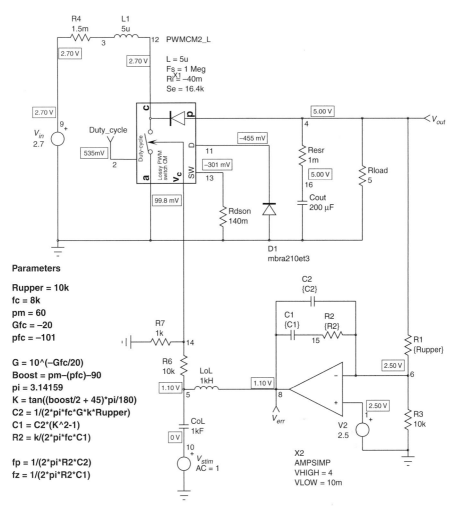

FIGURE 5-10a The CCM boost built around the Lossy model is compensated via a type 2 amplifier.

To provide enough swing on the error amplifier, we have installed a divider by 9 such that the upper current limit (2.5 A) corresponds to 900 mV output voltage.

After the ac sweep, the schematic reveals the operating dc points at the lowest input voltage. Given the various losses, the duty cycle exceeds the theoretical calculations (53% versus 46%), but semiconductor drops agree quite well with the data sheet predictions. Figure 5-10b portrays the ac sweep for both input voltages. The subharmonic peaking clearly appears at 500 kHz: damping is necessary.

Rather than apply the traditional ramp compensation method (50 to 75% of the inductor downslope), we will use Ridley's derivation, described in Eq. (2-182):

$$Q = \frac{1}{\pi(m_c D_0' - 0.5)} \qquad (5\text{-}78)$$

where

$$m_c = 1 + \frac{S_a}{S_1} \qquad (5\text{-}79)$$

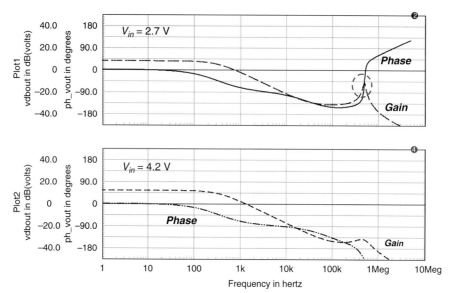

FIGURE 5-10b Ac small-signal response showing a peak when the input voltage is minimum.

This equation qualifies the double-pole quality coefficient observed at one-half the switching frequency. To calm unwanted oscillations, the designer must reduce the coefficient to 1. Analytically, this implies

$$m_c = \frac{\frac{1}{\pi} + 0.5}{D'_0} \qquad (5\text{-}80)$$

In all the above equations, the following notation applies:

D'_0 represents the steady-state off dc duty cycle (extracted from Fig. 5-10a dc bias).

S_a is the compensation ramp we are looking for, also named S_e in the literature.

S_1 and S_2 are, respectively, the on and off slopes, also denoted by S_n and S_f in the literature.

Given the boost configuration, we can easily derive the inductor on slope. It is (neglecting ohmic drops)

$$S_1 = \frac{V_{in}}{L} = \frac{2.7}{5u} = 540 \text{ kV/}\mu\text{s} \qquad (5\text{-}81)$$

Applying Eq. (5-80) leads to a parameter m_c equal to

$$m_c = \frac{\frac{1}{\pi} + 0.5}{D'_0} = \frac{818m}{464m} = 1.76 \qquad (5\text{-}82)$$

From Eq. (5-79) we extract S_a:

$$S_a = (m_c - 1)S_1 = 0.76 \times \frac{2.7}{5u} = 410 \text{ kV/s} \qquad (5\text{-}83)$$

When brought to the current-sense resistor, it becomes

$$S_a = 410k \times 40m = 16.4 \text{ kV/s} \qquad (5\text{-}84)$$

SIMULATIONS AND PRACTICAL DESIGNS OF NONISOLATED CONVERTERS   **457**

The Bode plot also reveals the presence of the main pole, defined in Chap. 2, App. 2A:

$$f_p = \frac{\frac{2}{R_{load}} + \frac{T_{sw}}{LM^3}\left(1 + \frac{S_a}{S_1}\right)}{2\pi C_{out}} = \frac{\frac{2}{5} + \frac{1u}{5u \times 6.35} \times 1.76}{6.28 \times 200u} = 362\,\text{Hz} \qquad (5\text{-}85)$$

We can then update the ac model with this ramp and see the final open-loop Bode plot (Fig. 5-10c).

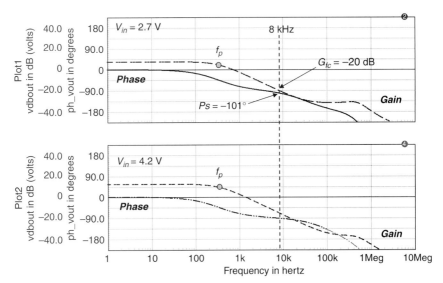

FIGURE 5-10c Ramp-compensated Bode plot of the CCM current-mode boost at both input voltage values.

Dealing with a CCM current-mode converter system leads us to implement a type 2 compensation circuit. To stabilize the circuit via a type 2 circuit, we use the k factor, which gives adequate results with first order systems. If the result does not suit us, we can still use manual placement as described in Chap. 2.

1. For a 5 V output from a 2.5 V reference voltage, we select a 250 µA bridge current. The lower and upper resistor values are therefore

$$R_{lower} = \frac{2.5}{250u} = 10\,\text{k}\Omega$$

$$R_{upper} = \frac{5 - 2.5}{250u} = 10\,\text{k}\Omega$$

2. Open-loop sweep the voltage-mode buck at the two input levels (Fig. 5-10c results).
3. From the Bode plot, we can see that the required gain at 8 kHz is -20 dB worst case.
4. The k factor macro computes a zero placed at 630 Hz and a pole at 101 kHz ($k = 12$) for a phase margin of 60°. The values passed to the simulator are the following:

$R_2 = 100\,\text{k}\Omega$
$C_1 = 2.5\,\text{nF}$
$C_2 = 15.6\,\text{pF}$

Once these elements are plugged into the simulator, a final run can be performed at both input levels. Results appear in Fig. 5-10d and confirm a phase margin above 60° together with a gain margin around 20 dB.

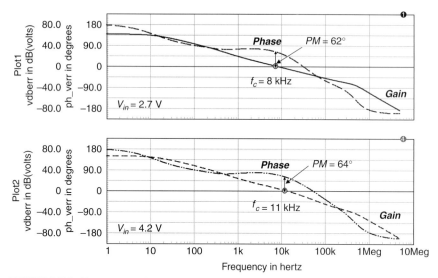

FIGURE 5-10d The compensated Bode plot shows a good phase and gain margins in all input voltage cases.

The transient test uses the averaged model together with a load evolving between 100 mA and 1 A. Figure 5-10e shows the results and confirms the stability. The drop stays within 70 mV, thus confirming the design hypothesis.

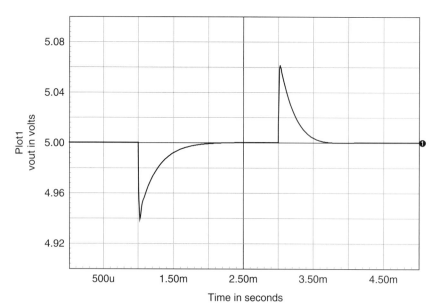

FIGURE 5-10e The transient response of the CCM boost when loaded by a 100 mA/1 A load (V_{in} = 2.7 V).

5.2.6 Transient Analysis

Figure 5-11a represents the transient simulation template using the generic current model PWMCM2. The controller parameters are the following:

$PERIOD = 1u$	Switching period of 1 μs (1 MHz)
$DUTYMAX = 0.9$	Maximum duty cycle up to 90%
$DUTYMIN = 0.01$	Minimum duty cycle down to 1%
$REF = 2.5$	Internal reference voltage of 2.5 V
$RAMP = 2.5$	Internal oscillator sawtooth amplitude, here 0 to 2.5 V
$VOL = 100m$	Error amplifier output when low
$VOH = 1.5$	Error amplifier output when high
$RATIO = 0.1$	Feedback signal is internally divided by 10 (1.5/10 = 150 mV)

We know that ramp compensation must be applied to this CCM boost. Its value has even been calculated: it corresponds to 76% of the on slope [Eq. (5-84)]. Capitalizing on what we derived in the buck converter operated in current mode, we can use the same set of equations:

$S'_1 = 21.6 \, \text{mV}/\mu\text{s}$ [Eq. (5-81) over the 40 mΩ sense resistor]

$S_{ramp} = 2.5 \, \text{V}/\mu\text{s}$ (2.5 V amplitude over a 1 μs period)

$R_{current} = 1 \, \text{k}\Omega$ (arbitrarily fixed)

$M_r = 76\%$ as discussed above (Eq. 5-82).

Given these numbers, Eq. (5-39) therefore gives

$$R_{ramp} = \frac{S_{ramp}}{M_r S'_1} R_{current} = \frac{2.5}{0.76 \times 21.6m} 1k = 152 \, \text{k}\Omega \quad (5\text{-}86)$$

Once this is plugged into the application circuit, we can run the simulation. Steady-state waveforms appear in Fig. 5-11b. The circuit is stable, without any subharmonic oscillations. Equation (5-73)

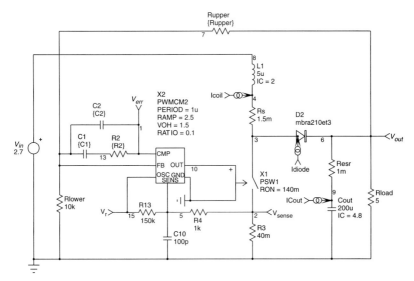

FIGURE 5-11a The CCM boost switching circuit featuring ramp compensation brought by R_{13}.

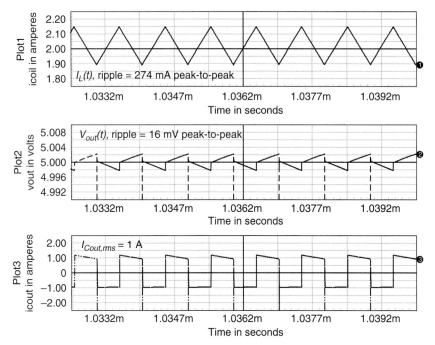

FIGURE 5-11b Steady-state waveforms collected further to a 1 ms simulation for $V_{in} = 2.7$ V.

is validated, and the output peak-to-peak ripple comfortably fits the original specifications. The inductor ripple slightly exceeds the specification because the original duty cycle calculations did not include the various ohmic losses (diode, MOSFET) which naturally induce a larger operating duty cycle. The efficiency is found to be around 91.7% at low line and falls to 90.5% at high line. This can be explained by the weight of each drop: in the first case, the diode drop is weighted by D' which is reduced at low line. At high line, as D goes down, the conductive portion of the diode (hence its losses) goes up, slightly hampering the total efficiency.

Measurements at low input voltage show a diode average power of 470 mW whereas the MOSFET dissipates 500 mW. As usual, these numbers are purely indicative, and bench measurements must be undertaken to obtain the final numbers.

As Fig. 5-11c confirms, the transient response to an output step is perfectly stable at both input levels.

5.2.7 Input Filter

For this boost, we have a requirement on the input current ac content, and it should stay below 1 mA peak to peak once the filter is installed. We have already shown how to calculate the elements of this filter in the buck converter example:

1. We display the input current signature and low input voltage and obtain its FFT. The peak current at 1 MHz reaches 101 mA (Fig. 5-12a).
2. The specification limits the peak input ripple to 1 mA. Therefore, the required attenuation needs to be

$$A_{filter} < \frac{1m}{101m} < 10m$$

or better than a 40 dB attenuation.

SIMULATIONS AND PRACTICAL DESIGNS OF NONISOLATED CONVERTERS **461**

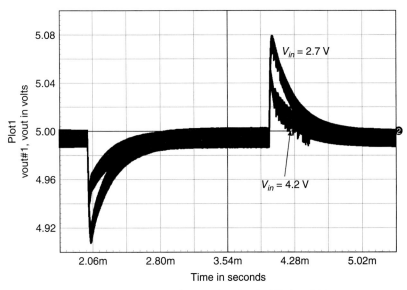

FIGURE 5-11c Transient response at two input levels, 2.7 and 4.2 V. $\Delta I = 0.9$ A.

3. Position the cutoff frequency of the LC filter using Eq. (1-198):

$$f_0 < \sqrt{0.01} \times F_{sw} < 100 \text{ kHz}$$

Let us select $f_0 = 100$ kHz.

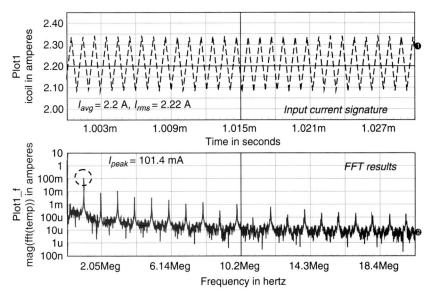

FIGURE 5-12a The input current signature of the CCM boost converter.

4. From simulation results, we can evaluate the rms current circulating in the filtering capacitor. This is actually the rms ac portion of the input current signature. The result represents one of the key selection criteria for filtering capacitors:

$$I_{ac} = \sqrt{I_{rms}^2 - I_{dc}^2} = \sqrt{2.22^2 - 2.2^2} = 297 \text{ mA rms}$$

which is a moderate value given the low input ripple. This is typical of a CCM boost converter where the input current characteristic form is nonpulsating.

5. From the result of step 3, we have two variables to play with: L and C. Most of the time, what matters is the total combined volume brought by the two elements. Given the frequency, let us choose a 1 µH inductor. Thus we can quickly determine the capacitor value via

$$C = \frac{1}{4\pi^2 f_0^2 L} = 2.53 \text{ µF}$$

or 3.3 µF for the next normalized value. We checked the data sheet of an SMD type and found that a TDK C2012JB1A335K [8] part could do the job. This part has a ripple current capability of 2 Arms at 1 MHz which gives us plenty of margin. It features an ESR of 5 mΩ.

6. From the vendor's catalog, the data sheet unveils the parasitic components of L and C: $r_{Cf} = 5$ mΩ (R_2) and $r_{Lf} = 3$ mΩ (R_1).

7. We now plug these parasitic resistors into Eq. (1-196) to check if the final attenuation remains within the limit, e.g., below 10m, despite the addition of the parasitic elements.

$$L = 10 \text{ µH} \qquad R_1 = 3 \text{ mΩ} \qquad C = 330 \text{ nF} \qquad R_2 = 5 \text{ mΩ}$$

$$\left\|\frac{I_{in}}{I_{out}}\right\| = \sqrt{\frac{0.005^2 + \frac{1}{(20.72)^2}}{(8m)^2 + \frac{1}{(20.72)^2} - \frac{2u}{3.3u} + (6.28)^2}} = 7.8m$$

which is below the 10m specification (see item 2).

8. Applying Eq. (1-185) gives the maximum filter output impedance:

$$\|Z_{outFILTER}\|_{max} = \sqrt{\frac{(3.3u \times 0.005^2 + 1u)(3.3u \times 0.003^2 + 1u)}{3.3u^2 \times 0.008^2}} = 37.8 \text{ Ω or } 31.6 \text{ dBΩ}$$

9. The converter static input resistance is classically obtained by using Eq. (1-181b): 1.31 Ω or 2.3 dBΩ at the lowest input voltage. Looking at the above result, where the output impedance peaks to 37.8 Ω, we will obviously find an overlapping area between the filter output impedance and the power supply input impedance: a damping action seems unavoidable in this case to ensure stability. Figure 5-12b offers a way to plot the input impedance without disturbing the operating point calculation. Figure 5-12c gathers both the boost input impedance results and the filter output impedances (undamped and damped).

10. The damping element, a series resistor, needs to be placed across the output capacitor to damp the peak. To avoid permanent heat dissipation, a capacitor appears in series with the resistor to block the dc component. To evaluate the series resistor value, you can either use Eq. (1-193) or place an arbitrary 1 Ω resistor in series with a capacitor of 3 to 4 times the filter capacitor. Then sweep the resistor value until the peak dampens. This what we did in Fig. 5-12c. A combination made of a 2.2 Ω resistor and a 10 µF capacitor gave adequate results.

11. Once the filter is properly damped, you need to recheck the Fig. 5-10d curves and verify that the phase margin does not suffer from the filter presence (Fig. 5-12d).

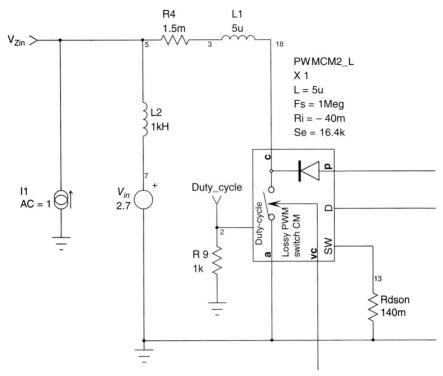

FIGURE 5-12b You can install an inductor in series with the source voltage to get the right dc point and then ac sweep by a current source to obtain the input impedance.

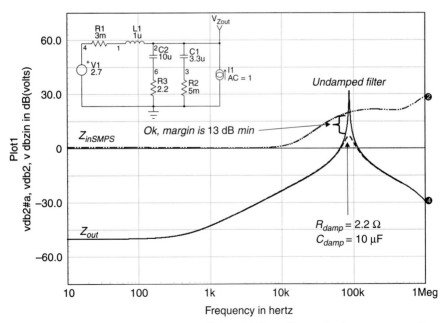

FIGURE 5-12c Because of the high quality coefficient of the input filter, overlapping areas are created . . . and compensated via a simple *RC* network.

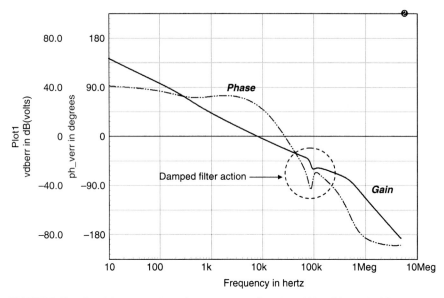

FIGURE 5-12d Once it is properly damped, an ac sweep confirms the stability of the whole CCM boost.

12. Final check: Install the *RLC* filter in Fig. 5-11a and see whether the input current amplitude appears to be within specifications. The result appears in Fig. 5-12e; we read $I_{in,peak}$ = 785 μA, so we are within the initial requirements. A transient load step can be undertaken as a final check if needed.

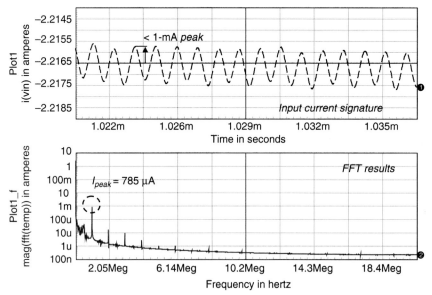

FIGURE 5-12e Once the filter is installed, we observe a low input current, compliant with the initial specification (V_{in} = 2.7 V).

5.2.8 Component Constraints for the Boost Converter

To end the boost design examples, we have gathered the constraints seen by the key elements used in this converter. These data should help you select adequate breakdown voltages of the diode and the power switch. All formulas relate to the CCM and DCM operation.

MOSFET		
$BV_{DSS} > V_{out}$	Breakdown voltage	
$I_{D,max} > I_{in} + \dfrac{\Delta I_L}{2}$	Maximum peak current	
$I_{D,rms} = I_{out}\sqrt{\left[\dfrac{D_{max}}{(1-D_{max})^2} + \dfrac{1}{3}\left(\dfrac{1}{2\tau_L}\right)^2 D_{max}{}^3 (1-D_{max})^3\right]}$		CCM
$I_{D,rms} = I_{out}\dfrac{\sqrt{1+2D_{max}{}^2/\tau_L}-1}{\sqrt{3D_{max}}}$		DCM

Diode	
$V_{RRM} > V_{out}$	Peak repetitive reverse voltage
$I_{F,avg} = I_{out}$	Continuous current

Capacitor		
$I_{C_{out},rms} = I_{out}\sqrt{\dfrac{D_{max}}{1-D_{max}} + \dfrac{D_{max}}{12}\left(\dfrac{1-D_{max}}{\tau_L}\right)^2}$		CCM
$I_{C_{out},rms} = I_{out}\sqrt{\dfrac{2}{3}\left(\dfrac{\sqrt{1+2D_{max}{}^2/\tau_L}-1}{D_{max}}\right)-1}$		DCM
with $\tau_L = \dfrac{L}{R_{load}T_{sw}}$		

5.3 THE BUCK-BOOST CONVERTER

The buck-boost converter combines both functions to either increase or decrease the input voltage. Unfortunately, the delivered output voltage polarity is negative with respect to ground. This is often considered the drawback of this converter topology. Nevertheless, in some cases, it does not present a problem to implement this structure. We will see this in the following design example.

5.3.1 A Voltage-Mode 12 V, 2 A Buck-Boost Converter Powered from a Car Battery

This application powers the negative rail of a dual-supply audio car amplifier. Symmetric supplies in audio amplifiers serve the purpose of getting rid of a dc blocking capacitor, usually found in mono supply architectures. Here, we suppose the amplifiers need ± 12 V to properly operate. The positive section can come from another regulator whereas the negative bias is provided by the buck-boost circuit. The specifications are as follows:

$V_{in,min} = 10$ V
$V_{in,max} = 15$ V

$V_{out} = -12$ V
$V_{ripple} = \Delta V = 250$ mV
V_{out} drop $= 250$ mV maximum from $I_{out} = 0.2$ to 2 A in 10 μs
$I_{out,max} = 2$ A
$F_{sw} = 100$ kHz

We start, as usual, by defining the duty cycle excursion during the converter operation at both input levels, full load. Since it is in continuous conduction mode, we can apply Eq. (1-135) where we simply consider V_{out} as a positive value to get rid of the minus sign:

$$D_{min} = \frac{V_{out}}{V_{in,max} + V_{out}} = \frac{12}{15 + 12} = 0.44 \qquad (5\text{-}87)$$

$$D_{max} = \frac{V_{out}}{V_{in,min} + V_{out}} = \frac{12}{10 + 12} = 0.545 \qquad (5\text{-}88)$$

To determine the needed inductor value for the buck-boost design, we look at Fig. 5-13 which depicts its current variations. We analytically define the ripple variation to select an

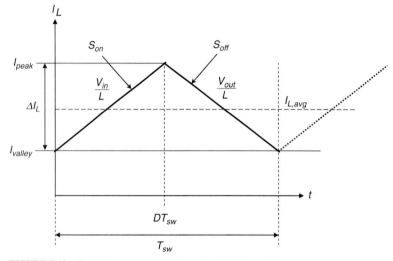

FIGURE 5-13 The inductor current waveform in a buck-boost design.

inductor value for maintaining this variation within acceptable limits. The peak current is defined by the excursion from the average inductor current to one-half the excursion ΔI_L. Note that, unlike with the boost or the buck converters, the average current in the inductor does not equal that of the input or the output current level.

From Fig. 5-13, it appears that

$$I_{peak} = I_{L,avg} + \frac{\Delta I_L}{2} \qquad (5\text{-}89)$$

where ΔI_L represents the inductor current excursion from the valley to the peak. This excursion occurs during the on time. Hence

$$\Delta I_L = \frac{V_{in} D T_{sw}}{L} \qquad (5\text{-}90)$$

Substitution of the duty cycle definition into Eq. (5-90) gives an updated ripple definition:

$$\Delta I_L = \frac{V_{in} T_{sw}}{L} \frac{V_{out}}{V_{in} + V_{out}} \qquad (5\text{-}91)$$

If we now link the inductor peak-to-peak current to the average inductor current by a ripple value called δI_r, we can rewrite Eq. (5-91):

$$\frac{\Delta I_L}{I_{L,avg}} = \delta I_r = \frac{V_{in} T_{sw}}{L} \frac{V_{out}}{V_{in} + V_{out}} \frac{1}{I_{L,avg}} \qquad (5\text{-}92)$$

In the buck-boost converter, the average inductor current is actually the sum of the average input current and the average output current (see Fig. 1-39a and b). Hence, Eq. (5-92) can be further updated:

$$\frac{\Delta I_L}{I_{L,avg}} = \delta I_r = \frac{V_{in} T_{sw}}{L} \frac{V_{out}}{V_{in} + V_{out}} \frac{1}{I_{in,avg} + I_{out,avg}} \qquad (5\text{-}93)$$

We know by definition that

$$\frac{P_{out}}{\eta} = V_{in} I_{in,avg} \qquad (5\text{-}94)$$

and

$$P_{out} = I_{out,avg} V_{out} \qquad (5\text{-}95)$$

Combining these power definitions into Eq. (5-93) then yields

$$\frac{\Delta I_L}{I_{L,avg}} = \delta I_r = \frac{V_{in} T_{sw}}{L} \frac{V_{out}}{V_{in} + V_{out}} \frac{1}{\frac{P_{out}}{V_{in}\eta} + \frac{P_{out}}{V_{out}}} \qquad (5\text{-}96)$$

From this equation, solving for L gives the design equation for the inductance value:

$$L = \frac{\eta V_{in}^2 V_{out}^2}{\delta I_r F_{sw} P_{out} (V_{in} + V_{out})(V_{out} + \eta V_{in})} \qquad (5\text{-}97)$$

In previous designs, we have set the ripple to around 10% of the average inductor current. Here, to purposely reduce the size of the inductor (but also its LI^2 storage capability), we will significantly increase the current ripple. Reducing the size of the inductor also helps to push the RHP zero to a higher frequency where it becomes less problematic for the loop control. If we select a 100% ripple (meaning ±50% above and below the average) together with a 90% efficiency, we need an inductance value of

$$L = \frac{0.9 \times 15^2 \times 12^2}{0.1 \times 100k \times 24 \times (15 + 12)(12 + 0.9 \times 15)} = 17.6 \,\mu\text{H} \qquad (5\text{-}98)$$

The inductor average current at $V_{in} = V_{in,min}$ will be

$$I_{L,avg} = I_{in,avg} + I_{out,avg} = P_{out}\left(\frac{1}{\eta V_{in}} + \frac{1}{V_{out}}\right) = 24 \times \left(\frac{1}{0.9 \times 10} + \frac{1}{12}\right) = 4.66 \,\text{A} \qquad (5\text{-}99)$$

Finally, using Eq. (5-91), we find the maximum peak current ($V_{in} = V_{in,min}$):

$$I_{peak} = I_{L,avg} + \frac{V_{in} T_{sw}}{2L} \frac{V_{out}}{V_{in} + V_{out}} = 4.66 + \frac{10 \times 10u}{35.2u} \times \frac{12}{10 + 12} = 6.2 \,\text{A} \qquad (5\text{-}100)$$

The output ripple equation for the buck-boost converter follows the same definition as for the boost converter. Therefore, we can still use Eq. (1-128b) or (1-154):

$$C_{out} \geq \frac{D_{min}V_{out}}{F_{sw}R_{load}\Delta V} \geq \frac{0.44 \times 12}{100k \times 6 \times 0.25} \geq 35.2\,\mu F \quad (5\text{-}101)$$

We now need to check the capacitor's rms current, as we know that this parameter actually guides us in the final capacitor selection. The rms current flowing into the output capacitor follows the same definition as from the boost converter. Thus

$$I_{C_{out},rms} = I_{out}\sqrt{\frac{D}{D'} + \frac{D}{12}\left(\frac{D'}{\tau_L}\right)^2} = 2\sqrt{\frac{0.545}{0.455} + \frac{0.545}{12}\left(\frac{0.455}{262m}\right)^2} = 2.3\,A \quad (5\text{-}102)$$

where $\tau_L = \frac{L}{R_{load}T_{sw}}$ as in Chap. 1. This is a fairly large value, and a single 35.2 µF capacitor has no chance of handling this current level! We have selected ultra low-impedance capacitors from Rybicon [1], the ZA series. As the maximum allowed current from a 470 µF capacitor is 1.7 A at an ambient temperature of 105 °C, we will put two of them in parallel:

$C = 470\,\mu F$
$I_{C,rms} = 1740\,mA$ @ $T_A = 105\,°C$
$R_{ESR,low} = 0.025\,\Omega$ @ $T_A = 20\,°C$, 100 kHz
$R_{ESR,high} = 0.05\,\Omega$ @ $T_A = -40\,°C$, 100 kHz
ZL series, 16 V

We then end up with an effective 940 µF capacitor featuring an ESR of 12.5 mΩ at 20 °C (25 mΩ at −40 °C).

As in any indirect energy transfer type of converter (such as the boost), we suffer from a RHP zero which hampers the available bandwidth. Let us first find the lowest-frequency location of this RHPZ and see if the available bandwidth together with the load capacitance satisfies the dropout specification:

$$f_{z2} = \frac{(1 - D_{max})^2 R_{load}}{2\pi D_{max} L} = \frac{207m \times 6}{6.28 \times 0.545 \times 17.6u} = 20.6\,kHz \quad (5\text{-}103)$$

Hence, choosing 20 to 25% of this value brings a possible crossover frequency of 5 kHz. Applying the dropout equation will tell us if this is enough, given the capacitor of 940 µF we already have:

$$\Delta V_{out} = \frac{\Delta I_{out}}{2\pi f_c C_{out}} = \frac{1.8}{6.28 \times 5k \times 940u} = 61\,mV \quad (5\text{-}104)$$

Is $25\,m\Omega \leq Z_{Cout}$ @ 5 kHz (34 mΩ)?? Yes

We are well within the original specification of 250 mV maximum drop, so we can go ahead with the ac simulation of this design.

5.3.2 Ac Analysis

Figure 5-14a represents the buck-boost converter wired with the PWM switch model. As the output voltage is negative, a simple −1 gain block reverses it for proper regulation. The PWM gain features a sawtooth with a 2 V amplitude, hence its 0.5 gain (−6 dB). As expected, the CCM buck-boost will resonate at different frequencies linked to the duty cycle:

$$f_{0,low} = \frac{1 - D_{max}}{2\pi\sqrt{LC}} = \frac{1 - 0.545}{6.28 \times \sqrt{17.6u \times 940u}} = 563\,Hz \quad (5\text{-}105a)$$

SIMULATIONS AND PRACTICAL DESIGNS OF NONISOLATED CONVERTERS   469

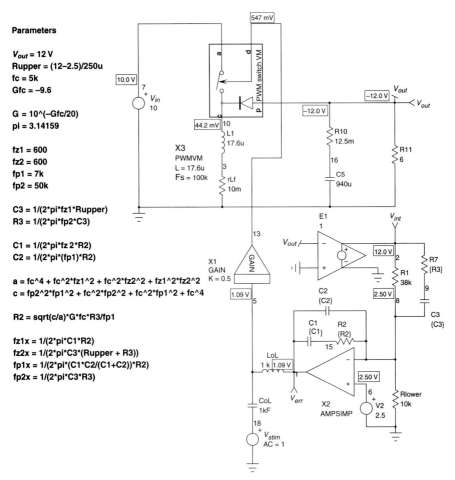

Parameters

V_{out} = 12 V
Rupper = (12−2.5)/250u
fc = 5k
Gfc = −9.6

G = 10^(−Gfc/20)
pi = 3.14159

fz1 = 600
fz2 = 600
fp1 = 7k
fp2 = 50k

C3 = 1/(2*pi*fz1*Rupper)
R3 = 1/(2*pi*fp2*C3)

C1 = 1/(2*pi*fz 2*R2)
C2 = 1/(2*pi*(fp1)*R2)

a = fc^4 + fc^2*fz1^2 + fc^2*fz2^2 + fz1^2*fz2^2
c = fp2^2*fp1^2 + fc^2*fp2^2 + fc^2*fp1^2 + fc^4

R2 = sqrt(c/a)*G*fc*R3/fp1

fz1x = 1/(2*pi*C1*R2)
fz2x = 1/(2*pi*C3*(Rupper + R3))
fp1x = 1/(2*pi*(C1*C2/(C1+C2))*R2)
fp2x = 1/(2*pi*C3*R3)

FIGURE 5-14a   The CCM *buck-boost* converter requires an inverting gain block (E_1) to reverse the output voltage before the error amplifier.

$$f_{0,high} = \frac{1 - D_{min}}{2\pi \sqrt{LC}} = \frac{1 - 0.44}{6.28 \times \sqrt{17.6u \times 940u}} = 693 \, \text{Hz} \qquad (5\text{-}105b)$$

Given the output capacitor and its ESR, we should see a zero moving between two positions at different temperatures:

$$f_{z,low} = \frac{1}{2\pi R_{ESR,high} C_{out}} = \frac{1}{6.28 \times 25m \times 940u} = 6.7 \, \text{kHz} \qquad (5\text{-}106)$$

$$f_{z,high} = \frac{1}{2\pi R_{ESR,low} C_{out}} = \frac{1}{6.28 \times 12.5m \times 940u} = 13.5 \, \text{kHz} \qquad (5\text{-}107)$$

Figure 5-14b depicts two Bode plots representing both input cases (10 and 15 V) plus the two ESR variations. From these two graphs, we can now proceed with the compensation calculations.

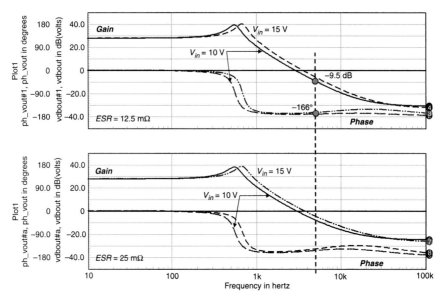

FIGURE 5-14b Bode plot results with different ESR and input conditions.

Dealing with a second-order system imposes the need to implement a type 3 compensation, and this what appears on the schematic. To stabilize the circuit via a type 3 circuit, we use the manual pole and zero method described in Chap. 3 and already implemented with the buck and boost design examples:

1. For a 12 V output from a 2.5 V reference voltage, we again select a 250 μA bridge current. The lower and upper resistor values are therefore

$$R_{lower} = \frac{2.5}{250u} = 10\,k\Omega$$

$$R_{upper} = \frac{12 - 2.5}{250u} = 38\,k\Omega$$

2. Open-loop sweep the voltage-mode buck-boost at the two input levels (Fig. 5-14b results).
3. From the Bode plot, we can see that the required gain at 5 kHz is around −10 dB worst case.
4. To cancel the *LC* filter peaking, place a double zero close to the resonant frequency, 600 Hz.
5. Since the zero occurs after the crossover frequency, we can place a first pole at 7 kHz.
6. Place a second pole at one-half of the switching frequency, to force the gain to further roll off, 50 kHz.
7. Using the manual placement method described in Chap. 3, evaluate all the compensator elements.

   $R_2 = 18.6\,k\Omega$
   $R_3 = 456\,\Omega$
   $C_1 = 15\,nF$
   $C_2 = 1.3\,nF$
   $C_3 = 7\,nF$

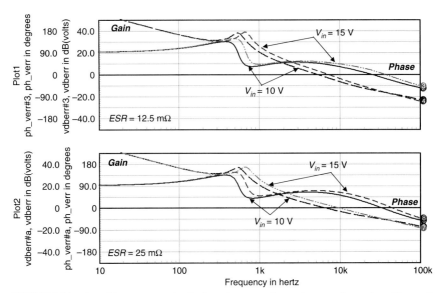

FIGURE 5-14c Once it is compensated, a final test shows a moderate phase margin together with enough gain margin.

Figure 5-14c unveils a 5 kHz crossover frequency featuring over 45° phase margin and −15 dB of gain margin. On the ac plot, we can clearly see the benefit of the capacitor zero coming in when its ESR reaches 25 mΩ. The bandwidth increases around 6.6 kHz at 15 V input, and the phase margin improves. The 45° phase margin obtained in the low ESR condition is not a panacea, and a different placement should improve the situation. For instance, the first pole is too close to the crossover frequency and can be farther to 15 kHz. Once an ac simulation is performed again, the phase margin increases to 61° (low ESR condition).

Using average model, an output step load can be performed to check the dropout transient response. Figure 5-14d shows the results at the lowest input level—we are out of specifications! What is the reason for this? Well, we have a fairly low resonant frequency of 560 Hz. To compensate the severe phase drop at this point, we inserted a double zero at 600 Hz. If you looked at App. 2B, you have seen that in presence of a large open-loop gain, the zeros you have placed in the compensation network now become poles of the closed-loop transfer function denominator (roots of the denominator). In other words, when these zeros occur at low frequency, they obviously hamper the system response time when a sudden change in output load occurs.

Given the RHPZ limitation, we cannot increase the crossover frequency. Thus, we increase the output capacitor value to 3.3 mF (7 × 470 μF capacitors in parallel, ESR minimum and maximum values drop to 3.6 and 7 mΩ). After proper compensation (a double zero placed at the new resonant frequency of 300 Hz), the transient response surely improves (Fig. 5-14e). But still, the result is not excellent as we would expect. Perhaps we could try to displace the double zeros? This is what we did: to show the effect of these zeros, we have gathered different responses as the double zeros move up the frequency axis. Thanks to the SPICE macro, it is child's play to change these values on the fly. Transient results appear in Fig. 5-14f with corresponding open-loop Bode plots in Fig. 5-14g. We can see how the transient response improves as the zeros go up, but this is detrimental to the phase margin at some point in the spectrum. Finally, conditional stability occurs with all the potential instability dangers it can bring (in particular if the system is subject to large gain changes as with voltage-mode operation). For instance, positioning the double zero at 500 Hz seems to meet the specifications and

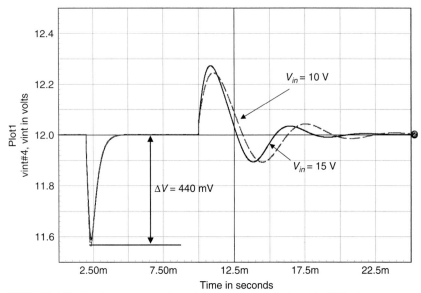

FIGURE 5-14d Transient response on the buck-boost converter with the original 940 μF capacitor.

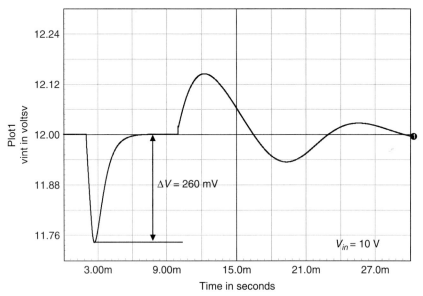

FIGURE 5-14e Updated transient response of the buck-boost converter with the output capacitor increased to 3.3 mF.

SIMULATIONS AND PRACTICAL DESIGNS OF NONISOLATED CONVERTERS 473

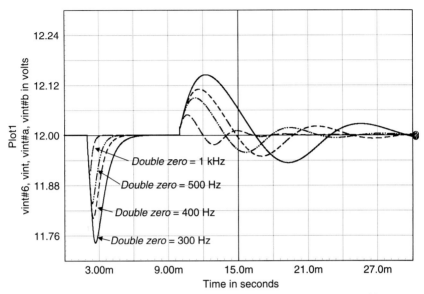

FIGURE 5-14f The transient response, the settling time, is affected by the double-zero positions.

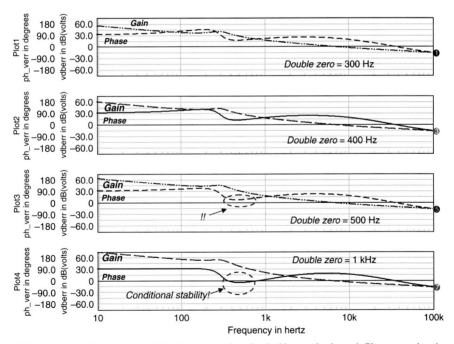

FIGURE 5-14g The open-loop Bode plots move when the double zero is changed. Please note that the crossover frequency remains the same.

offers a gain margin of more than 25 dB when the phase approaches 20°. Note that the crossover frequency remains unchanged to 5 kHz during all these trials: in other words, solely focusing on the crossover frequency is simply not enough when we are talking about feedback!

5.3.3 Transient Analysis

The transient section uses the generic model whose implementation in the buck-boost appears in Fig. 5-15a.

The sense resistor obeys Eq. (5-100) and offers an excursion up to 7 A. With a 30 mΩ resistor selection, the maximum peak voltage is found to be 210 mV (7 × 0.03). The behavioral source B_1 performs the voltage inversion to cope with the negative output. The generic controller features the following parameter changes:

$PERIOD = 10u$	Switching period of 10 μs (100 kHz)
$DUTYMAX = 0.9$	Maximum duty cycle up to 90%
$DUTYMIN = 0.01$	Minimum duty cycle down to 1%
$REF = 2.5$	Internal reference voltage of 2.5 V
$IMAX = 0.21$	Reset occurs when voltage on $IMAX$ pin exceeds 5 V
$VOL = 100m$	Error amplifier output when low
$VOH = 2.5$	Error amplifier output when high

Once the simulation is finished (around 60 s for an 8 ms run, Pentium 3 GHz), we can observe the results. They appear in Fig. 5-15b. The output voltage ripple stays within the limits (250 mV specified), but we see a lot of noise spikes. They are due to the sudden blocking action of the diode and to the presence of the ESR. Because of the pulsating nature of the output current, the presence of the ESR further degrades the output ripple. In these applications, the designer often adds a small second-stage LC filter to get rid of these spikes and to deliver a cleaner dc signal. If we install a 2.2 μH, 100 μF output filter, we can see the improvement in

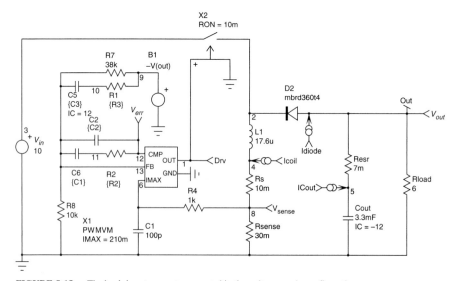

FIGURE 5-15a The buck-boost converter operated in the voltage-mode configuration.

SIMULATIONS AND PRACTICAL DESIGNS OF NONISOLATED CONVERTERS 475

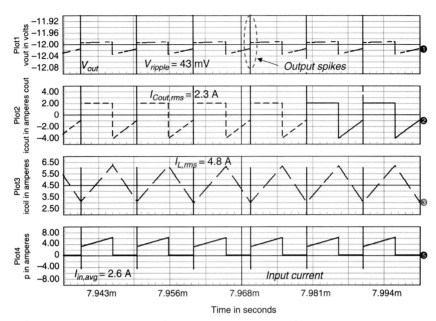

FIGURE 5-15b Simulation results of the buck-boost converter example.

Fig. 5-15c. Make sure the resonant frequency of the added filter is at least ten times above the crossover frequency to avoid stability issues. Unfortunately, the input current also suffers from the presence of noise spikes. This is the drawback of the buck-boost converters (including the flyback) where the input current is pulsating (as in the buck) but the output current is also (like the boost).

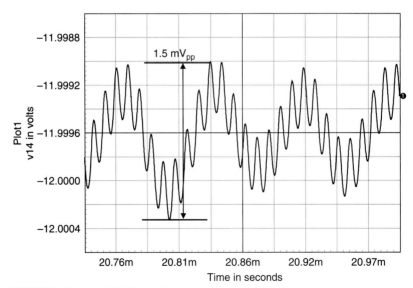

FIGURE 5-15c A small LC filter helps to reduce the output ripple.

5.3.4 A Discontinuous Current-Mode 12 V, 2 A Buck-Boost Converter Operating from a Car Battery

In this last example of the chapter, we have purposely used the same specification as in the previous design, but we now choose a current-mode converter operating in the discontinuous mode of operation. We will therefore be able to compare performance results and the various component stresses.

To design a discontinuous converter, we must first derive a formula to obtain the inductor value. This can be done by observing Fig. 5-16 which portrays the converter input current in CCM and DCM.

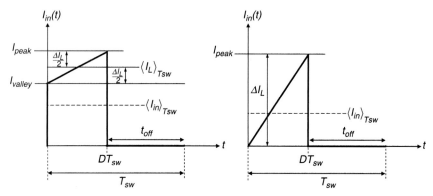

FIGURE 5-16 The input current of the buck-boost converter operated in CCM or DCM.

To derive a general equation, we use the CCM curve. From the two inductor current values I_{peak} and I_{valley}, we can write

$$I_{in,avg} = I_{L,avg}D = \frac{I_{peak} + I_{valley}}{2}D \quad (5\text{-}108)$$

$$I_{peak} - I_{valley} = \Delta I_L \quad (5\text{-}109)$$

The inductor current excursion depends on the input voltage and the on time:

$$\Delta I_L = I_{peak} - I_{valley} = \frac{V_{in}}{L}DT_{sw} \quad (5\text{-}110)$$

From the above equation, we can extract the duty cycle

$$\frac{(I_{peak} - I_{valley})LF_{sw}}{V_{in}} = D \quad (5\text{-}111)$$

Now replacing D in Eq. (5-108) gives

$$I_{in,avg} = \frac{(I_{peak} + I_{valley})(I_{peak} - I_{valley})LF_{sw}}{2V_{in}} \quad (5\text{-}112)$$

Applying Eq. (5-94), we finally obtain the input power definition for the CCM case:

$$\frac{P_{out}}{\eta V_{in}} = \frac{(I_{peak}^2 - I_{valley}^2)LF_{sw}}{2V_{in}} \quad (5\text{-}113)$$

$$P_{out} = \frac{1}{2}(I_{peak}^2 - I_{valley}^2)LF_{sw}\eta \quad (5\text{-}114)$$

SIMULATIONS AND PRACTICAL DESIGNS OF NONISOLATED CONVERTERS

In DCM, the valley point reaches zero and Eq. (5-114) simplifies to

$$P_{out} = \frac{1}{2} I_{peak}^2 L F_{sw} \eta \qquad (5\text{-}115)$$

We can now extract the inductor value from this equation:

$$L = \frac{2 P_{out}}{I_{peak}^2 F_{sw} \eta} \qquad (5\text{-}116)$$

Now, as we have two variables to pick up, either I_{peak} or L, we can identify another helpful equation defined in Chap. 1. Equation (1-151d) describes the critical inductance value at which the mode transition occurs. We therefore have the ability to equate Eqs. (5-116) and (1-151d) and solve for the peak current. Then reinserting the peak current value we found into Eq. (5-116) will give us the inductor value ensuring DCM at the lowest input voltage and maximum output current:

$$\frac{R_{load} \eta}{2 F_{sw}} \left(\frac{V_{in}}{V_{in} + V_{out}} \right)^2 = \frac{2 V_{out}^2}{I_{peak}^2 F_{sw} R_{load} \eta} \qquad (5\text{-}117)$$

$$I_{peak} = \frac{2(V_{in} + V_{out}) V_{out}}{\eta V_{in} R_{load}} = 9.3 \text{ A} \qquad (5\text{-}118)$$

Now substituting this value into Eq. (5-116) delivers the inductor value we need for this DCM operation:

$$L = \frac{2 V_{out}^2}{I_{peak}^2 F_{sw} \eta R_{load}} = \frac{2 \times 144}{9.3^2 \times 100k \times 0.9 \times 6} = 6.2 \, \mu\text{H} \qquad (5\text{-}119)$$

Now that we have the inductor value and the peak current value, let us now find the capacitor value. In DCM, the voltage ripple equation differs from that in CCM. Therefore, we must rederive a new voltage ripple equation that accounts for the inductor current falling to zero within a switching cycle.

Figure 5-17 depicts several interesting variables necessary to derive the output ripple: the diode current, the capacitor current, and the capacitive voltage.

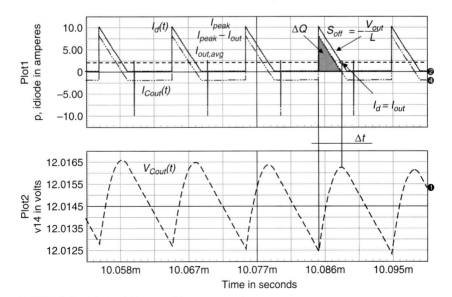

FIGURE 5-17 DCM variables around the output capacitor.

The diode current is made up of a dc portion (the output current) and an ac portion that flows into the capacitor. As long as the inductor current remains above the output current value, the capacitor will be recharged. When the inductor current has fallen to a near-zero value, the capacitor starts to deplete. The time at which the capacitor voltage starts to drop is shown in Fig. 5-17. Let us first calculate the time duration Δt during which the capacitor is recharged. The diode current starts from I_{peak} and drops to I_{out} with a slope that depends on the output voltage value and the inductor:

$$I_{peak} - \Delta t \frac{V_{out}}{L} = I_{out} \tag{5-120}$$

Extracting Δt gives

$$\Delta t = \frac{(I_{peak} - I_{out})L}{V_{out}} \tag{5-121}$$

The capacitor receives a certain amount of charge during Δt only. By integrating the positive area on the capacitor waveform, we can calculate the charge quantity it stores:

$$\Delta Q = \frac{I_{peak} - I_{out}}{2} \Delta t \tag{5-122}$$

Substituting Eq. (5-121) into (5-122) leads to the stored charge

$$\Delta Q = \frac{(I_{peak} - I_{out})^2 L}{2V_{out}} \tag{5-123}$$

Applying $\Delta Q = \Delta V \cdot C$, we reach the ripple voltage amount immediately:

$$\Delta V = \frac{(I_{peak} - I_{out})^2 L}{2V_{out}C_{out}} \tag{5-124}$$

We could also use the peak current definition from Eq. (5-118) and obtain a more general definition:

$$\Delta V = \frac{2L}{V_{out}C_{out}} \left(\frac{V_{in} + V_{out}}{\eta} \frac{V_{out}}{V_{in}R_{load}} - \frac{I_{out}}{2} \right)^2 \tag{5-125}$$

From this expression, we can calculate the amount of capacitance needed to meet the ripple voltage specification (250 mV):

$$C_{out} \geq \frac{(I_{peak} - I_{out})^2 L}{2V_{out}\Delta V} = \frac{(9.3 - 2)^2 \times 6.2u}{2 \times 12 \times 0.25} = 55\,\mu\text{F} \tag{5-126}$$

However, we know the role of the capacitor ESR in the output ripple whose contribution adds up to that of the capacitor. Thus, considering the capacitive contribution to be zero, the selected capacitor should have an ESR lower than

$$\text{ESR} \leq \frac{\Delta V}{I_{peak}} \leq \frac{0.25}{9.2} \leq 27\,\text{m}\Omega \tag{5-127}$$

Final step is the rms current. The equation below gives the rms current flowing in the capacitor in DCM.

$$I_{C_{out},rms} = I_{out}\sqrt{\frac{2}{3}\left(\frac{\sqrt{1 + 2D^2/\tau} - 1}{D}\right) - 1} \tag{5-128}$$

where $\tau_L = \dfrac{L}{R_{load}T_{sw}}$ as defined in Chap. 1. For a DCM buck-boost converter, the duty cycle can be computed by using Eq. (1-143):

$$D_{max} = \frac{V_{out}}{V_{in,min}}\sqrt{2\tau_L} = \frac{12}{10}\sqrt{2 \times 103m} = 0.545 \quad (5\text{-}129)$$

We can now compute the maximum rms current in the output capacitor:

$$I_{C_{out},rms} = I_{out}\sqrt{\frac{2}{3}\left(\frac{\sqrt{1+2D_{max}^2/\tau_L}-1}{D_{max}}\right)-1}$$

$$= 2\sqrt{666m\,\frac{\sqrt{1+1.47/103m}-1}{0.545}-1} = 3.2\,\text{A} \quad (5\text{-}130)$$

Based on the capacitor selection we made earlier, we need four capacitors in parallel to satisfy the ripple definition, the ESR limit, and the rms current stress. Thus the final equivalent capacitor features the following values:

$C_{eq} = 1880\ \mu\text{F}$
$I_{C,rms} = 7\,\text{A}\ @\ T_A = 105\ °\text{C}$
$R_{ESR,low} = 6.25\ \text{m}\Omega\ @\ T_A = 20\ °\text{C},\ 100\ \text{kHz}$
$R_{ESR,high} = 12.5\ \text{m}\Omega\ @\ T_A = -40\ °\text{C},\ 100\ \text{kHz}$
ZL series, 16 V → 4 × 470 µF capacitors in parallel

Now that we have all the part values calculated, we can go on to the ac analysis of this design.

5.3.5 Ac Analysis

In DCM current mode, we do not need any ramp compensation. This is the first advantage. The second one relates to the order of the converter that decreases to 1. The LC peaking seen before is gone, hence an easier compensation is possible now. The RHPZ also fades away, at least in the low-frequency portion and is no longer a design stability problem. However, we still have the ESR zeros placed at the following locations:

$$f_{z,low} = \frac{1}{2\pi R_{ESR,high}C_{out}} = \frac{1}{6.28 \times 12.5m \times 1.88m} = 6.8\,\text{kHz} \quad (5\text{-}131)$$

$$f_{z,high} = \frac{1}{2\pi R_{ESR,low}C_{out}} = \frac{1}{6.28 \times 6.25m \times 1.88m} = 13.5\,\text{kHz} \quad (5\text{-}132)$$

The main output pole depends on the load and the capacitor. In DCM current mode, it simplifies to

$$f_{p1} = \frac{2}{2\pi RC} = \frac{1}{\pi R_{load}C_{out}} = \frac{1}{3.14 \times 6 \times 1.88m} = 28.2\,\text{Hz} \quad (5\text{-}133)$$

A high-frequency pole also exists and its definition appears in App. 2A. It usually does not bother the designer as its phase lag is not perceived in the lower-frequency portion.

Figure 5-18a represents the ac circuit template using the PWM switch model. The sense resistor is found to let the current fly around 10 A [Eq. (5-118)]. With a chosen 100 mV drop,

480 CHAPTER FIVE

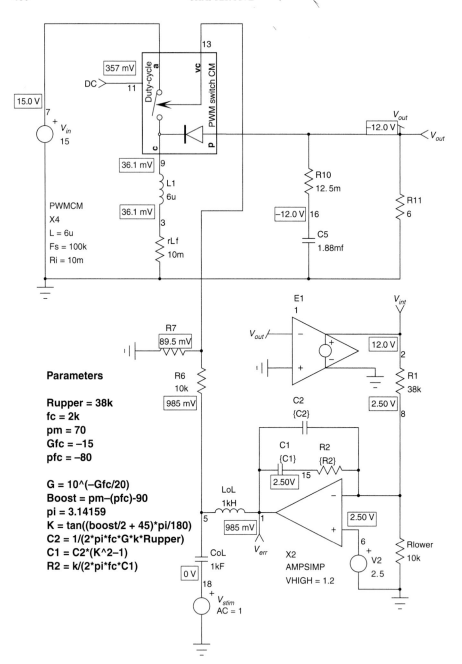

FIGURE 5-18a The DCM buck-boost ac analysis template. Note the type 2 compensation.

its value is

$$R_{sense} = \frac{100m}{10} = 10\,\text{m}\Omega \qquad (5\text{-}134)$$

To obtain a decent swing on the error amplifier, we will add an internal divide-by-10 via two external resistors. In a real controller, the excursion on the divider would be precisely clamped (e.g., to 1 V in a UC384X controller) to fix the maximum peak current in fault conditions. Here, for the sake of simplicity, we just clamp the error amplifier excursion to 1 V. In Fig. 5-18b, what bandwidth do we need? Equation (5-104) is still helpful:

$$f_c = \frac{\Delta I_{out}}{2\pi_c \Delta V_{out} C_{out}} = \frac{1.8}{6.28 \times 0.25 \times 1.88m} = 610\,\text{Hz} \qquad (5\text{-}135)$$

We can adopt 2 kHz for the purpose of the experiments, and it allows some margin. Here are the steps to follow for a DCM current-mode stability compensation:

1. The bridge divider does not change: $R_{upper} = 38\,\text{k}\Omega$ and $R_{lower} = 10\,\text{k}\Omega$.
2. Open-loop sweep the current-mode buck at the two input levels (Fig. 5-18b results).
3. From the Bode plot, we can see that the required gain at 2 kHz is around -15 dB worst case. The phase lag at this point is $-80°$.
4. We can place a first pole at the origin to benefit from the large open-loop gain of the op amp.
5. The zero must be placed below the crossover frequency to benefit from its phase boost as the closed-loop gain approaches 0 dB. We will place it around 20% of 2 kHz, so 400 Hz represents a good choice.

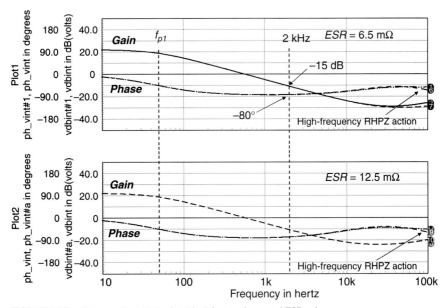

FIGURE 5-18b The open-loop Bode plot at both input voltages and ESR values.

6. The pole must cancel the zero ESR (if within the band of interest); otherwise place it at one-half the switching frequency.
7. The k factor gives good results for the DCM compensation. Its recommendations are the following for a 2 kHz bandwidth and a 70° phase margin:

$R_2 = 142\ \text{k}\Omega$

$C_1 = 3.2\ \text{nF}$

$C_2 = 101\ \text{pF}$

$f_z = \dfrac{1}{2\pi R_2 C_1} = 350\ \text{Hz}$

$f_p = \dfrac{1}{2\pi R_2 C_2} = 11\ \text{kHz}$

Once applied, Fig. 5-18c shows the compensated gain curve.

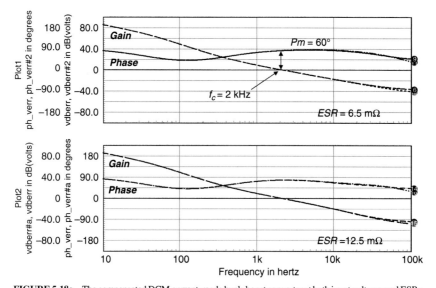

FIGURE 5-18c The compensated DCM current-mode buck-boost converter at both input voltages and ESRs.

Phase and gain margins are excellent and should ensure a rock-solid control stability. We can see that the Bode plot is almost unaffected by the input voltage variations or the ESR positions.

The transient response to an output step can now be performed and results appear in Fig. 5-18d.

We had some problems in obtaining convergence during this step, so we added a 100 nF capacitor on the PWM switch "c" node to ground and checked that it did not affect the frequency response. As one can see, we have a very stable response with a drop better than before (150 mV)! At the end we will discuss the differences between both designs, but stability is clearly in favor of DCM.

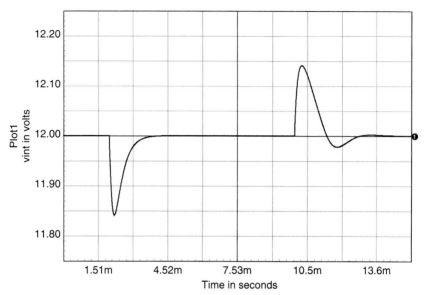

FIGURE 5-18d Output step response for the DCM current-mode boost converter. It was tested at 10 V, but the response does not significantly change for different input and ESR combinations.

5.3.6 Transient Analysis

The simulated circuit appears in Fig. 5-19a and does not differ much from that of the voltage-mode approach. The compensation network follows what we discussed above and imposes a 2 kHz bandwidth.

The difference in the passed parameters lies in the ratio that divides the error voltage by 10 to ensure enough voltage swing on the op amp:

$RATIO = 0.1$ Division ratio between op amp and current-sense set point
$VOH = 1.2$ V   Maximum op amp output voltage excursion

To ensure a true discontinuous mode, we have purposely reduced the inductor to 5 μH as Eq. (5-119) defines a value for boundary mode only (at the boundary between CCM and DCM).

Simulation results appear in Fig. 5-19b and show that the total output ripple is within specification. Equation (5-124) predicts a capacitive ripple of 11 mV, and we measure around 10.8 mV over the capacitor only (voltage on node 5). The ESR contribution lies around 65 mV as expected. The current circulating in the sense resistor actually looks like the inductor's as they are in the same path! That is something you will not see in the flyback converter as the off current circulates in the secondary side. The sense resistor alone dissipates 310 mW.

The peak inductor current is obviously larger than what was suggested by Eq. (5-118) as we selected a smaller inductor. We read 10.1 A associated with an rms current of 5.6 A. Given these results, the efficiency reaches 92% at low line, which is not bad at all. Now that we have two projects delivering the same amount of power, using exactly the same components but operating in different modes, we can compare the stresses and losses between both designs. Thanks to the simulation speed of the generic models, a few minutes are enough to extract key parameters:

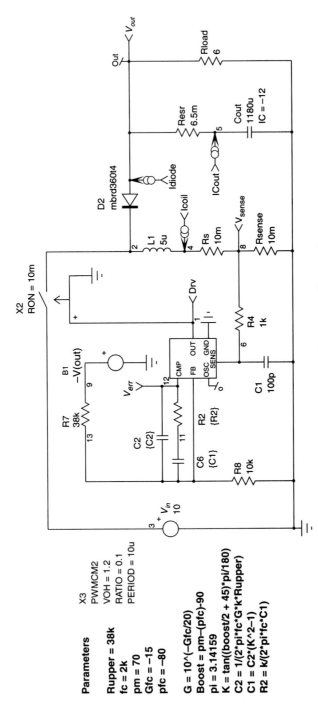

FIGURE 5-19a The current-mode switched model shows a similar arrangement to the voltage-mode approach.

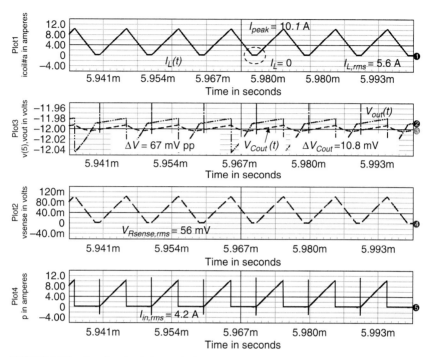

FIGURE 5-19b Transient simulation results for the DCM buck-boost converter operated at a 10 V input voltage.

Parameter	DCM	CCM
Power switch rms current (A)	4.2	3.5
Inductor rms current (A)	5.6	4.6
Inductor peak current (A)	10.1	6.1
Output capacitor rms current (A)	3.1	2.3
Output capacitor value (μF)	1880	3300
Output drop (mV)	150	200[1]
Output diode rms current (A)	3.7	3
Available bandwidth (kHz)	25	5
Efficiency low line (%)	92	93

Note: Double zero placed at 300 Hz.

From the above array, a few comments can be made:

- Circulating rms currents are larger in DCM than in CCM. This can be explained by the larger ac excursion straddling all dc currents.
- Peak currents are higher in DCM.
- The available bandwidth is greatly extended in DCM because the RHPZ is relegated to high frequencies (around 100 kHz in this example, see App. 2A for location details).
- Given the flexibility for the crossover frequency selection, the DCM output drop is better than in CCM and requires a smaller capacitor.

- The efficiency is about the same in this typical example, but as rms currents increase in DCM, they would force you to adopt lower $R_{DS(on)}$ components to cope with conduction losses.
- The output capacitor is smaller in DCM than in CCM, and there is no need for a double zero at low frequency that hampers the closed-loop gain: the DCM transient response is better than that of CCM! This can be easily understood as the current in the inductor cannot grow faster than V/L, a smaller L as in DCM leads to a faster system.

5.3.7 Component Constraints for the Buck-Boost Converter

To end the buck-boost design examples, we have gathered the constraints seen by the key elements used in this configuration. These data should help you select adequate breakdown voltages of the diode and the power switch. All formulas relate to the CCM and DCM operations.

MOSFET		
$BV_{DSS} > V_{out} + V_{in,max}$		Breakdown voltage
$I_{D,max} > I_{in} + \dfrac{\Delta I_L}{2}$		Maximum peak current [see Eqs. (5-100) and (5-118)]
$I_{D,rms} = I_{out} \sqrt{\left[\dfrac{D_{max}}{(1-D_{max})^2} + \dfrac{1}{3}\left(\dfrac{1}{2\tau_L}\right)^2 D_{max}{}^3(1-D_{max})^3\right]}$		CCM
$I_{D,rms} = I_{out} \dfrac{\sqrt{1 + 2D_{max}{}^2/\tau_L} - 1}{\sqrt{3D_{max}}}$		DCM
Diode		
$V_{RRM} > V_{out} + V_{in,max}$		Peak repetitive reverse voltage
$I_{F,avg} = I_{out}$		Continuous current
Capacitor		
$I_{C_{out},rms} = I_{out} \sqrt{\dfrac{D_{max}}{1-D_{max}} + \dfrac{D_{max}}{12}\left(\dfrac{1-D_{max}}{\tau_L}\right)^2}$		CCM
$I_{C_{out},rms} = I_{out} \sqrt{\dfrac{2}{3}\left(\dfrac{\sqrt{1+2D_{max}{}^2/\tau_L}-1}{D_{max}}\right) - 1}$		DCM
with $\tau_L = \dfrac{L}{R_{load}T_{sw}}$		

REFERENCES

1. http://www.rubycon.co.jp/en/
2. R. Erickson and D. Maksimovic, *Fundamentals of Power Electronic*, Kluwers Academic Press, 2001.
3. http://www.intusoft.com/nlpdf/nl32.pdf
4. http://www.vishay.com/docs/42030/731p.pdf
5. http://www.saga-sanyo.co.jp/oscon/cgi-bin/e_sizecode.cgi?id=SEQP

6. R. Severns and G. Bloom, *Modern dc-to-dc Switchmode Power Converter Circuits*, Reprint edition available at: www.ejbloom.com, book selection 101.
7. http://www.onsemi.com/pub/Collateral/NCP1013-D.PDF
8. http://www.component.tdk.com/tvcl_ircd.php#
9. http://www.onsemi.com

APPENDIX 5A THE BOOST IN DISCONTINUOUS MODE, DESIGN EQUATIONS

Although the buck converter only rarely operates in DCM, except in light-load conditions, the boost converter is often designed to work in this region. We have seen with the buck-boost design example that DCM brings several advantages such as a wider possible bandwidth (the RHPZ moves to higher frequencies) and the loss of a double pole at a moving resonant frequency (for voltage mode). However, you pay for these advantages with a higher rms current circulating in the inductor, the power MOSFET, and the output capacitor. Without going through a complete design here, we derive the basic equations to help you design a boost converter in DCM.

5A.1 Input Current

As we did with the buck-boost converter, we can draw the input current of the boost operated in CCM and DCM. Figure 5A-1 portrays both cases.

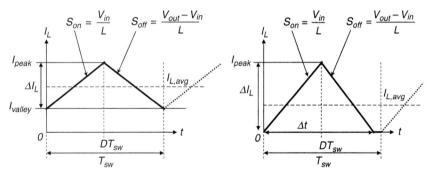

FIGURE 5A-1 The boost input current operated in both conduction modes: CCM on the left and DCM on the right.

From the DCM figure, we calculate the average input current (which is the average inductor current) by evaluating the area of the curve during the time Δt (magnetic activity in the inductor). This period of time is actually the sum of the on time and the downslope duration of the inductor current. Thus, we can write

$$\Delta t = \frac{I_{peak}}{S_{on}} + \frac{I_{peak}}{S_{off}} = \frac{I_{peak}L}{V_{in}} + \frac{I_{peak}L}{V_{out} - V_{in}} \tag{5A-1}$$

Rearranging this equation gives

$$\Delta t = LI_{peak}\left(\frac{1}{V_{in}} + \frac{1}{V_{out} - V_{in}}\right) = LI_{peak}\left[\frac{V_{out}}{V_{in}(V_{out} - V_{in})}\right] \tag{5A-2}$$

The average current during the magnetic activity (Δt duration) is found to be

$$I_{in,avg} = \frac{I_{peak}}{2} F_{sw} \Delta t = F_{sw} L I_{peak}^2 \left[\frac{V_{out}}{2V_{in}(V_{out} - V_{in})} \right] \quad (5A\text{-}3)$$

We know that P_{out} links to the input current via

$$I_{in,avg} = \frac{P_{out}}{\eta V_{in}} \quad (5A\text{-}4)$$

Then equating Eqs. (5A-3) and (5A-4) leads us to a definition for the inductor operated in DCM:

$$\frac{V_{out} I_{out}}{\eta V_{in}} = F_{sw} L I_{peak}^2 \left[\frac{V_{out}}{2V_{in}(V_{out} - V_{in})} \right] \quad (5A\text{-}5)$$

From this equation, we can extract the inductor value.

$$L = \frac{2 I_{out}(V_{out} - V_{in})}{\eta F_{sw} I_{peak}^2} \quad (5A\text{-}6)$$

The unknown in this equation is still the peak current. The easiest way to obtain this peak value is to set the boost converter at the border between CCM and DCM. In borderline operation [borderline conduction mode (BCM)], the peak current follows a shape as depicted by Fig. 5A-2.

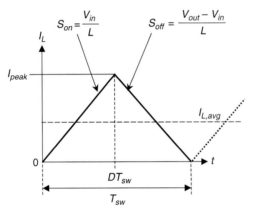

FIGURE 5A-2 The boost input current shape when operated in borderline conditions.

Computing the average value in BCM represents an easy exercise. If we write the area of the double triangle and average it over the switching period, we have

$$I_{in,avg} = \frac{I_{peak} T_{sw}}{2 T_{sw}} = \frac{I_{peak}}{2} \quad (5A\text{-}7)$$

Again, using Eq. (5A-4), we can extract the peak current definition in BCM:

$$\frac{P_{out}}{\eta V_{in}} = \frac{I_{peak}}{2} \quad (5A\text{-}8)$$

From this equation, the peak current definition comes quickly:

$$I_{peak} = \frac{2 P_{out}}{\eta V_{in}} = \frac{2 V_{out} I_{out}}{\eta V_{in}} \quad (5A\text{-}9)$$

Given the peak current definition, we substitute Eq. (5A-8) into (5A-6), and we obtain the borderline inductor value

$$L = \frac{V_{in}^2(V_{out} - V_{in})\eta}{2I_{out}V_{out}^2 F_{sw}} \tag{5A-10}$$

Suppose we have the following specifications for the DCM boost converter:

$V_{in,min} = 10$ V
$V_{in,max} = 15$ V
$V_{out} = 12$ V
$I_{out,max} = 2$ A
$F_{sw} = 100$ kHz
$\eta = 0.9$

Equation (5A-10) gives the inductor value to be in borderline at the minimum input voltage:

$$L = \frac{V_{in,min}^2(V_{out} - V_{in,min})\eta}{2I_{out}V_{out}^2 F_{sw}} = \frac{100 \times (12 - 10) \times 0.9}{4 \times 12 \times 12 \times 100k} = 3.13\,\mu H \tag{5A-11}$$

If we want to ensure true DCM and not borderline in the worst case (minimum input voltage, maximum load current), we can slightly reduce the inductor value by 10%, for instance, which gives us around 2.8 µH. The new peak current can be calculated by observing Fig. 5A-1:

$$I_{peak} = \frac{DV_{in}}{F_{sw}L} \tag{5A-12}$$

The maximum duty cycle in DCM derives from Eq. (1-111):

$$D_{max} = \frac{\sqrt{2LI_{out}F_{sw}(V_{out} - V_{in,min})}}{V_{in,min}} = \frac{\sqrt{2 \times 2.8u \times 2 \times 100k \times (12 - 10)}}{10} = 0.15 \tag{5A-13}$$

Substituted into Eq. (5A-12), it delivers the final peak current corresponding to the boost converter operated in DCM with a 2.8 µH inductor:

$$I_{peak} = \sqrt{\frac{2I_{out}(V_{out} - V_{in,min})}{LF_{sw}}} = \sqrt{\frac{2 \times 2 \times (12 - 10)}{2.8u \times 100k}} = 5.35\,A \tag{5A-14}$$

Now by using Eq. (1-119b), the converter will leave the DCM mode for an output load of

$$R_{critical} = \frac{2F_{sw}LV_{out}^2}{\left(1 - \frac{V_{in,min}}{V_{out}}\right)V_{in,min}^2} = \frac{2 \times 100k \times 2.8u \times 12 \times 12}{\left(1 - \frac{10}{12}\right) \times 100} = 4.84\,\Omega \tag{5A-15}$$

It corresponds to an output current of 2.5 A.

5A.2 Output Ripple Voltage

The calculation of the output ripple for the boost operated in DCM does not differ from that of the buck-boost converter also working in DCM. The idea consists of identifying the amount of charges brought to the capacitor during the off time. To help us, Fig. 5A-3 portrays the pertinent waveforms, including the diode and the capacitor currents. The positive shaded area on the capacitor signal highlights the period of time during which the diode current serves to recharge the capacitor (its voltage increases) and also powers the load. When the diode current passes below the load current, the capacitor takes over and starts to deplete.

The first algebra line derives the Δt duration. It is the time needed by the diode current to decrease and reach the output current:

$$I_{peak} - \frac{V_{out} - V_{in}}{L}\Delta t = I_{out} \tag{5A-16}$$

Extracting Δt gives the time definition:

$$\Delta t = \frac{(I_{peak} - I_{out})L}{V_{out} - V_{in}} \tag{5A-17}$$

Now, the stored charge corresponds to the area highlighted in Fig. 5A-3.

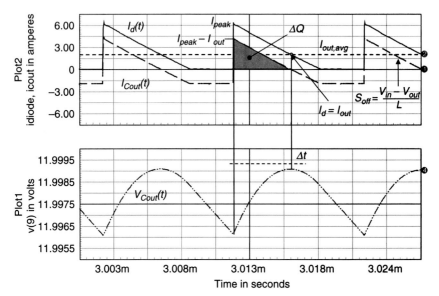

FIGURE 5A-3 The capacitive voltage output ripple in the boost discontinuous mode with currents circulating in the diode and the capacitor.

Given the simple triangle, we have

$$\Delta Q = \frac{1}{2}(I_{peak} - I_{out})\Delta t \tag{5A-18}$$

Updating Eq. (5A-18) with Eq. (5A-17), we have

$$\Delta Q = \frac{L}{2}\frac{(I_{peak} - I_{out})^2}{V_{out} - V_{in}} \tag{5A-19}$$

Knowing $\Delta Q = C \cdot \Delta V$, we finally obtain the output voltage ripple expression

$$\Delta V = \frac{L(I_{peak} - I_{out})^2}{2C_{out}(V_{out} - V_{in})} \tag{5A-20}$$

In this expression, the peak current can be replaced by Eq. (5A-14). We then obtain a more general definition of the boost output voltage ripple voltage in DCM:

$$\Delta V = \frac{\left(\sqrt{\frac{2I_{out}(V_{out} - V_{in})}{LF_{sw}}} - I_{out}\right)^2 L}{2C_{out}(V_{out} - V_{in})} \tag{5A-21}$$

CHAPTER 6
SIMULATIONS AND PRACTICAL DESIGNS OF OFF-LINE CONVERTERS—THE FRONT END

In this chapter, we investigate the front-end section, common to any ac–dc power supply. It can be a simple full-wave rectifier or a more complex power factor correction circuit. As rectification represents an important topic in mains connected topologies, we begin there.

6.1 THE RECTIFIER BRIDGE

An off-line power supply is nothing else than a dc–dc converter (a flyback or a forward, for instance) fed by a dc voltage obtained by rectifying the input line. This continuous voltage is produced by rectifying the sinusoidal alternating input voltage, whose polarity changes 50 or 60 times per second, depending on the region of the world. Usually, full-wave mode is employed, as single-wave rectification remains confined to low-power applications only (a few watts). The input voltage, the mains, follows the expression

$$V_{in}(t) = V_{peak}\sin(\omega t) \qquad (6\text{-}1)$$

where $\omega = 2\pi F_{line}$, with F_{line} the mains frequency, either 50 or 60 Hz. This frequency can increase up to 400 Hz in military and commercial aircraft applications. Here V_{peak} represents the sine wave peak voltage.

Figure 6-1a depicts a full-wave rectifier, the so-called Graetz bridge, implemented with four diodes in a single-phase off-line application.

The rectified voltage, usually labeled V_{bulk}, supplies a downstream converter, a flyback in this example. For the sake of analysis, this closed-loop converter can be replaced by a current source forcing a depletion current, defined as

$$I_{eq}(t) = \frac{P_{out}}{\eta V_{bulk}(t)} \qquad (6\text{-}2)$$

where P_{out} describes the downstream converter power delivered to its load and brought back to the rectifying bridge via the converter efficiency η.

Depending on the input polarity, two diodes are always conducting at the same time, dropping two V_f at a given current. Figure 6-1b and c shows how the current circulates as the source

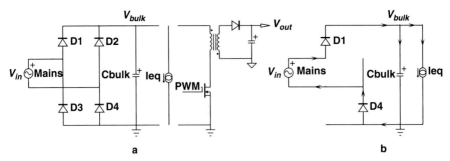

FIGURE 6-1a and b The full-wave rectifying bridge and its different conducting stages. The equivalent current source illustrates the converter loading the rectifier. Note the sine wave polarity on drawings b and c.

polarity changes: either D_1 and D_4 or D_3 and D_2. Thus, as soon as the mains voltage exceeds the bulk capacitor voltage by two forward voltage drops, the bulk capacitor is recharged until the sinusoidal crest is reached. The point t at which the recharge occurs is

$$V_{peak} \sin(\omega t) \geq V_{min} + 2V_f \qquad (6\text{-}3)$$

At this moment, a current pulse is generated and lasts until Eq. (6-3) is no longer satisfied. We will see below how to calculate the peak amplitude of this pulse. When the input voltage falls and blocks the two series diodes, the capacitor supplies the charge alone and its voltage drops: a voltage ripple takes place on the bulk rail (Fig. 6-1d). As you can see from these explanations, the bridge diodes conduct current only during a small portion of time over the entire mains period. Therefore, the total energy supplied by the source stays confined during these small events, bringing large peak and rms input currents. We will come back to this typical characteristic in a few paragraphs.

Figure 6-2a depicts the full-wave rectifier simulation sketch. We assumed a 45 W power supply affected by an optimistic 90% efficiency, further fed by a 100 µF capacitor. Note the initial condition on the capacitor which avoids a SPICE division by zero when the capacitor is totally discharged at power on. Simulation results include the various currents and voltage present on the circuit. They appear in Fig. 6-2b.

The capacitor acts as a receptor during the diode conduction time t_c and becomes a generator during the discharge time t_d when it supplies the load, the flyback converter. As soon as the input voltage reaches the capacitor residual voltage V_{min} at the end of t_d, two diodes conduct and a narrow current spike occurs with the sign depending on the mains polarity. These

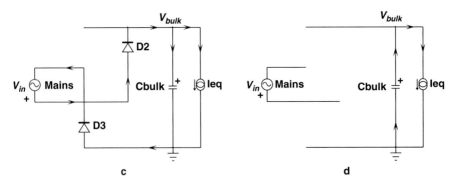

FIGURE 6-1c and d When all diodes are blocked ($V_{in,peak}$ is below the bulk voltage), the capacitor supplies the load on its own.

SIMULATIONS AND PRACTICAL DESIGNS OF OFF-LINE CONVERTERS—THE FRONT END

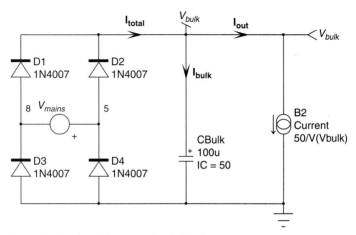

FIGURE 6-2a The full-wave rectifier in SPICE.

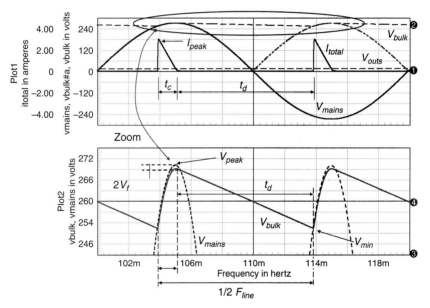

FIGURE 6-2b Simulation results of the full-wave rectifier. The lower portion of the picture zooms on the amplitudes in play during the capacitor recharge time t_c and the capacitor discharge time t_d.

spikes are seen positive only (thanks to the rectification work) within the total current I_{total} delivered to both the load and the capacitor. The dc (average) portion flows in the load, whereas the ac portion flows in the capacitor only.

6.1.1 Capacitor Selection

The dc–dc converter connected to the bulk capacitor operates within a certain input voltage range. In other words, if the input voltage drops below a certain level, the converter might overheat or enter an uncontrolled erratic mode of operation. The presence of the bulk capacitor helps to sustain the voltage between two line peaks (10 ms in Europe or 8.3 ms in the United States) and avoids a brutal drop on the dc–dc input. However, given the size and cost of the capacitor, it cannot be selected to permanently maintain the peak level: the bulk voltage will drop to V_{min}, a value that the designer has to fix. Usually, this value lies around 25 to 30% of the minimum peak line voltage. For instance, if the converter operates down to 85 Vrms, then V_{min} can be chosen around $(85 \times \sqrt{2}) \times 0.75 = 90$ Vdc.

To calculate the needed capacitor for the selected minimum bulk voltage V_{min}, we first need an equivalent schematic. We can update Fig. 6-1 during the off time or t_d, where all diodes are blocked. [Equation (6-3) is no longer satisfied.] During this time, the capacitor remains the only source delivering energy to the converter.

In Fig. 6-3, the capacitor acts as a generator, leading to voltage and current arrows pointing to a similar direction. Analytically, during the diode off time, the circulating current delivered by the capacitor can be again formulated as

$$I(t) = \frac{P_{out}}{\eta V_{bulk}(t)} \quad (6\text{-}4a)$$

We also know another equation to express this current:

$$I(t) = -C_{bulk}\frac{dV_{bulk}(t)}{dt} \quad (6\text{-}4b)$$

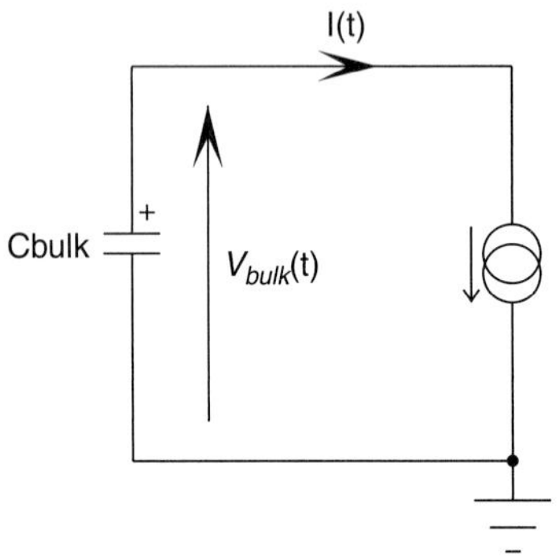

FIGURE 6-3 The capacitor supplies energy to the converter when all diodes are blocked.

Equating both equations gives

$$-\frac{P_{out}}{\eta V_{bulk}(t)} = C_{bulk}\frac{dV_{bulk}(t)}{dt} \quad (6\text{-}5a)$$

Rearranging this equality to define the power leads to

$$-\frac{P_{out}}{\eta} = C_{bulk}\frac{dV_{bulk}(t)}{dt}V_{bulk}(t) \quad (6\text{-}5b)$$

To solve this equation, we can integrate both sides from $t = 0$ (at the input peak) to $t = t_d$, just before refueling occurs (see Fig. 6-2b zoom):

$$\int_0^{t_d} -\frac{P_{out}}{\eta}dt = \int_0^{t_d} C_{bulk}\frac{dV_{bulk}(t)}{dt}V_{bulk}(t)\,dt \quad (6\text{-}6)$$

After simple manipulations, we obtain the following result:

$$-\frac{P_{out}}{\eta}t_d = \frac{C_{bulk}}{2} \cdot V_{bulk}^2 \Big|_0^{t_d} \quad (6\text{-}7)$$

The voltage at $t = 0$ is nothing else than the peak voltage V_{peak} (neglecting the voltage drop in the diodes), and for $t = t_d$, the bulk voltage reaches the minimum value V_{min}. Thus, replacing V_{bulk} by its values, we have

$$-\frac{P_{out}}{\eta}t_d = \frac{C_{bulk}}{2}\left(V_{min}^2 - V_{peak}^2\right) \quad (6\text{-}8)$$

Now rearranging to extract the capacitor value, we obtain

$$C_{bulk} = \frac{2P_{out}}{\eta\left(V_{peak}^2 - V_{min}^2\right)}t_d \quad (6\text{-}9a)$$

SIMULATIONS AND PRACTICAL DESIGNS OF OFF-LINE CONVERTERS—THE FRONT END **495**

This equation could also be derived by using the energy W released by the capacitor from V_{peak} to V_{min} during t_d:

$$W = W_{V_{peak}} - W_{V_{min}} = \frac{1}{2}C_{bulk}\left(V_{peak}^2 - V_{min}^2\right) \tag{6-9b}$$

As energy is power multiplied by time, we can write

$$\frac{P_{out}}{\eta}t_d = \frac{1}{2}C_{bulk}\left(V_{peak}^2 - V_{min}^2\right) \tag{6-9c}$$

and solving for C_{bulk} leads to the same result as Eq. (6-9a).

6.1.2 Diode Conduction Time

In Eq. (6-9a), we need to compute the discharge time to finalize the capacitor expression. Figure 6-4 focuses on the timings in play during a mains period.

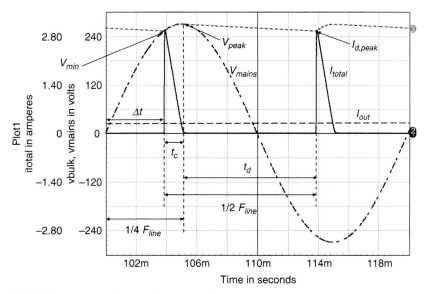

FIGURE 6-4 Updated timings during a complete input period.

The discharge time actually represents one-half of the total mains period, reduced by the diode conduction time t_c. Looking closely, we can see that the diode conduction time itself represents one-quarter of a mains period minus the time needed by the mains signal to reach V_{min}. This is Δt on the graph. Mathematically, it can be formulated as

$$V_{peak}\sin\omega\Delta t = V_{min} \tag{6-10}$$

Dividing both sides by V_{peak} yields

$$\sin\omega\Delta t = \frac{V_{min}}{V_{peak}} \tag{6-11}$$

Extracting the variable t from Eq. (6-11) gives us what we are looking for:

$$\Delta t = \frac{\sin^{-1}\left(\frac{V_{min}}{V_{peak}}\right)}{2\pi F_{line}} \qquad (6\text{-}12a)$$

Finally, we can now derive in a snapshot both the diode's conducting time t_c and the discharge time t_d.

$$t_c = \frac{1}{4F_{line}} - \Delta t = \frac{1}{4F_{line}} - \frac{\sin^{-1}\left(\frac{V_{min}}{V_{peak}}\right)}{2\pi F_{line}} \qquad (6\text{-}12b)$$

$$t_d = \frac{1}{2F_{line}} - t_c = \frac{1}{4F_{line}} + \frac{\sin^{-1}\left(\frac{V_{min}}{V_{peak}}\right)}{2\pi F_{line}} \qquad (6\text{-}12c)$$

Thanks to Eq. (6-13), the initial capacitor Eq. (6-9a) becomes

$$C_{bulk} = \frac{2P_{out}\left[\frac{1}{4F_{line}} + \frac{\sin^{-1}\left(\frac{V_{min}}{V_{peak}}\right)}{2\pi F_{line}}\right]}{\eta(V_{peak}^2 - V_{min}^2)} \qquad (6\text{-}13)$$

6.1.3 Rms Current in the Capacitor

We have seen in previous chapters that despite a precise capacitor value calculation, the final capacitor selection is always based on the rms current flowing through it. The bulk selection does not derogate from this rule. From Fig. 6-4, considering the origin of time when the recharge starts (at the end of (Δt)) and approximating the current decrease as a straight line, we have the following expression:

$$I_{total}(t) = I_{d,peak}\frac{t_c - t}{t_c} \qquad (6\text{-}14)$$

From this we can calculate the rms expression:

$$I_{total,rms} = \sqrt{2F_{line}\int_0^{t_c}\left(I_{d,peak}\frac{t_c - t}{t_c}\right)^2 dt} = I_{d,peak}\sqrt{\frac{2t_cF_{line}}{3}} \qquad (6\text{-}15)$$

The averaged value of I_{total} corresponds to the dc output current I_{out} supplied to the converter. Thus, we can write

$$I_{out} = \frac{I_{d,peak}t_c}{2}2F_{line} = F_{line}I_{d,peak}t_c \qquad (6\text{-}16)$$

From this the diode's peak current comes easily:

$$I_{d,peak} = \frac{I_{out}}{F_{line}t_c} \qquad (6\text{-}17)$$

It is important to note that the above equation is somewhat imprecise as it assumes that the diode conduction time equals the capacitor refueling time. However, that is not exactly true as the two series diodes conduct a current equal to the bulk refueling current plus the current consumed by the load. At a certain point, the diodes will block, and 100% of the load current will end up on the bulk capacitor alone. This occurs exactly at the point where the bulk capacitor current equals $-I_{out}$. This is the reason why a difference exists between the peak value expressed by Eq. (6-17) and the result you would obtain from simulation. The capacitor current can be precisely computed by deriving its voltage expression during the recharge time:

$$I_{C_{bulk}}(t) = C_{bulk}\frac{dV_{C_{bulk}}(t)}{dt} = C_{bulk}\frac{d(V_{peak}\sin(\omega t))}{dt} = 2\pi F_{line}C_{bulk}V_{peak}\cos(2\pi F_{line}t) \quad (6\text{-}18)$$

for $t \in [\Delta t, \Delta t + t_c]$. From this expression, it is possible to extract the capacitor peak current by replacing t by Δt as defined by Eq. (6-12a). It defines the exact point at which the sine wave reaches V_{min} and the capacitor recharge starts:

$$I_{C_{bulk},peak} = 2\pi F_{line}C_{bulk}V_{peak}\cos(2\pi F_{line}\Delta t) \quad (6\text{-}19a)$$

Once we know the capacitor peak current value, Eq. (6-17) can be improved by the following observation:

$$I_{d,peak} = I_{out} + I_{C_{bulk},peak} \quad (6\text{-}19b)$$

Comparisons between Eqs. (6-17) and (6-19b) have led to a conclusion whereby the first approximation gives a slightly higher rms current than what is actually obtained from simulation. It naturally improves the safety margin at the end. For the sake of simplicity, however, we will keep Eq. (6-17) for the design example.

We know that the total current I_{total} is made up of a dc portion and an ac portion. The dc portion flows in the load whereas the ac part goes through the capacitor. The link between the dc and ac components is as follows:

$$I_{total,rms} = \sqrt{I_{C_{bulk},rms}^2 + I_{out}^2} \quad (6\text{-}20)$$

The bulk capacitor rms current is thus obtained by manipulating Eq. (6-20):

$$I_{C_{bulk},rms} = \sqrt{I_{total,rms}^2 - I_{out}^2} \quad (6\text{-}21)$$

Updating Eq. (6-21) with Eqs. (6-15) and (6-17), we obtain

$$I_{C_{bulk},rms} = I_{out}\sqrt{\frac{2}{3F_{line}t_c} - 1} \quad (6\text{-}22a)$$

where t_c represents the diode conduction time as defined via Eq. (6-12b) and I_{out} is the converter average input current.

As the bulk capacitor supplies the downstream dc–dc converter, it also "sees" the converter high-frequency pulses typical of its input current signature. However, the ac switching component ($I_{smps,ac}$) only crosses the capacitor, as its dc value equals that of the dc current present in I_{total}. The rms current definition given by Eq. (6-22a) must thus account for the presence of this component. The final definition is therefore

$$I_{C_{bulk},rms,total} = \sqrt{I_{C_{bulk},rms}^2 + I_{smps,ac}^2} \quad (6\text{-}22b)$$

Reference 1 offers a different formula which more precisely accounts for the additional losses brought by these pulses. Of course, the mode of operation (DCM or CCM) will influence the final rms current used for the capacitor selection. The conclusion of the paper highlights an increase of 20% between results obtained with Eq. (6-22b) and a more accurate derivation accounting for the converter signature.

6.1.4 Current in the Diodes

The diodes see a current spike equal to the peak of Fig. 6-4. However, the time between each pulse becomes the input line period, compared to one-half of this period for the total current I_{total}. Thus, the diode rms current equals

$$I_{d,rms} = \sqrt{F_{line} \int_0^{t_c} \left(I_{d,peak} \frac{t_c - t}{t_c}\right)^2 dt} = I_{d,peak}\sqrt{\frac{t_c F_{line}}{3}} \qquad (6\text{-}23)$$

Replacing $I_{d,peak}$ by its definition [Eq. (6-17)] and simplifying, we obtain the rms current circulating in the diodes

$$I_{d,rms} = \frac{I_{out}}{\sqrt{3 F_{line} t_c}} \qquad (6\text{-}24)$$

where t_c represents the diode conduction time as defined in Eq. (6-12b).

The average diode current can be obtained via the following equation:

$$I_{d,avg} = \frac{F_{line} I_{d,peak} t_c}{2} = \frac{I_{out}}{2} \qquad (6\text{-}25)$$

If we neglect the diode dynamic resistance, each diode will dissipate:

$$P_d \approx V_f I_{d,avg} \approx V_f \frac{I_{out}}{2} \qquad (6\text{-}26)$$

6.1.5 Input Power Factor

The input rms current does not differ from that calculated by Eq. (6-15) and equals

$$I_{in,rms} = \frac{\sqrt{2} I_{out}}{\sqrt{3 F_{line} t_c}} \qquad (6\text{-}27)$$

again, where I_{out} depicts the average current drawn by the converter.

If we now calculate the average input power (in watts) and divide it by the apparent input power (VA), we obtain the rectifier power factor:

$$PF = \frac{W}{VA} = \frac{P_{in,avg}}{V_{in,rms} I_{in,rms}} \qquad (6\text{-}28)$$

Simply stated, the power factor gives information about the input energy spread over the mains period. With narrow input spikes, the instantaneous energy remains confined in the vicinity of the peak, inducing a high rms and peak current. The power factor exhibits a poor value, typical of a full-wave rectifier. To the opposite, if we are able to force the input current to spread over the entire line period, the power factor improves. This spread can be forced via a passive or an active solution, as described later in this chapter.

Capitalizing on the previous calculations, we can estimate the power factor of the full-wave rectifier. If we consider 100% efficiency and a constant converter input current (I_{out} = constant), then we can write

$$P_{in,avg} \approx I_{out} V_{bulk,avg} \qquad (6\text{-}29)$$

where the average bulk voltage is

$$V_{bulk,avg} = \frac{V_{peak} + V_{min}}{2} \quad (6\text{-}30)$$

Using Eq. (6-27), we can define the apparent input power as

$$P_{in,VA} = V_{in,rms} I_{in,rms} = V_{in,rms} \frac{\sqrt{2} I_{out}}{\sqrt{3 F_{line} t_c}} \quad (6\text{-}31)$$

Dividing Eq. (6-29) by Eq. (6-31) and rearranging, we obtain the power factor definition:

$$PF = \frac{V_{bulk,avg}}{V_{in,rms}} \sqrt{\frac{3}{2} F_{line} t_c} \quad (6\text{-}32)$$

Before we describe a real-world example, it is important to note that all the above equations are derived assuming a perfectly sinusoidal input voltage source, featuring a low output impedance. In reality, the mains is often distorted and because of the EMI filter presence, the output impedance varies quite a bit. Do not be surprised if the final results captured on the bench differ from the above calculations.

We are now ready to design the front-end stage.

6.1.6 A 100 W Rectifier Operated on Universal Mains

The power supply, a flyback, delivers 100 W (P_{conv}) to a given load with an 85% operating efficiency. The converter is told to operate on a universal mains input (85 Vrms to 275 Vrms, 47 Hz to 63 Hz) and requires a minimum operating voltage of 80 Vdc. The ambient temperature will be around 45 °C maximum. Let us start with the diode conduction time, the longest at the lowest line input:

$$t_c = \frac{1}{4 \times 60} - \frac{\sin^{-1}\left(\frac{80}{85 \times \sqrt{2}}\right)}{2 \times 3.14 \times 60} = 2.2 \text{ ms} \quad (6\text{-}33)$$

If your calculator is in degrees, replace the 6.28 term (2π) by 360.

The discharge time is immediately evaluated by

$$t_d = \frac{1}{2F_{line}} - t_c = \frac{1}{120} - 2.2m = 6.1 \text{ ms} \quad (6\text{-}34)$$

The bulk capacitor can therefore be calculated using Eq. (6-9a):

$$C_{bulk} \geq \frac{2 P_{conv}}{\eta (V_{peak}^2 - V_{min}^2)} t_d = \frac{2 \times 100}{0.85 \times (120^2 - 80^2)} 6.1m = 180 \, \mu\text{F} \quad (6\text{-}35)$$

Before the capacitor final selection, we need to assess its rms current:

$$I_{C_{bulk},rms} = \frac{P_{conv}}{\eta V_{bulk,avg}} \sqrt{\frac{2}{3 F_{line} t_c} - 1} = \frac{100}{0.85 \left(\frac{120 + 80}{2}\right)} \sqrt{\frac{2}{3 \times 60 \times 2.2m} - 1}$$

$$= 2.34 \text{ A} \quad (6\text{-}36)$$

From the original specifications, the capacitor needs to permanently sustain a steady-state voltage of 275 × 1.414 = 388 V. Hence, a 400 V type is mandatory. Given the moderate operating temperature of 45 °C, the capacitor will accept a higher ripple current compared to its maximum rating given at an ambient of 105 °C. In this particular case, the capacitor data sheet indicates a ripple multiplier above 2. An Illinois Capacitor [2] featuring the following characteristics has been chosen:

220 μF/400 V
I_{ripple} @ 120 Hz and 105 °C = 1.25 A
Ripple multiplier for T_A = 45 °C → 2.4
R_{ESR} = 1.1 Ω at 120 Hz and 20 °C
Reference: 227LMX400M2CH

Now that we have selected a 220 μF capacitor, the various times also need to be updated. From Eq. (6-35), we can compute the updated V_{min} given the new capacitor. The discharge time t_d is kept to its original value as (1) it will not change significantly with the capacitor increase and (2) an equation defining V_{min} without t_d would lead to an extremely complicated formula:

$$V_{min} = \sqrt{\frac{\eta C_{bulk} V_{peak}^2 - 2P_{conv} t_d}{C_{bulk} \eta}} = \sqrt{\frac{0.85 \times 220u \times 120^2 - 2 \times 100 \times 6.1m}{200u \times 0.85}} = 87 \text{ V} \quad (6\text{-}37)$$

If we now plug in 87 V as the minimum voltage into the above equations, we obtain

$$t_c = \frac{1}{4 \times 60} - \frac{\sin^{-1}\left(\frac{87}{85 \times \sqrt{2}}\right)}{2 \times 3.14 \times 60} = 2 \text{ ms} \quad (6\text{-}38a)$$

$$t_d = \frac{1}{2F_{line}} - t_c = \frac{1}{120} - 2m = 6.3 \text{ ms} \quad (6\text{-}38b)$$

The capacitor peak current can be evaluated by using Eq. (6-19a):

$$I_{C_{bulk}, peak} = 2\pi F_{line} C_{bulk} V_{peak} \cos(2\pi F_{line} \Delta t) = 6.28 \times 60 \times 220u \times 120$$
$$\times \cos(6.28 \times 60 \times 2.16m) = 6.7 \text{ A} \quad (6\text{-}39)$$

where Δt was evaluated to 2.16 ms using Eq. (6-12a). From this result, it is possible to evaluate the diode peak current by using Eq. (6-17) or Eq. (6-19b):

$$I_{d, peak} = I_{out} + I_{C_{bulk}, peak} = \frac{P_{out}}{\eta} \frac{2}{V_{min} + V_{peak}} + I_{C_{bulk}, peak}$$

$$= \frac{100}{0.85} \times \frac{2}{120 + 87} + 6.7 = 7.83 \text{ A} \quad (6\text{-}40a)$$

$$I_{d, peak} = \frac{I_{out}}{F_{line} t_c} = \frac{P_{out}}{\eta (V_{min} + V_{peak}) F_{line} t_c} = \frac{100}{0.85} \times \frac{2}{(120 + 87) \times 60 \times 2m} = 9.47 \text{ A}$$
$$(6\text{-}40b)$$

As expected, Eq. (6-40b) gives more pessimistic results than Eq. (6-40a). Simulation will later confirm the Eq. (6-40a) result.

Given the 220 μF selection, the bulk rms current slightly increases to 2.4 A which still gives some margin (maximum ripple equals 3 A given the ripple multiplier indicated by the manufacturer).

The diode currents are computed to the following values:

$$I_{d,rms} = \frac{P_{conv}}{\eta V_{bulk,avg} \sqrt{3F_{line}t_c}} = \frac{100}{0.85 \times \left(\frac{120 + 87}{2}\right)\sqrt{3 \times 60 \times 2m}} = 1.9\,\text{A} \quad (6\text{-}41\text{a})$$

$$I_{d,avg} = \frac{P_{conv}}{\eta V_{bulk,avg}^2} = 0.56\,\text{A} \quad (6\text{-}41\text{b})$$

Diodes such as the ON Semiconductor 1N5406 (600 V/3 A) are well suited for this application. The rms input current also needs to be known to select the right fuse:

$$I_{in,rms} = \frac{\sqrt{2}P_{conv}}{\eta V_{bulk,avg}\sqrt{3F_{line}t_c}} = \frac{\sqrt{2} \times 100}{0.85 \times \frac{(120+87)}{2}\sqrt{3 \times 60 \times 2m}} = 2.7\,\text{A} \quad (6\text{-}42\text{a})$$

A 250 V/4 A delayed type will do the job. A delayed type is necessary because of the large in-rush current at power on (see dedicated section below).

Finally, we can compute the power factor as imposed by this front-end section:

$$\text{PF} = \frac{V_{bulk,avg}}{V_{in,rms}}\sqrt{\frac{3}{2}F_{line}t_c} = \frac{104}{85}\sqrt{\frac{3}{2} \times 60 \times 2m} = 0.517 \quad (6\text{-}43)$$

Based on the above result, we could also evaluate the input rms current in a quicker way than via Eq. (6-42a):

$$I_{in,rms} = \frac{P_{out}}{\eta V_{in,min}\text{PF}} = \frac{100}{0.85 \times 85 \times 0.517} = 2.68\,\text{A} \quad (6\text{-}42\text{b})$$

Using Fig. 6-2a as a simulation fixture, we can plug in the diode and capacitor values and look for the various stresses. The simulation results appear in Fig. 6-5. As we can read, the simulated results give slightly more optimistic results than the analytical calculations, but they are very close to each other.

To compute the power factor, just measure the rms input current (2.5 A) and multiply it by the input voltage, 85 Vrms. This gives us the VA. The average input power is found to be around 120 W [multiply $V_{mains}(t)$ by $I_{in}(t)$ and average it over one period, you obtain watts]. The power factor reaches 0.57, a (poor) value typical of a full-wave application.

6.1.7 Hold-Up Time

The hold-up time defines the time during which the power supply can still deliver its nominal power as the mains input has disappeared (because of a transient power failure, for instance). The test consists of interrupting the input line when it crosses zero and watching the bulk level go down. At a given time, the power supply can no longer regulate, and its output falls to zero. Some protection circuit also exists which prevents the power supply from being overloaded in low-mains situations. This circuit is called a "brownout" detector, and we suppose that the designer did set it to 60 V in this example. In the simulation, as we do not observe any output voltage, we can program the current source B_2 to stop generating current when the bulk voltage passes below the above threshold:

```
B2 vbulk 0 V(Vbulk) > 60 ?117/(V(Vbulk)+1) : 0 ; IsSpice 117 W
; power consumption
G2 vbulk 0 Value = {IF (V(Vbulk) > 60, 117/(V(Vbulk)+1), 0 )};
;  PSpice 117 W power consumption
```

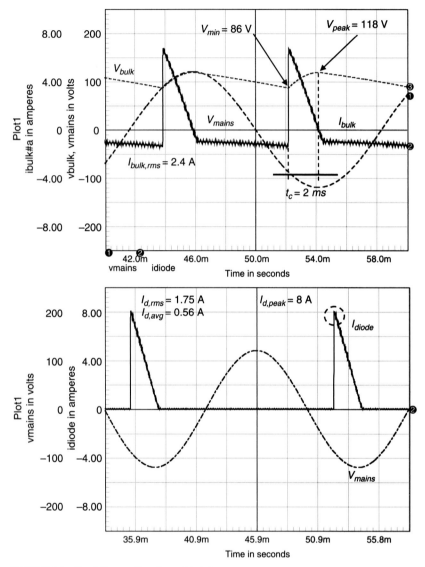

FIGURE 6-5 Simulation results and associated stresses.

Figure 6-6 shows the result for the rectifier which exhibits a hold-up time of 6 ms. This is a rather mediocre performance that can be enhanced by increasing the size of the bulk capacitor. If necessary, Eq. (6-9a) can be used to calculate the amount of capacitance to hold the bulk level long enough in case the mains were to go away. The parameter t_d would then be replaced by the targeted hold-up time.

6.1.8 Waveforms and Line Impedance

The above waveforms are idealized in the sense that the mains impedance is null: The waveform does not get distorted when the current spike occurs to recharge the bulk capacitor. In

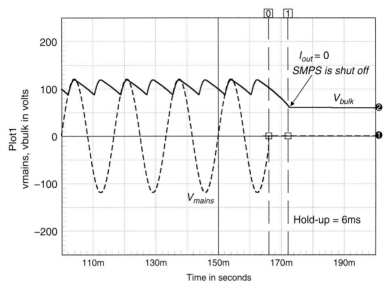

FIGURE 6-6 The bulk rail collapses as soon as the mains disappears. The downstream converter cannot maintain itself longer than 6 ms.

reality, the mains outlet into which you plug the rectifier features a particular output impedance. Unfortunately, the impedance of this local mains depends upon various elements such as the types of connected equipment: motors, light ballasts, and refrigerators. Thus, trying to predict it represents a difficult exercise. Furthermore, the EMI filter inserted in series with the converter also affects the impedance driving the bridge. In a newsletter [3], Intusoft proposes a simple representation of its local mains, made of a resistor and a low-value inductor. The updated rectifier now appears in Fig. 6-7a.

As one can imagine, when the charge current occurs, the peak current collapses the line voltage and engenders a so-called sag. Figure 6-7b displays the results where the distorted line now appears. You can also notice the current waveform which really differs from the previous assumptions.

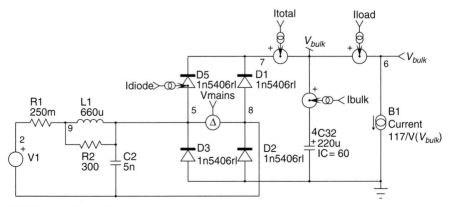

FIGURE 6-7a The rectifier now includes the mains impedance.

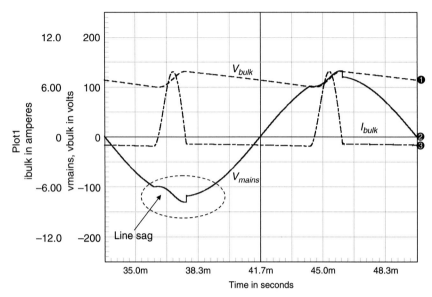

FIGURE 6-7b After line impedance inclusion, the waveform really changes and shows distortion.

This kind of waveform can also be obtained when the input line EMI filter artificially increases the line impedance. It tends to naturally reduce the peak current and thus the rms bulk current.

To check these simulation results versus reality, we have wired a rectifier bridge together with a 250 μF capacitor delivering 298 W to a resistive load. The first connection was made to a low-impedance ac source, and the variables in play were captured. They appear in Fig. 6-8a.

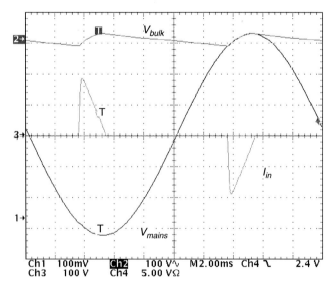

FIGURE 6-8a A full-wave rectifier loaded by 298 W and powered from a low-impedance source.

SIMULATIONS AND PRACTICAL DESIGNS OF OFF-LINE CONVERTERS—THE FRONT END 505

The triangular shape really resembles that of the simulated waveform. Also, note the absence of sag, attesting to the low impedance of the source. Now, we have connected the diode bridge to an adjustable autotransformer connected to the mains. The results appear in Fig. 6-8b.

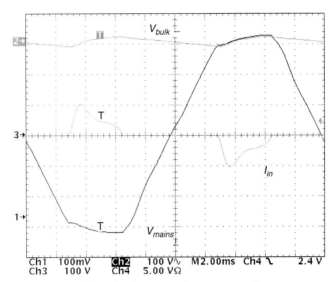

FIGURE 6-8b   The ac signal now comes from an autotransformer.

An autotransformer features a large series impedance which acts as a smoothing inductor. This results in a collapse of the ac signal, leading to a larger conduction time. The peak current is now reduced and leads to a smaller rms current: The stress on the bulk capacitor is reduced!

Finally, we have connected the rectifier to the mains, without any intermediate circuit. The waveforms appear in Fig. 6-8c.

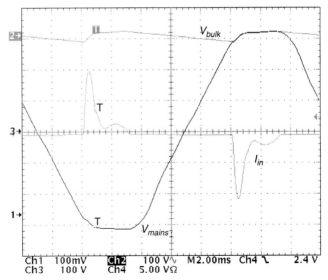

FIGURE 6-8c   When operated directly on the mains, the ac signal loses its purity.

The signal delivered by Electricité de France (EDF) looks like a smooth wave and shows almost no sag as the current peaks. Rather, it flattens on the peaks which creates a rather distorted current. During all these experiments, we have recorded several parameters. They are gathered below:

	Low-impedance ac	Autotransformer ac	Mains
$I_{in,rms}$ (A)	2.53	1.8	2.35
PF	0.51	0.69	0.52

The best power factor is obtained through the autotransformer! This makes sense as the inductive impedance smoothes the current and reduces its peak. The worst case actually occurs in the presence of the pure ac source featuring a low impedance. In other words, the calculations described in the above lines really represent the worst case and offer a good margin when the rectifier evolves on the mains. When the rectifier is finally built, make sure all parameters are measured either directly on the mains or via an ac source: Avoid the autotransformer which gives too optimistic values!

This experiment calls for another important comment: When you are running ac efficiency measurements on a board featuring a front-end EMI filter, always check and specify the source output impedance used for these measurements. If you use a low-impedance source, the lower PF will increase the circulating rms currents and worsen all conduction losses (common-mode chokes, in-rush resistors, etc.) leading to a degraded efficiency. On the other hand, if you take the same board and operate it from a variable transformer, as the PF improves, the rms current gets lower and the reduced conduction losses make your efficiency look better! We have experimentally tested a converter showing 84% efficiency on a low-impedance ac source and improving to 86% when connected directly to the distorted mains. Be sure your customer understands this concept as well!

6.1.9 In-Rush Current

At power on, when you first plug in the rectifier, the bulk capacitor is completely discharged. It results in a large charging current as the diode bridge is temporarily short-circuited by this discharged capacitor. With some really large bulk capacitor, this spike can actually trigger the mains breaker and destroy the rectifier diodes. Some manufacturers actually impose a maximum in-rush current when the user plugs in the converter right at the mains peak. Figure 6-9a represents a simple way to simulate an in-rush current. Of course, you can imagine the role of the input source impedance in this test fixture. Again, in-rush measurements shall not be evaluated with the autotransformer!

In Fig. 6-9b, a large input current appears as the switch closes. In this case, the capacitor ESR helps to reduce the current excursion. Still, the peak reaches 53 A. Of course, it crosses the diodes, and their selection must also account for this nonrepetitive spike. The parameter to look for is I_{FSM}. It equals 200 A for the 1N5406. Reference 4 discusses how to model input filter inductors and see how the saturation affects the peak current value.

The bulk capacitor (but the fuse as well) can suffer from these surges which clearly affect its lifetime. Several ways exist to limit the in-rush current, and a few known techniques are gathered in Fig. 6-10 for reference only. A simple relay short-circuiting a resistor, a negative temperature coefficient (NTC), and an active SCR represent possible solutions implemented in the industry.

Once the power supply is designed and built, including the front-end rectifying section, it is good practice to test it in a controlled-temperature chamber, operated at the highest ambient stated by the specification (for instance, 50 °C). Then load the power supply with its maximum current capability and power it via a relay to the maximum input voltage. The relay will be controlled by a square wave generator delivering pulses at a low pace: The on time must let the output

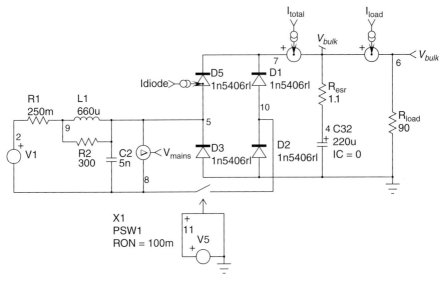

FIGURE 6-9a A switch inserted in series with the source simulates a sudden power on sequence.

reach regulation and stay there for a few hundred milliseconds. The off time should allow a complete discharge of the bulk capacitor before any new start-up. Select a low-impedance ac source to run these tests and avoid the autotransformers. Its inductive impedance will limit the in-rush pulse, naturally protecting the diode bridge and the bulk capacitor, unfortunately far from real life. Run this test for a few hours, including low-line conditions. Endurance tests,

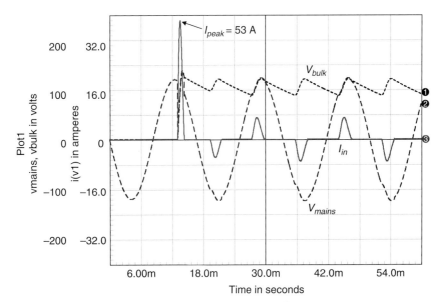

FIGURE 6-9b The in-rush current reaches 53 A in this particular example where the ac source is modeled according to Fig. 6-9a.

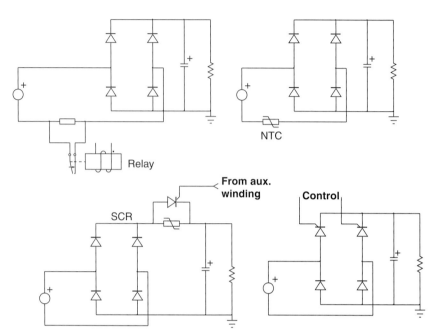

FIGURE 6-10 Various solutions exist to limit the surge current: a relay, a simple NTC, an SCR, an SCR-based rectifying bridge (a MOSFET can also do the job). In some cases, the NTC is even short-circuited by a relay to increase the efficiency.

also called burn-in tests, where the converter is operated for more than 100 h (168 h is typical) in the same conditions without relay this time (permanent voltage on the input), should also be performed to detect any design flaw. If the power supply under test survives these simple exercises, it is a good indication that the designer has adopted adequate margins.

6.1.10 Voltage Doubler

The voltage doubler was often used in the 1970s when power supplies capable of operating on a wider range (so-called universal input) were still not available. A voltage selector was used to select a 117 Vrms level or 230 Vrms. When operated on a 117 Vrms or a 230 Vrms input voltage, the doubler output voltage does not vary too much, hence the benefit of this configuration. Needless to say, a wrong selector position could lead to catastrophic failures. These days, doublers still find applications in low-line environments such as in the United States or Japan where they help to lift up the rectified dc voltage, hence reducing the average/rms current burden on the converter. Figure 6-11a shows how to build a doubler.

We have run a simulation to highlight the particular waveforms of this doubler. They appear in Fig. 6-11d.

The doubler configuration actually only uses two diodes of the bridge. And this makes sense when you are looking at the schematic. If D_2 or D_4 were to conduct in doubler mode, they would, respectively, short-circuit C_1 and C_2! In Fig. 6-11b, the capacitors alternatively charge to the peak input. It is interesting to note that the total voltage (V_{bulk}) passes through a minimum (V_{min}) when one capacitor also reaches its valley voltage (C_1, for instance, in Fig. 6-11b), but the other one stays halfway between its peak and valley. As both capacitors are in series, the minimum bulk voltage can be evaluated to

$$V_{min} = V_{C_1,min} + V_{C_2,avg} = V_{C_1,min} + \frac{V_{C_2,peak} + V_{C_2,min}}{2} \qquad (6\text{-}44)$$

SIMULATIONS AND PRACTICAL DESIGNS OF OFF-LINE CONVERTERS—THE FRONT END   509

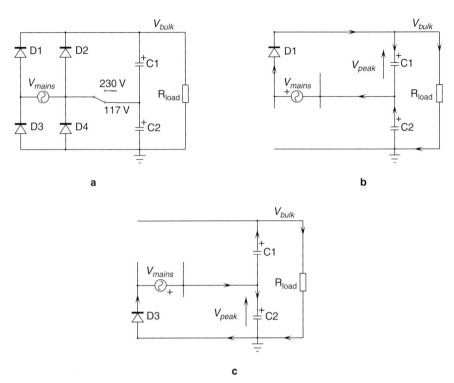

FIGURE 6-11a, b, c   A voltage doubler helps to increase the 117 Vrms line to reduce the burden on the downstream converter. In doubler operation, only two diodes are active.

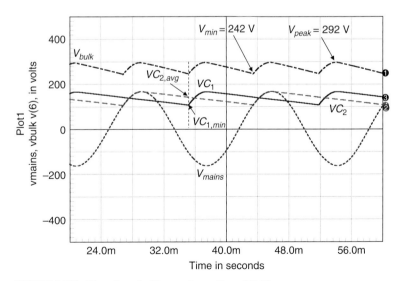

FIGURE 6-11d   The voltage doubler waveforms, $P_{out} = 117$ W.

Since both peaks and minima for the capacitors are equal, Eq. (6-44) leads to

$$V_{min} = \frac{3V_{C,min} + V_{C,peak}}{2} \qquad (6\text{-}45)$$

The design will fix the minimum bulk voltage as a target parameter; therefore, we need to evaluate the allowable $V_{C,min}$ for the capacitors:

$$V_{C,min} = \frac{2V_{min} - V_{C,peak}}{3} \qquad (6\text{-}46)$$

Rather than apply Eq. (6-4), we can write the energy difference between one peak and a valley for a single capacitor, knowing that this capacitor only supplies one-half of the required energy:

$$\frac{1}{2}CV_{C,peak}^2 - \frac{1}{2}CV_{C,min}^2 = \frac{P_{out}t_d}{2\eta} \qquad (6\text{-}47)$$

where t_d represents the discharge time of the capacitor. Solving for the capacitor value leads to

$$C_1 = C_2 = \frac{P_{out}}{\eta(V_{C,peak}^2 - V_{C,min}^2)}t_d \qquad (6\text{-}48)$$

The discharge time remains very close to that of Eq. (6-12c), but each capacitor now discharges over an entire mains period. Thus

$$t_d = \frac{1}{F_{line}} - t_c = \frac{3}{4F_{line}} + \frac{\sin^{-1}\left(\dfrac{V_{C,min}}{V_{C,peak}}\right)}{2\pi F_{line}} \qquad (6\text{-}49)$$

More information on the doubler can be found in Ref. 4.

6.2 POWER FACTOR CORRECTION

This section introduces the concept of passive and active power factor correction. This is a vast subject which would require an entire book of its own to thoroughly cover it. Further to explaining the reason for power factor correction, we will only discuss in detail the design of one of the most popular topologies, the borderline boost PFC. As usual, we have identified pertinent references available at the end of this chapter to enable you to strengthen your knowledge in this interesting domain of power electronics.

As observed in the previous lines, the presence of the capacitor on the full-wave rectifier induces input current spikes located in the vicinity of the ac source peaks. The load draws energy from the source only during a short time, needed to quickly recharge the bulk capacitor. Calculations show the circulation of a large rms current, mainly due to the presence of this high and narrow peak current. Figure 6-7a actually portrays the classical full-wave rectifier that we designed to power a 100 W load. Further to a simulation, we have collected the input variables presented below:

Case 1:

$V_{in,} = 85$ Vrms
$I_{in,peak} = 8$ A

$$I_{in,rms} = 2.5 \text{ A}$$
$$P_{in,avg} = 119 \text{ W}$$

Now we have removed from Fig. 6-7a the bulk capacitor and replaced the current source with a resistor drawing 119 W ($R = 60 \, \Omega$) from the input. As the rectified voltage now goes back to zero, the current source will generate a divide by 0 error, stopping the simulator, hence its replacement by a resistive load. We obtained the following results:

Case 2:

$$V_{in} = 85 \text{ Vrms}$$
$$I_{in,peak} = 2 \text{ A}$$
$$I_{in,rms} = 1.4 \text{ A}$$
$$P_{in,avg} = 119 \text{ W}$$

Figure 6-12 compares the input variable waveforms in both cases. The upper portion represents the resistive loading. In this configuration, the instantaneous power follows the sinusoidal signal and spreads over the entire half-period.

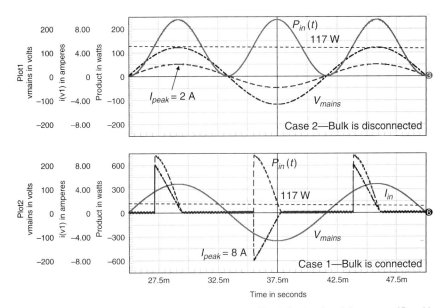

FIGURE 6-12 Typical input waveforms for a full-wave rectifier delivering dc and the same rectifier without the bulk capacitor—same average input power.

The lower case illustrates the classical rectifier associated with the bulk capacitor. As expected, the power drawn from the source stays confined in the vicinity of the peak input. The peak power $I_{in}(t)V_{in}(t)$ reaches 716 W whereas it was only 236 W in the previous case. For the same work, e.g., powering a load to produce light or sound, we have an rms current of 2.5 A in the first case and 1.4 A in the second one. In case 1, we have an excess of current circulating in the wires which contributes to overheating them. This excess of current does not contribute to the actual work; it just flows and increases the mains burden together with the distribution network. As an example, consider a European mains outlet which accepts up to 16 A rms of continuous current drawn from it. Case 1 would allow us to connect a maximum of six pieces of equipment (16/2.5), whereas case 2 would let us operate up to eleven devices (16/1.4).

6.2.1 Definition of Power Factor

The average input power drawn by a load actually corresponds to the instantaneous power $P_{in}(t)$ averaged over one cycle. This is expressed by

$$P_{in,avg} = \frac{1}{T}\int_0^T I_{in}(t)V_{in}(t)\,dt \quad \text{(watts, W)} \tag{6-50}$$

If we have in-phase sinusoidal signals, we can update Eq. (6-50) and solve the integral:

$$P_{in,avg} = \frac{1}{T}\int_0^T I_{in,rms}\sqrt{2}\sin(\omega t)V_{in,rms}\sqrt{2}\sin(\omega t)\,dt = I_{in,rms}V_{in,rms} \tag{6-51}$$

Keeping the same signals, we can also evaluate the product of both rms variables which defines the apparent power expressed in volt-amperes:

$$P_{in,apparent} = I_{in,rms}V_{in,rms} \quad \text{(volt-amperes, VA)} \tag{6-52}$$

These results show that a resistive loading brings an average input power equal to the product of the input rms variables. The ratio of Eq. (6-51) to Eq. (6-52) is called the power factor, denoted by PF:

$$PF = \frac{P_{in,avg}(W)}{I_{in,rms}V_{in,rms}(VA)} \tag{6-53}$$

The power factor reaches unity for a resistive loading and stays below 1 for other types of loads: In the previous example, the PF reaches 0.56 in case 1 and equals 1 in case 2. A low PF brings a higher rms current, as confirmed by the measurements.

Now, we can rework Eq. (6-51) by introducing a phase difference φ between the current and the voltage. The analytical expression for the average power becomes

$$P_{in,avg} = \frac{1}{T}\int_0^T I_{in,rms}\sqrt{2}\sin(\omega t + \varphi)V_{in,rms}\sqrt{2}\sin(\omega t)\,dt = V_{in,rms}I_{in,rms}\cos\varphi \tag{6-54}$$

If we apply the definition of Eq. (6-53), thus dividing Eq. (6-54) by Eq. (6-52), we obtain the power factor definition for sinusoidal signals

$$PF = \cos\varphi \tag{6-55}$$

where φ illustrates the phase difference between the current and the voltage.

6.2.2 Nonsinusoidal Signals

Equation (6-53) is always valid, whatever the voltage and current shapes. Equation (6-55) works as long as both signals' voltage and current signals are sinusoidal. The following lines detail how we can reformulate the power factor expression in the presence of a sinusoidal ac source (such as in Fig. 6-12) but resulting in a distorted current.

The definition of an rms signal implies the quadratic sum of its dc value with the fundamental and all the remaining harmonics. This is what Eqs. (6-56a) and (6-56b) detail:

$$I_{rms} = \sqrt{I_0^2 + \sum_{n=1}^{\infty} I_{n,rms}^2} \tag{6-56a}$$

$$V_{rms} = \sqrt{V_0^2 + \sum_{m=1}^{\infty} V_{m,rms}^2} \tag{6-56b}$$

From this equation, we can see how the presence of harmonics always contributes to the increase in the rms value. Now, if we update Eq. (6-54) via Eqs. (6-56a) and (6-56b), we can immediately simplify the resulting expression via two observations:

- When we have cross-product terms of different frequencies, in other words when $m \neq n$, the average power linked to these terms cancels. For example, if one of the terms multiplies the third and fifth harmonics, we have

$$\frac{1}{T}\int_0^T I_3 \sin(3\omega t + \varphi) V_5 \sin(5\omega t) dt = 0 \tag{6-57a}$$

Using SPICE, generate a sine wave of a frequency F_1 and another one at a frequency $F_2 = 3 \times F_1$. Multiply both waveforms together and take the average: You read zero.

- As we assumed we are dealing with a pure sinusoidal source, the input voltage does not contain any harmonics. Hence, all products implying current harmonics ($n = 2$ to $n = \infty$) are gone since their voltage counterpart does not exist. Only fundamentals of both signals carry the real power:

$$P_{in,avg} = \frac{1}{T}\int_0^T I_1 \sin(\omega t + \varphi) V_1 \sin \omega t$$

$$+ \underbrace{I_2 \sin(2\omega t + \varphi_2) V_2 \sin 2\omega t}_{=0} + \cdots + \underbrace{I_n \sin(n\omega t + \varphi_n) V_n \sin n\omega t\, dt}_{=0} \tag{6-57b}$$

Capitalizing on the above points, we can define the average power through the formula

$$P_{in,avg} = V_{1,rms} I_{1,rms} \cos \varphi \tag{6-58}$$

where φ represents the phase difference between both current and voltage fundamentals and V_1 and I_1 are the respective fundamental values of the input voltage and current.

Regarding the apparent power, the simple multiplication of both rms values now implies all the cross-products between the voltage fundamental and the current harmonics. Thus

$$P_{in,apparent} = V_{1,rms} \sqrt{\sum_{n=1}^{\infty} I_n^2} = V_{1,rms} I_{rms} \tag{6-59}$$

Note that the dc terms in Eqs. (6-56a) and (6-56b) have gone since the average values of both voltage and current are null. Finally, if we divide Eq. (6-58) by Eq. (6-59), we have an updated power factor definition now applying to a sinusoidal source dealing with a distorted current

$$PF = \frac{V_{1,rms} I_{1,rms}}{V_{1,rms} I_{rms}} \cos \varphi = \frac{I_{1,rms}}{I_{rms}} \cos \varphi = k_d k_\varphi \tag{6-60}$$

where $k_\varphi = \cos \varphi$ is called the *displacement factor*, φ represents the displacement angle between voltage and current fundamentals, and $k_d = \dfrac{I_{1,rms}}{I_{rms}}$ is called the *distortion factor*.

6.2.3 A Link to the Distortion

The total harmonic distortion (THD) of a signal is defined by the rms values brought by the harmonics excluding the fundamental ($n = 2$ to $n = \infty$) divided by the rms value of the fundamental alone:

$$\text{THD} = \frac{\sqrt{\sum_{n=2}^{\infty} I_{n,rms}^2}}{I_{1,rms}} = \frac{I_{rms}(dist)}{I_{1,rms}} \tag{6-61}$$

The harmonic content can be easily extracted from the total rms current via a few simple equations. We know that

$$I_{rms}^2 = I_0^2 + I_{1,rms}^2 + I_{rms}(dist)^2 \tag{6-62}$$

In this case, the current average value is null; the term I_0 thus fades away:

$$I_{rms}(dist) = \sqrt{I_{rms}^2 - I_{1,rms}^2} \tag{6-63}$$

Introducing Eq. (6-63) into Eq. (6-61), we have

$$\text{THD} = \frac{\sqrt{I_{rms}^2 - I_{1,rms}^2}}{I_{1,rms}} \tag{6-64}$$

If we now bring the denominator squared into the square root term, we can update Eq. (6-64):

$$\text{THD} = \sqrt{\left(\frac{I_{rms}}{I_{1,rms}}\right)^2 - 1} \tag{6-65}$$

Identifying the current division term to be the inverse of k_d, we have

$$\text{THD} = \sqrt{\frac{1}{k_d^2} - 1} \tag{6-66}$$

This definition offers a way to redefine the distortion factor in relation to the total harmonic distortion. Extracting k_d from Eq. (6-67) leads to its updated definition:

$$k_d = \frac{1}{\sqrt{1 + (\text{THD})^2}} \tag{6-67a}$$

If the THD is expressed as a percentage, the above equation becomes

$$k_d = \frac{1}{\sqrt{1 + \left(\dfrac{\text{THD}}{100}\right)^2}} \tag{6-67b}$$

In cases where both current and voltage fundamentals are in phase, otherwise stated as $k_\varphi = \cos \varphi = 1$, the power factor definition simplifies to

$$\text{PF} = k_d = \frac{1}{\sqrt{1 + \left(\dfrac{\text{THD}}{100}\right)^2}} \tag{6-68}$$

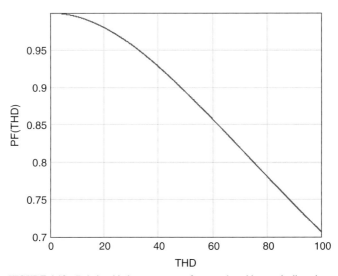

FIGURE 6-13 Relationship between power factor and total harmonic distortion.

Applying Eq. (6-68), Fig. 6-13 represents the power factor variations as the THD changes. We can see that a power factor of 0.95 gives more than 30% of THD. To reach less than 10% THD, the PF has to exceed 0.995!

Equation (6-60) offers an interesting way to visualize the power factor and see what parameter affects it (distortion, phase difference, or both). Figure 6-14 represents several examples where these numbers go through various values.

Note that the lower left case brings a null *pf*, since we have a 90° phase difference between current and voltage. This would be the case if a capacitor were directly connected to the mains.

To refresh your mind on ac signals, Ref. 6 spells out all you need to know about reactive, apparent, and true power. It is a really interesting document to download.

6.2.4 Why Power Factor Correction?

Equation (6-60) splits the power factor into two terms, distortion and phase angle. Let us see why a need for regulation exists in some countries.

- *Displacement factor*: This is the phase difference between both input signals, commonly named the *cosine phi*. If you look at plaques on your washing machine, light ballast, and so on, you will often see a statement signaling a cos φ greater than a certain value, 0.8 in France, for example. We have seen that a low power factor, or a mediocre cos φ, leads to the circulation of an rms current greater than what is really necessary to perform the work. This excess of rms current flows in the wires and forces the utility companies to oversize their distribution networks. In some countries, these companies even charge their end users in the event that the annual amount of reactive power exceeds a certain value.

- *Distortion factor*: We have seen a direct link between the distortion factor and the power factor. In the case of full-wave rectifiers, the displacement factor almost reaches 1, but the highly distorted current degrades the PF. Given the rich harmonic content, these distorted currents—imagine hundreds of uncorrected personal computers in a building—induce resonances which can perturb motors, generate noise, and bother sensitive equipment. In a

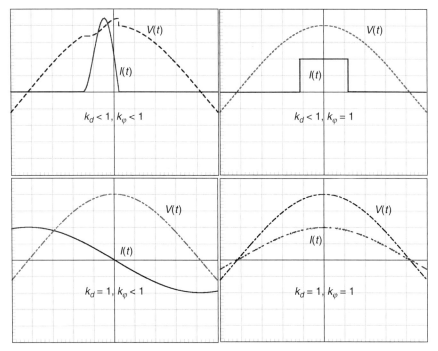

FIGURE 6-14 Various cases showing different k_d and k_φ values.

three-phase configuration, distortion can lead to an overheating of the neutral wire which normally does not carry any current.

An international standard, the IEC 1000-3-2 (EN 61000-3-2), defines the rules for the current distortion affecting mains-powered devices. Note that the standard does not discuss the power factor directly, but fixes current harmonic limits. The table below depicts the four classes in which the IEC 1000-3-2 classifies the equipment connected to the mains. It accounts for a recent amendment (A14, effective January 2001) which removed the "special wave shape" format for class D and reformulates the split between classes.

Class A	Balanced three-phase equipment, single-phase equipment not in other classes: • Household appliances except those identified as class D • Nonportable tools • Dimmers for incandescent lamps only • Anything not otherwise classified
Class B	Portable power tools
Class C	All lighting equipment, except incandescent lamp dimmers
Class D	Single phase, below 600 W, personal computers, television sets, PC monitors

Now, all switch-mode power supplies below 600 W (ac–dc adapters, TV supplies, and PC monitors) must conform to class D. Until now, given the mandatory harmonic levels for power levels beyond 75 W, the implementation of a PFC front-end stage represented the best option to meet the standard. This applies to Europe, Japan, and now China. The United States does not require the presence of a PFC. Reference 7 explores in detail the content of the standard and its origins.

TABLE 6-1 Harmonic Limits for All Four Classes

Harmonic number n	Class A Max. current, Arms	Class B Max. current, Arms	Class C Percent of fundamental	Class D 75 W < P < 600 W mArms/W	Class D Absolute limits Arms
3	2.3	3.45	30 . PF	3.4	2.3
5	1.14	1.71	10	1.9	1.14
7	0.77	1.155	7	1.0	0.77
9	0.4	0.6	5	0.5	0.4
11	0.33	0.495	3	0.35	0.33
13	0.21	0.315	3	0.296	0.21
$15 \leq n \leq 39$	2.25/n	3.375/n	3	3.85/n	2.25/n
2	1.08	1.62	2		
4	0.43	0.645			
6	0.3	0.45			
$8 \leq n \leq 40$	1.84/n	2.76/n			

6.2.5 Harmonic Limits

The international commission has established a list of harmonic amplitude limits for all four classes. We have assembled these limits in Table 6-1, reflecting where the standard considers individual harmonic levels. Thus, if the THD gives an idea of the harmonic content, the individual harmonic assessment up to the 39th remains the only measurement method to use to check whether the equipment passes or fails the test. Again, you can note the absence of a reference to the displacement factor as only the harmonic levels matter (a reference to the PF, however, appears in the class C definition).

SPICE can help us to unveil the harmonic content of a given signal via its FFT capabilities. Suppose we needed to assess the harmonic content of the signal in Fig. 6-12. We would run the simulation and then display the input current. Finally, invoking the FFT algorithm displays the total harmonic content. However, care must be taken to deal with a sufficient number of data points. The transient duration gives the resolution bandwidth or the step width between each FFT frequency point. For instance, if we select a transient duration of 40 ms, we will obtain a resolution of 1/40m or 25 Hz, leading to an imprecise graph. In the example below, we purposely increased the transient simulation to 300 ms, giving a resolution of 3 Hz. Figure 6-15 displays the results.

Note that the SPICE FFT engine displays peak current values whereas the standard talks about rms levels. A translation has to be made in either direction to keep similar units. This is what we did in the upper right corner of Fig. 6-15, transforming rms limits into peak ones. Applying Eq. (6-61), we can now manually compute the harmonic distortion by collecting the individual harmonic peak amplitudes. For the sake of the example, we went up to the 11th:

$$\text{THD} = \frac{\sqrt{1.87^2 + 1.32^2 + 0.79^2 + 0.51^2 + 0.48^2}}{2.22} = \frac{2.52}{2.22} = 1.13 \text{ or } 113\% \quad (6\text{-}69)$$

We can also run a discrete Fourier analysis, telling SPICE to observe and analyze the input source current. The result is very close to the above.

```
.FOUR 60 I(V1)

Fourier analysis for v1#branch:
No. Harmonics: 10, THD: 112.799%, Gridsize: 200, Interpolation
   Degree: 1
```

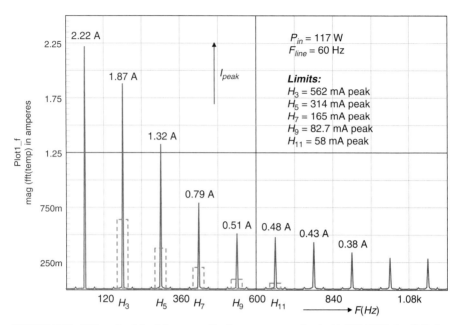

FIGURE 6-15 FFT results of the full-wave rectifier input current associated with IEC 61000 class D limits. $V_{in,peak} = 120$ V, $P_{out} = 117$ W.

Harmonic	Frequency	Magnitude	Phase	Norm. mag	Norm. phase
0	0	-8.8362e - 007	0	0	0
1	60	2.17646	-81.968	1	0
2	120	6.11649e - 005	-151.02	2.81029e - 005	-69.055
3	180	1.84776	-64.754	0.848974	17.2138
4	240	0.000113169	-124.84	5.19966e - 005	-42.87
5	300	1.31518	-43.162	0.604274	38.8054
6	360	0.000147133	-98.522	6.76019e - 005	-16.554
7	420	0.790208	-10.05	0.36307	71.9181
8	480	0.000156342	-72.26	7.18331e - 005	9.70806
9	540	0.508707	44.1788	0.233731	126.147

If distortion offers a way to discuss the input current purity, the standard actually considers harmonics up to the 39th position. Each harmonic amplitude should be individually checked by a power analyzer and compared to a stored value, representative of the standard limit. A signal lets you know if the equipment passes or fails the test.

Having the distortion on hand, we can now derive the power factor via Eq. (6-68):

$$\text{PF} = \frac{1}{\sqrt{1 + 1.13^2}} = 0.66 \quad (6\text{-}70)$$

6.2.6 A Need for Storage

Figure 6-16 represents the sinusoidal signals in play when the mains supplies a resistive load. The instantaneous power $P_{out}(t)$ depicts a squared sinusoid whose average value actually

SIMULATIONS AND PRACTICAL DESIGNS OF OFF-LINE CONVERTERS—THE FRONT END

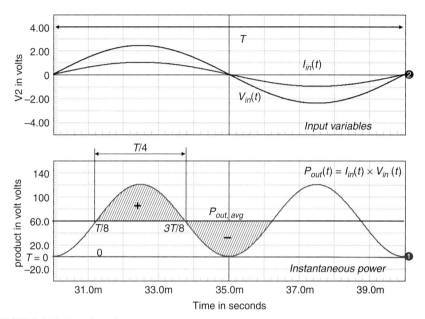

FIGURE 6-16 Waveforms in the presence of a resistive load leading to a PF of 1.

corresponds to the transmitted average power $P_{out,avg}$. Observing the waveform $P_{out}(t)$, one can see that

- The frequency of the instantaneous power equals twice the mains frequency. Actually $P_{out}(t)$ is a squared sinusoid.
- During a quarter of the period, the instantaneous power exceeds the average value ("plus" sign), whereas in the second portion, the signal passes below the dc output power ("minus" sign).

As pointed out, these signals represent a resistive load connected to the mains. In other words, if we would like to modify a circuit originally drawing a distorted current (for instance, the full-wave rectifier from Fig. 6-2a) in order to reach a PF of 1 (a resistive behavior), we should find a way to store and release energy as shown by Fig. 6-16, respectively, on the + and the − areas.

How much energy shall we store to offer a PF of 1, similar to that of Fig. 6-16? We can easily calculate the energy delivered by the source in the + region. This region lasts one-quarter of the mains frequency. Hence, the instantaneous energy delivered by the source can be evaluated by integrating the power from $T/8$ up to $3T/8$ if we take the origin of time from the left corner:

$$W_{source} = \int_{\frac{T}{8}}^{\frac{3T}{8}} 2P_{out} \sin^2(\omega t)\, dt = P_{out}\left(\frac{T}{4} + \frac{T}{2\pi}\right) \quad (6\text{-}71)$$

Within the + area, the load consumes an average power of P_{out}. Expressed in energy, it becomes

$$W_{load} = P_{out}\frac{T}{4} \quad (6\text{-}72)$$

The excess stored energy is simply the subtraction of Eq. (6-72) from Eq. (6-71):

$$W_{stored} = W_{source} - W_{load} = \frac{P_{out}}{\omega} \quad (6\text{-}73)$$

This result shows that active or passive power factor correction means storing and releasing energy: If the instantaneous input power $P_{in(t)}$ is above the average level, storage occurs. Conversely, if $P_{in(t)}$ is below the average level, energy release takes place. The squared sine wave will come back again as a capacitive output ripple when we discuss active correction techniques.

6.2.7 Passive PFC

A lot of papers describe the arrangement of an inductor and a capacitor to correct the power factor of a full-wave rectifier [8–13]. Possible configurations include Fig. 6-17a and b, among a variety of architectures documented in Ref. 10.

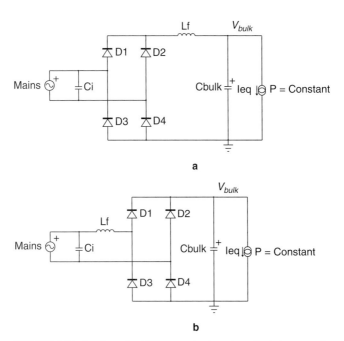

FIGURE 6-17a, b A passive PFC approach where the series inductor can be either dc or ac located.

On the above figures, the inductor can be placed on the dc side (Fig. 6-17a) or on the ac side (Fig. 6-17b). The inductor L_f modifies the harmonic content of the input current, whereas the capacitor C_i plays on the displacement factor. Hence, if the designer only cares about current distortion, the capacitor can be omitted.

Depending on the current flowing through the inductor, three different operating modes can be qualified: DCMI, DCMII, and CCM. Depending on where the inductor current cancelation occurs, either before or after one-half of the period, the rectifier operates in two different DCM modes. Figure 6-18 portrays these particular configurations and highlights the modes.

The exercise now consists of finding the proper LC configuration to obtain a good power factor and an acceptable power transfer. But before deriving useful equations, we need to make a few assumptions:

SIMULATIONS AND PRACTICAL DESIGNS OF OFF-LINE CONVERTERS—THE FRONT END

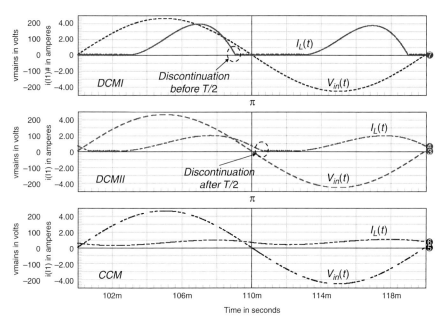

FIGURE 6-18 Three different operating modes, DCMI, DCMII, and CCM. The point where DCM occurs qualifies for a DCMI or DCMII type of discontinuous mode.

- C_{bulk} is large enough to keep V_{bulk} constant and ripple free.
- The ac source is ideal, and its output impedance is zero.
- There are no losses in the passive elements (L_f, C_{bulk}, and the diode bridge).

Then let us agree on some definitions:

$$V_{ref} = V_{in,rms} \quad (6\text{-}74)$$

This is the reference input voltage delivered by the ac source.

$$I_{ref} = \frac{V_{in,rms}}{Z_{L_f}} = \frac{V_{in,rms}}{\omega L_f} \quad (6\text{-}75)$$

The reference current illustrates the current seen by the ac source in case the rectifier input impedance becomes the series inductor impedance Z_{Lf}.

Finally,

$$m = \frac{V_{bulk}}{V_{in,peak}} \quad (6\text{-}76)$$

To understand this variable, we can glance at Fig. 6-17a, where we have an *LC* filter made up of L_f and C_{bulk}. If *L* is small, we are close to the classical full-wave operation and V_{bulk} reaches the input source peak value. In this case, $m = 1$ according to Eq. (6-76). Now, if *L* is very large, the output voltage V_{bulk} will no longer reach the peak but the average value of the full-wave signal. In other words,

$$V_{bulk} = \frac{2V_{in,peak}}{\pi} \quad (6\text{-}77)$$

or

$$m_{min} = \frac{V_{bulk}}{V_{in,peak}} = \frac{2}{\pi} \qquad (6\text{-}78)$$

Finally, the average power in watts can classically be defined as

$$P_{in} = P_{out} = \frac{1}{T}\int_0^T I_{in}(t)V_{in}(t)\,dt \qquad (6\text{-}79)$$

where $\omega = 2\pi F_{line}$. Based on the above expressions, a normalized power P_n can now be specified by the ratio of the average power to the so-called reference power:

$$P_n = \frac{P_{out}}{I_{ref}V_{ref}} = \frac{P_{out}}{\dfrac{V_{in,rms}^2}{\omega L_f}} = \frac{P_{out}\omega L_f}{V_{in,rms}^2} \qquad (6\text{-}80)$$

In Eq. (6-80), considering P_{out} constant, P_n presents the dimension of an inductance. The experiment followed by S. B. Dewan [8] consisted of sweeping values of L_f and collecting the resulting measured values for m and the PF plotted versus the P_n coefficient. After gathering these numbers on a chart, the author obtained a set of curves where peaks and valleys occurred. Careful examination of the graphic revealed a point in the discontinuous region where m reaches an optimum ($m = 0.79$) and the power factor peaks to 0.763. This value is obtained for $P_n = 0.104$. Thanks to Eq. (6-80), it now becomes possible to extract the optimum inductor value:

$$L_f = \frac{V_{in,rms}^2 P_n}{P_{out}\omega} \qquad (6\text{-}81)$$

Going back to the 120 Vrms 117 W full-wave rectifier and applying Eq. (6-80), we obtain an inductor value of

$$L_f = \frac{120^2 \times 0.104}{117 \times 6.28 \times 60} = 34\,\text{mH} \qquad (6\text{-}82)$$

In Dewan's paper, the optimum bulk capacitor value is given and equals

$$C_{bulk} = \frac{20 P_n}{m^2 \omega^2 L_f} \qquad (6\text{-}83)$$

Using the recommended values for m and P_n, respectively, 0.79 and 0.104, we obtain a capacitor of

$$C_{bulk} = \frac{5 \times 0.104}{0.79^2 \times 377^2 \times 34m} = 172\,\mu\text{F} \qquad (6\text{-}84)$$

Of course, a lower value can be used, but with a degradation of the peak-to-peak output ripple. As usual, the final capacitor selection depends on the rms current flowing through it. The updated simulation result is shown in Fig. 6-19a.

Simulation results appear in Fig. 6-19b and reveal the input current together with the output voltage. Individual measurements give

$$I_{in,rms} = 1.28\,\text{A} \qquad (6\text{-}85)$$

$$\text{PF} = \frac{117}{120 \times 1.28} = 0.76 \qquad (6\text{-}86)$$

SIMULATIONS AND PRACTICAL DESIGNS OF OFF-LINE CONVERTERS—THE FRONT END 523

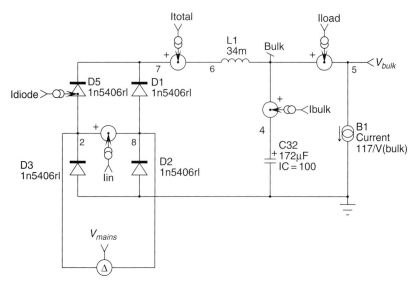

FIGURE 6-19a The 117 W rectifier equipped with the optimally designed *LC* corrector.

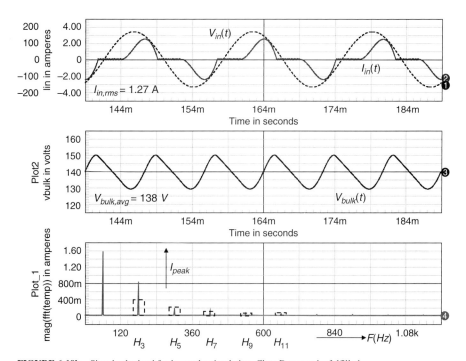

FIGURE 6-19b Signals obtained further to the simulation. Class D cannot be fulfilled.

$$V_{bulk,avg} = 138 \text{ V} \quad (6\text{-}87)$$

$$m = \frac{138}{170} = 0.81 \quad (6\text{-}88)$$

They pretty much agree with the predictions formulated by the author, in particular the power factor and the ratio of average output voltage to peak input voltage (m). However, the third and seventh harmonics do not allow the configuration to pass the class D recommendations. The SPICE Fourier analysis computes a THD of 55%, which is a good improvement, however, compared to the initial 112%!

6.2.8 Improving the Harmonic Content

Reference 12 points to a paper by Jovanović and Crow published in 1996 which focuses on the harmonic reduction via an LC filter optimized to purposely respect the IEC 1000-3-2 limits. The authors used a formula close to Eq. (6-81) to recommend an inductor value [9] of

$$L_f = \frac{V_{in,rms}^2 L_{ON}}{P_{out} F_{line}} \quad (6\text{-}89)$$

Using this formula, Jovanović and Crow calculated L_f with an L_{ON} coefficient ranging from 1m to 1 and then simulated Fig. 6-17a. Sticking to Eq. (6-60), they collected the displacement factor (denoted by K_d) and the resulting current distortion denoted by K_p (purity factor). Figure 6-20 reproduces the graph proposed in Ref. 12 which shows how K_d and K_p evolve with L_{ON}. Note the presence of the three operating modes as already described via Fig. 6-18.

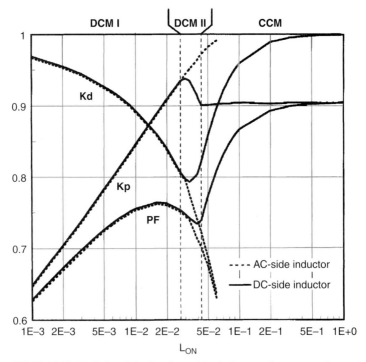

FIGURE 6-20 Evolution of the distortion and the displacement factors versus L_{ON}.

SIMULATIONS AND PRACTICAL DESIGNS OF OFF-LINE CONVERTERS—THE FRONT END   525

TABLE 6-2 Recommended Values for the Eq. (6-92) Coefficient

Rms input voltage V_{in}	L_{ON} range
220	0.006–0.03
230	0.004–0.03
240	0.003–0.03

Further to numerous simulations and experiments, Jovanović and Crow determined a range of recommended values for L_{ON}, summarized in Table 6-2, which allow the LC configuration to pass the class D specifications. The standard limits need to be met at the nominal operating voltage and the rated full power. The lowest L_{ON} value brings the smallest inductor size, whereas the upper L_{ON} range favors a better THD. The coefficient selection and inductor calculation take place at high line, but the physical size of the device will be determined by the operation at the lowest line, where the rms current is the highest.

Sticking to the 117 W rectifier but operated on a European mains this time (Fig. 6-17a with a 230 Vrms input), we tried to minimize the distortion with an L_{ON} of 0.024. Equation (6-89) suggests an inductor value of

$$L_f = \frac{230^2 \times 0.024}{117 \times 50} = 217 \text{ mH} \tag{6-90}$$

Figure 6-21 details the simulation results and the input current FFT.

After running the simulation, we obtained the following results:

$$m = \frac{241}{325} = 0.74 \tag{6-91}$$

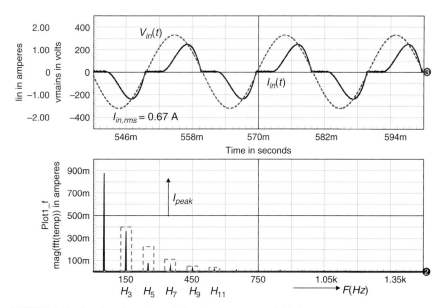

FIGURE 6-21 Resultant distortion analysis using the recommended inductor.

$$I_{in,rms} = 0.67 \text{ A} \tag{6-92}$$

$$PF = \frac{117}{230 \times 0.67} = 0.76 \tag{6-93}$$

The PF does not look good at first glance, but the high displacement factor brought by the large inductor explains it. The good news relates to the harmonic levels which are below class D limits as demonstrated by the dashed boxes in Fig. 6-21. In this particular example, the THD reaches 43%.

This type of passive filter often appears on low-cost ATX power supplies: This is the bulky inductor attached to the metallic chassis which makes the supply heavier than the normal power supply. There is one drawback, however, inherent to the inductor presence: In a normal configuration without the inductor, an input overvoltage encounters a low-impedance path via the bulk capacitor which resists the fast input change, limiting the diode voltage. When an inductor appears in the rectifier, the diode bridge loading is increased by the presence of the inductor. If voltage spikes occur on the mains, they no longer see a low impedance as the bulk capacitor is fed via the inductor; thus, the diode bridge output voltage increases, and in some cases the spikes avalanche the rectifiers and destroy them.

If the designer sees a need to correct the displacement factor, she or he can add a capacitor C_i in front of the diode bridge (Fig. 6-17a and b). Reference 12 suggests the following formula to compensate the input current displacement, when needed:

$$C_i = 0.12 \frac{P_{out}}{V_{ac,rms}^2 F_{line}} \tag{6-94}$$

Equation (6-94) leads to a 5 μF capacitor for the 230 Vrms 117 W full-wave rectifier. Once installed, the PF was increased up to 90.2%.

6.2.9 The Valley-Fill Passive Corrector

The valley fill offers a possible way to get rid of the bulky inductor found in passive correctors. This circuit was introduced by Jim Spangler and A. Behara [14] and further refined by K. Kit Sum in a 1997 publication [15]. Figure 6-22 shows the electric circuit consisting of two capacitors and a bunch of diodes. Simply put, the capacitors are charged in series (hence a smaller bulk value during the refueling time) and discharged in parallel via D_7 and D_5 when the discharge cycle occurs. D_6 simply prevents C_2 from discharging C_1.

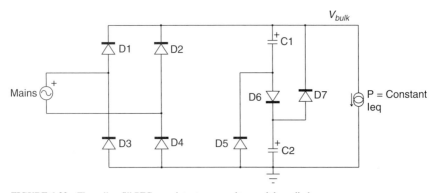

FIGURE 6-22 The valley-fill PFC associates two capacitors and three diodes.

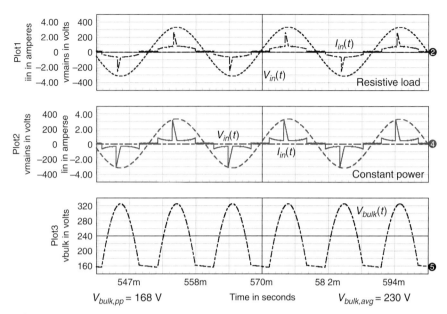

FIGURE 6-23 The valley-fill input signature depends on the nature of the load.

The main drawback of this corrector lies in the presence of spikes in the input current, as shown in Fig. 6-23. Also, as already found through experiment in Ref. 16, the circuit does not accommodate very well with a constant power load such as a closed-loop converter. You can see in Fig. 6-23 the input signature with loads of different nature.

This passive corrector finds applications in low-cost electronic ballasts, but its high output ripple can induce a high output current crest factor, which is a detrimental characteristic to the lamp lifetime.

6.2.10 Active Power Factor Correction

If the passive power factor correction sometimes offers an economical way to fulfill the IEC 1000-3-2 standard, it clearly suffers from several drawbacks as noted below:

- A bulky inductance increases the final weight and size.
- Despite a reduced harmonic content, the current still lags the voltage and degrades the power factor.
- The rectified output voltage is reduced compared to a standard full-wave rectifier and moves up or down as the input changes.

Active power factor correction, on the other hand, maintains a good power factor with both k_d and k_φ closer to 1. Furthermore, it delivers a regulated output voltage, naturally reducing (or eliminating) the wide input range burden of the downstream converter. The term *preregulator* perfectly applies as the PFC stage must be combined with a dc–dc converter that brings galvanic isolation (when necessary) and regulation speed. The single-stage approach, where the preconverter and the dc–dc converter are merged, starts to gain popularity as specialized controllers appear on the market (ON Semiconductor NCP1651, for instance). Active power factor correction might be one of the most popular subjects if you look at monthly publications such as IEEE magazines and others. Given the wide market this field represents, the interest in PFC is not going to fade away!

We have seen via Eq. (6-73) the need for storage. In an active PFC, storage is mostly made through a capacitor (the bulk capacitor) which stores and releases energy. Figure 6-24 sketches a classical configuration associating the preconverter with the dc–dc, here a flyback power supply:

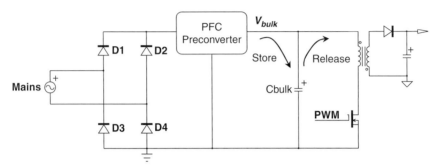

FIGURE 6-24 A corrected flyback converter shown here with its preconverter stage.

6.2.11 Different Techniques

Among all available topologies, the boost converter is by far the most popular structure used for active power correction. Competing solutions include the buck, the SEPIC, and the flyback, but they have difficulties competing with the boost, given its simplicity and ease of implementation. PFC circuits can be classified in several areas as follows:

1. *Constant on-time borderline mode (BCM)*: Associated with a boost converter operated in BCM (borderline conduction mode) and working in voltage mode or current mode, it represents one of the most popular topologies in the consumer market. The output voltage regulation occurs via a low-bandwidth closed-loop system which drives the peak current set point to ensure a near-unity power factor. The current loop imposes a sinusoidal peak inductor current, without actively tracking the average value of this current. Found in ac–dc notebook adapters or in light ballast applications, the market splits between many pin-compatible controllers among which are the MC33262 and the new NCP1606 from ON Semiconductor. The constant on-time BCM technique also suits flyback-based PFC.

   The BCM configuration can be used up to 300 W, although converters operated around the kilowatt exist. The main disadvantage of the BCM technique lies in the higher rms current (large ac swing) circulating in the circuit. However, the boost diode naturally turns off thanks to the inductor current returning to zero at the end of each switching cycle. For this reason, a cheap "lazy" diode can be used in these configurations since recovery losses do not exist.

2. *Fixed-frequency continuous mode (CCM)*: This category splits into two subsections:
   - *Average current control*: In this implementation, an amplifying chain featuring gain and bandwidth permanently tracks the average inductor current and makes it follow a sinusoidal reference. Dc regulation is then obtained by scaling up and down the sinusoidal reference. In this scheme, the high dc gain current error amplifier strives to perfectly shape the current envelope, and an excellent power factor is reached. The UC1854 from TI (former Unitrode) pioneered this technique, nowadays also implemented by the NCP1650 and the more recent NCP1653.

      In the constant t_{on} technique, the circuit imposes a shape to the peak inductor envelope, assuming the average value will naturally be sinusoidal. In average mode, the tracking between the set point and the resulting average current exists and guarantees a low distortion. Average control works without ramp compensation.

Average-mode control suffers from cusp distortion when the input voltage lies in the vicinity of the 0 V area. In this region, the available slew rate is too weak, and the inductor current will lag behind the set point for a short time, causing distortion.

- *Peak current control*: Similarly to the first structure we described, peak current control imposes a sinusoidal waveform on the inductor current but keeps it continuous this time. The technique requires proper adjustment of the ramp compensation; otherwise severe instability occurs near the input zero transition. The addition of the ramp degrades the resulting distortion, but solutions via the inclusion of offsets give adequate results. CCM peak current-mode control is not very popular, and only a few controllers exist (ML4812 from Fairchild, for instance).

The CCM mode allows the implementation of PFCs beyond the kilowatt. However, the boost diode recovery time heavily contributes to the power switch dissipation, and a snubber on both devices has to be designed.

3. *Analytical control law*: Without sensing the rectified envelope as the above would do, the controller internally elaborates a control law which forces power factor correction. This is the case for the NCP1601 (fixed-frequency DCM) or the NCP1653 (CCM) from ON Semiconductor. ICE1PCS01/2 from Infineon or the IR1150 from International-Rectifier also works with a control law.

Rather than an in-depth description of each possible scheme, this chapter offers a quick guided tour of the existing techniques and concentrates, at the end, on a BCM PFC design.

6.2.12 Constant On-Time Borderline Operation

When we studied BCM in Chap. 2, Eq. (2-222) taught us that the average value of a BCM current was equal to its peak divided by 2 (Fig. 2-99):

$$\langle I_L(t) \rangle_{T_{sw}} = \frac{I_{L,peak}}{2} \tag{6-95}$$

A BCM boost preconverter follows the architecture depicted by Fig. 6-25, here in a current-mode structure. Without the traditional bulk capacitor, the diode bridge delivers a full-wave

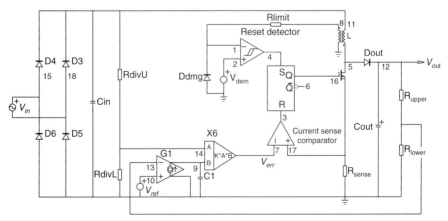

FIGURE 6-25 A borderline PFC architecture operated in current mode: A multiplier sets the peak current set point via high-voltage sensing.

rectified sinusoidal signal. Mostly C_{in} plays the role of an EMI filter but also avoids a real voltage decrease down to zero, saving some switching losses by restricting the duty cycle excursion. A portion of the rectified signal brought by R_{divU} and R_{divL} feeds a multiplier X_6 whose other input receives the loop error signal from the transconductance amplifier G_1 (a traditional op amp could also do the job). A capacitor C_1 rolls off the bandwidth to a low level (around 10 Hz), avoiding the natural output ripple to be sensed by the loop. Therefore, at steady state, the error amplifier outputs a flat dc signal. Why a low bandwidth? Remember, a PFC must store and release energy, implying the presence of a low-frequency ripple on the output (Fig. 6-19b). If the loop detects it via the sensing network R_{upper}/R_{lower}, it will lock onto it and a so-called tail chasing can occur. A strong bandwidth rolloff thus prevents this from happening. Unfortunately, it turns the PFC into a slowly reacting preconverter: It yields an extremely poor load transient response, the PFC Achille's heel.

Borderline operation requires inductor current sensing, and this is done here, via a few turns wound over the main inductor L. When the flux reaches zero (core reset), the voltage induced on this auxiliary winding collapses and initiates a new switching cycle. The time needed by the current to ramp down from the peak to zero depends on input and output conditions. Hence, the frequency constantly changes, a typical characteristic of BCM converters.

Elaborated with the half-wave input voltage image via the multiplier, the error voltage V_{err} imposes a sinusoidally varying cycle-by-cycle set point to the inductor current. In other words, the peak current set point changes as the rectified voltage moves up and down

$$I_{L,peak}(t) = \frac{V_{in}(t)}{L} t_{on} \quad (6\text{-}96)$$

where $V_{in}(t)$ represents the full-wave rectified input voltage.

The instantaneous input power $P_{in}(t)$ is obtained by multiplying the instantaneous input current and input voltage together. The instantaneous PFC input current is actually a time-continuous function of the individual inductor current averaged values over a switching cycle. As the average current value obeys Eq. (6-95), $P_{in}(t)$ follows Eq. (6-97):

$$P_{in}(t) = \frac{I_{L,peak}(t)}{2} V_{in}(t) \quad (6\text{-}97)$$

Extracting the peak current from the above and setting it equal to that of Eq. (6-96), we have

$$\frac{2P_{in}(t)}{V_{in}(t)} = \frac{V_{in}(t)}{L} t_{on} \quad (6\text{-}98)$$

From this equation, the instantaneous on time easily comes out and is equal to

$$t_{on}(t) = \frac{2P_{in}(t)L}{V_{in}(t)^2} \quad (6\text{-}99)$$

Now, we know the instantaneous input power follows a squared sine law [Eq. (6-71), assuming 100% efficiency]. Introducing the rms term V_{ac} into the input voltage definition, we can update Eq. (6-99):

$$t_{on} = \frac{4P_{out}\sin^2(\omega t)L}{[V_{ac}\sqrt{2}\sin(\omega t)]^2} = \frac{2LP_{out}}{V_{ac}^2} \quad (6\text{-}100)$$

where $V_{ac} = V_{in,rms}$.

In light of Eq. (6-100), we can see that the on time is kept constant over a line cycle when operating the boost in BCM. This is true not only for current mode, as we just saw, but also for voltage mode as the on time naturally keeps constant via the error amplifier delivering a continuous voltage at steady state (no ripple passes through as the gain has been rolled off). Finally, according to these steps, the input current follows Eq. (6-101):

$$I_{in}(t) = \frac{I_{L,peak}}{2}(t) = \frac{V_{ac}t_{on}}{2L}\sin(\omega t) = k\sin(\omega t) \quad (6\text{-}101)$$

where $k = \frac{V_{ac}t_{on}}{2L}$ remains constant. Equation (6-101) confirms that the input current follows a sinusoidal envelope: We perform power factor correction! Figure 6-26 portrays the typical waveforms obtained in a PFC converter operated in borderline mode.

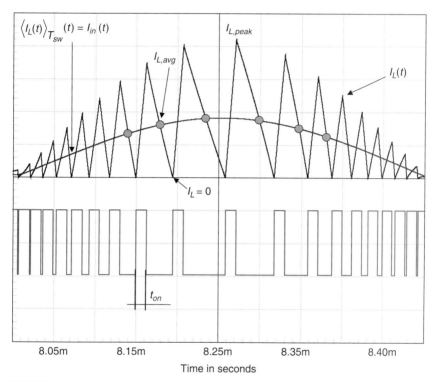

FIGURE 6-26 Typical waveforms obtained on a BCM PFC.

6.2.13 Frequency Variations in BCM

During the on time, the switch closes and the current ramps up to the peak imposed by the envelope at which point the MOSFET opens. The inductor current ramps down until it reaches 0: The inductor is told to be "reset." The controller detects this fact and initiates a new cycle. Looking at Fig. 6-26, you can observe a wide frequency variation thanks to the off-time modulation (t_{on} is constant). Again a few analytical steps will let us assess the frequency changes across a complete line cycle.

The switching period equals the sum of the on and off times:

$$T_{sw} = t_{on} + t_{off} \quad (6\text{-}102)$$

In a boost operated in BCM, the off time represents the time needed for the inductor current to fall from its peak to zero. Note that the equations below are a function of time t since the concerned instantaneous value depends on the sinusoidal input level:

$$t_{off}(t) = \frac{LI_{L,peak}}{V_{out} - V_{peak}|\sin(\omega t)|} \quad (6\text{-}103)$$

If we apply the Eq. (6-96) definition with Eq. (6-103), we have

$$t_{off}(t) = \frac{L \frac{V_{peak}|\sin(\omega t)|}{L} t_{on}}{V_{out} - V_{peak}|\sin(\omega t)|} = \left(\frac{V_{peak}|\sin(\omega t)|}{V_{out} - V_{peak}|\sin(\omega t)|} \right) t_{on} \quad (6\text{-}104)$$

Following Eq. (6-102), we finally can express the switching period:

$$T_{sw}(t) = \left(\frac{V_{peak}|\sin(\omega t)|}{V_{out} - V_{peak}|\sin(\omega t)|} \right) t_{on} + t_{on} = t_{on} \left(\frac{1}{1 - \frac{V_{peak}|\sin(\omega t)|}{V_{out}}} \right) \quad (6\text{-}105)$$

The switching frequency evolves as

$$F_{sw}(t) = \frac{1}{t_{on}} \left(1 - \frac{V_{peak}|\sin(\omega t)|}{V_{out}} \right) = \frac{V_{ac}^2}{2LP_{out}} \left(1 - \frac{V_{peak}|\sin(\omega t)|}{V_{out}} \right) \quad (6\text{-}106)$$

In the above lines, V_{peak} represents the peak input voltage. The absolute value expresses the results of the full-wave rectification. (The ripple is always positive.)

Here lies the main drawback of the BCM boost converter: its wide frequency excursion over a line cycle. It brings several other disadvantages listed below:

- Switching losses incur greater heat dissipation as the frequency increases.
- In light-load conditions, the BCM PFC switching frequency dramatically goes up. Its standby performance suffers.
- An internal clamp must limit the frequency excursion for the above reasons and slightly degrades the power factor at the end.

We have drawn the frequency excursion corresponding to Fig. 6-26 signals (Fig. 6-27). You can see the frequency span between the input peak and valley points.

As long as we keep the on time constant, we have seen that power factor correction is ensured with a BCM boost. A voltage-mode converter, having a fixed output voltage, also keeps its on time constant despite changes in its input. Well, in that case, using a BCM boost in voltage mode no longer requires sensing the sinusoidal envelope to produce a high power factor. Figure 6-28 shows the implementation: The high-voltage-sensing network has gone; a simple reset detector does the job of detecting the inductor reset, and we are all set! This technique is implemented in voltage-mode PFCs such as the MC33260.

6.2.14 Averaged Modeling of the BCM Boost

The BCM average model developed in Chap. 2 lends itself very well to simulating a boost converter employed as a power factor corrector. Figure 6-29a depicts the model arrangement when a classical input bridge is used. This example represents the MC33262 data sheet application circuit which delivers 160 W of output power.

SIMULATIONS AND PRACTICAL DESIGNS OF OFF-LINE CONVERTERS—THE FRONT END **533**

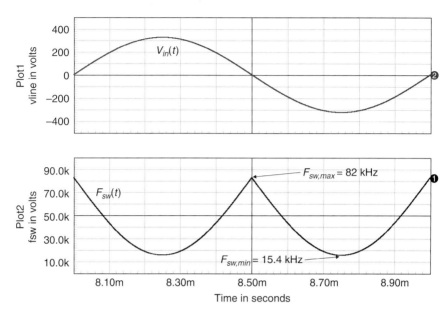

FIGURE 6-27 The frequency evolution within an input voltage cycle ($L = 2$ mH, $V_{in} = 230$ Vrms, $P_{out} = 160$ W).

Needless to say, the simulation time is extremely short compared to that of the cycle-by-cycle model. Here, a 1.2 s simulation was completed in 8 s with a 50 Hz input frequency! We can thus explore transient response, output ripple amplitude, and stability. Figure 6-29b shows the typical signals obtained from this example. The output ripple stays around 5 V, and the power factor reaches 0.997 at high line, a rather optimistic number, however. The bottom plot shows the switching frequency evolution over an entire input cycle. A discontinuity clearly occurs around the 0 V input because the duty cycle was pushed to the limits. The presence of C_{in} prevents

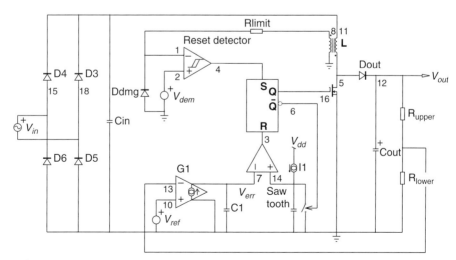

FIGURE 6-28 Voltage-mode operation of the boost in BCM also produces power factor correction.

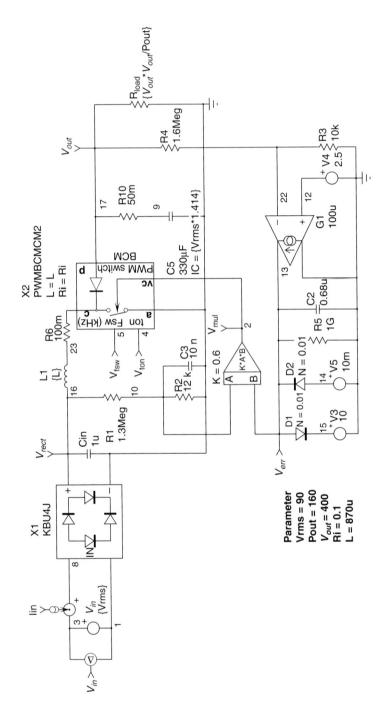

FIGURE 6-29a An averaged model can be used to simulate a PFC circuit. Here a MC33262-like configuration uses the BCM current-mode subcircuit. Note that we have used PWMBCM2 which prevents the use of a negative sense resistor (required for the current-mode PWM switch in the boost configuration).

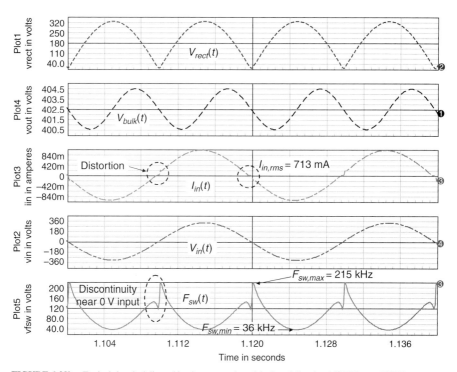

FIGURE 6-29b Typical signals delivered by the averaged model when delivering 160 W from a 230 Vrms source.

a complete collapse of $|V_{in}(t)|$ near the zero and helps keep the switching frequency excursion within reasonable limits.

Because a PFC is a closed-loop system in essence, its stability must be assessed. Thanks to the average models, we can quickly draw a Bode plot to check bandwidth and phase margin. Figure 6-30a displays how to wire the model to reveal its small-signal signature. The input voltage is replaced by a dc source, and stability exploration consists of assessing the various margins at different input voltages. For this example, we have selected a value of the rectified mains corresponding to 210 Vdc. Since the bias points are reflected on the schematic, they confirm the good operating point computed by the simulator.

If we now observe the Bode plot in Fig. 6-30b, the lack of phase margin pops up. Powering this circuit would probably lead to an unstable output. By adding a 22 nF capacitor across R_4, we create a zero in the vicinity of 5 Hz and a phase boost occurs. The updated plot looks much better, now offering a phase margin of 61°. We will come back to the BCM boost PFC in the design example.

6.2.15 Fixed-Frequency Average Current-Mode Control

Average mode control performs excellent power factor correction by tracking the average value of the inductor current now kept continuous over a switching cycle (CCM operation). We still have a sinusoidal set point as in the BCM boost, but a dedicated op amp makes sure that the cycle-by-cycle average inductor current precisely follows what the set point imposes. Figure 6-31 displays the typical curves obtained with a PFC operated in the CCM mode. On this curve, the inductor current operates in the continuous conduction mode over most of the mains cycle. However, at the beginning of the mains cycle, the inductor current cannot

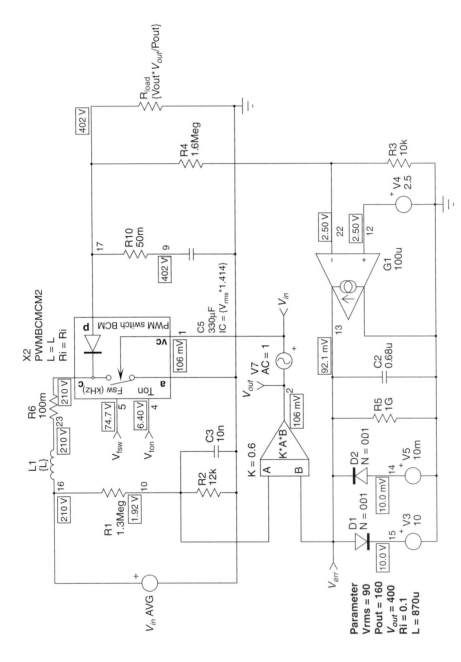

FIGURE 6-30a The BCM model used to obtain the open-loop Bode plot.

SIMULATIONS AND PRACTICAL DESIGNS OF OFF-LINE CONVERTERS—THE FRONT END   537

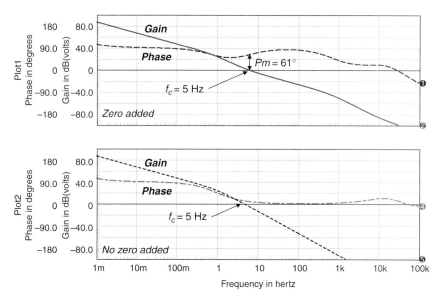

FIGURE 6-30b The Bode plot without compensation shows a weak phase margin. A local zero quickly boosts it around 5 Hz and improves the situation.

immediately jump to the set point as the available voltage across the inductor is too low: The averaged inductor value lags from the imposed set point and so-called cusp distortion occurs. The larger the inductance, the greater the cusp contribution.

Figure 6-32 displays the typical arrangement of such a preconverter implementing negative current sensing.

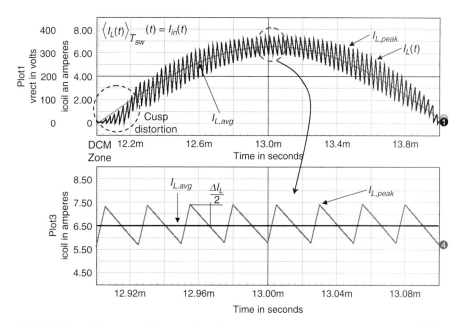

FIGURE 6-31 CCM-operated PFC curves. Distortion always occurs near 0 V as the converter enters DCM.

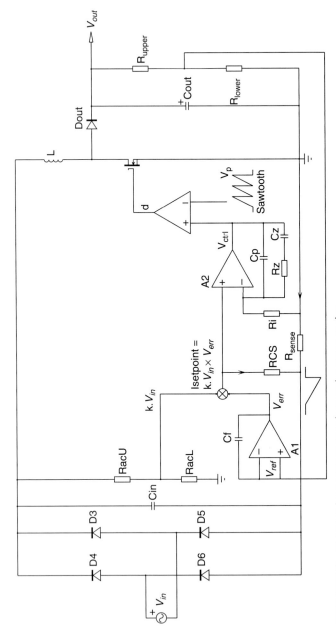

FIGURE 6-32 An average boost preconverter using negative current sensing.

An error amplifier A_1 adjusts the output voltage through the resistive divider R_{upper} and R_{lower}. C_f performs the compensation and rolls off the gain to avoid tail chasing on the output voltage ripple. (The error chain reacts to the output ripple.) A multiplier then generates a current which is a product of the rectified line voltage ($kV_{in}(t)$) and the error amplifier output level V_{err}. This current generator then feeds an offset resistor R_{CS} connected to the sense element R_{sense}, in series with the dc low-side input terminal. R_{CS} fixes the maximum peak inductor current allowed by the multiplier. For instance, suppose the multiplier delivers 300 μA and R_{CS} equals 2 kΩ; then the maximum voltage across R_{sense} reaches 600 mV.

This negative sensing technique often appears on high-power PFC controllers because of three reasons. (1) IC designers teach you how simple it is to multiply currents rather than voltages. (2) Since R_{sense} is low-side referenced, its reading simplifies the architecture. (3) As R_{sense} senses the total loop current involving C_{out}, the in-rush current safely blocks the controller until it recovers to an acceptable level.

The sinusoidal set point now feeds another error amplifier A_2. Its function consists of "amplifying" the error between the set point and the actual sensed signal. Given its filtering configuration as a type 2 circuit, A_2 delivers a voltage proportional to the cycle-by-cycle inductor average current. This is the key point in average mode control where the set point voltage now fixes the "instantaneous" averaged inductor current, unlike the peak current as in BCM. Figure 6-33a shows a close-up on this section, actually the heart of the average mode control PFC.

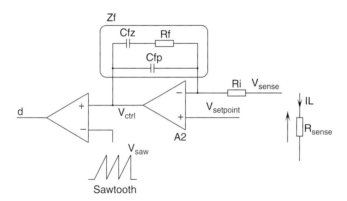

FIGURE 6-33a The error amplifier A_2 controls the duty cycle to shape the sinusoidal average inductor current.

The control voltage delivered by this structure can be expressed in the following way:

$$V_{ctrl}(s) = V_{set\,point}(s)[G(s) + 1] - V_{sense}(s)G(s) = V_{set\,point}(s) + G(s)[V_{set\,point}(s) - V_{sense}(s)]$$
(6-107)

where $G(s)$ represents the gain resulting from Z_f and R_i, $G(s) = \dfrac{Z_f(s)}{R_i}$. From this expression, we can see that the error amplifier actually delivers the set point waveform affected by a compensation term. This term, either subtracted or added depending on its sign, represents the amplified difference between the set point and the sensed inductor current. If the ramp features a peak amplitude of V_{saw}, then the duty cycle expression is found to be [17]

$$d(s) = \frac{V_{ctrl}(s)}{V_{saw}} = \frac{V_{setpoint}(s) + G(s)[V_{setpoint}(s) - V_{sense}(s)]}{V_{saw}}$$
(6-108)

The analysis of the amplifier block brings the position of the following poles and zero:

$$G(s) = \frac{k_c(1 + s/\omega_z)}{s(1 + s/\omega_p)} \qquad (6\text{-}109)$$

with

$$k_c = \frac{1}{R_f(C_{fz} + C_{fp})} \qquad (6\text{-}110)$$

$$\omega_z = \frac{1}{R_f C_{fz}} \qquad (6\text{-}111)$$

$$\omega_p = \frac{C_{fz} + C_{fp}}{R_f C_{fz} C_{fp}} \qquad (6\text{-}112)$$

As the expression shows, the origin pole provides high dc gain and helps to minimize the current tracking errors. Unfortunately, the above equations do not clearly show the midband gain value brought by this type 2 configuration: App. 3C derived a simpler expression where this midband immediately pops up. The literature recommends to place the pole and the zero to force the current loop crossing around 10 kHz [18 20]. This value offers good dynamic performance and filters the high-frequency switching noise inherently present in the sensed signal. Reference 18, in particular, discusses and documents the results obtained with different pole-zero placements and how they affect the distortion.

6.2.16 Shaping the Current

In average mode control, the set point no longer fixes the peak inductor current value, but the "instantaneous" average inductor current value. The inductor current is observed by the sense resistor R_{sense} which enters the error amplifier A_2. The type 2 filter eliminates the switching ripple, and an error voltage appears, which is an amplified version of the difference between the inductor average value and the set point. The duty cycle continuously adjusts to track what the set point imposes, hence the necessity for a 10 kHz bandwidth in the current loop. It corresponds to an order of magnitude lower than the switching frequency, 100 kHz in this example. Since the set point is made up of the rectified input voltage multiplied by the error level, we have

$$R_{sense}\langle I_L(t)\rangle_{T_{sw}} = kV_{in}(t)V_{err} \qquad (6\text{-}113)$$

The instantaneous average inductor current is nothing other than the input current; therefore,

$$R_{sense} I_{in}(t) = kV_{err} V_{peak} \sin \omega t \qquad (6\text{-}114)$$

Solving for $I_{in}(t)$, we have

$$I_{in}(t) = \frac{kV_{err} V_{peak}}{R_{sense}} \sin \omega t \qquad (6\text{-}115)$$

which is a sinusoidal waveform, considering the error voltage constant over a line cycle.

Figure 6-33b depicts the voltage-mode average model wired in an average mode control PFC. The arrangement directly implements what Fig. 6-32 suggested. For a 1 kW PFC, the sense resistor is designed as follows:

$$I_{in,rms} = \frac{P_{out}}{\eta V_{in,rms}} = \frac{1000}{0.95 \times 90} = 11.7 \text{ A rms} \qquad (6\text{-}116)$$

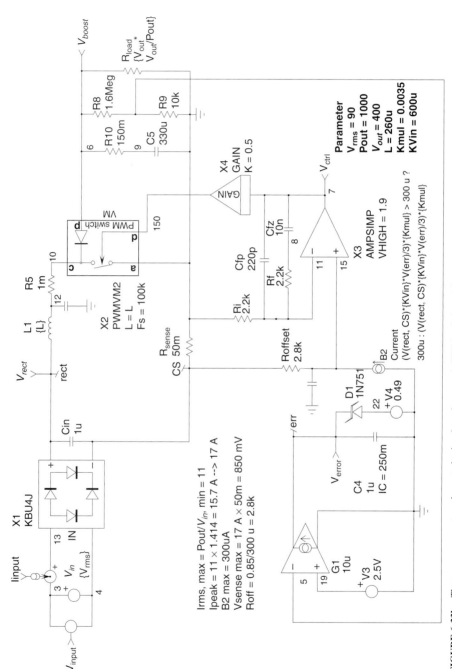

FIGURE 6-33b The average current-mode control using the voltage-mode PWM switch model.

To satisfy the above equation, the peak inductor current will need to reach

$$I_{L,peak} = I_{in,rms} \sqrt{2} = 11.7 \times 1.414 = 16.5 \text{ A} \quad (6\text{-}117)$$

Let's assume the multiplier delivers up to 300 μA maximum current. Power dissipation and manufacturer considerations lead us to pick up a sense resistor of 50 mΩ. Therefore, by selecting a maximum peak current of 17 A, the offset resistor R_{CS} is found to be

$$R_{CS} = \frac{R_{sense} I_{L,peak}}{I_{MUL}} = \frac{50m \times 17}{300u} = 2.8 \text{ k}\Omega \quad (6\text{-}118)$$

Figure 6-33b shows the error amplifier arrangement based on a transconductance amplifier. D_1 safely clamps its upper excursion to 5 V, whereas V_4 makes sure the amplifier output does not swing negative. The ABM source B_2 multiplies the error voltage by the rectified input voltage (via the coefficient KV_{in} in the parameter list) and transforms the result into a current clamped to 300 μA. X_3 performs the current loop amplification before driving the voltage-mode model via the PWM gain modulator (a 2 V ramp brings a −6 dB gain).

As with any average model configuration, Fig. 6-33b can lend itself to ac analysis or transient experiments. For the ac simulations, the input voltage is replaced by a dc source whose value equals low- and high-line input conditions, or any intermediate values. This helps to check the stability over the full input span. Figure 6-33c displays the corrected current loop transfer function obtained by inserting the ac sweep source in series with the 2.8 kΩ offset resistor. We can see around 10 kHz bandwidth together with a good phase margin.

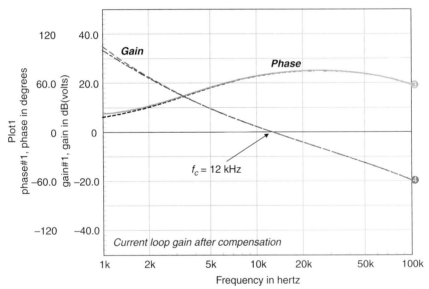

FIGURE 6-33c Current loop gains and phase at two different input levels, 90 Vrms and 230 Vrms: The crossover frequency remains unaffected. Note the phase polarity reversal due to the negative sensing method.

Despite the input voltage changes, the crossover frequency keeps constant. This is one of the benefits brought by average mode control.

Now it becomes possible to run a transient simulation and see what kind of power factor we obtain at both low and high line. Figure 6-33d portrays the current shape obtained at two different input levels. The distortion remains extremely low in both cases.

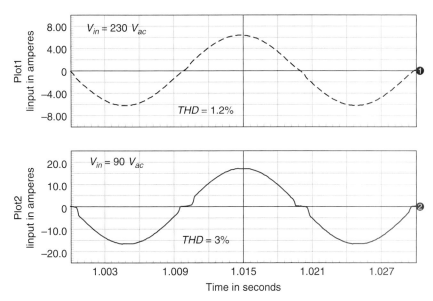

FIGURE 6-33d Input current signature at two input levels showing a THD below 5% in both cases.

6.2.17 Fixed-Frequency Peak Current-Mode Control

Peak current-mode control combines a boost converter also operated in CCM, but the internal set point now programs the peak inductor current rather than its average. Figure 6-34 depicts the internal circuitry found in a peak current-mode control CCM PFC implementing negative current sensing.

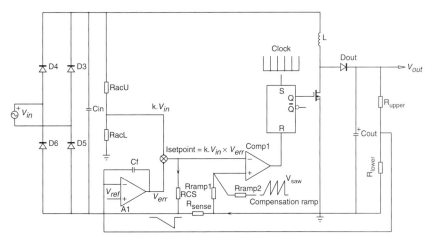

FIGURE 6-34 A peak current-mode control power factor correction controller.

The analysis of such a CCM configuration shows duty cycle variations above 50%, naturally implying ramp compensation to stabilize the converter. The maximum duty cycle occurs near the 0 V input where the controller strives to maintain CCM. Inadequate compensation level induces subharmonic oscillations as Fig. 6-35 displays.

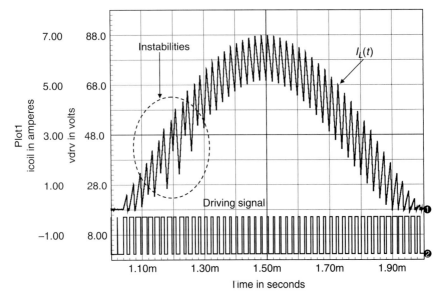

FIGURE 6-35 As the duty cycle increases above 50%, undercompensating the current loop brings subharmonic instabilities.

A few simple equations show that the peak current-mode control technique also produces near-unity power factor correction. The multiplier output sets the peak current set point via the offset resistor R_{CS}. Thus, we have

$$I_{L,peak} = \frac{I_{MUL}R_{CS}}{R_{sense}} = \frac{k|V_{in}(t)|V_{err}R_{CS}}{R_{sense}} \tag{6-119}$$

Replacing $V_{in}(t)$ with its rectified sinusoidal value, we have

$$I_{L,peak} = \frac{kV_{err}R_{CS}V_{peak}}{R_{sense}}|\sin \omega t| \tag{6-120}$$

where $V_{in}(t) = V_{peak}\sin(\omega t)$

If we consider the error amplifier output constant over a line cycle (slowly varying loop), then Eq. (6-120) simplifies to

$$I_{L,peak} = k_1|\sin \omega t| \tag{6-121}$$

with $k_1 = \frac{kV_{err}R_{CS}V_{peak}}{R_{sense}}$. According to Eq. (6-121), the scheme forces a sinusoidal input current.

6.2.18 Compensating the Peak Current-Mode Control PFC

As seen several times in Chap. 2, we need to inject a compensation ramp S_e to stabilize the current loop. For a boost converter, it can be 50% of the inductor downslope, as shown in Chap. 2. As the input voltage continuously varies, the amount of optimum ramp compensation permanently changes. The worst case arises when the line input crosses zero. In this particular case, the inductor off slope becomes

$$S_{L,off} = \frac{V_{out}}{L} \tag{6-122}$$

Reference 18 explains the effects of the ramp compensation on the input current waveform. At the time the mains crosses zero, little voltage exists across the inductor to force CCM. In other words, the inductor enters DCM, and its peak value diverges from the reference, causing distortion. When ramp compensation takes place, the duration of the DCM zone even extends as the ramp naturally opposes the duty cycle expansion, bringing more severe distortion.

Via a few lines of algebra, it is possible to show the effect of ramp compensation on the input line current. Let's look back to Chap. 2, Fig. 2-62, where we see how the final inductor current deviates from the initial set point because of the compensation ramp level S_a or S_e. Applying this concept to the PFC peak inductor current, we have

$$I_{L,peak}(t) = kV_{in}(t)V_{err} - S_e t_{on} = kV_{err}V_{peak}\sin\omega t - S_e t_{on} \tag{6-123}$$

where k represents the scaling factor brought by R_{acU}/R_{acL} in Fig. 6-34 and V_{err} is the error voltage delivered by the op amp A_1.

The CCM PFC boost converter transfer ratio obeys the following law:

$$\frac{V_{out}}{V_{in}(t)} = \frac{1}{1 - d(t)} \tag{6-124}$$

As V_{in} continuously changes, we can update Eq. (6-124) by

$$\frac{V_{out}}{V_{peak}\sin\omega t} = \frac{1}{1 - t_{on}(t)/T_{sw}} \tag{6-125}$$

from which the extraction of the on time comes easily:

$$t_{on}(t) = T_{sw}\left(\frac{V_{out} - V_{peak}\sin\omega t}{V_{out}}\right) \tag{6-126}$$

It may seem odd to see the term t within the on-time expression, but it really reflects its variations as the position on the sine wave evolves within the mains period.

Now, replacing t_{on} with its definition in Eq. (6-123) gives

$$I_{L,peak}(t) = \sin\omega t\left(kV_{peak}V_{err} + \frac{S_e T_{sw} V_{peak}}{V_{out}}\right) - S_e T_{sw} \tag{6-127}$$

A link exists between the inductor peak current and its average value: the total ripple current expressed as ΔI. Equation (6-128) defines this relationship:

$$I_{L,avg} = I_{L,peak} - \frac{\Delta I_L}{2} \tag{6-128}$$

For the boost converter, the inductor ripple current is defined as

$$\Delta I_L = \frac{V_{in}}{L}t_{on} = \frac{V_{peak}\sin\omega t}{L}t_{on}(t) \tag{6-129}$$

Using Eqs. (6-126) and (6-129), we can update Eq. (6-128):

$$I_{L,avg} = I_{L,peak} - T_{sw}\left(\frac{V_{peak}\sin\omega t}{2L}\right)\left(\frac{V_{out} - V_{peak}\sin\omega t}{V_{out}}\right) \tag{6-130}$$

Finally, by replacing the inductor peak current definition of Eq. (6-127) into Eq. (6-130), the average inductor current shows up [18]:

$$I_{L,avg} = \left(kV_{peak}V_{err} + \frac{S_e T_{sw} V_{peak}}{V_{out}} - \frac{T_{sw} V_{peak}}{2L}\right)\sin\omega t + \frac{T_{sw} V_{peak}^2}{2LV_{out}}\sin^2\omega t - S_e T_{sw} \tag{6-131}$$

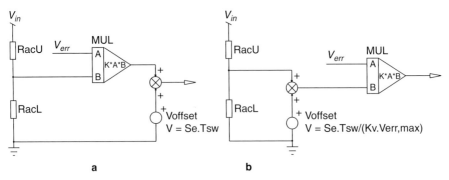

FIGURE 6-36 Solutions to fight the compensation ramp: (a) An offset is placed in series with the multiplier but causes trouble in light-load conditions. (b) The offset comes in series with the input voltage image, but although it works in light load, it does not completely fix the ramp problem.

In this long expression, almost all terms are constant, except the error voltage V_{err} which can contain some ripple. The inductance value also changes with the current and introduces distortion at the end. The subtracted final term represents the compensation ramp of amplitude $S_e T_{sw}$ which degrades the inductor average current. Some solutions fortunately exist to get rid of this term, for instance, via the insertion of a fixed offset in series with the multiplier. If this offset equals the compensation ramp peak value, then the terms cancel and its contribution is minimized. Although this technique offers good results at full load, the feedback loses the converter control in light-load conditions where the offset would impose a residual current.

Another solution consists of inserting this offset in series with the rectified voltage $kV_{in}(t)$. Figure 6-36 explains this solution where the offset no longer hampers the converter operation in light-load conditions. As V_{err} reduces in case the load current goes down, so does the multiplier output despite the offset presence. K_V is a coefficient linking the minimum and nominal ac input voltages. Reference 18 gives more details and shows results brought by this solution.

6.2.19 Average Modeling of the Peak Current-Mode PFC

We have successfully tested the current-mode PWM switch in a peak current-mode power factor correction circuit. Figure 6-37a displays the ac test fixture, whereas Fig. 6-38a represents the transient test circuit.

In this ac configuration, the bias levels are evidence of well-calculated dc operating points ($V_{out} = 402$ V), and the duty cycle reaches 30% at an input level of 230 Vrms. The optimum ramp compensation calculation uses Eq. (6-122):

$$S_{L,off} = \frac{V_{out}}{L} = \frac{400}{260u} = 1.53 \text{ A}/\mu s \tag{6-132}$$

Once reflected over the 50 mΩ sense resistor, it becomes a voltage slope.

$$S'_{L,off} = 1.53 \times 50m = 77 \text{ mV}/\mu s \tag{6-133}$$

If we take 50% of it, the ramp compensation level applied to the model becomes:

$$S_e = \frac{77m}{2} = 38 \text{ mV}/\mu s \text{ or } 38 \text{ kV/s} \tag{6-134}$$

The ramp compensation brings damping on the double pole as Fig. 6-37b details the plot of the control-to-inductor current transfer function.

Using Fig. 6-38a, we have purposely run a few simulations where the input current shape was explored. Figure 6-38b gathers the resulting signals. On the upper trace, the current delivered

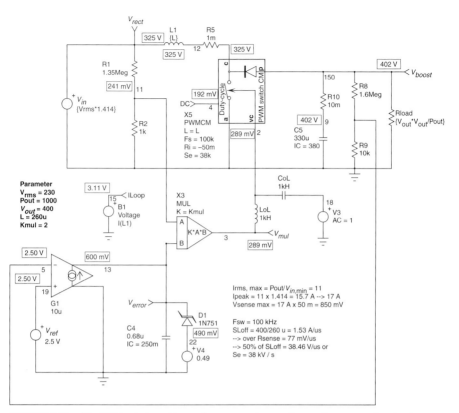

FIGURE 6-37a The PWM switch used in a peak current-mode control power factor correction circuit.

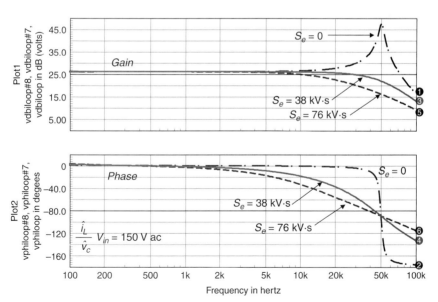

FIGURE 6-37b Injecting ramp compensation damps the peaking in the control-to-inductor current transfer function. The input voltage was set to 150 Vrms.

547

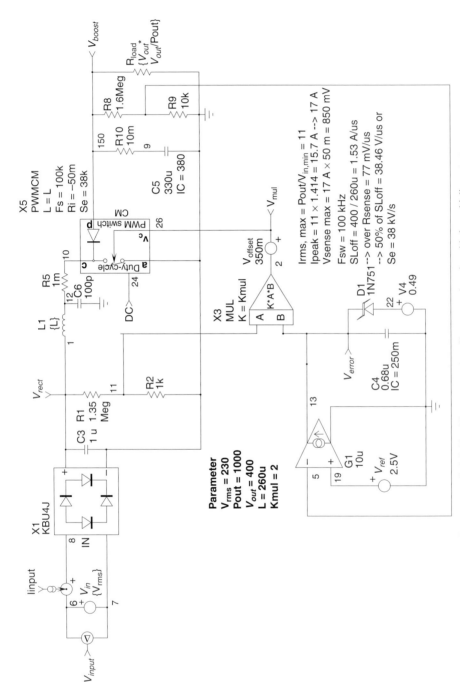

FIGURE 6-38a The transient test fixture uses a diode bridge. The compensation offset appears in series with the multiplier output.

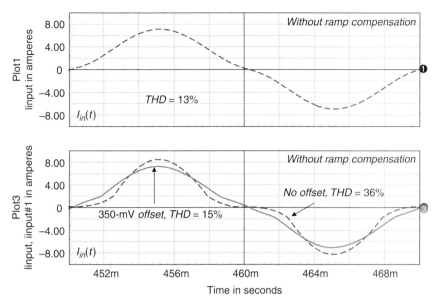

FIGURE 6-38b Input current delivered by the PFC operated in peak current-mode control either uncorrected (upper trace) or corrected via the addition of a series offset.

by the PFC without ramp compensation gives a 13% THD. Of course, the system enters instability at low line and cannot properly work. Once the ramp is injected [Eq. (6-134)], the stability improves but the distortion increases to 36%. Finally, a 350 mV offset enhances the situation and reduces the THD to 15%.

6.2.20 Hysteretic Power Factor Correction

Figure 6-39 displays the architecture employed in a hysteretic PFC converter. Unlike in the previous architectures, there is no internal clock. The controller fixes peak and valley currents through the activation of the power switch: The switch stays on until the inductor peak is reached where it opens. The current falls and once it touches the valley set point, a new on time occurs. To precisely follow the rectified envelope, the inductor current swing (hence the peak and valley levels) permanently changes. Unfortunately, in the vicinity of the 0 V region, the swing reduction tremendously increases the hysteretic frequency, hence switching losses. Despite its inherent simplicity, the wide frequency variations prevented the hysteretic converter from being successful. Cherry Semiconductor, in the 1990s, had a controller, the CS3810, but it has been abandoned now.

Several options are available to change the peak and valley on the fly. Figure 6-39 depicts one possible way through a voltage-driven hysteresis comparator. A second option consists of having two signals bringing different sinusoidal levels, toggled at the comparator pace: The highest amplitude fixes the peak, whereas the lowest fixes the valley. In our case, a variable hysteresis comparator can be built through a few SPICE lines:

IsSpice:

```
.SUBCKT COMPARHYSV NINV INV OUT VAR params: VHIGH=5 VLOW=100m
Rdum VAR 0 10Meg
B2 HYS NINV V = V(OUT) > {(VHIGH+VLOW)/2} ? V(VAR) : 0
```

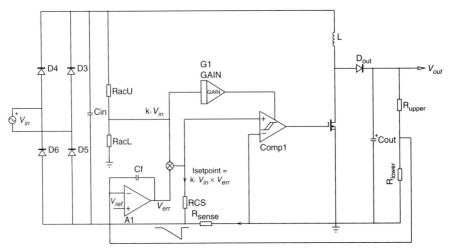

FIGURE 6-39 The hysteretic converter can be made with two sinusoidal set points, or via a variable hysteresis comparator as shown here.

```
B1 4 0 V = V(HYS,INV) > 0 ? {VHIGH} : {VLOW}
RO 4 OUT 10
CO OUT 0 10PF
.ENDS
```

PSpice:

```
.SUBCKT COMPARHYSV NINV INV OUT VAR params: VHIGH=5 VLOW=100m
Rdum VAR 0 10Meg
E2 HYS NINV Value = { IF ( V(OUT) > {(VHIGH+VLOW)/2}, V(VAR),
+ 0 ) }
E1 4 0 Value = { IF ( V(HYS,INV) > 0, {VHIGH}, {VLOW} ) }
RO 4 OUT 10
CO OUT 0 10PF
.ENDS
```

The complete application schematic appears in Fig. 6-40a and exactly implements the previous recommendations. Figure 6-40b collects all pertinent curves for the PFC analysis. The inductor current envelope nicely tracks the set point. This technique probably offers the best input current waveform, compared to the other approaches we covered.

6.2.21 Fixed-Frequency DCM Boost

The boost operated in constant on time and fixed frequency also performs power factor correction but with a greater distortion than its BCM counterpart. Figure 6-41 depicts a classical voltage-mode configuration which does not require high-voltage sensing since the converter operates in constant t_{on}. We purposely omitted the peak current sensing for the sake of simplicity, but such a circuit often appears to reset the main latch in an overcurrent situation. To demonstrate the power factor correction brought by this circuit, several lines of algebra are necessary. Let us start by observing the inductor current and voltage, as Fig. 6-42 proposes.

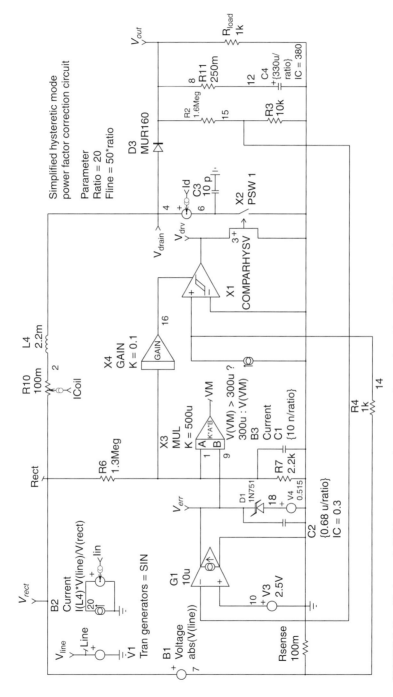

FIGURE 6-40a The hysteretic PFC implements negative current sensing together with a variable hysteresis comparator.

552 CHAPTER SIX

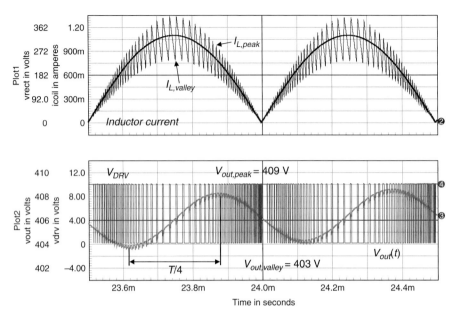

FIGURE 6-40b Simulation results of the PFC in hysteretic mode. Note the lack of distortion near the zero crossing region.

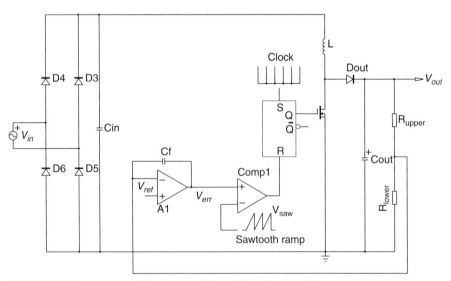

FIGURE 6-41 The boost operated in constant on time and discontinuous mode.

The average inductor current can be quickly derived by summing the areas under the on and off triangles:

$$I_{L,avg} = \left(\frac{I_{L,peak}}{2}t_{on} + \frac{I_{L,peak}}{2}t_{off}\right)\frac{1}{T_{sw}} = \frac{I_{L,peak}}{T_{sw}}\left(\frac{t_{on}}{2} + \frac{t_{off}}{2}\right) \quad (6\text{-}135)$$

First, we will need to derive a value for the off time, since the on time stays constant. To help us, Fig. 6-42 plots the instantaneous voltage across the inductor. Applying the inductor volt-second balance law, we have

$$V_{in}t_{on} - (V_{out} - V_{in})t_{off} = 0 \quad (6\text{-}136)$$

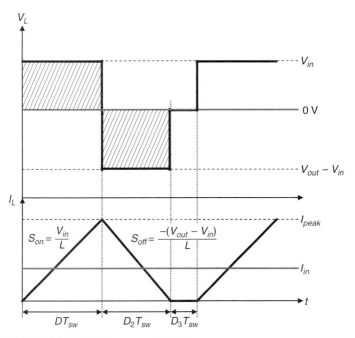

FIGURE 6-42 The inductor current in DCM.

from which the off-time duration comes easily:

$$t_{off} = \frac{V_{in}t_{on}}{V_{out} - V_{in}} \quad (6\text{-}137)$$

Now replacing t_{off} in Eq. (6-135) gives

$$I_{L,avg} = \frac{I_{L,peak}}{T_{sw}}\left(\frac{t_{on}}{2} + \frac{V_{in}t_{on}}{2(V_{out} - V_{in})}\right) = \frac{I_{L,peak}t_{on}}{T_{sw}}\left(\frac{V_{out}}{2(V_{out} - V_{in})}\right) \quad (6\text{-}138)$$

The inductor peak current is classically defined by

$$I_{L,peak} = \frac{V_{in}}{L}t_{on} \quad (6\text{-}139)$$

Updating Eq. (6-138) by the above, we obtain

$$I_{L,avg} = \frac{V_{in} t_{on}^2}{LT_{sw}} \frac{V_{out}}{2(V_{out} - V_{in})} = \frac{V_{in} t_{on}^2}{2LT_{sw}} \frac{1}{1 - \frac{V_{in}}{V_{out}}} \qquad (6\text{-}140)$$

Finally, the "instantaneous" average inductor current is thus

$$I_{L,avg}(t) = \frac{V_{peak} t_{on}^2}{2LT_{sw}} |\sin \omega t| \frac{1}{1 - \frac{V_{peak} |\sin \omega t|}{V_{out}}} \qquad (6\text{-}141)$$

We can rearrange the previous equation to unveil the emulated input resistance. After all, a PFC strives to shape the input current so that the preconverter looks like a resistive load. This is what Eq. (6-142) shows you:

$$I_{L,avg}(t) = \frac{V_{peak} |\sin \omega t|}{R_{in}} \frac{1}{1 - \frac{V_{peak} |\sin \omega t|}{V_{out}}} \qquad (6\text{-}142)$$

where R_{in} represents the emulated input resistance of the converter

$$R_{in} = \frac{2LT_{sw}}{t_{on}^2} = \frac{2LT_{sw}}{D^2 T_{sw}^2} = \frac{2L}{D^2 T_{sw}} \qquad (6\text{-}143)$$

In this expression, the term $\dfrac{1}{1 - \dfrac{V_{peak}|\sin \omega t|}{V_{out}}}$ induces distortion in the DCM boost operation.
It explains why the DCM boost converter performs power factor correction, but without the performance of the previous structures.

Introducing this DCM boost preconverter, we assumed the on time to remain constant. Unfortunately, deriving an analytical value for this parameter represents a difficult exercise as the following line will teach you. Whatever the input signal shapes, the boost converter instantaneous input power $P_{in}(t)$ always satisfies the equation

$$P_{in}(t) = I_{L,avg}(t) V_{in}(t) \qquad (6\text{-}144)$$

Given a 100% efficiency, the average input power is obtained from

$$P_{out} = \frac{1}{T_{line}} \int_0^{T_{line}} P_{in}(t) \cdot dt \qquad (6\text{-}145)$$

Replacing the average inductor current with its definition, we have

$$P_{out} = \frac{2}{T_{line}} \int_0^{\frac{T_{line}}{2}} \frac{V_{peak} t_{on}^2 \sin \omega t}{2LT_{sw}} \frac{1}{1 - \frac{V_{peak} \sin \omega t}{V_{out}}} V_{peak} \sin \omega t \cdot dt \qquad (6\text{-}146)$$

Note that we integrate over one-half of a line cycle to keep the sine term strictly positive: It saves us from manipulating absolute values. Rearranging a little, we finally have

$$P_{out} = \frac{V_{peak} t_{on}^2}{LT_{sw}T_{line}} \int_0^{\frac{T_{line}}{2}} \frac{\sin^2 \omega t}{1 - \frac{V_{peak} \sin \omega t}{V_{out}}} \cdot dt \qquad (6\text{-}147)$$

Unfortunately, this integral does not have a simple analytical solution: We need to adopt a numerical method to obtain the result and extract the on-time value. Possible solutions range from Excel and Mathcad to the autotoggling PWM switch model simulated with SPICE. Figure 6-43a depicts a way to wire it in an example running the DCM boost converter. Further to a dc point calculation, the model gives a duty cycle of 62%. It corresponds to an on time of 6.2 µs with a boost operated at a 100 kHz switching frequency. If we now run a transient simulation, Fig. 6-43b shows the resulting input current waveform. You can see the distorted shape which brings a THD of 10.3% when respecting DCM operation. If the converter entered CCM, the current would suddenly diverge from a quasi-sinusoidal envelope and distortion would greatly increase.

Discontinuous-mode fixed-frequency boost PFCs exhibit quite a large distortion compared to borderline types. However, the fixed-frequency operation offers better spectrum control and can help when synchronization between the preconverter and the downstream converter is a must.

6.2.22 Flyback Converter

The flyback topology covers probably 80% of power supplies designed in the world, especially in the consumer market. It is really one of the most popular structures, as Chap. 7 will let you discover. Implemented in PFC circuits, it brings the natural transformer galvanic isolation and can be used to deliver a loosely regulated dc voltage. A nonisolated downstream buck can then perform the dc–dc function for adequate response speed. The simplest technique to make PFC with a flyback uses fixed-frequency discontinuous operation with constant on time, such as voltage-mode control. Figure 6-44 shows this type of flyback power supply, operated without sensing the high-voltage rail. C_{in} is of low value and helps to filter out the differential noise generated by the PFC.

To check whether this configuration brings good power factor, we need to go through the equations once again. However, before proceeding, we look at the typical flyback input current waveform as depicted by Fig. 6-45. During the on time, the switch closes and applies the input voltage across the primary inductor L_p. The peak current thus evolves according to Eq. (6-148):

$$I_{L_p,peak} = \frac{V_{in}}{L_p} t_{on} \qquad (6\text{-}148)$$

From Fig. 6-45, the average inductor current is easily found:

$$I_{L_p,avg} = \frac{I_{L_p,peak}}{2} \frac{t_{on}}{T_{sw}} \qquad (6\text{-}149)$$

If we now substitute the peak current definition into Eq. (6-149), we have the instantaneous average input current relationship:

$$I_{L_p,avg}(t) = \frac{V_{peak}|\sin \omega t| t_{on}^2}{2L_p T_{sw}} = k|\sin \omega t| \qquad (6\text{-}150)$$

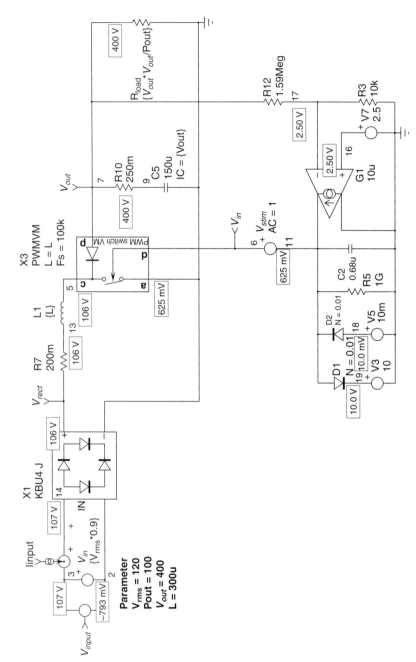

FIGURE 6-43a The voltage-mode PWM switch in a 400 V, 100 W low-line DCM boost converter.

SIMULATIONS AND PRACTICAL DESIGNS OF OFF-LINE CONVERTERS—THE FRONT END 557

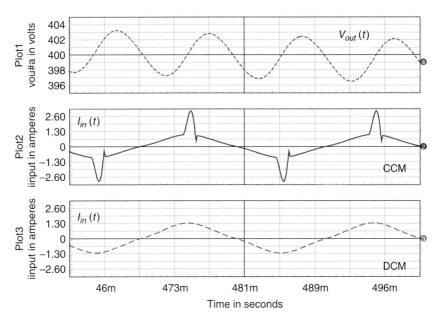

FIGURE 6-43b Fixed-frequency boost waveforms when running in DCM or CCM.

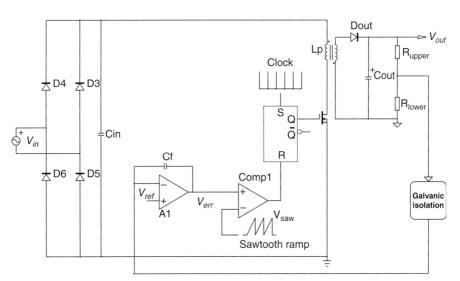

FIGURE 6-44 A flyback converter used in a DCM PFC preconverter.

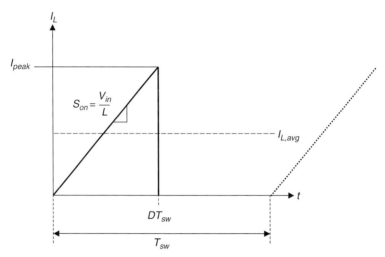

FIGURE 6-45 The input current of a DCM flyback converter.

where $k = \dfrac{V_{peak} \, t_{on}^2}{2L_p \, T_{sw}}$. If the on time t_{on} keeps constant, so does k and the instantaneous average current follows a sinusoidal envelope.

As we will see in Chap. 7, a discontinuous flyback converter draws an input power related to the inductor peak current squared. Thus, considering a 100% efficient converter, we have

$$P_{in}(t) = \frac{1}{2} L_p \left[I_{L_p, peak}(t) \right]^2 F_{sw} = \frac{1}{2} L_p \frac{V_{peak}^2 \sin^2 \omega t}{L_p^2} t_{on}^2 F_{sw} \qquad (6\text{-}151)$$

Rearranging the above equation gives

$$P_{in}(t) = \frac{V_{peak}^2 t_{on}^2 F_{sw}}{2L_p} \sin^2 \omega t \qquad (6\text{-}152)$$

If we now average the instantaneous input power over a line cycle, we should obtain the output power, given 100% efficiency. Based on Eq. (6-152), we then have

$$P_{out} = F_{line} \int_0^{T_{line}} \frac{V_{peak}^2 t_{on}^2 F_{sw}}{2L_p} \sin^2 \omega t \, dt \qquad (6\text{-}153)$$

Collecting the constant terms, we obtain

$$P_{out} = \frac{V_{peak}^2 t_{on}^2 F_{sw}}{2L_p} \langle \sin^2 \omega t \rangle_{T_{line}} \qquad (6\text{-}154)$$

The $\sin(x)$ function moves between -1 and 1, thus bounding $\sin(x)^2$ between 0 and 1. Its average value therefore equals 0.5:

$$P_{out} = \frac{V_{peak}^2 t_{on}^2 F_{sw}}{4L_p} \qquad (6\text{-}155)$$

Extracting t_{on} from the above gives

$$t_{on}^2 = \frac{4L_p P_{out}}{V_{peak}^2 F_{sw}} \tag{6-156}$$

Now, replacing V_{peak}^2 by $V_{peak}^2 = 2V_{ac}^2$, where V_{ac} is the rms value of the input voltage, we reach the final definition:

$$t_{on} = \frac{1}{V_{ac}}\sqrt{\frac{2L_p P_{out}}{F_{sw}}} \tag{6-157}$$

6.2.23 Testing the Flyback PFC

The easiest and fastest way to verify if the flyback converter used as a PFC does a good job is to use an average model: Simulation time is fast, and you can quickly explore its stability. Figure 6-46a portrays the average model wired in a transient configuration featuring an input diode bridge. The error amplifier uses a simple voltage-controlled current source G_1 associated with clipping diodes. The gain subcircuit X_2 emulates a 2 V ramp generator whose output directly feeds the PWM switch duty cycle input. The output voltage is stabilized to 48 V, and a small capacitor C_3 across the upper sensing resistor introduces a zero around the crossover frequency. Figure 6-46b depicts the input current waveforms at low and high line, confirming good distortion factors. The output ripple stays constant at 2 V peak to peak despite input voltage changes. Improvement on this parameter could come from an increase in the output capacitor.

So-called single-stage PFCs strive to combine a flyback together with power factor correction operation. Several solutions exist on the market among which the NCP1651 from ON Semiconductor operates in average current-mode control and lets the flyback operate in CCM.

This introduction to active power correction should give you a taste of the available solutions you can find in the silicon market. Of course, different options could be considered like the buck or the SEPIC, but they will lack the simplicity inherent in a boost converter. To capitalize on what we covered, two design examples, a boost PFC operated in BCM and in CCM, are presented now.

6.3 DESIGNING A BCM BOOST PFC

We have spent quite a long introduction on the PFC techniques, and it is now time to capitalize on the accumulated knowledge and see how to design a borderline boost power factor correction circuit. These PFC circuits are often found in high-power notebook adapters and converters delivering less than 300 W typically. The specifications are the following:

V_{in} = 85 Vrms to 275 Vrms—input voltage
V_{out} = 400 Vdc—output voltage
P_{out} = 150 W—output power
F_{line} = 50 Hz—mains frequency
η = 90%—efficiency
Hold-up time = 20 ms, minimum bulk voltage is 330 Vdc ($V_{out,min}$ imposed by downstream converter)
Output voltage ripple = 5%

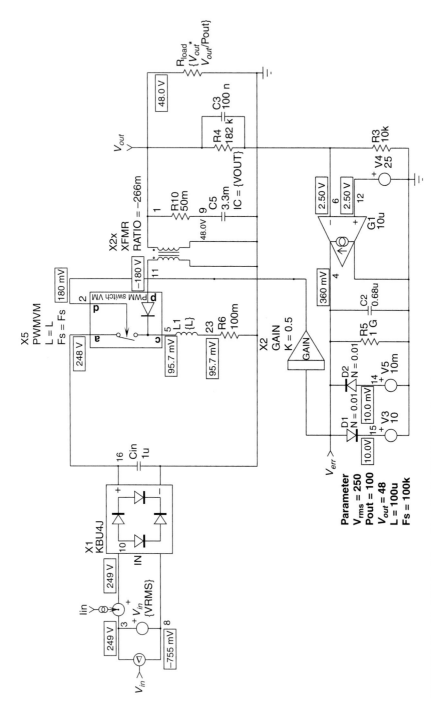

FIGURE 6-46a The PWM switch model used in the PFC flyback configuration.

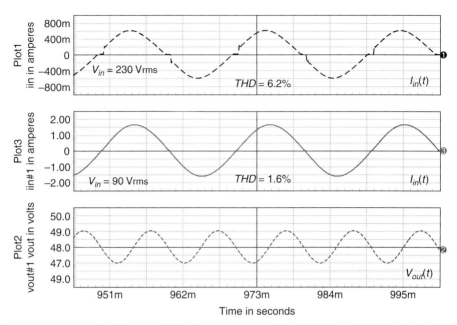

FIGURE 6-46b Transient simulation results of the DCM flyback used as a power factor correction circuit.

For cost reasons, we have selected an MC33262 from ON Semiconductor: The device is inexpensive and robust while performing near-unity power factor. From the above specification values, we can already calculate the inductor current: We know that Eq. (6-95) links its peak and its average values by

$$I_{L,peak} = \frac{2\sqrt{2}P_{out}}{\eta V_{in,min}} = \frac{2.83 \times 150}{0.9 \times 85} = 5.6 \text{ A} \qquad (6\text{-}158)$$

where η represents the PFC converter efficiency.

The MC33262 specifies a clamped current set point varying between 1.3 and 1.8 V. In the minimum of the specification, Eq. (6-158) must still be satisfied. The sense resistor can therefore be evaluated via the following simple equation:

$$R_{sense} = \frac{V_{th,max}}{I_{L,peak}} = \frac{1.3}{5.6} = 232 \text{ m}\Omega \qquad (6\text{-}159)$$

Unfortunately, given the inherent silicon wafer lot-to-lot variations, a certain dispersion exists on the maximum current set point. If we select a normalized 0.22 Ω resistor, the inductor peak current could increase up to

$$I_{L,max} = \frac{V_{th,max}}{220m} = \frac{1.8}{220m} = 8.2 \text{ A} \qquad (6\text{-}160)$$

Further, it would be wise to adopt a maximum peak current of 9 A for the inductor design. Such a number includes the necessary security margin, accounting for the inductor value drop as its core temperature rises. The power dissipated by this element follows Eq. (6-161) as derived in Ref. 21:

$$P_{R_{sense}} = \frac{4}{3} R_{sense} \left(\frac{P_{out}}{\eta V_{in,min}}\right)^2 \left[1 - \left(\frac{8\sqrt{2}V_{in,min}}{3\pi V_{out}}\right)\right] \tag{6-161}$$

This leads to a power dissipation of 0.9 W: Select a 0.22 Ω, 2 W resistor.

A BCM-operated PFC exhibits a widely varying switching frequency over the mains period. The minimum operating frequency is reached at the sinusoidal peak [the sine term in Eq. (6-106) equals 1]. To prevent acoustic noise in the inductor, care must be taken to avoid entering the audible range at the highest peak current conditions. The highest current peak is reached at low line, but the lowest switching frequency usually occurs at high line. However, in the latter case, since the peak current remains low, acoustic noise generation by the inductor does not cause any concern. Therefore, the inductor selection equation deals with the minimum acceptable switching frequency at low line. If we extract the inductor from Eq. (6-106), we have

$$L \leq \frac{\eta V_{in,min}^2}{2P_{out}F_{sw,min}} \left(1 - \frac{V_{peak,min}}{V_{out}}\right) \tag{6-162}$$

If we enter the design elements together with a minimum switching frequency of 20 kHz, we end up with an inductor whose value should not exceed 842 µH. Its rms current can be calculated via Eq. (6-163) [21]:

$$I_{L,rms} = \frac{2}{\sqrt{3}} \frac{P_{out}}{\eta V_{in,min}} = \frac{2}{\sqrt{3}} \frac{150}{0.85 \times 85} = 2.3 \text{ A} \tag{6-163}$$

To perform power factor correction with the BCM boost, the on time must remain constant, as we already demonstrated. Equation (6-164) helps us to compute the maximum and minimum on-time values. However, a discontinuity exists around the mains zero crossing event where t_{on} will strive to force current despite the absence of voltage. The average model will show this a few figures below.

$$t_{on,max} = \frac{2LP_{out}}{V_{in,min}^2} = \frac{2 \times 840u \times 150}{85^2} = 35 \text{ µs} \tag{6-164}$$

This value drops to 3.3 µs at the highest input level (275 Vrms).

The MC33262 includes a multiplier whose inputs receive an image of the bulk voltage and the output of the error amplifier, modified by a fixed offset to allow a real zero output. The resulting signal (a full-wave rectified sinusoidal voltage of output power–dependent amplitude) sets the peak current set point. The data sheet guidelines recommend to calculate the input divider network so that the ac multiplier input (labeled MUL) reaches 3 V at the highest line (Fig. 6-47). First, let us fix the resistive bridge current in R_1/R_2 to 250 µA to obtain an acceptable noise immunity. In low standby power applications, a lower current can also be selected, but the layout must be done with care. Equations are rather straightforward then.

$$R_2 = \frac{3}{250u} = 12 \text{ k}\Omega \tag{6-165}$$

The highest voltage on C_{in} corresponds to the peak of the maximum input voltage. Hence, R_3 can be quickly derived given the bridge current of 250 µA:

$$R_1 = \frac{275 \times \sqrt{2} - 3}{250u} = 1.6 \text{ M}\Omega \tag{6-166}$$

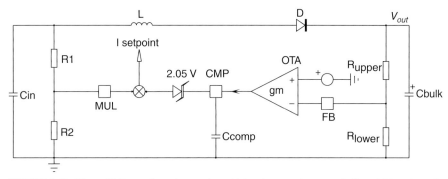

FIGURE 6-47 The multiplier configuration requires a high-voltage sensing network (R_1 and R_2) to shape the peak current set point.

Since the ac amplitude also participates to the peak current set point, the above divider network must ensure enough voltage on the multiplier output to allow a proper start-up at the lowest input/full power: A tweak might be necessary to properly start up while ensuring acceptable distortion at high line.

The output capacitor can be calculated via several possibilities. The first consists of satisfying the storage definition of Eq. (6-73):

$$W_s = \frac{P_{out}}{\omega} \tag{6-73}$$

In this case, the stored energy on the output capacitor directly relates to the output ripple we want. Let us adopt a ripple of 5% of the output voltage, thus 40 V peak to peak. The output peaks to

$$V_{out,peak} = 400 \times 1.05 = 420 \text{ V} \tag{6-167}$$

and it decreases to

$$V_{out,valley} = 400 \times 0.95 = 380 \text{ V} \tag{6-168}$$

The stored energy is then the difference between these values squared, multiplied by one-half the bulk capacitor value we are looking for:

$$W_s = \frac{1}{2}(V_{out,peak}^2 - V_{out,valley}^2)C_{bulk} \tag{6-169}$$

If combine Eqs. (6-73) and (6-169), the bulk capacitor definition emerges easily:

$$C_{bulk} = \frac{P_{out}}{\pi F_{line}(V_{out,peak}^2 - V_{out,valley}^2)} = \frac{150}{3.14 \times 50 \times (420^2 - 380^2)} = 30\,\mu\text{F} \tag{6-170}$$

The normalized value would be 47 μF/450 Vdc.

The second option, often seen in the literature, links the capacitor value to the hold-up time. Equation (6-9a) was already derived at the beginning of this chapter; we now update it as Eq. (6-9d) with the above PFC definitions:

$$C_{bulk} = \frac{2P_{out}}{(V_{out,peak}^2 - V_{out,min}^2)}t_d \tag{6-9d}$$

Here t_d is simply the hold-up time of 20 ms, and $V_{out,min}$ represents the minimum voltage accepted by the downstream converter (330 Vdc according to the starting specification). The updated bulk capacitor value thus becomes

$$C_{bulk} = \frac{2 \times 150}{(420^2 - 330^2)} 20m = 89 \, \mu F \tag{6-171}$$

The closest normalized value is 150 μF/450 V. The resulting ripple ΔV obtained with a 150 μF capacitor can be derived by applying a soft massage to Eq. (6-170):

$$C_{bulk} = \frac{P_{out}}{\pi F_{line}([V_{out} + \Delta V/2]^2 - [V_{out} - \Delta V/2]^2)} \tag{6-172}$$

Considering a ripple ΔV centered on the dc output voltage V_{out}, we can rearrange the above equation as

$$C_{bulk} = \frac{P_{out}}{\pi F_{line}(a+b)(a-b)} = \frac{P_{out}}{\pi F_{line} 2 V_{out} \Delta V} = \frac{P_{out}}{\omega V_{out} \Delta V} \tag{6-173}$$

The peak-to-peak ripple ΔV can now be obtained:

$$\Delta V = \frac{P_{out}}{\omega V_{out} C_{bulk}} = \frac{150}{314 \times 400 \times 150u} = 8 \, V \tag{6-174}$$

It is always important to verify that a sufficient margin exists between the final worst-case ripple and the overvoltage protection (OVP) set by the controller (in case a fault appeared, for instance). With the MC33262, the OVP is set to 8% above the selected output voltage (432 V with a 400 V output). Equation (6-174) teaches us that the peak voltage in regulation will reach 404 V roughly, giving enough margin for a reliable operation.

As usual, the rms current in the capacitor dictates the final choice. Without going into the details, the circulating rms current in the capacitor depends on the nature of the load. If a switching power supply loads the PFC, then its signature depends on the operating mode. But what actually matters is the operating frequency of the converter compared to that of the PFC controller. Depending on how they switch (synchronously, asynchronously, out of phase), the rms calculation really changes and can become a tedious exercise. Below appears the rms current definition for a resistive loading, as brilliantly derived by Monsieur Turchi [21]:

$$I_{C_{bulk},rms} = \sqrt{\left[\frac{1.6 P_{out}^2}{\eta^2 V_{in,min} V_{out}}\right] - \left(\frac{P_{out}}{V_{out}}\right)^2} = \sqrt{\frac{1.6 \times 150^2}{0.81 \times 85 \times 400} - \left(\frac{150}{400}\right)^2} = 1.08 \, A \tag{6-175}$$

Typical capacitor types are snap-in or radial and can be arranged in parallel to accept higher rms currents. The AXF series from Rubycon might represent a possible choice.

The catch diode "sees" both the ac capacitor current and the load current [21]. Equation (6-176) defines the rms current flowing through the diode, the average being the dc output current:

$$I_{d,rms} = \frac{1.26 P_{out}}{\eta \sqrt{V_{in,min} V_{out}}} = \frac{1.26 \times 150}{0.9 \sqrt{85 \times 400}} = 1.13 \, A \tag{6-176}$$

$$I_{d,avg} = \frac{P_{out}}{V_{out}} = 375 \, mA \tag{6-177}$$

From the above numbers, the diode conduction losses can be estimated by

$$P_{d,cond} = I_{d,rms}^2 R_d + I_{d,avg} V_f \tag{6-178}$$

where R_d represents the diode dynamic resistance at its operating point ($R_d = \dfrac{dV_f}{dI_d}@I_{d,avg}$) and V_f is the forward drop at the average current condition (375 mA). These numbers can easily be extracted from the V_f versus I_d data sheet curves. The BCM operation favors the implementation of inexpensive diodes since recovery losses do not hamper the power MOSFET dissipation budget: The diode naturally turns off when the inductor is fully depleted at the end of a switching cycle. This is not the case for a CCM PFC (or any CCM converter) where the diode turns off because the MOSFET suddenly conducts, brutally pulling the anode to ground, hence imposing V_{out} as a reverse voltage across the diode. When struggling to recover its blocking effect, the diode behaves as a short circuit, directly routing the output voltage to the MOSFET: losses occur on both the MOSFET and the diode. CCM PFCs require serious snubbing actions to reduce these losses to an acceptable level. For this 150 W BCM application, a diode such as a 1N4937 (1 A, 600 V) could easily do the job. However, beware of the in-rush current whose peak amplitude could easily damage the diode at power on. If this is a concern, select a more robust device such as the MUR460 or bypass the catch diode via an inexpensive but bigger diode (see the design example to learn how to install this diode).

The power MOSFET choice depends on the maximum allowable drain voltage that the quality department imposes. Usually, deratings of 15% of the maximum BV_{DSS} are common practice. In the case of a boost, having a 7 V upper ripple as in our case, the maximum voltage on the drain reaches 406 V. In fault conditions, e.g., in the presence of a defective feedback path, the OVP level peaks to 432 V. Hence, a 500 V MOSFET cannot fit the aforementioned bill, and we need a 600 V type. Some designers could, however, shift the nominal voltage down to 385 Vdc and then improve the margin to adopt a 500 V MOSFET.

The power dissipation budget of the MOSFET is principally made of several factors such as switching losses and conduction losses. The first ones require an evaluation prototype to measure the important transition times, and we recommend that you look at Ref. 21 to learn more about switching loss calculations. The conduction losses can be estimated given the $R_{DS(on)}$ of the MOSFET when its junction temperature reaches 100 °C. If rms currents are rather easy to evaluate in the presence of a fixed-frequency switching converter, the exercise brings more fun with a system for which both the frequency and the input voltage change! Let us see it for the sake of the exercise.

Figure 6-45 showed the current flowing in the MOSFET during the on time for a flyback converter; however, the waveform does not differ for the boost. The time-varying expression for the MOSFET current is easily derived, considering V_{in} constant:

$$I_D(t) = I_{L,peak}\dfrac{t}{t_{on}} \tag{6-179}$$

The peak current is defined by

$$I_{L,peak} = \dfrac{V_{in}}{L}t_{on} \tag{6-180}$$

Replacing the peak definition in Eq. (6-179) gives

$$I_D(t) = \dfrac{V_{in}}{L}t \tag{6-181}$$

To obtain the squared rms MOSFET current, raise Eq. (6-181) to the power 2 and integrate over a switching period:

$$I_{D,rms}{}^2 = \dfrac{1}{T_{sw}}\int_0^{t_{on}}\dfrac{V_{in}{}^2}{L^2}t^2 \cdot dt = \dfrac{V_{in}{}^2}{L^2 T_{sw}}\dfrac{t_{on}{}^3}{3} \tag{6-182}$$

Rearranging the above equation helps to get rid of the switching frequency:

$$I_{D,rms}^2 = \frac{1}{3}\left[\frac{V_{in}}{L}t_{on}\right]^2 \frac{t_{on}}{T_{sw}} = \frac{1}{3}I_{L,peak}^2 d \qquad (6\text{-}183)$$

where d represents the static boost duty cycle, defined as

$$d = \frac{V_{out} - V_{in}}{V_{out}} \qquad (6\text{-}184)$$

Thanks to the power factor operation, we know that the inductor peak current follows a sinusoidal envelope. The relationship between the peak envelope and the average input current is given by Eq. (6-95). We can slightly update it by plugging the peak input current

$$I_{L,peak}(t) = 2\sqrt{2}\frac{P_{out}}{\eta V_{ac}}|\sin(\omega t)| \qquad (6\text{-}185)$$

where $V_{ac} = V_{in,rms}$. Equation (6-183) can now be upgraded to become a time (the line period)-related definition:

$$I_{D,rms}^2(t) = \frac{8}{3}\left(\frac{P_{out}}{\eta V_{ac}}\right)^2 \sin^2(\omega t)\frac{V_{out} - \sqrt{2}V_{ac}|\sin(\omega t)|}{V_{out}} \qquad (6\text{-}186a)$$

A final integration over a half-line period helps to get rid of the absolute value and delivers the result we are looking for:

$$\langle I_{D,rms}^2(t)\rangle_{T_{line}} = 2F_{line}\int_0^{\frac{1}{2F_{line}}} \frac{8}{3}\left(\frac{P_{out}}{\eta V_{ac}}\right)^2 \sin^2(\omega t)\frac{V_{out} - \sqrt{2}V_{ac}|\sin(\omega t)|}{V_{out}} \cdot dt \qquad (6\text{-}186b)$$

After calculation, it becomes

$$\langle I_{D,rms}^2(t)\rangle_{T_{line}} = \left[\frac{2P_{out}}{\eta 3V_{ac}}\right]^2 \frac{3V_{out} - \frac{8\sqrt{2}V_{ac}}{\pi}}{V_{out}} = \left[\frac{2P_{out}}{\eta\sqrt{3}V_{ac}}\right]^2\left(1 - \frac{8\sqrt{2}V_{ac}}{3\pi V_{out}}\right) \qquad (6\text{-}187)$$

If we select a MOSFET featuring 150 mΩ at a junction temperature of 25 °C, the rule of thumb says that this value is likely to roughly double at a junction temperature of 100 °C. The conduction losses are then calculated by using Eq. (6-187):

$$P_{cond} = \langle I_{D,rms}^2(t)\rangle_{T_{line}} R_{DS(on)} = 3.8 \times 300m = 1.14\text{ W} \qquad (6\text{-}188)$$

Most nonisolated TO-220 packages roughly feature a thermal resistance junction-to-air $R_{\theta J-A}$ of 60 °C/W. In free-air conditions, without any heat sink, if we limit the die temperature excursion to 110 °C in a 70 °C ambient temperature, the package can then dissipate a maximum power of

$$P_{max} = \frac{T_{j,max} - T_{A,max}}{R_{\theta J-A}} = \frac{110 - 70}{60} = 660\text{ mW} \qquad (6\text{-}189)$$

From Eq. (6-188) (to which we need to add the averaged switching losses), the MOSFET will need a heat sink.

The error amplifier section implements an *operational transconductance amplifier* (OTA) whose output appears on pin 2 and features a transconductance (gm) of 100 μS.

We know that the PFC bandwidth must be rolled off around 20 Hz to avoid reacting on the 100 Hz output ripple. To roll off the gain, a simple capacitor connects from pin 2 to ground. Its value is found via Eq. (6-190):

$$C_{comp} = \frac{gm}{2\pi f_c} = \frac{100u}{6.28 \times 20} = 800 \, \text{nF} \tag{6-190}$$

Experience shows that a 0.68 µF capacitor can do the job. As mentioned in the BCM PFC analysis, a zero helps to improve the phase margin in the vicinity of the crossover frequency. In the BCM description, we offered to put a capacitor in parallel with the upper resistor. Another option is to place a resistor in series with the compensation capacitor. For a zero located at 10 Hz, the resistor value is

$$R_{comp} = \frac{1}{2\pi f_z C_{comp}} = \frac{1}{6.28 \times 0.68u \times 10} = 23.5 \, \text{k}\Omega \tag{6-191}$$

We are now all set; it is time to test the PFC behavior with the average model first, as the simulation time is the shortest.

6.3.1 Average Simulations

Figure 6-48a shows the average circuit using the borderline conduction subcircuit. You can see most of the element values that we discussed earlier. The OTA section has been improved with a B element (B_1) which clamps the maximum output current to 10 µA, in compliance with the data sheet. The two diodes D_1 and D_2 clamp the error amplifier excursion as in the real MC33262. Actually B_4 includes the offset deduced from the error amplifier output which leads to a real 0 V on the multiplier input when the output level exceeds the target. The first experiment consists of checking that the adopted configuration together with the selected element values performs power factor correction. Figure 6-48b depicts the input signals at nominal high line (230 Vrms) and nominal low line (100 Vrms) while the PFC front-end stage delivers its 150 W.

A quick stability check can be run via a simple step load from 75 to 150 W, just after the start-up sequence. A real ac measurement could also be undertaken, but we already showed how to do it in the borderline introduction section. Figure 6-48c describes the resulting waveforms and confirms the good stability of the circuit. The slightly better bandwidth at high line leads to a small voltage undershoot. Without the 23 kΩ zero resistor, we would observe a strong ringing on the output voltage.

The distortion offers a decent picture at low line but dramatically increases at high line. Why? This is so because the zero we inserted via R_8 lets the ripple come out from the error chain. Straddling the dc error signal, it becomes at high line a significant part of the multiplier output which naturally distorts the peak current set point. If we reduce the zero resistor to a few ohms, the distortion comes back to a better level (1.5%). Unfortunately, the start-up sequence and any load step will now suffer from an extremely oscillatory output, as we mentioned above. Figure 6-48d depicts the waveforms obtained at start-up with and without the insertion of the zero. Without the zero, the distortion is at its best, but the output voltage peaks to 440 V at power on. In reality, with an MC33262, an internal comparator would decapitate this excursion, possibly lethal to the downstream dc–dc converter. However, the oscillation presence should be fought by increasing the phase margin at the crossover frequency. You can do it through R_8 but to the detriment of the distortion. A balance then needs to be found between a reasonable transient response and a good distortion level.

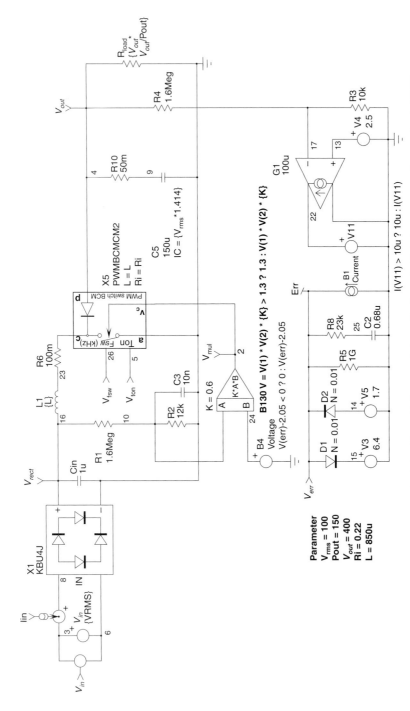

FIGURE 6-48a The 150 W PFC in the average simulation. The multiplier output is internally clamped to limit the current set point excursion to $1.3V/R_i$.

SIMULATIONS AND PRACTICAL DESIGNS OF OFF-LINE CONVERTERS—THE FRONT END **569**

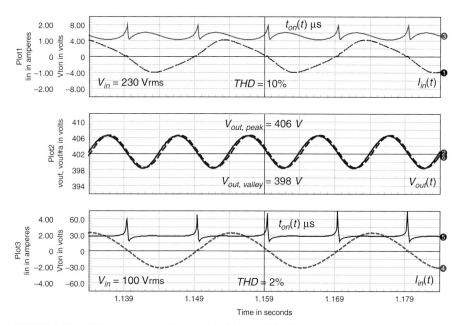

FIGURE 6-48b PFC waveforms at full power in both low- and high-line cases.

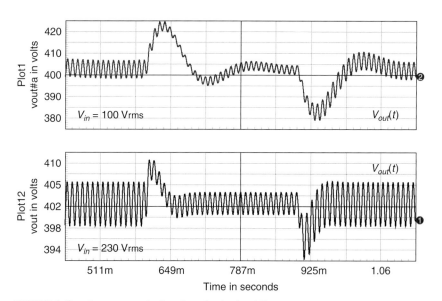

FIGURE 6-48c An output step load confirms the circuit stability.

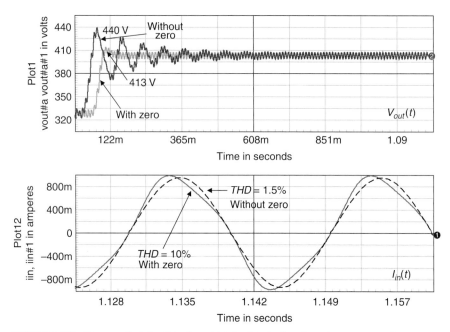

FIGURE 6-48d   The zero affects the distortion at high line but clearly reduces the start-up overshoot.

6.3.2 Reducing the Simulation Time

Average models simulate quickly because there is no switching component. To the opposite, the cycle-by-cycle simulation of PFC circuits represents a really time-consuming exercise, even with a fast computer. Reference 22 describes a way to speed up simulations by artificially altering some of the key element values by a fixed ratio, passed as a parameter to the simulation engine:

- $F_{line} = F_{line} \times ratio$
- $C_{bulk} = \dfrac{C_{bulk}}{ratio}$
- $C_{comp} = \dfrac{C_{comp}}{ratio}$

This is what Fig. 6-50a shows where we purposely selected a ratio of 10, operating the PFC at a 500 Hz line frequency. As the waveforms do not change in shape, we can use these simulation results to check the stress calculations for the key components. In the above sketches, we used a traditional diode bridge followed by a capacitor C_{in}.

To further simplify the configuration, Ref. 23 offers to replace these passive elements by two in-line equations following the suggestions in Fig. 6-49. As expected, the rectified sinusoidal voltage can be modeled by taking the absolute value of a sine wave source (labeled V_{in}) through the B_2 element. To display the real input current (and not the inductor current), a simple definition will reveal it via Eq. (6-192):

$$I_{in}(t) = I_L(t) \dfrac{V_{in}(t)}{|V_{in}(t)|} = I_{in,peak} |\sin(\omega t)| \dfrac{V_{peak} \sin(\omega t)}{V_{peak} |\sin(\omega t)|} = I_{in,peak} \sin(\omega t) \qquad (6\text{-}192)$$

This is what the B element current source B_3 performs.

SIMULATIONS AND PRACTICAL DESIGNS OF OFF-LINE CONVERTERS—THE FRONT END 571

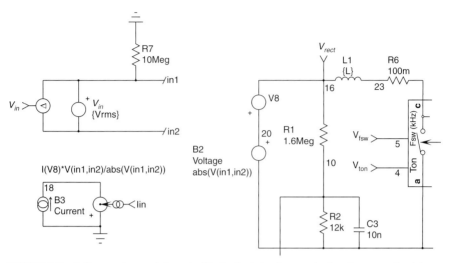

FIGURE 6-49 In-line equations can help to simplify the simulation fixture and reduce the computational time.

6.3.3 Cycle-by-Cycle Simulation

The example makes use of the BCM FreeRunDT model, actually a current-mode borderline controller already used in previous chapters.

We encapsulated the multiplier and the error amplifier of the MC33262 to simplify the drawing. A special section monitors the output voltage in case it ran into a start-up voltage condition. The controller is disabled by pulling the demagnetization pin high. This protection is absolutely needed as a boost converter, in essence, cannot run unloaded. As the output goes out of control, the controller could only enter into a minimum t_{on} situation where energy is always stored and released from the inductor: The output keeps going up until smoke comes out! Should you try it (for instance, with the first MC33261), the MOSFET would quickly blow up by exceeding its BV_{DSS}. Motorola solved this issue a while ago, by adding a comparator observing the feedback pin, as Fig. 6-50a X_5 does: It was the MC33262. New controllers such as the NCP1601 or NCP1653 include a dedicated OVP circuitry. Please note that a more comprehensive model of this MC33262 is available for download on the ON Semiconductor Web site (www.onsemi.com).

Figure 6-50b shows the simulation results obtained in 4 min on the 2.8 GHz computer. Most numbers agree with the theoretical calculations. The auxiliary V_{cc} is derived with a few turns across the inductor. Given the configuration, expect large variations on the controller V_{cc} pin as the PFC operates on universal mains. If the circuit does not clamp the gate-source voltage, it would be wise to safely clamp the V_{cc} to 15 V to (1) avoid the gate-source oxide breakdown and (2) limit $C_{iss}V_{GS}^2$ dissipation. The MC33262 accepts up to 30 V and safely clamps the V_{GS} to 15 V, so relax.

The input capacitor C_{in} plays a role for both EMI and the PFC operation as Ref. 24 highlights. As such, this capacitor should be placed after the diode bridge and not before. Acting as a reservoir, it avoids the boost input voltage to drop close to zero as the mains changes polarity. It thereby limits the duty cycle/switching frequency discontinuity. However, if it is increased too much, it opposes the smooth takeover between both mains positive and negative cycles, and the power factor suffers. The rule of thumb suggested by the above reference is 3 μF/kW. MKP types of capacitor (e.g., from WIMA [25]) are good candidates for these devices.

As we stated above, a 1 A diode such as a 1N4937 can suffer at power on, especially if the bulk capacitor has been increased for hold-up time reasons. If this is the case, you can use a bigger diode

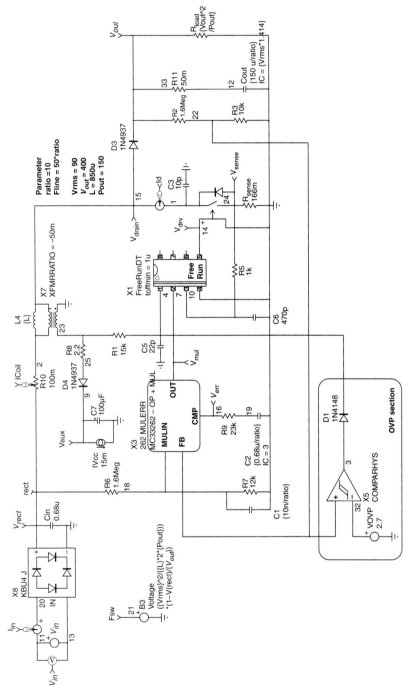

FIGURE 6-50a The cycle-by-cycle simulation of the MC33262-based 150 W PFC using the diode bridge.

SIMULATIONS AND PRACTICAL DESIGNS OF OFF-LINE CONVERTERS—THE FRONT END **573**

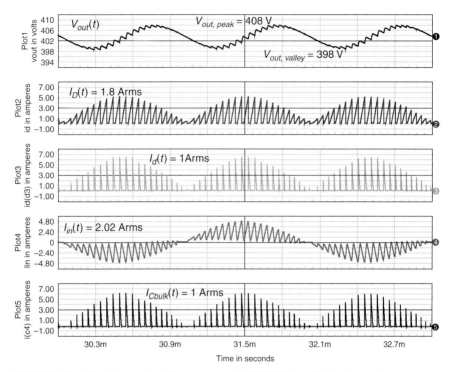

FIGURE 6-50b Simulation results of the cycle-by-cycle 150 W PFC for a 90 Vrms input voltage.

such as the MUR460 (4 A, 600 V) or install a bypass diode. This bypass element D_1 actually plays a role at start-up, but then remains silent the rest of the time. It absorbs the vast majority of the in-rush current and prevents the catch diode D_2 from undergoing current surges. Figure 6-50c shows how to install this external component. An MR756 (6 A, 600 V) can be selected.

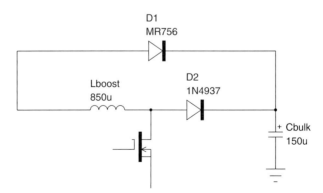

FIGURE 6-50c The bypass diode routes the in-rush current away from the catch diode.

6.3.4 The Follow-Boost Technique

For any linear or switching converter, performing the reduction or the elevation of the input voltage generates losses. It can be shown that the amount of generating losses greatly depends on the difference between the input and output levels: The smaller the difference, the better the efficiency. We have seen in Chap. 1 how true it was for linear regulators; it does not change significantly for switching converters. For a boost PFC, the efficiency really suffers at low line where the rectified peak of 130 Vdc (90 Vrms) must be raised to 400 Vdc. At the other extreme, the smaller step at high line (325 peak—230 Vrms—to 400 Vdc) eases the PFC burden. A certain category of downstream converters can accept operation over a wide input range (flybacks, for instance), but some do not accept it at all (forwards or resonant converters). For the first types, why not change the output voltage set point on the fly as the input mains varies? Several solutions are possible such as monitoring the input voltage and acting upon the feedback pin. ON Semiconductor purposely departed from this option with the MC33260 by driving the on time in relation to the output level squared (thus with respect to the output power). Let us see how it is performed.

In a boost PFC, the inductor peak current can be defined as follows:

$$I_{L,peak} = \frac{V_{ac}\sqrt{2}}{L} t_{on} \tag{6-193}$$

Thanks to the BCM operation, a straightforward relationship exists between the peak inductor current and the peak ac input current

$$I_{L,peak} = 2\sqrt{2} I_{ac} \tag{6-194}$$

where $I_{ac} = I_{in,rms}$. Combining the two above equations and extracting the on time give

$$t_{on} = \frac{2\sqrt{2} I_{ac} L}{\sqrt{2} V_{ac}} = \frac{2 I_{ac} L}{V_{ac}} \tag{6-195}$$

If now we redefine the ac current via the output power, we can update the equation:

$$t_{on} = \frac{2 P_{out} L}{\eta V_{ac} V_{ac}} = \frac{2 P_{out} L}{\eta V_{ac}^2} \tag{6-196}$$

The MC33260 operates in voltage mode and senses the output voltage via a current feedback technique. It thus uses a single resistor R_{FB} connected from a pin to the bulk capacitor. Based on its internal circuitry, the on time computed by the controller is expressed as

$$t_{on} = \frac{C_t R_{FB}^2}{k_{osc} V_{out}^2} \tag{6-197}$$

where k_{osc} and C_t, respectively, represent an internal constant linked to the circuit and an external timing capacitor.

If we now combine Eqs. (6-196) and (6-197), we obtain the MC33260 follower-boost equation:

$$V_{out} = R_{FB} V_{ac} \sqrt{\frac{C_t \eta}{2 L P_{out} k_{osc}}} \tag{6-198}$$

This equation shows that the output voltage linearly varies with the input voltage as the delivered power remains constant. If the output power changes, the voltage will automatically adjust to another level satisfying Eq. (6-198). Implemented in a boost PFC built with an MC33260, this particular equation will induce a behavior depicted by Fig. 6-51. We can see a linear increase of the output voltage at constant power. The slope at which the output voltage

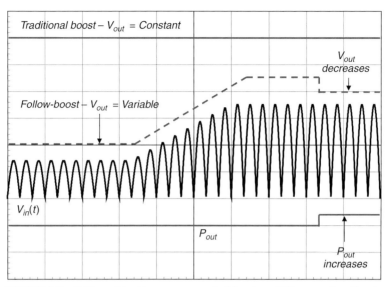

FIGURE 6-51 Unlike traditional boost converters, the follow-boost PFC output voltage linearly varies with the input voltage.

increases can be adjusted via the internal timer capacitor [the k_{osc} parameter in Eq. (6-197)]. By changing its value, we can force the PFC to enter regulation (the output voltage excursion is clamped) at a particular input level.

Thanks to a reduced differential between the input and output voltages, the inductor reset time increases. With a given peak current, a larger t_{off} implies a lower duty cycle, hence a reduced rms current. In other words, the follow-boost technique engenders less power dissipation in the MOSFET compared to the traditional boost PFC. Finally, we have seen that the inductor selection in a traditional BCM boost is mainly dictated by the minimum operating frequency [Eq. (6-162)]. Again, a benefit of the follow-boost mode lies in the expansion of the off time via a decrease in the reset voltage ($V_{reset} = V_{out} - V_{in}$). Thus, for a given frequency limit, a smaller inductor can be implemented. Combined, the above criteria lead to a lower cost PFC featuring a better efficiency!

WHAT I SHOULD RETAIN FROM CHAP. 6

The front-end stage of an off-line power supply represents an important portion that must not be neglected. We have often seen the bulk capacitor either oversized or selected on a cost basis only, without accounting for the rms current flowing into it. Needless to say, despite a good dc–dc design, the total lifetime suffers from this poor engineering practice. Here are the essentials you must keep in mind regarding the front-end stage:

1. You calculate the bulk capacitor value for a certain ripple amplitude definition, but the final selection is altered by the required rms current capability. The hold-up time also plays a role in this selection.
2. The peak rectification process generates a harmonic-rich input current which increases its rms content.

3. The high peak currents can distort the mains wave shape and affect the efficiency measurements as the network impedance varies.
4. Passive *LC* filters represent a solution to remove the main polluting harmonics from the current signal. This solution, however, hampers the power supply by making it heavier. It does not regulate the rectified output voltage.
5. Active solutions include PFC controllers driving a smaller inductor energized at a high switching frequency.
6. These PFCs can be made from various topologies operated in different modes: BCM/DCM for power levels up to 300 W and CCM above. In the consumer market, the borderline conduction mode boost power factor is presently one of the most popular structures.
7. To further improve the efficiency at low line, new PFCs implement a technique called the follow-boost. It automatically adjusts the output voltage to maintain a quasi-constant gap between the mains level and the regulated output voltage. High-power notebook adapters often implement this particular solution to offer excellent efficiency figures.

REFERENCES

1. M. Jovanović and L. Huber, "Evaluation of Flyback Topologies for Notebook AC–DC Adapters/Chargers Applications," *HFPC Proceedings*, May 1995, pp. 284–294.
2. http://www.illinoiscapacitor.com/
3. http://www.intusoft.com/nlpdf/nl29.pdf
4. S. M. Sandler, *Switchmode Power Supply Simulation with PSpice and SPICE3*, McGraw-Hill, New York, 2005.
5. Unitrode Power Supply Design Seminars SEM500, 1986.
6. http://www.ibiblio.org/obp/electricCircuits/AC/AC.pdf
7. http://www.boseresearch.com/RIGA_paper_27_JULY_04.pdf
8. F. C. Schwarz, "A Time Domain Analysis of the Power Factor for a Rectifier Filter System with Over and Subcritical Inductance," *IEEE Transactions on Industrial Electronics and Control Instrumentation*, vol. IECI-20, no. 2, pp. 61–68, May 1973.
9. S. B. Dewan, "Optimum Input and Output Filters for a Single-Phase Rectifier Power Supply," *IEEE Transactions on Industry Applications*, vol. IA-17, no. 3, pp. 282–288, May/June 1981.
10. A. W. Kelley and W. F. Yadusky, "Rectifier Design for Minimum Line Current Harmonics and Maximum Power Factor," *IEEE Applied Power Electronics Conference (APEC) Proceedings*, 1989, pp. 13–22.
11. R. Redl and L. Balogh, "Power Factor Correction in Bridge and Voltage-Doubler Rectifier Circuits with Inductors and Capacitors," *IEEE APEC Proceedings*, 1995, pp. 466–472.
12. M. Jovanović and D. Crow, "Merits and Limitations of Full-Bridge Rectifier with LC Filter in Meeting IEC 1000-3-2 Harmonic-Limit Specifications," *IEEE APEC Proceedings*, 1996. http://www.deltartp.com/dpel/conferencepapersdpel.html
13. R. Redl, "Low-Cost Line-Harmonic Reduction Techniques," tutorial course, HFPC '94, San Jose, California.
14. J. Spangler and A. K. Behara, "Electronic Fluorescent Ballast Using a Power Factor Correction Technique for Loads Greater Than 300 Watts," *APEC Proceedings*, 1991, pp. 393–399.
15. K. Sum, "An Improved Valley-Fill Passive Current Shaper," *Power System World*, 1997.
16. http://www.intusoft.com/nlpdf/nl68.pdf
17. Y. W. Lu et al., "A Large Signal Dynamic Model for DC-to-DC Converters with Average Current Mode Control," *APEC Proceedings*, 2004, pp. 797–803. http://www.ece.queensu.ca/hpages/labs/power/Publications.html

18. M. Jovanović and C. Zhou, "Design Trade-Offs in Continuous Current-Mode Controlled Boost Power-Factor Corrections Circuit," *HFPC Proceedings*, May 1992, p. 209. http://www.deltartp.com/dpel/conferencepapersdpel.html
19. L. Dixon, "Average Current-Mode Control of Switching Power Supplies," *Unitrode Switching Regulated Power Supply Design Seminar Manual*, SEM-700, 1990.
20. W. Tang, F. C. Lee, and R. B. Ridley, "Small-Signal Modeling of Average Current-Mode Control," *APEC Proceedings*, 1992, pp. 747–755.
21. J. Turchi, "Power Factor Correction Stages Operated in Critical Conduction Mode," Application note AND8123/D, www.onsemi.com
22. J. Turchi, "Simulating Circuits for Power Factor Correction," *PCIM Proceedings,* 2005, Nuremberg.
23. S. Ben-Yaakov and I. Zeltser, "Computer Aided Analysis and Design of Single Phase APFC Stages," *IEEE Applied Power Electronics Conference Professional Seminars,* 2003. http://www.ee.bgu.ac.il/~pel/public.htm
24. S. Maniktala, *Switching Power Supply Design and Optimization*, McGraw-Hill, New York, 2004.
25. www.wima.com
26. www.onsemi.com/pub/Collateral/HBD853-D.PDF

CHAPTER 7
SIMULATIONS AND PRACTICAL DESIGNS OF FLYBACK CONVERTERS

The flyback converter probably represents the most popular structure found on the market. The vast majority of consumer products make use of this converter: notebook adapters, DVD players, set-top boxes, satellite receivers, CRT TVs, LCD monitors, and so on. Three traits can justify this success: simplicity, ease of design, and low cost. For many designers, the flyback converter remains synonymous with poor EMI signature, large output ripple, and large-size transformers. After this fair warning, let us first discover how a flyback works before we go through several design examples.

7.1 AN ISOLATED BUCK-BOOST

If you remember the buck-boost converter (Fig. 1-38), you will certainly notice that the flyback takes its inspiration from this arrangement (Fig. 7-1).

The buck-boost delivers a negative voltage referenced to the input ground, without isolation. By swapping the inductor and the power switch, a similar arrangement is kept, but this time referenced to the input rail. Finally, coupling an inductor to the main one via a core, we obtain an isolated flyback converter. The secondary diode can be in the ground path (as shown), or in the positive wire, as more commonly encountered.

As with the nonisolated buck-boost, the energy is first stored from the input source during the on time. At the switch opening, the inductor voltage reverses and forward biases the catch diode, routing the inductor current to the output capacitor and the load. However, as the inductor and the load share the same ground, the output voltage is negative. On the flyback, a coupled-inductor configuration helps to adopt the needed polarity by playing on the winding dot positions and the diode orientation: the output voltage can be either positive or negative, above or below the input voltage, by adjusting the turns ratio. Physically separating the windings also brings galvanic isolation, needed for any mains-connected power supplies. *Galvanic?* Yes, this term relates to the physicist Luigi Galvani (1737–1798, Italy), who discovered the action of electric currents on nerves and muscles. So if you do not want to reiterate his experiments, you'd better watch the ability of the selected transformer to block any leakage current!

Further to this historical digression, let us purposely separate both the on and off events to see which exactly plays a role. Figure 7-2a portrays a parasitic-elements-free flyback when the power switch is closed. During this time, the voltage across the primary inductance L_p is equal

580 CHAPTER SEVEN

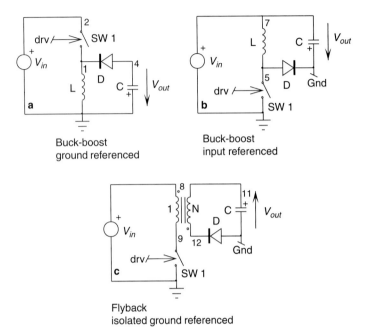

FIGURE 7-1 Rotating the buck-boost inductor leads to the flyback converter.

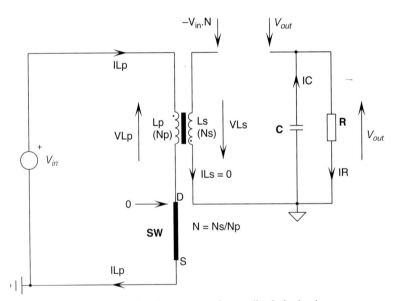

FIGURE 7-2a During the on time, the output capacitor supplies the load on its own.

to the input voltage (if we neglect the switch voltage drop). The current in this inductor increases at a rate defined by

$$S_{on} = \frac{V_{in}}{L_p} \tag{7-1}$$

Associating the on time t_{on} and the valley current, Eq. (7-1) can be updated to define the peak current value

$$I_{peak} = I_{valley} + \frac{V_{in}}{L_p} t_{on} \tag{7-2a}$$

Here I_{valley} represents the initial inductor current condition for $t = 0$, at the beginning of the switching cycle (if we are in CCM). If it is zero, as for the DCM case, we have

$$I_{peak} = \frac{V_{in}}{L_p} t_{on} \tag{7-2b}$$

During this time, there is no current flowing in the secondary side inductor. Why? Because of the winding dot configuration: the current enters on the primary side by the dot and should leave the secondary side by its dot as well. However, the diode presence prevents this current from circulating in this direction. During the on time, the dot arrangement on the transformer makes the diode anode swing negative, thus blocking it. As its cathode keeps to V_{out} thanks to the capacitor presence, the *peak inverse voltage* (PIV) undergone by the rectifier is simply

$$\text{PIV} = V_{in} N + V_{out} \tag{7-3}$$

where N is the turns ratio linking both inductors, equal to

$$N = \frac{N_s}{N_p} \tag{7-4}$$

We purposely adopted this definition for N to stay in line with the SPICE transformer representation where the ratio is normalized to the primary. An N ratio of 0.1 simply means we have a 1-to-0.1 turns relationship between the primary and the secondary, for instance, 20 turns on the primary and 2 turns on the secondary.

Because current does not circulate in both primary and secondary inductors at the same time, the term *transformer* is improper for a flyback converter but is often used in the literature (here as well), probably for the sake of clarity. A true transformer operation implies the simultaneous circulation of currents in the primary and secondary sides. For this reason, it is more rigorous to talk about coupled inductors. Actually, a flyback "transformer" is designed as an inductor, following Eq. (4A-20) recommendations.

When the PWM controller instructs the power switch to turn off, the voltage across the primary inductor suddenly reverses, in an attempt to keep the ampere-turns constant. The voltage developed across L_p now appears in series with the input voltage, forcing the upper switch terminal voltage (the drain for a MOSFET) to quickly jump to

$$V_{DS,off} = V_{in} + V_{L_p} \tag{7-5a}$$

However, as the secondary diode now senses a positive voltage on its anode (or a current circulating in the proper direction), it can conduct. Neglecting the diode forward drop V_f, the secondary-side-transformer terminal is now biased to the output voltage V_{out}. As a matter of fact, since both inductors are coupled, this voltage, translated by the turns ratio $1/N$, also appears on the primary side, across the primary inductor L_p. We say the voltage "flies" back across the transformer during the off time, as shown in Fig. 7-2b, hence the name *flyback*.

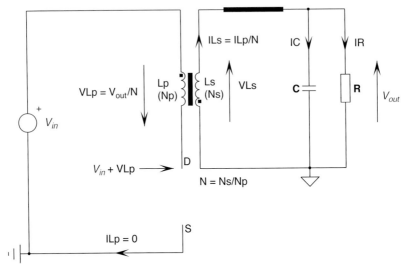

FIGURE 7-2b During the off time, the switch voltage jumps to the input voltage plus the primary inductor voltage.

The switch voltage at the opening is then

$$V_{DS,off} = V_{in} + V_{out}\frac{N_p}{N_s} = V_{in} + \frac{V_{out}}{N} = V_{in} + V_r \quad (7\text{-}5b)$$

where V_r is called the *reflected* voltage:

$$V_r = \frac{V_{out}}{N} \quad (7\text{-}5c)$$

The voltage applied across the inductor now being negative (with respect to the on-time sequence), it contributes to reset the core magnetic activity. The rate at which reset occurs is given by

$$S_{off} = -\frac{V_{out}}{NL_p} \quad (7\text{-}6)$$

By introducing the off-time definition, Eq. (7-6) can be updated to extract the valley current if the flyback operates in CCM:

$$I_{valley} = I_{peak} - \frac{V_{out}}{NL_p}t_{off} \quad (7\text{-}7a)$$

For the DCM case, where the inductor current goes back to zero, we have

$$I_{peak} = \frac{V_{out}}{NL_p}t_{off} \quad (7\text{-}7b)$$

If we now combine Eqs. (7-2a) and (7-7a), we can extract the dc transfer function of the CCM flyback controller:

$$I_{valley} = I_{peak} - \frac{V_{out}}{NL_p}t_{off} = I_{valley} + \frac{V_{in}}{L_p}t_{on} - \frac{V_{out}}{NL_p}t_{off} \quad (7\text{-}8)$$

Rearranging the above equations, we obtain

$$\frac{V_{out}}{NL_p}t_{off} = \frac{V_{in}}{L_p}t_{on} \quad (7\text{-}9)$$

Expressing the duty cycle D in relation to the on and off times, we reach the final definition for CCM:

$$\frac{V_{out}}{V_{in}} = \frac{Nt_{on}}{t_{off}} = \frac{NDT_{sw}}{(1-D)T_{sw}} = \frac{ND}{1-D} \quad (7\text{-}10)$$

It is similar to that of the buck-boost converter except that the turns ratio now takes place in the expression. Please note that inductor volt–seconds balance would have led us to the result in a much quicker way!

7.2 FLYBACK WAVEFORMS, NO PARASITIC ELEMENTS

To understand how typical signals evolve in a flyback converter, a simple simulation can be of great help. This is what Fig. 7-3a proposes. In this example, the transformer implements the simple T model described in App. 4-B. Figure 7-3b portrays the waveforms delivered by the simulation engine. If we now observe the input current, it is pulsating, exactly as a buck or a buck-boost input waveform. Looking at the diode current, we also observe a pulsating nature, confirming the bad-output-signature reputation of the flyback converter. The diode current appears at the end of the on time and jumps to a value depending on the turns ratio. These

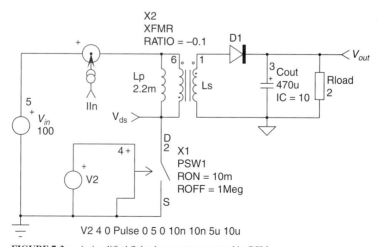

FIGURE 7-3a A simplified flyback converter operated in CCM.

584 CHAPTER SEVEN

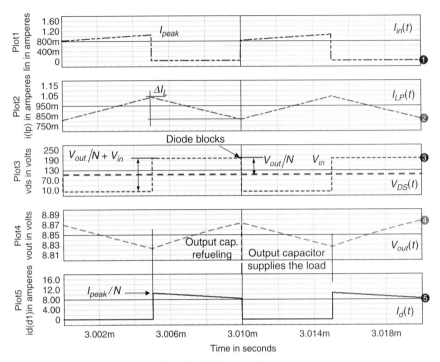

FIGURE 7-3b The simulation waveforms in CCM, no parasitic elements.

current discontinuities will naturally induce some output ripple on the capacitor, often further degraded by the presence of its ESR.

At the end of the off time, the upper switch terminal swings to a plateau level equal to that defined by Eq. (7-5b). Here, without a parasitic capacitor, the voltage transition is immediate. Given the 100 V input source, the switch jumps to around 200 V and stays there until the next on-time event. The rest of the waveforms are similar to those obtained with a buck-boost converter operated in CCM.

Calculating the average value of the input current will lead us to a very useful design equation, fortunately already derived in Chap. 5, in the buck-boost section, Eqs. (5-108) to (5-114). We can, however, derive a similar equation by observing the energy mechanisms. As the switch turns on, the inductor current is already at the valley point. The energy initially stored is thus

$$E_{L_p,valley} = \frac{1}{2}L_p I_{valley}^2 \tag{7-11}$$

At the end of the on time, the current has reached its peak value. The newly stored energy in the inductor then becomes

$$E_{L_p,peak} = \frac{1}{2}L_p I_{peak}^2 \tag{7-12}$$

Finally, the total energy accumulated by the inductor is found by subtracting these two equations:

$$E_{L_p,accu} = \frac{1}{2}L_p I_{peak}^2 - \frac{1}{2}L_p I_{valley}^2 = \frac{1}{2}L_p\left(I_{peak}^2 - I_{valley}^2\right) \tag{7-13}$$

Power (watts) is known to be energy (joules) averaged over a switching cycle. Given the converter efficiency η, the transmitted power is simply

$$P_{out} = \frac{1}{2}\left(I_{peak}^2 - I_{valley}^2\right)L_p F_{sw} \eta \qquad (7\text{-}14)$$

If during operation the load is reduced, the converter enters DCM, allowing the valley current to reach zero. We have simulated this behavior, and the updated set of curves is seen in Fig. 7-3c. The peak drain voltage remains the same, but as the primary-inductance-stored energy drops to zero, the secondary diode suddenly blocks. At this point, the core is said to be reset, fully demagnetized. Without diode conduction, the primary reflected voltage disappears, and the drain goes back to the input voltage. A third state prolongs the off time, where no current circulates in the primary branch. This is the dead time portion, already discussed in the small-signal chapter. In this case, Eq. (7-14) can be updated by setting the valley current to zero:

$$P_{out} = \frac{1}{2} I_{peak}^2 L_p F_{sw} \eta \qquad (7\text{-}15)$$

In both CCM and DCM plots, the second upper curve depicts the primary (or magnetizing) inductor current. This inductor stores energy during the on time and dumps its charge to the secondary side during the off time. On a real flyback circuit, it is impossible to observe this current as the magnetizing inductance cannot be accessed alone. You will thus use the input current to characterize the transformer rms and peak current stresses.

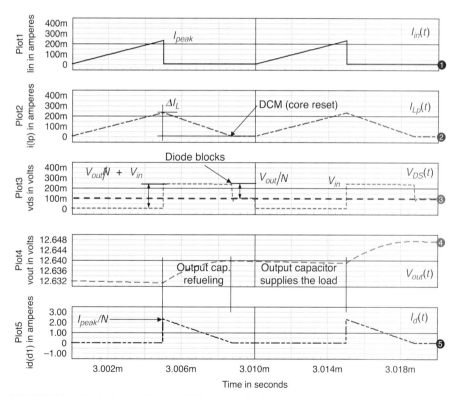

FIGURE 7-3c Simulation waveforms in DCM, no parasitic elements. As one can see on the output voltage, the converter is not fully stabilized at the time we captured the shot.

In the above examples, we ran the converters open loop, as exemplified by Fig. 7-3a. Despite similar duty cycles, the output level naturally varies as we operate in two different modes. Equation (7-10) defines the relationship between the output and the input voltage for the CCM mode. The link between V_{out} and V_{in} in DCM mode appears either in the buck-boost section [Eq. (1-143), just replace L by L_p and R_{load} by R_{load}/N^2] or in the DCM voltage-mode section, a few lines away.

7.3 FLYBACK WAVEFORMS WITH PARASITIC ELEMENTS

As perfection is not of this world, all our devices are affected by parasitic elements:

- The transformer features various capacitors, split between the windings and the primary inductance. We can lump all the capacitors in a single capacitor C_{lump} connected from the drain node (node 8) to ground.
- The coupling between the primary side and the secondary side is imperfect. Not all the stored energy in the primary flies to the secondary side. The symbol of such a loose coupling is the perfidious leakage inductance, a real plague of all flyback designs. In the example, we arbitrarily selected it to be 2% of the primary inductance, a rather poor transformer construction.
- The secondary diode also includes a certain amount of capacitance, especially if you use a Schottky. Otherwise, in CCM, the t_{rr} of a standard PN diode acts as a brief short-circuit seen from the primary side as a sharp current spike. The Schottky capacitance is reflected to the primary and included in C_{lump}.

Let us update Fig. 7-3a with the aforementioned parameters. Figure 7-4a shows the new converter. The simulation results are shown in Fig. 7-4b. We can see a lot of parasitic oscillations, playing

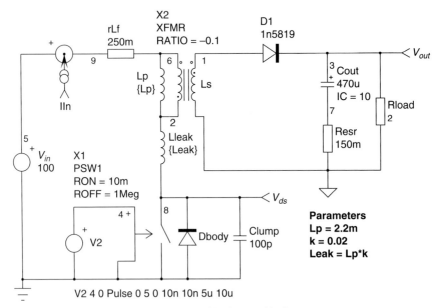

FIGURE 7-4a The updated flyback converter with a few parasitic elements.

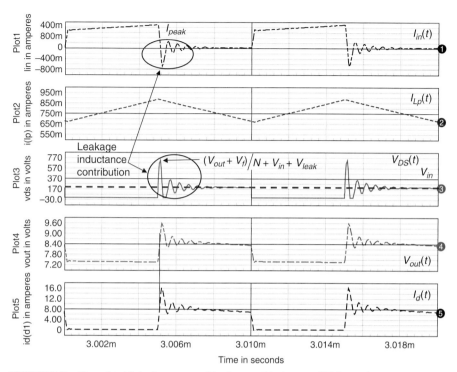

FIGURE 7-4b The updated flyback converter with a few parasitic elements, CCM operation.

an obvious role in the radiated EMI signature. Depending on their amplitude, these ringing waveforms often need to be attenuated by dampers. The output voltage now includes the capacitor ESR contribution, an effect of the current discontinuity at the switch opening. The reflected voltage includes the diode forward drop, as it sums up with the output voltage on the secondary side upper terminal. When the PWM controller biases the switch, the primary current also flows in the leakage inductance where energy is stored. When the PWM controller blocks the switch, this current needs to flow somewhere until this leakage inductance gets reset: it flows in the lump capacitor C_{lump} and makes the voltage very high. If no precautions are taken, the power switch (usually a MOSFET) can be destroyed by voltage breakdown. As we combine an inductor (the leakage term) and a capacitor, we obtain an oscillating LC network. This is the ringing you can see superimposed on all the internal signals. Very often, you need to clamp the voltage excursion, and also the oscillating wave amplitude, by an external damper. The leakage also delays the current flow into the output; this is seen as slowed-down edges on the diode current.

By increasing the load resistance and slightly shifting down the switching frequency, we can force a DCM operation. Figure 7-4c depicts the signals. They are very similar to those in the previous shot. As the peak current decreases, the voltage excursion on the switch is also reduced. Once the leakage ringing has gone, we enter the plateau region. During this time, as the core demagnetization has started, the primary inductance current drops. When it reaches zero, the secondary diode naturally blocks ($I_d = 0$). An oscillation takes place, involving the lump capacitor and the primary inductance L_p. The drain voltage freely rings from peaks to valleys, and when the switch closes again, the voltage brutally goes back to ground. If we wait to be in a valley before activating the switch, the converter is told to operate in "valley switching" mode. This is also called *borderline control*, which has several drawbacks such as frequency variations and noise inherent to valley jumping hesitations.

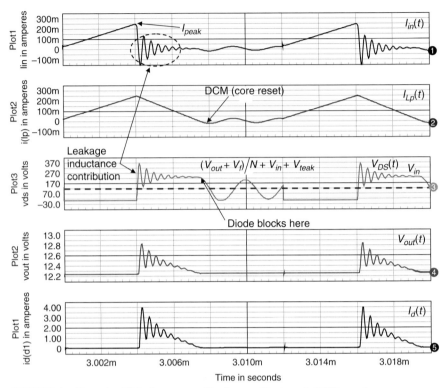

FIGURE 7-4c The updated flyback converter with a few parasitic elements, DCM operation.

7.4 OBSERVING THE DRAIN SIGNAL, NO CLAMPING ACTION

As shown in Fig. 7-5a, when a current flows in the primary inductance, it also circulates in the leakage inductance. During the on time, the current builds up in both terms until the PWM controller stops the switch: the primary current has reached its value I_{peak}. The secondary side circuit does not appear as the diode is blocked.

Immediately after the switch opening, Fig. 7-5b shows a simplified equivalent network. Both inductors are energized, and we assume here that the secondary diode has immediately started to conduct. The primary inductor is therefore replaced by a current source, in parallel with a voltage source whose value equals the reflected level from the secondary side. The leakage inductance is modeled by a source of I_{peak} value, routing its current through the lump capacitor. The capacitor upper terminal voltage quickly rises at a slope depending on I_{peak} and the capacitor value. On the rising edge of the signal, the slope is classically defined by

$$\frac{dV_{DS}(t)}{dt} = \frac{I_{peak}}{C_{lump}} \qquad (7\text{-}16)$$

The maximum voltage at which the capacitor voltage will ring depends on the LC network characteristic impedance Z_0. This term is made of the leakage inductance and the lump capacitor. Measured from the ground node, the drain will thus undergo the capacitor voltage in series

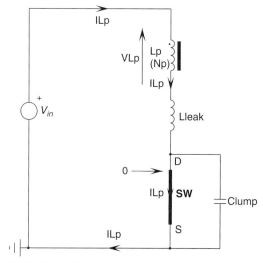

FIGURE 7-5a When the switch closes, the current flows through both the magnetizing inductor and its associated leakage term.

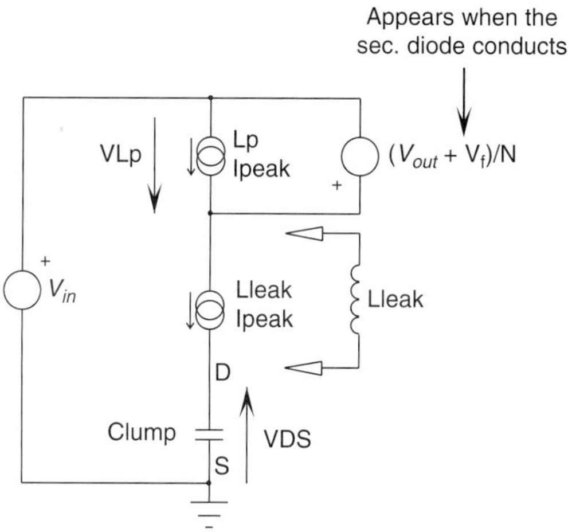

FIGURE 7–5b The leakage inductance charges the lump capacitor and makes the drain voltage abruptly go up.

with the sum of the reflected voltage (V_r, now considering the diode V_f) and the input source level V_{in}:

$$V_{DS,max} = V_{in} + V_r + I_{peak} Z_0 = V_{in} + \frac{V_{out} + V_f}{N} + I_{peak}\sqrt{\frac{L_{leak}}{C_{lump}}} \quad (7\text{-}17)$$

If no precaution is taken, risks exist to avalanche the power MOSFET by exceeding its BV_{DSS} and destroy it through an excess of dissipated heat. Artificially increasing the lump capacitor offers a simple way to limit the excursion at the switch opening and to protect the semiconductor (see the third design example). To illustrate this method, Fig. 7-5c depicts a flyback converter

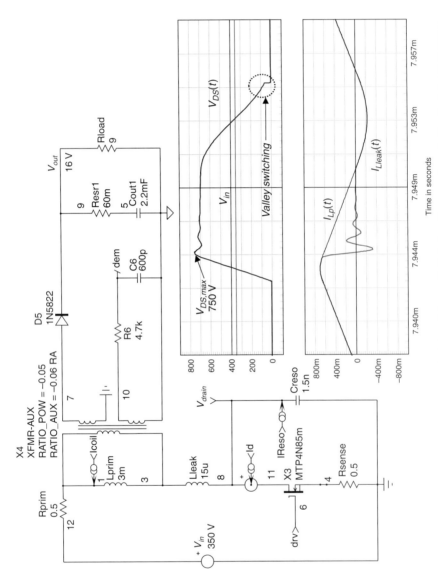

FIGURE 7-5c When a large capacitor connects to the drain node, it naturally limits the voltage excursion and softens the drain–source voltage.

operated in quasi-resonance (QR), also called borderline or valley switching operation. In this mode, the controller detects the valley in the drain voltage waveform and turns on the MOSFET, right in the minimum. Thus all capacitive losses are greatly diminished (if not removed) since a natural discharge of the lump capacitor occurs. Artificially increasing the capacitive term, here up to 1.5 nF, brings several advantages: (1) it slows down the rising voltage for a weakly radiating signal (a soft wave, low dV/dt) and (2) it limits the maximum excursion, allowing the use of an 800 V MOSFET at high line. As the graphs demonstrate, the drain voltage peaks to 751 V, as Eq. (7-17) predicted:

$$V_{DS,max} = 350 + \frac{16.6}{0.05} + 0.695\sqrt{\frac{15u}{1.5n}} = 751 \text{ V} \qquad (7\text{-}18)$$

A 1.5 nF capacitor connected to the drain does not come for free since it must accommodate high-voltage pulses and accept a certain level of rms current. Furthermore, if valley switching is not properly ensured by the controller, unacceptable CV^2 losses will degrade the converter efficiency.

In low-power applications, some designers adopt this method to limit the voltage excursion and do not use a standard clamping network. The solution, in certain conditions, offers a cost advantage, but care must be taken to respect the MOSFET breakdown limit, especially in high-surge input conditions. Low-cost cell phone travel chargers that typically run less than $1 do not really care about it.

Another solution is to use a more traditional clamping network.

7.5 CLAMPING THE DRAIN EXCURSION

The clamping network safely limits the drain amplitude by using a low-impedance voltage source V_{clamp}, hooked via a fast diode to the high-voltage input voltage rail V_{in} (Fig. 7-6a). In practice, the low-impedance voltage source is made by an *RCD* clamping network or a *transient voltage suppressor* (TVS). For the sake of the current explanation, we will consider the V_{clamp} source as a simple voltage source.

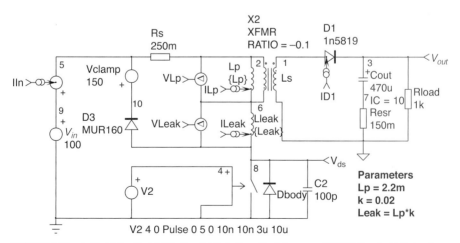

FIGURE 7-6a A clamping network (D_3, V_{clamp}) efficiently protects the power MOSFET against lethal voltage excursions.

When the switch opens, the voltage sharply rises, solely curbed by the drain lump capacitor. When the drain reaches a voltage equal to the sum of the input voltage and the clamping source, the series diode conducts and sharply cuts the drain-source excursion. Given a carefully selected clamping source level, the maximum excursion stays below the MOSFET breakdown voltage and ensures a reliable operation in worst-case conditions.

When the diode conducts, the equivalent schematic looks like that of Fig. 7-6b. Figure 7-6b is a more complex representation than Fig. 7-6a, as we consider the secondary diode not immediately conducting at the switch turn-off time. When the switch opens, all the magnetizing current also flows through the leakage inductance and equals I_{peak}. Both primary and leakage voltages reverse, in an attempt to keep the ampere-turns constant: the voltage across the drain starts to quickly rise, solely limited in growth speed by the lump capacitor [Eq. (7-16)]. On the secondary side, the voltage on the rectifying diode anode transitions from a negative voltage (the input voltage reversed and translated by the turns ratio N) to a positive voltage. When the drain exceeds the input voltage by the flyback voltage V_r, the secondary diode is forward biased. However, as drawn in Fig. 7-6b, the leakage current subtracts from the primary inductor current. The secondary side diode, despite its forward bias, cannot instantaneously attain a current equal to I_{peak}/N, since the initial primary current subtraction leads to zero ($I_{leak} = I_{peak}$). The current on this secondary side diode can only increase at a rate imposed by the leakage inductance and the voltage across it (Fig. 7-6c). Following its rise, the drain voltage quickly reaches a level theoretically equal to $V_{in} + V_{clamp}$ (neglecting the clamp diode forward drop and its associated overshoot): diode D_3 conducts and blocks any further excursion. The lump capacitor current goes to zero as its upper terminal voltage (the drain) is now fixed. The graph update appears in Fig. 7-6d. A reset voltage now applies across the leakage inductor terminals. Considering all voltages constant, we have

$$V_{reset} = V_{clamp} - \frac{V_{out} + V_f}{N} \qquad (7\text{-}19)$$

The current flowing through the leakage inductor immediately starts to decay from its peak value at a rate given by

$$S_{L_{leak}} = \frac{V_{reset}}{L_{leak}} = \left[V_{clamp} - \frac{V_{out} + V_f}{N} \right] \frac{1}{L_{leak}} \qquad (7\text{-}20)$$

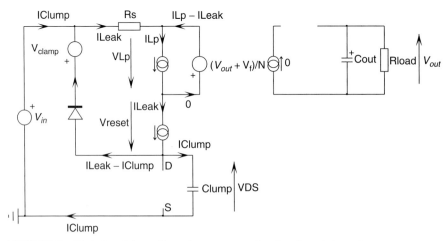

FIGURE 7-6b The leakage inductor diverts the current away from the primary inductor L_p.

SIMULATIONS AND PRACTICAL DESIGNS OF FLYBACK CONVERTERS 593

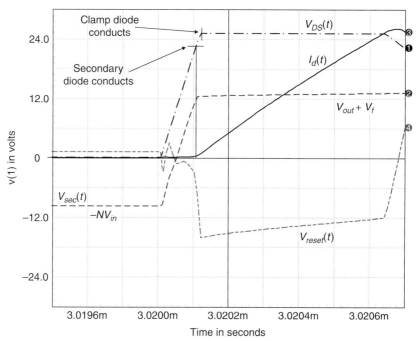

FIGURE 7-6c At the switch opening, the secondary voltage rises at a similar pace to that of the drain. The secondary diode is forward biased, and the secondary current starts to circulate.

The secondary current is no longer null and follows the equation

$$I_{sec}(t) = \frac{I_{L_p}(t) - I_{leak}(t)}{N} \tag{7-21}$$

When the leakage inductance current reaches zero, the clamp diode blocks: the secondary current is at its maximum, slightly below the theoretical value at the switching opening. Why?

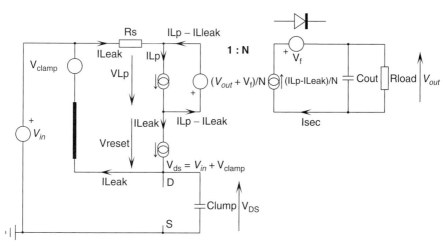

FIGURE 7-6d When the clamp diode conducts, a voltage is applied over the leakage inductor and resets it.

Because as long as the leakage inductor energy decays, it diverts current from the primary inductor. Figure 7-6e shows a representation of the currents in play, precisely at the switch turn-off.

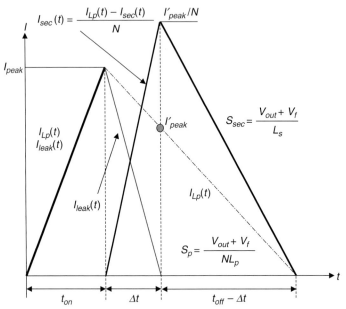

FIGURE 7-6e At turn-off, the leakage inductance delays the occurrence of the secondary side current.

In other words, the secondary diode current does not peak to I_{peak}/N, as it should with a zero leakage term, but to another value I'_{peak} reached when the leakage inductor is reset. This current value, on the primary side, can be calculated as follows, where S_p is the primary reset slope:

$$I'_{peak} = I_{peak} - S_p \Delta t = I_{peak} - \frac{V_{out} + V_f}{NL_p} \Delta t \qquad (7\text{-}22)$$

Now Δt corresponds to the time needed by the leakage inductor current to fall from I_{peak} to zero. Thus, we have

$$\Delta t = \frac{I_{peak}}{S_{L_{leak}}} = I_{peak} \frac{L_{leak}}{V_{reset}} = \frac{NL_{leak} I_{peak}}{NV_{clamp} - (V_{out} + V_f)} \qquad (7\text{-}23)$$

If we substitute Eq. (7-23) into (7-22), we obtain

$$I'_{peak} = I_{peak} - \frac{V_{out} + V_f}{NL_p} \frac{NL_{leak} I_{peak}}{NV_{clamp} - (V_{out} + V_f)} = I_{peak} \left[1 - \frac{L_{leak}}{L_p} \frac{1}{\frac{NV_{clamp}}{V_{out} + V_f} - 1} \right] \qquad (7\text{-}24)$$

A good design indication can be obtained from the ratio I'_{peak}/I_{peak}:

$$\frac{I'_{peak}}{I_{peak}} = \left[1 - \frac{L_{leak}}{L_p}\frac{1}{\frac{NV_{clamp}}{V_{out}+V_f}-1}\right] \quad (7\text{-}25)$$

To obtain a ratio close to 1, the obvious way is to reduce the leakage inductance term in comparison with the primary inductor. We can also speed up the leakage reset by satisfying the following equation:

$$V_{clamp} > \frac{V_{out}+V_f}{N} \quad (7\text{-}26)$$

Thus, by allowing a higher drain-source voltage excursion, within safe limits of course, you will naturally (1) reset the leakage inductor more quickly and (2) reduce the stress on the clamping network, both results leading to an overall better efficiency.

Following Eq. (7-24), the peak current seen by the diode on the secondary side is simply

$$I_{sec} = \frac{I'_{peak}}{N} = \frac{I_{peak}}{N}\left[1 - \frac{L_{leak}}{L_p}\frac{1}{\frac{NV_{clamp}}{V_{out}+V_f}-1}\right] \quad (7\text{-}27)$$

To test our analytical study, Fig. 7-6f portrays the simulated results of Fig. 7-6a. The peak current at the switch opening reaches 236 mA, and the secondary current at the end of the

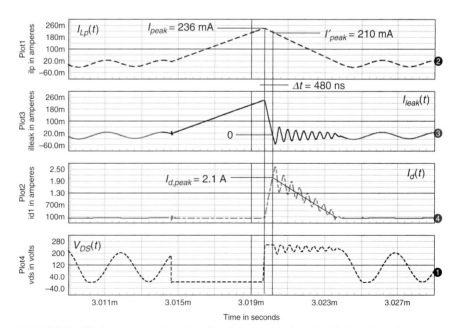

FIGURE 7-6f The simulation involving the leakage inductance confirms the delay brought by this term.

leakage reset time is approximately 2.1 A. Without the leakage term, the theoretical secondary current would have been 2.36 A. The reduction in current is thus around 11%, or 1.2% in power (I_{peak}^2), compensated by the feedback loop driving a slightly larger duty cycle. The converter efficiency obviously suffers from this situation.

Using Eq. (7-23), we can calculate the delay introduced by the leakage inductance, or its reset time:

$$\Delta t = \frac{NL_{leak}I_{peak}}{NV_{clamp} - (V_{out} + V_f)} = \frac{0.1 \times 44u \times 235m}{0.1 \times 150 - 13} = 520 \, ns \quad (7\text{-}28)$$

At the leakage inductor reset point, the primary current has fallen to

$$I'_{peak} = I_{peak}\left[1 - \frac{L_{leak}}{L_p}\frac{1}{\frac{NV_{clamp}}{V_{out} + V_f} - 1}\right] = 0.235 \times \left[1 - \frac{44u}{2.2m}\frac{1}{\frac{15}{13} - 1}\right] = 204 \, mA \quad (7\text{-}29)$$

The results delivered by the simulator are in good agreement with our calculations, despite slight discrepancies. They are mainly due to the clamp diode recovery time and the nonconstant reset voltage over the leakage inductor (reflected output ripple). If we reduce the clamping voltage to 140 V, the reset time increases to 800 ns, confirming the prediction of Eq. (7-26).

When the clamp diode blocks, the leakage inductor is completely reset. The new schematic becomes as shown in Fig. 7-7. At the diode turn-off point, the drain should normally drop to the reflected voltage plus the input voltage. However, the lump capacitor being charged to the clamp voltage, it cannot instantaneously go back to the plateau level. An oscillating energy

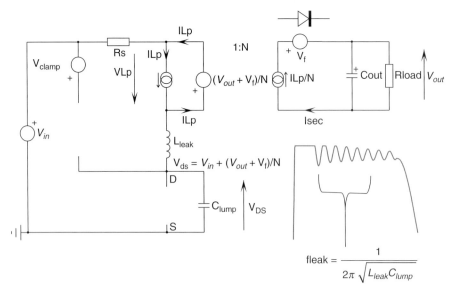

FIGURE 7-7 When the clamp diode blocks, a resonance occurs between the leakage inductor and the lump capacitor.

exchange takes place between L_{leak} and C_{lump}, making the drain freely oscillating at a frequency imposed by the network:

$$f_{leak} = \frac{1}{2\pi\sqrt{L_{leak}C_{lump}}} \tag{7-30}$$

Given the values indicated in Fig. 7-6a, the network rings at 2.4 MHz. Thanks to the ohm losses in the oscillating path, we have an exponentially decaying waveform which can radiate EMI. By inserting a resistor in the clamping network, we have a chance to damp these oscillations, as we will see later. Sometimes, the ringing is so severe that it falls below ground and forward biases the MOSFET body diode, resulting in additional losses. An RC network must then be installed across the primary side to damp these oscillations.

7.6 DCM, LOOKING FOR VALLEYS

As long as the secondary diode conducts, a reflected voltage appears across the primary inductor and brings its current down. If a new switching cycle occurs, whereas current still flows in the secondary, the primary inductor enters CCM and the rectifying diode gets brutally blocked as the switch closes. Depending on the technology of this diode, the sudden blocking engenders losses such as those related to t_{rr}. If the diode is a PN junction type, it is seen as a short-circuit reflected to the primary, with a big spike on the current-sense element (hence the need for a leading edge blanking (LEB) circuitry — see Chap. 4 for this), until it recovers its blocking capabilities. If this is a Schottky, there is no recovery time but there is an equivalent nonlinear large capacitor across the diode terminals. This capacitor also brings primary losses but to a lesser extent. As an aside, you have read that a Schottky diode does not feature t_{rr} but sometimes data sheets still specify it for these diodes. How can it be? In this case, t_{rr} actually occurs because the so-called guard rings, placed to avoid arcing on the diode die corners, are activated when the Schottky forward drop reaches the guard-ring equivalent PN forward voltage. In that case, these rings get activated, and this is the t_{rr} you see on the scope. It usually occurs at a high forward current.

Now, if the flyback enters DCM, the secondary diode will naturally turn off, without significant losses. At that point, the reflected voltage on the primary side disappears. Figure 7-8 depicts the new circuit. The lump capacitor is left charged to the reflected voltage plus the input voltage. However, we again have a resonating network made up of the inductors $L = L_p + L_{leak}$ and the lump capacitor. A decaying oscillation thus takes place between C_{lump} and L and lasts until all the stored energy has been dissipated in the inductor and capacitor ohmic terms (the resistor R_s in the circuit). The waveform temporal evolution can be described by using Eq. (1-171a). In our case, the drain node "falls" from the lump capacitor voltage and converges toward the input voltage:

$$V_{DS}(t) \approx V_{in} + V_r \frac{e^{-\zeta\omega_0 t}}{\sqrt{1-\zeta^2}}\cos(\omega_0 t) \tag{7-31a}$$

where

$$\omega_0 = \frac{1}{\sqrt{(L_p + L_{leak})C_{lump}}} \tag{7-31b}$$

is the undamped natural oscillation and

$$\zeta = R_s\sqrt{\frac{C_{lump}}{4(L_p + L_{leak})}} \tag{7-31c}$$

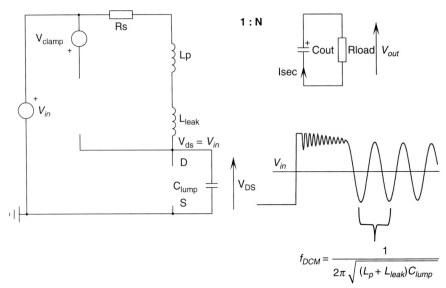

FIGURE 7-8 As DCM occurs, the drain rings to a frequency depending on the primary inductor in series with the leakage term and the lump capacitor.

represents the damping factor. V_r is the reflected voltage already seen and equals

$$V_r = \frac{V_{out} + V_f}{N} \tag{7-31d}$$

If we replace zeta in Eq. (7-31a) by its definition via Eq. (7-31c), then considering the damping factor small enough compared to 1 in the numerator, we can rewrite Eq. 7-31a in a more readable form:

$$V_{DS}(t) \approx V_{in} + V_r e^{-\frac{R_s}{2L}t} \cos(\omega_0 t) \tag{7-32}$$

with $L = L_p + L_{leak}$. The oscillating frequency comprises the leakage term plus the primary inductor as both devices now appear in series:

$$f_{DCM} = \frac{\omega_0}{2\pi} = \frac{1}{2\pi \sqrt{(L_p + L_{leak})C_{lump}}} \tag{7-33}$$

Figure 7-9 depicts the drain waveform in a more comprehensive form. We can see the leakage effect, with its associated ringing signal, followed by the plateau region, which is evidence for the secondary diode conduction. As the diode blocks, implying the core reset, an exponentially decaying oscillation takes place, alternating peaks and valleys. In quasi-resonance (QR) operation, the controller is able to detect the occurrence of these valleys. When in such valley, the drain level is minimum. If by design the reflected voltage equals the input voltage, then the waveform rings down to the ground. Assuming the PWM controller turns the MOSFET on right at this moment, *zero voltage switching* (ZVS) is ensured, removing all drain-related capacitive losses. This is what Fig. 7-5c already showed. The point at which the first valley occurs can be easily predicted by observing a −1 polarity of

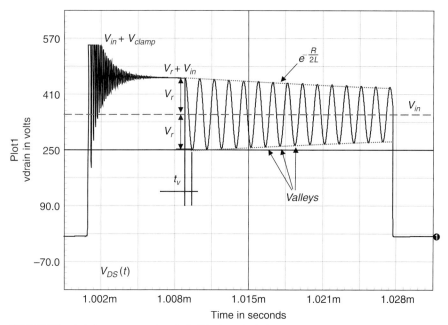

FIGURE 7-9 A drain-source waveform in DCM where the leakage ringing clearly shows up at the switch opening.

the cosine term in Eq. (7-32):

$$\cos(\omega_0 t_v) = \cos(2\pi f_0 t_v) = -1 \tag{7-34}$$

Solving this equation implies that $2\pi f_0 t_v = \pi$. Extracting t_v, gives

$$t_v = \frac{1}{2f_0} = \frac{2\pi\sqrt{(L_p + L_{leak})C_{lump}}}{2} = \pi\sqrt{(L_p + L_{leak})C_{lump}} \tag{7-35}$$

Thus, in a quasi-square wave resonant design, if a delay of t_v μs is inserted at the DCM detection point, the controller will turn the MOSFET on right in the minimum of its drain-source voltage.

By observing these parasitic ringings, it is possible to derive the transformer main element values. Appendix 7A shows how to do it.

7.7 DESIGNING THE CLAMPING NETWORK

The clamping network role is to prevent the drain signal from exceeding a certain level. This level has been selected at the beginning of the design stage, given the type of MOSFET you are going to use and the derating your quality department (or yourself!) imposes. Figure 7-10a depicts a waveform observed on the drain node of a MOSFET used in a flyback converter. The voltage sharply rises at the switch opening and follows a slope already defined by Eq. (7-16).

600 CHAPTER SEVEN

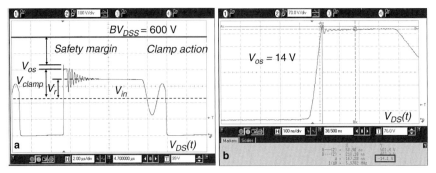

FIGURE 7-10 A typical DCM drain voltage showing the various voltage levels and the required safety margin.

In this particular example, the total drain node capacitor was measured at 130 pF, and the peak current at turn-off was 1.2 A. Equation (7-16) predicts a slope of

$$\frac{dV_{DS}(t)}{dt} = \frac{I_{peak}}{C_{lump}} = \frac{1.2}{130p} = 9.2 \text{ kV/}\mu\text{s} \tag{7-36}$$

This is exactly what is shown in Fig. 7-10b with a voltage swing of nearly 500 V in 50 ns. The clamp diode needs a certain amount of time before it actually conducts and blocks the drain excursion. It is interesting to note that the turn-on time of a PN diode can be extremely fast, whatever its type. A few nanoseconds is a typical number. What differs a lot between PN diodes is the blocking mechanism that can take several hundreds of nanoseconds (even microseconds for the 1N400X series) to let the crystal recover its electrical neutrality. (The recovery time is denoted by t_{rr}.) Here, zooming on the top of the waveform reveals an overshoot V_{os} of only 14 V, obtained with an MUR160. Later we will see different overshoots resulting from slower diodes such as the 1N4937 or even the 1N4007.

Accounting for this spike, we can now select the clamping voltage. Suppose we selected a MOSFET affected by a 600 V BV_{DSS} that is going to be used in a flyback converter supplied by a 375 Vdc (265 Vrms maximum) input source. Applying a derating factor k_D of 15% to the acceptable drain-source voltage (see App. 7B), we have the following set of equations, actually based on Fig. 7-10a indications.

- $BV_{DSS}k_D = 600 \times 0.85 = 510$ V   maximum allowable voltage in worst case
- $V_{os} = 20$ V   estimated overshoot attributed to the diode
- $V_{clamp} = BV_{DSS}k_D - V_{os} - V_{in} = 115$ V   selected clamping voltage (7-37a)

The 115 V clamping level is fixed and must be respected to keep the MOSFET operating in a safe area.

Now, let us define the transformer turns ratio. Equation (7-26) instructs to keep the clamping level well above the reflected voltage to authorize the quickest leakage term reset. In practice, satisfying this equation would imply high drain excursions (hence a more expensive 800 V component) or very low reflected voltages. Experience shows that a clamping voltage selected to be between 30 and 100% (the parameter k_c varying between 1.3 to 2) of the reflected voltage gives good results on efficiency figures. Let us assume we have a 12 V output adapter together with a rectifying diode featuring a V_f of 1 V and a clamp coefficient k_c of 2 (100%). To carry on with the design example, the transformer turns ratio N would simply be

$$V_{clamp} = k_c \frac{V_{out} + V_f}{N} \tag{7-37b}$$

Solving for N yields

$$N = \frac{k_c(V_{out} + V_f)}{V_{clamp}} = \frac{2(12 + 1)}{115} = 0.226 \qquad (7\text{-}38)$$

We now have an idea of the clamping voltage and the turns ratio affecting the transformer. Building of a low-impedance source connected to the bulk capacitor can be achieved via several possibilities, among which the *RCD* clamp represents the most popular solution.

7.7.1 The *RCD* Configuration

Figure 7-11 depicts the popular *RCD* clamping network. The idea behind such a network is to create a low-impedance voltage source of V_{clamp} value hooked to the bulk capacitor. This is a familiar structure, already depicted by Fig. 7-6a. The resistor R_{clp} dissipates power linked to the stored leakage energy, whereas the capacitor C_{clp} ensures a low ripple equivalent dc source. The clamping network dissipated power will increase as V_{clamp} approaches the reflected voltage. It is thus important to carefully select the clamping factor k_c, as you will see below.

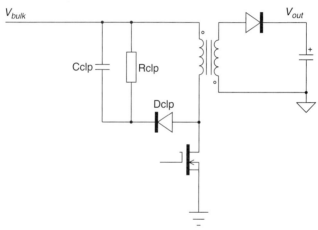

FIGURE 7-11 A typical clamping network made of a diode and two passive elements.

Looking back to Fig. 7-6d and 7-6e, we see the current circulating in the diode D_{clp} when it conducts (therefore when the drain voltage hits the clamp) can be expressed as

$$I_d(t) = I_{peak} \frac{\Delta t - t}{\Delta t} \qquad (7\text{-}39)$$

This current circulates until the leakage inductance is fully reset. The time needed for the reset is denoted by Δt and was already described by Eq. (7-23). If we now consider the clamping voltage constant (actually our dc source V_{clamp} previously used) during Δt, the average power dissipated by the clamping source is

$$P_{V_{clamp},avg} = F_{sw} \int_0^{\Delta t} V_{clamp} I_d(t) \cdot dt = F_{sw} V_{clamp} \int_0^{\Delta t} I_{peak} \frac{\Delta t - t}{\Delta t} \cdot dt \qquad (7\text{-}40)$$

Solving this integral gives us the power dissipated over a switching cycle

$$P_{V_{clamp},avg} = \frac{1}{2} F_{sw} V_{clamp} I_{peak} \Delta t \tag{7-41}$$

Now using Eq. (7-23) and substituting it into Eq. (7-41), we obtain a more general formula showing the impact of the reflecting voltage and the clamp level:

$$P_{V_{clamp},avg} = \frac{1}{2} F_{sw} L_{leak} I_{peak}^2 \frac{V_{clamp}}{V_{clamp} - \frac{V_{out} + V_f}{N}} = \frac{1}{2} F_{sw} L_{leak} I_{peak}^2 \frac{k_c}{k_c - 1} \tag{7-42}$$

Our dc source is made of a resistor and a capacitor in parallel. As all the average power is dissipated in heat by the resistor, we have the following equality:

$$\frac{V_{clamp}^2}{R_{clp}} = \frac{1}{2} F_{sw} L_{leak} I_{peak}^2 \frac{V_{clamp}}{V_{clamp} - \frac{V_{out} + V_f}{N}} \tag{7-43}$$

Extracting the clamp resistor from the above line leads us to the final result:

$$R_{clp} = \frac{2V_{clamp}\left[V_{clamp} - \frac{V_{out} + V_f}{N}\right]}{F_{sw} L_{leak} I_{peak}^2} \tag{7-44}$$

The capacitor ensures a low-ripple ΔV across the *RCD* network. Figure 7-12 depicts the capacitor waveforms, its current, and voltage. If we assume all the peak current flows in the capacitor

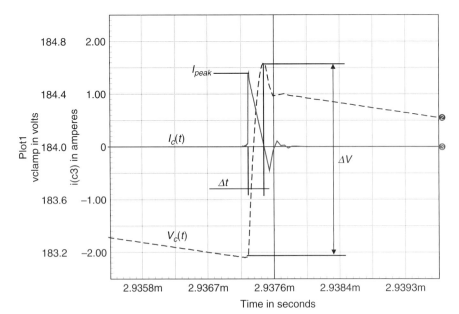

FIGURE 7-12 The clamp capacitor voltage and current waveforms.

during the reset time, the voltage developed across its terminal is obtained by first deriving a charge equation

$$Q_{C_{clp}} = I_{peak} \frac{\Delta t}{2} \qquad (7\text{-}45)$$

Then the voltage ripple seen by the clamping network is simply

$$I_{peak} \frac{\Delta t}{2} = C_{clp} \Delta V \qquad (7\text{-}46)$$

Solving for C_{clp} and replacing Δt by its definition [Eq. (7-23)], we obtain an equation for the clamp capacitor:

$$C_{clp} = \frac{I_{peak}^2 L_{leak}}{2\Delta V(V_{clamp} - V_r)} \qquad (7\text{-}47)$$

If now extract L_{leak} from Eq. (7-44) and replace it in Eq. (7-47), we reach a simpler expression:

$$C_{clp} = \frac{V_{clamp}}{R_{clp} F_{sw} \Delta V} \qquad (7\text{-}48)$$

Since this capacitor will handle current pulses, it is necessary to evaluate its rms content to help select the right component. The rms current can be evaluated by solving the following integral:

$$I_{clp,rms} = \sqrt{\frac{1}{T_{sw}} \int_0^{\Delta t} \left[I_{peak} \frac{\Delta t - t}{\Delta t} \right]^2 dt} = I_{peak} \sqrt{\frac{\Delta t}{3T_{sw}}} \qquad (7\text{-}49)$$

Based on the above example, we have the following design elements:

$I_{peak,max}$ = 2.5 A—maximum current the controller allows in fault or overload situations
F_{sw} = 65 kHz
L_{leak} = 12 µH
V_{clamp} = 115 V

From Eq. (7-44), we calculate the clamp resistor value

$$R_{clp} = \frac{2 \times 115 \left[115 - \frac{12+1}{0.226} \right]}{65k \times 12u \times 6.25} = 2.7\,k\Omega \qquad (7\text{-}50)$$

The power dissipated by the resistor reaches 4.9 W at full power, as delivered by the left term in Eq. (7-43). If we select a voltage ripple of roughly 20% of the clamp voltage, Eq. (7-48) leads to a capacitor value of

$$C_{clp} = \frac{V_{clamp}}{R_{clp} F_{sw} \Delta V} = \frac{115}{2.7k \times 65k \times 0.2 \times 115} = 28\,nF \qquad (7\text{-}51)$$

According to Eq. (7-49), the rms current reaches 266 mA.

7.7.2 Selecting k_c

Equation (7-42) lets us open the discussion regarding k_c. Its selection depends on design choices. The author recommends to keep it in the vicinity of 1.3 to 1.5 times the reflected voltage, in order to give greater margin on the turns ratio calculation. As indicated by Eq. (7-3), lower N ratios let the designer pick up lower V_{RRM} secondary diodes. Schottky diodes are common in either 100 V or 200 V V_{RRM} values, but 150 or 250 V is starting to appear also. Adopting diodes featuring high-reverse-voltage capabilities affects the efficiency: the V_f usually increases (sensitive to the average current) as well as the dynamic resistance R_d (sensitive to the rms current).

On the other hand, the choice of k_c depends on the amount of leakage inductance brought by the transformer. Poorly designed flyback transformers exhibit leakage terms in the vicinity of 2 to 3% of the primary inductance. Excellent designs bring less than 1%. The greater the leakage inductance, the larger the difference between the clamp level and the reflected voltage, implying k_c coefficients close to twice the reflected voltage or above. If you keep k_c low (1.3 to 1.5) in spite of a large leakage term, you will pay for it through a higher power dissipation on the clamping network. In case the transformer still suffers from high leakage despite several redesigns, bring k_c to 2 and select higher V_{RRM} secondary diodes. This is what was done in the design example given the leakage term value (1.5% of L_p).

Based on the example, we have plotted the clamp resistor dissipated power versus the clamp coefficient k_c. As the clamp level approaches the reflected voltage, we can see a dramatic increase in the clamp dissipated power (Fig. 7-13).

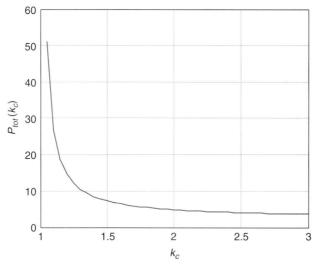

FIGURE 7-13 The dissipated power in the clamp resistor dramatically increases as the reflected voltage approaches the clamp voltage.

This plot stresses the need to push for a low leakage inductance, especially at high power levels. In the example, the 12 µH leakage term corresponds to 1.5% of the primary inductance. Several calculations were performed while assuming various transformer coupling qualities and trying to arbitrarily keep the average power on the clamp below 4 W. (The MOSFET breakdown voltage is fixed to 600 V.) Results are presented in the table below and show how other parameters are affected.

L_{leak} (% of L_p)	0.5	0.8	1	2
k_c	1.3	1.5	1.7	2
N	0.15	0.17	0.19	0.23
PIV (V)	67	76	84	97
V_{RRM} (V)	150	150	200	200
P_{clamp} (W)	3.4	3.8	3.8	6

In summary, if the leakage term is low, you can accept lower k_c factors (1.3) and thus reduce the voltage stress on the secondary diode, as explained. When the leakage inductance increases, keeping a reasonable dissipated power on the clamp implies an increase of N and the selection of higher voltage diodes with the associated penalties.

7.7.3 Curing the Leakage Ringing

Figure 7-10 showed some ringing appearing when the diode abruptly blocks. These oscillations find their roots in the presence of stray elements such as the leakage inductance, the lump capacitor, and all associated parasitic elements. Damping the network consists of artificially increasing the ohmic losses in the oscillating path. The damping resistor value can be found through a few simple equations pertinent to RLC circuits. The quality coefficient of a series RLC network is defined by

$$Q = \frac{\omega_0 L_{leak}}{R_{damp}} \tag{7-52}$$

To damp the oscillations, a coefficient of 1 can be the goal, implying that the damping resistor equals the leakage inductor impedance at the resonant frequency:

$$\omega_0 L_{leak} = R_{damp} \tag{7-53}$$

In our case, on the power supply prototype, we measured a ringing frequency f_{leak} of 3.92 MHz together with a leakage inductance of 12 μH. Thus, the damping resistor value must be

$$R_{damp} = 12u \times 6.28 \times 3.92 \text{Meg} = 295\ \Omega \tag{7-54}$$

A first possibility exists to dampen the RCD clamp itself as shown on Figure 7-14. This solution offers a simple nonpermanent dissipative way to help reducing the oscillations at the diode opening: once the diode D_{clp} is blocked (at the leakage inductor reset), this resistor no longer undergoes a current circulation. Unfortunately, the addition of this resistor affects the voltage peak associated with the current value at turn-off. This voltage increase ΔV simply equals

$$\Delta V = I_{peak} R_{damp} \tag{7-55}$$

Given the above result, care must be taken to not degrade the original diode overshoot ΔV by the insertion of the damping resistor. Start on the prototype with values around 10 ohms and increase the resistor to find a point where the overshoot and the ringing are acceptable. After several tweaks, a 47 Ω resistor brought the needed improvement. Figures 7-15a and b shows the result after the installation of the damping resistor. We can see in Fig. 7-15b that the ringing amplitude has been slightly reduced while the overshoot remains acceptable. Please note that R_{damp} acts on the oscillating network $L_{leak} C_{lump}$ through the diode recovery capacitance C_{rr}.

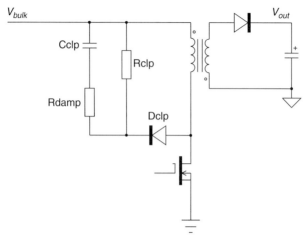

FIGURE 7-14 A resistor in series with the clamping capacitor helps to damp the parasitic oscillations.

Given its small value, the damping is less efficient than the solution described below. However, thanks to its effect on the overshoot itself, industrial applications, such as notebook adapters, make an extensive use of this technique which improves the radiated EMI signature.

Another possibility consists of damping the transformer primary side alone. It no longer touches the clamping network, but as it directly connects to the drain, it can impact the efficiency. Figure 7-16 represents this different option. The design procedure remains similar, and Eq. (7-53) still applies. The differences lies in the series capacitor, placed here to avoid a big resistive power dissipation as the switch closes. Reference 1 recommends a capacitor impedance equal to the resistor value at the resonant frequency of concern:

$$C_{damp} = \frac{1}{2\pi f_{leak} R_{damp}} \qquad (7\text{-}56)$$

Again, you might want to adjust this value a little to avoid overdissipation on the damping network. Figure 7-17a gathers some waveforms obtained on a flyback featuring a 22 μH leakage

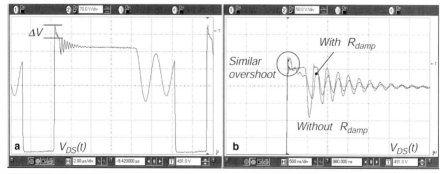

FIGURE 7-15 Installing the adequate damping resistor keeps the overshoot almost constant but reduces the ringing amplitude.

SIMULATIONS AND PRACTICAL DESIGNS OF FLYBACK CONVERTERS 607

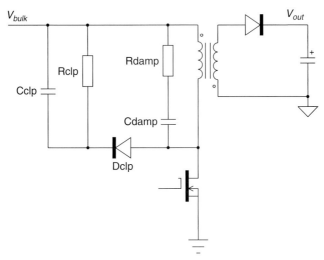

FIGURE 7-16 The RC damper connects on the transformer primary rather than on the clamping network.

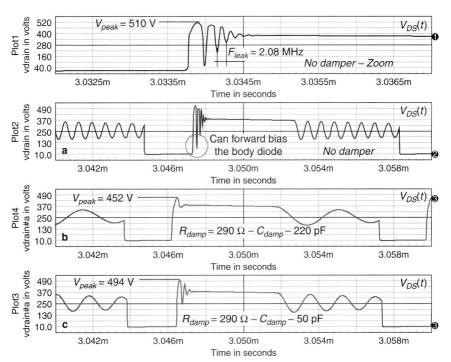

FIGURE 7-17a Damping the primary side also brings nice-looking waveforms!

inductor. The upper curves (a) describe the waveforms obtained without damping at all. The ringing becomes so severe that the body diode of the MOSFET can be forward biased, engendering further losses. If a discrete MOSFET can easily handle that, we do not recommend this forward biasing on a monolithic switcher where substrate injection could occur and damage the part through erratic behavior. A damper is mandatory in this case. Curve b in Fig. 7-17a depicts the fully damped waveform obtained from a resistor of 295 Ω and a capacitor of 220 pF, respectively, recommended by Eqs. (7-54) and (7-56). Despite the nice overall shape, where all the ringing has gone, the total loss budget has increased by 1.25 W. To reduce the heat, we decreased the damping capacitor down to 50 pF, and curve (c) appeared, still showing some ringing, but less severe than in the original waveform. The efficiency was almost left unaffected by this change.

The power dissipated by the resistor depends on the voltage stored by the damping capacitor during the switching events. During the on time, the capacitor charges to the input voltage V_{in}. During the off time, the capacitor jumps to the flyback voltage and stays there until the primary inductance resets (in DCM). Figure 7-17b depicts these events and shows the energy in play. If we suppose the capacitor is discharged to 0 at the beginning of the on time, then the needed energy to bring it up to the input voltage is (event 1)

$$E_{on} = \frac{1}{2} C_{damp} V_{in}^2 \qquad (7\text{-}57a)$$

Then, to charge the damping capacitor to the reflected voltage at the switch opening, you need to first bring the same energetic level as described by Eq. (7-57a) (discharge the capacitor down to zero, event 2) to which you add another jump equal to (event 3)

$$E_{off} = \frac{1}{2} C_{damp} \left[\frac{V_{out} + V_f}{N} \right]^2 \qquad (7\text{-}57b)$$

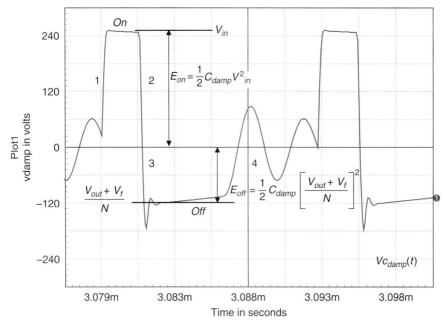

FIGURE 7-17b The damping capacitor voltage during the switching events.

Then, in DCM, the capacitor voltage rings and goes back to zero, releasing the energy also described by Eq. (7-57b) (event 4). As the capacitive current flows through the damping resistor, the total dissipation seen by this element is simply

$$P_{R_{damp}} = 2\left(\frac{1}{2}C_{damp}V_{in}^2 + \frac{1}{2}C_{damp}\left[\frac{V_{out}+V_f}{N}\right]^2\right)F_{sw} = C_{damp}\left[V_{in}^2 + \left(\frac{V_{out}+V_f}{N}\right)^2\right]F_{sw} \quad (7\text{-}57c)$$

In the above example, the input voltage was 250 V, the reflected voltage was 120 V, and there was a 71 kHz switching frequency. This leads to a theoretical dissipated power of

$$P_{R_{damp}} = C_{damp}\left[V_{in}^2 + \left(\frac{V_{out}+V_f}{N}\right)^2\right]F_{sw} = 220p \times [250^2 + 120^2] \times 71k = 1.2 \text{ W} \quad (7\text{-}57d)$$

Simulation gave 1.15 W. This damping technique, compared to the previous one, does hamper the light/no-load efficiency and might not represent a good option for power-sensitive projects.

When you install dampers as in the above example, a simple diagram such as Fig. 7-17b helps to figure out the power dissipated in the damping resistor.

7.7.4 Which Diode to Select?

The above example showed curves obtained with an ultrafast diode, the MUR160. (A UF4006 would have given similar results.) This diode being rather abrupt when it turns off, the equivalent $L_{leak}C_{lump}$ network rings a lot and requires a damping action as we just described. Some slower diodes can also be used, such as the 1N4937 or even the 1N4007. Yes, you have read it correctly, the 1N4007! This diode actually presents a rather lossy blocking mechanism which heavily damps the leakage ringing and just makes it disappear. As Fig. 7-18 shows, it

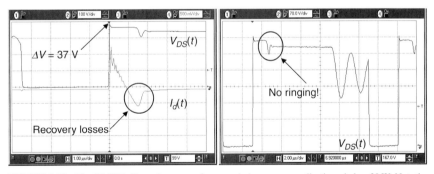

FIGURE 7-18 The 1N4007 offers adequate performance in low-power applications, below 20 W. Note the lack of ringing given the lossy mechanism.

gives excellent results, especially in radiated EMI. The 4007 has the reputation of being a slow diode, which is true when it starts to block. However, you can see the small overshoot of 37 V only, given the slope of 9.2 V/ns. As the diode recovers, it remains a short-circuit until it has fully swept away the minority carriers: this occurs at the negative peak. At this point, the diode recovers its blocking capability, and the current goes down slowly, naturally damping the leakage network. I have seen 1N4007 diodes used in high-power designs, well above 20 W (!), something I would absolutely not recommend.

There are other solutions known to help damp parasitic waveforms. They include lossy ferrite beads, widely used in the industry.

7.7.5 Beware of Voltage Variations

The *RCD* clamp network is unfortunately subject to voltage variations as the peak current changes. As a result, it is important to check that worst-case conditions do not jeopardize the MOSFET. Worst-case conditions are as follows:

1. *Start-up sequence, highest input level, full load*: Monitor $V_{DS}(t)$ by either directly observing the waveform or synchronizing the oscilloscope on the V_{cc} pin. The crucial point lies precisely when the feedback starts to take control. This is where the output voltage is at its peak and the primary peak current has not yet been reduced by the feedback loop.
2. *Highest input voltage, short-circuit*: Place a short-circuit on the secondary side (a real small piece of metal and not a long wire or the electronic load) and start the power supply. If the protection operates well, the converter should enter hiccup mode, trying to start up. As the auxiliary voltage does not show up (because of the output short-circuit), the controller quickly detects an *undervoltage lockout* (UVLO) and stops driving pulses. However, in the short period during which it drives the MOSFET, the peak current set point is pushed to its maximum limit and induces a large leakage kickback voltage. Make sure $V_{DS}(t)$ stays under control.

Figure 7-19 is a plot of the clamp voltage variations as the peak current changes. The design variable for Eq. (7-44) shall be the maximum peak current that the primary inductance can see. This is the maximum sensed voltage the controller can authorize over a given resistor, affected

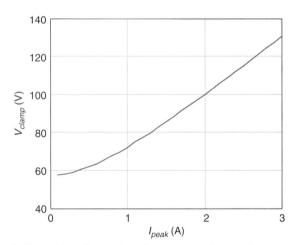

FIGURE 7-19 Clamp voltage changes due to primary peak current variations.

by the propagation delay t_{prop}. This propagation delay actually corrupts the current level imposed by the controller in short-circuit or in start-up conditions. Why? Because in either case, the feedback is lost and the peak current limit solely relies on the internal maximum set point. On a UC384X member, the maximum set point is 1.1 V. That is, when the current-sense comparator detects a voltage of 1.1 V, it immediately instructs the latch to reset and shuts the driver off. Unfortunately, it takes time for the turn-off order to propagate to the drive output and block the MOSFET. The propagation delay therefore includes the controller itself (internal logic delays) but also the driving chain to the MOSFET gate. Until it actually occurs, the primary current keeps growing and it overshoots. The overshoot depends on the primary slope and the propagation delay. Figure 7-20 illustrates this phenomenon.

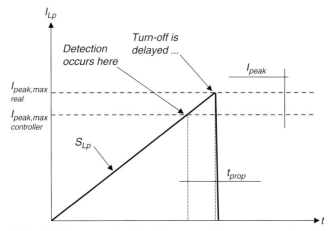

FIGURE 7-20 Propagation delay effects create primary current overshoot.

Suppose we have a primary inductance L_p of 250 µH and a maximum peak current set point of 1.1 V. The sense resistor equals 0.44 Ω, and the total propagation delay t_{prop} reaches 190 ns. Of course, if you insert a resistor in series with the drive, in an attempt to alter the EMI signature, you naturally introduce further delays which degrade the controller reaction time. The final peak current is obtained at the highest input voltage, 375 Vdc in this example:

$$I_{peak,max} = \frac{V_{sense,max}}{R_{sense}} + \frac{V_{in,max}}{L_p} t_{prop} = \frac{1.1}{0.44} + \frac{375}{250u} 190n = 2.8 \text{ A} \qquad (7\text{-}58)$$

In this example, the overshoot reaches almost 200 mA at high line. Suppose you have selected the clamp resistor based on 1.1 V developed over the 0.44 Ω sense element (2.5 A) to reach a clamp level of 115 V. You end up with a 124 V clamping level (8% more). But the worst is yet to come.

The worst situation arises when you have a really low primary inductance and a minimum t_{on} clamped down to 350 ns, for instance. The minimum t_{on} is made of the leading edge blanking (LEB) duration (if the controller features such circuitry, of course) added to the total propagation delay. If the controller features a 250 ns LEB with a 120 ns propagation delay, it will be impossible to reduce the on time below 370 ns. Why? Simply because at the beginning of each driving pulse, the controller is blind for 250 ns (to avoid a false tripping due to spurious pulses—diodes t_{rr} for instance) and takes another 120 ns to finally interrupt the current flow. During this time, the gate is held high and the MOSFET conducts. In short-circuit situations, not only does the primary current overshoot as explained above, but also it does not decrease during the off time: the converter enters a deep CCM mode because of the lack of reflected voltage. Actually, there is a reflected voltage, the diode forward drop divided by the transformer turns ratio:

$$V_r = \frac{V_f}{N} \qquad (7\text{-}59)$$

Since this reflected voltage is so low, it cannot bring the primary current back to its level at the beginning of the on time and so the current builds up at each pulse. In a short-circuit situation, either the clamp voltage runs away and the MOSFET quickly blows up, or the transformer saturates, leading to the same explosion! Figure 7-21 describes the situation.

In a start-up sequence, before the current reaches a high value, the output voltage rises and the reflected voltage starts to demagnetize the primary inductance. However, until a sufficient voltage imposes a proper downslope on the primary inductance, the current envelope peaks as described above. As predicted by Fig. 7-19, the clamping level goes out of control and the drain voltage dangerously increases. This is one of the major causes of power supply destruction at start-up. Figure 7-22 shows this behavior which must be seriously monitored when you design a high-power converter with a low primary inductance. In the presence of such difficulty, the solution lies in increasing the primary inductance, a choice which naturally limits peak current overshoots at high line. Another option consists of selecting a transient voltage suppressor.

7.7.6 TVS Clamp

A transient voltage suppressor (TVS) is nothing other than an avalanche diode (remember, zener effect occurs below 6.2 V; above, this is called the avalanche effect) able to take high-power pulses thanks to its large die size. The connection to the drain remains similar to that of the *RCD* clamp, as Fig. 7-23a shows. The TVS will clamp the voltage excursion of the drain, dissipating all the power as the clamp resistor would do. The TVS dynamic resistance being

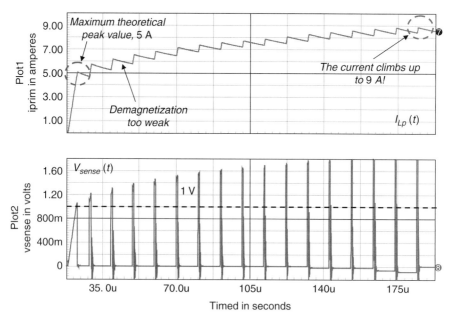

FIGURE 7-21 In short-circuit conditions, the current escapes from the PWM chip vigilance and runs out of control.

SIMULATIONS AND PRACTICAL DESIGNS OF FLYBACK CONVERTERS 613

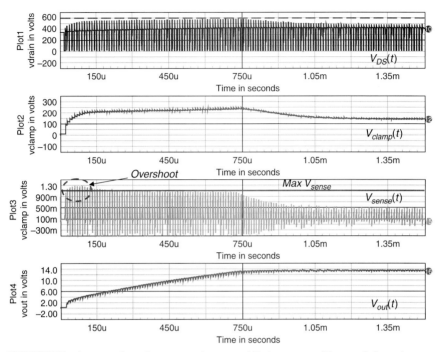

FIGURE 7-22 At start-up the peak current envelope can quickly increase, possibly engendering an increase of the clamp voltage. Carefully check this out once the design is finalized!

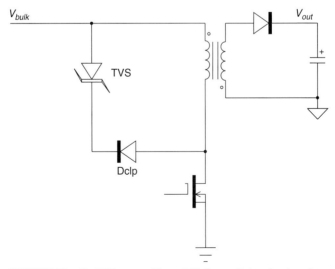

FIGURE 7-23a The TVS connects like an *RCD* clamp, offering a low-impedance clamping effect, which is less sensitive to current variations.

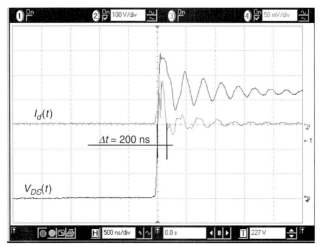

FIGURE 7-23b The TVS signature reveals an extremely brief conduction time, potentially radiating EMI noise.

small, its clamping voltage keeps rather constant despite large current variations. The dissipated power remains an important parameter to assess when the voltage selection is done. Capitalizing on the previously derived equations, the power dissipated by the TVS can be expressed as

$$P_{TVS} = \frac{1}{2} F_{sw} L_{leak} I_{peak}^2 \frac{V_z}{V_z - \frac{V_{out} + V_f}{N}} \qquad (7\text{-}60a)$$

where V_z represents the TVS breakdown voltage. Figure 7-23b shows a TVS typical signature, on the same converter we used for the Fig. 7-18 shots. As one can see, the TVS sharply clamps for a short time, 200 ns in the example. This narrow pulse can often radiate a wide spectrum of noise, and this is a reason (besides its cost) why the TVS is not very popular in high power applications. Make sure all connections are kept short, and place the TVS and its diode (here an MUR160) close to the transformer and the MOSFET. In rather leaky designs, some designers even place a 10 nF in parallel with the TVS to help absorb the current pulses.

The TVS offers one big advantage: in standby or in light-load conditions, the peak current remains low and the leakage kick is often not sufficient to reach the TVS breakdown voltage. It thus stays transparent, and the converter efficiency benefits from this operation. Without TVS, as the converter enters light load via skip cycle (a very common technique these days), the capacitor in the *RCD* network cannot keep up the clamp voltage as the recurrence between switching bursts goes down. Hence, the capacitor being discharged at each new switching bunch of pulses, it enters into action and dissipates a bit of power: the no-load standby power suffers from the situation.

7.8 TWO-SWITCH FLYBACK

One of the major limits in power delivery that the flyback suffers from can be linked to the presence of the leakage inductor. We have seen that classical single-switch solutions deal with this problem by routing the leakage energy to an external network: the energy is lost in heat, and the efficiency suffers. To use the flyback in higher power configurations, the two-switch structure might represent a possible solution. Figure 7-24a shows the application circuit. The architecture

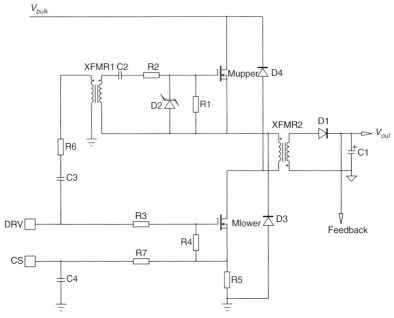

FIGURE 7-24a A two-switch flyback converter recycles the leakage energy at the opening of switches.

now uses two high-voltage MOSFETs but of smaller BV_{DSS}, compared to the single-switch approach. For instance, on a 400 V rail (assuming a PFC front-end stage), 500 V types can be implemented, implying a slightly better $R_{DS(on)}$ than their 600 V counterparts. The MOSFETs are turned on and off at the same time (same control voltage applied on the gate, the upper side floating with respect to ground). When both switches are conducting, the primary winding "sees" the bulk voltage. As the primary current reaches the peak limit, the controller classically instructs the switches to open. The current keeps circulating in the same direction and finds a path through the freewheel diodes D_3 and D_4. The transformer primary inductor immediately clamps to the reflected output voltage, and the leakage inductance resets with the following slope:

$$S_{leak} = \frac{V_{bulk} - \frac{V_{out} + V_f}{N}}{L_{leak}} \qquad (7\text{-}60b)$$

If you carefully observe Fig. 7-24b, the current circulates via the bulk capacitor, naturally recycling the leakage energy: the efficiency clearly benefits from this fact. The secondary-side diode current ramps up at a pace imposed by the leakage reset, rather slow given the longer leakage reset time inherent to the structure. Yes, you have guessed it; if you reflect more voltage than the input voltage, your colleagues are going to applaud at the first power-on!

Figure 7-24c offers a way to simulate the two-switch flyback using a dedicated current-mode controller. You could also try to reproduce the Fig. 7-24a transformer-based driving circuit, but it would take a longer simulation time. By the way, best practice would be to use a transformer made of two secondary windings for a perfect propagation delay match between the two transistors. If you understand the sentence "cost down!" then you understand why it becomes a single winding based. A bootstrapped solution could be used, but at the expense of a small refresh circuitry for the capacitor. We will come back to this with the two-switch forward example. Figure 7-24d collects all pertinent waveforms obtained from the simulator. As you can see on

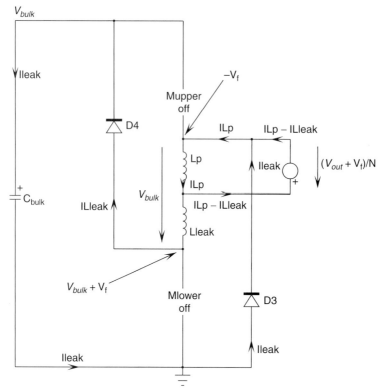

FIGURE 7-24b At the switch opening, the leakage current circulates through the freewheeling diodes, giving energy back to the bulk capacitor.

the upper trace, the leakage reset is rather smooth, something immediately seen on the secondary side diode current. However, this time, the reset duration does not generate losses dissipated in heat, but it brings the energy back to the bulk capacitor (see the positive jump on the source current, trace 10).

Despite its numerous benefits, such as leakage recycling, the two-switch flyback application has less success in the power supply industry than its two-switch forward counterpart.

7.9 ACTIVE CLAMP

The flyback with active clamp is currently gaining in popularity thanks to the ATX world requiring more efficient power supplies at a lower cost. The single-switch flyback with active clamp offers a possible alternative to the classical two-switch forward which can have difficulties in meeting the new ATX efficiency requirements: the total power supply efficiency is not allowed to drop below 80% for a loading ranging from 20 to 100% of the nominal power, thus its name, the 80+ initiative. The next step is to increase from 80 to 85%.

The principle behind the idea of active clamp still implies a capacitor storing the leakage energy at turn-off. However, rather than being simply dissipated in heat, the stored energy

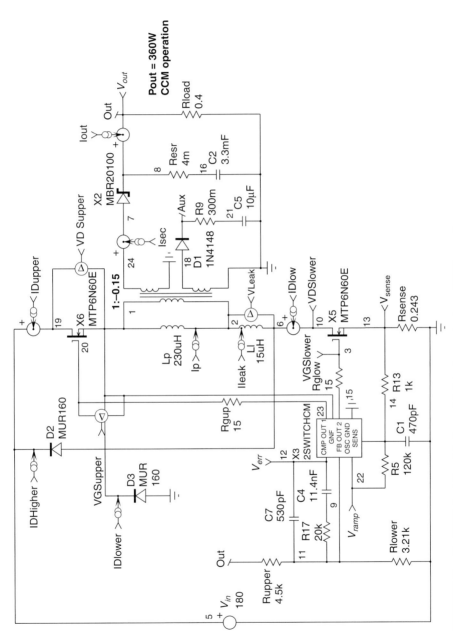

FIGURE 7-24c The SPICE simulation of the two-switch flyback converter, here a 360 W example.

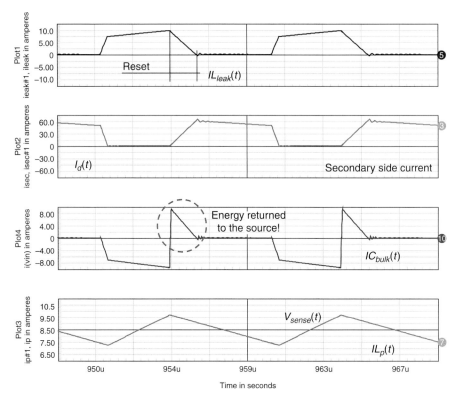

FIGURE 7-24d The simulation waveforms show the reset time slowing down the secondary side current.

is recycled to bring the drain voltage down to zero, naturally ensuring ZVS operation. This helps to (1) expand the use of the single-switch flyback beyond 150 W without paying switching losses incurred by the *RCD* clamp technique and (2) significantly increase the operating switching frequency, leading to the selection of lower size magnetic elements. This topic has been the subject of many comprehensive papers [3, 4, 5, and 6], and we will only scratch the surface here.

Figure 7-25 shows the flyback converter to which an active clamp circuit has been added. The system associates a capacitor (connected either to the bulk rail or to the ground) to a bidirectional switch made of SW and the diode D_{body}. This role can be played by a MOSFET, an N- or a P-channel, depending on the adopted reset scheme. To understand the operation, it is necessary to segment the various events in separate sketches and time intervals. Figure 7-26a and b gathers all these sketches, we comment on them one by one with illustrating graphics at the end.

In Fig. 7-26a, sketch a, the power switch has been turned on, as in any flyback operation. The current rises linearly with a slope dependent upon the input voltage and the combination of the leakage element with the primary inductance:

$$S_{on} = \frac{V_{in}}{L_p + L_{leak}} \qquad (7\text{-}61)$$

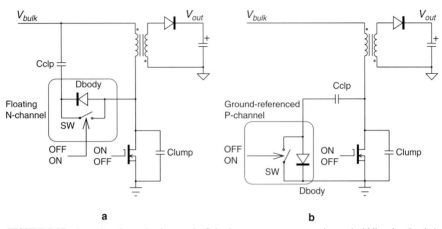

FIGURE 7-25 An active clamp circuit around a flyback converter uses a capacitor and a bidirectional switch.

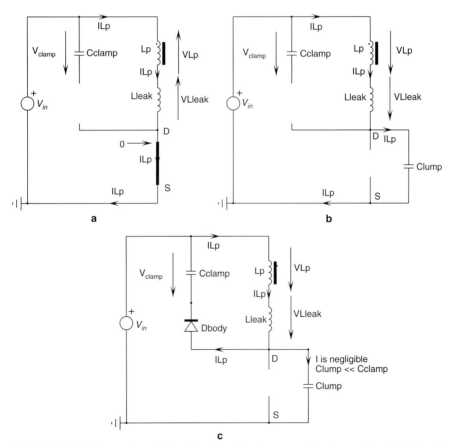

FIGURE 7-26a Different phases of the active clamp flyback converter. Here the secondary side diode still does not conduct.

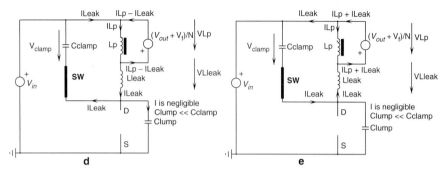

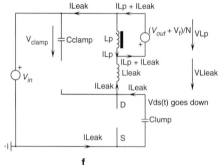

The leakage current has reached zero and now it reverses ---->

FIGURE 7-26b As the secondary diode conducts, the voltage across the primary inductance L_p is fixed to reflected voltage.

During this time, the input voltage splits between both terms as follows:

$$V_{L_p} = V_{in} \frac{L_p}{L_p + L_{leak}} \qquad (7\text{-}62)$$

$$V_{L_{leak}} = V_{in} \frac{L_{leak}}{L_p + L_{leak}} \qquad (7\text{-}63)$$

When the peak current reaches the set point imposed by the feedback loop ($I_{L,peak} = I_{peak}$), the power switch opens and the magnetizing current charges the lump capacitor. We are observing sketch b. The drain voltage immediately rises at a rate defined by Eq. (7-64):

$$\frac{dV_{DS}(t)}{dt} = \frac{I_{peak}}{C_{lump}} \qquad (7\text{-}64)$$

The voltage increases until it reaches the level imposed by the stored voltage present across the clamping capacitor V_{clamp} plus the input voltage V_{in}. At this time, the upper-switch body diode starts to conduct, and we are looking at sketch c. As the lump capacitor is much smaller than the clamp capacitor, it no longer diverts current and we can consider all the magnetizing current flowing through C_{clamp}. Note that we already had a resonant transition at the switch opening between C_{lump} and the total inductance made up of $L_{leak} + L_p$, but since its duration is short, you almost observe a linear waveform.

The clamp capacitor now fixes the voltage across both inductors $L_{leak} + L_p$ which act as a voltage divider. The voltage splits as follows:

$$V_{L_p} = -V_{clamp} \frac{L_p}{L_p + L_{leak}} \quad (7\text{-}65)$$

When the primary voltage has fully reversed to a point where the secondary diode conducts, we are now observing Fig. 7-26b, sketch d. Since the body diode is conducting ($V_{DS} = 0$), we can safely turn on the upper switch SW to benefit from zero voltage conditions across it. A small delay is thus necessary to let this body diode first conduct before activating the upper-side switch. The primary current classically decays at a slope imposed by the reflected voltage:

$$S_{off} = -\frac{V_{out} + V_f}{NL_p} \quad (7\text{-}66)$$

The leakage inductor now resonates with the clamp capacitor, and a sinusoidal waveform arises. The resonant frequency involves the leakage inductance and the clamp capacitor (neglecting the lump capacitor):

$$f_{reso1} = \frac{1}{2\pi \sqrt{L_{leak} C_{clamp}}} \quad (7\text{-}67)$$

The secondary side current is made up of the difference between the linear current delivered by the primary inductance [Eq. (7-66)] and the sinusoidal waveform drawn by the leakage term:

$$I_d(t) = \frac{I_{L_p}(t) - I_{L_{leak}}(t)}{N} \quad (7\text{-}68)$$

In a classical flyback, this period of time, where the leakage inductance diverts current, would be kept small as we want to reset the leakage term as quickly as possible. With the active clamp technique, on the contrary, the leakage term diverts current during the whole off time and actually turns on the secondary diode very smoothly.

When the resonant current waveform reaches zero, the current reverses and starts to flow in the other direction. It can do so thanks to the presence of the upper-side switch, still kept closed for the event. We are looking at sketch e. At a certain time, the leakage inductor current will reach a negative peak. The stored energy at this moment is simply

$$E_{L_{leak}} = \frac{1}{2} L_{leak} I_{peak}^2 \quad (7\text{-}69)$$

If we now open the upper-side switch, the leakage inductance current still flows in the same direction but now returns via the lump capacitor, creating a new resonance:

$$f_{reso2} = \frac{1}{2\pi \sqrt{L_{leak} C_{lump}}} \quad (7\text{-}70)$$

The current flowing through the lump capacitor now from the ground contributes to discharge it. The minimum of the voltage is reached at one-quarter of the period. Hence, if a delay is added to let the voltage swing to its valley, we can ensure zero voltage switching on the power MOSFET. This delay should be adjusted to

$$t_{del} = \frac{2\pi \sqrt{L_{leak} C_{lump}}}{4} = \frac{\pi}{2} \sqrt{L_{leak} C_{lump}} \quad (7\text{-}71)$$

The necessary condition to bring the lump capacitor down to zero implies that the energy stored in the leakage inductance at the switch opening [Eq. (7-69)] equals or exceeds the energy stored in the lump capacitor. Otherwise stated,

$$\frac{1}{2}L_{leak}I_{peak}^2 \geq \frac{1}{2}C_{lump}\left[V_{in} + \frac{V_{out} + V_f}{N}\right]^2 \qquad (7\text{-}72)$$

Capitalizing on this equation, we can extract a design law to obtain the resonant/leakage inductor value:

$$L_{leak} \geq \frac{C_{lump}\left(V_{in} + \frac{V_{out} + V_f}{N}\right)^2}{I_{peak}^2} \qquad (7\text{-}73)$$

In the above equation, the peak current at the switch opening is approximated by the peak current imposed by the controller at the end of the on time. In practice, given the damping action (ohmic losses), the final value slightly differs from this value. If the current at the upper-side switch opening moment is too low, the lump capacitor cannot properly discharge down to zero because Eq. (7-72) will not be satisfied. To solve this problem, you will need to either artificially increase the leakage term by inserting an inductor connected in series with the transformer primary or weaken the coupling between both primary and secondary windings, for instance, by using a different bobbin architecture. The peak current in Eq. (7-73) can be derived using equations already seen in Chap. 5 in the buck-boost section [Eqs. (5-99) and (5-100)]. Combining them leads to the definition of this peak current level

$$I_{peak} \approx I_{L,avg} + \frac{\Delta I_{L_p}}{2} = P_{out}\left(\frac{1}{\eta V_{in}} + \frac{N}{V_{out}}\right) + \frac{V_{in}D}{2L_pF_{sw}} \qquad (7\text{-}74)$$

Since the leakage inductance is defined, we can calculate the value of the clamp capacitor. Its value depends on the off-time duration at high line such that one-half of the resonant period always remains larger than the largest off-time duration. Otherwise, the negative peak in the resonance waveform might no longer correspond to the upper-side switch opening event, losing the relationship described by Eq. (7-72). Hence, the design obeys the following equation:

$$\frac{t_{res1}}{2} \geq (1 - D_{min})T_{sw} \qquad (7\text{-}75)$$

Replacing t_{res1} with its definition [Eq. (7-67)] and extracting the clamp capacitor value, we have

$$\pi\sqrt{L_{leak}C_{clamp}} \geq (1 - D_{min})T_{sw} \qquad (7\text{-}76)$$

$$C_{clamp} \geq \frac{(1 - D_{min})^2}{F_{sw}^2 \pi^2 L_{leak}} \qquad (7\text{-}77)$$

7.9.1 Design Example

To capitalize on the above equations, it is time to verify our assumptions with a simulated example. We can start from an initial flyback design to which we apply the active clamp technique. The design procedure difference between a flyback operated with or without active clamp is minor as long as we consider the leakage term negligible compared to the magnetizing inductor.

$$L_{leak} \ll L_p \tag{7-78}$$

Reference 6 calculates the duty cycle on a nonactive clamp flyback and compares it to an active clamp-operated converter. The difference remains about a few percent. What changes, however, concerns the maximum drain excursion at the switch opening. In a normal flyback operated with a classical *RCD* clamping network, the drain remains theoretically blocked below

$$V_{DS} = V_{in} + \frac{V_{out} + V_f}{N} + V_{clamp} \tag{7-79}$$

With an active clamp, the voltage developed across the leakage inductance during its resonating phase comes in series with the two first terms of Eq. (7-79). The excursion thus follows Eq. (7-80) [6]:

$$V_{DS} \approx V_{in} + \frac{V_{out} + V_f}{N} + \frac{2L_{leak}P_{out}}{\eta V_{in,max}D_{min}(1 - D_{min})} \tag{7-80}$$

The normal procedure would be to first fix the drain-source excursion desired (given the selected MOSFET) and then calculate the remaining elements. Unfortunately, the leakage term already appears in Eq. (7-80). However, as expressed by Eq. (7-78), we can arbitrarily pick up a value around 10% of the magnetizing inductance and see what turns ratio is authorized in relation to the MOSFET breakdown voltage. Later on, once the final leakage term is calculated, you have the possibility of rechecking the result of Eq. (7-80) and going for another pass if necessary.

The flyback we want to use features the following parameters:

$L_p = 770$ μH
$L_{leak} = 12$ μH
$1:N = 1:0.166$
$V_{in,max} = 370$ Vdc
$V_{in,min} = 100$ Vdc
$V_{out} = 19$ V
$I_{out,max} = 4$ A
$P_{out,max} = 76$ W
$F_{sw} = 65$ kHz
$C_{lump} = 220$ pF (measured according to App. 7A principles)
$\eta = 85\%$ (considered constant at low and high line for simplicity)

The duty cycle variations are computed first:

$$D_{max} = \frac{V_{out}}{V_{out} + NV_{in,min}} = \frac{19}{19 + 0.166 \times 100} = 0.534 \tag{7-81}$$

$$D_{min} = \frac{V_{out}}{V_{out} + NV_{in,max}} = \frac{19}{19 + 0.166 \times 370} = 0.236 \tag{7-82}$$

The high- and low-line peak currents reach [Eq. (7-74)]

$$I_{peak,high-line} \approx 76 \times \left(\frac{1}{0.85 \times 370} + \frac{0.166}{19}\right) + \frac{370}{2 \times 770u \times 65k} \times 0.236 = 1.8\,\text{A} \quad (7\text{-}83)$$

$$I_{peak,low-line} \approx 76 \times \left(\frac{1}{0.85 \times 100} + \frac{0.166}{19}\right) + \frac{100}{2 \times 770u \times 65k} \times 0.534 = 2.1\,\text{A} \quad (7\text{-}84)$$

With this information in hand, let us determine the needed resonating inductor:

$$L_{leak} \geq \frac{C_{lump}\left(V_{in} + \frac{V_{out} + V_f}{N}\right)^2}{I_{peak}^2} \geq \frac{220p \times \left(370 + \frac{19+1}{0.166}\right)^2}{1.8^2} = 16.3\,\mu\text{H} \quad (7\text{-}85)$$

Allowing a bit of margin, we add a small inductor of 8 µH in series with the transformer since the original leakage term already totals 12 µH (L_{leak} total = 20 µH). The rms current flowing into the added leakage inductor combines the main magnetizing current and the circulating clamp current. Reference 6 gives the following formula, reaching its maximum at low line:

$$I_{L_{leak},rms} = \sqrt{\frac{\left(\frac{P_{out}}{\eta V_{in,min}D_{max}}\right)^2(2D_{max}+1) + \frac{P_{out}}{\eta L_p F_{sw}}(1-D_{max}) + \frac{1}{4}\left(\frac{V_{in,min}D_{max}}{L_p F_{sw}}\right)^2}{3}} = 1.52\,\text{A} \quad (7\text{-}86)$$

Its maximum peak current is of course similar to that of Eqs. (7-83) and (7-84). Equation (7-87) now helps us to calculate the clamping capacitor value.

$$C_{clamp} \geq \frac{(1-D_{min})^2}{F_{sw}^2 \pi^2 L_{leak}} \geq \frac{(1-0.236)^2}{65k^2 \times 3.14^2 \times 20u} \geq 699\,\text{nF} \quad (7\text{-}87)$$

The voltage rating for this capacitor must exceed the reflected voltage plus the voltage developed across the leakage inductor. Using Eq. 7-80, we simply remove the first term:

$$V_{C_{clamp},max} \approx \frac{V_{out} + V_f}{N} + \frac{2L_{leak}P_{out}}{\eta V_{in,max}D_{min}(1-D_{min})}$$

$$= \frac{19+1}{0.166} + \frac{2 \times 20u \times 76}{0.85 \times 370 \times 0.236 \times (0.764)} = 120\,\text{V} \quad (7\text{-}88)$$

Finally, the ripple current circulates during the off time and obeys Eq. (7-89) [6]. The worst case occurs at low line:

$$I_{C_{clamp},rms} = I_{peak,max}\sqrt{\frac{1-D_{max}}{3}} = 2.1\sqrt{\frac{1-0.534}{3}} = 0.83\,\text{A} \quad (7\text{-}89)$$

As a final check, Eq. (7-80) predicts a voltage excursion of 490 V. Our MOSFET breaks at 600 V, so we are safe.

SIMULATIONS AND PRACTICAL DESIGNS OF FLYBACK CONVERTERS 625

Now that we have all resonant elements on hand, we can evaluate the upper-switch opening delay necessary to obtain ZVS on the MOSFET drain:

$$t_{del} = \frac{\pi}{2}\sqrt{L_{leak}C_{lump}} = 1.57 \times \sqrt{20u \times 220p} = 104\,\text{ns} \qquad (7\text{-}90)$$

7.9.2 Simulation Circuit

The circuit we used for the simulation example appears in Fig. 7-27. It implements the generic controller already developed for a synchronous rectification application. Output 1 drives the main switch whereas output 2 goes to the upper-side switch. Since its source floats, a real application would require a transformer or a high-voltage high-side driver such as the NCP5181 from ON Semiconductor. We removed any isolation circuit to reduce the complexity.

The controller switches at 65 kHz, and a 100 ns delay is inserted between both outputs. In theory, the first delay should be independently adjusted to let the body diode ensure ZVS on the upper-side switch. The second delay should then match the result of Eq. (7-90). Practically, a similar dead time is inserted without known operating problems. Figure 7-28a gathers a series of operating plots at high line. In this figure, we can see on the upper plot that both magnetizing current and resonant current are following the same shape at turn-on, but they diverge during the off time: the magnetizing current decreases linearly whereas the leakage current resonates with the clamp capacitor. Note that the secondary current smoothly

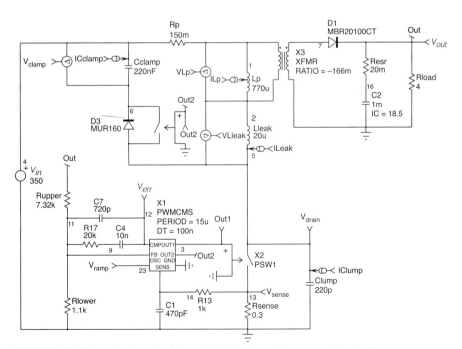

FIGURE 7-27 The active clamp simulation circuit using the synchronous generic controller.

626 CHAPTER SEVEN

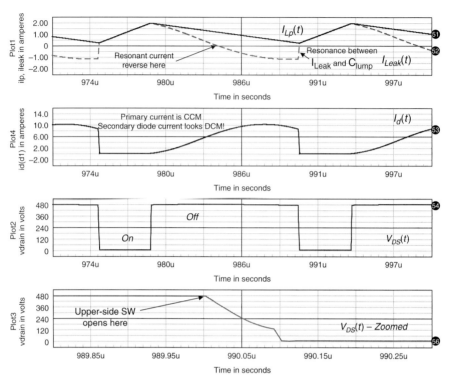

FIGURE 7-28a Typical signals of the active clamp flyback converter with an input voltage of 350 Vdc.

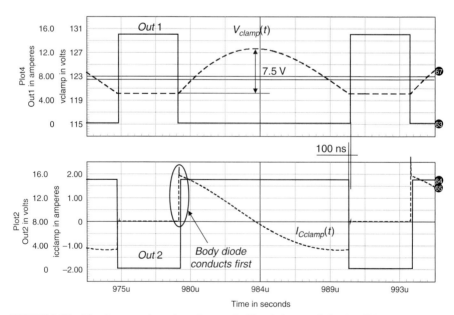

FIGURE 7-28b The clamp capacitor voltage does not significantly increase during the off time.

increases in the diode and looks as if it was DCM despite CCM on the primary side! This implies a higher peak current compared to a similar CCM flyback without active clamp, but switching losses on the diode are greatly reduced. The middle and last plots depict the drain voltage which stays almost flat on the plateau. On the zoom, you can clearly see the valley jump as soon as the upper-side switch opens. The delay seems well adjusted on this simulation. We should probably be able to force the wave farther down by increasing the leakage term.

Figure 7-28b focuses on the clamp voltage and shows that its voltage stays relatively constant during the transformer core reset. The bottom plot represents the clamp capacitor current and shows that it starts to flow before output 2 biases the upper-side switch: ZVS is ensured.

Figure 7-28c gathers plots where the clamp capacitor has been changed. Swept values are 220 nF, 680 nF, and 1 μF. When the capacitor no longer satisfies Eq. (7-77), $C_{clamp} = 220$ nF, the valley turn-on is lost on the main power MOSFET. However, the drain voltage excursion does not suffer from these variations.

The small-signal characteristic of the flyback converter operated with an active clamp does not significantly change compared to a traditional converter, as long as the clamp capacitor remains small compared to the reflected output capacitor.

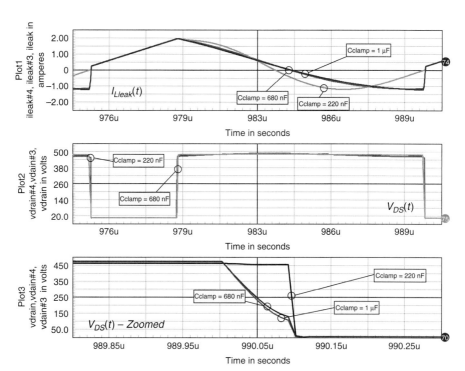

FIGURE 7-28c Three different clamp capacitor values have been swept: 220 nF, 680 nF, and 1 μF. If the maximum voltage excursion on the drain remains almost unchanged, the ZVS can be easily lost at low clamp capacitor values.

7.10 SMALL-SIGNAL RESPONSE OF THE FLYBACK TOPOLOGY

The flyback converter small-signal response does not differ from that of the buck-boost converter. Its dynamic behavior is affected by the operating mode, CCM or DCM, as well as the way the duty cycle is elaborated (voltage mode or current mode). Using the autotoggling models of the PWM switch represents an interesting exercise to explore the flyback small-signal response when operated in these various modes. Figures 7-29 and 7-30 depict how to wire the model in a single-output configuration. The converter delivers 19 V to a 6 Ω load (3.2 A), and we changed its primary inductance depending on the needed mode (CCM or DCM). The value at which the operating mode changes, given a certain input voltage (here 200 V), is called the *critical inductance*. Above this value, the converter delivering its nominal power will work in CCM. Below it, the converter will enter DCM. The critical inductance value can be derived by using formulas already seen in Chap. 1, the buck-boost section.

$$L_{p,crit} = \frac{R_{load}}{N^2 2 F_{sw}} \left(\frac{1}{1 + \frac{V_{out}}{NV_{in}}} \right)^2 = \frac{6}{0.166^2 \times 2 \times 65k} \left(\frac{1}{1 + \frac{19}{0.166 \times 200}} \right)^2 = 677\,\mu H$$

(7-91)

For the DCM converter, L_p will be fixed to 450 μH and for the CCM version, 800 μH. Let us now plot the Bode plot responses of the voltage-mode converter in both DCM and CCM (Fig. 7-29b). In DCM, the converter behaves as a first-order system in the low-frequency portion of the graph. The phase starts to drop to reach $-90°$, but the zero introduced by the output capacitor ESR kicks in and strives to pull the phase back to zero. However, as the frequency increases, the combined action of the high-frequency RHPZ and pole further degrades the phase. Their action is usually ignored as the cutoff frequency is often chosen way below their acting point. Note that the very first generation of models was unable to predict their presence.

In voltage-mode CCM, we can observe the peaking linked to the equivalent inductor L_e resonating with the output capacitor. We have a second-order system, degrading the phase at the resonating peak. The crossover frequency f_c must therefore be selected to be at least three times above the resonant frequency to avoid compensation difficulties when the gain crosses over the 0 dB axis. It must also stay below the worst-case RHP zero frequency to prevent further phase stress.

Figure 7-30a portrays the same converter arranged in a current-mode configuration. The modulator gain block is gone, and the loop is opened in a similar manner as before. The ac results are given in Fig. 7-30b and show that, in the low-frequency portion, the DCM and CCM operations do not differ that much in terms of Bode plots: they both exhibit first-order behavior. The CCM converter, however, turns into a third-order system after the appearance of the double pole placed at one-half of the switching frequency. Care must be taken to damp these subharmonic poles; otherwise instability can occur for a duty cycle greater than 50%. In this example, there is no ramp compensation.

7.10.1 DCM Voltage Mode

In light of the above curves, it appears obvious that stabilizing a voltage-mode converter operated in DCM looks simpler than stabilizing the same in CCM. This explains why mode transition in voltage mode often causes stability problems: a converter stabilized for only DCM cannot properly work in CCM. On the contrary, a converter stabilized for CCM might behave poorly in DCM because of an excessive compensation. Below is a summary of the voltage-mode small-signal flyback parameters pertinent to the error-to-output path.

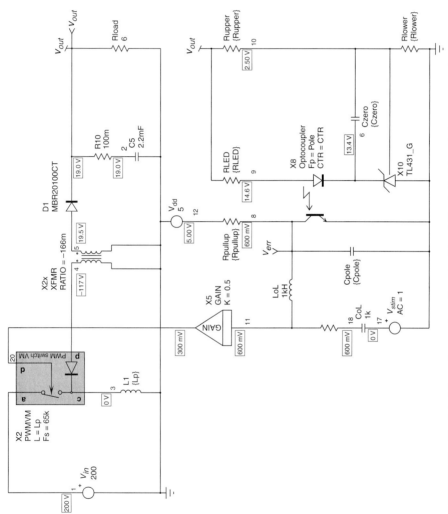

FIGURE 7-29a The flyback in a voltage-mode configuration. The −6 dB gain block brought by X_5 accounts for a 2 V peak amplitude sawtooth, and bias points are obtained after a DCM run.

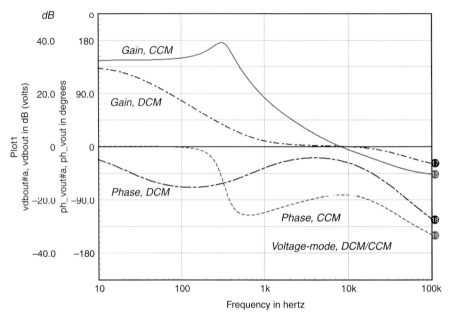

FIGURE 7-29b Bode plots of the voltage-mode converter operated in DCM or CCM. Under the same operating conditions, L_p is adjusted to change the mode. In CCM, the voltage-mode flyback behaves as a second-order system, which drops to a first-order in DCM.

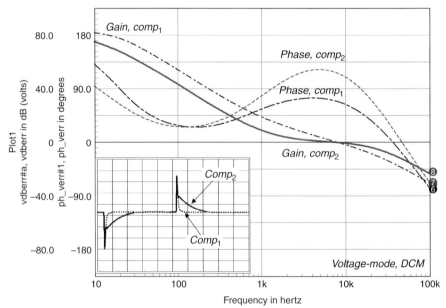

FIGURE 7-29c After compensation, despite a similar crossover frequency, the compensation featuring the zero positioned in the lower portion of the spectrum gives the slowest response (output current from 3 to 3.5 A).

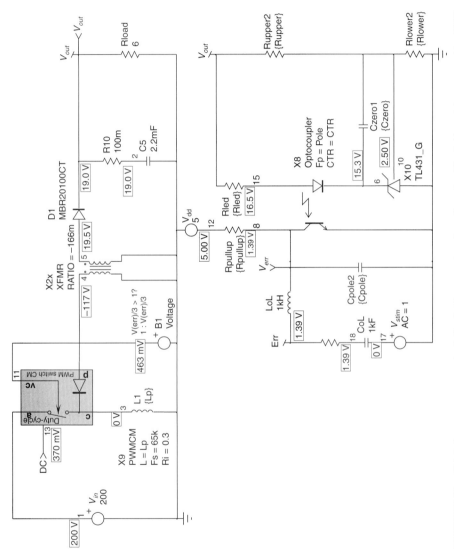

FIGURE 7-30a This is the same arrangement as before but in a current-mode configuration. The dc bias points correspond to a CCM simulation.

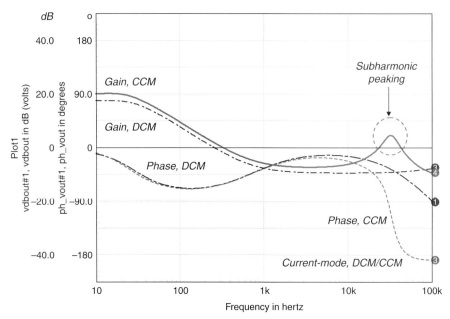

FIGURE 7-30b Bode plots of the current-mode converter operated in DCM or CCM. As previously indicated, L_p is adjusted to modify the mode.

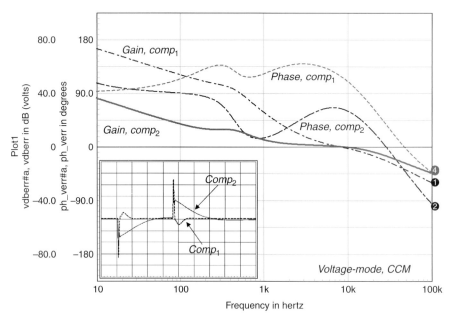

FIGURE 7-30c By changing the position of poles and zeros, conditional stability appears which can jeopardize the converter stability in some cases.

Voltage-Mode Control

	DCM	CCM
First-order pole	$\dfrac{1}{\pi R_{load} C_{out}}$	—
Second-order pole	$\dfrac{F_{sw}}{\pi}\left(\dfrac{1/D}{1+\dfrac{NV_{in}}{V_{out}}}\right)^2$	$\dfrac{1-D}{2\pi N \sqrt{L_p C_{out}}}$ (double)
Left half-plane zero	$\dfrac{1}{2\pi R_{ESR} C_{out}}$	$\dfrac{1}{2\pi R_{ESR} C_{out}}$
Right half-plane zero	$\dfrac{R_{load}}{2\pi N \dfrac{V_{out}}{V_{in}}\left(1+\dfrac{V_{out}}{NV_{in}}\right)L_p}$	$\dfrac{(1-D)^2 R_{load}}{2\pi D L_p N^2}$
V_{out}/V_{in} dc gain	$\dfrac{V_{out}}{V_{in}}$	$\dfrac{V_{out}}{V_{in}}$
V_{out}/V_{error} dc gain	$\dfrac{V_{in}}{V_{peak}}\sqrt{\dfrac{R_{load}}{2L_p F_{sw}}}$	$\dfrac{NV_{in}}{(1-D)^2 V_{peak}}$
Duty cycle D	$\dfrac{V_{out}}{V_{in}}\sqrt{\dfrac{2L_p F_{sw}}{R_{load}}}$	$\dfrac{V_{out}}{V_{out}+NV_{in}}$

where D = duty cycle
F_{sw} = switching frequency
R_{ESR} = output capacitor equivalent series resistance
M = conversion ratio = $M = \dfrac{V_{out}}{NV_{in}}$
L_p = primary inductance
R_{load} = output load
$N = \dfrac{N_s}{N_p}$ = transformer turns ratio
V_{peak} = PWM sawtooth amplitude

We have stabilized all converters with an 8 kHz crossover frequency and a phase margin of 70°. The DCM voltage-mode stabilization is a rather simple exercise considering the input voltage constant to 200 V and a load varying between 5 Ω (3.5 A) and 6 Ω (3 A). First, we need to calculate the position of the pole and zero inherent to this operating mode:

$$f_{p1,min} = \frac{1}{\pi R_{load,min} C_{out}} = \frac{1}{3.14 \times 5 \times 2.2m} = 29\,\text{Hz} \tag{7-92}$$

$$f_{p1,max} = \frac{1}{\pi R_{load,max} C_{out}} = \frac{1}{3.14 \times 6 \times 2.2m} = 24\,\text{Hz} \tag{7-93}$$

We now assume capacitor ESRs varying between 50 and 100 mΩ:

$$f_{z1,max} = \frac{1}{2\pi R_{ESR,min} C_{out}} = \frac{1}{6.28 \times 50m \times 2.2m} = 1.447\,\text{kHz} \tag{7-94}$$

$$f_{z1,min} = \frac{1}{2\pi R_{ESR,max} C_{out}} = \frac{1}{6.28 \times 100m \times 2.2m} = 723\,\text{Hz} \tag{7-95}$$

The high-frequency DCM RHPZ zero is evaluated; thus,

$$f_{z2} = \frac{R_{load}}{2\pi N \frac{V_{out}}{V_{in}}\left(1+\frac{V_{out}}{NV_{in}}\right)L_p} = \frac{6}{6.28 \times 0.166 \times \frac{19}{200}\left(1+\frac{19}{0.166 \times 200}\right)450u} = 85.6\,\text{kHz} \quad (7\text{-}96)$$

The second pole position depends on the DCM duty cycle. Thanks to Eq. (5-129), we can quickly obtain its value:

$$D = \frac{V_{out}}{NV_{in}}\sqrt{2\tau_L} = \frac{19}{0.166 \times 200}\sqrt{2 \times 0.138} = 0.3 \quad (7\text{-}97)$$

with

$$\tau_L = \frac{L_p N^2}{R_{load}T_{sw}} = \frac{450u \times 0.166^2}{6 \times 15u} = 0.138 \quad (7\text{-}98)$$

$$f_{p2} = \frac{F_{sw}}{\pi}\left(\frac{1/D}{1+\frac{NV_{in}}{V_{out}}}\right)^2 = \frac{65k}{3.14}\left(\frac{1/0.3}{1+\frac{0.166 \times 200}{19}}\right)^2 = 30.5\,\text{kHz} \quad (7\text{-}99)$$

The above formulas actually use results derived for a buck-boost converter but now include the transformer turns ratio for reflections on either side. The dc point in Fig. 7-29a confirms this result.

The dc gain linking the PWM input to the output stage can be obtained, given the PWM sawtooth peak amplitude (2 V in this example):

$$G_0 = 20\log_{10}\left[\frac{V_{in}}{V_{peak}}\sqrt{\frac{R_{load}}{2L_p F_{sw}}}\right] = 20\log_{10}\left[\frac{200}{2}\sqrt{\frac{6}{2 \times 450u \times 65k}}\right] = 30\,\text{dB} \quad (7\text{-}100)$$

This value is also confirmed by the Bode plot shown in Fig. 7-29b. Now that we have all in place, how do we compensate the whole thing? We can either use the *k* factor which gives adequate results in DCM or decide to place the corrective poles and zeros ourselves. Looking at Fig. 7-29b, we can see that an 8 kHz bandwidth requires no gain at all at the crossover point. The phase lag at this place is around −25°. A candidate for this compensator could be a simple type 1. The type 1 can work as long as the phase lag does not exceed −40° at the crossover point since no phase boost is provided by this structure. Otherwise, a type 2 can also do the job by placing the following poles and zeros:

- One pole at this origin, giving a high dc gain, thus a low dc output impedance and a good dc input rejection.
- A zero placed below the crossover frequency to bring the necessary phase boost (usually 1/5 of the crossover value gives good results).
- A pole situated at the capacitor ESR frequency (f_{z1}) or at one-half the switching frequency if the ESR value is too low. In this example, we placed it around 20 kHz.

The *k* factor placed a zero at 7 kHz and a pole at 8.7 kHz. As we have seen, spreading these poles and zeros apart (pushing the zero farther down the spectrum) will increase the phase boost, but the transient response might eventually suffer. Remember, the compensation zeros become the poles of your system in closed loop. To illustrate this, Fig. 7-29c depicts the compensated Bode plot with the following placement:

1. Compensation 1 places a zero at 7 kHz and a pole at 8.7 kHz.
2. Compensation 2 places a zero at 2 kHz and a pole at 20 kHz.

In both cases, the crossover frequency is 8 kHz, but compensation 1 gives a phase margin of 70° whereas compensation 2 increases it to 115°. The transient response in Fig. 7-29c speaks for itself; this is the slowest response. In conclusion, do not overcompensate the loop to obtain the largest phase margin. Keep it in the vicinity of 70° to 80° and always avoid going below 45° in the worst case. Of course, you also need to sweep all the parasitic parameters (ESR, for instance) and the operating conditions (input voltage, output load) to see how they affect the phase margin. Then, a compensation is required to remain stable in the worst case.

7.10.2 CCM Voltage Mode

By raising the primary inductance to 800 μH, we ensure the converter enters in CCM, as confirmed by Eq. (7-91). As we did before, we have selected a crossover frequency of 8 kHz. However, on a CCM buck-boost, boost, or flyback, what limits the bandwidth is the perfidious RHPZ. As we explained, this zero gives a boost in gain but stresses the phase rather than boosting it as a traditional left half-plane (LHP) zero would do. In other words, stabilizing the converter at a crossover frequency where the RHPZ starts to kick in represents a perilous, if not impossible, exercise. For this reason, it is advised to calculate the lowest RHPZ position and select 20 to 30% of it for the crossover frequency. In our example, let us run the exercise of assessing all pole and zero positions, as we did for the DCM case.

We first compute the duty cycle at a 200 V input voltage:

$$D = \frac{V_{out}}{V_{out} + NV_{in}} = \frac{19}{19 + 0.166 \times 200} = 0.36 \quad (7\text{-}101)$$

The right half-plane zero position is derived using the following equation:

$$f_{z2} = \frac{(1 - D)^2 R_{load}}{2\pi D L_p N^2} = \frac{(1 - 0.36)^2 \times 6}{6.28 \times 0.36 \times 800u \times 0.166^2} = 49.3 \text{ kHz} \quad (7\text{-}102)$$

In this particular case, 20% of this value is 10 kHz. We thus have a sufficient margin with the 8 kHz choice.

We know that the flyback converter operated in CCM peaks as any second-order converter would do. The double pole introduced by the resonance of L_s and C_{out} is placed at

$$f_{p2} = \frac{1 - D}{2\pi N \sqrt{L_p C_{out}}} = \frac{1 - 0.36}{6.28 \times 0.166 \times \sqrt{800u \times 2.2m}} = 462.8 \text{ Hz} \quad (7\text{-}103)$$

To have an idea how the phase will drop as we approach this point, the quality coefficient Q can also be calculated:

$$Q = \frac{(1 - D) R_{load}}{N} \sqrt{\frac{C_{out}}{L_p}} = \frac{(1 - 0.36) \times 6}{0.166} \times \sqrt{\frac{2.2m}{800u}} = 38.4 \text{ or } 31.6 \text{ dB} \quad (7\text{-}104)$$

However, given the damping brought by the diode dynamic resistance and the capacitor equivalent series resistor, the quality coefficient is reduced to 4 dB, as shown in Fig. 7-29b.

The output capacitor still brings a zero whose position is similar to its DCM counterpart as described by Eqs. (7-94) and (7-95). The dc gain of the control to output chain is obtained by applying the following formula:

$$G_0 = 20 \log_{10}\left[\frac{NV_{in}}{(1-D)^2 V_{peak}}\right] = 20 \log_{10}\left[\frac{0.166 \times 200}{(1-0.37)^2 \times 2}\right] = 32.4\,\text{dB} \quad (7\text{-}105)$$

The compensation of a CCM flyback converter operated in voltage mode requires a type 3 compensation network. The double pole present at the resonant frequency locally stresses the phase and requires the placement of a double zero right at the worst-case resonance. A possible placement could be as follows:

- One pole at this origin, giving a high dc gain, thus a low dc output impedance and a good dc input rejection.
- A double zero placed at the resonant frequency given by Eq. (7-103).
- A pole situated at the capacitor ESR frequency (f_{z1}) or at the RHPZ or at one-half the switching frequency. In our case, the RHPZ appears at 48 kHz, so placing a pole at one-half the switching frequency represents a possible choice.
- This third pole can be installed as the above definition explains. In this example, we will also place it at one-half the switching frequency.

Again, we can compare the results with the k factor, known to be less efficient in CCM. The k factor recommended to place the double zero around 3.7 kHz and a double pole at 18 kHz. It is thus very likely that the k factor proposal leads to a faster response given the higher position of the double zeros. In summary,

1. We have an 8 kHz crossover frequency.
2. Compensation 1 places a double zero at 3.7 kHz and a double pole at 18 kHz (k factor).
3. Compensation 2 places a double zero at the resonant frequency and a double pole at one-half the switching frequency.

Figure 7-30c illustrates the obtained compensated Bode plots, and the lower window shows the transient response. The k factor is faster, as we had already seen it in Chap. 3. However, look at the wide conditional stability area around 1 kHz. It is true that the gain margin in this zone is still about 20 dB, but some customers would not accept conditional zones at all. The placement of the double zero right at the resonant frequency gives a generous phase boost but unfortunately degrades the gain in the lower portion of the spectrum. If you now compare Figs. 7-30c and 7-29c, they offer very similar transient response despite different operating modes. The CCM with the second compensation option recovers slightly more slowly than its DCM counterpart, however.

In this example, we have used a type 3 compensation circuit based on a TL431. For flexibility purposes, we do not recommend using the TL431 in this type of configuration voltage-mode CCM. Because the LED series resistor plays a role in the gain and other pole-zero placement, it becomes difficult to combine the pole-zero positions and the right bias current when needed.

7.10.3 DCM Current Mode

The flyback operated in current mode is probably the most widely used converter. Thanks to its first-order behavior and its intrinsic cycle-by-cycle current protection, the converter lends itself very well to the design of rugged and easy-to-stabilize power supplies. The compensation is simple (first-order behavior in the low-frequency portion), and the converter naturally excels in input line rejection performance. As in voltage mode, the position of poles and zeros changes as the converter transitions from one mode to the other. This table summarizes their positions.

Current-Mode Control

	DCM	CCM
First-order pole	$\dfrac{1}{\pi R_{load} C_{out}}$	$\dfrac{\dfrac{D'^3}{\tau_L}\left(1 + 2\dfrac{S_e}{S_n}\right) + 1 + D}{2\pi R_{load} C_{out}}$
Second-order pole	$\dfrac{F_{sw}}{\pi}\left(\dfrac{1/D}{1 + \dfrac{NV_{in}}{V_{out}}}\right)^2$	—
Left half-plane zero	$\dfrac{1}{2\pi R_{ESR} C_{out}}$	$\dfrac{1}{2\pi R_{ESR} C_{out}}$
Right half-plane zero	$\dfrac{R_{load}}{2\pi N \dfrac{V_{out}}{V_{in}}\left(1 + \dfrac{V_{out}}{NV_{in}}\right) L_p}$	$\dfrac{(1 - D)^2 R_{load}}{2\pi DL_p N^2}$
V_{out}/V_{in} dc gain	—	$MN \dfrac{\dfrac{D'^2}{\tau_L}\left(M - 2\dfrac{S_e}{S_n}\right) - M}{\dfrac{D'^2}{\tau_L}\left(1 + 2\dfrac{S_e}{S_n}\right) + 2M + 1}$
V_{out}/V_{error} dc gain	$V_{in}\sqrt{\dfrac{R_{load} F_{sw}}{2L_p}}\dfrac{1}{S_e + S_n}$	$\dfrac{R_{load}}{R_i N}\dfrac{1}{\dfrac{D'^2}{\tau_L}\left(1 + 2\dfrac{S_e}{S_n}\right) + 2M + 1}$
Duty cycle D	$\dfrac{V_{out}}{V_{in}}\sqrt{\dfrac{2L_p F_{sw}}{R_{load}}}$	$\dfrac{V_{out}}{V_{out} + NV_{in}}$

where D = duty cycle
F_{sw} = switching frequency
R_{ESR} = output capacitor equivalent series resistance
$M = \dfrac{V_{out}}{NV_{in}}$ = conversion ratio
L_p = primary inductance
R_{load} = output load
$N = \dfrac{N_s}{N_p}$ = transformer turns ratio
$\tau_L = \dfrac{2L_p N^2}{R_{load} T_{sw}}$
$S_n = \dfrac{V_{in}}{L_p} R_{sense}$ = on-time slope, V/s
S_e = external compensation ramp slope, V/s
R_i = primary sense resistor

Our DCM power supply features the same values as above, except that it now operates in current mode with a sense resistor R_i of 300 mΩ. The low-frequency poles and zeros occupy the positions of the DCM voltage-mode converter. Equations (7-92) to (7-99) are thus still valid. The dc gain, however, changes and obeys the following equation:

$$G_0 = 20\log_{10}\left[V_{in}\sqrt{\frac{R_{load}F_{sw}}{2L_p}}\frac{1}{S_e + S_n}\right]$$

$$= 20\log_{10}\left[200\sqrt{\frac{6 \times 65k}{2 \times 450u}} \times \frac{1}{0 + \frac{200}{450u} \times 0.3}\right] = 30\,\text{dB} \qquad (7\text{-}106)$$

Again, the control-to-output curve looks like that of the voltage-mode DCM which requires a type 1 or type 2 amplifier, depending on the ESR help around the crossover frequency. The gain difference between Eq. 7-106 and the simulated gain relates to the internal structure of the current-mode controller. Figure 7-31a details the internal structure of a UC384X where

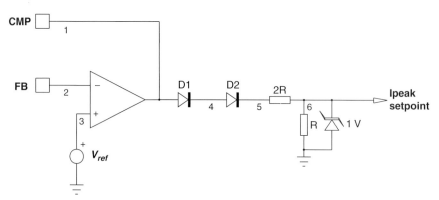

FIGURE 7-31a   The internal UC384X structure includes a 1 V clamp and a divider by 3.

a divider by 3 exists after two diodes in series. This divider brings a –9.5 dB insertion loss seen on Fig. 7-30b. In an NCP120X series member, the feedback undergoes a division by 4. The two series diodes make sure the set point falls to zero, even if the op amp low excursion does not reach ground. It also ensures the same null set point when an optocoupler directly connects to the CMP pin and exhibits a high saturation voltage (see Fig. 7-31b). The compensation follows the DCM voltage-mode guidelines. We have placed the compensation zero and pole at a similar location, respectively, 2 kHz and 20 kHz. Figure 7-32 shows the compensated Bode plot with its transient response: it is difficult to distinguish it from the voltage-mode response.

7.10.4   CCM Current Mode

In current-mode CCM, the right half-plane zero really hampers the available bandwidth as in CCM voltage mode. Equation (7-102) remains valid with an RHPZ located around 50 kHz. However, this time, there is no need to place a double zero thanks to the lack of resonant frequency.

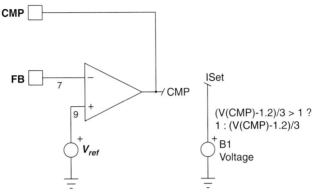

FIGURE 7-31b A possible SPICE implementation of the UC384X controller feedback section.

The main pole, whose position moves in relation to the compensation ramp, occurs at the following position:

$$f_{p1} = \frac{\frac{D'^3}{\tau_L}\left(1 + 2\frac{S_e}{S_n}\right) + 1 + D}{2\pi R_{load} C_{out}} \tag{7-107}$$

In this expression, the duty cycle is defined by Eq. (7-101) and is limited to 36%. Because of the CCM operation and the presence of subharmonic poles, it might be necessary to inject ramp

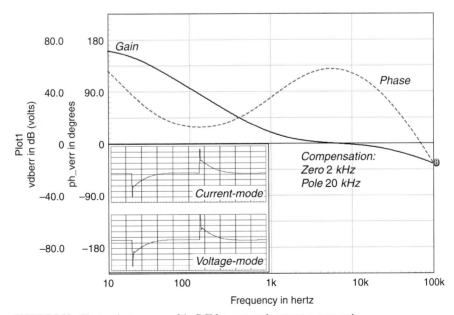

FIGURE 7-32 The transient response of the DCM current-mode converter, same scale.

compensation. A simple operation lets us assess the quality coefficient of the double poles located at one-half the switching frequency. If we go back to Chap. 2, without compensation ramp ($S_e = 0$) we have

$$Q = \frac{1}{\pi\left(D'\frac{S_e}{S_n} + \frac{1}{2} - D\right)} = \frac{1}{3.14 \times (0.5 - 0.36)} = 2.3 \qquad (7\text{-}108)$$

This result suggests that we damp the subharmonic poles to bring the quality coefficient below 1. Extracting the S_e parameter leads to

$$S_e = \frac{S_n}{D'}\left(\frac{1}{\pi} - 0.5 + D\right) = \frac{V_{in}R_i}{L_p D'}\left(\frac{1}{\pi} - 0.5 + D\right)$$

$$= \frac{200 \times 0.3}{850u \times (1 - 0.36)}\left(\frac{1}{3.14} - 0.5 + 0.36\right) = 19.6\,\text{kV/s} \qquad (7\text{-}109)$$

How much shall we compensate the converter? Is the suggested 19 kV/s slope enough? Well, sweeping the ramp amplitude reveals various responses, as Fig. 7-33 shows. Without any compensation at all, we can see the peaking above the 0 dB axis. This peaking is going to hurt the gain margin after compensation if it is not properly cured. When we add the external ramp, the peaking lowers and the situation improves when injecting more ramp. However, keep in mind that overcompensating the converter is not a panacea either, as its bevavior would approach that of a voltage mode.

A 19 kV/s slope means a ramp starting from 0 at the beginning of the on time and reaching 285 mV after 15 μs. We will see how to design such ramp in the next examples. In our model, we just set S_e to 19 kV and the compensation is done. Figure 7-34 gathers the compensated small-signal Bode plot and the step response it engenders. The compensation was similar to

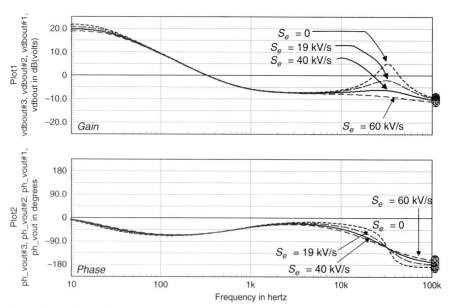

FIGURE 7-33 Effects of the ramp compensation on the CCM current-mode flyback.

SIMULATIONS AND PRACTICAL DESIGNS OF FLYBACK CONVERTERS

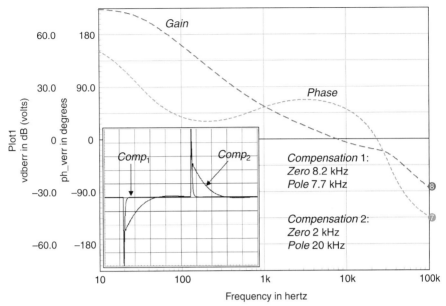

FIGURE 7-34 The CCM compensated flyback in current mode delivers almost the same transient response as that of the DCM version.

that of the DCM current-mode converter: a zero at 2 kHz and a pole placed at 20 kHz (compensation 2). Despite CCM, the step response is very close to that of the DCM version. The lack of double zero in the low-frequency spectrum clearly works in our favor. Compensation 1 is recommended by the *k* factor and places a zero and a pole at a similar location, actually canceling each other. The circuit becomes a type 1 compensator, simply placing a pole at the origin. The compensated result is the displayed Bode plot in the Fig. 7-34.

Further to these small-signal descriptions, let us summarize our observations:

- A CCM flyback converter requires a larger primary inductor compared to its DCM counterpart.
- CCM flyback converters operating in voltage mode or in current mode are subject to the same RHP zero which limits the available crossover frequency.
- A CCM voltage-mode flyback converter requires the placement of a double zero to fight the phase lag at the resonating point. The transient response is clearly affected.
- A voltage-mode CCM flyback can be stabilized using a type 3 compensation amplifier. But the classical TL431 configuration does not lend itself very well to its implementation.
- On the contrary, compensating a CCM current-mode flyback via a type 2 amplifier is an easy step.
- A DCM compensated voltage-mode flyback converter is likely to oscillate when entering CCM. Provided the RHP zero is far enough from the crossover frequency, the DCM current-mode flyback would be less sensitive to this mode transition.
- A CCM current-mode flyback converter whose duty cycle approaches 50% (or exceeds it) requires ramp compensation to avoid subharmonic oscillations. However, in certain cases, ramp compensation becomes mandatory at duty cycles as low as 35%: always compute the quality coefficient of the double to check for the necessity of compensation.

7.11 PRACTICAL CONSIDERATIONS ABOUT THE FLYBACK

Before showing how to design some flybacks converters, we need to cover a few other practical aspects, such as the start-up or the auxiliary supply of the controller.

7.11.1 Start-Up of the Controller

When you plug the power supply into the wall outlet, the bulk capacitor immediately charges to the peak of the input voltage (producing the so-called in-rush current). Since the controller is being operated from a low voltage (usually below 20 Vdc), it cannot be directly powered from the bulk, and a start-up circuitry must be installed. Figure 7-35 shows various solutions you could find on the market.

In Fig. 7-35, some initial energy is provided by the V_{cc} capacitor, charged by a start-up resistor. At power-on, when the capacitor is fully discharged, the controller consumption is zero and does not deliver any driving pulses. As V_{cc} increases, the consumed current remains below a guaranteed limit until the voltage on the capacitor reaches a certain level. This level, often called V_{CCon} or $UVLO_{high}$ (under voltage lockout) depending on the manufacturer, fixes the point at which the controller starts to deliver pulses to the power MOSFET. At this point, the consumption suddenly increases, and the capacitor depletes since it is the only energy reservoir. Its voltage thus falls until an external voltage source, the so-called auxiliary winding, takes over and self-supplies the controller. The capacitor stored energy must thus be calculated to feed the controller long enough that the auxiliary circuit takes over in time. If the capacitor fails to maintain V_{cc} high enough (because the capacitor has been set too small or the auxiliary voltage does not come), its voltage drops to a level called $UVLO_{low}$ or $V_{CC(min)}$. At this point, the controller considers its supply voltage too low and stops all operations. This safety level ensures that (1) the MOSFET receives pulses of sufficient amplitude to guarantee a good $R_{DS(on)}$ and (2) the controller internal logic operates under reliable conditions. When reaching the $UVLO_{low}$ level, the controller goes back to its original low consumption mode and, thanks to the start-up resistor, makes another restart attempt. If no auxiliary voltage ever comes, e.g., there is a broken diode or an output short-circuit, the power supply enters hiccup mode (or auto restart) where it pulses for a few milliseconds (the time the capacitor drops from $UVLO_{high}$ to $UVLO_{low}$), waits until the capacitor refuels, makes a new attempt, and so on. During the start-up event, the peak current is usually limited to a maximum value, very often 1 V developed across the sense resistor, and a brief noise can be heard in the transformer every time the controller pulses. This noise comes from the mechanical resonance of the magnetic material and the wire excitation induced by the high current pulses (remember the Laplace law).

Figure 7-36a depicts an oscilloscope shot of a power supply start-up sequence. You can see the voltage going up until it reaches the controller operating level. At this point, the voltage goes down as expected until the auxiliary voltage takes over. Figure 7-36b shows the same curve in

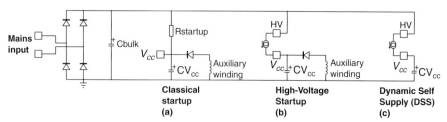

FIGURE 7-35 Several solutions are available to start up the power supply.

SIMULATIONS AND PRACTICAL DESIGNS OF FLYBACK CONVERTERS **643**

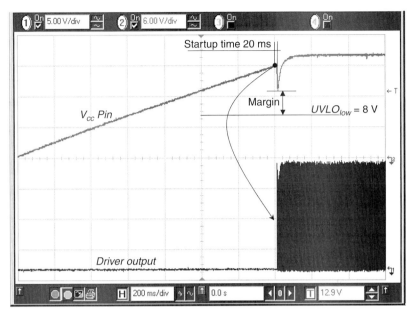

FIGURE 7-36a Start-up sequence where the auxiliary winding takes over in roughly 20 ms.

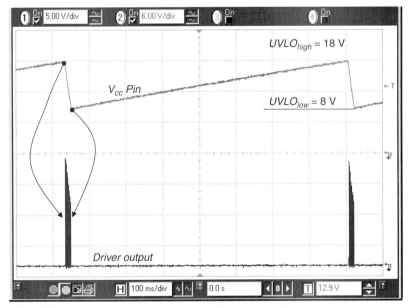

FIGURE 7-36b Same power supply where the auxiliary winding is lost at start-up.

the absence of auxiliary voltage: the controller cannot maintain its own V_{cc} and the hiccup takes place. The worst-case start-up conditions occur, as you can imagine, at the lowest input source (where the ripple on the bulk capacitor is maximum) and the highest load. A sufficient margin must exist between the point where the auxiliary winding takes over and the $UVLO_{low}$ point.

7.11.2 Start-Up Resistor Design Example

Figure 7-37 describes the current split between the start-up resistor and the V_{cc} capacitor at different events. Figure 7-37d details the associated timing diagrams.

1. At the beginning of t_1, the user plugs in the power supply: the current delivered by the start-up resistor charges the capacitor and supplies the controller. The chip is supposed to be in a complete off mode and draws a current I_{CC} equal to I_2. Actually, an internal comparator and a reference voltage are alive and observe the V_{cc} pin; hence, there is some current flowing in the chip. This current, depending on the technology, can range from 1 mA, for a UC384X bipolar controller, to a few hundred microamperes for a recent CMOS-based circuit. Always look for the maximum start-up current given in the data sheet and ignore the typical value. Make sure you have both extreme operating temperatures covered as well. (Flee these 25 °C typical values only!) Let us stick to 1 mA in this example and consider this value constant as V_{cc} elevates.

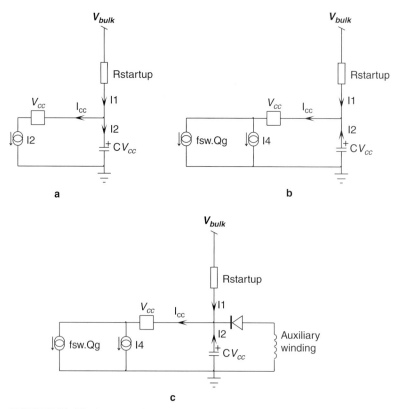

FIGURE 7-37 The three states the start-up sequence is made of.

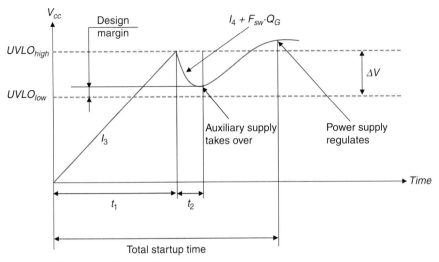

FIGURE 7-37d   Timing diagram of a start-up sequence.

2. When the voltage on the V_{cc} capacitor reaches $UVLO_{high}$, the controller wakes up all its internal circuitry (the bandgap, bias currents ...) and starts to deliver driving pulses: t_2 begins. The consumption increases to a current I_4, made of the chip natural consumption plus the current drawn by the MOSFET driving pulses (neglecting the driver switching losses):

$$I_{CC} = I_4 + Q_G F_{sw} \qquad (7\text{-}110)$$

where:

I_4 is the consumption current once the controller operates.
Q_G represents the maximum MOSFET gate charge (in nC).
F_{sw} is the controller switching frequency.

For a UC34X, I_4 is pretty high and goes up to 17 mA! For the sake of comparison, a NCP1200 from ON Semiconductor will only consume 1 mA at turn-on .... The parameter Q_G depends on your MOSFET type. Let us assume Q_G equals 35 nC which roughly corresponds to a 6 A/600 V MOSFET. Presuming a switching frequency of 100 kHz, the total current drawn by the controller during t_2 is simply:

$$I_{CC} = I_4 + Q_G F_{sw} = 17m + 35n \times 100k = 20.5 \text{ mA} \qquad (7\text{-}111)$$

3. During this t_2 time, neglecting the start-up current brought by $R_{start\text{-}up}$, the V_{cc} capacitor supplies the total controller current on its own (I_2 now reversed on Fig. 7-37b). If the capacitor is too small, the chip low operating limit $UVLO_{low}$ will precociously be touched and the power supply will fail to start-up in one clean shot. It may then start through several attempts or not start at all. Thus, what matters now is the time t_2 taken by the auxiliary winding to take over and self-supply the controller. Experience shows that a duration of 5 ms is a reasonable number to start with. This fixes the minimum value of the V_{cc} capacitor, involving Eq. 7-111, the time t_2, and the maximum voltage swing (V over the capacitor (6 V for the UC384X):

$$C_{Vcc} \geq \frac{I_{CC} t_2}{\Delta V} \geq \frac{20.5m \times 5m}{6} = 17 \, \mu\text{F} \qquad (7\text{-}112)$$

Given the unavoidable dispersion on the component value, select a 33 μF capacitor whose voltage rating depends on your auxiliary winding excursion.

4. Next step, the start-up time. Let us assume the following specifications for our power supply:

$V_{in,min} = 85$ Vrms, $V_{bulk,min} = 120$ Vdc (no ripple before the power supply starts-up)
$V_{in,max} = 265$ Vrms, $V_{bulk,max} = 375$ Vdc
$F_{sw} = 100$ kHz
Maximum start-up time = 2.5 s

If we consider t_1 as being the major contributor to the total start-up time, then we can calculate the necessary current to reach the $UVLO_{high}$ in less than 2.4 s, including a 100 ms safety margin:

$$IC_{V_{cc}} \geq \frac{UVLO_{high} C_{V_{cc}}}{2.4} = \frac{17 \times 33u}{2.4} = 234 \, \mu A \qquad (7\text{-}113)$$

In Eq. 7-113, take the highest $UVLO_{high}$ from the data sheet again to make sure all distribution cases are covered. From this equation, we need to add the controller fixed start-up current I_3 of 1 mA. Hence, the total current the start-up resistor must deliver reaches 1.3 mA. Knowing a minimum input line of 85 Vrms, the resistor is simply defined by:

$$R_{startup} = \frac{V_{bulk,min} - UVLO_{high}}{1.3m} = \frac{120 - 17}{1.3m} = 79.3 \, k\Omega \qquad (7\text{-}114)$$

Select a 82 kΩ resistor. Unfortunately, once the supply has started, the controller no longer needs the resistor. It just wastes power in heat and contributes to significantly degrade the efficiency in light load conditions. In this case, the resistor dissipates at high line (neglecting the bulk ripple):

$$P_{R_{startup}} = \frac{(V_{bulk,max} - V_{aux})^2}{R_{startup}} = \frac{(375 - 15)^2}{82k} = 1.6 \, W \qquad (7\text{-}115)$$

You might need to put three 27 kΩ/1 W resistors in series to dissipate the heat and sustain the voltage.

5. Assemble the power supply with the calculated element values and connect a probe to the controller V_{cc} pin. Supply the system with the lowest ac input (85 Vrms here) and load the converter with the maximum current per your specification. Now observe the start-up sequence; you should see something close to what Fig. 7-37d illustrates. Make sure you have enough margin at the point where the auxiliary voltage takes off and the $UVLO_{low}$ level.

1.6 W represents quite a bit of power to dissipate. Furthermore, placed too close to electrolytic capacitors, they might locally increase their operating temperature and reduce their lifetime. Is there a way to reduce the consumption or get rid of this start-up resistor? The answer is yes on both cases!

7.11.3 Half-Wave Connection

To reduce the start-up resistor consumption, we can change its upper terminal connection from the bulk capacitor to the diode bridge input. Figure 7-38 shows you the way:

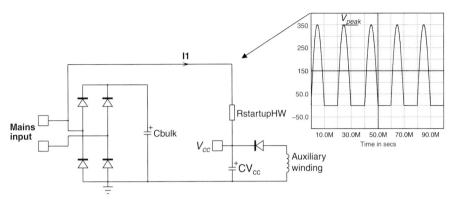

FIGURE 7-38 Connecting the start-up resistor to the diode bridge input reduces the power dissipation.

Thanks to the bridge, a half-wave rectified waveform appears on the upper terminal of $R_{start-upHW}$. The average voltage seen by the resistor:

$$V_{R_{startupHW},avg} = \frac{V_{ac,peak}}{\pi} - V_{cc} \qquad (7\text{-}116)$$

where $V_{ac,peak}$ represents the peak of the ac input voltage, $265\sqrt{2}$ at high line.

The charging current I_1 at V_{cc} close to $UVLO_{high}$ then becomes:

$$I_1 = \frac{\dfrac{V_{ac,peak}}{\pi} - UVLO_{high}}{R_{startupHW}} \qquad (7\text{-}117)$$

Equation 7-114, which corresponds to a traditional bulk connection, can be rewritten highlighting the charging current definition:

$$I_1 = \frac{V_{ac,peak} - UVLO_{high}}{R_{startup}} \qquad (7\text{-}118)$$

To obtain the same charging current via Fig. 7-38 suggestion, let us equate expressions 7-117 and 7-118 and extract the relationship between $R_{start-upHW}$ and $R_{start-up}$. For the sake of simplicity, the peak voltage being much larger than the V_{cc}, we will neglect the $UVLO$ term present in both equations. Calculation thus gives us:

$$R_{startupHW} = \frac{R_{startup}}{\pi} \qquad (7\text{-}119)$$

Following our study, we shall now evaluate the power dissipated in $R_{start-upHW}$ and see how much power the half-wave connection helps us to save. Again, we need to evaluate the power in both cases (half-wave and bulk connection) and compare the results as we did in the above lines. When the resistor connects directly to the bulk, again neglecting the V_{cc} term and the bulk ripple, the resistor dissipates:

$$P_{R_{startup}} = \frac{V_{ac,peak}^2}{R_{startup}} \qquad (7\text{-}120)$$

When the resistor now connects to the half-wave signal, the resistor undergoes sinusoidal signals. Therefore:

$$P_{R_{startpHW}} = \frac{1}{T}\int_0^{T/2} I_{R_{startupHW}}(t) \cdot V_{R_{startupHW}}(t) \cdot dt = \frac{1}{T}\int_0^{T/2} \frac{V_{ac,peak}}{R_{startupHW}} \sin\omega t \cdot V_{ac,peak} \sin\omega t \cdot dt \quad (7\text{-}121)$$

Rearranging Eq. 7-121 lets us solve the integral quicker since the second term is nothing else than the rms input voltage squared, but defined over one-half the period only:

$$P_{R_{startpHW}} = \frac{1}{R_{startupHW}} \frac{1}{T}\int_0^{T/2} (V_{ac,peak} \sin\omega t)^2 \cdot dt = \frac{1}{R_{startupHW}} \frac{(V_{ac,rms})^2}{2} \quad (7\text{-}122)$$

Introducing the peak value in Eq. 7-123 definition, we finally have:

$$P_{R_{startpHW}} = \frac{1}{R_{startupHW}} \frac{\left(\frac{V_{ac,peak}}{\sqrt{2}}\right)^2}{2} = \frac{V_{ac,peak}^2}{4R_{startupHW}} \quad (7\text{-}123)$$

Now, it is interesting to compare the relationship between both dissipation budgets (half-wave and bulk connection) by simply dividing Eq. 7-120 by Eq. 7-123:

$$\frac{P_{R_{startup}}}{P_{R_{startupHW}}} = \frac{(V_{ac,peak})^2}{R_{startup}} \frac{4R_{startupHW}}{(V_{ac,peak})^2} = \frac{4R_{startupHW}}{R_{startup}} \quad (7\text{-}124)$$

Now using Eq. 7-119, the relationship becomes:

$$\frac{P_{R_{startup}}}{P_{R_{startupHW}}} = \frac{4R_{startup}}{\pi R_{startup}} = \frac{4}{\pi} \approx 1.27 \quad (7\text{-}125)$$

As a conclusion, if we apply the half-wave connection to our original bulk-connected start-up scheme, we gain 21% in power dissipation. Let us apply this result to our design example:

$$R_{startupHW} = \frac{R_{startup}}{\pi} = \frac{82k}{\pi} = 26\,\text{k}\Omega \quad (7\text{-}126a)$$

from Eq. 7-123:

$$P_{R_{startpHW}} = \frac{V_{ac,peak}^2}{4R_{startupHW}} = \frac{(265 \times \sqrt{2})^2}{4 \times 26k} = 1.35\,\text{W} \quad (7\text{-}126b)$$

In that case, two 13 kΩ/1 W resistors in series would be ok, compared to three 27 kΩ/1W in the direct connection case.

Well, we certainly reduced the dissipation lost in heat, but the best would be to get rid of it.

7.11.4 Good Riddance, Start-up Resistor!

Figure 7-39 depicts a solution involving either a bipolar or a MOSFET in an active current source. Actually, the transistor acts as a ballast whose current is limited by the same resistor calculated through Eq. 7-114. When the auxiliary voltage builds up, it simply reverse biases the transistor which disconnects the start-up resistor. D_1 in the bipolar option prevents from an

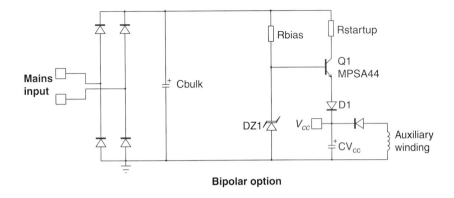

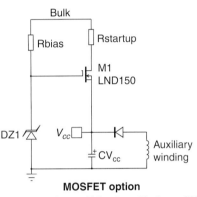

FIGURE 7-39 A simple ballast based on a high-voltage bipolar or MOSFET cancels the power dissipated by the start-up resistor.

excessive base-emitter reverse voltage. It can be omitted for the MOSFET option. In both cases, the zener voltage is selected to be at least 2 V above the maximum start-up level $UVLO_{high}$ of the concerned PWM controller: 1 V for the V_{be} drop at low temperatures and 1 V for the margin. With an UC384X featuring a 17 V start-up (UC3842/44), select a 20 V zener.

7.11.5 High-Voltage Current Source

Some semiconductor manufacturers have introduced a technology capable of a direct connection to the bulk. Known as the *very high voltage integrated circuit* (VHVIC) at ON Semiconductor, the process accepts up to 700 V and, as such, is well suited to build high-voltage current sources. Figure 7-40 shows the internal high-level circuitry of a chip including such approach (a member of the NCP120X series for instance).

At power-up, the reference level equals the $UVLO_{high}$ level, e.g. 12 V, and the source delivers a certain current (usually a few mA). When the V_{cc} voltage reaches the upper level, the comparator senses it and turns the current source off. As in the start-up resistor case, the capacitor stays on its own to supply the controller. The auxiliary winding is then supposed to take over before reaching the second level, $UVLO_{low}$. If not, the current source turns *on* and *off* at a pace imposed by the V_{cc} capacitor: this is hiccup mode.

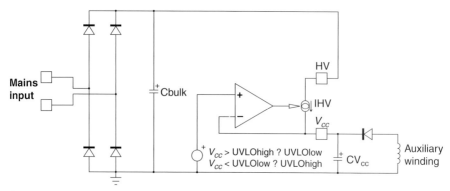

FIGURE 7-40 Thanks to a proprietary technology, the NCP120X circuits can directly connect to the bulk capacitor and saves some dissipated power.

When both *UVLO* levels are selected to be close to each others, let us say 12 V and 10 V, the start-up current source turns into a dynamic self-supply (DSS), delivering an average level of 11 V. You can thus forget the auxiliary winding! If you connect a probe to the V_{cc} pin of a NCP1200 or NCP1216, you might observe a signal as on Fig. 7-41. The HV pin is pulsating from the nominal current source value (here 4 mA) to almost zero when the source is off (the leakage lies around 30-40 μA). The regulation type of the DSS is hysteretic. In other words, the on-time duration will automatically adjust depending on the consumed current by the controller. If the controller consumes 2 mA and the source peak is 4 mA, the duty cycle will be established to 50%. The DSS type of a controller self-supply suits low-power circuits. For instance, a NCP1200 driving a 2 A MOSFET. Remember that Eq. 7-110 still holds and the

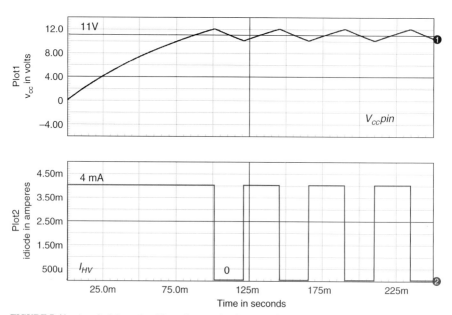

FIGURE 7-41 A typical dynamic self-supply operation from an NCP1216 controller.

operating frequency associated to the MOSFET Q_G will quickly limit the operations by either exceeding the DSS capabilities or inducing too much heat on the die. This average power dissipation limit constrains the DSS usage to driving low Q_G MOSFETs. If we assume a 15 nC Q_G driven at a 65 kHz switching frequency, a 1 mA controller consumption, then the total DSS dissipated power reaches:

$$P_{DSS} = I_{CC}V_{bulk,max} = (I_4 + Q_G F_{sw}) = (1m + 65k \times 15n) \times 375 = 740 \, \text{mW} \quad (7\text{-}127)$$

Given the DIP8 or DIP14 power dissipation capabilities (below 1 W with a large copper area around them), the DSS usage is naturally limited to the operation of low gate-charge MOSFETs.

7.11.6 The Auxiliary Winding

The auxiliary winding is an important part of the supply since it powers the controller after the start-up sequence. Figure 7-42a represents a typical arrangement where the start-up resistor appears. A small resistor like R_1 can sometimes be inserted to limit the effect of the leakage inductance. We will see the role it plays in protecting the converter against short-circuits but also how it can defeat the auxiliary supply in standby conditions.

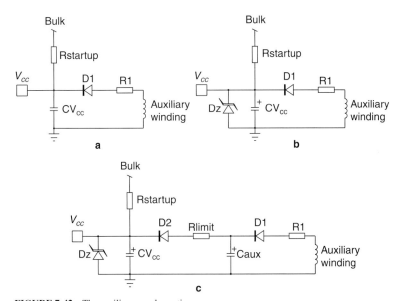

FIGURE 7-42 The auxiliary supply section.

Most of designers select a 1N4148 for the auxiliary circuit. Watch out for the reverse voltage seen by this diode:

$$\text{PIV} = \frac{N_{aux}}{N_p} V_{in} + V_{cc} \quad (7\text{-}128)$$

where V_{cc} represents the dc cathode voltage after rectification. The 1N4148 accepts up to 100 V of repetitive reverse voltage. Other good candidates are the 1N4935 or the BAV20 which accept up to 200 V.

The voltage on the auxiliary winding depends on the controller and the selected MOSFET. Most of these devices do not accept V_{GS} levels above 20 V, though some recent components accept up to ± 30 V. If you look at a MOSFET data sheet, you will see that the $R_{DS(on)}$ is specified at 10 V. If you increase the driving voltage to 15 V, you might gain a few percent in conduction losses, but not a lot. However, the power dissipated by the driver will increase. Also, the MOSFET lifetime depends on several parameters among which the driving voltage plays a significant role, so these are two good reasons to not overdrive the MOSFET! So what value then? Well, around 15 V looks like a good number, given that the effective driving voltage also depends on the sense resistor drop. Since the sense resistor appears in series with the source, when it develops 1 V, this voltage subtracts from the driving voltage.

If the driver includes a driving clamp, naturally limiting the gate-source voltage below 15 V, then there is no harm to let the driver supply swing to a higher level as long as it can sustain it (usually 20 Vdc for a CMOS-based process). On the other hand, if the auxiliary winding is subject to large variations, you might need a clamp. Figure 7-42b offers a simple means to clamp the voltage via the limiting resistor R_1. This works fine as long as you can increase the resistor to a point where the zener power dissipation stays under control. Problems start to arise in standby, where only a few pulses are present. During their presence, the capacitor C_{Vcc} must be fully replenished; otherwise the controller will stop operation. The amount of charge needed to refuel the capacitor depends on the peak current flowing in the capacitor. If you limit it via a series resistor, you store less charge and the voltage drops.

Increasing the V_{cc} capacitor remains a possible option, but it clearly hampers the start-up time. Figure 7-42c shows the so-called split supply solution where a big capacitor stores the necessary amount of charge in standby but stays isolated from the V_{cc} pin at start-up. The V_{cc} capacitor still obeys the design equations we have derived, but the auxiliary capacitor C_{aux} can be freely increased to sustain the standby: it will not delay the start-up time thanks to D_2. If necessary, the series resistor R_1 can be omitted in case of a long time difference between pulses in standby. Figure 7-43 shows how skip-cycle (see below what it means) slices a continuous PWM pattern into short pulses. In this case, the capacitor must keep the auxiliary level during 30 ms at least.

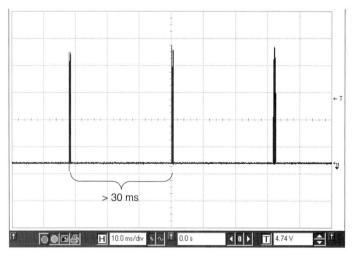

FIGURE 7-43 Typical drive pulses of a controller featuring skip-cycle. Please note the time distance between two bursts.

7.11.7 Short-Circuit Protection

A true output short-circuit will be seen by the controller as a sudden loss of feedback information. Yes, during the start-up sequence, the system also runs open loop since no regulation has yet occurred. Circuits permanently monitoring the feedback loop have thus a means to detect this kind of event. On the market, however, some controllers do not include short-circuit protection. They rely on the collapse of the auxiliary winding when the output is short-circuited to enter a protective burst operation. Well, on paper, the idea looks great. Unfortunately, the perfidious leakage inductance sabotages the coupling between the power winding and the auxiliary winding. Figure 7-44 depicts a typical shot obtained on the anode of an auxiliary winding

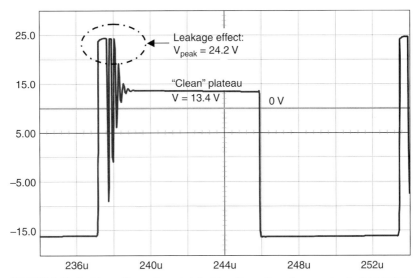

FIGURE 7-44 In this auxiliary voltage shot, the big spike confirms the presence of a large leakage inductance between the power and the auxiliary windings.

diode. As you can imagine, with the diode acting as a peak rectifier, the auxiliary voltage will go to 23.5 V whereas the plateau is at 12.7 V, considering a V_f of 0.7 V for the auxiliary diode. The plateau represents the output voltage, scaled down via the appropriate turns ratio between the power and auxiliary windings. Even if the plateau falls to a few volts because of the output short-circuit, the auxiliary supply will not collapse: the controller will keep driving the MOSFET, and the output current will probably destroy the output diode after a few minutes.

To avoid this problem, the best idea is to sense the feedback loop and make a decision regardless of the auxiliary conditions. For inexpensive controllers not implementing this method, you need to damp the ringing waveform by installing an inductor and its associated damper. A resistor can also help, but you will experience troubles in standby (V_{cc} capacitor refueling deficiency). Figure 7-45 shows how to install these passive elements on the auxiliary winding, just before the diode. (Typical values are shown here.) If you do not damp and install the inductor alone, depending on its value, it might ring so much that you could destroy the auxiliary diode by exceeding its maximum reverse biasing possibilities. It did happen. Figure 7-46a and b shows the auxiliary signals obtained without and with the LC filter.

This technique will help detect a real short-circuit on the secondary-side output. True overload detection is another story.

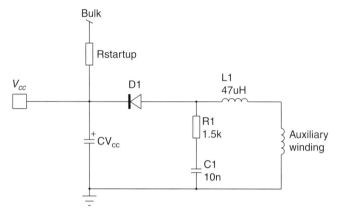

FIGURE 7-45 An *LC* network helps to tame the leakage spike at turn-off.

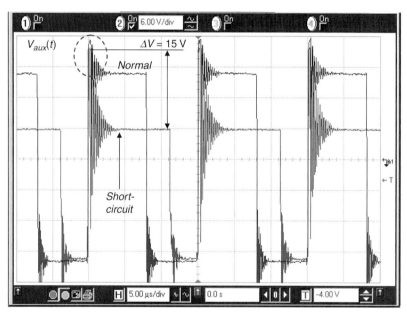

FIGURE 7-46a The auxiliary signal without a damping network (connection to the diode anode). Note the peak voltage amplitude; the auxiliary supply does not collapse.

7.11.8 Observing the Feedback Pin

An overload detection scheme could be implemented in different ways:

- The controller permanently observes the feedback signal, knowing that it should be within a certain range for regulation. If the signal escapes this range, there must be a problem: a flag is asserted.
- Rather than look at the feedback, the controller observes the current-sense pin and looks for overshoots beyond the maximum limit. If the current exceeds the limit, a flag is asserted.

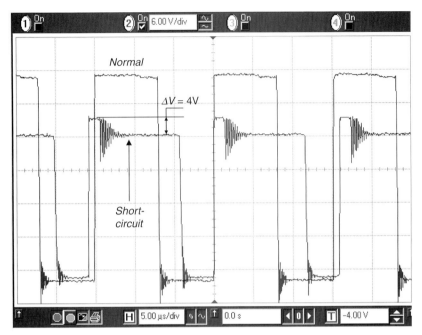

FIGURE 7-46b The auxiliary signal with a damping network (connection to the diode anode).

The first option is already implemented in a variety of controllers, such as the NCP1230 from ON Semiconductor. When the flag is asserted, a timer is started to delay the reaction to a fault. After all, a start-up sequence is also seen as a fault since no feedback signal appears prior to regulation. The timer is there to give a sufficient time for starting up. Typical values are in the range of 50 to 100 ms. Figure 7-47 describes a possible logic arrangement observing the feedback. To deliver the maximum peak current (1 V over the sense resistor), the feedback pin must be in the vicinity of 3 V. If it is above, close to the internal V_{dd} level, the PWM chip no longer has control over the converter. When this situation is detected by comparator CMP2, the switch *SW* opens and the external capacitor can charge. If the fault lasts long enough, the timer capacitor voltage will reach the V_{timer} reference level, confirming the presence of a fault. The controller can then react in different ways: go into autorecovery hiccup mode or simply totally latch off the circuit. If the fault lasts too short a time, the capacitor is reset and waits for another event. Some schemes do not fully reset the timer capacitor, but accumulate the events. This technique offers the best fault detection performance.

The precision of the overload detection depends on several factors, among which the TL431 bias plays a role. If you go back to Fig. 3-45, you can see how a badly biased TL431 degrades the squareness of the *I/V* characteristic. Hence, if you need a precise detection point, you should also pay attention to the secondary side circuitry.

In DSS-based controllers (NCP1200 or NCP1216), the error flag is tested as the V_{cc} voltage reaches 10 V. You thus have a natural short-circuit protection without the auxiliary winding problems.

7.11.9 Compensating the Propagation Delay

If you go back to Fig. 7-20, you understand how the primary slope affects the maximum current limit. This phenomenon will make your flyback converter deliver more power at high line

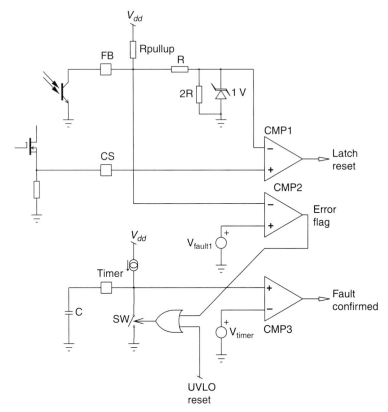

FIGURE 7-47 Monitoring the feedback brings excellent short-circuit/overload protection, regardless of the auxiliary V_{cc} state.

than at low line, explaining the difficulty to precisely react to an overload situation. To compensate this phenomenon, you need an *overpower protection* (OPP) circuit. The idea behind this solution is to offset the current-sense information to cheat the PWM controller. By increasing the voltage floor as the bulk level goes up, the current ramp reaches the maximum peak limit sooner. This, in effect, reduces the peak current excursion and clamps down the maximum available power (Fig. 7-48).

Figure 7-49a to c depicts most of the popular solutions found on the market. Figure 7-49a illustrates the simplest solution, but it clearly affects the current consumption on the bulk rail and requires high-voltage sensing. Figure 7-49b forces rotation of the diode in the auxiliary winding ground, and it can sometimes change the EMI signature. However, it does not burden the bulk rail with a permanent dc consumption. Figure 7-49c offers another option by creating an offset *ex nihilo* from a forward winding. As such, you create a voltage source solely dependent on the bulk rail.

The compensation level depends on various parameters such as the total propagation delay which also includes the driver turn-off capability: even if the controller reacts within a few hundred nanoseconds, if the driver cannot pull the gate down quickly enough, the total reaction time suffers and the current keeps increasing. A small PNP can help strengthen the pull-down action, as shown a few lines below.

The calculation of the OPP resistor requires a few lines of algebra, especially if we involve some ramp compensation. To help us in this task, Fig. 7-50 shows a simplified source arrangement

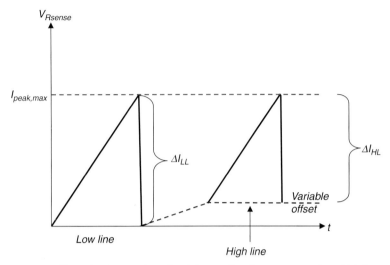

FIGURE 7-48 Offsetting the voltage floor helps to reduce the current excursion at high line.

where the ramp generator finds its place. The easiest way to solve this system is to apply the superposition theorem:

V_{ramp} and V_{bulk} grounded: $V_{CS} = \dfrac{R_{ramp}||R_{OPP}}{R_{ramp}||R_{OPP} + R_{comp}} V_{sense}$ (7-129)

V_{ramp} and V_{sense} grounded: $V_{CS} = \dfrac{R_{ramp}||R_{comp}}{R_{ramp}||R_{comp} + R_{OPP}} V_{bulk}$ (7-130)

V_{bulk} and V_{sense} grounded: $V_{CS} = \dfrac{R_{OPP}||R_{comp}}{R_{OPP}||R_{comp} + R_{ramp}} V_{ramp}$ (7-131)

From these equations, the total sense voltage V_{CS} is obtained by summing Eqs. (7-129) through (7-131). We can then extract a value for the compensation resistor we are looking for:

$$R_{OPP} = \dfrac{(V_{CS} - V_{bulk})R_{ramp}R_{comp}}{V_{ramp}R_{comp} + V_{sense}R_{ramp} - V_{CS}(R_{ramp} + R_{comp})}$$ (7-132)

To compensate a given converter, place it in the operating conditions under which you would like it to enter hiccup. For instance, our flyback delivers 3 A for the nominal power, and the maximum of the specification is 4 A. Supply the converter with the highest input level and load it with 3.7 A, including some safety margin. Now, collect the information needed to feed Eq. (7-132) with. Let us assume we have measured the following data:

$V_{bulk} = 370$ V (maximum value in this example)
$S_{ramp} = 133$ mV/μs

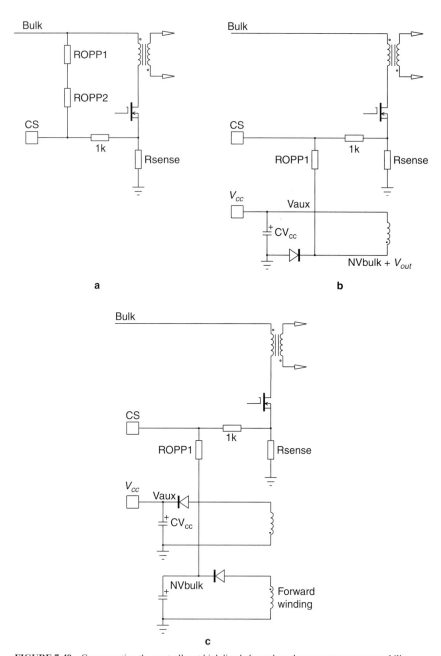

FIGURE 7-49 Compensating the controller at high line helps reduce the converter power capability.

SIMULATIONS AND PRACTICAL DESIGNS OF FLYBACK CONVERTERS 659

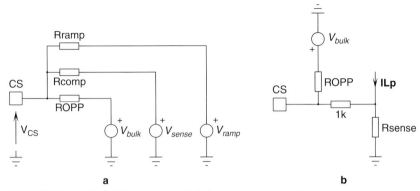

FIGURE 7-50 A simplified representation including a ramp generator if present.

$V_{sense} = 830$ mV
$F_{sw} = 65$ kHz
$t_{on} = 3.5$ μs
$R_{ramp} = 20$ kΩ
$R_{comp} = 1$ kΩ
$V_{CS,max} = 1.1$ V—from the controller data sheet, maximum current limit

V_{sense} is the peak voltage measured across the sense resistor, t_{on} is the operating on time, and S_{ramp} is the ramp compensation slope.

Beyond $P_{out,max}$, we would like to shut down the power supply. What overpower resistor must we install to follow the suggestions of Fig. 7-49a? Applying Eq. (7-132) gives

$$R_{OPP} = \frac{(V_{CS} - V_{bulk})R_{ramp}R_{comp}}{S_{ramp}t_{on}R_{comp} + V_{sense}R_{ramp} - V_{CS}(R_{ramp} + R_{comp})}$$

$$= \frac{(1.1 - 370) \times 20k \times 1k}{0.133 \times 3.5 \times 1k + 830m \times 20k - 1.1 \times (21k)} = 1.22 \, M\Omega \quad (7\text{-}133)$$

In some cases, there is no ramp compensation at all. The compensation resistor formula simplifies to

$$R_{OPP} = R_{comp}\frac{V_{bulk} - V_{CS}}{V_{CS} - V_{sense}} = 1k \times \frac{370 - 1.1}{1.1 - 0.83} = 1.36 \, M\Omega \quad (7\text{-}134)$$

To plot the effects of the OPP compensation from Fig. 7-50b, we can derive the expression of the peak current (no external ramp in this example):

$$I_{L_p}(V_{bulk}) = \frac{R_{OPP} + R_{comp}}{R_{sense}R_{OPP}}\left(V_{CS} - V_{bulk}\frac{R_{comp}}{R_{comp} + R_{OPP}}\right) + \frac{V_{bulk}}{L_p}t_{prop} \quad (7\text{-}135)$$

Results of compensated and noncompensated converters appear in Fig. 7-51.

Despite this technique, given the dispersion on the current-sense limit, the primary inductance, and the sense elements, precisely clamping the secondary side current represents a difficult exercise. Another solution lies in directly monitoring the secondary side current and instructing the controller to shut down.

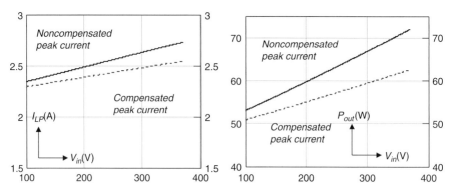

FIGURE 7-51 Comparison of a noncompensated and a compensated flyback converter operated in DCM.

7.11.10 Sensing the Secondary Side Current

In an isolated flyback, the controller uses an optocoupler to sense the output voltage and reacts accordingly depending on the power demand. Why not also use this loop to pass some information pertaining to the output current? Figure 7-52 offers a solution associating a TL431 and a bipolar transistor.

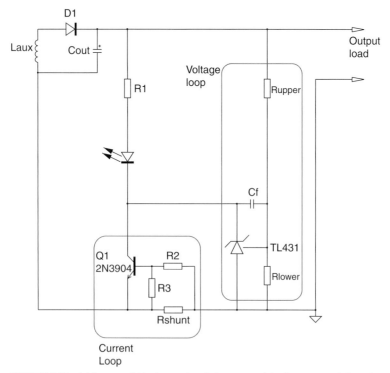

FIGURE 7-52 Adding a small bipolar transistor helps to control the direct current delivered to the load.

When the current flowing through R_{shunt} produces a voltage across R_3 below Q_1 threshold voltage (≈ 650 mV at $T_j = 25$ °C), the TL431 controls the loop alone. When the voltage drop on R_3 reaches 650 mV, then Q_1 starts to conduct and takes over the TL431: more current flows into R_1 and the controller reduces its duty cycle. In a short-circuit condition, the TL431 leaves the picture and Q_1 makes the converter work as a constant-current generator. Figure 7-53 shows the kind of output characteristic you can get from this arrangement. This

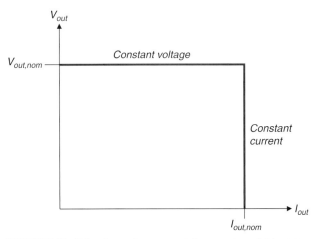

FIGURE 7-53 Below the maximum current, the converter maintains a constant output voltage. If the load consumes more current, the current loop takes over and controls the converter.

is called a *constant-current constant-voltage* (CC-CV) operation. The divider made around R_2 and R_3 helps to add a capacitor across R_3 in case a delay is necessary (for printer applications, for instance, where the load is made of short high-current bursts). If you set R_3 to 10 kΩ and R_2 to 1 kΩ, then triggering on a 2 A output current means a shunt resistor of the following value:

$$R_{shunt} = \frac{V_{be}(R_3 + R_2)}{I_{out} R_3} = \frac{650m(10k + 1k)}{10k \times 2} = 357 \, m\Omega \tag{7-136}$$

Unfortunately, the power dissipated by the shunt reaches

$$P_{shunt} = I_{out}^2 R_{shunt} = 4 \times 0.357 = 1.43 \, W \tag{7-137}$$

This power dissipation can be explained by the voltage drop needed across the shunt to trip the circuit. It is true that a germanium transistor would offer a better alternative, but who remembers the AC127 found in our old car radios (ahem, and the OC70)?

Also, as explained, despite a rather good distribution of the transistor V_{be}, its value changes with the junction temperature with a slope of -2.2 mV/°C. It might not represent a problem with a loose output current limit specification, say, $\pm 15\%$, but in some cases, it might not be acceptable. A way to circumvent this problem is to select a dedicated controller such as the MC33341 or the more recent NCP4300 from ON Semiconductor ($V_{drop} = 200$ mV).

7.11.11 Improving the Drive Capability

The transistor turn-off time also takes its share of the total propagation delay. When you are using a large Q_G MOSFET, pulling down the gate requires a certain amount of current that the controller can sometimes have problem absorbing. In that case, a low-cost 2N2907 can help to speed up the gate discharge. Figure 7-54a depicts the way to wire it; Fig. 7-54d shows the improvement.

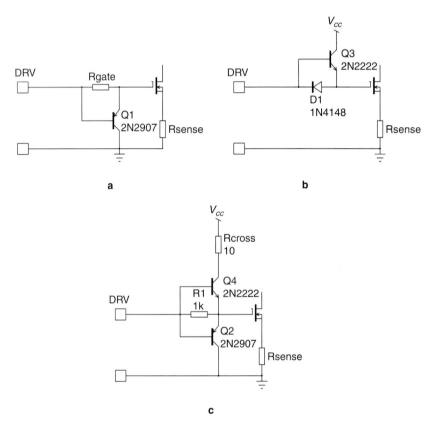

FIGURE 7-54a, b, and c Adding low-cost bipolars can dramatically improve the driving performance and ease the controller burden in presence of large Q_G MOSFETs.

We have seen through Eq. (7-110) that the controller dissipation can reach a few hundred milliwatts in the presence of important V_{cc} voltages and big MOSFETs. To remove the driving burden from the controller, a simple external NPN transistor wired as an emitter follower can reduce the package temperature (Fig. 7-54b). The diode D_1 ensures a fast discharge of the gate. Finally, if you need both fast charge and fast discharge performance, Fig. 7-54c might represent a solution. The resistor in series with V_{cc} limits the cross-conduction current, a heavy noise generator that can often disturb the controller. If this low-cost buffer is not enough in terms of current, you can always select a dedicated dual driver such as the MC33151/152, which is able to deliver 1.5 A peak.

By the way, why does the UC384X output stage always deliver more than the new CMOS-based controllers found on the market? Well, this is mainly due to the bipolar nature of the UC384X driver which behaves as a real current source. When the upper or lower transistor

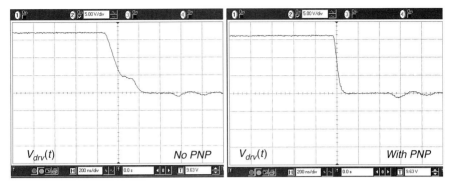

FIGURE 7-54d This picture shows the difference in turn-off times when a small PNP is added as recommended in the text.

closes, to either bias or discharge the gate, the transistor becomes a current source whose value stays relatively independent of its V_{ce} (see Early effect). With a output stage made of MOSFETs, when closed, the upper-side MOSFET offers a non linear resistive path. The current available to drive the flyback transistor gate thus depends on the gate-source level itself.

$$I_{gate}(t) = \frac{V_{cc} - V_{GS}(t)}{R_{DS(on)}} \tag{7-138}$$

In this formula, the resistive term in the denominator behaves in a nonlinear manner, and its value depends on the die temperature: when a CMOS-based controller heats up, its driving capability reduces. It can explain why the EMI signature changes after a warm-up sequence or why the switching noise on the controller also improves after this period: as both lower and upper side $R_{DS(on)}$ have increased, the shoot-through current has diminished.

7.11.12 Overvoltage Protection

An overvoltage protection (OVP) circuit protects the load, and the converter itself, when the loop control is lost, for instance, when the optocoupler breaks or a wrong solder joint is made in production. It the controller sometimes hosts its own OVP circuitry, it often does not and you need to design it as a separate circuit. Depending on your end customer, since an OVP detection is considered a dangerous event, most of designs are latched and remain off when such a situation arises. The reset occurs when the user unplugs the converter from the wall outlet. Figure 7-55a and b offers two possibilities to detect these OVPs: you can either observe the secondary voltage image on the auxiliary winding or detect the information via the output voltage directly. In the latter, you need an optocoupler to transport the secondary side information to the primary side: the cost is obviously higher, but it does not suffer from coupling problems (see Fig. 7-44 for leakage problems). Observing the auxiliary voltage always refers to the leakage term. In Fig. 7-55a, the monitored voltage is physically separated from the V_{cc} line. Thus, we can easily install a filter without interacting with the controller supply. Then a voltage divider made up of R_9 and R_{10} selects the level at which the circuit must latch. When the voltage over R_9 reaches ≈ 0.65 V at $T_j = 25$ °C, then the discrete *silicon-controlled rectifier* (SCR) latches and pulls V_{cc} down to ground. Make sure an adequate margin exists, especially on transient load removals or start-up sequences, to avoid false tripping of the latch. A 100 nF (C_5) capacitor helps to improve the noise immunity. As the controller V_{cc} rail can be of rather low impedance, it is necessary to locally degrade it via resistor R_4, set to 47 Ω in this example; otherwise the SCR could suffer from an overcurrent problem.

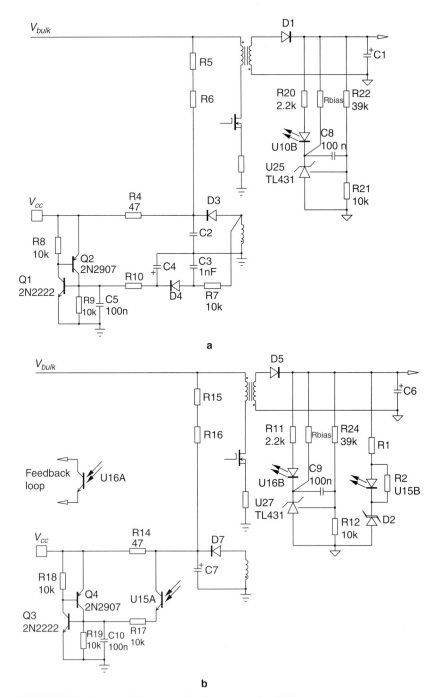

FIGURE 7-55 A discrete SCR latches off in the presence of an OVP event.

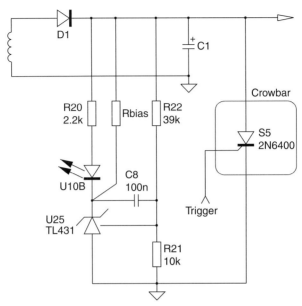

FIGURE 7-56 A crowbar circuit immediately short-circuits the converter output in case of loop failure. The trigger signal can come from a dedicated controller such as the MC3423.

Figure 7-55b does not differ that much, except that the order to latch comes from the secondary side. The optocoupler LED will pass current when the output voltage reaches the zener voltage (D_2) plus a 1 V drop from the LED itself. At this moment, the U_{15A} emitter will go up and bias the SCR to protect the converter.

In case the load cannot suffer any voltage runaway, the crowbar option might represent a possible choice (Fig. 7-56). The circuit detects the OVP as Fig. 7-55b does, but it triggers a powerful thyristor to actually short-circuit the converter supply. If the power supply features a short-circuit protection, hiccup mode will be entered until the supply is stopped. Readers interested in this device can have a look at the MC3423, a dedicated crowbar driver from ON Semiconductor.

7.12 STANDBY POWER OF CONVERTERS

Since the days when Thomas Edison powered the first lightbulb in 1879, electricity has been considered a gift, and our modern consumer lifestyles show this to be true. Unfortunately, freedom always comes at a price. International agencies have now been set up to regulate energy production amid concerns that atmospheric emissions from power stations could be the cause of many weather-related problems. Solutions need to quickly emerge which reduce energy bills while helping us consume electricity in a clever way. Working closely with technical committees such as the International Energy Agency (IEA) [7], some semiconductor companies offer new, ready-to-use integrated solutions to tackle the standby power problem. This paragraph reviews the roots of the standby losses and presents currently available solutions to help designers easily meet new coming standards or so-called codes of conduct.

7.12.1 What Is Standby Power?

Let us start by defining standby power so as to avoid any of the common misunderstandings between designers and final users. We can distinguish different kinds of standby power depending on the type of equipment concerned:

1. *When the active work of an apparatus connected to the mains has come to an end, the power drawn by this apparatus should ideally tend to* 0 W. By *active work*, we mean a function for which the apparatus is designed, for instance, charging a battery. When the user disconnects a cell phone from the charger and leaves it plugged in the mains outlet, the charger should become inert, drawing nearly zero power. If you want to check for this parameter, just touch a charger or an ac adapter case (your notebook's, for example) and feel its temperature through your hand. You will be surprised how hot some of these can be, clearly revealing their poor standby power performance!

2. *When the active work of an apparatus connected to the mains has been temporarily deactivated, either automatically or through a user demand, the power drawn from this apparatus shall be the smallest possible.* Again, by active work, we can take the example of a TV set left in standby via a remote control order, but whose circuitry (including the shining front LED!) shall be kept alive to respond to a wake-up signal as soon as the user wishes it. It is the designer's duty to keep the internal consumption very low (using low-power µPs, high-efficiency LEDs, etc.). However the final element remains the switch-mode power supply (SMPS) connected to the ac supply. Unfortunately, most of current SMPS efficiencies drop to a few tens of percent when operated well below their nominal power. If you have a 25% efficiency at 500 mW output power, then you consume close to 2 W. A lot of electronic apparatus spend most of their time in this mode. So the power consumed represents a significant fraction of the domestic power budget at around 5%.

Documents and various links pertinent to standby power can be found on the Web site via Refs. 7 and 8.

7.12.2 The Origins of Losses

Figure 7-57 portrays the typical arrangement of the flyback converter, one of today's most popular converter topologies used in consumer products. Components are symbolically represented for a better understanding of the process. Suppose that the power supply switches at 100 kHz and

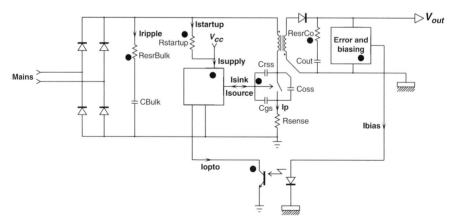

FIGURE 7-57 The various losses that the converter is the seat of. Every spot on the picture is a source of loss.

delivers 12 V nominal dc output voltage. With no load on its output, i.e., in a standby mode, the controller will naturally reduce the duty cycle to ensure the right output level. Let us now try to identify the loss contributors. If we start from the right side, we first have a feedback network whose function is to ensure that the output level stays within given specifications. A TL431-based network requires at least 1 mA to properly operate plus a few microamperes flowing through the sensing network. You also need to bias the optocoupler LED in order to instruct the primary side controller to reduce the power transfer. Including all these secondary losses, we come to an output power of around 24 mW in no-load situation if we consider a total secondary side current of 2 mA.

A few other milliwants is circulating through various other resistive elements (secondary diode, capacitor ESRs, etc.) but since we assume that the rms current is low, they can be neglected. On the primary side, every time the power MOSFET closes, it discharges the drain node parasitic capacitance consisting of MOSFET C_{oss} and C_{rss}, transformer stray capacitance, etc. Each capacitor C is charged to a voltage V and discharged at a given switching frequency F_{sw}. The average losses generated by these elements are defined by

$$P_{loss} = 0.5CV^2F_{sw} \qquad (7\text{-}139)$$

Here C and V are fixed elements and thus difficult to modify. However, since the switching frequency can be selected, it represents a first possible trail to follow. The MOSFET itself presents ohmic losses and dissipates an average power given by (conduction losses in DCM)

$$P_{MOSFET,cond} = \frac{1}{3}I_{peak}^2 DR_{DS(on)} \qquad (7\text{-}140)$$

The pulse width modulator (PWM) controller, the heart of the SMPS, needs power to beat. If the device is a bipolar UC384X based, it is likely to consume around 20 to 25 mA in total, which from a 12 V source gives 240 mW best case. The driving current also represents an important part of the budget. Operating a 50 nC MOSFET at 100 kHz requires a 5 mA average current [Eq. (7-110)] directly consumed from the controller V_{cc}, whatever the duty cycle. Another 60 mW of power... The biggest portion of all? The naughty start-up resistor which ensures at least 1 mA of start-up current needed for these bipolar controllers! On a 230 Vrms mains, another 300 mW is wasted in heat.

If we sum up all contributors, including front-stage losses (diode, bulk capacitor ESR, etc.) and drain clamp losses, we can easily end up with a no-load standby power of around 1 W.

7.12.3 Skipping Unwanted Cycles

We have seen that the switching frequency plays a significant role in the power loss process. Since we do not transmit any power in standby (or just a little depending on the application), why bother the MOSFET with a continuous flow of pulses? Why not just transmit the needed power via a burst of pulses and stay quiet the rest of the time? We would surely introduce a bit of output ripple because of this quiet period, but a lot of switching losses would fade away! This is the principle of skip-cycle regulation: the controller skips unnecessary switching cycles when entering the standby area. Figure 7-58 depicts how skip cycle takes place in low standby controllers from ON Semiconductor (NCP120X series). The controller waits until the output power demand goes low and then starts skipping cycles. Skipping a cycle means that some switching cycles are simply ignored for a certain time. Figure 7-59 gives the basic circuitry needed for this technique. When the feedback voltage on the FB pin passes below the skip source V_{skip}, the comparator CMP1 resets the internal latch. As all pulses are now stopped, the output voltage starts to drop, leading to a movement of the feedback voltage. When the skip comparator detects that the FB voltage has gone above the skip source V_{skip}, pulses are released, bringing V_{out} toward the target. At this moment, the FB voltage falls again below the skip source value and a hysteretic regulation takes place. If V_{skip} equals 1 V and the FB pin can swing up to 3 V (current-sense limit

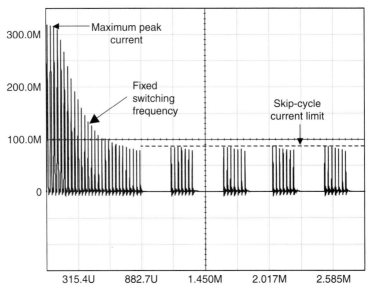

FIGURE 7-58 Skip-cycle takes place at low peak currents which guarantees noise-free operation.

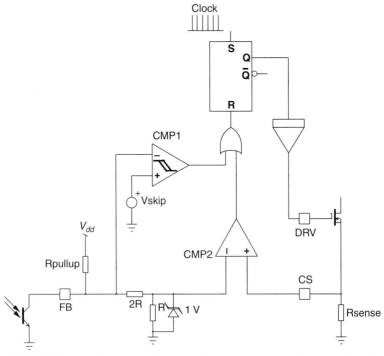

FIGURE 7-59 A simple comparator observing the feedback loop is enough to implement skip-cycle.

of 1 V), then the skip operation occurs at 30% of the maximum authorized peak current. When the power demand increases again, the pulse bunches come closer to each other until the controller goes back to a full PWM pattern with a variable peak current.

Given the hysteretic nature of the regulation in skip mode, you have no control over the way pulse packets are arranged by the controller. What matters is the loop bandwidth, the load level, and the hysteresis on the skip-cycle comparator. The best standby is obtained when only a few pulses occur, separated by several milliseconds of silence (see Fig. 7-43). Of course, it then becomes difficult to maintain the self-supply, and all losses must be chased. If the peak current level at which skip-cycle occurs is selected low enough, you will not have audible noise issues but standby will suffer. Noise comes from the mechanical resonance of the transformer (or the *RCD* clamp capacitor, especially disk types), excited by (1) the audible frequencies covered during the hysteretic regulation and (2) the sharp discontinuity associated with the pulse packet appearance. However, circuits such as the NCP120X series naturally minimize the noise generation because you can select the skip-mode peak current.

7.12.4 Skipping Cycles with a UC384X

Of course you can skip cycles with a UC384X! Using a low-cost LM393, Fig. 7-60a shows how to stop operations of the PWM controller by lifting up its current-sense pin. Thanks to the absence of an internal LEB, if the current-sense pin is pulled higher than 1 V, the circuit fully stops pulsing. Let us now have one of the comparator input observing the feedback voltage and the other receiving a portion of the reference level—our skip reference source V_{skip} in the previous example. Add a simple PNP transistor to the comparator output and the trick is done. As the comparator noninverting pin (FB node) reaches the voltage set by R_6, Q_1 base is pulled down and the controller stops and restarts the switching operations to a pace imposed by the feedback loop: skip-cycle is born. Figure 7-60b shows the captured waveforms at different output levels, P_{out1} being the lowest one. Thanks to this circuit, we were able to divide the no-load standby power by 2.

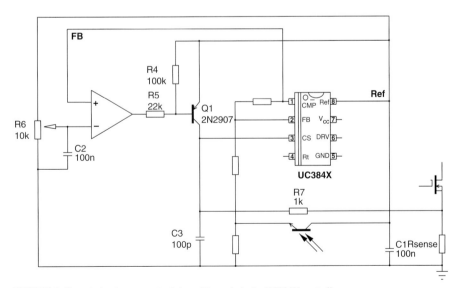

FIGURE 7-60a A simple comparator brings skip-cycle to the UC384X controller.

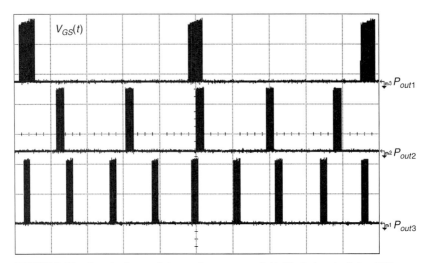

FIGURE 7-60b Once implemented, the standby power was divided by 2, dropping below 1 W.

7.12.5 Frequency Foldback

The frequency foldback technique offers another interesting alternative to the skip-cycle operation. Rather than artificially slice the switching pattern, a *voltage-controlled oscillator* (VCO) starts to act when the imposed peak current reaches a certain low point. As the controller does not allow this current to further decrease, the only way to reduce the transmitted power lies in a switching frequency reduction. This solution is usually implemented in partnership with a quasi-resonant flyback converter. As the frequency naturally increases when the load gets lighter, an internal oscillator monitors the minimum switching period. Usually, the designer forbids the frequency to exceed 70 kHz for EMI concerns. When the peak current freezes (around 30% of the maximum allowed peak current), the frequency can no longer increase as the feedback loses current control (the current is now fixed). The VCO thus takes over and linearly decreases the frequency to a few hundred hertz if necessary. Figure 7-61a shows the frequency versus output power plot whereas Fig. 7-61b plots a few operating waveforms from the NCP1205, a controller implementing this principle.

7.13 A 20 W, SINGLE-OUTPUT POWER SUPPLY

This design example describes how to calculate the various component values for a 20 W flyback converter operated on universal mains with the following specifications:

$V_{in,min}$ = 85 Vrms

$V_{bulk,min}$ = 90 Vdc (considering 25% ripple on the bulk capacitor)

$V_{in,max}$ = 265 Vrms

$V_{bulk,max}$ = 375 Vdc

V_{out} = 12 V

$V_{ripple} = \Delta V$ = 250 mV

V_{out} drop = 250 mV maximum from I_{out} = 0.2 to 2 A in 10 μs

SIMULATIONS AND PRACTICAL DESIGNS OF FLYBACK CONVERTERS

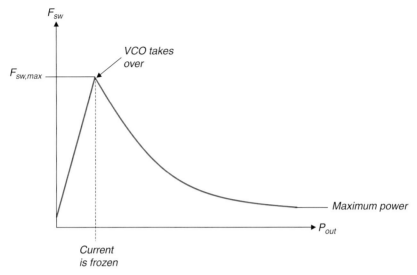

FIGURE 7-61a The QR flyback sees its frequency going up as the load current goes down. At a certain point, since the peak current cannot decrease anymore, the controller folds the frequency back.

$I_{out,max} = 1.66$ A
MOSFET derating factor $k_D = 0.85$
Diode derating factor $k_d = 0.5$
RCD clamp diode overshoot $V_{os} = 15$ V

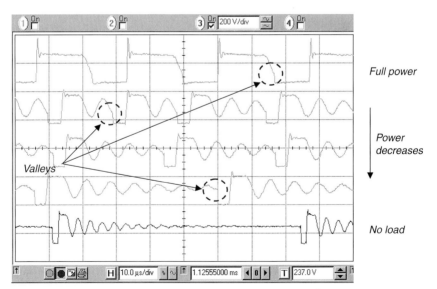

FIGURE 7-61b Here, several drain-source waveforms were captured at different output power levels. Please note the frequency reduction as explained and the valley switching even at light loads.

First, we are going to select peak current-mode control for several reasons:

1. The market offer in terms of controllers is rich.
2. Current-mode control inherently offers superior input voltage rejection.
3. The primary current is permanently monitored.
4. DCM-to-CCM mode transition is not a problem for this operating mode.

The switching frequency will be selected to be 65 kHz. It offers a good compromise among switching losses, magnetic size, and EMI signature. Given that the conducted EMI standard CISPR-22 specifies an analysis between 150 kHz and 30 MHz, having a 65 kHz switching frequency implies a second harmonic below 150 kHz and a third harmonic already reduced in amplitude. For your information, the vast majority of ac–dc adapters for notebooks are operated at 65 kHz.

Now, the operating mode: DCM or CCM? This is obviously the key point. Each mode has its advantages, and the points below offer a brief summary of pros and cons:

DCM

- Small inductor.
- No RHPZ in the low-frequency portion, higher crossover frequency achievable.
- First-order system, even in voltage mode, simple to stabilize.
- Simple low-cost secondary diode does not suffer from t_{rr} losses.
- No turn-on losses on the MOSFET—$I_D = 0$ at turn-on (not considering capacitive losses).
- Valley switching is possible in quasi-resonant mode.
- It is easier to implement synchronous rectification on the secondary side.
- It is not subject to subharmonic oscillations in current mode.
- Large ac ripple, inducing conduction losses on the MOSFET and other resistive paths (ESRs, copper wires).
- Bigger hysteresis losses on the ferrite material.

CCM

- Low ac ripple, smaller conduction losses compared to DCM.
- Low hysteresis losses due to operation on B-H minor loops.
- Low ripple on the output.
- t_{rr} related losses on both the secondary side diode and the primary side MOSFET.
- Requires fast diodes or Schottky to avoid excessive losses.
- Turn-on losses on the MOSFET—$I_D \neq 0$ at turn-on, overlap of $V_{DS}(t)$ and $I_D(t)$.
- Requires a compensation ramp in peak current-mode control when duty cycle is above 50%.
- It is more difficult to stabilize in voltage mode.
- RHPZ hampers the available bandwidth.
- Despite similar energy storage, the inductance increases in CCM and so does the transformer size.

Well, reading these notes, you probably see more arguments in favor of the DCM operation than the CCM. This is true for the low-power range where DCM represents the easiest way to go. Generally speaking, CCM concerns low output voltages with high current, e.g., 5 V at 10 A, whereas DCM would be suited for higher voltages and lower currents. Most of the cathode ray tube–based TVs are running a DCM flyback converter delivering 130–160 V/1 A or less. Why DCM? Because high-V_{RRM} inexpensive diodes are usually slow and CCM would induce a lot of switching losses.

Based on what is seen on the market, below 30 W, you can design DCM without any problems. Above, ac losses often become predominant in your design, especially if the converter size prevents you from using large aluminum cans featuring low ESRs. As such, adopting CCM helps reduce the stress on the secondary side capacitors and the transformer wires. If this is true for low line where CCM seems to be unavoidable, what about high line? Well, a good tradeoff consists of running CCM in low-line conditions and entering DCM in the upper voltage range, reducing turn-on losses, which is less problematic at low bulk voltages. This is what was demonstrated in Ref. 5 and is often implemented in high-power designs of 60 to 100 W ac–dc adapters. Nevertheless, some designers still prefer to operate in critical conduction mode (so-called boundary mode or quasi-square wave resonant mode), even in high-power designs because of (1) the reduced switching stress on semiconductors such as the secondary diode or the primary MOSFET and (2) the ease of implementation of synchronous rectification.

Ok, let us go all DCM for this small 20 W converter. As usual, a few steps will guide us through the design procedure. For a flyback, it is good practice to start with the transformer turns ratio.

We explained in this chapter that the turns ratio is intimately connected to the maximum allowable drain–source voltage of the MOSFET. In the industry, it is common practice to select 600 V MOSFETs for universal mains operations. It will be our choice here. Then, if we assume a good transformer featuring a leakage inductance below 1% of the primary inductance, a k_c of 1.5 seems to be a reasonable number. Combining Eqs. (7-37a) and (7-37b), we have

$$N = \frac{k_c(V_{out} + V_f)}{BV_{DSS}k_D - V_{os} - V_{bulk,max}} = \frac{1.5 \times (12 + 0.6)}{600 \times 0.85 - 15 - 375} = 0.157 \quad (7\text{-}141)$$

Let us pick a turns ratio of 0.166 or $1/N = 6$. In the above equation, we assumed the diode forward drop to be 0.6 V.

By using the equation derived in Chap. 5 [Eq. (5-118)] and tweaking it to account for the transformer turns ratio, we have an equation which delivers the peak current to be in DCM boundary mode at the lowest input voltage:

$$I_{peak} = \frac{2\left(V_{min} + \dfrac{V_{out} + V_f}{N}\right)(V_{out} + V_f)N}{\eta V_{bulk,min} R_{load}} \quad (7\text{-}142)$$

In this equation, the term $V_{bulk,min}$ relates to the minimum voltage seen on the bulk rail, also called V_{min} in Fig. 6-4. Assume our bulk capacitor obeys Chap. 6 recommendations and lets the maximum ripple be around 25% of the rectified peak. In that case, the minimum input voltage seen by the converter is

$$V_{min} = V_{bulk,min} = 0.75 \times 85 \times \sqrt{2} = 90 \text{ Vdc} \quad (7\text{-}143)$$

After Fig. 6-4 notations, the average low-line bulk voltage will be

$$V_{bulk,avg} = \frac{V_{peak} + V_{min}}{2} = \frac{85 \times \sqrt{2} + 90}{2} = 105 \text{ V} \quad (7\text{-}144)$$

Updating Eq. (7-142) with real values leads to a peak current of

$$I_{peak} = \frac{2 \times \left(90 + \dfrac{12 + 0.6}{0.166}\right) \times (12 + 0.6) \times 0.166}{0.85 \times 90 \times 7.2} = 1.26 \text{ A} \quad (7\text{-}145)$$

Being given the peak current at which we are going to work makes the inductor calculation an easy step:

$$L_p = \frac{2P_{out}}{I_{peak}^2 F_{sw} \eta} = \frac{2 \times 20}{1.26^2 \times 65k \times 0.85} = 456\,\mu H \tag{7-146}$$

Let us adopt a rounded value of 450 µH. The duty cycle variations can now be deduced as the peak current remains constant at high- and low-line conditions. The variable for the voltage names used below refers to Fig. 6-2b.

$$t_{on,max} = \frac{I_{peak} L_p}{V_{bulk,min}} = \frac{1.26 \times 450u}{90} = 6.3\,\mu s \tag{7-147}$$

$$D_{max} = \frac{t_{on,max}}{T_{sw}} = \frac{6.3u}{15u} = 0.42 \tag{7-148}$$

$$t_{on,min} = \frac{I_{peak} L_p}{V_{bulk,max}} = \frac{1.26 \times 450u}{375} = 1.5\,\mu s \tag{7-149}$$

$$D_{min} = \frac{t_{on,min}}{T_{sw}} = \frac{1.5u}{15u} = 0.10 \tag{7-150}$$

However, given the bulk ripple at low line, the duty cycle will also move between minimum and maximum values as given by Eqs. (7-147) to (7-150). To calculate the average conduction MOSFET losses at low line, we should calculate the squared rms current as a function of $D(t)$ and $V_{in}(t)$ since the ripple modulates these variables. Therefore, we should integrate this definition over a complete mains cycle to reach "averaged" power losses dissipated during a ripple cycle. To avoid this tedious calculation (the ripple is not exactly a ramp), we will calculate the rms content at the valley value as an extreme worst case:

$$I_{D,rms} = I_{peak}\sqrt{\frac{D_{max}}{3}} = 1.26 \times \sqrt{\frac{0.42}{3}} = 471\,mA \tag{7-151}$$

Since our MOSFET sustains 600 V, what $R_{DS(on)}$ must we select? A TO-220 package vertically mounted and operating on free air exhibits a thermal resistor junction-to-air $R_{\theta J\text{-}A}$ of roughly 62 °C/W. The maximum power this package can dissipate, without an added heat sink, depends on the surrounding ambient temperature. In this application, we assume the converter to operate in an ambient of 50 °C maximum. Choosing a maximum junction temperature for the MOSFET die of 110 °C, the maximum power accepted by the package is thus

$$P_{max} = \frac{T_{j,max} - T_A}{R_{\theta J-A}} = \frac{110 - 50}{62} = \frac{60}{62} = 0.96\,W \tag{7-152}$$

The conduction losses brought by the rms current circulation are

$$P_{cond} = I_{D,rms}^2 R_{DS(on)} @ T_j = 110\,°C \tag{7-153}$$

From Eqs. (7-152) and (7-153), the $R_{DS(on)}$ at 110 °C must be smaller than

$$R_{DS(on)} @ T_j = 110\,°C \leq \frac{P_{max}}{I_{D,rms}^2} \leq \frac{0.96}{0.471^2} \leq 4.3\,\Omega \tag{7-154}$$

The MOSFET on resistance stated at a 25 °C junction almost doubles when the junction heats up and reaches 110 to 120 °C. As such, a first indication shows that the selected MOSFET will need a 25 °C $R_{DS(on)}$, which is lower than 2.1 Ω. Of course, switching losses are needed to complete the selection. Unfortunately, as they involve numerous stray elements, trying to analytically predict them is an impossible exercise. A SPICE simulation with the right MOSFET and transformer model can give an indication, but bench measurements remain the only way to accurately estimate them. In DCM, turn-on losses should theoretically be null, but the lump capacitor present on the drain discharges through the MOSFET. It induces the following losses, depending where the turn-on occurs: right at the sinusoidal top (maximum losses) or in the valley of the waveform (minimum losses):

$$P_{SW,lump}{}^{max} = \frac{1}{2}C_{lump}\left(V_{bulk,max} + \frac{V_{out} + V_f}{N}\right)^2 F_{sw} \quad (7\text{-}155a)$$

$$P_{SW,lump}{}^{min} = \frac{1}{2}C_{lump}\left(V_{bulk,max} - \frac{V_{out} + V_f}{N}\right)^2 F_{sw} \quad (7\text{-}155b)$$

The lump capacitor value can be extracted following App. 7A. A typical turn-off sequence appears in Fig. 7-62a. The turn-off sequence depends on the environment around the drain node, such as the presence of a snubber capacitor. Figure 7-62b shows an oscilloscope shot at a switch opening event which favorably compared to the simulation data. Until the V_{GS} reaches the plateau level, nothing changes. The drain does not move, but the current starts to bend, depending on the MOSFET transconductance (gm). At the beginning of the plateau, the MOSFET starts to block, the drain voltage rises, and the current further bends. The current keeps circulating in the MOSFET (acting as a linear resistor) until the drain voltage reaches a level where another current path exists (the current must flow somewhere, right?). At this time, the current falls to zero as it has been fully diverted elsewhere. In this particular case (case 1),

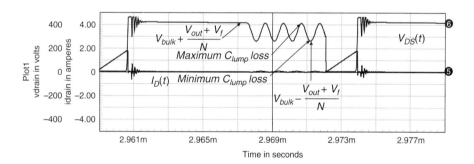

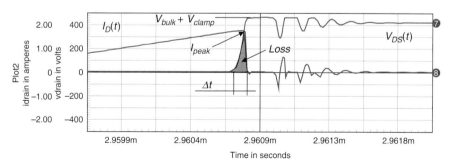

FIGURE 7-62a A typical turn-off sequence for a DCM flyback converter.

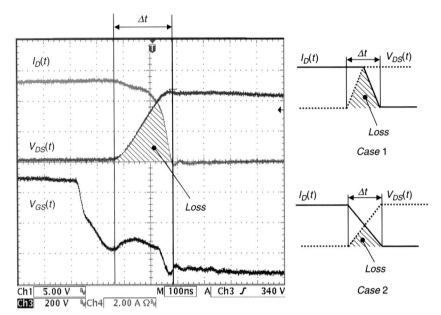

FIGURE 7-62b Worst case occurs when the drain voltage immediately rises up at the switch opening (case 1). If some snubber exists to slow the voltage rise, the idealized case 2 can happen with a more favorable loss budget.

considering the overlapping section as a triangle of Δt width, the average power dissipated by the MOSFET at turn-off is a simple triangle area.

$$P_{SW,off} = F_{sw} \int_0^{\Delta t} I_D(t) V_{DS}(t) \cdot dt = F_{sw} \left[\int_0^{\frac{\Delta t}{2}} I_{peak}(V_{bulk} + V_{clamp}) \frac{2t}{\Delta t} \cdot dt \right.$$

$$\left. + \int_{\frac{\Delta t}{2}}^{\Delta t} (V_{bulk} + V_{clamp}) I_{peak} \frac{2(\Delta t - t)}{\Delta t} \cdot dt \right]$$

$$P_{SW,off} = \frac{I_{peak}(V_{bulk} + V_{clamp}) \Delta t}{2} F_{sw} \quad (7\text{-}156a)$$

If some snubber exists, the drain voltage will rise later and the overlap becomes more favorable (case 2). In that operating mode, the power dissipation shrinks to become:

$$P_{SW,off} = F_{sw} \int_0^{\Delta t} I_D(t) V_{DS}(t) \cdot dt = F_{sw} \int_0^{\Delta t} I_{peak} \frac{\Delta t - t}{\Delta t} (V_{bulk} + V_{clamp}) \frac{t}{\Delta t} \cdot dt$$

$$= \frac{I_{peak}(V_{bulk} + V_{clamp}) \Delta t}{6} F_{sw} \quad (7\text{-}156b)$$

On the final prototype, observing both MOSFET voltage and current, you can either extract the overlap time Δt and all needed variables, or have the oscilloscope compute the loss contribution for you. Once all losses are identified, the total MOSFET power dissipation must be calculated at both low- and high-line values. At low line, conduction losses are predominant, whereas at high line, switching losses take the lead:

$$P_{MOSFET} = P_{cond} + P_{SW,lump} + P_{SW,off} \tag{7-157}$$

Depending on the selected MOSFET and the various contributions, you might need to either select a lower $R_{DS(on)}$ type or add a small heat sink if the total power exceeds Eq. 7-152 limit. Considering Eq. 7-154 as a worst case, a MOSFET like the IRFBC30A could be the right choice ($BV_{DSS} = 600$ V, $R_{DS(on)} = 2.2\ \Omega$).

Before closing down the MOSFET section, we can estimate the power dissipation burden on the controller driving stage. To fully turn the IRFBC30A on, a 23 nC electricity quantity (Q_G) needs to be brought to its gate-source space. If the controller operates from a 15 V auxiliary winding at a 65 kHz frequency, the dissipation on the chip is:

$$P_{drv} = F_{sw} Q_G V_{cc} = 65k \times 23n \times 15 = 22\ \text{mW} \tag{7-158}$$

Assuming our controller considers a current limit when the voltage image on its dedicated pin hits 1 V (a very popular value among controllers on the market), the sense resistor value (sometimes called the burden resistor) is evaluated by:

$$R_{sense} = \frac{1}{I_{peak}} \tag{7-159}$$

We need a peak current of 1.26 A. Considering a design margin of 10% ($I_{peak} = 1.38$ A), the sense resistor value equals:

$$R_{sense} = \frac{1}{1.26 \times 1.1} = 0.72\ \Omega \tag{7-160}$$

Unfortunately, this value does not fit the normalized resistor series E24/E48. Several solutions exist to cure this:

1. Put several resistors in parallel to reach the right value. In this case, two 1.4 Ω resistors would give 0.7 Ω.
2. Select a higher value and install a simple resistive divider. For instance, select a 0.82 Ω resistor. The needed peak current is 1.38 A. Once applied over the 0.82 Ω, it will develop 1.13 V. The divider ratio thus must be $\frac{1}{1.13} = 0.88$. By selecting a series resistor of 1 kΩ, the ground resistor is simply 7.3 kΩ. Figure 7-63 details the arrangement.

Using Eq. (7-151), the power dissipated by the sense resistor reaches

$$P_{R_{sense}} = I_{D,rms}^2 R_{sense} = 0.471^2 \times 0.82 = 181\ \text{mW} \tag{7-161}$$

Now, we need to protect the MOSFET against the leakage spikes. We have seen how to derive the RCD clamp values, so let us put the equations to work. We have slightly tweaked expressions (7-44) to reveal the clamp coefficient k_c:

$$R_{clp} = \frac{(k_c - 1)\left[2k_c(V_{out} + V_f)^2\right]}{N^2 F_{sw} L_{leak} I_{peak}^2} = \frac{(1.5 - 1)\left[2 \times 1.5 \times (12 + 0.6)^2\right]}{0.166^2 \times 65k \times \frac{450u}{100} \times 1.38^2} = 15.5\ \text{k}\Omega \tag{7-162}$$

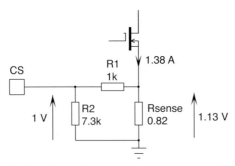

FIGURE 7-63 When a sense resistor requires a value difficult to find, it is always possible to install a divider which artificially increases the sensed voltage, at the cost of efficiency, however.

$$C_{clp} = \frac{k_c(V_{out} + V_f)}{NR_{clp}F_{sw}\Delta V} = \frac{1.5 \times 12.6}{0.166 \times 15.5k \times 65k \times 11} \approx 10\,\text{nF} \qquad (7\text{-}163)$$

In the above equations:

L_{leak} is the leakage inductance. We selected 1% of the primary inductance for the example. Of course, you would measure it on the transformer prototype, representative of the production series.

ΔV is the selected ripple in percentage of the clamp voltage; 11 V roughly corresponds to 10% of the clamp voltage ($\approx$114 V).

Finally, the power dissipated in the clamp resistor will guide us through the resistor selection:

$$P_{R_{clp}} = 0.5F_{sw}L_{leak}I_{peak}^2\frac{k_c}{k_c - 1} = 0.5 \times 65k \times 4.5u \times 1.38^2 \times \frac{1.5}{0.5} = 835\,\text{mW}$$
$$(7\text{-}164)$$

To dissipate 1 W, two 33 kΩ, 1 W resistors in parallel will do.

Having defined what we have on the primary side, let us look at the diode. Given N, we can calculate the secondary diode voltage stress:

$$\text{PIV} = NV_{bulk,max} + V_{out} = 0.166 \times 375 + 12 = 74\,\text{V} \qquad (7\text{-}165)$$

With a k_d coefficient of 0.5 (diode derating factor), select a diode featuring a 150 V V_{RRM} and accepting at least 4 A of continuous current. An MBRS4201T3 in an SMC package (surface mount) looks like the good choice:

$V_{RRM} = 200$ V
$I_{F,avg} = 4$ A
$V_f = 0.61$ V at $T_j = 150\,°\text{C}$ and $I_{F,avg} = 4$ A
$I_R = 800\,\mu\text{A}$ at $T_j = 150\,°\text{C}$ and $V_r = 74$ V

The total power dissipation endured by this component is related to its dynamic resistance R_d, its forward drop V_f, and its leakage current (in particular for a Schottky). Total losses are defined by

$$P_d = V_{T0}I_{d,avg} + R_d I_{d,rms}^2 + DI_R \text{PIV} \approx V_f I_{d,avg} \qquad (7\text{-}166)$$

In our case, with such a derating factor for the reverse voltage, we can neglect the leakage current contribution as it brings very little additional losses. Also, at these small rms currents, the dynamic resistor contributes to almost nothing. Hence, Eq. (7-166) simplifies to

$$P_d = V_f I_{d,avg} = 0.61 \times 1.6 \approx 1\,\text{W} \qquad (7\text{-}167)$$

Dissipating 1 W on an SMC package can represent a challenge, especially if you cannot benefit from a wide copper area on the board layout. If this is the case, a TO-220 diode such as the MBR20200 (industry standard) can easily dissipate 1 W in free-air conditions, without heat sink [Eq. (7-152)] and could be adopted instead. You could argue that a Schottky diode is not a necessity in this DCM example. A fast diode can also do the job, at the expense of a slightly higher forward drop.

As we have seen in the numerous design examples, we first calculated the capacitor value to obtain the right ripple value (considering the capacitive contribution alone), but the ESR always degraded the result. This time, being in DCM, we should directly evaluate the maximum ESR we can accept to pass the ripple condition of 250 mV. The secondary peak current can be known via the primary peak current and the transformer turns ratio:

$$I_{sec,peak} = \frac{I_{peak}}{N} = \frac{1.26}{0.166} = 7.6\,\text{A} \qquad (7\text{-}168)$$

Based on this number,

$$R_{ESR} \leq \frac{V_{ripple}}{I_{sec,peak}} \leq \frac{0.25}{7.6} \leq 33\,\text{m}\Omega \qquad (7\text{-}169)$$

Searching a capacitor manufacturer site (Rubycon, for example), we found the following reference:

680 μF − 16 V − YXG series
Radial type, 10 (φ) × 16 mm
$R_{ESR} = 60\,\text{m}\Omega$ at $T_A = 20\,°\text{C}$ and 100 kHz
$I_{C,rms} = 1.2\,\text{A}$ at 100 kHz

Associating several of these capacitors in parallel, we should reach the required equivalent series resistor. Now, keep in mind that the ESR increases as the temperature decreases. To keep the ripple at the right value at low temperatures, you might need to increase the total capacitor value, relax the original specification, or install a small output LC filter, as we will see later on. What is the rms current flowing through the parallel combination?

$$I_{C_{out,rms}}^2 = I_{sec,rms}^2 - I_{out,avg}^2 \qquad (7\text{-}170)$$

The secondary side current requires the use of Eq. (7-151) where the off time now enters the picture:

$$I_{sec,rms} = I_{sec,peak}\sqrt{\frac{1 - D_{max}}{3}} = 7.6 \times \sqrt{\frac{1 - 0.42}{3}} = 3.34\,\text{A} \qquad (7\text{-}171)$$

Substituting this result into Eq. (7-170) gives

$$I_{C_{out,rms}} = \sqrt{3.34^2 - 1.66^2} = 2.9\,\text{A} \qquad (7\text{-}172)$$

In other words, three 680 μF capacitors have to be put in parallel to accept the above rms current, assuming they share the current equally. The total ESR drops to

$$R_{ESR,total} = \frac{60m}{3} = 20\,\text{m}\Omega \tag{7-173}$$

The loss incurred by this resistive path amounts to

$$P_{C_{out}} = I_{C_{out},rms}{}^2 R_{ESR} = 2.9^2 \times 20m = 168\,\text{mW} \tag{7-174}$$

Given all these results, it is time to look at the small-signal response. Using our flyback current-mode template, we can feed it with the calculated values:

$L_p = 450\,\mu\text{H}$
$R_{sense} = 0.7\,\Omega$
$N = 0.166$
$C_{out} = 2040\,\mu\text{F}$
$R_{ESR} = 20\,\text{m}\Omega$
$R_{load} = 7.2\,\Omega$

The application circuit appears in Fig. 7-64 where you can see a TL431 arranged as a type 2 amplifier.

What bandwidth do we need to satisfy our 250 mV drop as expressed in the initial specification?

$$f_c \approx \frac{\Delta I_{out}}{2\pi \Delta V_{out} C_{out}} = \frac{1.8}{6.28 \times 0.25 \times 2040u} = 562\,\text{Hz} \tag{7-175}$$

Let us shoot for 1 kHz, a reasonable value to reach also giving us some margin. The steps to follow in order to stabilize a DCM current-mode appear below:

1. The bridge divider calculation assumes a 250 μA current and a 2.5 V reference (TL431). Thus

$$R_{lower} = \frac{2.5}{250u} = 10\,\text{k}\Omega \tag{7-176a}$$

$$R_{upper} = \frac{12 - 2.5}{250u} = 38\,\text{k}\Omega \tag{7-176b}$$

2. Open-loop sweeps the current-mode flyback at the lowest input level (90 Vdc). Make sure the optocoupler pole and its CTR are properly entered to compensate for its presence. The laboratory measurement gave 6 kHz with a CTR varying from 50% to 150%. Figure 7-65 shows the results. In this example, we considered an optocoupler pull-up resistor of 20 kΩ (as in the NCP1200 series, for instance).

3. From the Bode plot, we can see that the required gain at 1 kHz is around +20 dB worst case. The phase lag at this point is −88°.

4. The *k* factor gives good results for the DCM compensation. Its recommendations are the following for a 1 kHz bandwidth and a targeted 60° phase margin:

$R_{LED} = 3\,\text{k}\Omega$
$C_{pole} = 2.2\,\text{nF}$
$C_{zero} = 15\,\text{nF}$

Once applied, Fig. 7-66 shows the compensated gain curves at both input voltages. Further sweeps of ESRs and CTR do not show compensation weaknesses. Do not forget to remove the

SIMULATIONS AND PRACTICAL DESIGNS OF FLYBACK CONVERTERS

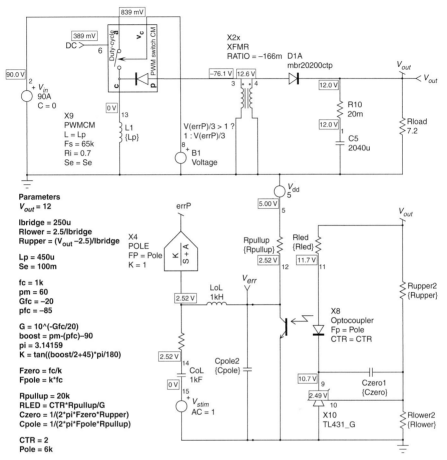

FIGURE 7-64 The ac configuration of the 20 W converter where the optocoupler pole takes place.

optocoupler pole X_4 before running the compensated ac sweeps (see Chap. 3 for more details)! When the board stability is confirmed via a load step on the average template, we can use the current-mode generic model already described in previous chapters. The cycle-by-cycle circuit appears in Fig. 7-67. The optocoupler is wired as an emitter follower, and the external voltage source mimics an internal 5 V V_{dd}. This is the most popular feedback implementation available on numerous controllers. Note the presence of a 300 mV source in series with the optocoupler collector to get rid of the saturation voltage occurring in light-load conditions (remember the two diodes in series in our UC34X representation, Fig. 7-31a). We purposely added an auxiliary winding, assuming the controller needs some self-supply. Given the turns ratio relationship, we assume the auxiliary voltage will reach around 13 V, neglecting the leakage inductance contribution. Figure 7-68 gathers all the pertinent waveforms collected at low line, full power. As you can see, some of the amplitudes confirm our theoretical calculations except the secondary variables. Why? Because a few equations [Eq. (7-171)] assumed a conduction at the boundary between CCM and DCM where $D' = 1 - D$. However, the inductor value and loading conditions impose a more pronounced DCM mode where the dead time appearance bothers Eq. (7-171).

Figure 7-69 depicts the obtained ripple at the lowest input voltage for a 20 W loading. A peak-to-peak measurement confirms a ripple amplitude of 174 mV, in line with the original specification.

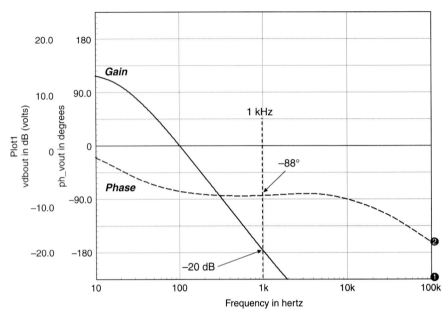

FIGURE 7-65 Open-loop Bode plots at the lowest input voltage, including the optocoupler pole and lowest CTR.

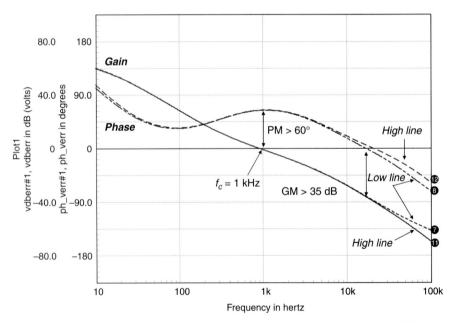

FIGURE 7-66 Once compensated, the Bode plot shows adequate bandwidth and phase margin at 1 kHz.

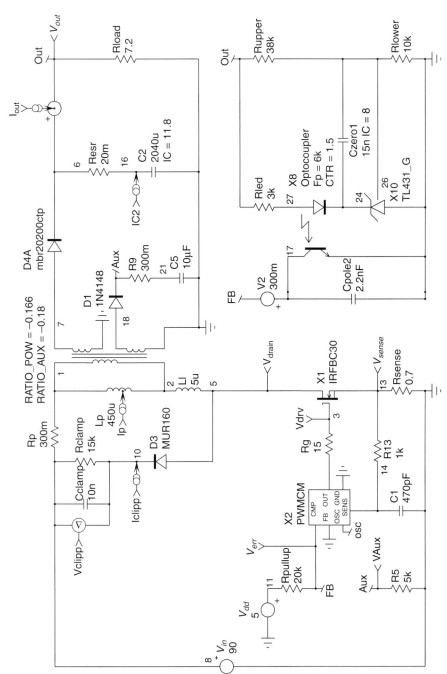

FIGURE 7-67 The cycle-by-cycle simulation template of the 20 W converter.

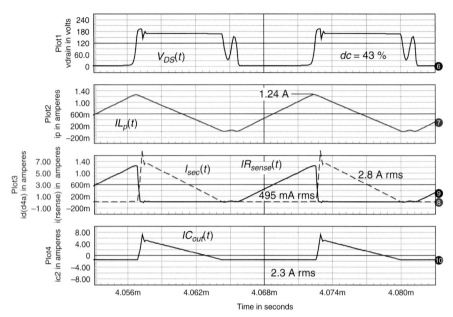

FIGURE 7-68 Cycle-by-cycle results on some pertinent waveforms (V_{in} = 90 Vdc).

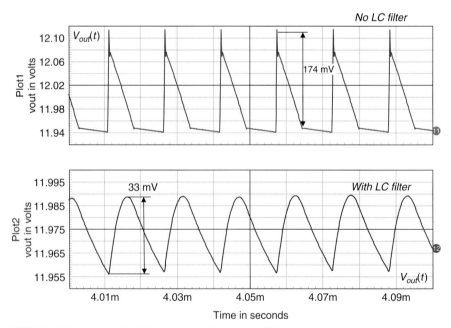

FIGURE 7-69 Output ripple full load, with and without a post *LC* filter.

However, there is almost no margin against unavoidable ESR variations. To improve the situation, we can install a small *LC* filter. The cutoff frequency of this filter must be well above the crossover frequency (at least 2 to 3 times) to avoid further stressing the phase at this point. Here, we put a 2.2 µH 100 µF filter, exhibiting a cutoff value of 10.7 kHz, 10 times above 1 kHz. The capacitor does not see much ac ripple as all is undergone by the front-end capacitor C_2. Figure 7-70 shows how to wire the TL431: the fast lane (see Chap. 3) goes before the *LC* filter, whereas the R_{upper}/R_{lower} network connection remains unchanged. Failure to keep the fast lane connection as suggested would induce oscillations, given the high-frequency gain shown by this path. Figure 7-69 (lower curve) also portrays the ripple with the *LC* filter installed and shows a 33 mV peak-to-peak amplitude.

Finally, a step load confirms the good behavior of our simulated template, in both ac and transient modes (Fig. 7-71).

What controller can we select to build this converter? There are many to chose from. Given

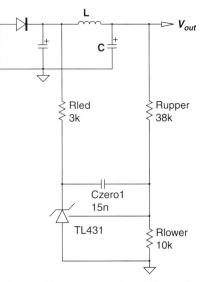

FIGURE 7-70 The insertion of an *LC* filter requires connection care with a TL431.

the low output power, an NCP1216 in DIP package (for improved thermal performance) from ON Semiconductor can easily do the job. Thanks to its high-voltage capability, there is no need for a transformer featuring an auxiliary winding: the controller is self-supplied by the high-voltage rail. The optocoupler directly connects to the feedback pin. Figure 7-72 portrays the application schematic. A few remarks regarding this implementation are appropriate:

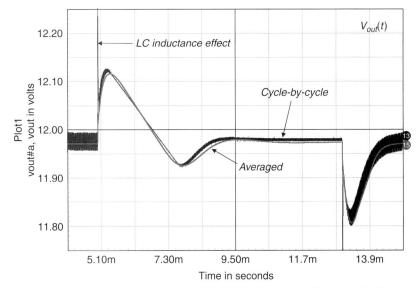

FIGURE 7-71 A 0.2 to 1.6 A load step test confirms the power supply stability at low line. See how the average model response superimposes on the cycle-by-cycle model response.

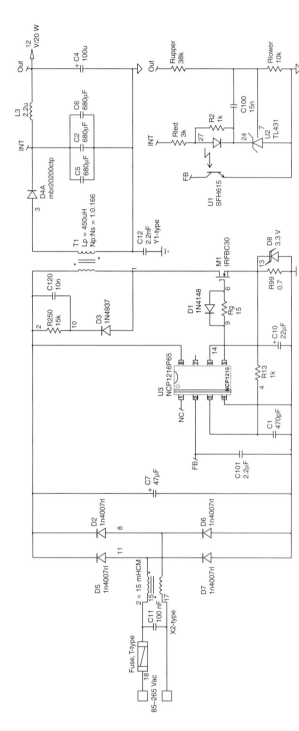

FIGURE 7-72 A typical implementation using a high-voltage controller from ON Semiconductor.

- The input filter uses the leakage inductance of the common-mode inductors to create a differential mode section with C_7.
- The zener diode D_8 is often found on large-volume consumer products. Its role is to limit the controller current-sense pin voltage excursion in case of a drain–source short-circuit on the main MOSFET. Sometimes, two 1N4007s in series are placed there for increased ruggedness. As the MOSFET source is clamped by the diodes during the fault, the controller does not see a lethal level until the fuse blows. Experience shows that the controller is often spared, thanks to this trick.
- R_2 transforms the optocoupler LED in a constant-current generator ($\approx$ 1 V/1 kΩ). It provides the bias to the TL431. If we prefer a resistor wired from the TL431 cathode to V_{out}, this solution also works well.
- All capacitors such as C_{101}, C_1, C_{10}, and R_{13} must be placed as close as possible to the controller in order to improve the noise immunity.

For the transformer, there are two options:

1. You select a transformer manufacturer (Coilcraft, Pulse Engineering, Coiltronix, Vogt, Delta Electronics, etc.) and provide the following data:

L_p = 450 µH
$I_{Lp,max}$ = 1.5 A
$I_{Lp,rms}$ = 500 mA
$I_{sec,rms}$ = 3 A
F_{sw} = 65 kHz
V_{in} = 100 to 375 Vdc
V_{out} = 12 V at 1.6 A
$N_p:N_s$ = 1:0.166

Based on this information, the manufacturer will be able to pick up the right core and discuss with you the winding arrangements, the pinout, etc. Ask for an impedance versus frequency plot for both the primary inductance and the leakage inductance. Watch for the inductance falloff as the temperature rises. It might hamper the power capability as the ambient temperature rises.

2. Take a look at App. 7C and follow the building instructions written by Charles Mullett.

7.14 A 90 W, SINGLE-OUTPUT POWER SUPPLY

This second design example describes a 90 W flyback converter operated on universal mains and featuring the following specifications:

$V_{in,min}$ = 85 Vrms
$V_{bulk,min}$ = 90 Vdc (considering 25% ripple on bulk capacitor)
$V_{in,max}$ = 265 Vrms
$V_{bulk,max}$ = 375 Vdc
V_{out} = 19 V
$V_{ripple} = \Delta V$ = 250 mV

V_{out} drop = 250 mV maximum from I_{out} = 0.5 to 5 A in 10 μs
$I_{out,max}$ = 5 A
T_A = 70 °C
MOSFET derating factor k_D = 0.85
Diode derating factor k_d = 0.5
RCD clamp diode overshoot V_{os} = 20 V

Given the power rating, we are going to design a converter operating in CCM, at least in the lower range of the input voltage. To design a CCM converter, we can either calculate the primary inductance to obtain a certain inductor ripple current at low line or select the voltage at which the converter leaves CCM to enter DCM. The second option offers greater flexibility to choose the operating mode at a selected input voltage. Before digging into the calculation details, we recall the inductor current when running in CCM. Figure 7-73 portrays the waveform.

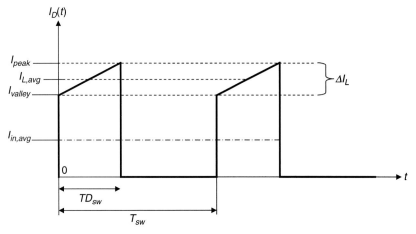

FIGURE 7-73 The inductor current in the continuous conduction mode.

Based on this signal, we derived a design equation, already introduced in Chap. 5 [Eq. (5-97)]:

$$L = \frac{\eta V_{bulk,min}^2 \left(\frac{V_{out} + V_f}{N}\right)^2}{\delta I_r F_{sw} P_{out} \left(V_{bulk,min} + \left(\frac{V_{out} + V_f}{N}\right)\right)\left(\left(\frac{V_{out} + V_f}{N}\right) + \eta V_{bulk,min}\right)} \quad (7\text{-}177)$$

In the above expression, the term δI_r defines the amount of ac peak-to-peak ripple across the inductor average current:

$$\delta I_r = \frac{\Delta I_L}{I_{L,avg}} \quad (7\text{-}178)$$

and V_{min} illustrates the lowest dc input voltage on the bulk rail (refer to Fig. 6-4). The bulk capacitor will be calculated to let the bulk voltage drop to 90 Vdc.

When equal to 2, the converter runs in full discontinuous mode. Different notations also exist in the literature, for instance, found under K_{RF} in Ref. 9. Despite a 2 in the denominator (to have K_{RF} equal to 1 in DCM), the expression does not basically change. Experience shows that depending on the input voltage range, some δI_r values are recommended to obtain the best design tradeoff:

- For a universal mains design (85 to 265 Vrms), choose a δI_r between 0.5 and 1.
- For a European input range (230 Vrms $\pm$ 15%), choose a δI_r between 0.8 and 1.6.

Let us choose $\delta I_r = 0.8$ in this example.

We explained in the previous design the intimate relationship between the turns ratio and the maximum allowable MOSFET drain–source voltage. Again, we will select a 600 V MOSFET for the universal mains operations, together with a derating factor k_D of 0.85. Assuming a good transformer coupling, a k_c of 1.5 represents a reasonable number. Thus, the turns ratio must be larger than

$$N = \frac{k_c(V_{out} + V_f)}{BV_{dss}k_D - V_{os} - V_{bulk,max}} = \frac{1.5 \times (19 + 0.6)}{600 \times 0.85 - 20 - 375} = 0.255 \quad (7\text{-}179)$$

Let us pick a turns ratio of 0.25 or $1/N = 4$. In the above equation, we assumed the diode forward drop to be 0.6 V. Applying Eq. (7-177), we find an inductor value of

$$L = \frac{0.85 \times 90^2 \times \left(\frac{19.6}{0.25}\right)^2}{0.8 \times 65k \times 90 \times \left[90 + \left(\frac{19.6}{0.25}\right)\right]\left[\left(\frac{19.6}{0.25}\right) + 0.85 \times 90\right]} \approx 320\,\mu\text{H} \quad (7\text{-}180)$$

From Fig. 7-73, the inductor average current relates to the average input current via the following formula:

$$I_{in,avg} = I_{L,avg} D \quad (7\text{-}181)$$

The maximum average input current depends on the lowest input voltage V_{min} and the delivered power:

$$I_{in,avg} = \frac{P_{out}}{\eta V_{min}} = \frac{90}{0.85 \times 90} = 1.18\,\text{A} \quad (7\text{-}182)$$

The duty cycle in Eq. (7-181) obeys the flyback relationship:

$$D_{max} = \frac{V_{out}}{V_{out} + NV_{min}} = \frac{19}{19 + 0.25 \times 90} = 0.46 \quad (7\text{-}183)$$

Introducing this result into Eq. (7-181) gives the average inductor current

$$I_{L,avg} = \frac{I_{in,avg}}{D_{max}} = \frac{1.18}{0.46} = 2.56\,\text{A} \quad (7\text{-}184)$$

Based on these results, the salient points of Fig. 7-73 are easily derived:

$$\Delta I_L = I_{L,avg}\delta I_r = 2.56 \times 0.85 = 2.18\,\text{A} \quad (7\text{-}185)$$

$$I_{peak} = I_{L,avg} + \frac{\Delta I_L}{2} = I_{L,avg}\left(1 + \frac{\delta I_r}{2}\right) = 2.56\left(1 + \frac{0.85}{2}\right) = 3.65 \text{ A} \qquad (7\text{-}186)$$

$$I_{valley} = I_{L,avg} - \frac{\Delta I_L}{2} = I_{L,avg}\left(1 - \frac{\delta I_r}{2}\right) = 2.56\left(1 - \frac{0.85}{2}\right) = 1.47 \text{ A} \qquad (7\text{-}187)$$

The primary rms current calculation requires the solution of a simple integral. As usual, the task consists of writing the time-dependent equation of the considered variable. In Fig. 7-73, the valley current ($t = 0$) is actually the peak current minus the ripple. At the end of the on time, the current reaches I_{peak}. The equations are thus

$$t = 0 \rightarrow I_{L_p}(t) = I_{valley} = I_{peak} - \Delta I_L \qquad (7\text{-}188)$$

$$t = DT_{sw} \rightarrow I_{L_p}(t) = I_{valley} + \Delta I_L \qquad (7\text{-}189)$$

$$I_{L_p}(t) = I_{valley} + \Delta I_L \frac{t}{DT_{sw}} = I_{peak} - \Delta I_L + \Delta I_L \frac{t}{DT_{sw}} \qquad (7\text{-}190)$$

After integration of Eq. (7-190), it becomes

$$I_{L,rms} = \sqrt{\frac{1}{T_{sw}} \int_0^{DT_{sw}} \left(\frac{\Delta I_L t}{DT_{sw}} + I_{peak} - \Delta I_L\right)^2 dt} = \sqrt{D\left(I_{peak}^2 - I_{peak}\Delta I_L + \frac{\Delta I_L^2}{3}\right)} \qquad (7\text{-}191)$$

Injecting Eqs. (7-183), (7-185), and (7-186) results in Eq. (7-191) gives the final rms current circulating in the transformer primary, the MOSFET, and the sense resistor:

$$I_{L,rms} = \sqrt{D_{max}\left(I_{peak}^2 - I_{peak}\Delta I_L + \frac{\Delta I_L^2}{3}\right)}$$

$$= \sqrt{0.46\left(3.58^2 - 3.58 \times 2.18 + \frac{2.18^2}{3}\right)} \approx 1.8 \text{ A} \qquad (7\text{-}192)$$

Given the rms current, we can try to find a suitable MOSFET. For a 90 W output power, there is no way a MOSFET without heat sink can survive in an ambient temperature of 70 °C. Try to find low-$R_{DS(on)}$ devices, such as the following:

Reference	Manufacturer	$R_{DS(on)}$	BV_{DSS}	Q_G
2SK2545	Toshiba	0.9 Ω	600 V	30 nC
2SK2483	Toshiba	0.54 Ω	600 V	45 nC
STB11NM60	ST	0.45 Ω	650 V	30 nC
STP10NK60Z	ST	0.65 Ω	650 V	70 nC
SPP11N60C3	Infineon	0.38 Ω	650 V	60 nC
SPP20N60C3	Infineon	0.19 Ω	650 V	114 nC

Pay attention to the package (an isolated full-pack version is simpler to mount on a heat sink than a nonisolated type) but also to the total gate charge. As the MOSFET becomes bigger (a lot of primary cells in parallel), the amount of gate charge significantly increases and so does the average driving current delivered by the controller. Let us pick an SPP11N60C3 from Infineon, a popular model in ac–dc adapters. The conduction losses at a junction temperature of 110 °C are

$$P_{cond} = I_{D,rms}{}^2 R_{DS(on)} @ T_j = 110 \text{ °C} = 1.8^2 \times 0.6 \approx 2 \text{ W} \tag{7-193}$$

The switching losses now account for the turn-on term made up of the lump capacitive discharge (already present in DCM) and the current–voltage overlap. This time, the current jumps to the valley term and no longer starts from zero. Figure 7-74a depicts the typical CCM waveforms.

As in the DCM example, a real-case waveform was captured in Fig. 7-74b. The gate-source voltage starts to increase until the plateau level is reached. At that time, the drain voltage falls and the current rises as the MOSFET operates in a linear manner. The driver signal has complete control over the signal steepness, and a resistor can be inserted to slow down this sequence on both variables. It was not the case in the turn-off event where the inductor acted as a constant-current source, imposing the drain voltage slope via the lump capacitor. In CCM, slowing down the turn-on event reduces the secondary side stress for the diode which is abruptly blocked. The resulting spike on the primary side reduces in amplitude and the radiated EMI greatly improves. If we consider a crossing point in the middle of the waveforms, Eq. (7-156b) still holds, except that we should consider the new variables in play at the switch closing time:

$V_{DS}(t)$ transitions from the plateau voltage to zero

$I_D(t)$ transitions from zero to the valley level

If we consider Δt the overlap time, switching losses at turn-on are thus

$$P_{SW,on} = \frac{I_{valley}\left(V_{bulk} + \dfrac{V_{out} + V_f}{N}\right)\Delta t}{6} F_{sw} \tag{7-194}$$

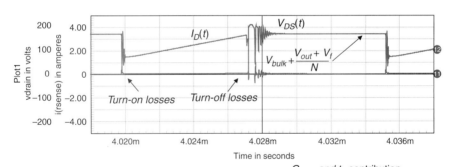

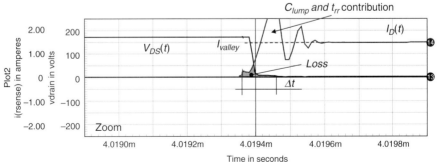

FIGURE 7-74a A simulated typical turn-on sequence in a CCM flyback converter.

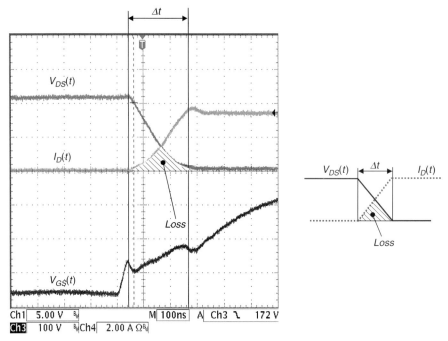

FIGURE 7-74b A real-case turn-on sequence on a CCM converter.

These losses consider a crossing time at the middle of both curves which, in reality, often differs from this ideal case. Again, bench measurements are mandatory to check the total power dissipation budget. Note that measurements will include the lump capacitor and the secondary diode t_{rr} contributions (if any).

The off-time losses still obey Eq. (7-156b). Finally, the MOSFET will dissipate.

$$P_{MOSFET} = P_{cond} + P_{SW,on} + P_{SW,off} \quad (7\text{-}195)$$

Based on Eq. (7-193), and without knowing the outcome of Eq. (7-195), our MOSFET will need a heat sink.

The total gate charge of the SPP11NC60C3 amounts to 60 nC. Therefore, the power dissipated by the driver (and not the MOSFET!) when supplied by a 15 Vdc source (the auxiliary V_{cc} voltage) reaches

$$P_{drv} = F_{sw} Q_G V_{cc} = 65k \times 60n \times 15 = 59 \text{ mW} \quad (7\text{-}196)$$

As for the DCM example, let us assume the controller considers a current limit when the voltage image of the primary current reaches 1 V. Therefore, the sense resistor value is evaluated by

$$R_{sense} = \frac{1}{I_{peak}} \quad (7\text{-}197)$$

We need a peak current of 3.6 A. Considering a design margin of 10% ($I_{peak} = 4$ A), the sense resistor value equals

$$R_{sense} = \frac{1}{4} = 0.25 \, \Omega \quad (7\text{-}198)$$

SIMULATIONS AND PRACTICAL DESIGNS OF FLYBACK CONVERTERS 693

Given this value and the rms current, the power dissipation of this element amounts to

$$P_{sense} = I_{D,rms}^2 R_{sense} = 1.8^2 \times 0.25 = 810 \text{ mW} \qquad (7\text{-}199)$$

Two 0.5 Ω, 1 W SMD types wired in parallel will do.

The MOSFET protection goes along with clamping elements calculations. The below expressions unveil the resistor and capacitor values by using familiar definitions:

$$R_{clp} = \frac{(k_c - 1)\left[2k_c(V_{out} + V_f)^2\right]}{N^2 F_{sw} L_{leak} I_{peak}^2} = \frac{(1.5 - 1)\left[2 \times 1.5 \times (19 + 0.6)^2\right]}{0.25^2 \times 65k \times \frac{320u}{100} \times 3.65^2} \approx 3.4 \text{ k}\Omega \qquad (7\text{-}200)$$

$$C_{clp} = \frac{k_c(V_{out} + V_f)}{NR_{clp} F_{sw} \Delta V} = \frac{1.5 \times 19.6}{0.25 \times 3.4k \times 65k \times 12} \approx 44.3 \text{ nF} \qquad (7\text{-}201)$$

In the above equations:

L_{leak} is the leakage inductance. We selected 1% of the primary inductance for the example. Of course, you would measure it on the transformer prototype, representative of the production series.

ΔV is the selected ripple in percentage of the clamp voltage; 12 V roughly corresponds to 10% of the clamp voltage ($\approx$117 V).

Finally, the power dissipated in the clamp resistor is

$$P_{R_{clp}} = 0.5 F_{sw} L_{leak} I_{peak}^2 \frac{k_c}{k_c - 1} = 0.5 \times 65k \times 3.2u \times 3.65^2 \times \frac{1.5}{0.5} \approx 4 \text{ W} \qquad (7\text{-}202)$$

To dissipate 4 W, three 10 kΩ, 2 W resistors in parallel will behave ok. Given the converter power, use an ultrafast device such as the MUR160 for the clamp diode rather than a slower diode.

Now, the secondary side diode. The peak inverse voltage needs to be assessed before a selection can be made:

$$\text{PIV} = NV_{bulk,max} + V_{out} = 0.25 \times 375 + 19 = 112.8 \text{ V} \qquad (7\text{-}203)$$

As usual, with a diode voltage derating factor k_d of 50%, select a diode featuring a 200 V V_{RRM} and accepting at least 10 A. (Its average current is nothing more than the direct output current.) An MBR20200CT in a TO-220 package could be a possible choice.

V_{RRM} = 200 V
$I_{F,avg}$ = 20 A (two diodes inside, each accepting 10 A)
V_f = 0.8 V maximum at T_j = 125 °C and $I_{F,avg}$ = 10 A
I_R = 800 μA at T_j = 150 °C and V_r = 80 V

The total power dissipation endured by this component is related to its dynamic resistance R_d, its forward drop V_f, and its leakage current (in particular for a Schottky). Total losses are defined by

$$P_d = V_{T0} I_{d,avg} + R_d I_{d,rms}^2 + DI_R \text{PIV} \approx V_f I_{d,avg} \qquad (7\text{-}204)$$

In our case, the leakage current does not bring significant losses, so we will neglect it. Hence, for each diode, Eq. (7-204) simplifies to

$$P_d = V_f I_{d,avg} = 0.8 \times 2.5 \approx 2 \text{ W} \qquad (7\text{-}205)$$

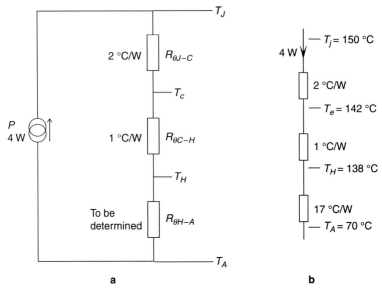

FIGURE 7-75a, b The power can be pictured as a current source feeding thermal resistances in series.

In the above equation, we assumed an equal current sharing between both diodes as they sit on the same die. The total power dissipation is thus twice what Eq. (7-205) states: 4 W. A heat sink is necessary. Figure 7-75a and b portrays the electrical analogy between thermal resistance and resistors. A thermal resistance of 5 °C/W means that the component temperature rises by 5 °C for each watt you put in. The equivalent ohm law still applies, and we can write

$$T_j - T_A = P(R_{\theta J-C} + R_{\theta C-H} + R_{\theta H-A}) \tag{7-206}$$

where T_j and T_A = junction and ambient temperatures, respectively
$R_{\theta J-C}$ = thermal resistance between junction and component case. It is usually around a few degrees Celsius per watt (2 °C/W here)
$R_{\theta C-H}$ = thermal resistance between case and heat sink. If you use a good isolator and grease, less than 1 °C/W is achievable
$R_{\theta H-A}$ = thermal resistance between heat sink and ambient air. This is what you are looking for to select the right type
P = power to be dissipated

The maximum junction temperature depends on several parameters among which the mold compound (the black powder the component is made of) plays a role. For an MBR20200 diode, you can safely limit the maximum temperature to 150 °C ($T_{j,max}$ for the MBR20200 is 175 °C). From Eq. (7-206), we can extract the thermal resistance our heat sink will need to exhibit to maintain the diode junction temperature below 150 °C when it is immersed into an ambient temperature of 70 °C:

$$R_{\theta H-A} = \frac{T_{j,max} - T_A}{P} - R_{\theta J-C} - R_{\theta C-H} = \frac{150 - 70}{4} - 2 - 1 = 17 \text{ °C/W} \tag{7-207}$$

Our diode junction when operated at a 70 °C ambient temperature will thus theoretically increase up to

$$T_j = T_A + PR_{\theta J-A} = 70 + 4 \times (17 + 1 + 2) = 150\,°C \tag{7-208}$$

Once the prototype is assembled, make sure all heat sink temperatures are controlled and comply with your QA recommendations. Some state that all heat sink temperatures must be below 100 °C at the maximum ambient temperature. As Fig. 7-75b shows, we would fail to meet this recommendation and a larger heat sink would be necessary (7.5 °C/W with a junction working at 88 °C).

If for various reasons (cost, reverse voltage, and so on) we could not select a Schottky diode, how would the reverse recovery time impact the diode dissipation budget? It is necessary to understand the diode blocking phenomenon via a simple drawing, as Fig. 7-76 offers [10]. In this figure, we have highlighted several timing intervals. Let us comment on them, one by one:

1. At t_1, the power MOSFET has been switched on, and the diode begins its blocking process. The slope at which the current decays is imposed by the external circuit, mainly the inductance in the mesh.
2. At the beginning of t_a, the diode behaves as a short-circuit, and the charge stored when the diode was conducting (more precisely, a remaining portion since recombination did evacuate

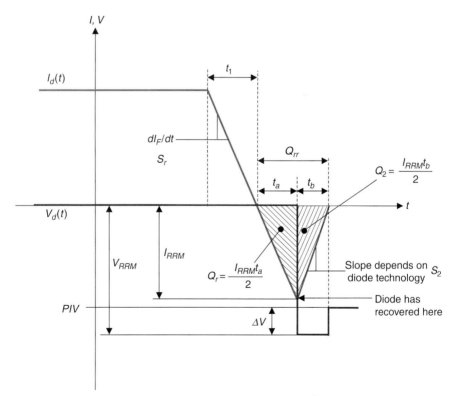

FIGURE 7-76 The diode in a blocking configuration.

a bit of it already) is evacuated via a negative current I_{RRM}. The current amplitude is linked to the blocking slope imposed by the external circuit: the steeper the imposed slope is, the more negative I_{RRM} swings.

$$S_r = -\frac{dI_F}{dt} = -\frac{PIV}{L} \qquad (7\text{-}209)$$

3. At the end of t_a, the charge is fully evacuated (Q_r on the drawing) and the diode is going to recover its blocking capability. It must bring a negative charge $-Q_2$ in order to reconstruct the internal barrier.
4. The current quickly decreases to zero via a slope S_2, and a voltage spike occurs across the diode terminals (sudden opening of the circuit). The amplitude of this spike depends on I_{RRM} and the speed at which the current goes back to zero (t_b duration, slope S_2). This speed depends on the technology. One talks about the diode "softness." The softness is actually defined by the ratio t_b/t_a. Diodes can recover in an abrupt, soft, or snappy way, generating more or fewer spikes, ringing, and other EMI unfriendly goodies.
5. Diode specifications state the reverse recovery time t_{rr}, the peak recovery current I_{RRM}, and the total charge Q_{rr} made of Q_r plus Q_2. All these parameters are defined at a given current slope dI_F/dt. It is interesting to note that Q_{rr} increases as the junction temperature goes up (the minority carrier lifetime increases). For instance, a 30EPH06 from International-Rectifier exhibits a Q_{rr} of 65 nC at $T_j = 25\ °C$ and climbs up to 345 nC for $T_j = 125\ °C$!

In the above picture, the power dissipated by the diode appears during the interval t_b:

$$P_{t_b} = F_{sw} V_{RRM} \frac{t_b I_{RRM}}{2} = F_{sw} V_{RRM} Q_2 \qquad (7\text{-}210)$$

The spike brought by the inductor in the circuit during t_b can be classically evaluated via

$$\Delta V = L \frac{dI_F}{dt} = L S_2 \qquad (7\text{-}211)$$

Based on Fig. 7-76, the total voltage excursion V_{RRM} can thus be rewritten

$$V_{RRM} = PIV + L S_2 \qquad (7\text{-}212)$$

If we divide all terms of Eq. (7-212) by the PIV, and by observing the inverse of S_r [Eq. (7-209)], we obtain

$$\frac{V_{RRM}}{PIV} = 1 + \frac{L S_2}{PIV} = 1 + \frac{S_2}{S_r} \qquad (7\text{-}213)$$

Now, if we define the slopes according to their respective variables, we have

$$\frac{V_{RRM}}{PIV} = 1 + \frac{\frac{I_{RRM}}{t_b}}{\frac{I_{RRM}}{t_a}} = 1 + \frac{I_{RRM}}{t_b} \frac{t_a}{I_{RRM}} = 1 + \frac{Q_r}{Q_2} \qquad (7\text{-}214)$$

From Eq. (7-214), we can extract V_{RRM} and replace it in Eq. (7-210):

$$P_{d,t_{rr}} = F_{sw} PIV \left(1 + \frac{Q_r}{Q_2}\right) Q_2 = F_{sw} PIV (Q_r + Q_2) = F_{sw} PIV Q_{rr} \qquad (7\text{-}215)$$

The t_{rr}-related losses are difficult to predict since they depend on the blocking slope, temperature, and diode technology itself. Also, few manufacturers give details on these parameters, making the exercise even more difficult. Final temperature assessment in the worst-case environment is thus mandatory to verify the junction stays within a reasonable value.

Now that we are done with the semiconductor portion, let us discuss the capacitor selection. As for the DCM design, we consider the ESR as the dominant term. The secondary peak current needs to be known before proceeding:

$$I_{sec,peak} = \frac{I_{peak}}{N} = \frac{3.6}{0.25} = 14.4 \text{ A} \tag{7-216}$$

Based on this number,

$$R_{ESR} \le \frac{V_{ripple}}{I_{sec,peak}} \le \frac{0.25}{14.4} \le 17 \text{ m}\Omega \tag{7-217}$$

Searching a capacitor manufacturer site (Vishay), we found the following reference:

2200 µF − 25 V − 135 RLI series
Radial type, 12.5 (φ) × 40 mm
$R_{ESR} = 44$ mΩ ay $T_A = 20$ °C and 100 kHz
$I_{C,rms} = 2$ A at 100 kHz and 105 °C

Placing several of these capacitors in parallel, we should reach the required equivalent series resistor. Now, keep in mind that the ESR increases as the temperature goes down. To keep the ripple at the right value at low temperatures, you might need to increase the total capacitor value, relax the original specification, or install a small output LC filter as we will see later on. What is the rms current flowing through the parallel combination?

$$I_{C_{out,rms}}^2 = I_{sec,rms}^2 - I_{out,avg}^2 \tag{7-218}$$

The secondary side current requires the use of Eq. (7-192) where the off time now matters:

$$I_{sec,rms} = \sqrt{(1 - D_{max})\left(I_{sec,peak}^2 - I_{sec,peak}\frac{\Delta I_L}{N} + \frac{\Delta I_L^2}{N^2 3}\right)}$$

$$= \sqrt{(1 - 0.46)\left(14.4^2 - 14.4 \times \frac{2.18}{0.25} + \frac{2.18^2}{0.25^2 \times 3}\right)} = 7.6 \text{ A} \tag{7-219}$$

Putting this result into Eq. (7-218) gives

$$I_{C_{out,rms}} = \sqrt{7.6^2 - 4.7^2} \approx 6 \text{ A} \tag{7-220}$$

Given the individual rms capability of each capacitor (2 A at 105 °C), we need to put three of them in parallel for a total capability of 6 A. A final bench measurement at the minimum input voltage and maximum current will indicate if the capacitor temperature is within safe limits or not. Given the association of capacitors, the equivalent series resistor drops to

$$R_{ESR,total} = \frac{44m}{3} = 14.6 \text{ m}\Omega \tag{7-221}$$

The total loss incurred by this resistive path amounts to

$$P_{C_{out}} = I_{C_{out,rms}}^2 R_{ESR} = 6^2 \times 14.6m = 525 \text{ mW} \tag{7-222}$$

Since all our elements are designed, it is time to look at the small-signal response of this converter. The current-mode template can be fed with the following data:

$L_p = 320 \, \mu H$
$R_{sense} = 0.25 \, \Omega$
$N = 0.25$
$C_{out} = 6600 \, \mu F$
$R_{ESR} = 14.6 \, m\Omega$
$R_{load} = 4 \, \Omega$

Figure 7-77 portrays the application schematic which does not differ from that of the DCM example since the model autotoggles between both modes. The display of operating points confirms the behavior of the simulated circuit ($V_{out} = 19$ V, duty cycle = 46.7%). For the sake of simplicity, we kept the same optocoupler parameters as in the DCM design.

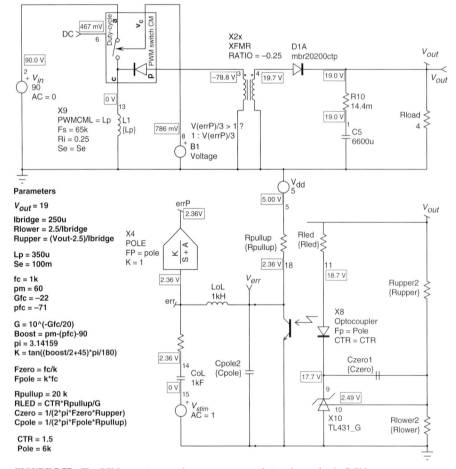

FIGURE 7-77 The CCM converter uses the same current-mode template as for the DCM case.

What bandwidth do we need to satisfy the 250 mV drop as expressed in the initial specification?

$$f_c \approx \frac{\Delta I_{out}}{2\pi_c \Delta V_{out} C_{out}} = \frac{4.5}{6.28 \times 0.25 \times 6600u} = 434\,\text{Hz} \quad (7\text{-}223)$$

Let us select 1 kHz, a reasonable value to reach, also giving us some margin within the recommendation of Eq. (7-223). Is the right half-plane zero far enough to avoid this associated phase lag?

$$f_{z_2} = \frac{(1-D)^2 R_{load}}{2\pi D L_p N^2} = \frac{(1-0.46)^2 \times 4}{6.28 \times 0.46 \times 320u \times 0.25^2} = 20.2\,\text{kHz} \quad (7\text{-}224)$$

The answer is yes, we are operating at a crossover frequency located below the 20% of the RHPZ location. What about the converter peaking, since we operate in CCM with a duty cycle close to 50%?

$$Q = \frac{1}{\pi\left(D'\frac{S_e}{S_n} + \frac{1}{2} - D\right)} = \frac{1}{3.14 \times (0.5 - 0.46)} = 8 \quad (7\text{-}225)$$

This result suggests that we damp the subharmonic poles to bring the quality coefficient below 1. Extracting the S_e parameter (the external ramp amplitude) leads to

$$S_e = \frac{S_n}{D'}\left(\frac{1}{\pi} - 0.5 + D\right) = \frac{V_{in} R_i}{L_p D'}\left(\frac{1}{\pi} - 0.5 + D\right)$$

$$= \frac{90 \times 0.25}{320u \times (1 - 0.46)}\left(\frac{1}{3.14} - 0.5 + 0.46\right) = 36\,\text{kV/s} \quad (7\text{-}226)$$

We will see the effects of this external ramp and how to practically generate it. The steps to follow to stabilize a CCM current-mode appear below:

1. The bridge divider calculation assumes a 250 µA current and a 2.5 V reference (TL431). Thus

$$R_{lower} = \frac{2.5}{250u} = 10\,\text{k}\Omega \quad (7\text{-}227a)$$

and

$$R_{upper} = \frac{19 - 2.5}{250u} = 66\,\text{k}\Omega \quad (7\text{-}227b)$$

2. Open loop sweeps the current-mode flyback at the lowest input level (90 Vdc). Make sure the optocoupler pole and its CTR are properly entered to compensate for its presence. The laboratory measurement gave 6 kHz with a CTR varying from 50 to 150%. Figure 7-78 unveils the results. In this example, we considered an optocoupler pull-up resistor of 20 kΩ (as in the NCP1200 series, for instance).
3. From the Bode plot, we can see that the required gain at 1 kHz is around +22 dB worst case. The phase lag at this point is −71°.
4. The k factor gives good results for the current-mode CCM compensation, generally speaking for first-order behaviors. Its recommendations are the following for a 1 kHz bandwidth and a targeted 80° phase margin:

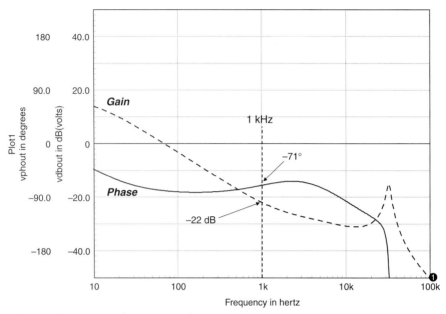

FIGURE 7-78 The ac sweep of the CCM current mode does not reveal any large phase discontinuity, except in the vicinity of the double subharmonic pole.

$R_{LED} = 2.4\ \text{k}\Omega$
$C_{pole} = 2.2\ \text{nF}$
$C_{zero} = 10\ \text{nF}$

Once applied, Fig. 7-79 shows the compensated gain curves at both input voltages. Further sweeps of ESRs and CTR do not show compensation weaknesses. As usual, do not forget to remove the optocoupler pole X_4 before running the compensated ac sweeps (see Chap. 3 for more details)! The simulation reveals a peaking due to the CCM operation and a duty cycle close to 50% at low line, but the associated gain margin seems reasonable. We have then added some ramp compensation from 10 to 30 kV/s, and results appear in the inset graph. A 20 kV compensation looks good enough for our converter. Increasing the ramp level too much could degrade the peak current capability and would prevent the power supply from delivering its full power at low line. Once compensation values are adopted, you can sweep the ESRs between the minimum and the maximum stated in the capacitor data sheets to check that the converter remains stable.

After the ac analysis comes the cycle-by-cycle simulation to verify our assumptions about the voltage and current variables and the stability. Figure 7-80 portrays the transient simulation template we have used. Note the presence of the secondary side LC filter which removes all the ESR-related spikes.

To add ramp compensation, we have implemented the Fig. 7-81 solution (introduced by Virginia Tech 20 years ago). It offers excellent noise immunity as it does not touch any oscillator section. You can select the various elements of the ramp generator as you need; we found that the RC values of 18 kΩ and 1 nF gave good results for 65 kHz operation. If we make the ramp resistance high enough, constant-current charge equations can thus be used with reasonable accuracy. Hence, referring to Fig. 7-81, we find

$$I_{C_1} = \frac{V_{drv,high}}{R_2} = \frac{15}{18k} = 800\ \mu\text{A} \qquad (7\text{-}228)$$

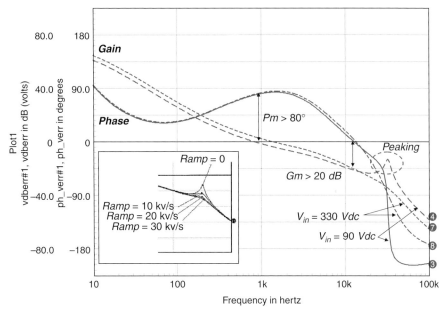

FIGURE 7-79 Once compensated, the converter exhibits a comfortable phase margin at both input levels. The lower left frame shows the effect of ramp compensation on the peaking. A value of 20 kV/s seems to suffice in this case.

For a 0.46 duty cycle and a 15.4 μs switching period, the on-time duration is 7 μs. Thus, the voltage across C_1 ramps up to

$$V_{C_1} = \frac{I_{C_1} t_{on,max}}{C_1} = \frac{800u \times 7u}{1n} = 5.6 \text{ V} \qquad (7\text{-}229)$$

Its voltage slope is then

$$S_{ramp} = \frac{V_{peak}}{t_{on,max}} = \frac{5.6}{7u} = 800 \text{ kV/s} \qquad (7\text{-}230)$$

Applying Eq. (5-39) with the above data leads to a ramp resistor of

$$R_{ramp} = \frac{S_{ramp}}{S_e} R_3 = \frac{800k}{20k} 1k = 80 \text{ k}\Omega \qquad (7\text{-}231)$$

This value is inserted into the Fig. 7-80 simulation. Except the ramp compensation circuit, the application file does not differ that much from the DCM example. Of course, you could replace the generic controller by a real model, but the simulation time would suffer. We recommend to test the whole configuration (check turns ratio, currents, and so on) using a fast model; then, once everything is within limits, you can try a more comprehensive model. Note the presence of the secondary LC filter, again inserted to reduce the high-frequency ripple.

As shown by Fig. 7-82, the valley and peak currents on the primary side have smaller values than the ones theoretically calculated. This can be explained by an overall better efficiency compared to the 85% we selected in the original calculation. On SPICE, the MOSFET $R_{DS(on)}$ stays constant despite a higher junction temperature. For instance, the efficiency measurement gives 91%, a rather good value. The MOSFET total loss amounts to 1.3 W, and the diode losses are 3 W.

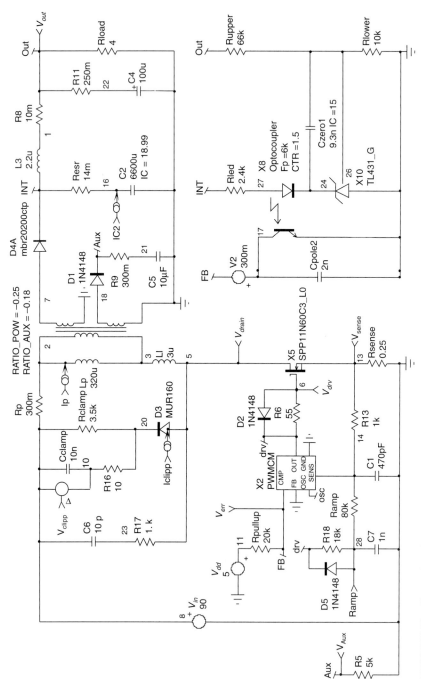

FIGURE 7-80 The transient simulation template for the CCM current-mode converter.

In this particular example, selecting the components based on the pure theoretical values gives a good design margin from the beginning. Be careful, though, when selecting capacitors with a small margin or without any margin at all, for this can quickly lead to catastrophic failures in production given the dispersions: take margins!

Figure 7-83 portrays the voltage excursion on the MOSFET drain for the highest input voltage (375 V) and shows us that the *RCD* clamp network was properly selected. In the laboratory, if you missed something in this calculation, the power-on sequence usually triggers an ovation in the surrounding audience. In SPICE, all stays quiet (no fireworks!) if you exceed the breakdown voltage and you can discretely readjust values to match your target!

Finally, a load step confirms the compensation calculations by showing an output voltage drop of 100 mV at 90 Vrms (Fig. 7-84).

The application schematic uses an NCP1230 from ON Semiconductor (Fig. 7-85). This controller includes a lot of interesting features such as a fault timer which offers excellent short-circuit protection, even if the coupling between the auxiliary and the power winding suffers (feedback observation, see Fig. 7-47). Also, given the 90 W output power, if you decide to

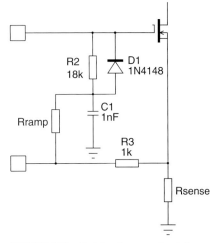

FIGURE 7-81 Deriving the ramp signal from a low-impedance path such as the driver pin is a good way to build the compensation waveform.

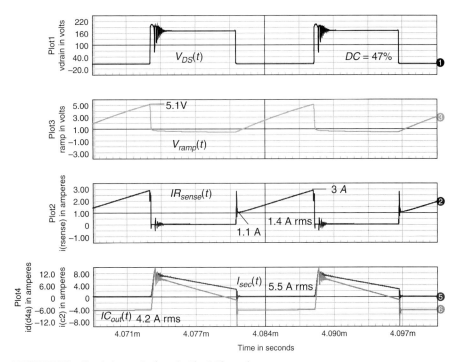

FIGURE 7-82 Simulation results from the Fig. 7-80 template.

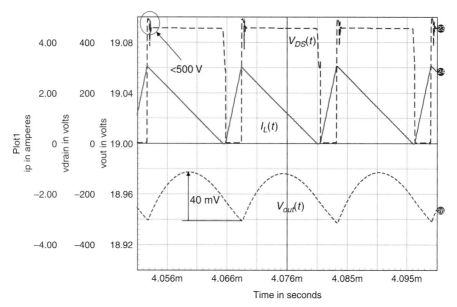

FIGURE 7-83 Drain-source voltage at 375 V of input voltage and maximum load. The diode overshoot does not appear, but as it usually corresponds to 15 to 30 V, there is a lot of margin until 600 V.

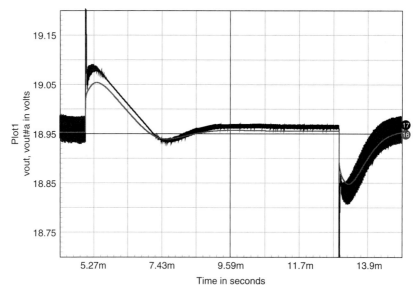

FIGURE 7-84 The transient response shows a converter under control, leading to a drop of roughly 100 mV.

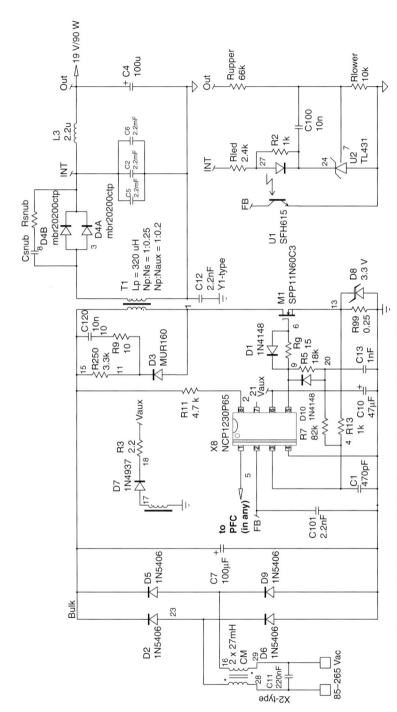

FIGURE 7-85 The final application schematic where the controller includes a V_{cc} for a PFC, if one is installed.

include a PFC front-end stage, then the 1230 can directly shut down the PFC in standby, further saving power. The power supply must still be designed to operate on wide mains (to start up when the PFC is off in low mains condition), but it must thermally be designed for high line only (when the PFC is on). The schematic includes provisions for a snubber installed across the secondary side diode, as ringing often occurs at this location.

The transformer design requires the knowledge of the following data that can be either sent to a transformer manufacturer or used to build the transformer yourself, sticking to the design example available in App. 7C.

$L_p = 320\ \mu H$
$I_{Lp,max} = 4\ A$
$I_{Lp,rms} = 1.8\ A$
$I_{sec,rms} = 8\ A$
$F_{sw} = 65\ kHz$
$V_{in} = 90\ to\ 375\ Vdc$
$V_{out} = 19\ V\ at\ 4.7\ A$
$N_p:N_s = 1:0.25$
$N_p:N_{aux} = 1:0.2$

7.15 A 35 W, MULTIOUTPUT POWER SUPPLY

This third design example describes a 35 W flyback converter operated on universal mains and featuring the following specifications:

$V_{in,min} = 85\ Vrms$
$V_{bulk,min} = 90\ Vdc$ (considering 25% ripple on bulk capacitor)
$V_{in,max} = 265\ Vrms$
$V_{bulk,max} = 375\ Vdc$
$V_{out1} = 5\ V \pm 5\%, I_{out,max} = 2\ A$
$V_{out2} = 12\ V \pm 10\%, I_{out,max} = 2\ A$
$V_{out3} = -12\ V \pm 10\%, I_{out,max} = 0.1\ A$
$V_{ripple} = \Delta V = 250\ mV$ on all outputs
$T_A = 50\ °C$
MOSFET derating factor $k_D = 0.85$
Diode derating factor $k_d = 0.5$

This power supply could suit a consumer product, such as a set-top box, a VCR, or a DVD recorder. Of course, more outputs would be necessary, but this example can be easily translated to other needs. In this application, we are going to use quasi-square wave resonance (so-called QR mode). Why? Because

1. The power supply always operates in DCM, so it is easier to stabilize.
2. You can thus select inexpensive "lazy" diodes, so there are no t_{rr}-related problems.
3. If you switch in the valley, you reduce C_{lump} losses and can sometimes purposely increase this capacitor to get rid of an expensive RCD clamp.
4. Secondary side synchronous rectification sees benefits from the guaranteed DCM operation.

On the other side, the drawbacks of the QR operation are as follows:

1. The operating frequency varies in relation to the input and output conditions.
2. The frequency increases in light-load conditions, as do switching losses.
3. A frequency clamp or an active circuitry must be installed to limit the frequency excursion in light-load conditions; otherwise standby losses can be extremely important.
4. DCM operation incurs higher rms currents compared to the CCM mode.
5. In the lowest line and heaviest loading conditions, the frequency can decrease and enter the audible range, causing acoustical noise troubles.

Despite these drawbacks, a lot of ac–dc adapters and set-top box makers use the QR mode to improve the overall efficiency, mainly thanks to synchronous rectification. To design a QR converter, we must first understand the various signals present in the flyback operated in this mode and derive an equation to design the primary inductor. As usual, the design starts with the turns ratio definition. In a QR design, you strive to reflect a large amount of voltage to the primary side to bring the wave valley (when the secondary diode blocks) as close as possible to the ground. Thus all losses linked to the drain lump capacitor are minimized, if not totally canceled ($V_{DS} = 0$). For this reason, designers often select an 800 BV_{DSS} type of MOSFET to allow maximum reflection.

In moderate power designs, such as this one, it is possible to connect an additional capacitor between drain and source to actually reduce the voltage excursion at the switch opening [see Eq. (7-17) and Fig. 7-5c]. If a sufficient margin exists, then goodbye costly *RCD* clamp! For this reason, the diode overshoot parameter (V_{os}) and k_c disappear from our turns ratio definition to the benefit of a new variable, V_{leak}. Variable V_{leak} corresponds to the voltage excursion brought by the leakage inductance and the capacitor connected between drain and source C_{DS}:

$$V_{leak} = I_{peak}\sqrt{\frac{L_{leak}}{C_{DS}}} \qquad (7\text{-}232)$$

Figure 7-86 depicts a typical signal captured on a QR converter where no *RCD* clamp has been wired.

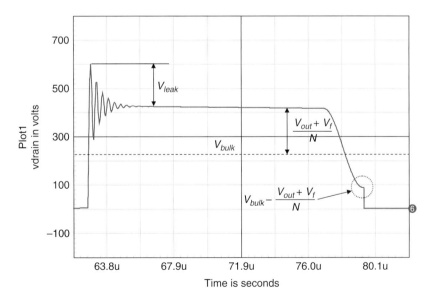

FIGURE 7-86 A QR waveform where the maximum excursion depends on the leakage inductor and the capacitor wired between drain and source.

To take into account this additional parameter, the turns ratio definition can be updated as follows:

$$N = \frac{V_{out} + V_f}{BV_{DSS}k_D - V_{bulk,max} - V_{leak}} \quad (7\text{-}233)$$

Let us look now at the primary inductor signals. Figure 7-87 depicts the current flowing in a flyback inductor when operated in QR or borderline mode. One interesting thing, already

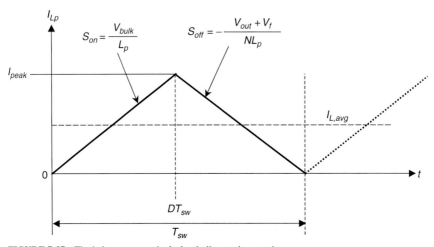

FIGURE 7-87 The inductor current in the borderline mode operation.

highlighted in the PFC section, relates to the inductor average current when QR is employed in the flyback converter:

$$I_{L,avg} = \frac{I_{peak}}{2} \quad (7\text{-}234)$$

Looking at the input current, Fig. 7-88 shows how it evolves with time. The average value delivered by the source becomes

$$I_{in,avg} = \frac{I_{peak}D}{2} \quad (7\text{-}235)$$

To derive the inductor equations, let us use the slope definitions in Fig. 7-87 to find the on- and off-time durations:

$$t_{on} = I_{peak}\frac{L_p}{V_{bulk}} \quad (7\text{-}236)$$

$$t_{off} = I_{peak}\frac{NL_p}{V_{out} + V_f} \quad (7\text{-}237)$$

As the converter operates in BCM, the DCM power conversion formula perfectly holds:

$$P_{out} = \frac{1}{2}L_p I_{peak}^2 F_{sw}\eta \quad (7\text{-}238)$$

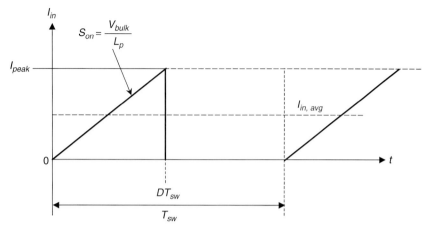

FIGURE 7-88 The input current (or drain current) in the borderline mode operation.

From this we can extract the peak current definition:

$$I_{peak} = \sqrt{\frac{2P_{out}}{F_{sw}L_p\eta}} \qquad (7\text{-}239)$$

If we sum up Eqs. (7-236) and (7-237) to obtain the switching frequency, further replacing the peak current by the Eq. (7-239) definition, we have

$$\frac{1}{F_{sw}} = t_{on} + t_{off} = \sqrt{\frac{2P_{out}}{F_{sw}L_p\eta}} L_p \left(\frac{1}{V_{bulk}} + \frac{N}{V_{out} + V_f}\right) \qquad (7\text{-}240)$$

As the minimum frequency occurs when the bulk voltage is minimum (low line), then extracting the switching frequency and the inductor leads to

$$L_p = \frac{\eta(V_{out} + V_f)^2 V_{bulk,min}^2}{2P_{out}F_{sw,min}(V_{out} + V_f + NV_{bulk,min})^2} \qquad (7\text{-}241)$$

$$F_{sw,min} = \frac{\eta(V_{out} + V_f)^2 V_{bulk,min}^2}{2P_{out}L_p(V_{out} + V_f + NV_{bulk,min})^2} \qquad (7\text{-}242)$$

where $F_{sw,min}$ represents the minimum switching frequency at low line ($V_{bulk,min}$) and full power. Equation (7-242) shows how the frequency varies:

- Full load, the frequency is minimum and can enter the audible range. Conduction losses are prominent over switching losses.
- Light load, the frequency quickly increases and degrades the efficiency by switching loss contribution.

Figure 7-89 portrays the typical frequency evolution versus load variations for a QR converter.
The next step consists of assessing the rms current, worst case again, low line and full power. The formula remains similar to the one used in the DCM design:

$$I_{D,rms} = I_{peak}\sqrt{\frac{D_{max}}{3}} \qquad (7\text{-}243)$$

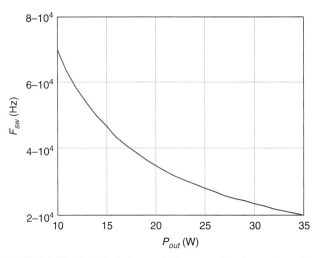

FIGURE 7-89 As the load changes, the frequency also adjusts to keep the DCM mode.

To derive the duty cycle, Eq. (7-235) can be tweaked:

$$\frac{P_{out}}{V_{bulk,min}\eta} = \frac{I_{peak}D_{max}}{2} \qquad (7\text{-}244)$$

From this D_{max} is extracted:

$$D_{max} = \frac{2P_{out}}{I_{peak}V_{bulk,min}\eta} \qquad (7\text{-}245)$$

Substituting this equation into Eq. (7-243) gives

$$I_{D,rms} = \sqrt{\frac{2I_{peak}P_{out}}{3\eta V_{bulk,min}}} \qquad (7\text{-}246)$$

Using Eq. (7-246), we can compute the conduction losses of the MOSFET:

$$P_{cond} = I_{D,rms}^2 R_{DS(on)} @ T_j = 110\ °C \qquad (7\text{-}247)$$

As we connect an additional capacitor from the drain to the source, it will result in additional switching losses. Assuming we switch the MOSFET on right in the wave valley (Fig. 7-86), the associated switching losses at the highest line input are

$$P_{sw} = 0.5\left(V_{bulk,max} - \frac{V_{out}+V_f}{N}\right)^2 C_{DS}F_{sw,max} \qquad (7\text{-}248)$$

where $V_{bulk,max}$ represents the bulk voltage at the highest line input (375 Vdc in this example). In that case, the converter is assumed to operate at a frequency defined by Eq. (7-242) where $V_{bulk,min}$ is replaced by $V_{bulk,max}$. Depending on the controller, this frequency can be achieved or clamped if it exceeds a certain value (usually less than 150 kHz for EMI issues).

The design now comes to a point where we need to select a turns ratio to derive the rest of the variables. To select this particular number N, we first need to list the constraints we have:

$F_{sw,min}$: This is the minimum switching frequency at full power and low line. If selected too low, it can bring the converter into the audible range and generate acoustical noise problems. Too high, it can quickly push the power supply into the upper range of switching frequencies, and switching losses will dominate. 40 kHz can be a number to start with.

$F_{sw,max}$: As emphasized above, we do not want a very high switching frequency because Eq. (7-248) will lead the MOSFET power dissipation budget. Also, for EMI reasons, keeping the frequency around 70 kHz can be a good point. Let us stick to this value for the maximum frequency.

L_{leak}: This is the leakage inductance. We assume it to be 1% of the primary inductance:

$$L_{leak} = \frac{L_p}{100}$$

V_{leak}: This is the key number that we need to optimize. If it is selected too low, the C_{DS} capacitor increases to keep the drain excursion within control, and again the MOSFET suffers from switching losses at high line. If it is too high, we reflect less voltage on the primary side and the conduction losses now become dominant at low line.

To help select the right V_{leak} value, we will plot both switching and conduction losses, respectively, at high- and low-line conditions, as a function of V_{leak}. We could also plot the total contribution $P_{sw} + P_{cond}$ at low- and high-line conditions, but we considered switching and conduction losses separately here for the sake of simpler equations. We then select the point at which both contributions equal, in order to obtain a balanced budget between switching and conduction losses at the two input voltage extremes. The iteration exercise also includes several values for the $R_{DS(on)}$ in order to dissipate, at the end, a reasonable amount of power on the MOSFET, at an acceptable cost. A few trials indicated that an $R_{DS(on)}$ of 1.5 Ω could be the right choice, leading to a dissipated power of 2 W. Based on this $R_{DS(on)}$ figure, Fig. 7-90 portrays the curves obtained through the definition of conduction and switching

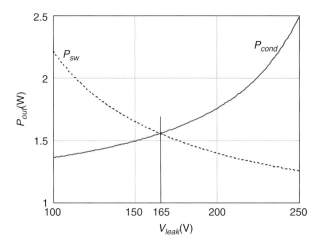

FIGURE 7-90 Curves showing the evolution of switching losses (high line) and conduction losses (low line) as a function of the selected drain voltage overshoot. The intersection corresponds to a point where individual contributions are equal. The point could be refined by plotting total losses at both input voltage extremes but at the expense of more complicated formulas.

losses as functions of V_{leak}:

$$P_{cond}(V_{leak}) = \frac{4P_{out}^2 R_{DS(on)}}{3(V_{out} + V_f)(\eta V_{bulk,min})^2}\left(V_{out} + V_f + \frac{(V_{out} + V_f)V_{bulk,min}}{BV_{DSS}\alpha - V_{bulk,max} - V_{leak}}\right) \quad (7\text{-}249)$$

$$P_{sw}(V_{leak}) = \frac{P_{out}F_{sw,max}(2V_{bulk,max} - BV_{DSS}\alpha + V_{leak})^2}{100 V_{leak}^2 F_{sw,min}\eta} \quad (7\text{-}250)$$

Reading Fig. 7-90 [or equating Eqs. (7-249) and (7-250)], we find a V_{leak} value of 165 V. We can now substitute this number into the above design equations to obtain the recommended values:

$$N = \frac{V_{out} + V_f}{BV_{dss}\alpha - V_{bulk,max} - V_{leak}} = \frac{5 + 0.8}{800 \times 0.85 - 375 - 165} = 0.041 \quad (7\text{-}251)$$

We will select a turns ratio of 25, or $N = 0.04$. Knowing the turns ratio, we can calculate the inductor value given the minimum operating frequency we have chosen (40 kHz):

$$L_p \leq \frac{\eta(V_{out} + V_f)^2 V_{bulk,min}^2}{2P_{out}F_{sw,min}(V_{out} + V_f + NV_{bulk,min})^2}$$

$$\leq \frac{0.8 \times (5 + 0.8)^2 \times 90^2}{2 \times 35 \times 40k \times (5 + 0.8 + 0.041 \times 90)^2} \leq 864\,\mu H \quad (7\text{-}252)$$

Selecting a 860 μH indicator; the maximum peak current at low line is found to be

$$I_{peak} = \sqrt{\frac{2P_{out}}{F_{sw,min}L_p\eta}} = \sqrt{\frac{2 \times 35}{0.8 \times 40k \times 860u}} = 1.6\,A \quad (7\text{-}253)$$

The corresponding rms current then reaches a value of

$$I_{D,rms} = \sqrt{\frac{2I_{peak}P_{out}}{3\eta V_{bulk,min}}} = \sqrt{\frac{2 \times 1.6 \times 35}{3 \times 0.8 \times 90}} = 0.72\,A \quad (7\text{-}254)$$

Based on our $R_{DS(on)}$ trial, we have selected an 800 V MOSFET, the STP7NK80Z from ST:

$BV_{DSS} = 800\,V$
$R_{DS(on)}$ at $T_j = 25\,°C = 1.5\,\Omega$
$Q_g = 56\,nC$

According to Eq. (7-254), the low-line conduction losses on the MOSFET amount to

$$P_{cond} = I_{D,rms}^2 R_{DS(on)} @ T_j = 110\,°C = 0.72^2 \times 3 = 1.55\,W \quad (7\text{-}255)$$

Assuming a leakage inductor being 1% of L_p (8.6 μH), then), we can calculate the capacitor to be installed between drain and source to satisfy Eq. (7-232):

$$C_{DS} = \left(\frac{I_{peak}}{V_{leak}}\right)^2 L_{leak} = \left(\frac{1.6}{165}\right)^2 \times 8.6u = 808\,pF \quad (7\text{-}256)$$

Select an 820 pF, 1 kV type.

The worst-case switching losses occur at high line, in the valley [Eq. (7-248)] for a properly tweaked converter, when the switching frequency reaches the selected limit, 70 kHz in this design example:

$$P_{sw} = 0.5\left(V_{bulk,max} - \frac{V_{out} + V_f}{N}\right)^2 C_{DS} F_{sw,max}$$

$$= 0.5 \times \left(375 - \frac{5.8}{0.04}\right)^2 \times 820p \times 70k = 1.52\,\text{W} \qquad (7\text{-}257)$$

With these numbers on hand, we can now assess the MOSFET dissipation at both input voltage extremes:

$$P_{tot}@V_{bulk,min} = P_{cond} + P_{SW} = 1.55 + 0 \approx 1.55\,\text{W} \qquad (7\text{-}258\text{a})$$

$$P_{tot}@V_{bulk,max} = P_{cond} + P_{SW} = 0.5 + 1.52 \approx 2\,\text{W} \qquad (7\text{-}258\text{b})$$

At low line, as the reflected voltage exceeds the bulk voltage, the MOSFET body diode conducts, and we have perfect zero-voltage switching: there are no turn-on switching losses. In high-line conditions, conduction losses go down but switching losses dominate. In both cases, we neglected turn-off losses, given the snubbing action of C_{DS} whose presence delays the rise of $V_{DS}(t)$. Bench measurements can confirm or deny this design assumption. For a dissipated power of 2 W, we need to choose a heat sink offering the following thermal resistance:

$$R_{\theta H-A} = \frac{T_{j,max} - T_A}{P} - R_{\theta J-C} - R_{\theta C-H} = \frac{110-50}{2} - 2 - 1 = 27\,°\text{C/W} \qquad (7\text{-}259)$$

We need a peak current of 1.6 A. The sense resistor can therefore be calculated, accounting for a 10% margin on the peak current selection:

$$R_{sense} = \frac{1}{I_{peak} \times 1.1} = \frac{1}{1.6 \times 1.1} = 0.57\,\Omega \qquad (7\text{-}260)$$

You might need to slightly decrease this value on the bench (or use a divider, Fig. 7-63). This is so because the minimum switching frequency does not account for the delay needed to switch right in the drain–source valley. This delay artificially decreases the switching frequency, and a higher peak current is needed to pass the power at low line.

The sense element will dissipate:

$$P_{R_{sense}} = I_{D,rms}^2 R_{sense} = 0.72^2 \times 0.57 = 295\,\text{mW} \qquad (7\text{-}261)$$

Select three 1.8 Ω, 0.25 W resistors wired in parallel.

We have three windings, and the first turns ratio for the 5 V output is already known. To obtain the other ratios, we have the choice between three options described by Fig. 7-91:

- The normal way imposes three distinct windings, asking a set of two connecting points on the secondary-side bobbin. Despite a good coupling between the windings, this is the least performing in terms of cross-regulation. Each wire is sized according to its own individual current requirement.
- The ac stack consists of stacking the winding before the rectifying diode. It can work on secondaries sharing the same ground and a similar polarity. The number of turns goes down, compared to the previous solution. The wire size of the lower output winding (5 V) must be sized according to the rms currents drawn from the above winding (12 V).
- The dc stack, on the other hand, connects the windings on the dc output, starting from the lowest voltage. In Fig. 7-91, we have stacked the 12 V output on top of the 5 V one. Experience shows that this solution offers superior cross-regulation performance compared to the ac stack. It is often found on multioutput printer power supplies. Given the arrangement, the 12 V diode sees a reduced voltage stress compared to the normal solution. We will come back to this in the secondary study.

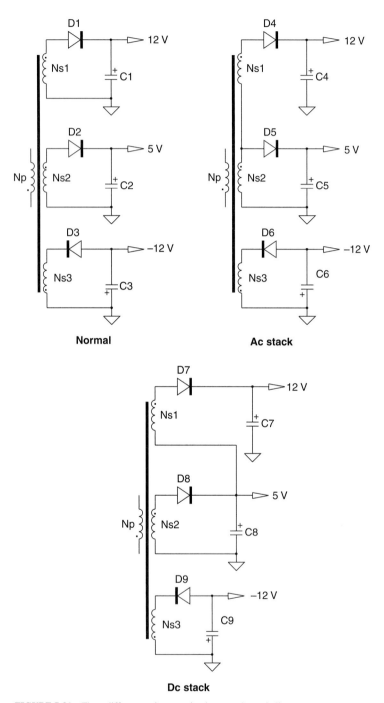

FIGURE 7-91 Three different options to wire the secondary windings.

Transformer constructions can become complicated using these techniques while ensuring compliance with the international safety standards. Reference 11 presents several solutions on how to arrange the various windings and how that affects performance. In this example, we will adopt the dc stack solution.

With an original turns ratio of 1:0.04 for the 5 V output, we can quickly derive the turns ratios of the other windings. The negative 12 V winding ratio is simply

$$N_p:N_{s,12neg} = 1:N_{s,5}\frac{12}{5} = 1:0.04 \times 2.4 = 1:0.096 \qquad (7\text{-}262)$$

The 12 V being stacked on top of the 5 V, it must deliver 7 V. Thus

$$N_p:N_{s,12pos} = 1:N_{s,5}\frac{7}{5} = 1:0.04 \times 1.4 = 1:0.056 \qquad (7\text{-}263)$$

What is the problem here? We do not account for the forward drop of each diode at the selected output current. Either we update Eqs. (7-262) and (7-263) to include this V_f information or we go through average simulations, using the selected diodes, and adjust the ratios to fit the targets.

Regarding these diodes, what are their respective stresses and losses? For the 5 V output, the peak inverse voltage is the following:

$$\text{PIV} = V_{bulk,max}N + V_{out} = 375 \times 0.04 + 5 = 20\,\text{V} \qquad (7\text{-}264)$$

Select a 40 V Schottky diode, sustaining 6 A at least. Remember, given the stacked configuration, the 5 V diode also sees the output current of the 12 V line ($I_{d,avg} = 2 + 2 = 4$ A)! An MBRF2060CT fits the bill:

MBRF2060CT

TO-220 full pack

V_f at [$T_j = 100\,°C$ and $I_d = 4$ A] = 0.5 V

Maximum junction temperature = 175 °C

Its associated conduction losses are then

$$P_{cond} = V_f I_{d,avg} = V_f I_{out} = 0.5 \times 4 = 2\,\text{W} \qquad (7\text{-}265)$$

For a 2 W dissipation need, a small heat sink must be added. It should have a thermal resistance of

$$R_{\theta H-A} = \frac{T_{j,max} - T_A}{P} - R_{\theta J-C} - R_{\theta C-H} = \frac{150 - 50}{4} - 2 - 1 = 22\,°C/W \qquad (7\text{-}266)$$

For the negative winding, -12 V, the reverse voltage reaches

$$\text{PIV} = V_{bulk,max}N + V_{out} = 375 \times 0.04 + 12 = 27\,\text{V} \qquad (7\text{-}267)$$

Select a 60 V, 1 A diode, fast switching type given the low output current (100 mA). We found that an MBR160LRG was a good choice:

MBR160T3G

SMA package

V_f at [$T_j = 100\,°C$ and $I_d = 0.1$ A] = 0.3 V

Its associated conduction losses are

$$P_{cond} = V_f I_{d,avg} = V_f I_{out} = 0.3 \times 0.1 = 30\,\text{mW} \qquad (7\text{-}268)$$

For the positive 12 V winding, the PIV slightly differs as the +5 V comes in series with the winding when it swings negative. Thus

$$\text{PIV} = V_{bulk,max} N - 5 + V_{out} = 375 \times 0.04 + 12 - 5 = 22 \text{ V} \tag{7-269}$$

For a 2 A output and a 40 V breakdown, the MBRS540T3G looks good (5 A diode):

MBRS540T3G

SMC package

V_f at $[T_j = 100\ °C$ and $I_d = 2\ A] = 0.35$ V

Its associated conduction losses are then

$$P_{cond} = V_f I_{d,avg} = V_f I_{out} = 0.35 \times 2 = 700 \text{ mW} \tag{7-270}$$

If we install the diode over a 3.2 cm² copper frame for each lead, the junction-to-ambient thermal resistance drops to 78 °C/W. The diode junction in a 50 °C ambient thus reaches

$$T_j = T_A + R_{\theta J-A} P_{cond} = 50 + 78 \times 0.7 = 104\ °C \tag{7-271}$$

The secondary side capacitors must now be chosen, given the ripple characteristics for each output. We usually calculate the capacitor constraints by estimating the peak (ESR-linked ripple) and the rms (losses) currents flowing through them. Unfortunately, on a multioutput flyback converter, it is extremely difficult to predict current wave shapes and derive their associated peak and rms values. The presence of multiple leakage inductances degrades the whole picture, and using the standard T model would lead to errors. Figure 7-92 shows the cantilever

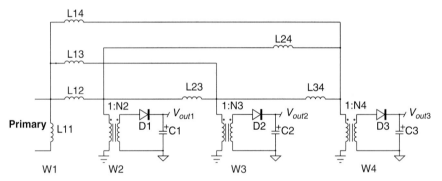

FIGURE 7-92 The upgraded cantilever model accounts for the multiple leakage inductors brought by the transformer construction.

model developed by CoPEC and presented in Ref. 12. This model leads to a good agreement between the simulated waveforms and the observed oscilloscope signals, but finding its parameters experimentally is a difficult and long exercise. The configuration and the way transitions occur depend on the installed clamping circuit. It is possible to show that at the switch opening event, the magnetizing current splits between the winding at a pace imposed by the leakage inductances L_{12}, L_{13}, and L_{14} and not by the individual output currents. As a result, the secondary side currents are sometimes far from the classical triangular wave shapes, and the simple equations that we derived no longer hold. On this particular design, given the low current on the third winding, we can try to find the rms and peak currents by using a simple two secondary winding equivalent arrangement, assuming truly triangular waveforms.

In other words, the results given below are approximate, and bench experiments are mandatory to individually check the assumptions.

We first need to calculate the maximum duty cycle at low line [Eq. (7-245)]:

$$D_{max} = \frac{2P_{out}}{I_{peak}V_{bulk,min}\eta} = \frac{2 \times 35}{1.6 \times 90 \times 0.8} = 0.61 \quad (7\text{-}272)$$

Now, as we operate in DCM, the current flowing in the 5 V diode, which also includes the 12 V, 2 A loading current as they are dc stacked, peaks to the following value, again assuming a plain triangular shape:

$$I_{sec,peak5V} \approx \frac{2(I_{out1} + I_{out2})}{1 - D_{max}} = \frac{2 \times 4}{1 - 0.61} = 20.5 \text{ A} \quad (7\text{-}273)$$

As both direct currents are equal on the 5 V and 12 V lines, we consider that they perfectly split in two, leading to a 10 A peak current on their respective diodes. Please keep in mind that this is just a rough approximation given the role of the leakage inductances of Fig. 7-92 which force different turn-off times on all diodes. Based on this result, for both 5 V and +12 V outputs, the capacitor ESRs must be smaller than

$$R_{ESR} \leq \frac{V_{ripple}}{I_{sec,peak}} \leq \frac{0.25}{10.5} \leq 24 \text{ m}\Omega \quad (7\text{-}274)$$

Searching a capacitor manufacturer's site (Rubycon), we found the following references:

5 V output:

2200 µF – 10 V – YXG series

Radial type, 12.5 (φ) × 20 mm

$R_{ESR} = 35$ mΩ at $T_A = 20$ °C and 100 kHz

$I_{C,rms} = 1.9$ A at 100 kHz

12-V output:

1500 µF – 16 V – YXG series

Radial type, 12.5 (φ) × 20 mm

$R_{ESR} = 35$ mΩ at $T_A = 20$ °C and 100 kHz

$I_{C,rms} = 1.9$ A at 100 kHz

Putting two capacitors in parallel for each output must give the ESR we are looking for. The rms current flowing in the 5 V line, which also includes the 12 V consumption, can be roughly evaluated via

$$I_{sec,rms5V} \approx I_{sec,peak5V}\sqrt{\frac{1 - D_{max}}{3}} = 20.5 \times \sqrt{\frac{1 - 0.61}{3}} = 7.4 \text{ A} \quad (7\text{-}275)$$

Assuming an equal split between currents, we can use Eq. (7-170) to calculate each capacitor rms current:

$$I_{C_{out,rms}} \approx \sqrt{\left(\frac{I_{sec,rms5V}}{2}\right)^2 - I_{out,avg}^2} = \sqrt{3.7^2 - 2^2} = 3.1 \text{ A} \quad (7\text{-}276)$$

In other words, the combination of two capacitors in parallel fits the rms current figure. The total ESR drops to

$$R_{ESR,total} = \frac{35m}{2} \approx 18\,m\Omega \tag{7-277}$$

The loss incurred by this resistive path amounts to

$$P_{C_{out}} = I_{C_{out},rms}^2 R_{ESR} = 3.1^2 \times 18m = 173\,mW \tag{7-278}$$

We are now set for the capacitors, at least for the two main outputs. Given the transformer leakage behavior, we cannot predict the rms and peak currents for the -12 V output. Therefore, a cycle-by-cycle simulation will bring us additional information. However, we will arbitrarily put a 470 µF capacitor for the average simulation.

For the average simulations, we are going to use the borderline model already tested in the PFC applications.

$L_p = 860\,\mu H$
$R_{sense} = 0.5\,\Omega$
$N_p:N_{s1} = 0.04$, 5 V/2 A, $R_{load} = 2.5\,\Omega$
$N_p:N_{s2} = 0.053$, 12 V/2 A, $R_{load} = 6\,\Omega$
$N_p:N_{s3} = 0.088$, -12 V/100 mA, $R_{load} = 120\,\Omega$
$C_{out1} = 4400\,\mu F$, $R_{ESR} = 18\,m\Omega$
$C_{out2} = 3000\,\mu F$, $R_{ESR} = 18\,m\Omega$
$C_{out2} = 470\,\mu F$, $R_{ESR} = 250\,m\Omega$ (arbitrarily selected)

In this design, given the multiple-output configuration, we want to implement weighted feedback. That is, the TL431 will observe several outputs, each affected by a certain weight depending on its required tolerance. Figure 7-93 shows how to wire these resistors on the 5 V and the +12 V outputs.

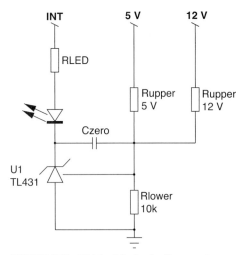

FIGURE 7-93 Weighted feedback offers superior performance in multioutput configurations.

The calculation is similar to the regular way, except that a coefficient—the weight—is affected to the upper-side resistors. In this example, the 5 V needs to be the most precise ($\pm 5\%$), but the 12 V cannot drop below 10.8 V. An initial trial where we assign 70% (W_1) to the 5 V and 30% (W_2) to the 12 V can be tested. The calculation is as follows:

1. Choose a bridge current. We usually select 250 μA, leading to a lower-side resistor of

$$R_{lower} = \frac{V_{ref}}{I_{bridge}} = \frac{2.5}{250u} = 10\,k\Omega \qquad (7\text{-}279)$$

2. The upper-side resistors follows the same procedure as for a single-output calculation, except that the final result is divided by the corresponding weight:

$$R_{upper5V} = \frac{V_{out1} - V_{ref}}{I_{bridge} W_1} = \frac{5 - 2.5}{250u \times 0.7} = 14.3\,k\Omega \qquad (7\text{-}280)$$

$$R_{upper12V} = \frac{V_{out2} - V_{ref}}{I_{bridge} W_2} = \frac{12 - 2.5}{250u \times 0.3} = 126.6\,k\Omega \qquad (7\text{-}281)$$

What bandwidth do we need, given the capacitor selection? Let us use the same formula which links the output drop (neglecting ESR effects) to the available bandwidth. In lack of specification for the output current changes, we will assume a step from 200 mA to 2 A on the 5 V and 12 V lines. We select the output featuring the lowest capacitance in this case.

$$f_c \approx \frac{\Delta I_{out}}{2\pi_c \Delta V_{out} C_{out}} = \frac{1.8}{6.28 \times 0.25 \times 3000u} \approx 382\,Hz \qquad (7\text{-}282)$$

We will adopt a 1 kHz bandwidth, giving us a comfortable margin. Here, we do not need to worry about an RHPZ as it does not exist (in the low-frequency domain at least), thanks to the permanent BCM operation. The complete ac circuit appears in Fig. 7-94 and requires some comments:

1. We assume an internal division by 3, as for the other controllers used in design examples 1 and 2. If you choose an NCP1207A or an NCP1337 from ON Semiconductor, this is the case.
2. As we use weighted feedback, we cannot observe a single output since two are used for the loop. We will thus observe the total result, made of the sum of the 5 and 12 V lines, affected by the chosen weight. The observed point is labeled OLW.
3. As for the previous design, we first sweep with the optocoupler in the loop path to account for its pole and the additional phase shift it brings. It is then further removed to check the final result. We kept the pole at 6 kHz with a CTR of 1.5. Of course, these parameters will change depending on the type of component you use here (see Chap. 3 for more details on the optocoupler).
4. The bias points show a switching frequency of 45 kHz and a peak current of 1.4 A (V_c/R_i).
5. Observing an on time of 13.4 μs with a 45 kHz switching gives a duty cycle of 60.3%, in agreement with Eq. (7-272).

The resulting Bode plot appears in Fig. 7-95. At a 1 kHz point, the deficit in gain is -32 dB, and the phase lags by 83°. The k factor method gives acceptable results for a first-order system like this QR flyback. Hence, thanks to the automated method, the compensation can be quickly made following the procedure recommendations (see other design examples and Chap. 3 for more details):

720 CHAPTER SEVEN

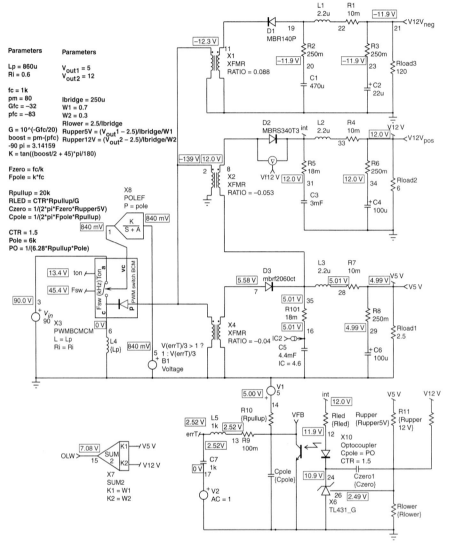

FIGURE 7-94 The complete ac circuit for the multioutput flyback converter where the optocoupler pole purposely appears in the chain.

1. We place a pole at 7 kHz, $C_{pole} = 1.2$ nF.
2. A zero is located at 150 Hz, $C_{zero} = 75$ nF.
3. The LED resistor is calculated to offer a 32 dB gain boost at 1 kHz. $R_{LED} = 750\Omega$.

For the above calculation, we have arbitrarily selected the 5 V output upper resistor to calculate the zero location. Since we have two loops, this placement will obviously affect the other loop (the 12 V path), but as long as the final result observed on the feedback pin gives adequate phase margin, we are safe. Figure 7-96 confirms this assumption by plotting the open-loop gain after

SIMULATIONS AND PRACTICAL DESIGNS OF FLYBACK CONVERTERS 721

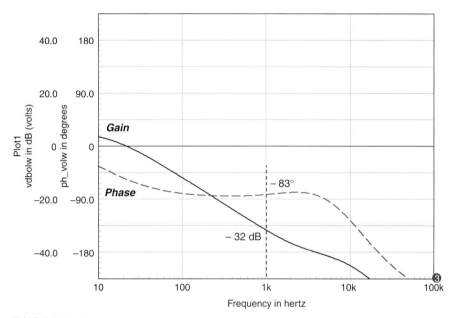

FIGURE 7-95 The open-loop small-signal response of the multiple-output flyback converter operated in QR.

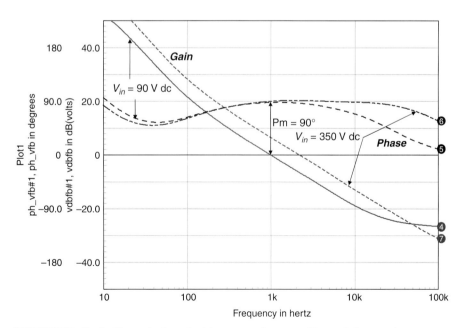

FIGURE 7-96 The final loop gain shows the right crossover frequency with a good phase margin.

proper compensation (observed on the optocoupler collector). As usual, all stray parameters such as ESRs must now be swept to check that they do not threaten the phase margin in any case.

We can now assemble the cycle-by-cycle model and verify a few key parameters such as rms currents and the peak voltage on the drain at the maximum line input. Figure 7-97 depicts the adopted schematic. The core reset detection is obtained via an auxiliary winding to which an RC delay is added (R_9C_9, on pin 1 of the controller). This delay helps to switch right in the minimum of the drain wave and slightly reduces the switching frequency, leading to a higher peak current.

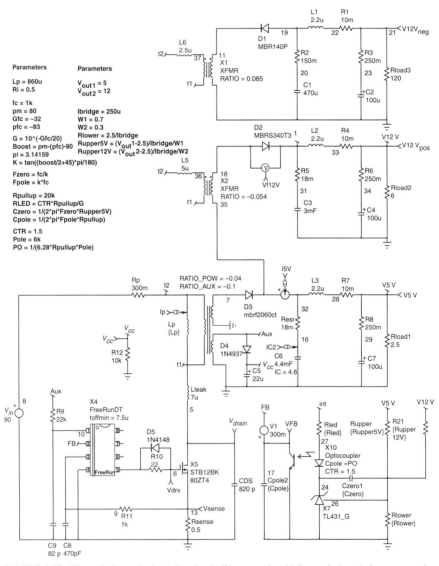

FIGURE 7-97 The cycle-by-cycle simulation circuit. We purposely added some leakage inductances on the transformer model, as described in App. 4B.

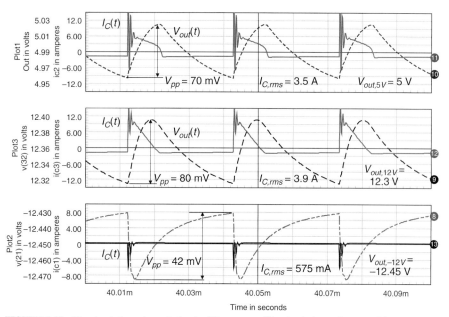

FIGURE 7-98 The simulation schematic for the QR converter. The simulation makes use of the FreerunDT model.

Simulation results appear in Fig. 7-98, showing the output voltages and their associated main capacitor ripples (before the LC filter). The capacitor rms currents are slightly above the analytical prediction, mainly because of the primary current split at the switch opening which induces a non triangular shape. The negative line requires a capacitor capable of accepting an rms current of around 600 mA, with a peak reaching 6 A. Note the short pulse duration on this particular winding. The primary current now climbs up to 1.7 A, for a 33 kHz operating frequency.

Figure 7-99 offers simulation results on the MOSFET drain signal. At low line, as we reflect more than the input voltage, the body diode is biased and the switch turns on at zero voltage: there are no switching losses at turn-on. At high line, the delay introduced by $R_9 C_9$ makes us switch right in the valley, limiting switching losses. The delay after DCM detection lies around 3 µs. Thanks to the capacitor installed between drain and source, the peak voltage reaches 580 V, well below the 800 V limit. Needless to say, your transformer manufacturer must not change the transformer construction once the final prototype and its associated leakage inductance have been qualified!

Finally, a low-line transient response on the 5 V line will let us know if our design is properly stabilized, at least on the computer! The 5 V output is stepped from 1 to 2 A in 10 µs. Figure 7-100 portrays the transient response delivered by the cycle-by-cycle simulation. All drops are within the specification limits. Given the simplified transformer model used here, bench validations are necessary on both small-signal and transient aspects.

Figure 7-101 depicts the final application schematic involving an NCP1207A or the more recent NCP1337. The auxiliary winding serves the purpose of core reset detection only, as the 1207A is self-supplied by its internal dynamic self-supply (DSS). Make sure, however, that the Q_G and the maximum switching frequency are compatible with the DSS current capability. To cope with the eventuality of a weak supply, a simple diode wired to the auxiliary winding can permanently supply the controller.

The following transformer information can then be passed to the manufacturer for the prototype definition. RMS results are obtained from a low-line simulation:

724 CHAPTER SEVEN

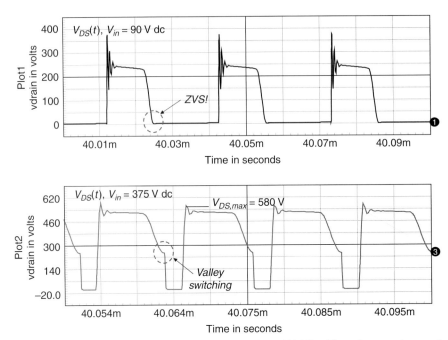

FIGURE 7-99 The drain-source voltage waveforms at low line and high line. The peak stays under control with the simple drain–source capacitor.

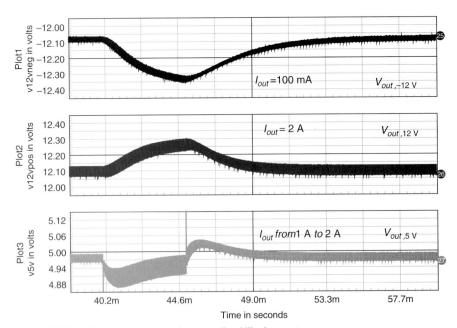

FIGURE 7-100 The transient response shows a well-stabilized converter.

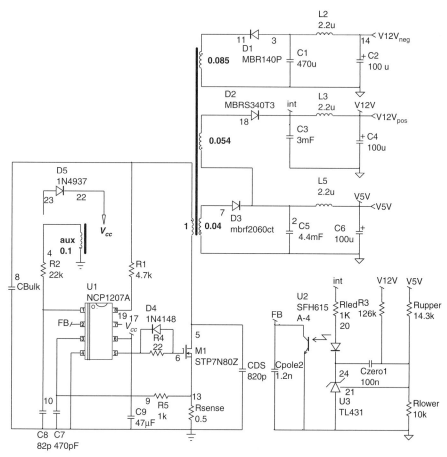

FIGURE 7-101 The final converter circuit using an NCP1207A. An option exists to wire the auxiliary winding as an auxiliary V_{cc} in case the DSS were too weak.

$L_p = 860 \ \mu H, I_{p,rms} = 800 \ mA, I_{peak} = 1.75 \ A$
$N_p{:}N_{s1} = 1{:}0.04, 5 \ V/2 \ A, I_{sec1,rms} = 8 \ A$
$N_p{:}N_{s2} = 1{:}0.053, 12 \ V/2 \ A - \text{dc stacked on } N_{s1}, I_{sec2,rms} = 4.3 \ A$
$N_p{:}N_{s3} = 1{:}0.088, -12 \ V/100 \ mA, I_{sec3,rms} = 580 \ mA$
$F_{sw} = 40 \ kHz$

7.16 COMPONENT CONSTRAINTS FOR THE FLYBACK CONVERTER

To complete the flyback design examples, we have gathered the constraints seen by the key elements used in this configuration. These data should help you select adequate breakdown voltages of the diode and the power switch. All formulas relate to the CCM and DCM operations.

MOSFET

$$BV_{DSS} > V_{in,max} + \frac{V_{out} + V_f}{N} + V_{clamp} + V_{OS} \quad \text{Breakdown voltage}$$

$$I_{D,rms} = \sqrt{D_{max}\left(I_{peak}^2 - I_{peak}\Delta I_L + \frac{\Delta I_L^2}{3}\right)} \quad \text{CCM}$$

$$I_{D,rms} = I_{peak}\sqrt{\frac{D}{3}} \quad \text{DCM}$$

Diode

$$V_{RRM} > V_{out} + NV_{in,max} \quad \text{Peak repetitive reverse voltage}$$

$$I_{F,avg} = I_{out} \quad \text{Continuous current}$$

Capacitor

$$I_{C_{out},rms} = I_{out}\sqrt{\frac{D_{max}}{1 - D_{max}} + \frac{D_{max}}{12}\left(\frac{1 - D_{max}}{\tau_L}\right)^2} \quad \text{CCM}$$

$$I_{C_{out},rms} = I_{out}\sqrt{\frac{2}{3}\left(\frac{\sqrt{1 + 2D_{max}^2/\tau_L} - 1}{D_{max}}\right) - 1} \quad \text{DCM}$$

$$\text{with } \tau_L = \frac{L_{sec}}{R_{load}T_{sw}} \text{ and } L_{sec} = L_p N^2$$

WHAT I SHOULD RETAIN FROM CHAP. 7

The flyback converter is certainly the most widespread converter on the consumer market. Given its simplicity of implementation, many designers adopt it for power levels up to 200 W. Throughout this chapter, we have reviewed a few important things to keep in mind when you are thinking about a flyback:

1. The leakage inductance plagues any transformer design. As power goes up, if your leakage term does not stay under control, your clamping network might dissipate too much power for a given printer circuit board (pcb) area. Make sure the manufacturer you have selected understands this and guarantees a constant low-leakage term.
2. Watch out for the maximum voltage seen by the semiconductors, and in particular, the primary MOSFET. Adopt a sufficient design margin from the beginning, or failures may occur in the field. The same applies for the secondary diode: do not hesitate to place dampers across diodes to calm dangerous oscillations.
3. Often overlooked is the rms current on the secondary side capacitors, as it greatly affects their lifetime. The problem becomes huge if you design in DCM or in quasi-resonant mode.
4. The small-signal study leads to an easier design for peak current-mode converters compared to voltage-mode ones. Transition from one mode to the other is well handled for current-mode control power supplies whereas it becomes a more difficult exercise for voltage-mode based converters.

5. Simulation does not lend itself very well to multioutput converters, unless you can plug a good transformer model. Thus, bench measurements are even more important to check the rms currents found in the various secondaries.
6. The standby power criterion is becoming an important topic these years. We recommend that your designs already include low standby power conversion techniques; that is, pick up less power-greedy controllers implementing skip-cycle or off-time expansion in light loads and so on. UC384X-based designs are really poor on the power-saving performance and should be left over for no-load operated power supplies permanently connected to the mains. So-called green controllers are generally slightly higher priced than the standard controllers, but the benefit for the distribution network is incomparable.

REFERENCES

1. http://www.ridleyengineering.com/snubber.htm
2. C. Nelson, LT1070 Design Manual, Linear Technology, 1986.
3. http://scholar.lib.vt.edu/theses/available/etd-71398-22552/
4. D. Dalal, "Design Considerations for Active Clamp and Reset Technique," Texas-Instruments Application Note, slup112.
5. M. Jovanovic and L. Huber, "Evaluation of Flyback Topologies for Notebook AC-DC Adapters/Chargers Applications," *HFPC Proceedings,* May 1995, pp. 284–294.
6. R. Watson, F. C. Lee, and G. C. Hua, "Utilization of an Active-Clamp Circuit to Achieve Soft Switching in Flyback Converters," *Power Electronics, IEEE Transactions,* vol. 11, Issue 1, pp. 162–169, January 1996.
7. http://www.iea.org/
8. http://www.eu-energystar.org/en/index.html
9. AN4137, www.fairchild.com
10. J.-P. Ferrieux and F. Forest, "Alimentation à Découpage Convertisseurs à Résonance," 2d ed., Masson, 2006.
11. AN4140, www.fairchild.com
12. B. Erickson and D. Maksimovic, "Cross Regulation Mechanisms in Multiple-Output Forward and Flyback Converters," http://ece.colorado.edu/~pwrelect/publications.html
13. http://www.mag-inc.com/pdf/2006_Ferrite_Catalog/2006_Design_Information.pdf
14. http://www.unitconversion.org/unit_converter/area-ex.html

APPENDIX 7A READING THE WAVEFORMS TO EXTRACT THE TRANSFORMER PARAMETERS

Observing the drain–source waveform of a flyback converter can teach the designer a lot. First, you can immediately see if the power supply operates in DCM or CCM: identifying some ringing at the end of off time as in Fig. 7A-1 unveils a DCM mode. Second, measuring the various slopes and frequencies in play leads to extracting the parameters pertinent to the transformer. Figure 7A-1 represents the signal captured on a flyback converter supplied from a 300 Vdc input source, the drain voltage, but also the primary current. To determine the transformer elements, we have several equations we can use:

The primary slope S_{L_p}, indirectly delivered by measuring the *di/dt* of the drain current, equals

$$S_{L_p} = \frac{V_{in}}{L_p + L_{leak}} \qquad (7A\text{-}1)$$

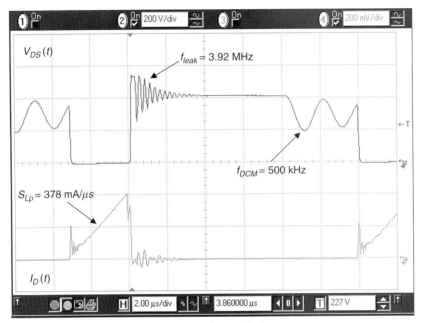

FIGURE 7A-1 A typical flyback wave obtained from a flyback converter.

The ringing frequency at the clamp diode opening involves the leakage inductance and the drain lump capacitor:

$$f_{leak} = \frac{1}{2\pi \sqrt{L_{leak} C_{lump}}} \qquad (7A\text{-}2)$$

Finally, the DCM frequency also implies the lump capacitor but associated with the sum of leakage and primary inductors:

$$f_{DCM} = \frac{1}{2\pi \sqrt{(L_{leak} + L_p) C_{lump}}} \qquad (7A\text{-}3)$$

Observing Fig. 7A-1, we found the following values:

$S_{Lp} = 378$ mA/μs
$f_{DCM} = 500$ kHz
$f_{leak} = 3.92$ MHz
$V_{in} = 300$ V

We have a set of three unknowns (L_p, L_{leak}, and C_{lump}) and three equations. Solving this system gives the individual transformer elements:

$$C_{lump} = \frac{S_{L_p}}{4 V_{in} f_{DCM}^2 \pi^2} = \frac{378k}{4 \times 300 \times 500k^2 \times 3.14^2} = 127\,\text{pF} \qquad (7A\text{-}4)$$

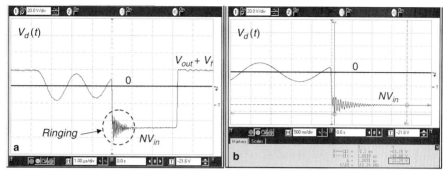

FIGURE 7A-2a and b Secondary side diode anode signal.

$$L_{leak} = \frac{1}{4\pi^2 f_{leak}^2 C_{lump}} = \frac{1}{4 \times 3.14^2 \times 3.92 Meg^2 \times 127p} = 13\,\mu H \qquad (7A\text{-}5)$$

$$L_p = \frac{V_{in} - S_{L_p} L_{leak}}{S_{L_p}} = \frac{300 - 378k \times 13u}{378k} = 780\,\mu H \qquad (7A\text{-}6)$$

Looking back to the real transformer data sheet, we have a primary inductor of 770 μH and a leakage term of 12 μH. Our results are within very good limits. Just a final comment: make sure the frequencies read by the oscilloscope do not change too much when you observe the ringing with the probe in close proximity of the drain point (no electrical contact). Otherwise, it would mean the probe capacitance changes the ringing frequency and should be accounted for in the final results.

To obtain the turns ratio of any secondary winding, hook your scope probe on the anode of the secondary side diode of concern and measure the voltage. You should see something like Fig. 7A-2. The voltage drops to the input voltage multiplied by the turns ratio N. As we measured around 50 V from a 300 V input voltage, the ratio is simply

$$N = \frac{50}{300} = 166m \qquad (7A\text{-}7)$$

Looking at Fig. 7A-2b triggers a very important comment. Equation (7-3) defines the peak inverse voltage of the secondary diode without considering any spikes. The figure speaks for itself: we have 13.3 V above what Eq. (7.3) would give us. Thus, always consider a safety margin when you are applying the theoretical formula and run a complete set of measurements on the final prototype to check this margin! Here, the worst-case condition would be the maximum input voltage. Adjust the loading current to check how the ringing amplitude evolves and chose the peak value. Then, apply an ×2 coefficient for the selected diode V_{RRM}.

APPENDIX 7B   THE STRESS

The voltage or current stress represents an important factor in a component's lifetime, passive or active. Without entering into the details of energy activation and Arrhenius' law (Svante Arrhenius, Swedish, 1859–1927), there are a few commonsense remarks we can express in this appendix.

7B.1 Voltage

First, the breakdown voltage. Have you ever observed a voltage on a component above what is stated as the maximum rating in the data sheet, for instance, 620 V on a 600 V MOSFET? Why is the component still alive? Maybe I have a good design since it does not fail? Well, this is so because the manufacturer installs some guard bands on the probing program. These guard bands ensure that the worst element found after the wafer diffusion, and accepted for release, satisfies the stated breakdown. For instance, a 600 V MOSFET will be released as long as its leakage current, worst case, at 600 V is less than a few tens of microamperes. Thus, thanks to an extensive characterization work carried at various junction temperatures and including process variations, the manufacturer knows that a permanent 600 V bias will be safe. However, as in the production chain, there can be strong candidates, those whose breakdown is above the ratings (e.g., 630 V) and those which are weak (e.g., breaking down close to 600 V). Do not rely on a higher breakdown voltage considering the safety margin installed by the manufacturer—never! Make sure you apply your own safety margin on top of this one. A lot of SMPS makers derate their own operating voltages and currents by 15% compared to what is stated. This the derating factor k_D used throughout Chap. 7. Hence, for a 600 V MOSFET, tailor your clamping network to stay within 510 V in worst-case conditions. Tough, no? Perhaps a 650 V MOSFET will help, in certain cases. Avoid the 800 V versions; their $R_{DS(on)}$ is larger and they are much more expensive than their 600 V counterparts. Note that BV_{DSS} of MOSFETs and diodes (for a breakdown above 6.2 V) exhibits a positive temperature coefficient.

On the same tone, another important recommendation: do not use the MOSFET body diode avalanche capability as a permanent transient voltage suppressor (TVS). This is the best way to quickly kill the component. First, MOSFET avalanche capabilities are usually stated with an energy obtained from a large inductor and a small current, which is not the reality. Most of the time, you have a small inductor and a large current (thus a strong I_{rms}^2 component) flowing in your circuit. The fuse effect is thus strengthened with a larger rms current. When the MOSFET finally avalanches, a few observations can be made:

- The current flows through the internal parasitic NPN transistor obtained by the vertical diffusion of the die. Its collector–base junction avalanches.
- As the current flows through this path, only a small percentage of the die is activated; hence, there is a resistance to the current flow much larger than if 100% of the die were activated.
- A voltage breakdown is like a lightening strike: you cannot precisely predict where the current will flow in the die. Bad luck might route the avalanche current to weaker portions of the silicon and smoke pops up.

As a summary, you can allow accidental avalanche on the MOSFET, for instance, in the presence of input surges. If this is true for vertical MOSFETs (who remembers the VMOS these days!), never go above the BV_{DSS} for lateral MOSFETs such as those used in monolithic switchers. They also have a body diode, but it cannot accept any avalanche energy. A simple low-energy pulse would thus be lethal to the component.

Similar recommendations apply to the diode. Some data sheets state a nonrepetitive peak reverse voltage (V_{RSM}), again meaning that only an accidental surge is allowed. The safety factor for a diode, denoted by k_d, given the presence of ringing, is 100%. Yes, if you calculate a peak inverse voltage (PIV) of 50 V, select a 100 V diode.

Regarding the MOSFET driving voltage, there is no need to exceed 15 V. Above a V_{GS} of 10 V, the $R_{DS(on)}$ decreases only by a few percent and you dramatically increase the driving losses while reducing the MOSFET lifetime. The approximate heat lost in the MOSFET driver (the PWM controller) can be found from

$$P_{driver} = V_{GS} Q_G F_{sw} \tag{7B-1}$$

where Q_G represents the total gate charge needed to fully turn on the MOSFET and F_{sw} is the switching frequency.

By the way, the gate charge represents an important value. It directly indicates the switching speed you can expect with a given drive output capability. If your data sheet indicates a total gate charge of 100 nC (a strong power MOSFET), and the driver can deliver a *constant* peak current of 1 A, you turn on in 100 ns!

$$t_{sw} = \frac{Q_G}{I_{peak}} \tag{7B-2}$$

In practice, CMOS-based drivers exhibit a resistive output impedance, and the driving current diminishes as $V_{GS}(t)$ increases. I know, it was not the case with bipolar-based output stages such as those from the UC384X series: the early effect was small enough to make them behave as almost ideal current sources despite V_{ce} variations.

Make sure no drain spikes bring the gate-source voltage above the gate oxide breakdown voltage (± 20 V or ± 30 V depending on the type of MOSFET). A gate-source zener often helps to reduce the associated risks with spikes coupled from the drain to the gate when high-voltage spikes show up. Unfortunately, given the sharpness of the zener, oscillations can sometimes occur and must be carefully observed, especially if you fear a large stray inductance (long wire to the MOSFET gate from the drive output, for instance). Inserting a small resistor close to the MOSFET gate (10 Ω) will help to damp the whole parasitic network.

7B.2 Current

Current stress relates to power dissipation and transient thermal impedance. Again the subject would need a complete book to be properly treated. Basically, the junction temperature of the device must stay within acceptable limits. What are these limits? With a diode or a transistor, for instance, the mold compound (the black plastic around the component leads) represents physical limits beyond which the junction cannot go. Why? Because the mold powder would melt down and deteriorate and contaminate the die. The die itself (at least the silicon) can reach temperatures as high as ≈ 250 °C, above which the silicon is said to be intrinsic. (It loses its semiconductor properties and becomes fully conducting—if it is not destroyed, it recovers when cooling down.) In TO3 packages, where there is no mold compound, you can allow high operating temperatures. For instance, ON Semiconductor 2N3055 states a maximum junction temperature as high as 200 °C! Put the same die in a traditional TO220 package, and the maximum rating would probably drop to around 150 °C.

A 20% derating on the maximum die temperature represents an acceptable tradeoff. For instance, a TO220 MOSFET accepts a junction temperature up to 150 °C, so make sure the worst-case die temperature stays below 110 to 120 °C. An MUR460 (4 A, 600 V ultrafast diode) accepts a junction temperature up to 175 °C. A 20% derating would lead to a maximum junction temperature of 140 °C. The junction temperature can have another side effect. Some SMPS manufacturers' quality departments (or their end customers) do not accept heat sink or component body temperature above 100 °C in the maximum ambient temperature. Make sure the heat sink calculation, involving the various thermal resistances, leads to the right case temperature. Sometimes, you find that a given MOSFET $R_{DS(on)}$ will pass the specification, but given the size constraints on the heat sink dimensions, you might need to adopt a much lower on resistance type to pass the case temperature requirement.

Another less known phenomenon relates to the so-called thermal fatigue. This occurs when you have important junction temperature excursions within a very short time. Because the energy duration is short, the heat does not have the time to propagate to the case and heat sink, and a local hot spot occurs. The temperature gradient can be so large that it induces a physical stress on the junction or the connection pads which often give up. For instance, a semiconductor operated at an average junction temperature of 90 °C with a 20 °C ripple on it will suffer less than a 60 °C operated junction subject to a positive 50 °C excursion every 10 ms.

APPENDIX 7C TRANSFORMER DESIGN FOR THE 90 W ADAPTER

Contributed by Charles E. Mullett

This appendix describes the procedure to design the transformer for the 90 W CCM flyback converter described in Chap. 7. Thanks to the derived equations and simulations, we have gathered the following electrical data pertaining to the transformer design:

$L_p = 320\ \mu H$	Primary inductor
$I_{Lp,max} = 4\ A$	Maximum peak current the transformer will accept without saturation
$I_{Lp,rms} = 1.8\ A$	Primary inductor rms current
$I_{sec,rms} = 8\ A$	Secondary side rms current
$F_{sw} = 65\ kHz$	Switching frequency
$V_{out} = 19\ V\ @\ 4.7\ A$	Output voltage and the dc output
$N_p:N_s = 1:0.25$	Primary to secondary turns ratio
$N_p:N_{aux} = 1:0.2$	Primary to the auxiliary winding turns ratio

7C.1 Core Selection

Based on the above data, several methods exist to determine the needed core size. In this appendix, we will use the area product definition. The first step therefore consists of finding the required area product $W_a A_c$, whose formula appears in the Magnetics Ferrite Core catalog available through the Ref. 1 link:

$$W_a A_c = \frac{P_{out}}{K_c K_t B_{max} F_{sw} J} 10^4 \qquad (7C\text{-}1)$$

where $W_a A_c$ = product of window area and core area, cm^4
P_{out} = output power, W
J = current density, A/cm^2
B_{max} = maximum flux density, T
F_{sw} = switching frequency, Hz
K_c = conversion constant for SI values (507)
K_t = topology constant (for a space factor of 0.4). This is the window fill factor. For the flyback converter, use $K_t = 0.00033$ for a single winding and $K_t = 0.00025$ for a multiwinding configuration.

Some judgment is required here, because the current density and flux density are left to the designer. A reasonable, conservative current density for a device of this power level, with natural convection cooling, is 400 A/cm^2.

The flux density at this frequency (65 kHz), for modern power ferrites such as Magnetics "P" material, is usually around 100 mT (1000 gauss). This can also be approached by choosing a loss factor of 100 mW/cm³ (a reasonable number for a temperature rise of 40 °C) and looking up the flux density on the manufacturer's core material loss data. Doing this for the Magnetics Kool Mµ® material yields a flux density of 45 mT (450 gauss). We will use this for the design.

Therefore we have

$$W_a A_c = \frac{P_{out}}{K_c K_t B_{max} F_{SW} J} 10^4 = \frac{90 \times 10000}{507 \times 0.00033 \times 0.045 \times 65000 \times 400} = 4.6 \, \text{cm}^4 \quad (7\text{C-}2)$$

Looking at the core-selection charts in the Magnetics Kool Mµ® catalog, we choose the DIN 42/20 E-core. It features an area product of 4.59 cm⁴, a very close match to the requirement. For future reference, it has a core area of $A_e = 2.37 \, \text{cm}^2$.

7C.2 Determining the Primary and Secondary Turns

Per the design information in the Magnetics Ferrite Core catalog, and based on Faraday's law, we have

$$N_p = \frac{V_{in,min} 10^4}{4 B A_e F_{sw}} \quad (7\text{C-}3)$$

where N_p = primary turns
 $V_{in,min}$ = minimum primary voltage
 B = flux density, T
 A_e = effective core area, cm²
 F_{sw} = switching frequency, Hz

This assumes a perfect square wave where the duty ratio D is 50%. In our case, the duty ratio will be limited to 0.46 and will occur at minimum input voltage, or 90 Vdc (85 Vrms considering 25% ripple on the bulk capacitor). One can either correct for these in the equation or use another version that expresses the same requirement in terms of the voltage at a given duty ratio.

$$N_p = \frac{V_{in,min} t_{on,max} 10^4}{2 B A_e} \quad (7\text{C-}4)$$

where $t_{on,max}$ = the maximum duration of the pulse applied to the winding. And, in the denominator, the 4 is replaced by a 2, due to the fact that the square wave actually applied the voltage during one-half of the period, so it was accompanied by another factor of 2.

In this case,

$$t_{on,max} = D_{max} T_{sw} = \frac{D_{max}}{F_{sw}} = \frac{0.46}{65k} = 7.07 \, \mu s \quad (7\text{C-}5)$$

Thus,

$$N_p = \frac{90 \times 7.07u \times 10^4}{2 \times 0.045 \times 2.37} = 30 \, \text{turns} \quad (7\text{C-}6)$$

Now, applying the required turns ratio to determine the secondary turns gives

$$N_{S1} = 0.25 N_p = 0.25 \times 30 = 7.5 \, \text{turns} \quad (7\text{C-}7)$$

The other secondary winding, the auxiliary winding, must have a turns ratio of 0.2, so the best fit of turns for these two secondaries is 10 turns (power) and 8 turns (auxiliary). We then round off the main secondary turns to 10 turns and recalculate the primary turns:

$$N_p = \frac{10}{0.25} = 40 \text{ turns} \qquad (7\text{C-8})$$

And the auxiliary winding will have 8 turns.

$$N_{S2} = 40 \times 0.2 = 8 \text{ turns} \qquad (7\text{C-9})$$

7C.3 Choosing the Primary and Secondary Wire Sizes

The wire sizes can now be determined, based on the previously chosen current density of 400 A/cm². In the case of the primary, the stated current is 1.8 A rms, so the required wire area will be

$$Aw(pri) = \frac{I_{Lp,\,rms}}{J} = \frac{1.8}{400} = 0.0045 \text{ cm}^2 = 0.45 \text{ mm}^2 \qquad (7\text{C-10})$$

This corresponds closely to a wire size of 21 AWG, which features an area of 0.4181 mm².

Fitting the windings into the bobbin usually requires some engineering judgment (and sometimes trial and error), as one would like to avoid fractional layers (they increase the leakage inductance, meaning that they result in poorer coupling and reduced efficiency) and also minimize skin effect by keeping the diameter below one skin depth. In this case, using two conductors of approximately one-half the area is a wise choice. This suggests two conductors of 24 AWG. The final choice will depend on the fit within the bobbin.

Interleaving the primary winding is also advantageous, as it significantly reduces the leakage inductance and also reduces the proximity losses. With this in mind, we will try to split the winding into two layers of one-half the turns, or 40/2 = 20 turns on each layer, with the secondary windings sandwiched in between the two series halves of the primary. With two conductors of no. 24 wire, the width of the 20-turn layer will be (assuming single-layer insulation on the wire) 0.575 mm dia. × 2 × 20, or 23 mm. The bobbin width is 27.43 mm, so this doesn't leave enough margin at the sides for safety regulation requirements. Reducing the wire size to no. 25 is the answer. The reduction in wire size is well justified, because a fractional layer would surely increase the winding losses.

The secondary current is 8 A rms. For a similar current density of 400 A/cm², the wire size will be

$$Aw(\text{sec}) = \frac{I_{sec,\,rms}}{J} = \frac{8}{400} = 0.02 \text{ cm}^2 = 2 \text{ mm}^2 \qquad (7\text{C-11})$$

The corresponding wire size is 14 AWG, which has an area of 2.0959 mm². To reduce the skin effect and product a thinner layer (and also spread it over a greater portion of the bobbin), we will use two conductors of no. 17 wire, which will have a total area of 2 × 1.0504 mm², or 2.1008 mm². Since the diameter of single-insulated no. 17 wire is 1.203 mm, the 10 turns of two conductors will occupy a width of 10 × 2 × 1.203, or 24.06 mm. This does not fit well (allowing for safety margins) within the bobbin width of 26.2 mm. Choose two conductors of no. 18 wire, and check later to see that the copper loss is acceptable.

7C.4 Choosing the Material, Based on the Desired Inductance, or Gapping the Core If Necessary

Based on the calculations detailed in Chap. 7, the desired primary inductance is 320 μH. The desired inductance factor A_L can now be determined:

$$A_L = \frac{L_p}{N^2} = \frac{320u}{40^2} = 200 \text{ nH/turn}^2 \quad (7C\text{-}12)$$

The ungapped A_L of the K4022-E060 core (60u material) is 194, so no gap is needed.

7C.5 Designs Using Intusoft Magnetic Designer

Intusoft's Magnetic Designer software is a powerful tool for designing transformers and inductors. It allows the designer to quickly arrive at a basic design, then optimize it by providing fast recalculations of losses, leakage inductance, etc., as the designer experiments with variations in the winding structure, core size and shape, and other design choices.

We have entered the design data into the software, as illustrated by the following two screens (Figs. 7C-1 and 7C-2). Note that the predicted temperature rise is 24.16 °C, and the core window is only 33.43% full. This suggests that a smaller core can be used, without exceeding the 40 °C maximum temperature rise. Figure 7C-2 shows the actual winding structure as computed by the software. The EE 42/15 core might be a good choice. It is available in 90u material and has an A_L of 217 nH/turn². To achieve the required inductance, raise the turns to 48, 12, and 10. (The 10 turns is hopefully close enough to 9.8.)

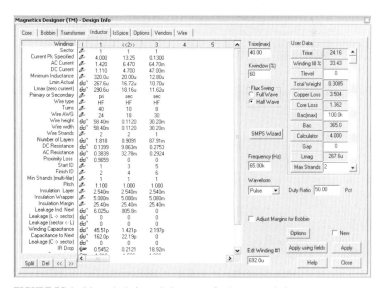

FIGURE 7C-1 Magnetic designer design screen for the present design.

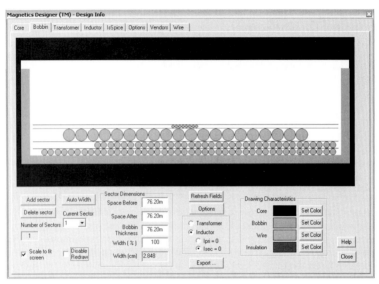

FIGURE 7C-2 The software proposes a winding structure based on the design data.

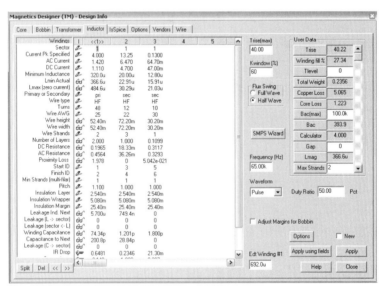

FIGURE 7C-3 Screen shot of design using the EE 42/15 core.

SIMULATIONS AND PRACTICAL DESIGNS OF FLYBACK CONVERTERS

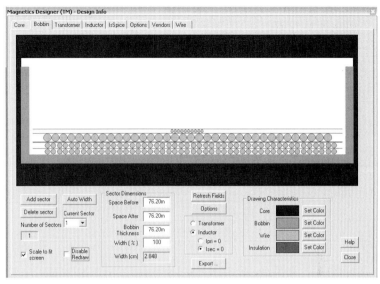

FIGURE 7C-4 Screen shot of the recommended wire arrangement.

Figure 7C-3 shows the design details, with a predicted temperature rise of 40.22 and a fill factor of 27.34%. The bobbin is still not very full, but several attempts led to the conclusion that to fill it further would lead to excessive copper loss. If a higher permeability material were available, a smaller core could be used. The final winding structure is shown in Fig. 7C-4.

CHAPTER 8
SIMULATIONS AND PRACTICAL DESIGNS OF FORWARD CONVERTERS

If the flyback converter excels in moderate output current applications, the forward is really well suited for low output voltages and strong currents. For instance, ATX power supplies, or so-called silver boxes, are made of a single-switch or two-switch forward converter delivering currents up to 50 A with output voltages ranging from 12 V down to 3.3 V. The forward converter belongs to the buck-derived family type of topologies. The difference between a forward and a buck lies in the transformer presence, made necessary by the galvanic isolation requirement. Let us discover how the forward works before we spend time on a real-life design example.

8.1 AN ISOLATED BUCK CONVERTER

Figure 8-1a represents the original buck converter with its floating switch situation. *Floating* means that one of its terminals is not ground-referenced. When the switch closes, it applies the input voltage on the freewheel diode cathode. You can see that input and output share a common ground, not really what we want for galvanic isolation. In Fig. 8-1b, a transformer appears, which isolates both grounds. When the primary power switch closes, so does its secondary side counterpart (we assume an isolated drive for this case), and a fraction of the input voltage appears on the freewheel diode just as buck does, except that the transformer imposes its ratio on the input voltage and brings isolation. The converter can thus increase or decrease the input voltage by selecting different turns ratios. However, most of the designs decrease the input level. Finally, Fig. 8-1c shows the final architecture, called a *single-switch forward*, where the power transistor becomes ground-referenced (simpler to drive) and the secondary switch becomes a diode. Note the transformer polarity which applies a positive voltage on the series diode at the switch closing.

Let us now look at the real configuration when the power switch closes. If we look back at the equivalent transformer model, we see the presence of an inductor on the primary side. This inductor, called L_p in a flyback or L_{mag} in a forward, simply relates to the inductance created by the primary turns over a magnetic material. In the case of a flyback, this inductance plays the role of an energy storage element. For the forward case, we also store energy in the magnetizing inductance but as it does not participate to the energy transfer between the primary and secondary sides, it becomes an undesirable part we have to deal with. A few sketches linked to the

740 CHAPTER EIGHT

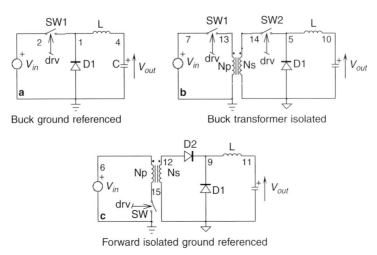

FIGURE 8-1 The forward transforms a buck topology into an isolated structure.

on and off events should shed some light on the problem to be solved. Figure 8-2 starts with the turn-on sequence. When the MOSFET closes, it applies the input voltage across the magnetizing inductor L_{mag}. The magnetizing current I_{mag} thus increases at a pace classically defined by

$$S_{mag,on} = \frac{V_{in}}{L_{mag}} \qquad (8\text{-}1)$$

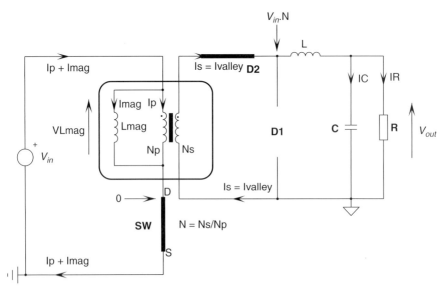

FIGURE 8-2 At turn-on, the transformer-translated input voltage appears on the secondary side, and the primary inductor is magnetized.

To avoid the transformer saturation, the magnetizing current will always be discontinuous. We will see why later. Using Eq. (8-1), we can derive a time-dependent definition for the magnetizing current:

$$I_{mag}(t) = \frac{V_{in}}{L_{mag}} t \qquad (8\text{-}2)$$

However, given the transformer dot arrangement, a positive voltage appears on the secondary side diode D_2 which conducts; the inductor current now circulates in the transformer secondary. Given the coupling between the primary and the secondary, this current now reflects on the primary side via the turns ratio N. The current seen by the source and the MOSFET drain becomes

$$I_D(t) = I_{mag}(t) + I_s(t)N \qquad (8\text{-}3)$$

where $N = \dfrac{N_s}{N_p}$.

The secondary current is nothing other than the rising inductor current supplying the load and the capacitor. The slope of this current depends on the voltage applied over the inductor L during the on time. Given the transformer ratio, we have

$$I_s(t) = I_{valley} + \frac{NV_{in} - V_{out}}{L} t \qquad (8\text{-}4)$$

where I_{valley} represents the initial current in the inductor at turn-on.

Updating Eq. (8-3) with Eqs. (8-2) and (8-4) brought to the primary side via the turns ratio N, we obtain

$$I_D(t) = NI_{valley} + t\left(\frac{V_{in}}{L_{mag}} + \frac{(NV_{in} - V_{out})N}{L}\right) \qquad (8\text{-}5)$$

Figure 8-3 displays the current waveforms, where we have separated the various signals for the sake of clarity. It is important to note that, in reality, you could not physically separate the magnetizing current from the total current. On the bench, inserting a current probe in series with the transformer primary would give the sum of the magnetizing current and the reflected secondary current. In Fig. 8-3, we have

- $I_D(t)$ represents the total primary current circulating in the MOSFET drain, inclusive of the reflected secondary side current $I_p(t)$ plus the magnetizing current.
- $I_p(t)$ depicts the secondary current reflected to the primary side, without the magnetizing current.
- In the above definitions, I_{valley} and I_{peak} respectively refer to the buck inductor L valley and peak currents.
- $I_{mag}(t)$ shows the theoretical magnetizing current circulating in the transformer.

The MOSFET drain current therefore rises until it reaches the peak limit imposed by the controller, in the case of a current-mode controller. In the case of a voltage-mode controller, the current stops its race when the pulse width circuit instructs the power switch to turn off. Now we are in a situation where we have two energized inductors: the secondary side inductor L and the magnetizing inductor L_{mag}. Figure 8-4 portrays the sketch when the primary MOSFET opens. On the secondary side, as for any buck converter and neglecting the leakage inductance, the freewheel diode immediately enters conduction and keeps the current circulating in the same direction. The junction point at both cathodes is considered to be null, as we neglect D_1 forward drop in that case. Thus the inductor downslope becomes

$$S_{L,off} = -\frac{V_{out}}{L} \qquad (8\text{-}6)$$

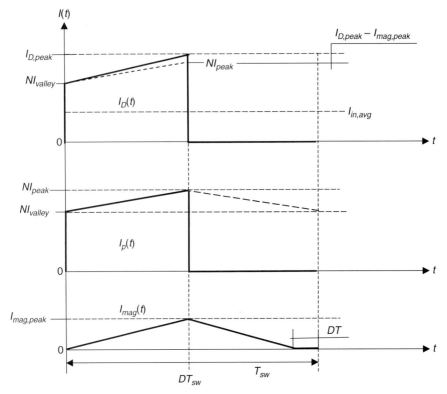

FIGURE 8-3 When the switch closes, a current made up of the reflected load current plus the magnetizing current circulates in the MOSFET and the source.

Unfortunately, on the primary side, we do not see any path to route the stored magnetizing energy which is of no use in the forward power transfer mechanism. Does it remind you of a particular situation? Yes, the leakage inductor in the flyback converter. In both cases, we have to deal with an unwanted energy that we can either dissipate in heat or try to recycle in some way.

Before exploring the need for a core reset, we can observe the voltage appearing at the cathode junction and note that we have a square wave signal toggling between NV_{in} (D_2 conducts, on time) and almost 0 (D_1 conducts, off time). Therefore, the forward output voltage follows a similar law as the buck operating in CCM whose input voltage is translated by the transformer turns ratio:

$$V_{out} = NV_{in}D \tag{8-7}$$

8.1.1 Need for a Complete Core Reset

Figure 8-5a represents a transformer built on a toroid. We see a primary winding made of N_p turns and a secondary winding made of N_s turns. When we apply a voltage across the primary winding, a current circulates in the primary side and also appears on the secondary side. Both primary and secondary currents are linked by the turns ratio N, as we already know. What is now interesting to observe is that both primary and secondary windings are "communicating" together via the flux φ circulating in the magnetic medium.

SIMULATIONS AND PRACTICAL DESIGNS OF FORWARD CONVERTERS 743

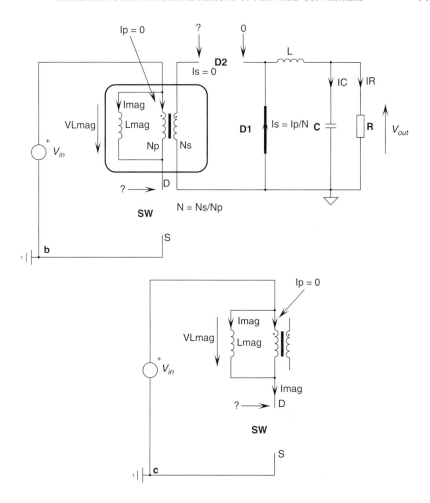

FIGURE 8-4 As the MOSFET opens, D_1 immediately conducts and ensures ampere-turns circulation for L. On the primary side, where does the magnetizing current flow?

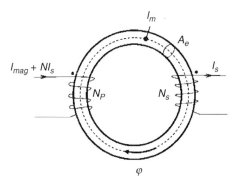

FIGURE 8-5a The flux links both windings when it circulates in the magnetic medium.

The flux circulation is actually obtained by deriving a portion of the primary current to generate this so-called magnetizing current I_{mag}, without which no electrical link would exist between both primary and secondary windings. The fact that we have wound turns on a magnetic medium, or in the air, creates a physical inductance and can be measured with an *LC* meter. This *magnetizing* inductor gets energized by the primary voltage and creates the *magnetizing current*. Figure 8-5b describes the situation via a simple drawing.

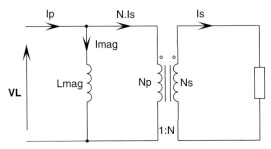

FIGURE 8-5b The magnetizing current is diverted from the primary current to create the linking flux between both windings.

We have seen in App. 4A the relationship between the applied primary volt-seconds $V_{in}t_{on}$ and the flux excursion it induces. By combining Eqs. (4A-17) and (4A-18), we have

$$\varphi(t) = \frac{1}{N_p} \int V_L(t) \cdot dt \qquad (8\text{-}8a)$$

The instantaneous flux density easily emerges as the core effective area A_e links both variables:

$$B(t) = \frac{1}{N_p A_e} \int V_L(t) \cdot dt \qquad (8\text{-}8b)$$

When the voltage step is applied over the transformer primary at turn-on, it forces a linear flux increase inside the transformer. The flux excursion depends not only on the time during which the voltage appears on the primary (the so-called volt-seconds) but also on the primary turns number N_p, as indicated by Eqs. (8-8a) and (8-8b). Figure 8-6a portrays the simplified representation of an excursion example. We can see the flux going up during the on time, cycle 1. This magnetizing excursion is denoted by ΔB_{on}. If the flux density does not return to its initial point, then the next cycle, denoted by 2, will start on top of the previous one and transformer saturation will quickly occur.

In Fig. 8-5c, note the presence of Roman numerals, pointing out the so-called operating quadrants. When the flux excursion is strictly positive, the transformer is said to operate in the first quadrant, denoted by I. The single-switch forward converter (or its two-switch variation) operates in the first quadrant. As the flux density starts from the remanent point B_r, the allowed positive excursion below the saturation level is somewhat limited. On the other hand, if the flux density starts from $-B_r$ prior to turn-on, the path to the positive saturation limit naturally increases, easing the transformer design to avoid saturation problems. Active clamp converters operate the transformer in the first and third quadrants, naturally improving the transformer utilization.

To avoid the saturation problem depicted by Fig. 8-5c, we need to bring the flux density back to its initial point at the beginning of each switching cycle. To achieve this, we can apply a reset voltage of opposite polarity across L_{mag} during the off time to bring down the core flux. However, the time during which the reset voltage occurs must be of sufficient duration to bring the flux density back to its starting point. In other words, the applied volt-seconds during the

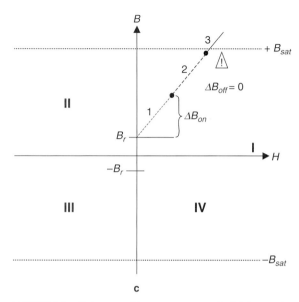

FIGURE 8-5c If the magnetizing current operates in CCM, saturation quickly occurs.

on time must equal the volt-seconds imposed on the magnetizing inductor during the off time. Otherwise stated:

$$V_{in}t_{on} = V_{reset}t_{off} \tag{8-9}$$

where V_{reset} illustrates the voltage applied over the magnetizing inductor during the off time.

Failure to do this, as shown in Fig. 8-5d, engenders transformer saturation after a few cycles. In this example, the off-time excursion denoted by ΔB_{off} cannot fully reset the core. This is often designated as a flux walk-away situation, and it is incumbent upon the designer to prevent this phenomenon from occurring. On the other hand, when a proper reset exists, as in Fig. 8-5e, the flux density comes back to its initial point and the operation is safe.

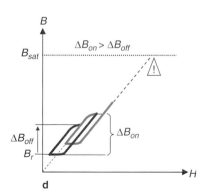

FIGURE 8-5d If an improper reset is made, saturation quickly occurs.

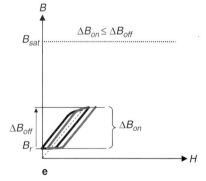

FIGURE 8-5e On the contrary, if equal amounts of volt-seconds are applied at both magnetizing and demagnetizing phases, the saturation problem disappears.

8.2 RESET SOLUTION 1, A THIRD WINDING

The most common solution to reset the core during the off time consists of adding another winding N_{sr} on the main transformer. Figure 8-6a shows the configuration. This winding operates in a flyback configuration, thanks to the winding dot positions.

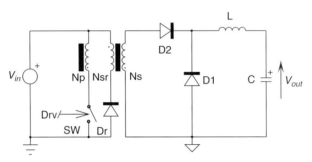

FIGURE 8-6a A third winding helps to reset the core during the off time.

During the on time, where SW is closed, the reset diode D_r is blocked as its cathode jumps to a voltage equal to

$$V_K = V_{in} + V_{in}N_r = V_{in}(1 + N_r) \qquad (8\text{-}10)$$

where $N_r = \dfrac{N_{sr}}{N_p}$.

Thus no current circulates in this reset winding, as shown by Fig. 8-6b, which depicts the situation. When the controller instructs the power switch to open, we come back to the situation of Fig. 8-4, now updated in Fig. 8-6c. The voltage across the magnetizing inductor reverses and forward biases the reset diode D_r: an image of the magnetizing current now flows in the reset

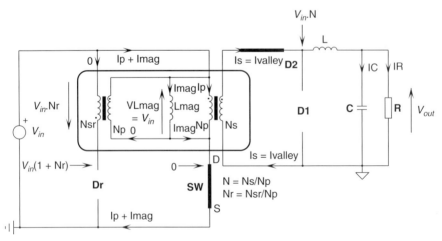

FIGURE 8-6b At the switch closing time, the reset diode cathode is biased to a level made up of the input level plus the reset winding reflected voltage.

SIMULATIONS AND PRACTICAL DESIGNS OF FORWARD CONVERTERS 747

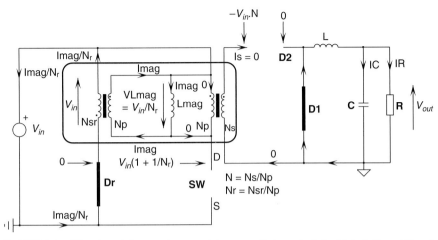

FIGURE 8-6c When the switch opens, the magnetizing current finds a path via the third winding and flows in the input source.

winding and circulates via the input source. We recycle the magnetizing current, to the benefit of the converter efficiency. Figure 8-6d zooms in on the primary portion, as the secondary side "disappears" from the picture (D_2 is blocked). As the reset diode D_r conducts, the input voltage also appears across the reset winding N_{sr}. Thanks to the transformer coupling, this voltage reflects across the magnetizing inductance and imposes a downslope equal to

$$S_{mag,off} = -\frac{V_{in}}{N_r L_{mag}} \qquad (8\text{-}11)$$

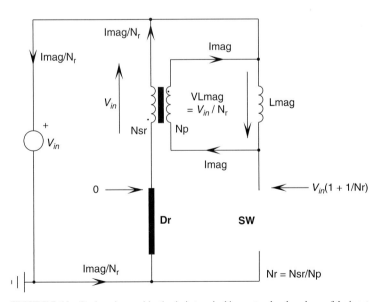

FIGURE 8-6d On the primary side, the drain terminal jumps to a level made up of the input voltage plus the reflected reset voltage.

Seen from the drain terminal, the reflected reset voltage becomes wired in series with the input source. The MOSFET thus sustains a peak voltage equal to

$$V_{DS} = V_{in} + \frac{V_{in}}{N_r} = V_{in}\left(1 + \frac{1}{N_r}\right) \tag{8-12}$$

If you remember the flux remarks, the transformer windings remain coupled as long as a core flux links them all. When the core is finally reset at the end of the off time, the flux has gone back to zero, Eq. (8-12) no longer holds and the drain goes back to the input voltage, as with the flyback case in DCM. By observing the signal $V_{DS}(t)$, we thus have an idea of the core reset point. This is what Fig. 8-6e shows you, where two different reset ratios are illustrated:

- $N_r = 1$: This is the most popular choice. The drain voltage jumps to twice the input voltage [Eq. (8-12)] and the duty cycle cannot exceed 50%. As reset voltages are equal (V_{in}) and Eq. (8-9) must still be satisfied, there is no other solution than limiting the duty cycle to below 50% where $t_{on} = t_{off}$. Usually, 45% is selected to include a safety margin via a dead time insertion between each switching cycle.
- $N_r < 1$: The designer accepts to demagnetize more quickly than he or she magnetizes, thus authorizing duty cycles above 50%. The penalty lies in the MOSFET BV_{DSS} selection that can quickly hamper the converter price (Eq. 8-12).

A typical forward application circuit appears in Fig. 8-7a. It pictures a 100 W, 28 V converter operating in current mode. Forward topologies are rarely operated in voltage mode.

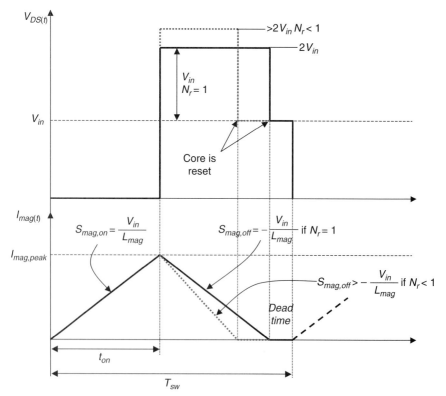

FIGURE 8-6e At the core reset occurrence, the drain voltage swings back to the input rail until the next on cycle.

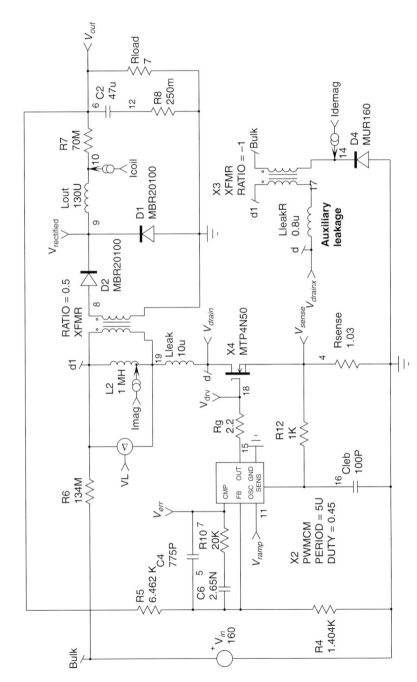

FIGURE 8-7a A typical simulation template for a forward converter using the current-mode generic model.

Given the second-order behavior of the system, the compensation becomes a more difficult exercise, compared to current mode. Also, building a type 3 error amplifier with a TL431 can be tricky, as already highlighted in Chap. 3. We will therefore concentrate on current-mode examples in this chapter. As their buck counterparts, forwards are designed to operate in CCM at full power and only enter DCM in light-load conditions.

In Fig. 8-7a, the demagnetization uses a 1:1 transformer ratio to which a leakage inductor has been added. As you can imagine, the spike associated with this inductor can affect the MOSFET drain-source maximal excursion, and care must be taken when specifying the transformer. It clearly appears in Fig. 8-7b details where all pertinent waveforms are gathered. The first signal shows the drain-source evolution. When the core resets, the drain node freely swings back to the input voltage and stays there until the next cycle. The voltage across the transformer must, of course, satisfy the inductor volt balance, and the average voltage at steady state is null. The etched area corresponds to the applied volt-seconds at turn-on and turn-off: they are equal. The third waveform is similar to that obtained with a classical buck converter. Regarding the primary current, as the secondary side inductor operates in CCM, the drain current jumps to the valley. Please note that the sense information can be affected by a sharp spike, linked to the buck diode t_{rr}. Filtering is sometimes necessary (LEB circuitry or a simple RC network as here) to avoid an erratic operation of the current-sense comparator. At the end of the on time, the peak current reaches the reflected buck current which adds to the magnetizing current. The compensation, in this example, uses a type 2 scheme built around the internal operational amplifier available in the model.

When you are looking at the drain-source waveform, something should trigger your interest: why isn't there any full wave ringing when the core reset is reached? After all, since the magnetizing current is becoming null, the drain is suddenly relaxed and is free to oscillate, just

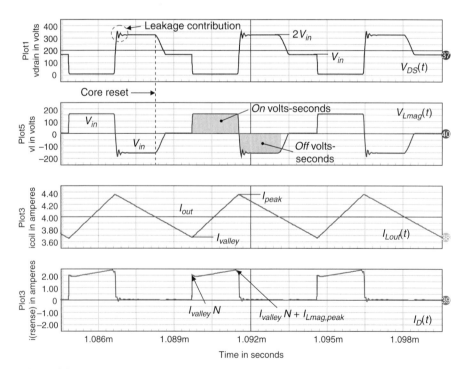

FIGURE 8-7b The pertinent waveform of the single-switch forward converter.

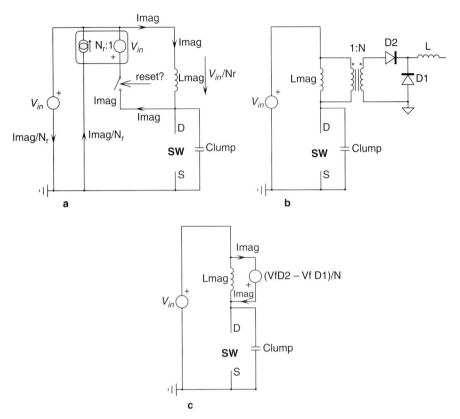

FIGURE 8-8 At the switch opening, the drain rings but is quickly clamped by the secondary diodes.

as in the DCM flyback case. Why isn't there any sinusoidal waveform on the MOSFET drain? Actually, there is one, but it is quickly clamped—by the secondary side diodes. Figure 8-8a shows a close-up on the primary path while resetting the core. The secondary side is absent, D_2 being blocked. When the magnetizing current has fallen to zero, V_{in} is no longer applied across the transformer primary and we have a resonating tank made of L_{mag} and C_{lump}, tuned at the following frequency:

$$f_{mag} = \frac{1}{2\pi\sqrt{L_{mag}C_{lump}}} \qquad (8\text{-}13)$$

At the core reset point, when diode D_r blocks (the switch opening on Fig. 8-8a), the drain voltage starts to resonate and its voltage falls from the $2V_{in}$ plateau (Fig. 8-8b). As a result, the voltage across the magnetizing inductor quickly diminishes toward 0, as shown in Fig. 8-7b. When it reaches zero, the drain is at V_{in}. Because of the resonance, the drain voltage would like to go farther down, as with a DCM flyback. However, as the magnetizing inductor voltage starts to increase, the secondary voltage develops and forward biases D_2 whose cathode is negative, since D_1 freewheels. If we consider both forward drops to be equal, we have a zero on the secondary side, reflected over L_{mag}: any further voltage excursion on the primary side is blocked (Fig. 8-8c). Also, thanks to the LC network, the magnetizing current was resonating below zero and is now pausing, because of the "short-circuit" imposed by D_2 and D_1.

752 CHAPTER EIGHT

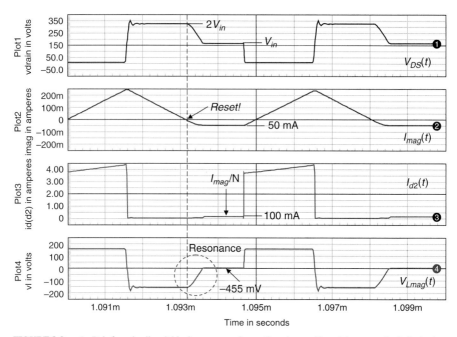

FIGURE 8-9a As D_1 is freewheeling, it blocks any excursion on the primary side and thus stops the drain ringing.

Figure 8-9a shows the plots obtained when we are looking carefully at the simulation signals. In reality, because of the freewheel action, the D_1 forward drop is stronger than that of D_2, leading to the reflection of a negative voltage over L_{mag}. The magnetizing current pauses and circulates as a current in D_2. This is clearly seen as a positive step prior to the on-time starting point on the $I_{d2}(t)$ curve. When the power switch turns on again, the magnetizing current starts from a negative value and increases toward its peak.

If the forward converter now operates in DCM, as the load becomes lighter, the short-circuit imposed by D_2 disappears as it blocks when the buck inductor L is reset. In this example, L gets reset before the magnetizing inductor. When the latter resets, the drain can freely ring as the previous clamping conditions have gone (Fig. 8-9b).

8.2.1 Leakage Inductance and Overlap

As described in the flyback section several times, imperfectly coupled windings can be represented by a perfect transformer ratio associated with a leakage inductor. The forward does not derogate to this rule for the coupling between the primary and the secondary windings. Figure 8-10a shows the equivalent circuit where the leakage inductance takes place in series with the secondary power winding. We are in the freewheel mode and the MOSFET just turned on: In this mode, the inductor current I_L circulates in D_1, but as D_2 starts to conduct, a current I_s appears in the secondary side. Given Kirchhoff's law, we can write

$$I_s(t) + I_{d_1}(t) = I_L(t) \tag{8-14}$$

If we neglect the ac ripple in inductor L (assume L is large), we can consider its current to be dc and equal to the output current I_{out}. As the inductor behaves as a constant-current source, if

SIMULATIONS AND PRACTICAL DESIGNS OF FORWARD CONVERTERS 753

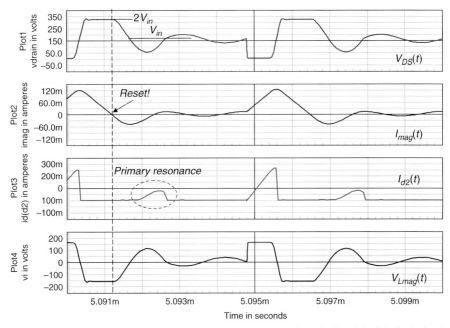

FIGURE 8-9b If the forward operates in DCM, D_1 no longer acts as a short-circuit and the drain freely rings!

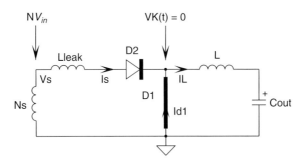

FIGURE 8-10a The leakage term delays the secondary side current rise time and forces conduction on both diodes.

the current I_s starts to increase as the primary MOSFET is turned on, the current in the freewheel diode must decrease at the same rate to satisfy Eq. (8-14): when I_s has reached I_{out}, $I_{d1} = 0$. The time t_1 needed for this transition depends on the secondary winding voltage $V_s(t)$ and the leakage inductor value. If we neglect the D_1 forward drop while freewheeling, we have

$$I_s(t) = \frac{NV_{in}}{L_{leak}}t \qquad (8\text{-}15)$$

and

$$I_{d_1}(t) \approx I_{out}\left(\frac{t_1 - t}{t_1}\right) \approx \frac{NV_{in}}{L_{leak}}(t_1 - t) \qquad (8\text{-}16)$$

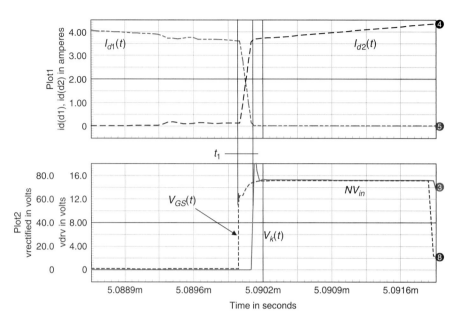

FIGURE 8-10b The overlap actually reduces the average voltage on the cathode junction and forces a wider duty cycle to compensate the loss.

During this period, both diodes are conducting and we have a so-called overlap. As the two diodes conduct, we have 0 V on average at the cathode junction until D_1 is fully blocked ($V_K = 0$). This occurs when $I_s(t) = I_{out}$ at $t = t_1$; thus

$$t_1 = \frac{L_{leak} I_{out}}{NV_{in}} \qquad (8\text{-}17)$$

This delay reduces the average voltage on left terminal L and forces the controller to increase the duty cycle. Figure 8-10b illustrates this phenomenon via curves obtained from the Fig. 8-7a simulation template. It occurs in a similar manner at turn-off. We will see in a dedicated section (8.8.1) that mag-amp circuits use this phenomenon to provide an efficient secondary side regulation technique.

During the overlap, the secondary is short-circuited by the diodes simultaneously conducting. This short-circuit is reflected on the primary side where the voltage across the magnetizing inductor is zero: the magnetizing current pauses and circulates in the secondary diodes, translated by the turns ratio.

The lost area shown on the lower portion of the graph occurs as the MOSFET turns on and lasts until D_1 is blocked. The average voltage loss can be evaluated through the etched-area integration, as highlighted by Fig. 8-10c.

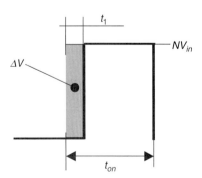

FIGURE 8-10c The average voltage lost during t_1 can be evaluated by integrating the etched area.

$$\Delta V = \frac{1}{T_{sw}} \int_0^{t_1} NV_{in} \cdot dt = F_{sw} NV_{in} t_1 \qquad (8\text{-}18a)$$

Replacing t_1 by its definition [Eq.(8-17)] gives

$$\Delta V = F_{sw} L_{leak} I_{out} \quad (8\text{-}18\text{b})$$

This equation states that the power converter behaves as a voltage source featuring an $L_{leak}F_{sw}$ output impedance during the overlap time. There are no particular losses linked to this event. However, the duty cycle must increase, compared to the theoretically computed value.

Figure 8-10d, e, and f shows oscilloscope shots captured on an ATX single-switch forward power supply. Figure 8-10d represents the drain (upper curve) and the anode of the secondary side diode $V_A(t)$. You believe the coupling is very good, but actually the converter uses a variation of the reset winding technique where a capacitor is inserted between the drain and the reset diode anode, when the reset winding is referenced to ground. This reduces losses and improves efficiency. Reference 1 describes this method. Figure 8-10e details the primary voltage at turn-on where the overlap zone is clear. Figure 8-10f portrays the overlap occurring at turn-off.

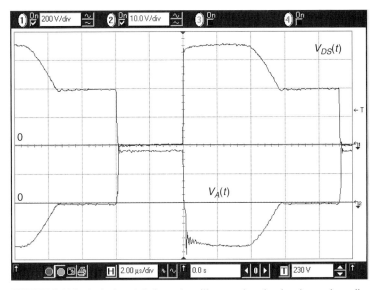

FIGURE 8-10d A single-switch forward oscilloscope shot showing the good coupling between the primary and the reset windings.

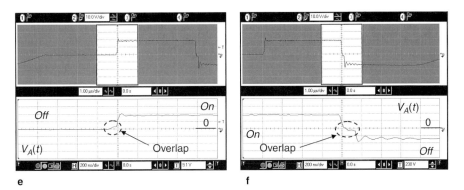

FIGURE 8-10e, f When zooming in on the secondary side anode, actually the primary voltage scaled down by N, the overlap area emerges clearly.

8.3 RESET SOLUTION 2, A TWO-SWITCH CONFIGURATION

Like the flyback converter with its leakage energy, the recycling of the magnetizing current in a single-switch configuration (also called singled-ended) imposes a voltage stress on the MOSFETs at turn-off. In our example, despite a 375 Vdc maximum input voltage, the 1:1 reset ratio forces the adoption of a MOSFET able to sustain at least twice this level. For safety, it is recommended to pick up a 900 V BV_{DSS} type, but we have found inexpensive ATX supplies operating with 800 V devices. Is it interesting to take such a risk, knowing that your downstream power electronic boards and disks are worth several hundred dollars? In this case, the two-switch forward type of power supply offers a more reliable way of dealing with the magnetizing current. If you remember the two-switch flyback, its forward counterpart does not significantly differ. Figure 8-11a portrays a typical application of a dual-ended forward converter. Both MOSFETs are now driven from a common gate-drive transformer, but it requires a 1:2 transformer ratio. The primary section of this transformer swings between $+\frac{V_{DRV}}{2}$ and $-\frac{V_{DRV}}{2}$, given the presence of the coupling capacitor (a dc block) C_2. If we want to select a 1:1 gate-drive transformer, we just need to insert another series capacitor with each secondary and add a way to restore the dc component. Usually a simple zener diode can do the job quite well. It was the solution presented in Fig. 7-24a, where a single-ended transformer drove the upper-side MOSFET only. In Fig. 8-11a, we departed slightly from this solution to suppress the secondary side dc-block capacitors. It is possible, thanks to the two bipolar transistors. They ensure a fast turn-off sequence and always maintain the gate in a low-impedance state when the DRV pin is low. This technique avoids problems when a controller capable of skipping

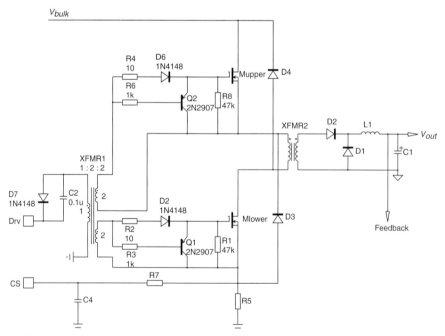

FIGURE 8-11a A two-switch forward converter uses two MOSFETs and a couple of diodes to recycle the magnetizing energy.

cycles (the NCP1217A, for instance) is implemented. When the driver suddenly stops switching as it enters into the skip mode, a resonance occurs between C_2 and the gate-drive transformer magnetizing inductor. This circulating current in the primary engenders an unwanted secondary voltage which can reactivate both MOSFETs without reaction from the controller as its output is off. Needless to say, this situation can be dangerous. Thanks to the bipolar presence, when the driver output goes low, diode D_7 fixes the gate-drive transformer primary to ground and Q_1/Q_2 firmly blocks both MOSFETs. Even if this event lasts for a while, the transformer is fully reset and both secondary windings act as a short-circuit, properly pulling down each MOSFET's gate. Tests of this typical circuitry have led to excellent results in skip-mode operation on consumer applications.

At turn-on, both power switches are conducting, and V_{bulk} appears across the primary transformer, exactly as in the single-ended configuration. As switches open (Fig. 8-11b) the voltage across the primary winding reverses and the magnetizing current finds its way through D_4 and D_3. Thanks to both diodes, the magnetizing current falls with a slope equal to

$$S_{mag,off} = -\frac{V_{bulk} + 2V_f}{L_{mag}} \tag{8-19}$$

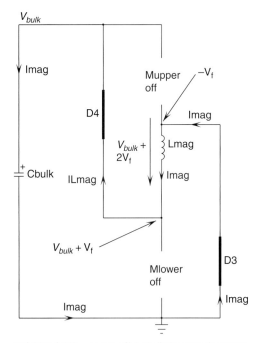

FIGURE 8-11b At turn-off, both diodes route the magnetizing current back to the source and improve the efficiency.

As with the single-switch version, the maximum duty cycle must remain below 50% or 45% including a safety margin. However, as both MOSFETs are wired in series, 500 V types can now be implemented.

Figure 8-12a portrays a simulation template using the PWMCM singled-ended controller biasing the bipolar-based gate-drive circuit. We have inserted a leakage inductor which should create overlap on the secondary side. This is what Fig. 8-12b confirms, where the magnetizing

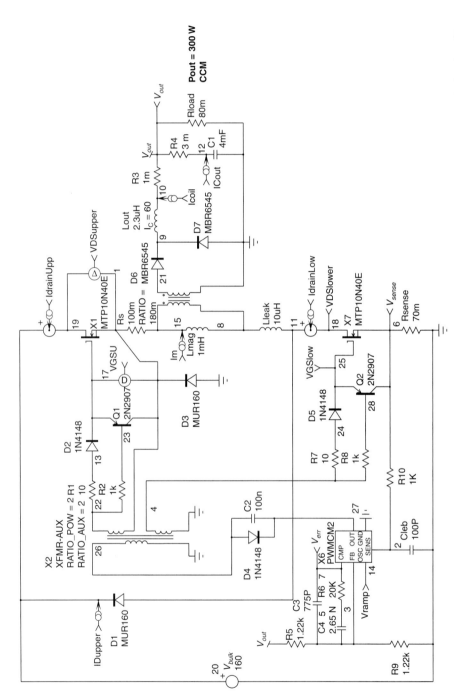

FIGURE 8-12a The two-switch forward simulation template using a 1:2:2 gate-drive transformer. The output is 5 V at 60 A. $V_K(t)$ is observed at the junction of D_6 and D_7 cathodes.

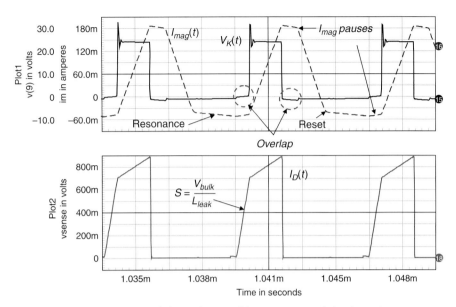

FIGURE 8-12b Given the leakage inductor, the magnetizing current pauses during the overlap.

current pauses as both secondary diodes conduct together. Of course, SPICE offers us a way to observe the magnetizing current, but this is not something you could check on the bench. Figure 8-12c depicts the voltages on both power MOSFETs: there is no ringing at the opening sequence, and the voltage peaks to the input level.

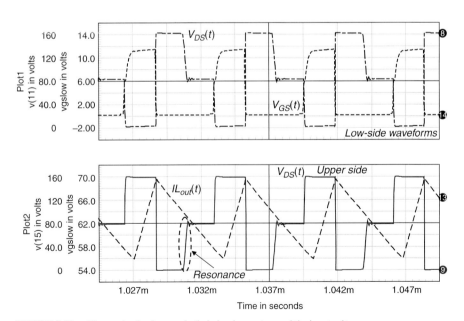

FIGURE 8-12c The maximal voltage on both drains does not exceed the input voltage.

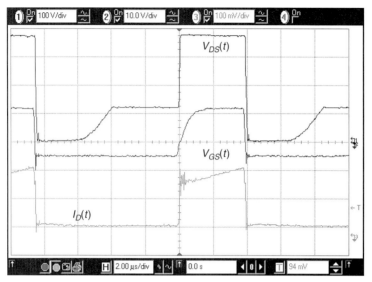

FIGURE 8-12d A two-switch forward converter has been built, and the above waveforms have been captured.

Figure 8-12d shows an oscilloscope shot captured on a real two-switch forward implementing a similar gate-drive circuit. Waveforms do not differ that much from the simulation. Reference 2 details a design procedure for a two-switch forward converter.

8.3.1 Two-Switch Forward and Half-Bridge Driver

In some cases, a gate-drive transformer can be considered too bulky, especially in dc-dc applications requiring a compact design, e.g., for brick converters. The half-bridge driver, as detailed in App. 8A, might therefore represent a possible solution to replace the transformer. Figure 8-13a presents the proposed sketch.

Unfortunately, the two-switch forward does not lend itself very well to the implementation of a half-bridge driver. Why? Because the lower C_{boot} terminal (node HB in Fig. 8-13a, U_2 pin 6) does not swing to the ground when M_{lower} turns on, as in a classical half-bridge configuration. On the contrary, it drops to $-V_f$ during the reset time, in other words, when both MOSFETs are off and the magnetizing current circulates in D_3/D_4. When the core is fully reset, both diodes stop conducting and the HB node turns to a high-impedance state (both MOSFET and the freewheel diodes are blocked): the refresh of the bootstrap capacitor C_{boot} prematurely disappears and the upper-side UVLO is quickly reached, disturbing the converter operation. The proposed circuit involving U_1 and M_1 actually builds an independent 1.5 μs pulse width occurring a few hundred nanoseconds after M_{lower} has been released. The pulse drives M_1 which independently pulls the HB pin to ground. No current circulates since M_{upper} is off. This technique has been successfully tested on a high-voltage two-switch forward converter used in an ATX power supply, originally working with a gate-drive transformer. Figure 8-13b portrays a typical shot showing the effect of the added circuitry.

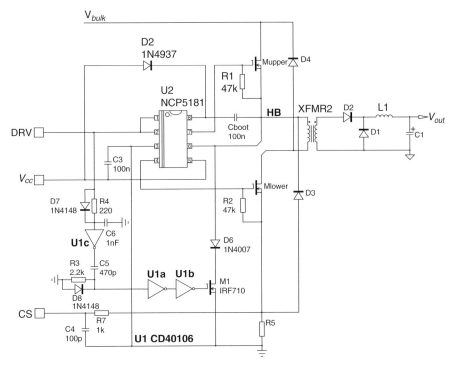

FIGURE 8-13a A half-bridge driver can be used in a two-switch forward application, but it requires a simple refresh circuit built around M_1 and U_1.

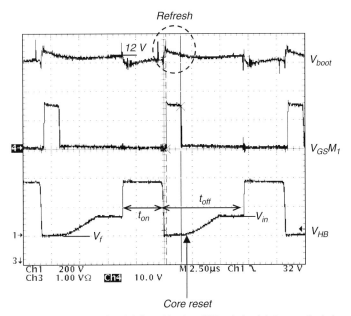

FIGURE 8-13b The signal delivered by the additional circuit helps to refresh the bootstrap capacitor despite light-load conditions.

8.4 RESET SOLUTION 3, THE RESONANT DEMAGNETIZATION

The resonant demagnetization technique offers another nondissipative way of resetting the core. This topology is often used in low-power dc-dc converters, switching at a high frequency. It consists of building a resonant circuit made up of the magnetizing inductor and the lump capacitor present on the drain node. The resonance brought by the network resets the core by using all parasitic elements and offers a way to expand the duty cycle excursion above 50% (Fig. 8-14a).

The lump capacitor placed between drain and source combines all capacitors either present on the primary side (the MOSFET C_{oss}, the transformer parasitic capacitor) or reflected from the secondary side (diode junction capacitors or from synchronous rectifiers). Figure 8-14b and c and Fig. 8-15a and b portray the resonant demagnetization forward converter at different stages. We purposely omitted the leakage inductor in this representation for the sake of simplicity.

- Figure 8-14b: The power switch is closed, and a current circulates in the MOSFET. This current is made up of the magnetizing inductor current plus the secondary side current reflected via the transformer turns ratio. The magnetizing current increases in a linear manner as in the traditional forward mode. The lump capacitor, here materialized as a component placed between the drain and source terminals, is discharged to the on-state MOSFET voltage drop (≈ 0).

- Figure 8-14c: The current flowing in the drain at the switch opening diverts into the lump capacitor. This current equals the reflected secondary side current I_s added to the magnetizing current. Considering the current constant during this short event, the pace at which the drain voltage increases is therefore given by

$$\frac{dV_{DS}(t)}{dt} = \frac{I_{D,peak}}{C_{lump}} \qquad (8\text{-}20)$$

where $I_{D,peak} = I_{mag,peak} + NI_{s,peak}$. Given the primary current value at the switch opening, it is usually a steep slope.

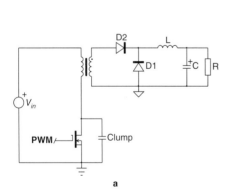

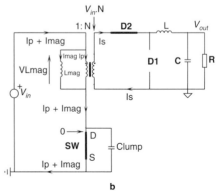

FIGURE 8-14a The demagnetization winding is removed, and the lump capacitor resonates with the magnetizing inductor.

FIGURE 8-14b At the switch closing, the input voltage appears across the magnetizing inductor and D_2 conducts.

SIMULATIONS AND PRACTICAL DESIGNS OF FORWARD CONVERTERS 763

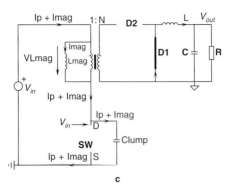

FIGURE 8-14c At the switch opening, the drain current is diverted in the lump capacitor, creating a steep voltage slope across the MOSFET drain-source terminals.

- Figure 8-15a: The rate will suddenly change when the drain terminal reaches the input voltage. At this point, diode D_2 starts to block and the freewheel portion via D_1 begins. The lump capacitor current falls to the magnetizing current level (Fig. 8-16) and the slope on the drain terminal transitions to a sinusoidal arch waveform. Its frequency depends on the resonating network created by the magnetizing inductor and the lump capacitor:

$$f_0 = \frac{1}{2\pi \sqrt{L_{mag} C_{lump}}} \qquad (8\text{-}21a)$$

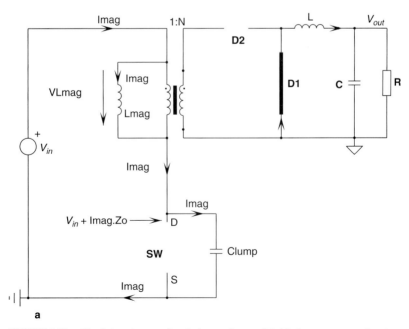

FIGURE 8-15a The drain voltage reaches the input voltage and D_2 blocks: a resonance takes place on the drain between L_{mag} and C_{lump}.

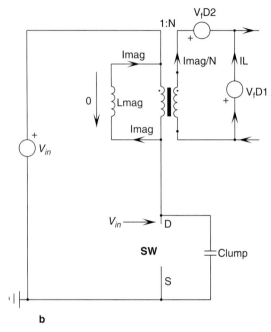

FIGURE 8-15b   The drain oscillation has fallen down to V_{in} and the magnetizing current now circulates in the secondary winding.

where the lump capacitor C_{lump} is

$$C_{lump} = C_{DS} + C_D N^2 + C_T \tag{8-21b}$$

where C_{DS} = total capacitor across MOSFET, C_{oss} plus external capacitor if any
C_D = total secondary side capacitance created by diode junctions, associated with another capacitor if installed
C_T = transformer equivalent primary capacitor obtained from measuring the resonating frequency of the device once a prototype exists

The peak voltage on the drain reaches a level now dictated by the following equation:

$$V_{DS,peak} = I_{mag,peak} \sqrt{\frac{L_{mag}}{C_{lump}}} + V_{in} \tag{8-22}$$

The voltage across the magnetizing inductor being sinusoidal, as is its current also, it crosses zero right at the drain-source top (90° phase shift). At this point, the core is reset. As the drain voltage continues its drop toward V_{in}, the magnetizing current reverses and pushes the B-H curve into the third quadrant.

The arch duration corresponds to one-half of the resonating period; hence

$$t_r = \pi \sqrt{L_{mag} C_{lump}} \tag{8-23}$$

- Figure 8-15b: The drain fall is suddenly stopped as it crosses V_{in}, starting the secondary diode D_2 conduction again (see Fig. 8-8c). The magnetizing current now circulates in the secondary winding, scaled by the turns ratio N, while pausing on the primary side as the magnetizing inductor voltage is null. When the MOSFET turns off again, a new cycle occurs and the current linearly ramps up again.

The simulated waveforms of such resonant forward converters appear in Figs. 8-16, 8-17, and 8-18. The drain peaks to a level defined by Eq. (8-22). At this point, we can observe a zero magnetizing current, and the core is reset. The current keeps going negative and stops when the drain voltage reaches the input level. In these simulations, we purposely wired a capacitor between drain and source. The resonating current therefore circulates in the sense resistor during the off time, as shown by the middle curve. At turn-on, the MOSFET must discharge the lump capacitor and it undergoes switching losses:

$$P_{SW,lump} = \frac{1}{2} V_{C_{lump}} V_{in}^2 F_{sw} \qquad (8\text{-}24)$$

In Fig. 8-18, we can observe the instantaneous voltage across the transformer primary terminals. As explained, the volt-seconds during the on time must equal the volt-seconds during the off time. Failure to maintain this relationship ends up in core saturation. If the off-time duration dictated by the controller does not allow the sinusoidal arch to return to V_{in}, core saturation will occur. This can be stated by the following equation:

$$\pi \sqrt{L_{mag} C_{lump}} \leq (1 - D_{max}) T_{sw} \qquad (8\text{-}25)$$

Therefore, when you design a converter, make sure a dead time exists during the start-up sequence where the duty cycle is pushed to the maximum. The resonant frequency depends on

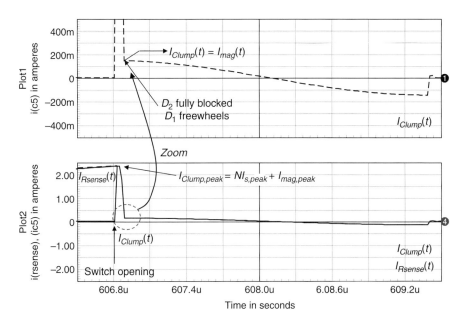

FIGURE 8-16 The lump capacitor current transitions from a value linked to the secondary side current plus the magnetizing current, to the magnetizing current alone when D_1 freewheels 100%.

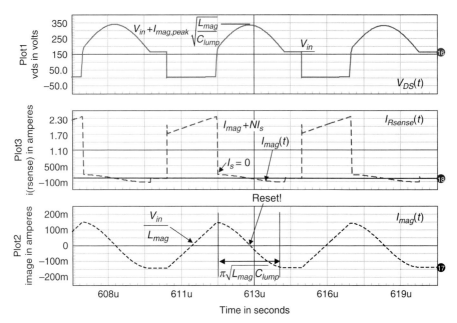

FIGURE 8-17 Simulated waveforms of the resonant forward converter.

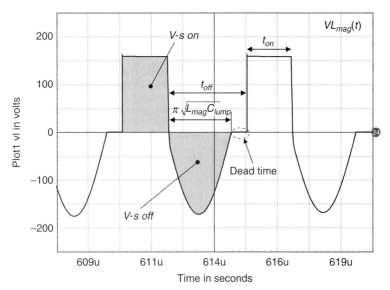

FIGURE 8-18 The voltage across the primary confirms the equality of the volt-seconds during the on and off states.

the magnetizing inductor and the parasitic capacitors among which the transformer stray elements also play a role. Once the transformer prototype is built, you can measure its own resonating frequency f_T, given by

$$f_T = \frac{1}{2\pi \sqrt{L_{mag} C_T}} \qquad (8\text{-}26)$$

Equation (8-25) defines the relationship between the minimum reset time (when $D = D_{max}$) and the resonating frequency involving the magnetizing inductor associated with all the parasitic capacitors seen on the drain node. If we consider the transformer capacitor contribution alone and multiply both equation ends by 2, we obtain

$$2\pi \sqrt{L_{mag} C_T} \ll 2T_{sw}(1 - D_{max}) \qquad (8\text{-}27)$$

Otherwise stated:

$$\frac{1}{2\pi \sqrt{L_{mag} C_T}} \ll \frac{1}{2T_{sw}(1 - D_{max})} \qquad (8\text{-}28)$$

Replacing the left term by Eq. (8-26), we have

$$f_T \gg \frac{F_{sw}}{2(1 - D_{max})} \qquad (8\text{-}29)$$

It is therefore not unusual to gap the core to reasonably bring the magnetizing inductor down and increase the transformer resonating frequency. Gapping the core also improves the magnetizing inductor distribution spread across production. You pay it by a larger magnetizing current.

By adjusting the peak drain voltage, you have a way to offer off-time volt-seconds larger than the on-time volt-seconds. This resonant technique thus allows you to operate at duty cycles above 50% but at the expense of a voltage stress sustained by the MOSFET. As indicated, the transformer now works in quadrants I and III, naturally leading to a better core utilization.

Design examples of resonant forward topologies for telecom dc-dc converters are described in Refs. 3 and 4.

8.5 RESET SOLUTION 4, THE RCD CLAMP

Resetting the forward transformer core consists of authorizing a drain swing of sufficient amplitude above V_{in} at the opening event. The swing excursion must be such that the reset voltage applied across the magnetizing inductor brings the flux back to its starting point. The traditional forward (and its two-switch counterpart) applies a reset voltage V_{in} across the primary (the drain peaks to twice V_{in} with a 1:1 reset winding ratio), naturally limiting the duty cycle excursion below 50%. The resonant forward structure lets the MOSFET drain oscillating in a smooth sinusoidal wave whose amplitude defines the maximum allowable duty cycle excursion. Operation beyond 50% is therefore possible but at the penalty of a higher voltage stress on the MOSFET. This is the reason why this technique remains confined to dc-dc telecom applications (V_{in} from 36 to 72 Vdc).

Another popular reset scheme uses the *RCD* clamp technique, like the one used to reset the leakage inductor in flyback topologies. This technique has gained a lot of popularity in single-switch forward converters for ATX power supplies. It is less expensive than using a tertiary winding, and as it allows duty cycle excursions above 50%, it greatly improves the behavior in half-cycle dropout tests.

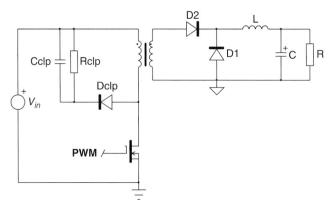

FIGURE 8-19 An *RCD* clamping network offers a path to the magnetizing current.

Figure 8-19 depicts a forward converter equipped with an *RCD* clamping network.

To explain how it works, we do not have any choice but to go through multiple drawings, dissecting each operating stage. Actually, the *RCD* operation does not really differ from that of the resonant reset technique, except that we purposely clamp the voltage excursion before the peak given by Eq. (8-22). In these sketches, the *RC* network has been replaced by a voltage source of a fixed value, representative of the steady-state clamp voltage. Figure 8-20a and b represents the first two intervals:

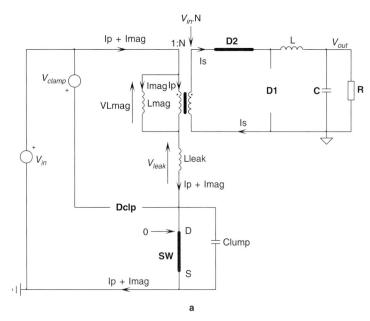

FIGURE 8-20a The turn-on phase is similar to that of traditional forward converters.

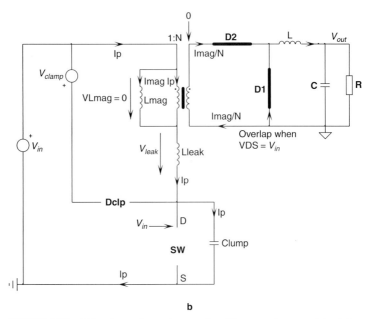

FIGURE 8-20b When the overlap on the secondary side occurs, the magnetizing inductance is short-circuited. This configuration occurs during a brief period of time.

- Fig. 8-20a: The MOSFET is on and the current ramps up in the magnetizing inductor. The secondary diode D_2 conducts; D_1 is blocked. The lump capacitor is discharged.
- Figure 8-20b: The MOSFET just opens; the drain voltage swings up to a rate imposed by the; total primary current now flowing into the lump capacitor (see Eq. 8-20).
- The drain-source voltage reaches the input voltage V_{in}; D_2 starts to block as D_1 begins to freewheel. As both diodes are conducting, a short-circuit appears across the magnetizing inductance on the primary side. The magnetizing current pauses and L_{mag} "leaves" the network during the brief overlap time. The new resonance now involves L_{leak} and C_{lump} as the magnetizing inductor is short-circuited:

$$f_{leak} = \frac{1}{2\pi \sqrt{L_{leak}C_{lump}}} \tag{8-30}$$

- Fig. 8-21a: The reflected secondary current is now zero since D_1 fully freewheels. The current in the leakage inductor drops to the magnetizing current level (see Fig. 8-16), and the resonating frequency goes to that defined by Eq. (8-31):

$$f_0 = \frac{1}{2\pi \sqrt{(L_{mag} + L_{leak})C_{lump}}} \tag{8-31}$$

where the lump capacitor is made of

$$C_{lump} = C_{DS} + C_D N^2 + C_T$$

Its value is similar to that pertaining to Eq. (8-21b).

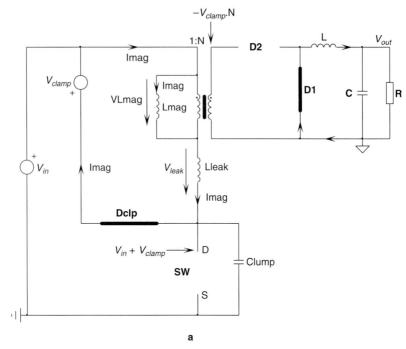

FIGURE 8-21a When the drain voltage reaches the clamp level, the associated diode D_{clp} conducts and stops the drain excursion.

- The drain voltage increases above V_{in} in a sinusoidal manner until D_{clp} starts to conduct. The drain excursion is fixed by the clamp voltage and the magnetizing current comes down at a pace defined by (neglecting L_{leak})

$$S_{mag,\,off} \approx -\frac{V_{clamp}}{L_{mag}} \qquad (8\text{-}32)$$

- When the magnetizing current crosses zero, the transformer core is reset, D_{clp} blocks and the drain resonates with the same network L_{mag}, L_{leak}, and C_{lump}, down to V_{in}. At this point, D_2 conducts and ensures a circulating path for the magnetizing current on the secondary side (Fig. 8-21b).

The simulated waveforms of the *RCD* forward converter appear in Fig. 8-22. At the switch opening, the voltage on the drain ramps up as on the resonant forward. At a certain time, the drain hits the clamp level and the associated diode starts to conduct. Conduction occurs until the magnetizing current reaches zero. At this point, D_{clp} blocks and an oscillation occurs. The current resonates in the negative portion and pushes the transformer in the third quadrant. When the drain reaches the input voltage, the secondary side diode D_2 enters conduction and clamps the primary winding to almost zero: the magnetizing current pauses until the MOSFET turns on again. Figure 8-22 displays the primary inductor voltage which reveals the brief oscillation described by Eq. (8-30) during the turn-off overlap. Figure 8-23 zooms in on the secondary diodes, confirming the small overlap, when both devices conduct at the same time.

SIMULATIONS AND PRACTICAL DESIGNS OF FORWARD CONVERTERS

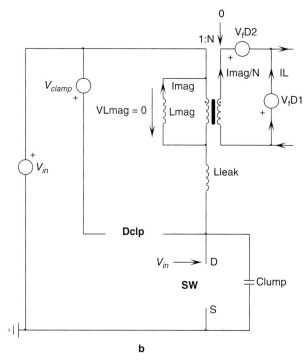

FIGURE 8-21b When the drain voltage falls down to the input voltage and tries to go further down, the secondary diodes conduct the magnetizing current.

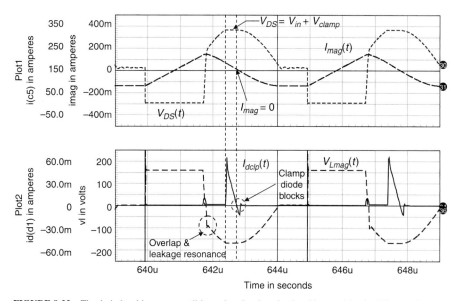

FIGURE 8-22 The drain level increases until it reaches the clamping level imposed by the *RC* network.

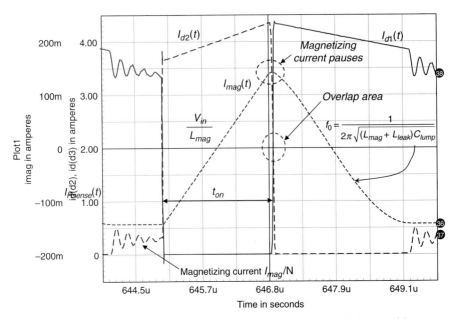

FIGURE 8-23 A close-up of the secondary diode currents shows the recirculation of the magnetizing current in D_2 before the MOSFET turns on again.

The way an $\overline{RCD}$ forward converter operates is quite complex given the various resonances in play. Reference 5 details the operation and describes a design procedure. Reference 6 analyzes the converter in a different way, including leakage inductance effects. For the sake of simplicity, these effects are neglected in the calculations below.

To determine the clamp element values, first we need to know the current circulating when the clamp diode conducts. The conduction point, denoted by I_{clp}, appears in Fig. 8-24.

To obtain the value I_{clp}, first we calculate the time needed by the drain-source voltage to reach the clamp voltage when starting from V_{in}. This time segment is denoted by Δt in Fig. 8-24. For the sake of simplicity, we purposely neglected the rise time from 0 to V_{in}. The peak of the waveform, when unclamped, has been given by Eq. (8-22). If we consider a sinusoidal evolution from the point at which $V_{DS}(t)$ crosses the input voltage to the clamp level, we can write

$$I_{mag,peak}\sqrt{\frac{L_{mag}}{C_{lump}}}\sin\omega\Delta t = V_{clamp} \qquad (8\text{-}33)$$

Extracting Δt, we have

$$\Delta t = \arcsin\left(\frac{V_{clamp}}{I_{mag,peak}\sqrt{\dfrac{L_{mag}}{C_{lump}}}}\right)\frac{1}{2\pi f_0} \qquad (8\text{-}34)$$

Now replacing f_0 by its definition [Eq. (8-31)] where the leakage term has been neglected gives

$$\Delta t = \arcsin\left(\frac{V_{clamp}}{I_{mag,peak}\sqrt{\dfrac{L_{mag}}{C_{lump}}}}\right)\sqrt{L_{mag}C_{lump}} \qquad (8\text{-}35)$$

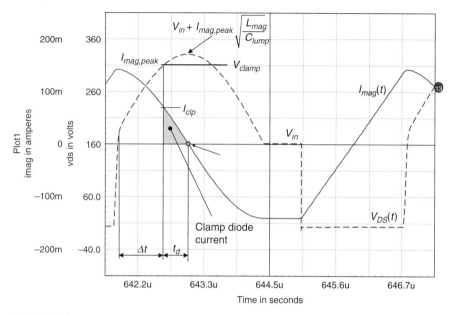

FIGURE 8-24 A study of the resonating waveform reveals when the clamp diode conducts.

If we consider the magnetizing current centered on zero (symmetric operation), then at steady state, it reaches a peak value obtained by the following definition:

$$I_{mag,peak} = \frac{V_{in}}{2L_{mag}} t_{on} \tag{8-36}$$

Given the resonance, the current falls until $V_{DS}(t)$ returns to V_{in}. The current reaches I_{clp} at Δt, where the clamp diode conducts. Thus

$$I_{clp} = I_{mag,peak} \cos(2\pi f_0 \Delta t) \tag{8-37}$$

The diode conduction time t_d is easily obtained, given the slope imposed on the magnetizing current defined by Eq. (8-32). The diode current peaks to I_{clp} and falls to zero, leading to a triangular shape:

$$t_d = I_{clp} \frac{L_{mag}}{V_{clamp}} \tag{8-38}$$

Well, we now have everything to calculate the *RCD* elements. The first element to determine is the minimum clamp voltage. Remember, in a forward, we deal with volt-seconds. Therefore, there must be sufficient voltage developed across the transformer primary during the off time; otherwise saturation will occur. In other words, if we clamp too low, the voltage excursion above V_{in} will not be high enough to reset the core during the minimum off time. To satisfy the volt-seconds during the resonance, let us glance at Fig. 8-25.

The volt-seconds applied during the on time is easy to compute:

$$\langle V_{L_{mag}} \rangle_{t_{on}} = V_{in} t_{on} \tag{8-39}$$

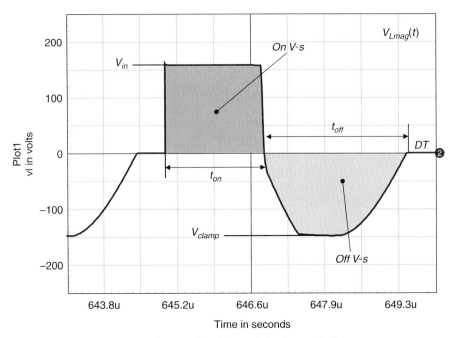

FIGURE 8-25 The voltage across the magnetizing inductor during the on and off times.

The volt-seconds during the off time requires a slightly larger effort. We need to write the time equation of $V_{Lmag}(t)$ during the off time and integrate it over the resonating cycle. We thus have

$$\langle V_{L_{mag}} \rangle_{toff} = V_{clamp} \int_0^{t_{off}} \sin\left(2\pi \frac{t}{2t_{off}}\right) \cdot dt = \frac{2V_{clamp}}{\pi} t_{off} \qquad (8\text{-}40)$$

There is a small error brought about by the clamping effect, but it can be considered negligible if the clamp is kept close to the peak. To demagnetize the core in all conditions, that is, when the off time is minimum, we need to ensure the following inequality:

$$V_{in} t_{on} \leq \frac{2V_{clamp}}{\pi} t_{off} \qquad (8\text{-}41)$$

Extracting the clamping voltage, we have

$$\frac{\pi V_{in} t_{on}}{2t_{off}} \leq V_{clamp} \qquad (8\text{-}42)$$

In worst-case conditions, low line and full power, in Fig. 8-25 the dead time vanishes and the off time shrinks to

$$t_{off,min} = (1 - D_{max})T_{sw} \qquad (8\text{-}43)$$

Based on the above expression and Eq. (8-7), we can update Eq. (8-42) to

$$V_{clamp} \geq \frac{\pi V_{out}}{2N(1 - D_{max})} \qquad (8\text{-}44)$$

Now, as we resonate during the off time, Eq. (8-27) still holds:

$$\pi \sqrt{L_{mag}C_{lump}} \leq (1 - D_{max})T_{sw} \qquad (8\text{-}45)$$

Extracting the magnetizing inductor value gives

$$L_{mag} \leq \frac{(1 - D_{max})^2}{F_{sw}^2 \pi^2 C_{lump}} \qquad (8\text{-}46)$$

Thus first C_{lump} must be estimated, and then L_{mag} which can require some adjustment via the introduction of a gap. Several iterations are needed to obtain the final result.

The clamp resistor R_{clp} can be calculated by assuming a buck-boost converter built around L_{mag}, R_{clp}, and C_{clp} and delivering a voltage V_{clamp}. As the current circulating in L_{mag} when D_{clp} conducts has been calculated by Eq. (8-37), we have

$$P_{clamp} = \frac{V_{clamp}^2}{R_{clp}} = \frac{1}{2}I_{clp}^2 F_{sw} L_{mag} \qquad (8\text{-}47)$$

Extracting R_{clp} leads to

$$R_{clp} = \frac{2V_{clamp}^2}{I_{clp}^2 F_{sw} L_{mag}} \qquad (8\text{-}48)$$

It is interesting to note that unlike with a flyback converter where the clamping resistor can be reduced to limit the voltage excursion within safe limits, we need excursion above V_{in} to reset the core. If R_{clp} is too low, reset will not occur. Equation (8-47) gives the dissipated power and will guide you through the right selection. Unlike the other reset techniques, the clamp voltage remains independent of the input voltage.

The clamp capacitor has to be calculated in order to keep the ΔV ripple within a certain range. If we look back at Fig. 8-24, the charge stored by the capacitor during the clamp diode conduction time is

$$Q_{C_{clp}} = I_{clp}\frac{t_d}{2} \qquad (8\text{-}49)$$

The voltage ripple seen by the clamping network is simply ($Q = VC$)

$$I_{clp}\frac{t_d}{2} = C_{clp}\Delta V \qquad (8\text{-}50)$$

Replacing t_d by its definition [Eq. (8-38)] and extracting C_{clp}, we obtain

$$C_{clp} = \frac{I_{clp}^2 L_{mag}}{V_{clamp}^2 \Delta V} \qquad (8\text{-}51)$$

The clamp capacitor also plays a role during transient events as it should not prevent V_{clamp} from going up, if necessary, to reset the core despite a duty cycle pushed to the maximum. Reference 7 discusses this topic in detail.

To verify the calculations, we have assembled a simulation circuit as presented by Fig. 8-26a. The converter delivers 28 V and features the following elements:

$L_{mag} = 1$ mH
$N = 0.51$
$C_{lump} = 700$ pF
$F_{sw} = 200$ kHz
$D = 0.35$
$V_{in} = 160$ Vdc
$f_0 = 190$ kHz

First, let us calculate the steady-state peak magnetizing current:

$$I_{mag,peak} = \frac{V_{in}D}{2L_{mag}F_{sw}} = \frac{160 \times 0.35}{2m \times 200k} = 140 \text{ mA} \qquad (8\text{-}52)$$

From this number, we can check the maximum peak voltage without a clamping network:

$$I_{mag,peak}\sqrt{\frac{L_{mag}}{C_{lump}}} + V_{in} = 0.140\sqrt{\frac{1m}{700p}} + 160 = 327 \text{ V} \qquad (8\text{-}53)$$

For the sake of the example, we are going to place the clamp level at 150 V, hence an excursion of 310 V, slightly below the maximum without clamp. From Eq. (8-35), we calculate the time at which the drain reaches the clamp level:

$$\Delta t = \arcsin\left(\frac{V_{clamp}}{I_{mag,peak}\sqrt{\frac{L_{mag}}{C_{lump}}}}\right)\sqrt{L_{mag}C_{lump}}$$

$$= \arcsin\left(\frac{150}{0.140 \times \sqrt{\frac{1m}{700p}}}\right)\sqrt{1m \times 700p} = 1.1 \times 836n = 920 \text{ ns} \qquad (8\text{-}54)$$

Again, make sure that the calculator is set in radians to obtain the above result. Let us now find the magnetizing current value flowing through the diode when $V_{DS}(t)$ reaches V_{clamp}:

$$I_{clp} = I_{mag,peak}\cos(2\pi f_0 \Delta t) = 0.14 \times \cos(6.28 \times 190k \times 920n)$$

$$= 0.14 \times 0.455 = 64 \text{ mA} \qquad (8\text{-}55)$$

We now have everything to calculate the right clamp resistor:

$$R_{clp} = \frac{2V_{clamp}^2}{I_{clp}^2 F_{sw} L_{mag}} = \frac{2 \times 150^2}{64m^2 \times 200k \times 1m} = 55 \text{ k}\Omega \qquad (8\text{-}56)$$

If we select a typical ripple of 5% of V_{clamp} (7 V_{pp}), then the clamp capacitor is found to be

$$C_{clp} = \frac{I_{clp}^2 L_{mag}}{V_{clamp} 2\Delta V} = \frac{0.064^2 \times 1m}{150 \times 2 \times 7} = 2 \text{ nF} \qquad (8\text{-}57)$$

Figure 8-26a portrays the simulation schematic adopted for the *RCD* forward converter.

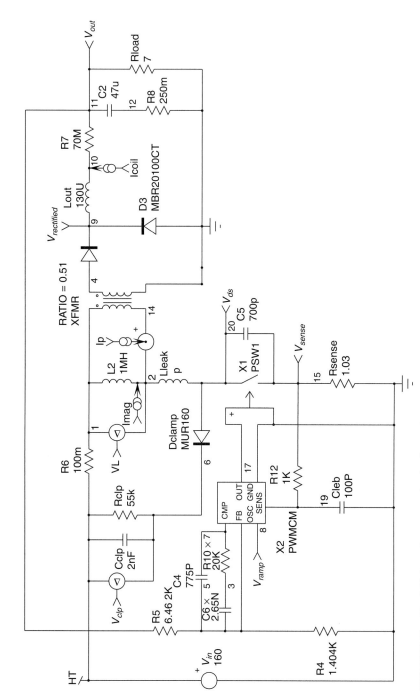

FIGURE 8-26a The *RCD* forward converter using the clamp value derived in the above lines.

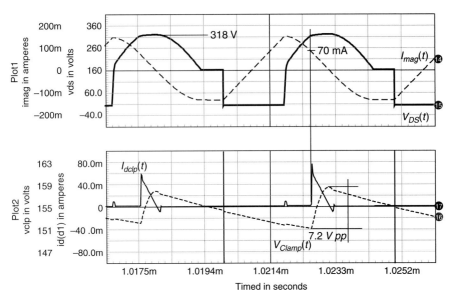

FIGURE 8-26b Simulation results of the *RCD* forward converter are in agreement with our calculations.

Figure 8-26b details the simulation results obtained after a few minutes.

In these forward converter designs, fully resonant or *RCD* clamp-based, the care brought to the transformer construction is a preponderant criterion. If for any reason the leakage term can no longer be neglected, because of a loose coupling, the above definitions must be revisited, as proposed by Ref. 6:

$$R_{clp} = 2V_{clamp}^2 \left[(V_{out}/N)^2 \frac{T_{sw}}{L_{mag}} - \frac{2V_{out}V_{clamp}}{N\sqrt{L_{mag}/C_{lump}}} + F_{sw}L_{leak}(I_{sec}/N)^2 \right]^{-1} \quad (8\text{-}58)$$

$$C_{clp} = \frac{D_{max}T_{sw}}{R_{clp}\Delta V/V_{clamp}} \quad (8\text{-}59)$$

We have resimulated the Fig. 8-26a circuit with a 1% leakage inductor. Results appear in Fig. 8-26c. To keep the same level on the drain, the clamp resistor was reduced to 4.6 kΩ and the clamp capacitor increased to 22 nF. In this example, the clamp resistor dissipates around 6 W, to compare with 410 mW when neglecting the leakage inductor. Facing this lost power increase, the designer must ask whether the *RCD* technique represents the best choice in this case.

8.6 RESET SOLUTION 5, THE ACTIVE CLAMP

Among the emerging topologies, the forward featuring active clamp starts to spread in off-line applications where good efficiency is needed. Unlike the other structures we already studied, the active clamp forward offers an elegant means to fully recycle the magnetizing current via zero-voltage switching operation on the main power MOSFET. Built around a floating upper-side

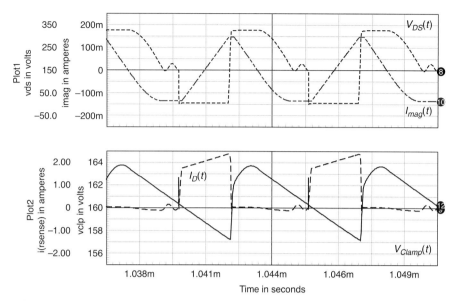

FIGURE 8-26c Same simulations as in Fig. 8-26b with a leakage term increased to 1%.

switch, the active clamp converter is seen in Fig. 8-27a. A possible connection to ground implies the use of a P channel, as proposed by Fig. 8-27b. The benefits of the active clamp converter are listed below:

1. The transformer reset lasts for the whole off-time event. When you are implementing synchronous rectification on the secondary side, this implies a full drive signal availability for the freewheel synchronous transistor. This differs from conventional forwards where the drive signal disappears as soon as core reset is reached, leaving the body diode alone to do the job.
2. As we will see, the transformer is driven in quadrants 1 and 3, leading to better transformer utilization. This naturally offers the possibility of pushing the duty cycle above 50%, naturally engendering the reduction of the transformer turns ratio. This, in turn, relieves the

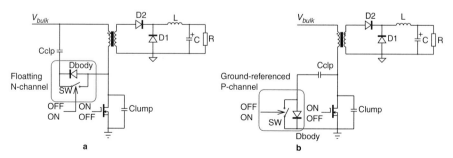

FIGURE 8-27 An active clamp forward converter requires a second switch authorizing the clamp current to circulate in both directions.

primary switch current stress and allows the selection of lower breakdown voltage diodes and MOSFETs on the secondary side.
3. The technique ensures the recycling of the energy stored in parasitic elements, such as the lump capacitor and the leakage inductor. If carefully designed, the forward active clamp can bring zero voltage switching on the power switch, naturally opening the door to higher switching frequency operations with all the physical benefits they bring, in particular smaller magnetic parts.
4. The voltage stress on the power switch is relatively constant and does not depend on the input voltage. However, some precautions must be observed to keep the clamp voltage under control in transient events such as power-off or load steps.

If you compare the active clamp forward converter to the active clamp flyback converter described in Chap. 7, the principle remains very close: how to reset the magnetizing current in a nondissipative way and ensure zero voltage switching (ZVS) on the drain when the power switch turns on again?

To reach this goal, the switch SW made of a MOSFET with its inherent body diode allows current to conduct in both directions: one spontaneous, via the body diode, and the other one, in reverse, through the MOSFET properly biased. This MOSFET must be opened at a certain time, forcing the peak inductive current driven by L_{mag} to find a way through the lump capacitor. The MOSFET can be either floating and hooked to the high-voltage rail (V_{bulk}) or referenced to the ground. In the first case, this would be an N channel and would require a way to bias it via the bootstrap technique or a gate-drive transformer. The second technique is built around a P channel and naturally simplifies the gate control. However, in off-line applications, the component needs to be a high-voltage type which dramatically increases its cost.

Let us take the various events step by step to describe the active clamp operation. Figure 8-28a and b shows the current circulation at the switch closing time. We assume to be at steady state, e.g., C_{clp} is already charged to V_{clamp}:

- Figure 8-28a: The MOSFET is on, and the current ramps up in the magnetizing inductor. The secondary diode D_2 conducts, and D_1 is blocked. The lump capacitor is discharged, and V_{DS} is almost zero. The leakage inductor is crossed by the reflected output inductor current which peaks to I_{mag} plus NI_s.
- Figure 8-28b: The MOSFET just opens, and the drain voltage swings up, mainly driven by the leakage inductor. The voltage quickly increases at a pace depending on the peak current and the lump capacitor. If we consider the leakage current constant during this short event, the slope is thus

$$\frac{dV_{DS}(t)}{dt} = \frac{I_{D,peak}}{C_{lump}} \qquad (8\text{-}60)$$

where $I_{D,peak} = I_{mag,peak} + NI_{s,peak}$.
- The drain voltage has now reached the input voltage. The secondary side current transfers from D_2 to D_1, actually short-circuiting the secondary winding and, by reflection, the primary inductor. The magnetizing inductor current pauses as L_{mag} is short-circuited. The drain-source voltage keeps increasing until it reaches $V_{in} + V_{clamp}$, where the clamp diode conducts. A current now circulates in the clamp capacitor, and V_{clamp} entirely appears across the leakage inductor: its reset has started. After a time Δt, the leakage inductor current has dropped to the magnetizing current level, ending the overlap sequence: D_2 is fully blocked. During that time, the clamp capacitor has undergone a sudden voltage jump, given the Q charge brought during Δt. On an active clamp converter, the clamp capacitor voltage does not significantly change at steady state. Therefore, in considering a leakage inductor constant

SIMULATIONS AND PRACTICAL DESIGNS OF FORWARD CONVERTERS 781

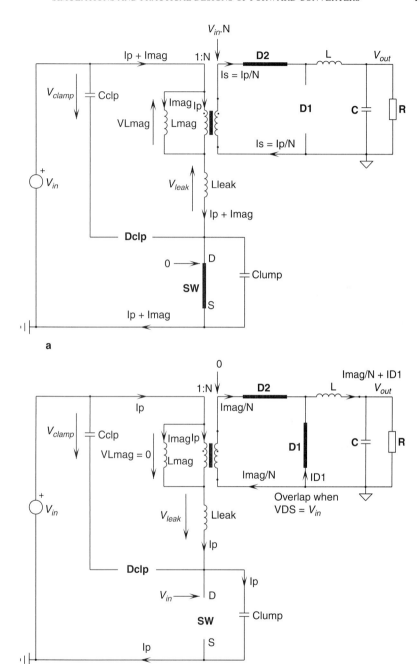

FIGURE 8-28 When the MOSFET closes, the primary current rises to a peak imposed by the reflected secondary side valley current. When it reaches the input voltage, the overlap sequence starts, short-circuiting the magnetizing inductor.

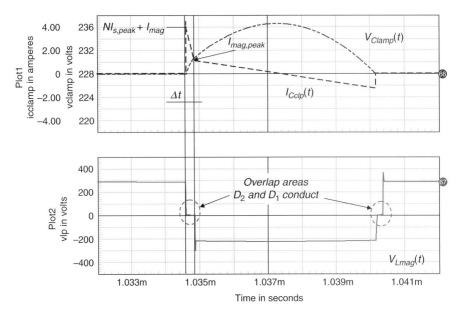

FIGURE 8-29 A close-up at the switch opening shows a quick voltage rise until the overlap sequence short-circuits the primary inductor.

voltage reset, we can approximate the overlap duration:

$$\Delta t = \frac{NI_s - I_{mag,peak}}{V_{clamp}} L_{leak} \quad (8\text{-}61)$$

Figure 8-29 offers a close-up of the overlap sequence where the leakage inductance partial reset clearly appears.

- As the clamp diode is now conducting, the controller can activate the upper-side switch (if we consider Fig. 8-27a) in zero-voltage condition. The controller must thus generate a phase-reversed signal slightly delayed compared to the main switch drive. This is exactly like a half-bridge arrangement including dead time control.
- Figure 8-30a: At the end of the overlap sequence, a resonance occurs between the magnetizing inductor associated with the leakage inductor and the capacitors present in the mesh: C_{clp} plus C_{lump}. The drain voltage slightly increases and reaches a peak when the magnetizing current crosses zero; at this time, the core is reset. Considering the clamp voltage constant (we assume C_{clp} is large enough to maintain a low ripple) and neglecting the leakage term, we see the magnetizing current falls at a rate defined by

$$S_{mag,off} = -\frac{V_{clamp}}{L_{mag}} \quad (8\text{-}62)$$

- Figure 8-30b: The magnetizing current crosses zero where it changes direction, thanks to the upper switch allowing conduction in both directions.
- Figure 8-31a: The clamp voltage decreases until the magnetizing current reaches its maximum negative value. At that time, the controller instructs the upper switch to open. The magnetizing current does not have another way other than circulating via the lump capacitor which starts to discharge: V_{DS} falls to V_{in}.

SIMULATIONS AND PRACTICAL DESIGNS OF FORWARD CONVERTERS 783

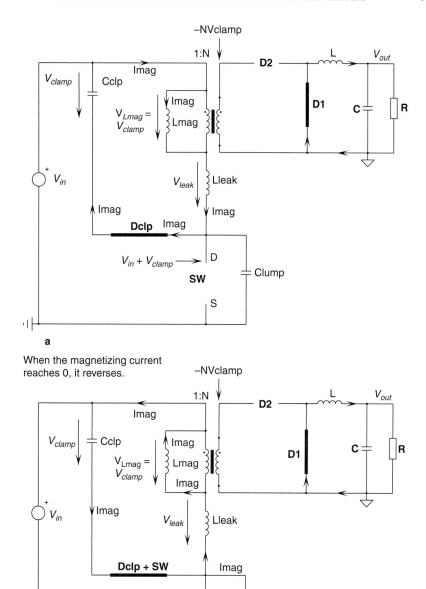

FIGURE 8-30 The drain voltage has reached the clamp level, and after the overlap duration, the magnetizing current finds a way through the clamp capacitor.

784 CHAPTER EIGHT

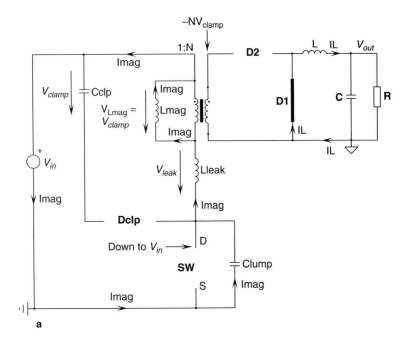

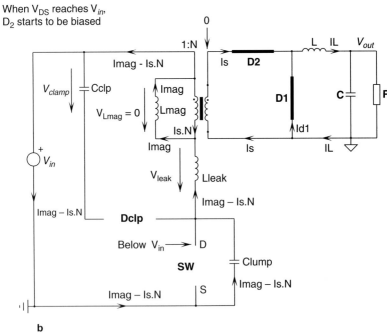

FIGURE 8-31 The clamp voltage pushes the magnetizing current toward its negative peak. At a certain point, the upper switch opens, and the magnetizing current does not have any other possibility than flowing through the lump capacitor.

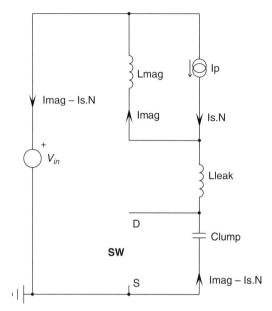

FIGURE 8-32 The reflected current corresponding to D_2 starting to conduct subtracts from the magnetizing current.

- Figure 8-31b: The drain voltage is now at V_{in}, and the primary voltage starts to be positive again, bringing D_2 into conduction. If D_2 conducts current in the secondary winding, it appears on the primary side as well. What is now important to understand is the way currents split in the primary side. Figure 8-32 zooms in on this particular configuration. The presence of the leakage inductor is important as it plays a favorable role in this case. The current circulating in the lump capacitor is expressed by

$$I_{C_{lump}}(t) = I_{mag}(t) - NI_s(t) \qquad (8\text{-}63a)$$

If the leakage inductor is null, the worst case occurs and $I_s(t)$ immediately jumps to I_{out}, when the output ripple is neglected (assuming L is large). Therefore, Eq. (8-63a) updates to

$$I_{C_{lump}} = I_{mag,peak-} - NI_{out} \qquad (8\text{-}63b)$$

where $I_{mag,peak-}$ represents the negative peak of the magnetizing current.

If the leakage inductor exists and really delays the occurrence of NI_s on the primary side, the best-case scenario leads to

$$I_{C_{lump}} = I_{mag,peak-} \qquad (8\text{-}63c)$$

It simply means that when the upper switch opens, 100% of the magnetizing current immediately reverses and circulates alone to discharge the lump capacitor. You have understood that, to satisfy the discharge of the lump capacitor, the magnetizing current must always be greater than the reflected secondary side current. Actually, to ensure the total reset of C_{lump}, we can use the energetic budget brought by Eq. (8-63b):

$$\frac{1}{2}L_{mag}(I_{mag,peak-} - NI_{out})^2 \geq \frac{1}{2}C_{lump}V_{in}^2 \qquad (8\text{-}64)$$

If a leakage inductor exists, less current will be diverted from L_{mag} and V_{DS} will be brought below ground, biasing the MOSFET body diode. We will thus ensure full zero-voltage switching! In this picture, we consider the energy stored by the leakage inductor to be much lower than that stored in the magnetizing inductor.

Figure 8-33a and b zooms in on this event and shows the effects with two different leakage inductors. In Fig. 8-33a, the reflected secondary side current quickly appears and subtracts from the lump capacitor discharge current before the drain touches zero. The ZVS cannot be obtained. To make it happen, you should increase the magnetizing negative peak current defined by

$$I_{mag,peak-} = \frac{V_{in}}{2L_{mag}} t_{on} \qquad (8\text{-}65)$$

As the controller fixes the duty cycle to maintain V_{out}, the only way to reach this goal is to gap the transformer. Yes, the circulating current increases and slightly degrades the conduction losses on the primary side, but the occurrence of ZVS brings a real benefit to the overall efficiency in off-line and high-frequency applications.

In Fig. 8-33b, the primary side leakage inductor delays the appearance of the reflecting secondary side current. The magnetizing current is thus almost alone in reducing the drain: as the figure confirms, we have full zero-voltage switching. However, if you increase the leakage inductor, the overlap sequence might artificially force a higher duty cycle and efficiency can suffer. Having the drain falling to around 100 V in an off-line application (as in Fig. 8-33a) already represents a tremendous improvement over a non-ZVS solution.

It is now time to see a simulation example where we want to deliver 5 V/50 A from a 300 Vdc source. The transformer turns ratio is 1:0.05, the converter operates at a 100 kHz switching

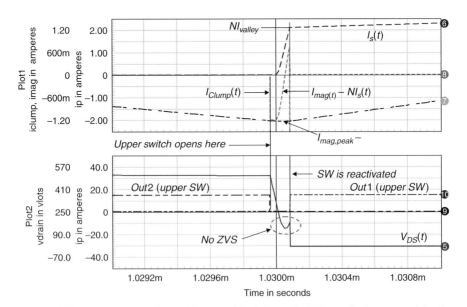

FIGURE 8-33a The leakage term is 1% of the magnetizing inductor; the ZVS operation is not ensured, but the drain has fallen to nearly 100 V before restart, already improving the switching loss budget.

SIMULATIONS AND PRACTICAL DESIGNS OF FORWARD CONVERTERS 787

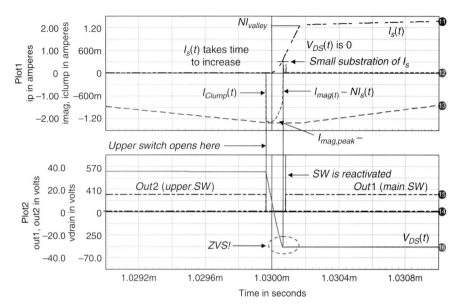

FIGURE 8-33b The leakage term is 4% of the magnetizing inductor, and the ZVS operation is fully ensured.

frequency, and the lump capacitor approaches 200 pF. The first question: What is the clamp voltage level? Well, in these resonant topologies, you do not choose a clamp voltage to protect the MOSFET as in a flyback circuit. The clamp level is actually imposed by the duty cycle excursion and the need to reset the core. To reset the core, Fig. 8-34 sketches the primary

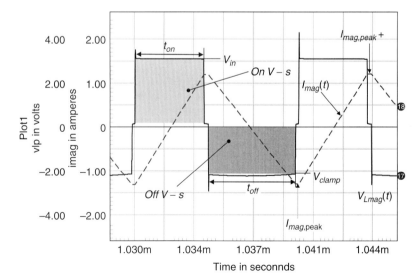

FIGURE 8-34 The inductor volt-seconds balance at steady state. The magnetizing current makes the transformer operate in quadrants I and III. Note the magnetizing inductor is in CCM.

inductor voltage at steady state and its associated magnetizing current: during the on time V_{Lmag} stays at V_{in}, and during the off time it stays stuck to V_{clamp}. We can therefore write the following volt-seconds balance relationship:

$$V_{in}t_{on} = V_{clamp}t_{off} \tag{8-66}$$

Introducing the duty cycle D and extracting the clamp voltage give

$$V_{clamp} = V_{in}\frac{D}{1-D} \tag{8-67a}$$

You have recognized that this formula depicts a buck-boost operated in CCM. The active clamp forward converter operates with a magnetizing current remaining continuous within one switching cycle.

From Eq. (8-7), we can extract an input voltage definition and replace V_{in} in the above expression. Thus

$$V_{clamp} = \frac{V_{out}}{N(1-D)} \tag{8-67b}$$

This equation confirms the relative insensitivity of the clamp voltage in relation to the input variations as long as the loop adjusts the duty cycle to regulate V_{out}. If we plot Eq. (8-67a) having an input voltage varying from 200 to 380 V, V_{clamp} (and thus the drain-source voltage excursion) stays relatively constant, as shown in Fig. 8-35a.

Well, what about the duty cycle limit for the controller, for instance, during start-up or an overload event where the loop is temporarily lost? Take a look at Fig. 8-35b which plots the clamp voltage excursion in case the duty cycle runs away. This curve teaches that your controller must exhibit a good precision on the maximum duty cycle limit; otherwise, it is likely that your colleagues will start to applaud at the first power-on sequence!

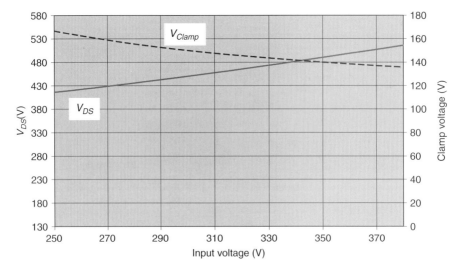

FIGURE 8-35a As long as the loop keeps control, the clamp voltage stays relatively constant despite input voltage variations. As a result, the voltage stress undergone by the MOSFET varies by around 100 V over the considered input range.

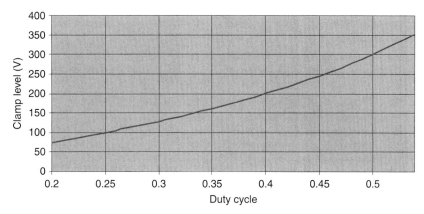

FIGURE 8-35b If the maximum duty cycle does not stay under control, the clamp voltage escapes! ($V_{in} = 300$ V)

Back to our design procedure. Suppose we have precisely fixed the maximum duty cycle to 45% by design. Equation (8-67a) thus indicates a clamp voltage at equilibrium of

$$V_{clamp} = V_{in}\frac{D}{1-D} = 300 \times \frac{0.45}{1-0.45} = 245 \text{ V} \tag{8-68}$$

With a 300 V maximum operating level (as an example here), we should pick a MOSFET capable of sustaining

$$BV_{DSS} \geq \frac{V_{in,max} + V_{clamp,max}}{k_d} = \frac{300 + 245}{0.85} \geq 641 \text{ V} \tag{8-69}$$

A 650 V type would probably do the job here. Again, pay attention to the maximum duty cycle excursion and make sure some margin exists in all cases.

The clamp capacitor selection requires knowledge of both the allowable ripple and the impact of its combination with the magnetizing inductor. Figure 8-36 depicts the voltage shape across the clamp capacitor when the upper-side diode conducts.

To assess the ripple amplitude, we must integrate the current flowing into the clamp capacitor. It has linear shape, ramping down from $I_{mag,peak+}$ to $I_{mag,peak-}$. The voltage discontinuity at the beginning is linked to the leakage inductor and will be neglected for the calculation. The ripple voltage can be expressed by

$$\Delta V = \frac{1}{C_{clp}}\int_0^{t_{off}/2} I_{C_{clp}}(t) \cdot dt = \frac{1}{C_{clp}}\int_0^{(1-D)T_{sw}/2} \frac{V_{in}}{2L_{mag}}DT_{sw}\frac{[(1-D)T_{sw}/2] - t}{(1-D)T_{sw}/2} \cdot dt \tag{8-70}$$

which gives

$$\Delta V = \frac{DV_{in}(1-D)T_{sw}^2}{8C_{clp}L_{mag}} \tag{8-71}$$

Extracting V_{in} from Eq. (8-67a) and accordingly updating Eq. (8-71), we have

$$\Delta V = \frac{V_{clamp}(1-D)^2 T_{sw}^2}{8C_{clp}L_{mag}} \tag{8-72}$$

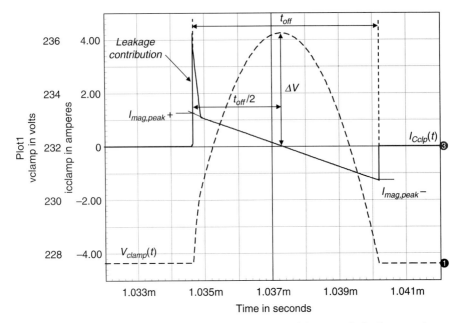

FIGURE 8-36 The upper-side body diode MOSFET routes the magnetizing current in the clamp capacitor.

As we are looking for a C_{clp} definition, we complete the derivation by extracting it from Eq. (8-72):

$$C_{clp} = \frac{V_{clamp}(1 - D)^2 T_{sw}^2}{8\Delta V L_{mag}} \quad (8\text{-}73)$$

The capacitor selection is made at the highest line level, corresponding to the worst operating condition relative to voltage stress. To continue the design, we must look at the magnetizing inductor. To aim for a ZVS operation (or near-ZVS, if decreasing V_{DS} to 100 V is acceptable as in Fig. 8-33a), we must satisfy Eq. (8-64) where we replaced the peak magnetizing current definition by Eq. (8-65). We can then derive the magnetizing inductor value:

$$L_{mag} = \frac{C_{lump}V_{in} + NI_{out}T_{sw}D + \sqrt{C_{lump}^2 V_{in}^2 + 2C_{lump}NV_{in}I_{out}T_{sw}D}}{2I_{out}^2 N^2} \quad (8\text{-}74)$$

The numerical application gives an inductor of 300 μH. To fit these requirements, the designer has to gap the core to increase the magnetizing current, a necessary condition for ZVS or near-ZVS operation. In the upcoming example, we adopted 500 μH, leading to a capacitor value of 200 nF (V_{clamp} = 245 V with ΔV = 15 V). Reference 8 offers some guidance for an active clamp design, as well as some small-signal indications for the loop compensation. Reference 9 goes into greater detail regarding the stability analysis, in particular, the relationship between the $L_{mag}C_{clp}$ interaction and the selected crossover frequency f_c. If we look at Eq. (8-72), it is exactly the same as for the CCM voltage-mode buck-boost converter where the resonant frequency is given by

$$f_0 = \frac{1 - D}{2\pi \sqrt{L_{mag}C_{clp}}} \quad (8\text{-}75)$$

As you can imagine, the resonance brought by the clamping network affects the Bode plot of the active clamp forward converter where it stresses the phase in the vicinity of the resonance. As a general recommendation, the crossover frequency must be kept far away from the lowest resonating point (occurring at the low input voltage); otherwise the phase shift might be extremely difficult to compensate. Some options also consider damping the clamping capacitor by a series resistor and capacitor.

There is no average model for the active clamp converter that we could present here. Reference 10 presents a way to use the PWM switch model but does not specifically describe an active clamp circuit as proposed by Fig. 8-27. Reference 11 describes the average switch modeling technique applied to an active clamp SEPIC converter. Fortunately, some software programs are able to predict the open-loop gain of switching converters such as PSIM or TRANSIM/Simplis [12]. Figure 8-37a portrays an active clamp converter drawn with the PSIM schematic capture software, using an example from Refs. 7 and 8 in Chap. 3. It shows an open-loop converter built with perfect elements and activated at a 100 kHz switching frequency. The ac response is delivered in a snapshot and appears in Fig. 8-37b. The resonating dip clearly appears and is located at 9.4 kHz, which, given an operating duty cycle of 0.38 simulated by PSIM, confirms the prediction of Eq. (8-75).

Figure 8-38a depicts the cycle-by-cycle active clamp converter we have built, without compensation as we were interested in the start-up sequence only. We selected PWMCMS, originally derived for synchronous converter simulations, but given its inherent deadtime generation, it lends itself very well to this simulation. Usually, the first delay should be adjusted to close the

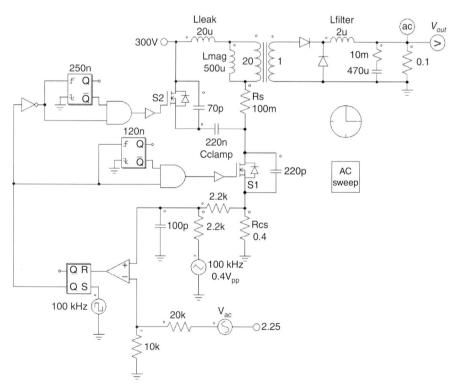

FIGURE 8-37a PSIM can work with switching circuits and extract the small-signal response after a few minutes.

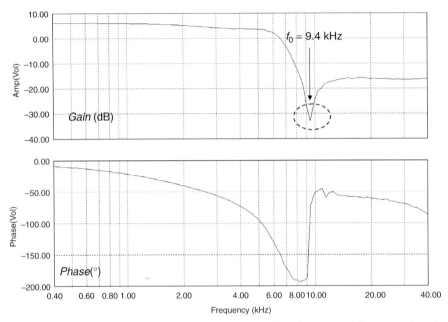

FIGURE 8-37b The small-signal response delivered by PSIM confirms the presence of the resonance brought by the clamping network.

upper-side switch just after the diode conduction to ensure ZVS. Its value is not that critical then. The main switch delay should, however, be carefully tweaked to restart the power switch right in the minimum of the drain-source valley. Reference 13 gives some hints on how to calculate these delays. Figure 8-38b portrays the start-up sequence: the clamp capacitor is fully discharged, and the drain-source voltage undulates as the output voltage rises. The magnetizing current becomes positively biased which induces additional losses on both the upper-switch body diode and the main power MOSFET (Fig. 8-38c). In normal operation, the diode naturally turns off when the magnetizing current reverses. In the presence of such heavy positive offset, the body diode is still conducting when the controller reactivates the main MOSFET. The dc bias on the magnetizing current can also occur at steady state, due to the predominance of parasitic terms such as the leakage inductor (negative bias) or excessive lump capacitor (positive bias). This topic is thoroughly discussed in Ref. 9.

Given the offset on the magnetizing current at startup, ZVS is lost on the main power MOSFET until steady state occurs. Figure 8-38d represents this event.

Some active clamp dedicated controllers exist on the market such as the LM5025 from National Semiconductor or the UCC2897 from Texas Instruments. ON Semiconductor has recently released the NCP1562A which features a soft-stop sequence. The soft stop consists of gradually reducing the duty cycle when the controller enters an undervoltage lockout or an overvoltage. Slowly reducing the duty cycle helps to properly discharge the clamp capacitor and prevents any excessive drain-source stress. Figure 8-38e shows three start-up sequences with different initial voltages on the clamp capacitor. The best case occurs when C_{clp} is fully discharged, it gives a maximum drain-source voltage of 520 V. The worst case occurs if C_{clp} is not discharged from the previous turn-off event: the drain peaks to 570 V.

This ends our discussion on the active clamp forward converter, a structure extremely popular in telecommunication markets. The technique starts to appear in off-line applications, and the upcoming high efficiency standards will probably make it a more popular structure in the coming years.

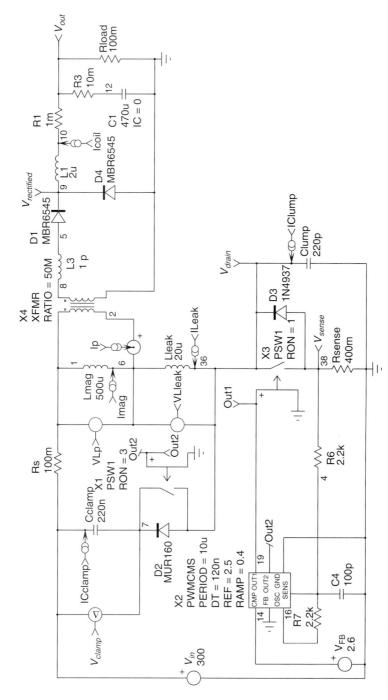

FIGURE 8-38a The PWMCMS generic controller features dead time control that can be used for active clamp simulation purposes.

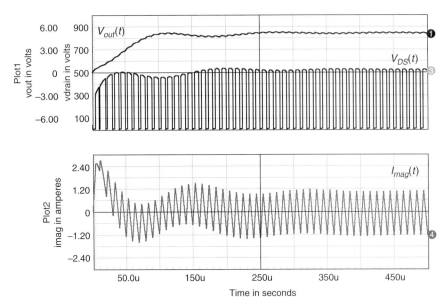

FIGURE 8-38b At start-up, the clamp capacitor charges up, and the magnetizing current becomes positively biased.

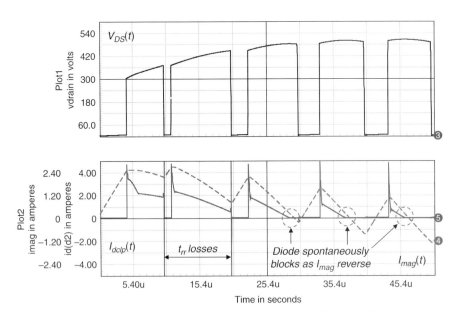

FIGURE 8-38c During transient events, such as the start-up sequence, the positive magnetizing current excursion cancels the natural upper-side diode blocking. It produces losses on both the body diode and the power MOSFET.

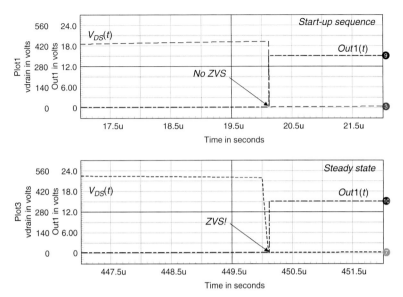

FIGURE 8-38d The upper section shows the drain-source signal during the start-up sequence where the ZVS is lost. The lower curve shows that the main output *out*1 drives the MOSFET after the drain has fallen to zero: this is ZVS.

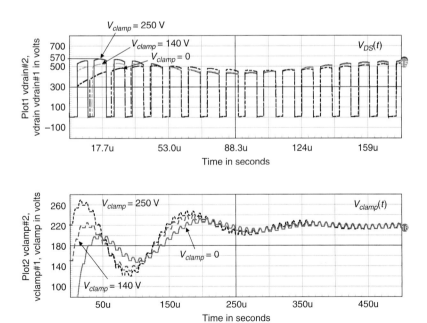

FIGURE 8-38e A power-on sequence with different initial charges on the clamp capacitor at the beginning of the operations. The NCP1562A with its soft-stop circuitry helps to reduce the stress associated with the clamp capacitor precharge.

8.7 SYNCHRONOUS RECTIFICATION

When output currents in excess of 10 A are flowing through the output diodes, a simple power dissipation calculation linked to the forward drops shows the necessity of finding an alternative rectification solution. For instance, assume the selected diodes feature a drop of 500 mV ($T_j = 150\,°C$) at 10 A; then on a forward converter operating at a 32% duty cycle, the conduction losses per diode would be (neglecting the rms contribution)

$$P_{D2} = I_{out}DV_f = 10 \times 0.32 \times 0.5 = 1.6\,\text{W} \qquad (8\text{-}76\text{a})$$

$$P_{D1} = I_{out}(1 - D)V_f = 10 \times 0.68 \times 0.5 = 3.4\,\text{W} \qquad (8\text{-}76\text{b})$$

$$P_{tot} = P_{D_2} + P_{D_1} = 1.6 + 3.4 = 5\,\text{W} \qquad (8\text{-}76\text{c})$$

where D_2 is the diode in series with the secondary winding and D_1 is the freewheel diode. To reduce this power dissipation budget, a solution called *synchronous rectification* consists of replacing the traditional diodes by low-$R_{DS(on)}$ MOSFETs self-driven either from the secondary winding or via a dedicated controller. The self-driven technique appears in Fig. 8-39a and represents a popular solution given its low-cost implementation.

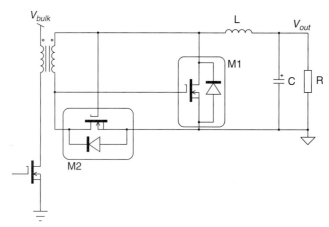

FIGURE 8-39a Self-driven synchronous MOSFETs in a forward converter.

The series diode played by M_2 has been rotated and now appears in the return path. When the primary MOSFET closes, the M_2 gate jumps to NV_{in} and ensures a proper bias during the on time. When the on time ends, M_1 ensures the freewheel portion and conducts as long as a voltage appears across its gate-source connection. In other words, when core reset occurs, the flux linking the secondary to the primary fades away and M_1 loses its bias. The current now flows in the M_1 body diode, not really known for its outstanding performance; the efficiency suffers. Figure 8-39a presents several drawbacks. First, both driving voltages move in relation to the input source. The transformer design and the MOSFET selection must thus account for the minimum driving voltage at low line and the maximum allowable V_{GS} at high line. Zener diodes can be used, but since the driving impedance on the secondary side is low,

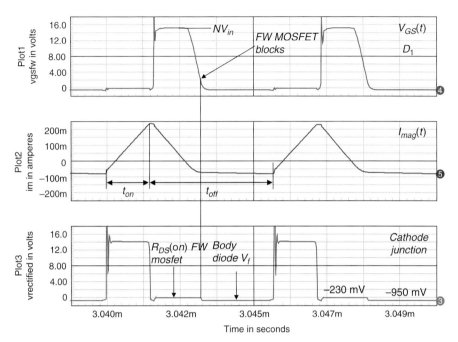

FIGURE 8-39b Typical self-driven MOSFET signals captured on a two-switch forward converter.

their use is difficult. In off-line applications, the PFC front-end stage becomes almost mandatory. Second, as highlighted by Eqs. (8-76a) and (8-76b), depending on the duty cycle, the burden on each diode changes: If you have a duty cycle of 45%, the burden is rather well spread between diodes. However, if the duty cycle drops to 30%, the freewheel diode will dissipate more since $1 - D$ becomes larger than D. Therefore, losing the driving voltage and transferring the burden to the body diode represent the major drawback for self-driven synchronous rectifiers.

Figure 8-39b depicts typical waveforms simulated on a 3.3 V, 10 A forward powered by a 48 V input. The transformer features a 1:0.31 turns ratio, the duty cycle establishes at 32%, and both synchronous MOSFETs feature an $R_{DS(on)}$ of about 20 mΩ at a high junction temperature. We can clearly see, as the freewheel MOSFET debiases (shortly after the magnetizing current slope flattens), that the voltage drop changes from a few hundred millivolts (MOSFET conducting) to a larger level, close to 1 V, linked to the body diode conducting the current. The power dissipated by each MOSFET can now be updated. If we neglect the ripple on the output inductor (the circulating current is a square wave of I_{out} plateau), then the series MOSFET losses can be defined as

$$P_{M2} = I_{out}^2 D R_{DS(on)} = 10^2 \times 0.32 \times 20m = 0.64 \text{ W} \tag{8-77}$$

The computation of the freewheel MOSFET losses requires a little bit more attention as they spread between the channel and the body diode. The MOSFET conducts during the time needed by the magnetizing current to drop from its peak value to zero. This time is exactly the on time, since the voltage applied during the on time equals the voltage applied to reset the core (in a two-switch or single-switch structure). The channel conduction losses are simply equal to that given by Eq. (8-77):

$$P_{M1,channel} = I_{out}^2 D R_{DS(on)} = 10^2 \times 0.32 \times 20m = 0.64 \text{ W} \tag{8-78a}$$

The body diode will conduct the remaining time which is the switching period minus 2 on times. Therefore,

$$P_{M1,body} = I_{out}(1 - 2D)V_f = 10 \times 0.36 \times 950m = 3.4 \text{ W} \quad (8\text{-}78b)$$

The total freewheeling MOSFET losses are thus the sum of the two above results which is 4 W. The power dissipated is larger than that when one is using Schottky diodes! Well, if we look at Fig. 8-39b, the body diode conduction time is longer than the synchronous action. This body diode dropping around 1 V, its power dissipation really destroys the benefit brought by the synchronous rectification. Further reducing the freewheeling MOSFET $R_{DS(on)}$ would not improve things since the body diode is the guilty party. To fight this problem, some designers add a Schottky diode in parallel with M_1, obviously increasing the cost.

For this reason, self-driven synchronous rectification does not engender tremendous interest in single-switch or dual-switch forward converters—unless you choose to go for a costly dedicated controller, independently driving both MOSFET gates which also solves the driving voltage levels as well. The equation changes when we consider the active clamp reset or a resonant reset structure. The strength of these topologies resides in a much longer demagnetizing time, offering a driving voltage of larger duration compared to the classical implementation. Actually, for the forward clamp, the freewheeling MOSFET V_{GS} exists for the whole off-time duration!

Figure 8-39c shows the driving voltage when implementing the synchronous rectification with a resonant reset forward. As can be seen, the body diode only conducts for a small portion of the off time, no longer hampering the efficiency. Of course, the resonant reset does not shield you from the dwell time appearing at high line, where the duty cycle reduction imposes a longer plateau, implying the freewheeling driving voltage loss. But the efficiency can be optimized at certain operating input conditions. Finally, the freewheeling gate-source voltage stays constant to the clamp voltage, scaled down by the transformer turns ratio. References 14 and 15 discuss synchronous rectification designs and provide interesting solutions to study.

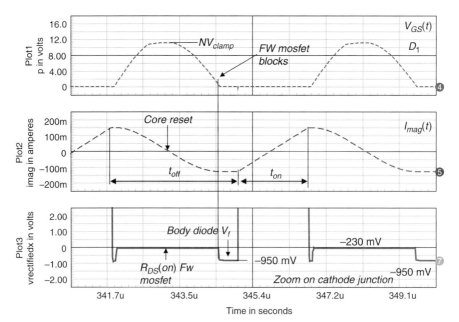

FIGURE 8-39c In a resonant reset forward, the freewheeling synchronous MOSFET receives a constant drive voltage. In some conditions, this MOSFET can stay on a longer time compared to the traditional reset scheme.

8.8 MULTIOUTPUT FORWARD CONVERTERS

The converters studied so far were delivering a single-output voltage. In applications requiring multiple outputs, the forward must be configured to deliver several distinct voltages, different from the main regulated one, for instance a 12 V, a 5 V, and a 3.3 V, if we are talking about an ATX converter. Several solutions exist on the market to build a reliable and efficient multioutput forward converter. Among these options, the magnetic amplifiers (the so-called mag-amps) have pioneered the postregulation technique. The postregulation technique uses leading edge modulation. The trailing edge modulation, classically implemented by PWM modulators, voltage mode, or current mode, interrupts the MOSFET gate-drive pulse when the goal is reached: the peak current is within the wanted window or the ramp has crossed the error voltage. Leading edge modulation, to the contrary, delays the occurrence of the MOSFET gate drive, its interruption event being fixed. Mag-amps use leading modulation, as described by Fig. 8-40.

8.8.1 Magnetic Amplifiers

The term *magnetic amplifiers* actually designates a saturable inductor connected to the main output winding and feeding a traditional buck arrangement (Fig. 8-41). A current I_{rst} provided by a regulation block adjusts the reset time and provides a means to regulate the output voltage. Mag-amp circuits are successful in high-power applications owing to their good efficiency and low implementation cost. Without entering into design details, let us try to understand how they work.

If you remember the overlap definition (Fig. 8-10a to c), the saturable inductor plays a role similar to the leakage term, by delaying the time at which the series diode conducts 100% of the freewheeling current. To explain this phenomenon differently, Fig. 8-42a shows a simple LR circuit where L represents the saturable inductor (the curved bar above the inductor indicates the saturation capability). When the series switch closes, as the inductor features a high permeability μ_r and thus a high inductance, the whole input voltage appears across its terminal. The current builds up in the coil until the core quickly saturates, bringing μ_r to 1: the inductor now looks like a short-circuit, and after a certain delay necessary to reach saturation, all the input voltage shows up across R_{load}. Figure 8-42b portrays the output voltage evolution and confirms the presence of the delay.

The delay comes from the time needed by the flux density B to build up from an initial stored value (at $t = 0$, B_0) to reach saturation where the inductance collapses. If we suppose a

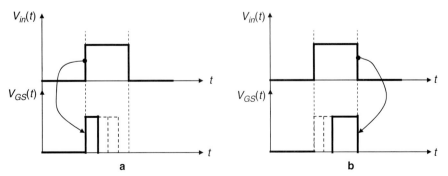

FIGURE 8-40 Trailing edge modulation on the left (a) is widely spread on PWM modulators. The leading edge technique (b) is implemented by mag-amps and other synchronous buck converters.

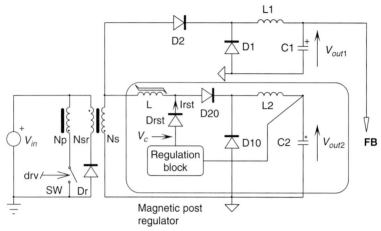

FIGURE 8-41 The saturable inductor is directly connected to the existing secondary winding to buck another voltage, lower than the main output.

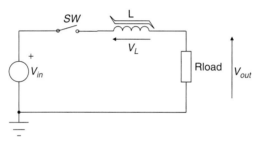

FIGURE 8-42a A saturable core placed in series with the input source saturates after a few microseconds, delaying the voltage appearance over the load resistor.

fully linear flux density growth, from Eq. (8-8b), and consider a constant voltage across the saturable inductor at the turn-on time (see the overlap section), we can write

$$B_0 + \frac{V_L t}{N_L A_e} = B_{sat} \tag{8-79a}$$

Extracting the time, we obtain

$$\Delta t = \frac{(B_{sat} - B_0) N_L A_e}{V_L} \tag{8-79b}$$

where B_{sat} = saturation flux density
B_0 = initial flux density at $t = 0$
N_L = number of turns on saturable inductor core
A_e = core cross-sectional area
V_L = inductor voltage equal to NV_{in} during on time

Equation (8-79b) reveals the principle of mag-amp regulation. By controlling the point B_0 from which the magnetizing operation starts, we have a means to act upon the delay time and

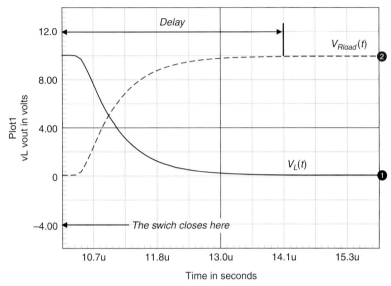

FIGURE 8-42b The resulting curves confirm the presence of delay, brought by the time the flux builds up in the core, and brings saturation.

thus an elegant way to control the postregulated output voltage. If we left the saturable inductor in place, without any control, the initial flux density point would be the remanent value B_r. The delay brought by the core to move its operating point from B_r to B_{sat} therefore represents an incompressible delay, inherent to the magnetic material.

Let us now look back at Fig. 8-41, which depicts the postregulated buck converter. In the sketch, we can see the saturable inductor connected in series with diode D_{20}, further biased by another diode called D_{rst}. During the primary on time, the secondary voltage jumps to NV_{in} and blocks the reset diode D_{rst}. The inductor saturates in a time Δt and NV_{in} appears on the D_{20} anode. The applied volt-seconds area (φ_{mag}) appears in Fig. 8-43a where the inductor voltage goes to zero when saturation is reached. The Δt delay shows up in Fig. 8-43b where we observe

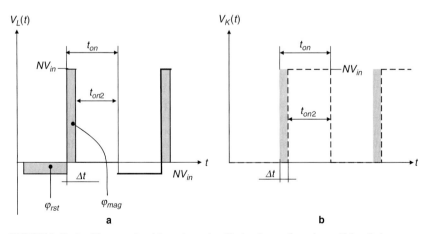

FIGURE 8-43a, b The saturation delay reduces the effective duty cycle on the rectifying diodes.

the resulting voltage at D_{20}/D_{21} cathodes. As expected, the effective on time t_{on2} is reduced compared to the original t_{on}, bringing the postregulated output down:

$$V_{out} = NV_{in}\frac{t_{on2}}{T_{sw}} = NV_{in}\frac{t_{on} - \Delta t}{T_{sw}} \quad (8\text{-}80)$$

We are neglecting both diodes' forward drops.

At the primary MOSFET turn-off time, the secondary voltage changes its polarity to $-NV_{in}$. D_{20} blocks and D_{21} freewheels. Now D_{rst} can conduct, and since it now imposes V_c on its anode, a current I_{rst} of opposite direction flows in the saturable inductor. The reset volt-seconds is depicted by the φ_{rst} area in Fig. 8-43a. This reset current, whose amplitude is controlled by the regulation block, serves the purpose of demagnetizing the core, sliding its operating point down the BH curve. At the end of the off time, the operating point is in D. This path is depicted by Fig. 8-44 which portrays a typical mag-amp hysteresis curve on the left, where point D is located. Once the demagnetization time is over (at the end of the off time), the inductor flux builds up, moving the magnetic operating point from D toward saturation: if D was left at a position close to saturation (B_r, for instance, in lack of control current during start-up), the saturation delay Δt is minimum. On the contrary, if D were pushed down to 0 T or even in the negative quadrant, point D', for instance, the delay time would lengthen. The complete path appears on the right side of the same figure, where during the magnetization phase, the operating point joins the first magnetization curve and moves up to saturation again. Thus, having a control circuit observing the output voltage and accordingly driving a current to demagnetize/remagnetize the inductor represents the basis of mag-amp control.

The material used for these mag-amp applications must exhibit a square $B\text{-}H$ curve. A so-called material "squareness" is defined by the ratio B_r/B_{sat} and is usually close to 1. To obtain square curves and minimize losses, cores are made of amorphous materials. Reference 16 describes 3R1 cores from Philips, whereas Ref. 17 points to Metglas, the site of an amorphous core manufacturer widely used in mag-amp designs.

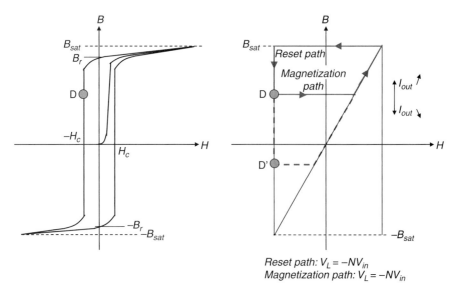

FIGURE 8-44 The operating point D is moving along the core $B\text{-}H$ curve.

Based on the above observations, we can write a few equations. The demagnetization area, or the reset volt-seconds, is obtained by finding the area φ_{rst}:

$$\varphi_{rst} = (NV_{in} - V_c)(1 - d)T_{sw} \qquad (8\text{-}81a)$$

The magnetic field in the core increases to a value given by

$$H_{rst} = \frac{NI_{rst}}{l_e} \qquad (8\text{-}81b)$$

where l_e represents the mean magnetic path length and I_{rst} the inductor current controlled by regulation loop. During the on time, the core flux increases to a level shown as φ_{mag} in Fig. 8-42a and described by

$$\varphi_{mag} = NV_{in}\Delta t \qquad (8\text{-}82a)$$

The magnetic field in the core increases to a value given by

$$H_{rst} = \frac{NI_s}{l_e} \qquad (8\text{-}82b)$$

where I_s represents the secondary side current circulating when the core is saturated.

At equilibrium, the volt-seconds balance law applies to the saturable inductor, implying that Eq. (8-81a) delivers a result equal to that of Eq. (8-82a):

$$NV_{in}\Delta t = (NV_{in} - V_c)(1 - d)T_{sw} \qquad (8\text{-}83)$$

If we extract the delay, we have

$$\Delta t = \frac{(NV_{in} - V_c)(1 - d)T_{sw}}{NV_{in}} = \left(1 - \frac{V_c}{NV_{in}}\right)(1 - d)T_{sw} \qquad (8\text{-}84)$$

Substituting this result into Eq. (8-80) gives the postregulated output voltage expression

$$V_{out} = NV_{in}\frac{dT_{sw} - \left[\left(1 - \frac{V_c}{NV_{in}}\right)(1 - d)T_{sw}\right]}{T_{sw}} = NV_{in}\left(d - \left(1 - \frac{V_c}{NV_{in}}\right)(1 - d)\right) \qquad (8\text{-}85)$$

If we consider a constant duty cycle at steady state, the output voltage can be adjusted via the control voltage V_c.

A simulation example is depicted by Fig. 8-45a where we can see a saturable core reset via a simple bipolar transistor, driven through a TL431. This is a series reset circuit since Q_1 appears in series with the saturable inductor. Other reset schemes exist; they are described in the references at the end of the chapter.

In this example, we generated a square wave signal whose polarity and amplitude are close to what the secondary winding of a forward would give. The output voltage delivers 5 V at 10 A. When the TL431 senses an excess of voltage on V_{out}, it pulls Q_1 base to ground and forces a reset current to circulate. This current induces a drop across R_4 and modulates the reset voltage amplitude on the saturable inductor:

$$V_L = I_{rst}R_4 - (NV_{in} + V_{out}) + V_{ce(sat)} \qquad (8\text{-}86)$$

By adjusting this current, the TL431 changes the position of point D in Fig. 8-44, depending on the power demand. Figure 8-45b describes some waveforms obtained from the simulation.

804 CHAPTER EIGHT

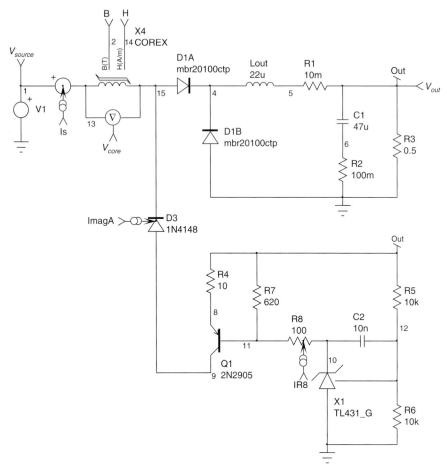

FIGURE 8-45a An example of a series reset mag-amp circuit built with a TL431 and a simple bipolar transistor.

The first current jump on the inductor is linked to the right horizontal move of point D or D' in Fig. 8-44. The second links to the flux density increase until saturation is obtained. At this time, the saturable inductor current equals the buck inductor current.

The design of mag-amp post regulators is described in several papers [16–22]. Small-signal modeling is tackled in Refs. 20 and 21, showing how a SPICE model can help compensate a mag-amp feedback loop.

8.8.2 Synchronous Postregulation

If the above mag-amps offer a reliable way to build low-cost postregulated outputs, they suffer from a few drawbacks among which core losses can be significant. Also, the overall performance can suffer if one wants to buck a high input voltage down to a low level, e.g., 15 to 3.3 V. Another solution starts to take off and consists of implementing another full buck circuit regulating on the plateau voltage of the secondary side winding. Figure 8-46 depicts the configuration made up of either one MOSFET and a diode or two MOSFETs as in a synchronous half-bridge association.

SIMULATIONS AND PRACTICAL DESIGNS OF FORWARD CONVERTERS 805

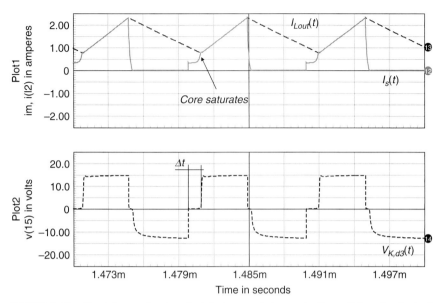

FIGURE 8-45b Simulation results where the inductor current shows saturation until it reaches the buck ripple current.

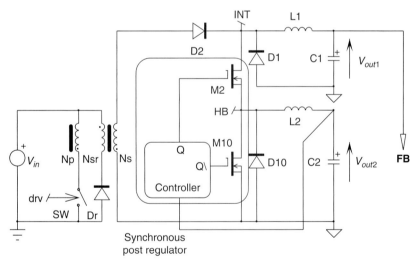

FIGURE 8-46 The synchronous postregulator controls the output voltage by using leading edge modulation, just as a mag-amp would.

The modulation scheme follows the signals shown in Fig. 8-47, where we can see an error signal decreased when the output is close to the output target and increased when above. The ramp is shown to be positive on this representation; however, integrated circuits released on the market for these postregulation purposes usually implement a negative slope ramp. As the PWM pattern on the HB node confirms, the regulation uses leading edge modulation: on the

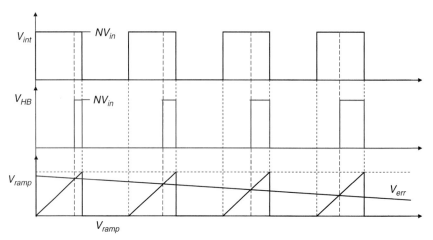

FIGURE 8-47 These signals confirm the leading edge regulation technique implemented in the buck postregulator.

left portion of the graphic, the load current is reduced, and the duty cycle on the HB node remains small. As the load current is increased, the error voltage drops. The crossing point between the error voltage and the ramp therefore happens sooner in the switching cycle, and the duty cycle at the HB node lengthens. At full load or during the start-up sequence, neglecting internal propagation delays, the HB node duplicates the voltage at node INT, something a mag-amp cannot do because of its B_r/B_{sat} inherent delay.

For the sake of illustration, we have built a simulation template portrayed in Fig. 8-48a. Two N channels are used plus an AND-based circuit that generates the half-bridge dead time. A TL431 appears in the feedback loop, but its signal needs to be reversed, given the error signal polarity. We wired two PNP transistors in a current-mirror configuration. The loop compensation is similar to that of a voltage-mode buck and requires a type 3 network. This is what the TL431 offers, following the guidelines outlined in Chap. 3. The start-up sequence immediately subjected to a load step appears in Fig. 8-48b and confirms the correct compensation of the circuit. Figure 8-48c zooms in on the step event and shows how the PWM pattern changes to accommodate the current variation.

There are several dedicated controllers on the market performing the above function: the NCP4330/31 from ON Semiconductor, the LM5115 from National Semiconductor, or the UCC2540 from Texas Instruments. This synchronous postregulating technique tries to make mag-amps an obsolete option for multioutput forward designers. However, mag-amp costs have dropped over the past years, and their simple circuitry still makes them an attractive solution.

8.8.3 Coupled Inductors

Multioutput forwards are obtained by adding several distinct secondary windings, each equipped with a buck inductor and a couple of diodes. If the feedback loop properly regulates one output, the second or third output suffers from cross-regulation problems linked to transformer coupling deficiencies, forward drop dispersions, and so on. Coupling the inductors on the same core represents a well-known and extremely efficient solution.

Let us look at a two-output forward as it appears in Fig. 8-49 in a simplified way, where the inductors are not coupled and operate in CCM. The first output delivers 5 V at 50 A whereas

SIMULATIONS AND PRACTICAL DESIGNS OF FORWARD CONVERTERS   807

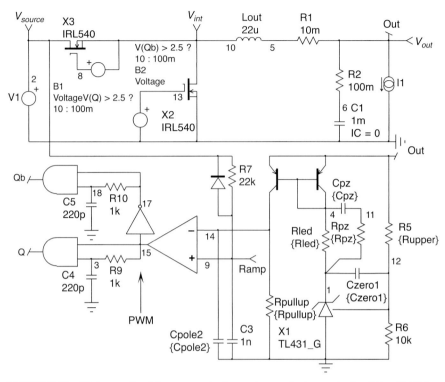

FIGURE 8-48a We have replaced the mag-amp by the synchronous buck converter to which a TL431 has been added.

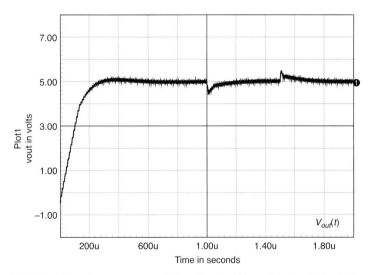

FIGURE 8-48b A start-up sequence followed by a load change from 5 to 10 A exhibits good stability.

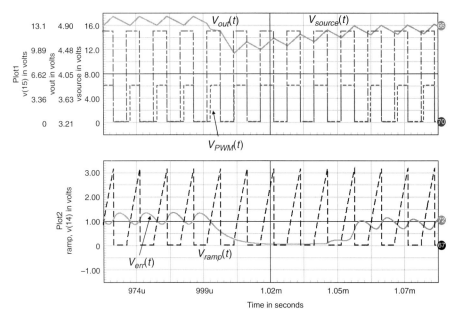

FIGURE 8-48c The response to a transient loading shows how the error voltage regulates via the leading edge modulation.

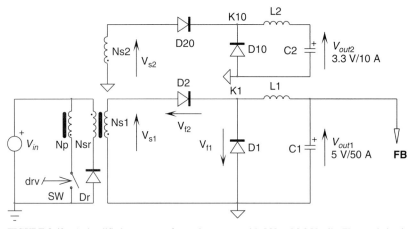

FIGURE 8-49 A simplified two-output forward converter with 5 V and 3.3 V rails. The regulation is made on the 5 V output.

the second output supplies 3.3 V at 10 A. The feedback loop observes the 5 V rail only. If we consider the average voltage across the buck inductor equal to zero, then the average voltage observed on node K_1 or K_{10} equals the respective dc output voltages. Looking at Fig. 8-50, we can write the following equation:

$$V_{out} = (V_{s,peak} - V_{f2})d - V_{f1}(1 - d) \tag{8-87}$$

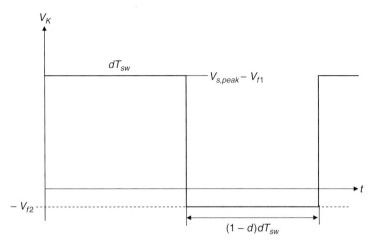

FIGURE 8-50 The voltage appearing on the cathode junction equals the output voltage at steady state.

where $V_{s,peak}$ = peak secondary voltage on considered winding
V_{f_2} = series diode forward drop
V_{f_1} = freewheeling diode forward drop

If we assume both forward drops to be equal to V_f, then Eq. (8-87) simplifies to

$$V_{out} = V_{s,peak}d - V_f \tag{8-88}$$

In the dual-output forward, we will assume that the design leads to the following numbers:

$n = \dfrac{N_{s_2}}{N_{s_1}} = 0.66$ secondary turns ratio between output 1 (5 V) and output 2 (3.3 V)
$V_{f_{10}} = V_{f_{20}} = 0.6\text{ V}$ diode forward drops for 3.3 V, 10 A output
$V_{f_1} = V_{f_2} = 0.75\text{ V}$ diode forward drops for 5 V, 50 A output
d = duty cycle = 30%

From Eq. 8-88, we can calculate the peak voltage delivered by the 5 V winding. This level is predictable since the feedback loop permanently monitors the corresponding dc output and ensures that the whole chain delivers 5 V on the output.

$$V_{s1,peak} = \dfrac{V_{out1} + V_{f1}}{d} = \dfrac{5 + 0.75}{0.3} = 19.17\text{ V} \tag{8-89}$$

Because the second winding is linked to the main one by the n ratio, we can estimate its peak voltage to be

$$V_{s2,peak} = V_{s1,peak}n = 19.17 \times 0.66 = 12.65\text{ V} \tag{8-90}$$

Applying Eq. (8-88), we then obtain the dc output voltage measured on this second winding:

$$V_{out2} = V_{s2,peak}d - V_{f10} = 12.65 \times 0.3 - 0.6 = 3.2\text{ V} \tag{8-91}$$

When the primary MOSFET conducts, a voltage develops across each inductor and forces the respective current to increase:

$$V_{L_1,on} = V_{s_1,peak} - V_{f1} - V_{out1} = 19.17 - 0.75 - 5 = 13.42 \text{ V} \quad (8\text{-}92a)$$

$$V_{L_2,on} = V_{s2,peak} - V_{f10} - V_{out2} = 12.65 - 0.6 - 3.2 = 8.85 \text{ V} \quad (8\text{-}93a)$$

During the freewheeling phase, the inductors are biased to a voltage equal to

$$V_{L_1,off} = -V_{f1} - V_{out1} = -0.75 - 5 = -5.75 \text{ V} \quad (8\text{-}92b)$$

$$V_{L_2,on} = -V_{f10} - V_{out2} = -0.6 - 3.2 = -3.8 \text{ V} \quad (8\text{-}93b)$$

As observed, despite a slight variation on the nonregulated rail (3.2 V versus a 3.3 V target), both L_1 and L_2 on and off voltages are exactly linked by the ratio n. Therefore, if we decide to wind both inductors on the same core, it is mandatory to keep the same n relationship between the transformer secondaries and their corresponding inductor windings. Failure to respect this rule will create a conflict among the above expressions, forcing the circulation of an unexpected large ripple current.

To understand how multiple winding-based circuits operate, we look at an equivalent sketch using reflected elements. Figure 8-51 shows a two-source arrangement following Eq. (8-88).

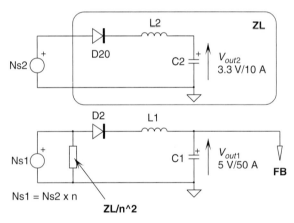

FIGURE 8-51 The two-winding transformer is replaced by two independent sources in series with one forward drop, as indicated by Eq. (8-88).

Thanks to the transformer configuration, it is possible to reflect the whole impedance loading the 3.3 V winding to the 5 V winding. This is obtained by following App. 8B guidelines:

- All impedances being divided by n^2, ESRs and inductors are indeed divided by n^2 but capacitors are multiplied by n^2.
- Voltage sources follow a similar rule and are divided by n. Currents, to the contrary, will be multiplied by n.

Figure 8-52a accounts for these changes. Note that the output voltage available on the reflected elements must also undergo a translation by $1/n$ to obtain the final value of 3.2 V. In Fig. 8-52a, as both the inductor and the diode are in series, we can slide the diode to the inductor right terminal and thus update the drawing to Fig. 8-52b. At this point, we can now couple both

SIMULATIONS AND PRACTICAL DESIGNS OF FORWARD CONVERTERS 811

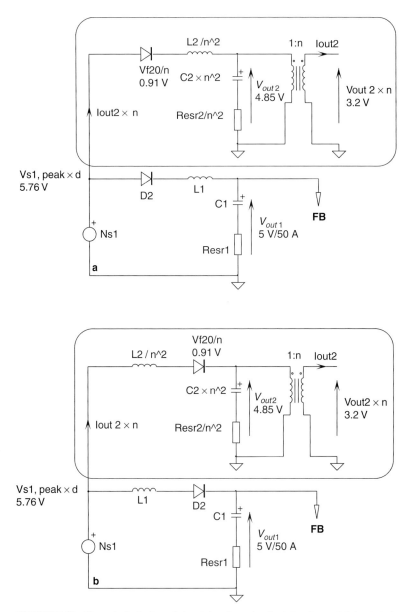

FIGURE 8-52 The equivalent schematic involving elements reflected on the regulated output via the turns ratio n. Capacitor ESRs have been added for a more precise representation. The two inductors are still not coupled.

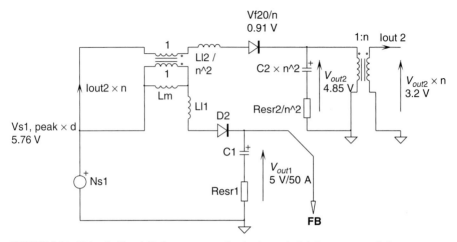

FIGURE 8-53 Using the T model helps to represent the circuit once both inductors are coupled.

inductors by using the equivalent transformer T model, already introduced in Chap. 4 (Fig. 4B-4, App. 4B). The result appears in Fig. 8-53, where L_{l1} and L_{l2} model the leakage terms inherent to the transformer construction. Of course, L_{l2} is reflected accounting for the squared secondary turns ratio n. Because the 3.3 V rail is normalized to the 5 V rail by the ratio n, all on and off inductor voltages are also linked by n: Eqs. (8-92a) and (8-93a), as well as Eqs. (8-92b) and (8-93b). In other words, the on and off voltages of Fig. 8-53 inductors being equal, the inductors are coupled with a 1:1 turns ratio.

To better understand the benefit brought by a 1:1 coupling, Fig. 8-54a represents the inductive section alone. Observing voltages on nodes 1, x and y, we can write

$$V_p = V_s \tag{8-94a}$$

$$V_1 - V_x = V_1 - V_y \tag{8-94b}$$

which implies

$$V_x = V_y \tag{8-94c}$$

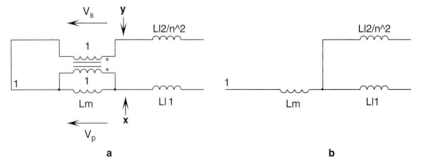

FIGURE 8-54 The T model with its leakage inductors simplifies to a simple connection linking both leakage terms after the magnetizing inductor.

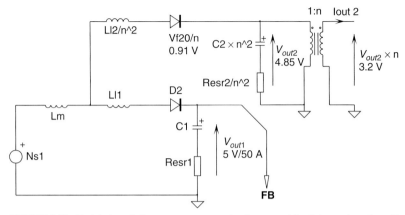

FIGURE 8-55 Both leakage inductors now connect after the magnetizing inductor, in series with the input source.

Based on this result, Fig. 8-54a can be updated to Fig. 8-54b. Finally, when dropped into the original schematic, the inductor arrangement leads to Fig. 8-55. To understand how the ripple current splits between the outputs, we can further tweak Fig. 8-55 in a more compact way. The result appears in Fig. 8-56 and offers the insight we are looking for.

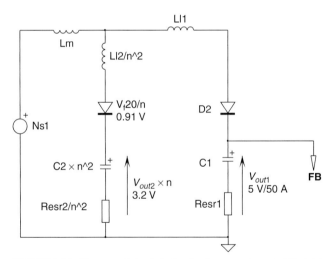

FIGURE 8-56 The magnetizing inductor clearly appears to control the total normalized ripple current before it splits in relation to the leakage inductor values.

In this figure, it now clearly appears that the total normalized current flows through L_m whose value actually determines the ripple amplitude. The ripple portion on both outputs now depends on the leakage terms: the lowest term gets the highest share of ripple. According to the capacitor type, it might be preferable to steer the major portion of the total ripple current into C_1, for instance. In that case, you should wind the inductor so that L_{l1} is made smaller than L_{l2} to favor a current circulation via the 5 V winding rather than the 3.3 V.

The simulation of a multioutput forward converter is possible and appears in Fig. 8-57a and b. In this circuit, we can clearly see the T model applied across both outputs, a 5 V, 50 A and a 3.3 V, 10 A rail. The feedback is made, thanks to a TL431 observing the 5 V output. We have optimized the compensation network with a type 2 circuit built with a TL431. The small-signal study of the multioutput forward will be covered in the stability analysis section.

Figure 8-58a depicts the ripple current in both output capacitors for coupled inductors, whose turns ratio is theoretically perfectly adapted to the transformer secondary side ratios (0%). As you can see, a small mismatch (here 1%) and ripple currents dramatically increase. For a 3% mismatch, the rms ripple current is multiplied by 2.5.

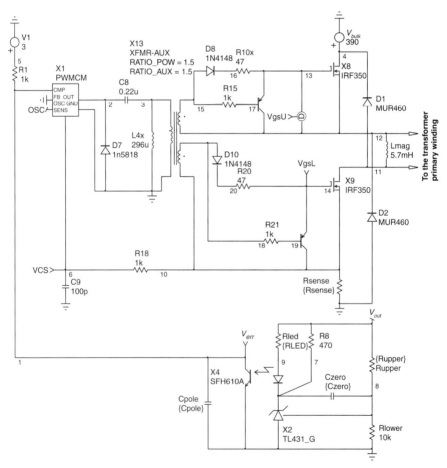

FIGURE 8-57a The primary portion of the multioutput forward converter uses the PWMCM generic controller. The feedback uses a TL431 to monitor the 5 V rail. The compensation network reproduces a type 2 compensation.

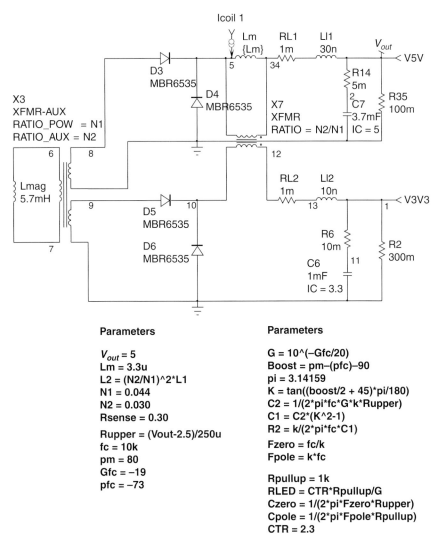

FIGURE 8-57b The secondary side implements the T model arrangement, but a coupling coefficient could also do the job.

Figure 8-58b shows the result of altering the leakage elements on both secondary sides. As expected, if we decrease L_{l1} on the 5 V rail and increase L_{l2} on the 3.3 V output, the ripple in the 5 V rail increases whereas it diminishes in the 3.3 V rail.

Finally, Fig. 8-58c depicts the step load response of the multioutput forward converter where the 3.3 V rail undergoes a 2 to 10 A variation while the 5 V output delivers a continuous 50 A current. You can note the very good result on the 3.3 V output (80 mV peak deviation only), despite a main regulation performed on the 5 V rail.

You can find additional details on multioutput converters in Refs. 23 and 24. The latter presents a way to predict crossregulation using the cantilever model via a three-output example.

816 CHAPTER EIGHT

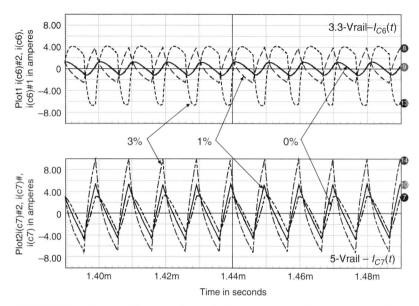

FIGURE 8-58a The turns ratio between the coupled inductors must match as closely as possible the turns ratio between the secondaries. A small mismatch leads to a higher ripple circulation.

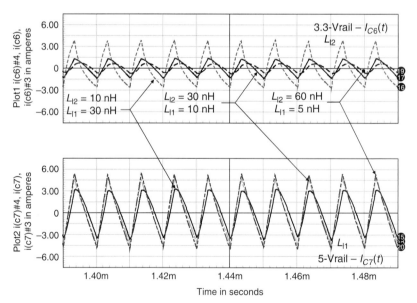

FIGURE 8-58b Ripple steering can be undertaken by increasing one particular leakage term rather than the other one.

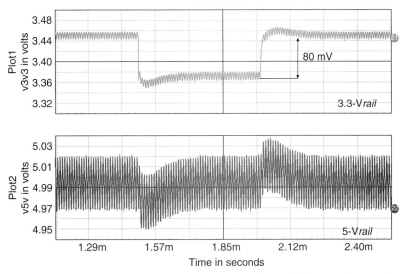

FIGURE 8-58c Thanks to inductor coupling, the multioutput forward delivers excellent transient results.

8.9 SMALL-SIGNAL RESPONSE OF THE FORWARD CONVERTER

In spite of a transformer addition, the small-signal response does not differ too much from that of a simple buck converter. As usual, the ac response changes in relation to the operating modes, DCM and CCM, and the vast majority of forward converters are designed to operate in the CCM mode. A lot of off-line forward converters use current mode for the ease of compensation. Voltage mode remains a possibility, but only if an op amp-based compensation is available. A TL431-based type 3 compensation circuit, especially at low output levels, does not really offer the required design flexibility.

8.9.1 Voltage Mode

Figure 8-59a represents a typical forward configuration in the voltage mode using the autotoggling model, PWMVM. The circuit delivers 12 V at 20 A from a 36 to 72 Vdc source. The TL431 ensures a type 3 compensation and we assume the PWM section features a 2.5 V sawtooth amplitude. Figure 8-59b displays the open-loop Bode plot obtained at a 36 V input level where the converter is operated in CCM or DCM. The loading point at which the converter toggles between both modes for a 36 V input can be found by using the equation already derived for the buck converter:

$$R_{critical} = 2F_{sw}L\frac{NV_{in}}{NV_{in} - V_{out}} = 2 \times 100k \times 25u\frac{0.91 \times 36}{0.91 \times 36 - 12} = 7.9 \, \Omega \quad (8\text{-}95)$$

Figure 8-59b shows a peaking linked to the presence of the output *LC* filter. This second-order network stresses the phase lag at the resonating point, and it must be compensated by

818 CHAPTER EIGHT

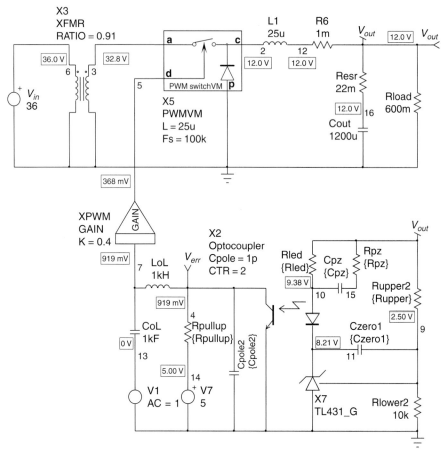

FIGURE 8-59a The autotoggling model in a voltage-mode forward converter.

a double-zero placement at the peaking frequency. In DCM, the open-loop gain quickly drops and confirms the absence of resonance since the inductor state variable has left the picture. The phase margin reaches 90° (first-order behavior) and starts to improve as the output capacitor and its associated zero begin to kick in. Considering a 10 kHz bandwidth, the manual pole-zero placement described in Chap. 3 gives the right crossover frequency associated with a good phase margin. Figure 8-59c confirms it.

The compensation was obtained by placing a double zero at the resonant frequency:

$$f_{z1} = f_{z2} = \frac{1}{2\pi \sqrt{LC_{out}}} = \frac{1}{6.28 \times \sqrt{25u \times 1.2m}} = 919 \, \text{Hz} \quad (8\text{-}96)$$

then one pole to compensate the ESR-related zero was positioned at

$$f_{p1} = \frac{1}{2\pi R_{ESR} C_{out}} = \frac{1}{6.28 \times 22m \times 1.2m} = 6 \, \text{kHz} \quad (8\text{-}97)$$

SIMULATIONS AND PRACTICAL DESIGNS OF FORWARD CONVERTERS   819

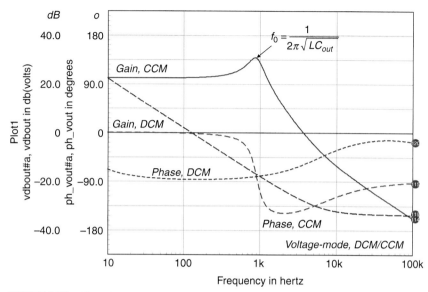

FIGURE 8-59b   The voltage-mode forward open-loop Bode plot response.

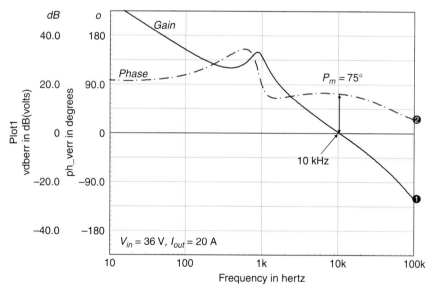

FIGURE 8-59c   After compensation, the forward converter exhibits a 10 kHz crossover frequency at a 36 V input voltage.

and finally a second pole was located at one-half the switching frequency, or 50 kHz. Given a pull-up resistor of 2.2 kΩ and an optocoupler CTR of 2, the spreadsheet calculated the following elements to cross over at 10 kHz:

$R_{LED} = 7$ kΩ
$C_{ZERO1} = 7$ nF
$C_{POLE2} = 1.5$ nF
$C_{PZ} = 33$ nF
$R_{PZ} = 785$ Ω

To test the stability, a load step was performed and results appear in Fig. 8-59d. There is no overshoot, and the drop stays within 3% of the output voltage.

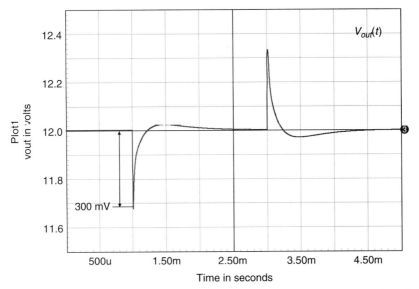

FIGURE 8-59d A load transient confirms the stability of the compensated voltage-mode forward converter.

As explained several times, the TL431 does not lend itself very well to the exercise of generating a type 3 compensation. The LED series resistor must be calculated to place the second pole-zero network, but it also needs to fulfill bias current conditions in any load situation. The problem worsens for low output voltages where the LED bias can become a real problem, especially if the collector current is rather high (low pull-up resistor, not enough headroom given the LED drop, and the TL431 operating level). In this case, several iterations can be necessary to reach the right compensation curve, in particular by playing on the pull-up value, the optocoupler CTR, and the adopted bandwidth (if you can). You might finally end up with a poor phase margin or a wrong crossover frequency and decide to go for an op amp-based compensation. The other option consists of choosing the current-mode topology in place of the voltage-mode topology!

The table below lists the small-signal voltage-mode forward parameters when operated in CCM. As a buck-derived converter is never designed to operate in DCM at nominal load, we have not included the small-signal elements pertinent to this mode. However, if needed, they can be easily retrieved from the buck section where V_{in} is simply replaced by NV_{in}.

SIMULATIONS AND PRACTICAL DESIGNS OF FORWARD CONVERTERS 821

TABLE 8-1

	CCM
First-order pole	—
Second-order pole	$\dfrac{1}{2\pi\sqrt{LC_{out}}}$
Left half-plane zero	$\dfrac{1}{2\pi R_{ESR}C_{out}}$
Right half-plane zero	—
V_{out}/V_{in} dc gain	ND
V_{out}/V_{error} dc gain	$\dfrac{NV_{in}}{V_{peak}}$
Duty cycle D	$\dfrac{V_{out}}{NV_{in}}$

where D = duty cycle
  F_{sw} = switching frequency
  R_{ESR} = output capacitor equivalent series resistance
  $M = \dfrac{V_{out}}{NV_{in}}$ = conversion ratio
  R_{load} = output load
  C_{out} = output capacitor
  L = secondary side buck inductor
  $N = \dfrac{N_s}{N_p}$ = transformer turns ratio
  V_{peak} = PWM sawtooth amplitude

8.9.2 Current Mode

Figure 8-60a depicts a similar forward converter, but this time it is operated in current mode. The circuit uses the autotoggling current-mode model whose control voltage comes from an in-line equation B_1. This source actually mimics a 1 V clamp action (as on a UC384X or NCP120X series) with a divider taking place between the feedback pin and the internal current-sense comparator (see Fig. 7-31a and b). The maximum peak current is therefore clamped to 1 V over the sense resistor. The model parameter R_i actually corresponds to the real circuit sense resistor R_{sense}, reflected after the transformer and scaled by its turns ratio N: 36.4 mΩ.

If we now observe the current circulating through the MOSFET drain of a single-switch forward converter, the shape looks like that of Fig. 8-3: the reflected secondary side current plus the magnetizing current. Equation (8-3) translates this statement into an analytical form. From this observation, the magnetizing current acts as a compensating ramp on forward converters. In other words, very often current-mode forward converters are stable because of the magnetizing current naturally compensating the current loop. However, it must be checked whether the ramp is large enough to make its compensating action or some more ramp is externally needed. In the other case that often arises, too much magnetizing current overcompensates the converter, and there is nothing you can do! Let us run the exercise for the present case:

$$S_{off} = \frac{V_{out}}{L} = \frac{12}{25u} = 480 \text{ kA/s} \qquad (8\text{-}98)$$

When it is reflected over R_i (already scaled by the turns ratio), we have a voltage slope of

$$S'_{off} = S_{off} R_i = 480k \times 36.4m = 17.5 \text{ mV/}\mu\text{s} \qquad (8\text{-}99)$$

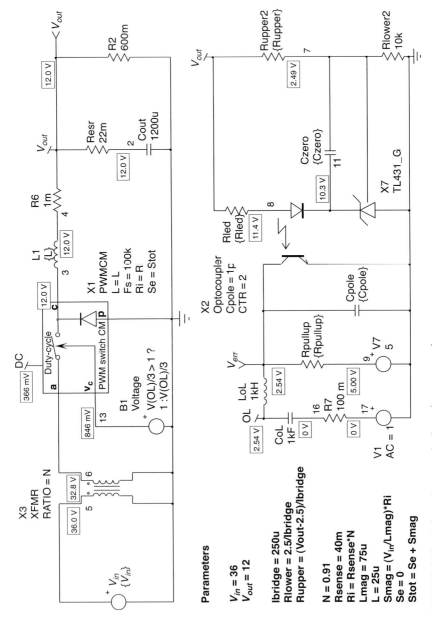

FIGURE 8-60a A forward converter implementing current mode.

The magnetizing inductor of the transformer establishes to 75 µH on the tested prototype. It generates a current at low line equal to

$$S_{mag} = \frac{V_{in,min}}{L_{mag}} = \frac{36}{75u} = 480 \text{ kA/s} \qquad (8\text{-}100)$$

Now transforming Eq. (8-100) into a sense voltage, we have

$$S'_{mag} = S_{mag} R_{sense} = 480k \times 40m = 19.2 \text{ m/V}\mu\text{s} \qquad (8\text{-}101)$$

In this particular case, the magnetizing ramp provides a nearly 110% compensation level. This is clearly an overcompensated design, but besides increasing the magnetizing inductor, there is almost nothing you can play with to change the situation.

An ac sweep of the configuration does not reveal a peaking of any kind, despite a CCM operation and a duty cycle approaching 40% (Fig. 8-60b).

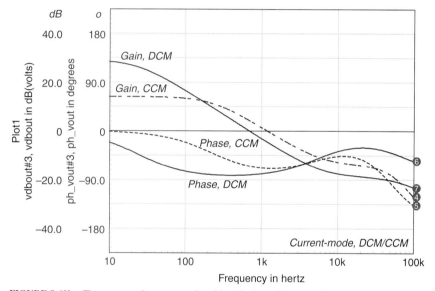

FIGURE 8-60b The ac sweep does not reveal peaking, thanks to the magnetizing current compensating the converter.

Given the first-order behavior and a 10 kHz crossover frequency (to match that of the voltage-mode forward), the k factor applied to a type 2 compensator recommends the following pole and zero placement:

$$f_{z1} = 7.2 \text{ kHz} \qquad (8\text{-}102)$$

$$f_{p1} = 14 \text{ kHz} \qquad (8\text{-}103)$$

Once this is calculated for the TL431 circuit, we have the following components:

$R_{PULL\text{-}UP} = 2.2 \text{ k}\Omega$
$R_{LED} = 1 \text{ k}\Omega$
$C_{ZERO} = 585 \text{ pF}/680 \text{ pF}$
$C_{POLE} = 5.2 \text{ nF}/6.8 \text{ nF}$

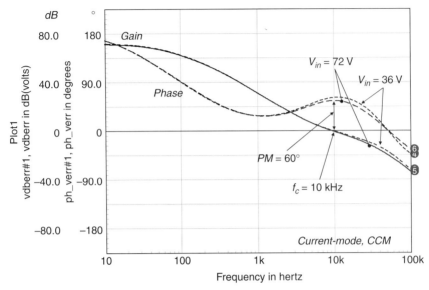

FIGURE 8-60c Once compensated, the forward converter exhibits excellent phase margin at both operating voltages (full load).

In Fig. 8-60c, the phase margin approaches 60° at both low- and high-line conditions, which is a solid design. Stepping the output confirms the stability as shown in Fig. 8-60d. Given the absence of zeros in the lower portion of the spectrum (as in voltage mode), note the faster recovery compared to the voltage-mode case.

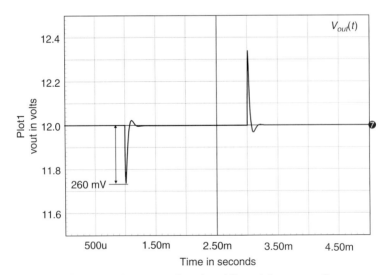

FIGURE 8-60d The load step test confirms the stability and shows an excellent recovery time, faster than the voltage-mode version.

TABLE 8-2

	CCM
First-order pole	$\dfrac{\dfrac{1}{R_{load}C_{out}} + \dfrac{T_{sw}}{LC_{out}}[m_c D' - 0.5]}{2\pi}$
Second-order pole	—
Left half-plane zero	$\dfrac{1}{2\pi R_{ESR} C_{out}}$
Right half-plane zero	—
V_{out}/V_{in} dc gain	$\dfrac{ND\left[m_c D' - \left(1 - \dfrac{D}{2}\right)\right]}{\dfrac{L}{RT_{sw}} + [m_c D' - 0.5]}$
V_{out}/V_{error} dc gain	$\dfrac{R_{load}}{R_i} \dfrac{1}{1 + \dfrac{R_{load}T_{sw}}{L}[m_c D' - 0.5]}$
Duty cycle D	$\dfrac{V_{out}}{NV_{in}}$

where D = duty cycle
 F_{sw} = switching frequency
 T_{sw} = switching period
 R_{ESR} = output capacitor equivalent series resistance
 $M = \dfrac{V_{out}}{NV_{in}}$ = conversion ratio
 L = secondary side inductance
 L_p = primary side or magnetizing inductance
 R_{load} = output load
 $N = \dfrac{N_s}{N_p}$ = transformer turns ratio
 $S_n = \dfrac{NV_{in} - V_{out}}{L} NR_{sense}$ this is the on-time secondary slope reflected to the primary
 S_e = external compensation ramp slope, V/s
 R_i = primary sense resistor R_{sense} reflected to secondary side ($R_i = NR_{sense}$)
 $m_c = 1 + \dfrac{S_e}{S_n}$

As with the voltage-mode case, Table 8-1 lists the small-signal elements related to the forward converter operated in current mode. Again, the CCM version only appears in this section.

8.9.3 Multioutput Forward

The autotoggling PWM switch model, in current mode or voltage mode, lends itself very well to the multioutput forward converter with coupled inductors. The trick consists of changing the transformer position, as already highlighted in Chap. 2, in Fig. 2-111c precisely. However, as we now deal with several inductors put on coupled secondaries, we need to calculate what inductor is actually "seen" by the PWM switch model on its "C" node. The easiest way is through the representation of the multioutput forward (two outputs in this case) and calculating the primary slope S_p imposed on the MOSFET during the on time. This slope will determine the equivalent inductor L_{eq} seen at the primary switch closing event. Figure 8-61 depicts the converter where the output loads have purposely been removed.

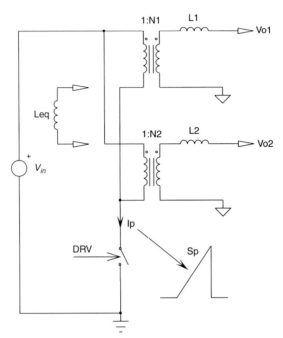

FIGURE 8-61 This simple model helps to determine the equivalent inductor L_{eq} seen by the PWM switch on the primary side of the multioutput forward converter.

Let us go through a few lines of algebra, considering an infinite magnetizing inductance. The primary slope S_p is actually made up of the sum of secondary slopes, each reflected to the primary by their respective turns ratios:

$$S_p = \frac{N_1 V_{in} - V_{o1}}{L_1} N_1 + \frac{N_2 V_{in} - V_{o2}}{L_2} N_2 \qquad (8\text{-}104)$$

Considering both outputs perfectly coupled via the transformer and the inductors, we have

$$V_{o2} = V_{o1} \frac{N_2}{N_1} \qquad (8\text{-}105)$$

Replacing V_{o2} in Eq. (8-104), we have

$$S_p = \frac{N_1 V_{in} - V_{o1}}{L_1} N_1 + \frac{N_2 V_{in} - V_{o1}\frac{N_2}{N_1}}{L_2} N_2 = \frac{N_1^2 V_{in} - V_{o1} N_1}{L_1} + \frac{N_2^2 V_{in} - V_{o1}\frac{N_2^2}{N_1}}{L_2} \qquad (8\text{-}106)$$

Dividing by N_1^2 in the first term and by N_2^2 in the second gives

$$S_p = \frac{V_{in} - V_{o1}/N_1}{L_1/N_1^2} + \frac{V_{in} - V_{o1}/N_1}{L_2/N_2^2} = (V_{in} - V_{o1}/N_1)\left(\frac{1}{L_1/N_1^2} + \frac{1}{L_2/N_2^2}\right) = \frac{V_{in} - V_{o1}/N_1}{L_{eq}}$$
(8-107)

where

$$L_{eq} = \frac{1}{\left(\dfrac{1}{L_1/N_1^2} + \dfrac{1}{L_2/N_2^2} \cdots + \dfrac{1}{L_n/N_n^2}\right)} = (L_1/N_1^2) \| (L_2/N_2^2) \cdots \| (L_n/N_n^2) \quad (8\text{-}108)$$

In the two-output case, provided both L_1 and L_2 are perfectly coupled, we have

$$L_2 = \left(\frac{N_2}{N_1}\right)^2 L_1$$
(8-109)

The equivalent inductor value feeding the PWM switch thus becomes

$$L_{eq} = \frac{1}{\dfrac{N_1^2}{L_1} + \dfrac{N_2^2}{L_2}} = \frac{1}{\dfrac{N_1^2}{L_1} + \dfrac{N_2^2}{L_1}\dfrac{N_1^2}{N_2^2}} = \frac{1}{\dfrac{N_1^2}{L_1} + \dfrac{N_1^2}{L_1}} = \frac{L_1}{2N_1^2}$$
(8-110)

Capitalizing on Figs. 8-57b, 8-60a, and 8-61, we can now build the averaged multioutput forward converter using coupled inductors. It is seen in Fig. 8-62a. We used a T model transformer arrangement, but a traditional SPICE k coupling coefficient would lead to similar results. As the subcircuit now directly connects to the input voltage, the sense resistor is kept to its normal value and a calculated result is attributed to the equivalent inductor passed to the model. The operating points appear on the schematic and reveal a duty cycle of 39%, together with a peak current of 3 A (B_1 output voltage over R_{sense}). The upper section of Fig. 8-62b depicts the open-loop Bode plot obtained for a nominal power of 280 W at an input voltage of 330 Vdc. As observed, there are no resonating peaks, and the system looks easy to compensate for a 10 kHz bandwidth. After adopting a type 2 compensation network, the final Bode plot appears below and shows a nice phase margin of 80°. To test the model's validity, we performed a step load on the unregulated 3.3 V output for both the average and cycle-by-cycle models. Figure 8-62c portrays the obtained results with both approaches, showing good agreement between the waveforms.

To confirm the validity of our approach, we actually built a two-switch current-mode converter featuring almost the same characteristics as the one tested here, except we had 12 V and 5 V outputs coupled together. The TL431 loop only included the 5 V rail, and the controller was an NCP1217A from ON Semiconductor. After the selection of a proper compensation network using the k factor (10 kHz bandwidth), the 12 V output was subjected to a load step of 8 A amplitude.

PSpice simulations and experimental results are gathered in Fig. 8-63 and show good agreement between the curves. The measured drop on the 12 V output almost matches the simulated voltage. It is evidence of not only the validity of the small-signal approach, but also the correct modeling of the feedback chain, made of the TL431 and the optocoupler models.

This small-signal section ends the description of the forward converter. It is now time to discover how to design such a converter using modern semiconductors.

828 CHAPTER EIGHT

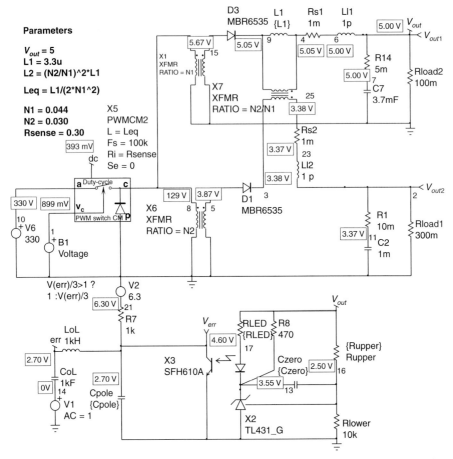

FIGURE 8-62a The multioutput forward converter using coupled inductors and the PWMCM average switch model.

8.10 A SINGLE-OUTPUT 12 V, 250 W FORWARD DESIGN EXAMPLE

This converter delivers a regulated 12 V, 22 A output used to power a game station or a multimedia computer. Given the amount of power, a preconverter (PFC stage) is mandatory, but its description will not be included here (Chap. 6 deals with this subject). The specifications are as follows:

$V_{bulk,min} = 350$ Vdc
$V_{bulk,max} = 400$ Vdc
$V_{out} = 12$ V
$V_{ripple} = \Delta V = 50$ mV
$\Delta V_{out} = 250$ mV maximum from $I_{out} = 10$ to 20 A in 10 μs

SIMULATIONS AND PRACTICAL DESIGNS OF FORWARD CONVERTERS 829

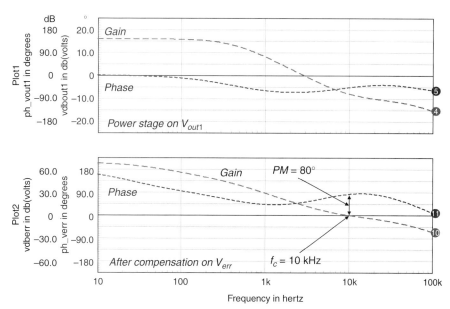

FIGURE 8-62b The open-loop Bode plot of the coupled forward converter reveals a friendly shaped ac response.

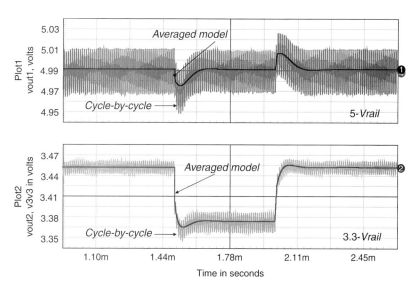

FIGURE 8-62c The transient step on the cycle-by-cycle model perfectly matches a similar step on the ac model.

830 CHAPTER EIGHT

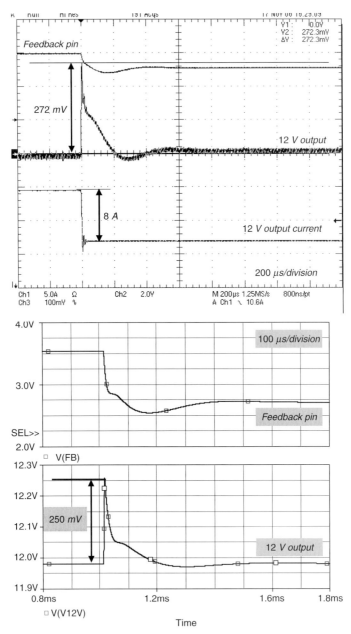

FIGURE 8-63 A 5 V/12 V two-switch forward converter was built and tested for a load transient. The obtained results match the simulation quite well in shape.

$I_{out,max} = 22$ A
$T_A = 70$ °C
$F_{SW} = 100$ kHz
MOSFET derating factor $k_D = 0.85$
Diode derating factor $k_d = 0.5$

Given the amount of power, a single-switch configuration could not be used. We are thus going to pick up a two-switch topology. First, let us calculate the necessary transformer turns ratio, given a maximum duty cycle D_{max} of 45% and a target efficiency of 90%. Based on Eq. (8-7), we have

$$V_{out} = \eta V_{bulk,min} D_{max} N \qquad (8\text{-}111)$$

Extracting the turns ratio, we obtain

$$N = \frac{V_{out}}{\eta V_{bulk,min} D_{max}} = \frac{12}{0.9 \times 350 \times 0.45} = 0.085 \qquad (8\text{-}112a)$$

Using this value in Eq. (8-111), we can estimate the minimum duty cycle at high line, changing the bulk voltage parameter:

$$D_{min} = \frac{V_{out}}{\eta V_{bulk,max} N} = \frac{12}{0.085 \times 0.9 \times 400} = 0.39 \qquad (8\text{-}112b)$$

The crossover frequency f_c will arbitrarily be selected at 10 kHz. Above, the converter could pick up switching noise and would require a more careful layout. Below, the stringent dropout specification would result in selecting a larger output capacitor. However, some iterations are always possible if a suitable capacitor value is not found at the beginning of the calculation steps. Considering a voltage drop mostly dictated by f_c, the output capacitance, and the step load current, we can derive a first capacitor value by using a formula already encountered:

$$C_{out} \geq \frac{\Delta I_{out}}{2\pi f_c \Delta V_{out}} \geq \frac{10}{6.28 \times 10k \times 0.25} \geq 636\,\mu F \qquad (8\text{-}113a)$$

The above case assumes an ESR much lower than the capacitor impedance at the crossover frequency:

$$R_{ESR} \leq \frac{1}{2\pi f_c C_{out}} \leq \frac{1}{6.28 \times 10k \times 636u} \leq 25\,m\Omega \qquad (8\text{-}113b)$$

We must therefore select a capacitor whose worst-case ESR remains below the capacitor impedance at the crossover frequency, in order to limit its contribution to the transient output drop. A 1000 µF ZL capacitor from Rubycon seems to be a good choice:

$C = 1000$ µF
$I_{C,rms} = 1820$ mA at $T_A = 105$ °C
$R_{ESR,low} = 23$ mΩ at $T_A = 20$ °C
$R_{ESR,high} = 69$ mΩ at $T_A = -10$ °C
ZL series, 16 V

Given a ΔI_{out} of 10 A, the above room temperature ESR component would, alone, generate an output voltage undershoot of

$$\Delta V_{out} = \Delta I_{out} R_{ESR} = 10 \times 23m = 230 \, \text{mV} \qquad (8\text{-}113c)$$

which is unacceptable given a specification of 250 mV! In other words, the selected capacitor will need, at least, to exhibit an ESR equal to one-half of what Eq. (8-113b) recommends (10 mΩ). We are going to parallel three ZL capacitors, to obtain the following equivalent component:

$C_{out} = 3000 \, \mu\text{F}$
$I_{C,rms} = 5.5$ A at $T_A = 105\,°\text{C}$
$R_{ESR,low} = 7.6$ mΩ at $T_A = 20\,°\text{C}$
$R_{ESR,high} = 23$ mΩ at $T_A = -10\,°\text{C}$
ZL series, 16 V

The undershoot might reach the limit at low temperatures, unless we do not believe the game station to operate below 0 °C. If there is a need to play inside an igloo, then the requirement for four or five capacitors becomes obvious! The final check will include the circulating rms current. However, given the nonpulsating nature of the buck output, we do not expect this current to be that high.

Given the output power level and the rather large output capacitor, we can consider the total ripple voltage contributed by the ESR term alone. Thus, if we adopt an ESR of 15 mΩ (approximate value at 0 °C), the maximum peak-to-peak output ripple current must be lower than

$$\Delta I_L \leq \frac{\Delta V}{R_{ESR,max}} \leq \frac{50m}{15m} \leq 3.3 \, \text{A} \qquad (8\text{-}114)$$

To obtain the inductor value, we write the buck ripple expression based on the off-time duration:

$$\Delta I_L = \frac{V_{out}}{L}(1 - D_{min})T_{sw} \qquad (8\text{-}115)$$

Using Eq. (8-114), we can derive a minimum inductor value for L:

$$L \geq \frac{V_{out}}{\Delta I_L}(1 - D_{min})T_{sw} \geq \frac{12}{3.3}(1 - 0.39)10u \geq 22 \, \mu\text{H} \qquad (8\text{-}116)$$

If we consider a 10% drop in the inductor value at high temperature and current, let us adopt a 25 μH output inductor. A 25 μH inductor forces the circulation of the following rms current in the capacitor:

$$I_{C_{out},rms} = I_{out}\frac{1 - D_{min}}{\sqrt{12\tau_L}} = 20 \times \frac{1 - 0.39}{3.5 \times 4.17} = 835 \, \text{mA} \qquad (8\text{-}117)$$

where $\tau_L = \frac{L}{R_{load}T_{sw}} = \frac{25u}{600m \times 10u} = 4.17$, as shown in Chap. 1. Given the equivalent capacitor current capability (5.5 A), there is no problem here.

The secondary side current will peak to a value given by

$$I_{s,peak} = I_{out} + \frac{\Delta I_L}{2} = 20 + 1.65 = 21.65 \, \text{A} \qquad (8\text{-}118)$$

On the primary side, this current reflects to

$$I_{p,peak} = I_{s,peak} N = 21.65 \times 0.085 = 1.84 \, \text{A} \tag{8-119a}$$

And the valley current reaches

$$I_{p,valley} = \left(I_{out} - \frac{\Delta I_L}{2}\right) N = (20 - 1.65) \times 0.085 = 1.56 \, \text{A} \tag{8-119b}$$

If we consider a controller featuring a 1 V maximum peak current, then the sense resistor is computed via the following expression, where a 10% margin appears:

$$R_{sense} = \frac{1}{I_{p,peak} \times 1.1} = \frac{1}{1.84 \times 1.1} \approx 500 \, \text{m}\Omega \tag{8-120}$$

The current waveform circulating in the MOSFETs, the transformer primary, and the sense resistor is similar to that of Fig. 7-73 (primary current of the 90 W CCM flyback example). The rms current expression slightly differs since it includes the reflection of the secondary side ripple to the primary side by the turns ratio N (the magnetizing current contribution is neglected here):

$$I_{p,rms} = \sqrt{D_{max}\left(I_{p,peak}^2 - I_{p,peak}\Delta I_L N + \frac{(\Delta I_L N)^2}{3}\right)} \tag{8-121a}$$

$$I_{p,rms} = \sqrt{0.45 \times \left(1.84^2 - 1.84 \times 3.3 \times 85m + \frac{(3.3 \times 85m)^2}{3}\right)} = 1.14 \, \text{A} \tag{8-121b}$$

The power dissipation of the sense resistor amounts to

$$P_{R_{sense}} = I_{p,rms}^2 R_{sense} = 1.14^2 \times 500m = 650 \, \text{mW} \tag{8-122}$$

It corresponds to 0.3% of the 250 W output power, but given the 90% efficiency requirement, you might want to opt for a current transformer to save some power. This type of device senses a current, usually via a single-turn primary winding, which reflects to the secondary side and circulates via a "burden resistor." A voltage is thus available to feed the current-sense input of the controller. Such a transformer acts as a high-pass filter, featuring an equivalent cutoff frequency since obviously no dc signal can be transmitted. The design of such a transformer goes beyond the scope of this design example, but readers interested in the topic can find a useful description of the current-sense transformer operation in Ref. 25.

8.10.1 MOSFET Selection

The MOSFETs are selected based on the maximum input voltage and the derating factor k_d of 0.85. If we choose 500 V devices (in a two-switch forward converter, the transistor stress is limited to the input voltage), the maximum high-voltage rail must be limited to

$$V_{bulk,max} = BV_{DSS} k_d = 500 \times 0.85 = 425 \, \text{V} \tag{8-123}$$

If the PFC does not include skip cycle in light-load operation, chances are that its output voltage will reach the overvoltage protection (OVP) level. The converter thus enters a kind of autorecovery hiccup mode. It is therefore important to check that one respects Eq. (8-123) despite the OVP detection. On an MC33262, as the OVP is set to 8% of the regulated output, it might be necessary to reduce the nominal bulk rail to 390 Vdc to keep the right safety margin on the MOSFET.

After browsing several manufacturer websites, we selected the IRFB16N50KPBF from International Rectifier. It features the following specifications:

IRFB16N50KPBF

TO220AB

$BV_{DSS} = 500$ V

$R_{DS(on)} = 770$ mΩ at $T_j = 110\ °C$

$Q_G = 90$ nC

$Q_{GD} = 45$ nC

Thanks to Eq. (8-121b), we can estimate its conduction losses as

$$P_{cond} = I_{D,rms}^2 R_{DS(on)} @ T_j = 110°C = 1.14^2 \times 0.77 \approx 1\ \text{W} \tag{8-124}$$

The turn-on switching losses require a measurement on the prototype, but they can roughly be estimated by (see Fig. 7-74b)

$$P_{SW,on} = \frac{I_{p,valley} V_{bulk} \Delta t}{6} F_{sw} \tag{8-125a}$$

If the MOSFET driver delivers a peak current of 1 A, then we could estimate the overlap time, where current and voltage cross with each other, to be in the vicinity of

$$\Delta t \approx \frac{Q_{GD}}{I_{DRV}} \approx \frac{45n}{1} \approx 45\ \text{ns} \tag{8-125b}$$

This is a very rough calculation intended to give an idea of a possible dissipation budget. Also, in this calculation, we consider a constant driving current, independent of the gate-source voltage. If this is true for bipolar drivers (constant-current generators), such as the UC384X family, the picture changes with CMOS-based output stages whose output current inherently relates to the V_{GS} evolution (resistive output behavior).

Calculation of Eq. (8-125b) involves the Miller charge Q_{GD} lower than the total gate charge. The $V_{DS}(t)$ and $I_D(t)$ transitions occur during this moment and bring losses, as shown in Fig. 7-74b. Once introduced into Eq. (8-125a), turn-on losses amount to

$$P_{SW,on} = \frac{I_{p,valley} V_{bulk,max} \Delta t}{6} F_{sw} = \frac{1.56 \times 400 \times 45n}{6} \times 100k \approx 470\ \text{mW} \tag{8-125c}$$

Turn-off losses are similar to those captured by Fig. 7-62b and are evaluated by

$$P_{SW,off} = \frac{I_{p,peak} V_{bulk} \Delta t}{2} F_{sw} \tag{8-125d}$$

These losses reach a level of

$$P_{SW,off} = \frac{I_{p,peak} V_{bulk,max} \Delta t}{2} F_{sw} = \frac{1.84 \times 400 \times 45n}{2} \times 100k = 1.65\ \text{W} \tag{8-125e}$$

Again, these numbers are to be manipulated with caution since they involved nonlinear parameters, but also are not necessarily well characterized such as the driving current capability. They are published here to show you the impact of these variables on the final loss budget. There is no need to insist on the necessity of measuring them on the final bench prototype.

8.10.2 Installing a Snubber

If we install a snubber across the MOSFET drain-source terminals, we have the ability to slightly delay the drain-source voltage rise. As a result, the crossing area where the voltage and current overlap can be significantly reduced. Such a snubber appears in Fig. 8-64a. Ignore R and D for the moment, and assume C connects directly across the MOSFET.

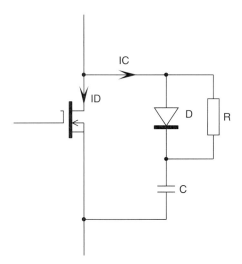

FIGURE 8-64a An *RCD* snubber helps to reduce turn-off losses.

When the driver instructs the MOSFET to open, the drain current falls at a pace imposed by the controller driving current. However, as the current needs to flow somewhere, it diverts into the capacitor C. The drain-source voltage thus increases a pace mainly fixed by this capacitor. It obeys the following law:

$$V_{DS}(t) = \frac{1}{C}\int_0^t I_C(t) \cdot dt \qquad (8\text{-}126)$$

The capacitor current looks like a rising ramp, starting from 0 (MOSFET closed) to the peak current value circulating at the switch opening: $I_{D,peak}$. Figure 8-64b portrays the typical waveforms associated with a snubbing action. If we call Δt the time needed by the primary current to transfer from the MOSFET drain to the snubber capacitor, we can derive the drain-source voltage equation:

$$V_{DS}(t) = \frac{1}{C}\int_0^t I_{D,peak}\frac{t}{\Delta t} \cdot dt = \frac{I_{D,peak}t^2}{2C\Delta t} \qquad (8\text{-}127)$$

Given its quadratic nature, the voltage shape will follow a parabolic curve. As shown in Fig. 8-64b, the switching losses reduce owing to the capacitor action. With the help of Eq. (8-127), we can derive the expression of these losses and use it as a guide for the capacitor selection:

$$P_{SW;off} = \frac{1}{T_{sw}}\int_0^{\Delta t} I_D(t)V_{DS}(t) \cdot dt = \frac{1}{T_{sw}}\int_0^{\Delta t}\frac{I_{D,peak}(\Delta t - t)}{\Delta t}\frac{I_{D,peak}t^2}{2C\Delta t} \cdot dt = \frac{(I_{D,peak}\Delta t)^2}{24T_{sw}C} \qquad (8\text{-}128)$$

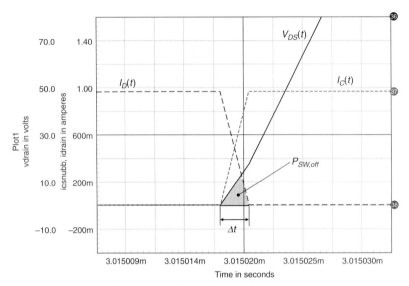

FIGURE 8-64b Turn-off waveforms with a snubbing capacitor.

Unfortunately, if we add a capacitor in parallel to the power switch, the additional switching losses at turn-on (the discharge of C in the switch) will ruin the effort to reduce the dissipated power on the MOSFET. These capacitive losses amount to

$$P_{SW, cap} = \frac{CV_{bulk}^2}{2T_{sw}} \qquad (8\text{-}129)$$

The above equation reveals a linear relationship between the capacitor value and the contributed losses. However, observing Eqs. (8-128) and (8-129), we see that a change in the capacitor value engenders opposite results: if C increases, the turn-off losses decrease but the capacitive loss increases. Therefore, a point exists where both curves cross, indicative of the optimum capacitor value bringing equal contributions from both loss mechanisms. This point can be obtained by deriving the total loss expression in relation to C and finding the C value at which it cancels:

$$P_{tot, SW} = P_{SW, off} + P_{SW, cap} = \frac{(I_{D, peak}\Delta t)^2}{24T_{sw}C} + \frac{CV_{bulk}^2}{2T_{sw}} \qquad (8\text{-}130)$$

$$\frac{d}{dC}\left(\frac{(I_{D, peak}\Delta t)^2}{24T_{sw}C} + \frac{CV_{bulk}^2}{2T_{sw}}\right) = \frac{V_{bulk}^2}{2T_{sw}} - \frac{(I_{D, peak}\Delta t)^2}{24T_{sw}C^2} \qquad (8\text{-}131)$$

$$\frac{V_{bulk}^2}{2T_{sw}} - \frac{(I_{D, peak}\Delta t)^2}{24T_{sw}C^2} = 0 \qquad (8\text{-}132)$$

Extracting C gives

$$C = \frac{\sqrt{3}}{6}\frac{I_{D, peak}\Delta t}{V_{bulk, max}} = 288m\frac{1.84 \times 45n}{400} = 60\,\text{pF} \qquad (8\text{-}133)$$

Back to our design example, we have entered the above equations in Mathcad and asked for a plot gathering both switching losses, as expressed by Eqs. (8-128) and (8-129). The result

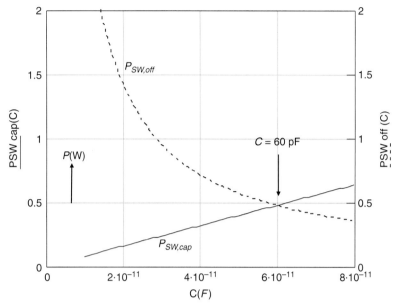

FIGURE 8-64c Plotting both switching loss curves on a similar chart shows the point at which they cross: this point designates the optimum snubbing capacitor value.

appears in Fig. 8-64c and confirms the analytical derivation. When a 60 pF snubbing capacitor is selected, the total loss budget is spread as below:

$$P_{SW,off} = P_{SW,cap} = 0.48 \text{ W} \tag{8-134}$$

This budget is entirely dissipated by the MOSFET at turn-on and turn-off.

Compared to 1.65 W, we have reduced the turn-off MOSFET loss which now amounts to nearly 1 W. Can we further relieve the MOSFET junction and keep it in a safer temperature area? Yes, we can dissipate the turn-on capacitive loss in an external resistor rather than in the MOSFET. The power dissipation budget will remain constant, but nearly one-half of it will go through the added resistor rather than the MOSFET. This resistor comes in series with C. However, if the resistor dissipates power, it now introduces an offset $I_{D,peak}R$ at the switch opening which could ruin the snubber benefits. To get rid of this series offset, a diode simply short-circuits the resistor during the capacitor charge (Fig. 8-64a). At turn-on, the minimum on time must ensure a full discharge of C via R; otherwise turn-off losses will increase as an offset builds up on the capacitor. An RC network is considered charged or discharged if its charging or discharging conditions are kept at least during 3τ, where $\tau = RC$. In our case, this condition translates to the following equation:

$$3RC < D_{min}T_{sw} \tag{8-135}$$

Extracting R and plugging already derived values, we have

$$R < \frac{D_{min}T_{sw}}{3C} < \frac{0.39 \times 10u}{3 \times 60p} < 21.6 \text{ k}\Omega \tag{8-136}$$

This resistor is chosen to sustain the Eq. (8-134) result (0.5 W); therefore a 1 W type can do the job or two 0.5 W elements in series, to allow high-voltage operation.

We can now compute the total MOSFET losses, inclusive of conduction and switching contributions:

$$P_{tot} = P_{cond} + P_{SW,on} + P_{SW,off} = 1 + 0.47 + 0.48 \approx 2\,\text{W} \qquad (8\text{-}137)$$

Given a TO-220 package, a heat sink becomes necessary. The design procedure has already been described in Chap. 7 design examples, section 7.14 for instance.

8.10.3 Diode Selection

The choice of the primary freewheeling diodes depends on the transformer magnetizing inductor. Usually, in off-line applications, this current remains low enough to accommodate 1 A diodes. In our case, let us assume the transformer exhibits a magnetizing inductor of 10 mH. At the maximum on time and minimum bulk voltage, the magnetizing current will peak to

$$I_{mag,peak} = \frac{V_{bulk,min}}{L_{mag}} D_{max} T_{sw} = \frac{350}{10m} \times 0.45 \times 10u = 157\,\text{mA} \qquad (8\text{-}138a)$$

If we consider both on and off slopes to be equal, the magnetizing current will reset at the following delay, after the MOSFET turn-off event:

$$t_{reset} = I_{mag,peak} \frac{L_{mag}}{V_{bulk,min}} = 157m \times \frac{10m}{350} = 4.48\,\mu\text{s} \qquad (8\text{-}138b)$$

Both on and reset times are of equal duration since reset voltages are similar (V_{bulk})! The average current can now be derived in a snapshot:

$$I_{mag,avg} = \frac{(t_{on} + t_{reset})I_{mag,peak}}{2T_{sw}} = \frac{(0.45 \times 10u + 4.48u) \times 157m}{2 \times 10u} = 70\,\text{mA} \qquad (8\text{-}138c)$$

For this application, diodes such as the MUR160 could do the job. Let us now take care of the secondary side.

In the forward converter, both secondary side diodes sustain a similar peak inverse voltage (PIV). Given a turns ratio of 0.085, the diodes have to sustain the following PIV:

$$\text{PIV} = NV_{bulk,max} = 0.085 \times 400 = 34\,\text{V} \qquad (8\text{-}139)$$

Then 100 V Schottky devices could be employed in this application. The MBR60H100CT, housed in a TO-220 package (or the MBR40H100WT, in TO-247), features a drop of 0.6 V for an average current of 20 A ($T_j = 150\,°\text{C}$). The series diode would then dissipate the following power in worst-case conditions (low line, for a maximum on time):

$$P_d = V_f I_{out} D_{max} = 0.6 \times 20 \times 0.45 = 5.4\,\text{W} \qquad (8\text{-}140a)$$

The freewheeling diode would dissipate slightly more as it conducts during the off time which worsens at high line (maximum off time):

$$P_d = V_f I_{out}(1 - D_{min}) = 0.6 \times 20 \times (1 - 0.39) = 7.3\,\text{W} \qquad (8\text{-}140b)$$

On average, these diodes would dissipate around 12 W or 5% of the total output power. It seems reasonable to adopt synchronous rectification to further reduce this budget. On a buck-derived converter, we can neglect the output ac ripple, making the synchronous rectification with MOSFETs an attractive solution. Why? Because in a flyback or in a boost converter, the resistive path of the MOSFET—its $R_{DS(on)}$—adds another dissipating factor, absent from a diode-based rectifying solution. In effect, the dynamic resistance R_d of a diode is usually small enough to neglect its contribution to the losses, even in the presence of a large ac ripple. The statement changes with a MOSFET where its $R_{DS(on)}$ becomes the main dissipative path, further aggravated by a large ac current. In a forward converter, the output ac ripple is small, confirmed by a capacitive current below 1 A [Eq. (8-117)]. Now, if we want a dissipated power of 2 W for each synchronous diode, we need the following $R_{DS(on)}$, neglecting the ac component on the total rms value:

$$R_{DS(on)} \leq \frac{P_D}{I_{out}^2} \leq \frac{2}{20^2} \leq 5\,\text{m}\Omega \tag{8-141}$$

Given the low resistor requirement, the application would require several MOSFETs operated in parallel to satisfy Eq. (8-141). The NTP75N06 presents a 19 mΩ resistance at $T_j = 150\,°C$ and a 60 V BV_{DSS}. To obtain an acceptable $R_{DS(on)}$ value, we would parallel three of them, making a total resistive path of 6.6 mΩ at $T_j = 150\,°C$. Considering the increased burden on the freewheeling rectifier, four of them would be needed to reach 5 mΩ. A discussion must now take place to balance the price of two Schotkky diodes plus their heat sink versus nine MOSFETs, however probably operated without any heat sink given the total power spread among them. Also, self-driven MOSFETs do not represent a panacea, mainly because of the driving voltage shortage when the core is reset. This shortage activates the freewheeling MOSFET body diode for a short time and further degrades the efficiency. A dedicated controller might help here, but at an increased expense.

8.10.4 Small-Signal Analysis

Figure 8-65a represents the current-mode forward converter using the PWMCM autotoggling ac model. All elements are fed with the value calculated above. The optocoupler was found to exhibit a pole located at 25 kHz, given the rather low pull-up resistor value. An ac sweep was performed (Fig. 8-65b), and the supply has been compensated for a 10 kHz bandwidth using the automated k factor tool or the Excel spreadsheet supplied on the CD-ROM. The computed values are the following:

$f_z = 2.7\,\text{kHz}$
$f_p = 37\,\text{kHz}$
$R_{pull\text{-}up} = 1\,\text{k}\Omega$
$R_{LED} = 100\,\Omega$
$R_{upper} = 38\,\text{k}\Omega$
$C_{zero} = 1.5\,\text{nF}$
$C_{pole} = 4.7\,\text{nF}$

Note the presence of the B_1 source modeling the internal divide-by-3 network, usually found on current-mode controllers (Fig. 7-31a and b). To test the validity of this compensating approach, we have also built a cycle-by-cycle model, featuring a simple set-reset scheme built on two transient sources V_{reset} and V_{set}. They are written as indicated below, whereas the schematic appears in Fig. 8-66a.

```
Vset 3 0 PULSE 0 10 0 1n 1n 50n {Tsw}
Vreset 16 0 PULSE 0 10 {Tsw*Dmax} 1n 1n 50n {Tsw}
```

T_{sw} and D_{max} are passed parameters and, respectively, change the switching frequency and the maximum duty cycle. If we look back to Chap. 2, the current-mode average model calculates

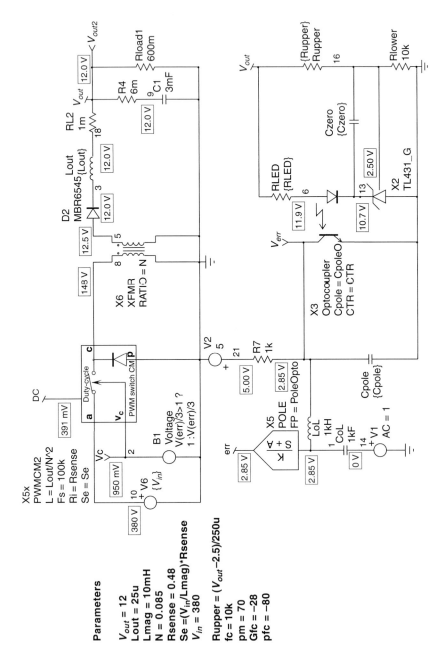

FIGURE 8-65a The small-signal model uses the PWMCM switch model where the transformer appears after the model.

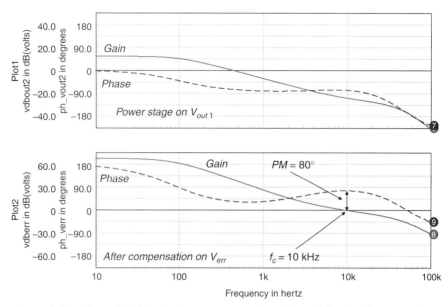

FIGURE 8-65b The resulting Bode plots in an uncompensated open loop and after the right compensation.

the peak current based on the control voltage and the ramp compensation level. The duty cycle is then further derived based on the switching period and the imposed peak current. In some cases, a conflicting situation may arise when the imposed peak current and thus its associated on time get clamped by the maximum duty cycle. In this option, the average model simply fails to give the right large-signal transient response. This is what you can see in Fig. 8-66b where we compare the responses given by the average model (no switching component), the cycle-by-cycle model with a maximum duty cycle set to 80%, and the same converter operated with a duty cycle of 45%. Note that the 80% case has no physical meaning since you could not operate a traditional forward converter in this configuration. It is displayed to show the duty cycle limit impact on the transient response.

The explanation lies in the on-time limit imposed by the duty cycle clamp which, in turn, clamps down the maximum primary peak current. The inductor rise time is therefore inherently limited and cannot increase at a higher pace, despite the feedback loop pushing for the maximum effort. This is what you can see in Fig. 8-66c, comparing both slopes: 312 mA/µs for the nonlimited version versus 80 mA/µs versus the clamped one. No wonder the transient response of the 45% version gives a larger voltage drop: 200 mV. However, it is still within the specification. If the drop had been larger than expected, there would be no need to improve the crossover frequency since the loop is already asking for the maximum power. You should increase the output capacitor and rework the compensation network.

8.10.5 Transient Results

We have updated the Fig. 8-66a switches with MOSFET models, and we did run a simulation again (Fig. 8-67a). The selected controller is PWMCMS whose second output directly drives the transformer, avoiding a dc coupling capacitor. First, the selected MOSFETs differ from the original choice; their SPICE models were lacking at that time. The simulation thus uses SPP12N50C3 which features an $R_{DS(on)}$ of 700 mΩ at 25 °C and a Miller capacitance of 26 nC.

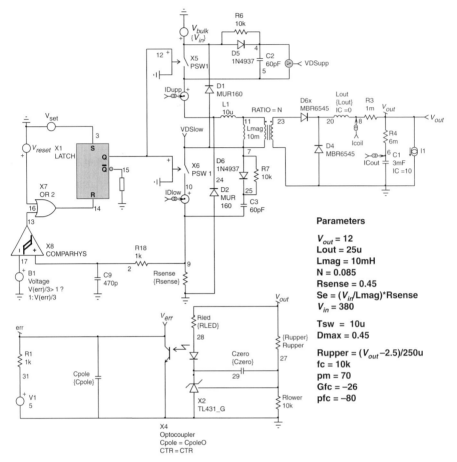

FIGURE 8-66a The cycle-by-cycle model using a simplified current-mode control circuitry featuring maximum duty cycle limit.

Figure 8-67b portrays a zoom in on the drain and current waveforms. Once we integrate (average) the multiplication of both signals, we obtain an average power of 1.3 W per MOSFET, slightly better than we calculated. However, keep in mind that most SPICE MOSFETs operate at a 25 °C $R_{DS(on)}$, dividing conduction losses by 2 ($R_{DS(on)}$ at room temperature usually doubles when the junction reaches 150 °C).

Figure 8-67c displays the driving signals, and the output voltage, showing a peak-to-peak ripple lower than 20 mV (specification states 50 mV). Also, the inductor peak and valley currents agree with the initial calculations.

To obtain the efficiency, compute the input power by measuring the average current flowing in the dc source, further multiplied by the source voltage to obtain P_{in}. The measurement of P_{out} over P_{in} gives 95%, of course without the inclusion of other losses, among which the transformer's play a significant role. This transformer could be built by a dedicated manufacturer provided you supply the data below. These data must include the maximum volt-seconds seen by the primary side. If you look at Eq. (4A-22), it states

$$NB_{sat}A_e > V_{in,max}t_{on,max} \tag{4A-22}$$

SIMULATIONS AND PRACTICAL DESIGNS OF FORWARD CONVERTERS

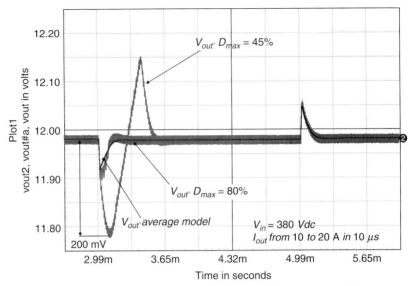

FIGURE 8-66b The transient response delivered by all configurations, average, cycle-by-cycle with 80% limit and 45%. The impact of the real 45% limit is clearly seen on the drop level.

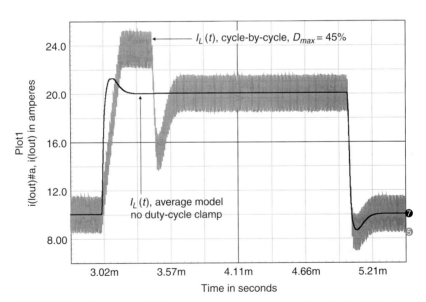

FIGURE 8-66c The duty cycle limit impact clearly appears when one is observing the inductor slope during the transient load change.

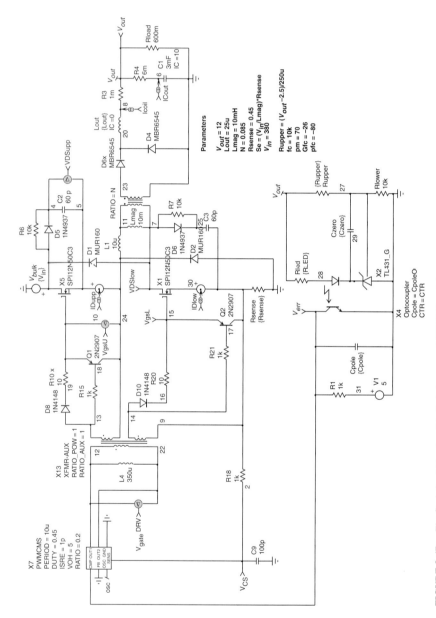

FIGURE 8-67a The final simulation template featuring power MOSFET models.

SIMULATIONS AND PRACTICAL DESIGNS OF FORWARD CONVERTERS 845

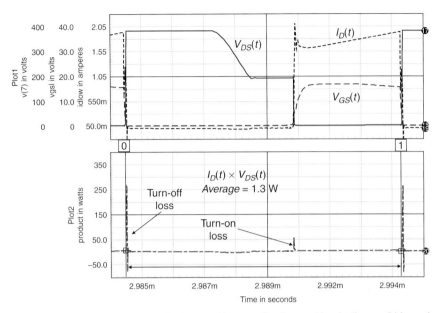

FIGURE 8-67b A zoom in on the MOSFET transitions reveals a fast transition, leading to a fairly good dissipation level per capita.

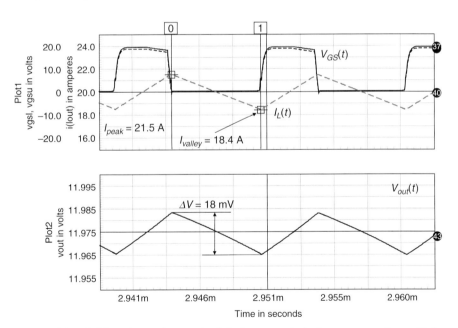

FIGURE 8-67c Driving voltage and output signal: the low output ripple shows great compliance with the starting specification of 50 mV, so we have enough margin.

This is the design condition for a forward transformer. If you do not respect it, the transformer will quickly saturate after a few switching cycles. In our case, given the input voltage and the duty cycle limit, we have the following volt-seconds maximum value:

$$V-s_{,max} = V_{in,max} T_{sw} D_{max} = 400 \times 10u \times 0.45 = 1.8m \tag{8-142}$$

The parameters passed to the manufacturer are the following ones:

$I_{Lp,max} = 2$ A
$I_{p,rms} = 1.1$ A
$I_{s,rms} = 12.6$ A
$F_{sw} = 100$ kHz
$V_{in} = 380$ to 400 Vdc
$V_{out} = 12$ V at 20 A
$N_p:N_s = 1:0.085$

The inductor also has its own set of specifications. They mainly relate to the continuous output current and the maximum peak current excursion.

$L = 25$ μH
$I_{p,max} = 25$ A
$I_{L,rms} = 20$ A
$F_{sw} = 100$ kHz
$\Delta I_L = 3$ A peak to peak

The final schematic appears in Fig. 8-68 and shows a control section made of an NCP1217A operated at a 100 kHz switching frequency. This controller features skip-cycle operation and will allow the converter to operate in no load without a runaway situation. This is the typical Achille's heel of controllers such as the UC384X that cannot go to 0% duty cycle. This characteristic often engenders output overvoltage situations as the minimum on time cannot maintain the dc target in no-load conditions. Skip cycle elegantly solves this problem.

The controller dc supply comes from an auxiliary circuit, a low-power flyback controller, for instance: a monolithic circuit such as a member of the NCP101X series or the NCP1027 from ON Semiconductor would be the perfect candidate for such a function.

8.10.6 Short-Circuit Protection

As it is presented in the sketch, there is no short-circuit protection. Unlike a flyback converter, an auxiliary winding wired on the primary side only delivers NV_{in} during the on time, making the converter blind to any secondary side problem. Unfortunately, there is no simple way to reflect the secondary voltage and take action in case of difficulties. One possibility lies in deriving an auxiliary signal from the output inductor, by winding a few turns over the existing core. The signal is then brought back on the primary side to be rectified. In case of short-circuit, this auxiliary V_{cc} collapses and protects the converter via a recurrent burst mode (the controller tries to restart). Figure 8-69a presents this solution. However, the creepage distance that needs to be respected and the safety burden on this "transformer" (it must sustain the 3 kV "hi-pot" test) make the solution unpractical and costly.

On the other hand, you can select controllers implementing the protection circuit shown in Fig. 7-47. The NCP1212 from ON Semiconductor is a possible solution in that case. For single-switch forward applications, it is possible to implement the DSS-based option described in Fig. 8-69b which uses an NCP1216A.

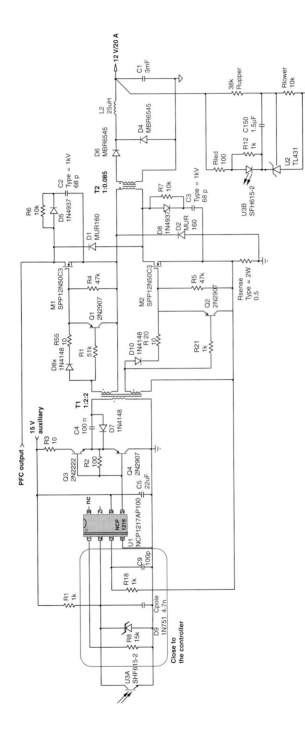

FIGURE 8-68 The final converter schematic shows an NCP1217A used as the main controller.

847

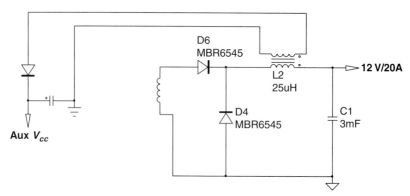

FIGURE 8-69a A possible solution to self-supply the controller from the secondary side. It, however, brings other constraints on the output inductor such as galvanic isolation and high-voltage dielectric strength.

This option uses the dynamic self-supply (DSS) which powers the controller via the high-voltage rail. However, as the controller cannot deliver the total driving current on its own (DSS capability limited to a few milliamperes only), especially if you drive a big MOSFET, you need the help of an auxiliary winding. The auxiliary winding in a forward converter delivers NV_{in} and exhibits wide voltage variations, especially in a universal mains application. The bipolar transistor shields the controller from these variations and makes the drive current flow through the winding, not the controller. The chip power dissipation is thus limited to the internal circuitry consumption multiplied by the high-voltage rail level. Please go through AND8069D available from the ON Semiconductor website for more details about these tricks.

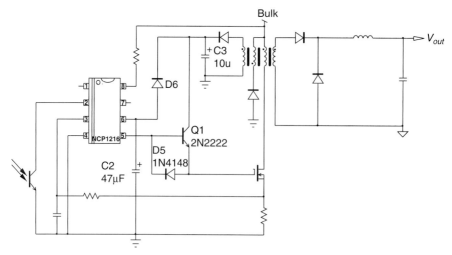

FIGURE 8-69b A DSS associated with an auxiliary winding helps self-power the controller without overloading its high-voltage section. Short-circuit detection is ensured by monitoring the feedback voltage.

8.11 COMPONENT CONSTRAINTS FOR THE FORWARD CONVERTER

To end our forward design example, we have gathered the constraints seen by the key elements used in this configuration. These data should help you to select adequate breakdown voltages of the diode and the power switch. All formulas relate to the CCM operation, as for the buck converter.

MOSFET		
$BV_{DSS} > 2V_{in,max}$	Breakdown voltage with a simple third winding reset scheme	
$I_{D,rms} = \sqrt{D_{max}\left(I_{peak}^2 - I_{peak}\Delta I_L + \frac{\Delta I_L^2}{3}\right)}$	CCM	Neglecting magnetizing current

Diode	
$V_{RRM} > NV_{in,max}$	Peak repetitive reverse voltage
$I_{F,avg} = D_{max}I_{out}$	Continuous current for the series diode
$I_{F,avg} = (1 - D_{min})I_{out}$	Continuous current for the freewheeling diode

Capacitor
$I_{C_{out},rms} = I_{out}\dfrac{1 - D_{min}}{\sqrt{12\tau_L}}$
with $\tau_L = \dfrac{L}{R_{load}T_{sw}}$ and N, the transformer turns ratio, N_s/N_p.

WHAT I SHOULD RETAIN FROM CHAP. 8

The forward converter is widely used in telecommunication power supplies (dc-dc) owing to its nonpulsating output current signature. ATX power supplies also make large use of this topology, in single- or dual-switch approaches, for applications ranging from 150 to 500 W. The forward converter excels in low output voltage and strong output current applications, e.g., a 5 V, 50 A power supply.

1. The forward topology belongs to the buck-derived family of converters. The energy transfer takes place at the switch closing event. The converter implements a real transformer, unlike the flyback which uses coupled inductors.

2. As in a buck converter, the forward output current is nonpulsating, naturally reducing the rms current circulating in the output capacitor. As a forward converter always operates in CCM at its maximum power, the rms current circulating in the power switch(es) is low, naturally reducing conduction losses.

3. When operated in voltage mode, the forward converter behaves as a second-order system and requires the adoption of a type 3 compensation network. Its current-mode version accommodates a type 1 or type 2 version and can easily transition from one operating mode to another.

4. The transformer magnetizing inductor must be operated in DCM, whatever the input or output conditions. Failure to ensure this condition leads to transformer saturation, quickly followed by a brief loud noise!
5. The most standard demagnetization circuit uses a third winding, recycling the magnetizing current via the input source. The duty cycle is thus naturally limited to 50%, and the transformer operates in quadrant I only.
6. The two-switch forward converter represents an excellent tradeoff between complexity and ease of energy recycling. However, the duty cycle is still clamped below 50%.
7. Numerous variations exist to extend the duty cycle limit beyond 50%. However, none of these solutions is a panacea, and they all require a careful design to cover all possible operating cases.
8. The active clamp gains in popularity as (1) it allows the recycling of the magnetizing current, (2) the duty cycle can exceed 50%, and (3) zero voltage switching is possible, allowing a higher switching frequency. Its complexity, in particular for off-line applications, does not make it an easy converter to design.
9. In multioutput power supplies, coupling the output inductors on the important outputs dramatically improves the crossregulation performance.
10. In the case of low output rails in multioutput converters, several postregulation solutions exist such as magnetic amplifiers or the more recent controlled synchronous rectification.

REFERENCES

1. N. Machin and J. Dekter, "New Lossless Clamp for Single-Ended Converter," www.rtp.com.au/papers/w1329.PDF
2. Ed Walker, "Design Review: A Step-by-Step Approach to AC Line-Powered Converters," SEM1600, Texas Instruments seminars, 2004 and 2005.
3. S. Hariharan and D. Schie, "Designing Single Switch Forward Converters," Power Electronics Technology, October 2005, www.powerelectronics.com
4. T. Huynh, "Designing a High-Frequency, Self Resonant Reset Forward DC-DC for Telecom Using Si9118/19," Vishay AN724.
5. C. D. Bridge, "Clamp Voltage Analysis for RCD Forward Converters," APEC 2000.
6. C. S. Leu, G. C. Hua, and F. C. Lee, "Analysis and Design of RCD Clamp Forward Converter," *VPEC Proceedings*, 1999.
7. M. Madigan and M. Dennis, "50 W Forward Converter With Synchronous Rectification and Secondary Side Control," SEM1300, TI seminars, 1999.
8. G. Stojcic, F. C. Lee, and S. Hiti, "Small-Signal Characterization of Active-Clamp PWM Converters," *Power Systems World*, 1996.
9. Q. Li, "Developing Modeling and Simulation Methodology for Virtual Prototype Power Supply System," Ph.D. dissertation, CPES, VPI&SU, March 1999.
10. I. Jitaru and S. Bîrcă-Gălăteanu, "Small-Signal Characterization of the Forward-Flyback Converters with Active Clamp," *IEEE APEC*, 1998, pp. 626–632.
11. P. Athalye, D. Maksimović, and B. Erickson, "Average Switch Modeling of Active Clamp Converters," IECON 2001.
12. http://www.transim.com/
13. D. Dalal, "Design Considerations for the Active Clamp and Reset Technique," TI Application Note SLUP112.
14. Philips, "25 Watt DC-DC Converter Using Integrated Planar Magnetics," Technical Note, Philips Magnetic Products.
15. C. Bridge, "The Implication of Synchronous Rectifiers to the Design of Isolated, Single-Ended Forward Converters," TI Application Note SLUP175.

16. http://www.elnamagnetics.com/library/square.pdf
17. http://metglas.com/tech/index.htm
18. "Mag-Amp Core and Materials," Bulletin SR4, http://www.mag-inc.com/pdf/sr-4.pdf
19. B. Mamano, "Magnetic Amplifier Control for Simple Low-Cost, Secondary Regulation," http://focus.ti.com/lit/ml/slup129/slup129.pdf
20. Sam Ben-Yaakov, "A SPICE Compatible Model of Mag-Amp Post Regulators," APEC, 1992.
21. M. Jovanović and L. Huber, "Small-Signal Modeling of Mag-Amp PWM Switch," PESC, 1997.
22. F. Cañizales, "Etat de l'art sur la post-regulation magnétique," *Electronique de Puissance*, no. 9.
23. L. Dixon, "Coupled Filter Inductors in Multi-Output Buck Regulators," SEM500, Unitrode seminars 1986.
24. D. Maksimović, R. Erickson, and C. Griesbach, "Modeling of Cross-Regulation in Converters Containing Coupled Inductor," APEC, 1998.
25. B. Mamano, "Current Sensing Solutions for Power Supply Designers," Texas Instruments Application Note SLUP114.

APPENDIX 8A HALF-BRIDGE DRIVERS USING THE BOOTSTRAP TECHNIQUE

Despite its inherent simplicity and reliability, the transformer-based driving circuit does not please designers when small and slim converters must be developed. A high-voltage integrated circuit implementing the bootstrap technique thus represents a possible solution, usually housed in a DIP8 or SO8 package. International Rectifier pioneered this technique back at the beginning of the 1980s with the IR2110. Several manufacturers now offer this kind of high-voltage solution able to swing up to 600 V.

Figure 8A-1 depicts a standard half-bridge (HB) schematic where a fast diode and a capacitor form the heart of the bootstrap architecture: when the lower-side MOSFET turns on, the HB pin drops near zero and the so-called bootstrap capacitor C_{boot} immediately charges up to $V_{cc} - V_f$ via D_2. During the next cycle, once M_{lower} has been blocked, the upper-side MOSFET turns on and brings the HB pin to the bulk level. The upper terminal of C_{boot} is thus immediately translated to $V_{bulk} + V_{cc} - V_f$ and provides a HB-referenced floating voltage bias to M_{upper}: C_{boot} starts to deplete as it delivers energy to the driver. When the next cycle takes place, M_{upper} opens and M_{lower} closes, immediately refreshing the capacitor voltage: a voltage ripple ΔV takes place across C_{boot}. To avoid erratic behavior at power-on, the driving logic usually first turns on M_{lower} so that C_{boot} provides a floating V_{cc} ready for the next upper-side driving event. Figure 8A-2a and b portrays this succession of events.

Sometimes, in self-oscillating circuits, this logic control does not exist, and an erratic spike occurs at power-on on the HB pin. This spike can be linked to the incomplete charge of the bootstrap capacitor which prematurely triggers the upper-side UVLO. To prevent this from happening, you can wire a resistor from the HB pin to the ground of a few hundreds kilo-ohms. This leak will provide a dc charging path to C_{boot} as soon as the user powers up the converter. Thus, before the circuit actually starts to pulse, the upper side is ready to operate and the spike disappears.

The bootstrap capacitor must be calculated to supply the upper-side MOSFET driving current without discharging too deeply. If this capacitor cannot sustain the energy need during the upper-side operation, the dedicated UVLO section trips and stops the bridge operation. Figure 8A-3 depicts the charge process over the bootstrap capacitor when the upper-side MOSFET is activated:

The first drop on Fig. 8A-3 is due to the charge transfer between the bootstrap capacitor and the upper-side MOSFET. This charge transfer equals the total gate charge Q_G of the concerned transistor. The second depletion zone labeled Q_{dc} is linked to the dc current delivered by the bootstrap capacitor during the high-side on time: it corresponds to the driver quiescent current I_{drv} (the internal bias consumption) and the pulldown resistor R_{PD} installed between the gate

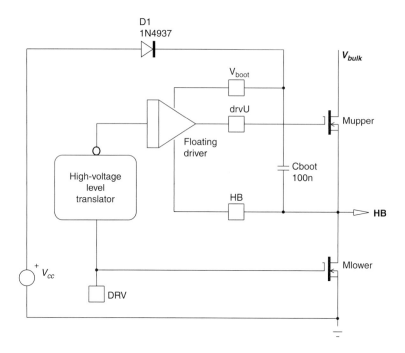

FIGURE 8A-1 A typical half-bridge driver using the bootstrap technique.

and the source of the upper-side transistor. The total gate charge taken from the capacitor during the high-side on time is therefore expressed by:

$$Q_{tot} = Q_G + Q_{dc} = Q_G + D_{max}T_{sw}\left(\frac{V_{cc} - V_f}{R_{PD}} + I_{drv}\right) \qquad (8A\text{-}1)$$

where Q_G represents the upper-side MOSFET total gate-charge, I_{drv} the current consumed by the high-side driver during its on time and R_{PD}, the gate-source resistor installed on the upper-side MOSFET if any. If we accept a voltage drop ΔV on the capacitor at the end of the on time, then C_{boot} can be calculated via Eq. 8A-2:

$$C_{boot} \geq \frac{Q_G + D_{max}T_{sw}\left(\frac{V_{cc} - V_f}{R_{PD}} + I_{drv}\right)}{\Delta V} \qquad (8A\text{-}2)$$

Suppose our half-bridge design features the following data:

$Q_G = 50$ nC
$F_{sw} = 100$ kHz
$D_{max} = 50\%$
$R_{PD} = 47$ kΩ
$I_{drv} = 600$ μA
$V_{cc} = 12$ V
$V_f = 0.8$ V

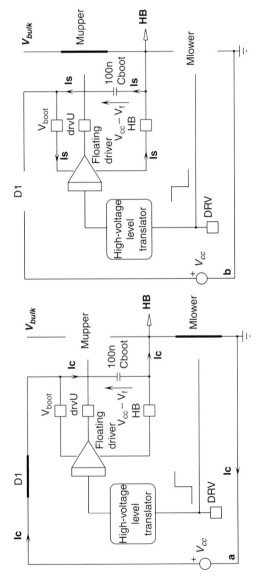

FIGURE 8A-2 The floating supply is obtained by charging the bootstrap capacitor as the lower-side switch turns on.

If we assume a half-bridge driver featuring a UVLO of 10 V, we cannot accept a voltage drop greater than:

$$\Delta V = V_{cc} - V_f - UVLO = 12 - 0.8 - 10 = 1.2V \quad (8A\text{-}3)$$

Selecting a 1 V voltage drop and based on the above numbers, we should pick-up a bootstrap capacitor of:

$$C_{boot} \geq \frac{Q_G + D_{max}T_{sw}\left(\frac{V_{cc} - V_f}{R_{PD}} + I_{drv}\right)}{\Delta V}$$

$$\geq \frac{50n + \left(\frac{(12 - 0.8)}{47k} + 600u\right) \times 0.5 \times 10u}{1} \geq 54nF \quad (8A\text{-}4)$$

To include some design margin, a 100 nF/50 V type of capacitor can be selected. In the above equation, we assumed DT_{sw} to be the time during which the upper side switch is activated. It should be replaced by $(1\text{-}D)T_{sw}$ in case the upper-side switch is turned on during the clock off time.

The recharge of the bootstrap capacitor can generate a sharp current spike given the nature of the circuit: a voltage source V_{cc} directly driving a capacitor. To limit the current and therefore the radiated noise, the designer can insert a 10 to 22 Ω resistor in series with the bootstrap diode. Care must then be taken to check that the smallest on time affecting M_{lower} is long enough to fully recharge C_{boot} despite the new series resistor (Fig. 8A-3). The diode must sustain the reverse voltage linked to V_{bulk} minus V_{cc}. A 1N4397 can do the job for most high-voltage applications. Figure 8A-4 shows a typical half-bridge configuration built around an NCP5181 and driven by a dead time generator. This generator reduces the shoot-through current in the MOSFETs and improves the efficiency.

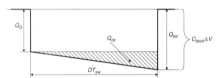

FIGURE 8A-3 The bootstrap capacitor depletion is brought by the upper-side MOSFET activation but also by the driving current consumed during the on time.

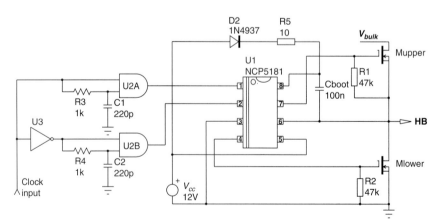

FIGURE 8A-4 A typical half-bridge circuit based on the NCP5181 from ON Semiconductor to which a dead time generator has been added.

As any integrated circuits do, the half-bridge driver presents a sensitivity to negative bias. When the lower-side body diode freewheels, the HB node can swing in the negative direction. Depending on the amount of electrons injected into the substrate (depends on the peak current and its duration—Coulombs then), an erratic behavior can happen, leading to an unexpected failure. To avoid this problem, limit the length of the loop involving M_{lower} and the controller ground. A resistor (a few ohms) can also be inserted between the HB node and pin 6 with a low-V_f high-voltage diode wired between pins 6 and 3, close to the controller. This resistor can also affect the driving speed of M_{upper}, so do not oversize it.

APPENDIX 8B IMPEDANCE REFLECTIONS

When a resistor R_L is connected to a transformer's secondary side, an equivalent resistor R_{eq} is "seen" from the primary side. Figure 8B-1 shows this typical example where the turns ratio is normalized to the primary, as used throughout all chapters:

If the transformer features an infinite magnetizing inductor and neglecting any leakage terms, we can write

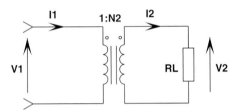

FIGURE 8B-1 Evaluating the equivalent resistance seen from the primary.

$$I_1 = N_2 I_2 \tag{8B-1}$$

but also

$$V_2 = V_1 N_2 \tag{8B-2}$$

By definition,

$$I_2 = \frac{V_2}{R_L} \tag{8B-3}$$

Substituting Eq. (8B-3) into (8B-1), we have

$$I_1 = \frac{V_1}{R_{eq}} = \frac{N_2 V_2}{R_L} \tag{8B-4}$$

Rearranging the above expression leads to

$$R_{eq} = \frac{V_1 R_L}{N_2 V_2} \tag{8B-5}$$

Owing to Eq. (8B-2), we have

$$R_{eq} = \frac{R_L}{N_2^2} \tag{8B-6}$$

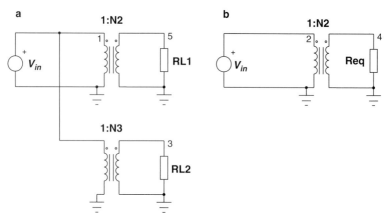

FIGURE 8B-2 When connected to an auxiliary winding, R_{L2} can be reflected over R_{L1}.

A situation can arise when a load is connected to a secondary winding. In that case, the transformer reduction to a single winding often simplifies the stability analysis. Figure 8B-2a and b shows a typical example.

We can first reflect R_{L2} on the primary side of its winding, applying Eq. (8B-6):

$$R_{eq1} = \frac{R_{L2}}{N_3^2} \qquad (8B\text{-}7)$$

Then we can "push" R_{eq1} over R_{L1} by still applying Eq. (8B-6) but reversed:

$$R_{eq2} = R_{eq1} N_2^2 = \frac{R_{L2} N_2^2}{N_3^2} \qquad (8B\text{-}8)$$

Finally, we can write the definition of R_{eq}, as represented by Fig. 8B-2b:

$$R_{eq2} = R_{L1} \,\|\, \frac{R_{L2} N_2^2}{N_3^2} \qquad (8B\text{-}9)$$

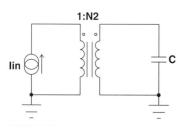

FIGURE 8B-3 A capacitor is now connected to the secondary side.

In lieu of a resistor, we can reflect a simple capacitor (Fig. 8B-3). In that case, for a sinusoidal excitation, the capacitor impedance is

$$Z_C = \frac{1}{jC\omega} \qquad (8B\text{-}10)$$

Eq. (8B-6) holds:

$$C_{eq} = C N_2^2 \qquad (8B\text{-}11)$$

In Fig. 8B-4a, we also reflect an impedance, but to another winding. Applying a similar methodology as for Fig. 8B-2a, we can show that the equivalent capacitor is

$$C_{eq} = C_{L2} \left(\frac{N_3}{N_2}\right)^2 \,\|\, C_{L1} \qquad (8B\text{-}12)$$

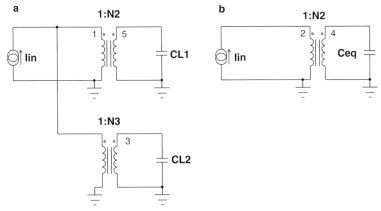

FIGURE 8B-4 When connected to an auxiliary winding, C_{L2} can be reflected over C_{L1}.

If now we speak about impedances, based on the above, we can write that the reflection of a given impedance to the primary side of a transformer featuring a $1{:}N_2$ turns ratio is governed by

$$Z_{eq} = \frac{Z_L}{N_2^2} \tag{8B-13a}$$

Generally speaking, if the impedance is a resistor, we have

$$Z_{eq} = \frac{R_L}{N_2^2} \tag{8B-13b}$$

For a capacitor, we would obtain

$$Z_{eq} = \frac{1}{2\pi C_L N_2^2 f} \tag{8B-13c}$$

and for an inductor

$$Z_{eq} = \frac{2\pi L_L f}{N_2^2} \tag{8B-13d}$$

where f is a sinusoidal signal.

As capacitors are always being associated with their equivalent series resistor (ESR), we can update Fig. 8B-3 as Fig. 8B-5a shows.

The association in series of a capacitor C and a resistor R leads to a complex admittance Y defined by:

$$Y = \frac{sC}{1 + sRC} = \frac{sC}{1 + s\tau} \tag{8B-14}$$

where $\tau = RC$, the network time constant. The impedance Z brought by the series network obeys the following expression:

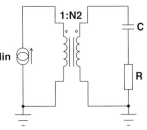

FIGURE 8B-5a A capacitor naturally exhibits an equivalent series resistor.

expression:

$$Z = \sqrt{R^2 + \left(\frac{1}{C\omega}\right)^2} \qquad (8\text{B-}15)$$

If we apply Eq. (8B-6), we find that the equivalent impedance seen from the primary is

$$Z_{eq} = \sqrt{\left(\frac{R}{N_2^2}\right)^2 + \left(\frac{1}{\omega C N_2^2}\right)^2} \qquad (8\text{B-}16)$$

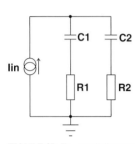

FIGURE 8B-5b Paralleling two impedances does not lead to paralleling resistors and adding capacitors.

As Fig. 8B-5b shows, paralleling complex impedances is also very common since this happens when paralleling two capacitors affected by individual ESRs. Unfortunately, the total impedance obtained via this element combination is not a simple expression. Let us derive admittance expressions, which are easier to manipulate when impedances are in parallel. The first case assumes $R_1 C_1 = R_2 C_2$.

If we call Y_1 the admittance of $R_1 C_1$ and Y_2 the admittance of $R_2 C_2$, both networks being in parallel, $Y_{tot} = Y_1 + Y_2$:

$$Y_{tot} = \frac{1}{R_1 + \frac{1}{sC_1}} + \frac{1}{R_2 + \frac{1}{sC_2}} = \frac{sC_1}{sR_1 C_1 + 1} + \frac{sC_2}{sR_2 C_2 + 1} = \frac{sC_1}{s\tau_1 + 1} + \frac{sC_2}{s\tau_2 + 1} \qquad (8\text{B-}17)$$

If $\tau_1 = \tau_2 = \tau$, then the final admittance simplifies to

$$Y_{tot} = \frac{s(C_1 + C_2)}{1 + s\tau} \qquad (8\text{B-}18\text{a})$$

which looks like Eq. (8B-14) where the capacitor C is the sum of both capacitors (as if they were paralleled) and the ESR is the value that once combined with $C_1 + C_2$ gives τ_1 or τ_2. This resistor is simply

$$R = \frac{\tau_1}{C_1 + C_2} = R_1 \| R_2 \qquad (8\text{B-}18\text{b})$$

It makes sense since for high switching frequencies, both capacitors are short-circuits.

- When two series RC networks $R_1 C_1$ and $R_2 C_2$ of similar time constants are paralleled, the resulting equivalent series RC network is made of $C = C_1 + C_2$ and $R = R_1 \| R_2$.

Now, let us select different time constants for $R_1 C_1$ and $R_2 C_2$.
Starting from Eq. (8B-17), we can combine after simplification by s and neglecting the 1 in the expression:

$$Y_{tot} = \frac{(C_1 + C_2) + s(\tau_1 C_2 + \tau_2 C_1)}{\left(s\frac{\tau_1 \tau_2}{(\tau_1 + \tau_2)} + 1\right)(\tau_1 + \tau_2)} \qquad (8\text{B-}19)$$

which clearly differs from Eq. (8B-18a).

- In conclusion, two paralleled RC networks of different time constants do not reduce to a single RC network!

APPENDIX 8C TRANSFORMER AND INDUCTOR DESIGNS FOR THE 250 W ADAPTER

Contributed by Charles E. Mullett

This appendix describes the procedure to design the transformer and output inductor for the 250 W forward converter described in Chap. 8. Thanks to the derived equations and simulations, we have gathered the following electrical data pertaining to the transformer design.

8C.1 Transformer Variables

$V - s_{,max} = V_{in,max} T_{sw} D_{max} = 400 \times 10u \times 0.45 = 1.8m$

$I_{Lp,max} = 2$ A

$I_{p,rms} = 1.1$ A

$I_{s,rms} = 12.6$ A

$F_{sw} = 100$ kHz

$V_{in} = 380$ to 400 Vdc

$V_{out} = 12$ V at 20 A

$N_p:N_s = 1:0.085$

8C.2 Transformer Core Selection

Based on the above data, several methods exist to determine the needed core size. In this appendix, we will use the area-product definition. The first step therefore consists of finding the required area product $W_a A_c$, whose formula appears in the Magnetics Ferrite Core catalog available through the Ref. 1 link:

$$W_a A_c = \frac{P_{out}}{K_c K_t B_{max} F_{sw} J} 10^4 \tag{8C-1}$$

where $W_a A_c$ = product of window area and core area, cm^4
P_{out} = output power, W
J = current density, A/cm^2
B_{max} = maximum flux density, T
F_{sw} = switching frequency, Hz
K_c = conversion constant for SI values (507)
K_t = topology constant (for a space factor of 0.4). This is the window fill factor. For the forward converter, use $K_t = 0.0005$.

Some judgment is required here, because the current density and flux density are left to the designer. A reasonable, conservative current density for a device of this power level, with natural convection cooling, is 400 A/cm^2.

The flux density at this frequency (100 kHz), for modern power ferrites such as Magnetics "P" material, is usually around 100 mT (1000 gauss). This can also be approached by choosing a loss factor of 100 mW/cm^3 (a reasonable number for a temperature rise of 40 °C) and looking up the flux density in the manufacturer's core material loss data. Doing this for the Magnetics P material yields a flux density of 110 mT. We will use this for the design.

Therefore, we have

$$W_d A_c = \frac{P_{out}}{K_c K_t B_{max} F_{sw} J} 10^4 = \frac{250 \times 10000}{507 \times 0.0005 \times 0.11 \times 100000 \times 400} = 2.24\, cm^4 \quad (8C\text{-}2)$$

Looking at the core selection charts in the Magnetics Ferrite Catalog, we choose the ETD-39 core. It features an area product of 2.21 cm⁴, a very close match to the requirement. For future reference, it offers a core area of $A_e = 1.23$ cm².

8C.3 Determining the Primary and Secondary Turns

Per the design information in the Magnetics Ferrite Core catalog, and based on Faraday's law,

$$N_p = \frac{V_{in,min} 10^4}{4 B A_e F_{sw}} \quad (8C\text{-}3)$$

where N_p = primary turns
$V_{in,min}$ = primary voltage
B = flux density, T
A_e = the effective core area, cm²
F_{sw} = the switching frequency, Hz

This assumes a perfect square wave where the duty ratio D is 0.5. In our case, the duty ratio will be limited to 0.45, and that will occur at minimum input voltage, or 380 Vdc. One can either correct for these in the equation [adjust the result by the factor (0.45/5) × (380/400)] or use another version that expresses the same requirement in terms of the voltage at a given duty ratio.

$$N_p = \frac{V_{in,min} t_{on,max} 10^4}{2 B_{max} A_e} \quad (8C\text{-}4)$$

where $t_{on,max}$ = the duration of the pulse applied to the winding. And, in the denominator, the 4 is replaced by a 2, due to the fact that the square wave actually applied the voltage during one-half of the period, so it was accompanied by another factor of 2.

In this case,

$$t_{on,max} = D_{max} T_{sw} = \frac{D_{max}}{F_{sw}} = \frac{0.45}{10^5} = 4.5\, \mu s \quad (8C\text{-}5)$$

Thus,

$$N_p = \frac{380 \times 4.5u \times 10^4}{2 \times 0.11 \times 1.23} = 63.2\, \text{turns} \quad (8C\text{-}6)$$

Now, applying the required turns ratio to determine the secondary turns, we have

$$N_s = 0.085 N_p = 0.085 \times 63.2 = 5.37\, \text{turns} \quad (8C\text{-}7)$$

We now round off the secondary turns to 5 turns, then recalculate the primary turns:

$$N_p = \frac{5}{0.085} = 58.8\, \text{turns} \quad (8C\text{-}8)$$

We can now round this off to

$$N_p = 58.8\, \text{turns} \quad (8C\text{-}9)$$

8C.4 Choosing the Primary and Secondary Wire Sizes

The wire sizes can now be determined, based on the previously chosen current density of 400 A/cm². In the case of the primary, the stated current is 1.1 A rms, so the required wire area will be

$$Aw(pri) = \frac{I_{p,rms}}{J} = \frac{1.1}{400} = 0.253 \text{ mm}^2 \tag{8C-10}$$

This corresponds to a wire size of 23 AWG, which has an area of 0.2638 mm².

Fitting the windings into the bobbin usually requires some engineering judgment (and sometimes trial and error), as one would like to avoid fractional layers (they increase the leakage inductance, meaning that they result in poorer coupling and reduced efficiency) and also minimize the skin effect by keeping the diameter below one skin depth. In this case, using two conductors of approximately one-half the area is a wise choice. This suggests two conductors of 26 AWG. The final choice will depend on the fit within the bobbin.

Interleaving the primary winding is also advantageous, as it significantly reduces the leakage inductance and also reduces the proximity losses. With this in mind, we will try to split the winding into two layers of one-half the turns, or $58/2 = 29$ turns on each layer, with the secondary winding sandwiched in between the two series halves of the primary. With two conductors of no. 26 wire, the width of the 29-turn layer will be (assuming single-layer insulation on the wire) 0.443 mm diameter $\times 2 \times 29$, or 25.69 mm. The bobbin width is 26.2 mm, so this doesn't leave enough margin at the sides for safety regulation requirements. Reducing the wire size to no. 27 is the answer. The reduction in wire size is well justified, because a fractional layer would surely increase the winding losses.

The secondary current is 12.6 A rms. For a current density of 400 A/cm², the wire size will be

$$Aw(sec) = \frac{I_{s,rms}}{J} = \frac{12.6}{400} = 3.15 \text{ mm}^2 \tag{8C-11}$$

The corresponding wire size is 12 AWG, which has an area of 3.32 mm². To reduce the skin effect and produce a thinner layer (and also spread it over a greater portion of the bobbin), we will use three conductors of no. 17 wire, which will have a total area of 3×1.05 mm² or 3.15 mm². Since the diameter of the single-insulated no. 17 wire is 1.203 mm, the five turns of three conductors will occupy a width of $5 \times 3 \times 1.203$, or 18.05 mm. This fits well within the bobbin width of 26.2 mm.

8C.5 Gapping the Core

To ensure enough primary magnetizing current to properly reset the core (drive the stray capacitance and allow the voltage across the winding to reverse), one must usually reduce the primary inductance from the core's ungapped value to one that will cause an adequate magnetizing current. A popular rule of thumb is to make the magnetizing current around 10% of the primary current. Since the primary current is 2 A peak, we will let the magnetizing current rise to 0.2 A. The desired primary inductance, then, with a primary voltage of 380 V and a pulse duration of 4.5 µs, is

$$L_p = \frac{V_{in,min}}{\frac{di}{dt}} = \frac{380}{\frac{0.2}{4.5u}} = 8.55 \text{ mH} \tag{8C-12}$$

The desired inductance factor A_L can now be determined:

$$A_L = \frac{L_p}{N^2} = \frac{8.55m}{58^2} = 2542 \text{ nH/turns}^2 \qquad (8\text{C-}13)$$

The ungapped A_L of the ETD-39 core with P material is 2420, so no gap is needed.

8C.6 Designs Using Intusoft Magnetic Designer

Intusoft's Magnetic Designer software is a powerful tool for designing transformers and inductors. It allows the designer to quickly arrive at a basic design, then optimize it by providing fast recalculations of losses, leakage inductance, etc., as the designer experiments with variations in the winding structure, core size and shape, and other design choices. To illustrate the iteration procedure, we have designed two different transformers. In both of these designs, shown in Figs. 8C-1 through 8C-4, the primary has been split into two layers in series and the secondary is interleaved in between them. This reduces the proximity losses significantly, since the mmf (magnetomotive force) induced in the inside layer of the primary is reduced by the opposing current in the secondary. Note the ac resistances of each of the layers and the winding losses.

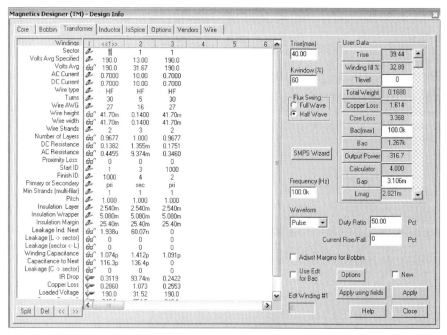

FIGURE 8C-1 Design alternative 1, using an ETD-39 core.

SIMULATIONS AND PRACTICAL DESIGNS OF FORWARD CONVERTERS

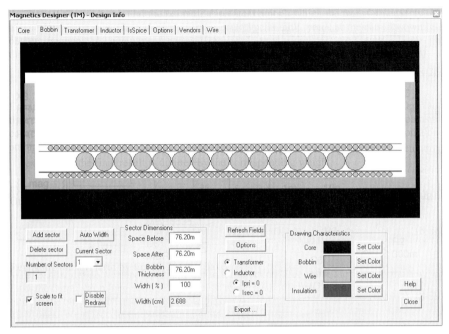

FIGURE 8C-2 Winding structure for design alternative 1.

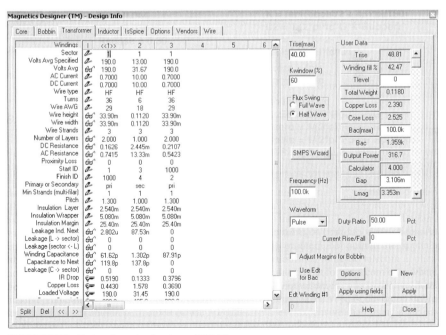

FIGURE 8C-3 Design alternative 2, using an ETD-34 core.

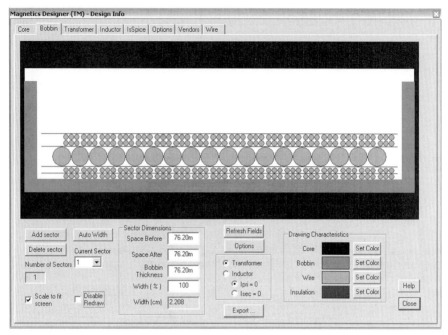

FIGURE 8C-4 Winding structure for design alternative 2.

Note the following:

Core type	ETD 39
Primary	60 turns, 2 × no. 27
Secondary	5 turns, 3 × no. 16
Copper loss	1.614 W
Core loss	3.368 W
Temperature rise	39.44 °C
Dc resistance, layers 1 + 2	0.3133 Ω
Ac resistance, layer 1	0.4455 Ω
Ac resistance, layer 3	0.3460 Ω

Note the following:

Core type	ETD 34
Primary	72 turns, 3 × no. 29
Secondary	6 turns, 3 × no. 18
Copper loss	2.390 W
Core loss	2.525 W
Temperature rise	48.81 °C
Dc resistance, layers 1 + 2	0.3733 Ω
Ac resistance, layer 1	0.7415 Ω
Ac resistance, layer 3	0.5423 Ω

Comparison of the two designs:

	Design 1	Design 2
Core type	ETD 39	ETD 34
Primary	60 turns, 2 × no. 27	72 turns, 3 × no. 29
Secondary	5 turns, 3 × no. 16	6 turns, 3 × no. 18
Copper loss	1.614 W	2.390 W
Core loss	3.368 W	2.525 W
Temperature rise	39.44 °C	48.81 °C
Dc resistance, layers 1 + 2	0.3133 Ω	0.3733 Ω
Ac resistance, layer 1	0.4455 Ω	0.7415 Ω
Ac resistance, layer 3	0.3460 Ω	0.5423 Ω

Note the much higher ac resistance in design 2, mostly because the windings are thicker. This is what raised the copper loss in design 2. In design 1, the smaller number of turns raised the flux density, resulting in greater core loss, but the lower ac resistance more than compensated for it and resulted in a lower total loss and lower temperature rise (part of the reason for the lower temperature rise was due to the larger structure, of course).

8C.7 Inductor Design

For simple, single-winding inductors such as this, there are many off-the-shelf products available from reputable suppliers throughout the world. Also, many of these suppliers have design aids posted on their websites and in their catalogs. These provide quick and easy pathways to the desired inductor.

From Chap. 8 calculation and simulation results, we have the following inductor data:

Inductor Variables

$L = 25$ μH

$I_{p,max} = 24$ A

$I_{L,rms} = 20$ A

$F_{sw} = 100$ kHz

Approaching the design independently begins with determining the core size, based on the stored energy, and choosing the core material based on the cost-performance tradeoffs, including considerations of the amount of ac flux versus dc flux. In this case, the inductor will be designed for smoothing the output current and will have relatively low ac flux density compared to the dc flux density (in flyback transformers and in many inductors used in power factor correction circuits, this is not the case). Given that the ac flux density will be low, one can choose the core material based on low cost and high-saturation flux density. Powdered iron toroids and E cores are popular in these applications, as the material costs are minimal.

As in the transformer design, the size can be estimated in terms of the area product of the inductor, and this is based on the power handling capacity of the inductor. One would like to be able to name the allowable loss and ultimate temperature rise in advance, and then be able to calculate the core size that will give these final results. But predicting the losses in advance and estimating the surface temperature are relatively complex tasks and will not be attempted here. We will, instead, choose a reasonable current density for the wire, based on practical experience, and then estimate the ultimate power loss and temperature rise in the finished inductor.

8C.8 Core Selection

One can find a number of expressions for the area product of the core, due to different considerations of the thermal properties and shapes. But the general basis for calculation of the area product is quite universal. The area product is the result of multiplying the core area by the window area. This makes sense, as the window area is related to the copper area, and hence the current capacity of the device, and the core area is related to the flux capacity, and the flux capacity is related to the volt-seconds applied to the core by the winding. At a given frequency, one can see that the area product will be determined by current and voltage, and this is obviously related to power. Hence, the power handling capacity of a magnetic component determines its area product, usually expressed in cm^4.

A popular expression for the area product is

$$W_a A_c = \frac{L I_{p,max} I_{L,rms}}{K_{cu} B_{max} J} 10^4 \quad cm^4 \tag{8C-14}$$

Several of the popular core materials, such as Magnetics Hi-Flux and the Micrometals powered iron materials, can handle flux densities as high as 1.0 T and are available in toroidal shapes. The permeability of these materials decreases with the flux density, long before saturation, so it will save design time and avoid iterations to pick a conservative number for the flux density, perhaps 0.6 T. For windings of one layer of wire, the fill factor (ratio of copper area to window area) K_{cu} should be about 0.3. Putting these into Eq. (8C-14), we have

$$W_a A_c = \frac{L I_{p,max} I_{L,rms}}{K_{cu} B_{max} J} 10^4 = \frac{25u \times 24 \times 20}{0.3 \times 0.6 \times 400} 10^4 = 1.67 \; cm^4 \tag{8C-15}$$

This suggests a core from Magnetics in the family of 34.3 mm outside-diameter cores (55588 series), available in several materials, or the T131 series from Micrometals.

Using the Magnetics Hi-Flux material, in the 125u material (core part #58585), the inductance factor is 79 nH/turn² = 0.079 μH/turn², and the magnetic path length, for future use, is $l_e = 8.95$ cm. The required turns, neglecting the drop in permeability due to saturation of the material, is 18 (rounded up from 17.8), as shown in Eq. (8C-16).

$$N = \sqrt{\frac{L}{A_L}} = \sqrt{\frac{25}{0.079}} = 17.8 \; turns \tag{8C-16}$$

This yields a magnetizing force of

$$H = \frac{0.4 \pi N I}{l_e} = \frac{0.4 \times \pi \times 18 \times 20}{8.95} = 50.5 \; oersteds = 0.505 \; ampere - turns \, per \, meter \tag{8C-17}$$

Referring to the characteristic curves for this material, one observes that the permeability decreases to 65% of its initial value with this magnetization. Increasing the turns to compensate for this will result in a new value for the turns of

$$N = \frac{18 \; turns}{0.65} = 28 \; turns \tag{8C-18}$$

Now, with 28 turns, the permeability is only 43% of its initial value, but because of the increase in turns, the inductance is still over the required 25 μH, as shown below:

$$L = N^2 A_L = (28)^2 \times 0.43 \times 0.079 = 26.6 \; \mu H \tag{8C-19}$$

8C.9 Choosing the Wire Size and Checking the DC Resistive Loss

The wire conductor area at a current density of 400 A/cm² and 20 A (ac component of current is negligible) is

$$A_w = \frac{I_{L,rms}}{J} = \frac{20}{400} = 0.05 \text{ cm}^2 \quad (8\text{C-20})$$

The nearest AWG size is 11, and according to the data on the core, a single layer of 26 turns of no. 11 wire will have a total resistance R_{dc} of 0.00348 Ω. Extrapolating to 28 turns gives an estimated resistance of 0.00375 Ω. Thus, the dc resistance loss is

$$P_{cu} = I_{L,rms}^2 R_{dc} = 20^2 \times 0.00375 = 1.5 \text{ W} \quad (8\text{C-21})$$

8C.10 Checking the Core Loss

The ac flux density is a result of the pulse waveform appearing across the winding of the inductor. For simplification, assume that the waveform is a square wave (a close enough approximation for this purpose) and that its peak amplitude is equal to the output voltage. Then, from Faraday's law (and including the core area $A_e = 0.454$ cm² $= 0.454 \times 10^{-4}$ m²)

$$B_{ac} = \frac{V_{out}}{4NA_e F_{sw}} = \frac{12}{4 \times 28 \times 0.454 \times 10^{-4} \times 100k} = 0.024 \text{ T} \quad (8\text{C-22})$$

Using the core loss density curves in the Magnetics catalog, the core loss density is about 150 mW/cm³. Multiplying this by the core volume of 4.06 cm³ results in a total core loss of

$$P_{core} = 150 \times 4.06 = 609 \text{ mW} \quad (8\text{C-23})$$

8C.11 Estimating the Temperature Rise

With a total loss of $P_{cu} + P_{core} = 1.5 + 0.609 = 2.109$ W, the temperature rise in normal convective cooling is estimated to be

$$\Delta T \text{ (°C)} = \left(\frac{P(\text{W})}{A_s(\text{in}^2)}\right)^{0.8} = \left(\frac{P(\text{W})}{A_s(\text{cm}^2)/(2.54^2)}\right)^{0.8} \times 100 \quad (8\text{C-24})$$

The wound inductor will have an approximate shape of 50 mm diameter × 29 mm high. Assuming one circular surface has no cooling, the area will be the circumference times the height, plus one circular area, or

$$A_s = \pi \times 5 \times 2.9 + \pi \left(\frac{5.0}{2}\right)^2 = 65.2 \text{ cm}^2 \quad (8\text{C-25})$$

Therefore,

$$\Delta T = \left(\frac{P(\text{W})}{65.2/2.54^2}\right)^{0.8} \cdot 100 = \left(\frac{2.109 \times 2.54^2}{65.2}\right)^{0.8} = 28.5 \text{ °C} \quad (8\text{C-26})$$

The inductor appears to fall within the desired specifications and should perform well in this application.

APPENDIX 8D CD-ROM CONTENT

The CD includes demonstration versions of numerous simulation software programs. They should give you a taste of how the product behaves and will perhaps guide you through an acquisition later on. Unfortunately, given the CD-ROM size limit, we were unable to include some of the largest files, and we recommend that the user download the targeted demonstration version directly from the editor website.

To let you simulate and learn from the book, I have included many of the chapter examples for OrCAD *PSpice* and Intusoft *IsSpice*. Some will work with the size-limited demonstrations version whereas others will require the full licensed version. The files appear in their respective directory on the CD-ROM. Thanks to the help of other software editors, numerous examples from the book running on their demonstration versions are also present on the disk. This is the case for Spectrum-Software *Micro-Cap*, DesignSoft *TINA*, Beige Bag *B2 SPICE*.

For those of you interested in ready-to-use industrial examples, more comprehensive models and application circuits are also available as a separate set of files translated for different simulator versions. Please check the author's website for more details (http://perso.orange.fr/cbasso/Spice.htm).

The CD-ROM contains the following demonstration versions and software:

Intusoft ICAP/4, www.intusoft.com

Cadence OrCAD, http://www.cadence.com/products/orcad/index.aspx (to be downloaded)

Spectrum Software Micro-Cap, http://www.spectrum-soft.com/index.shtm

Powersim PSIM, http://www.powersimtech.com/

Transim SIMetrix, http://www.transim.com/

National-Instruments Multisim, http://www.ni.com/multisim/

Beige Bag Software B2 Spice, http://www.beigebag.com/

R. Andresen 5SPICE, http://www.5spice.com/index.htm

Penzar TOPSPICE, http://penzar.com/topspice/topspice.htm

Ridley Engineering, http://www.ridleyengineering.com

The k factor spreadsheet has been updated and now includes independent pole-zero placement plus the TL431 in type 2 and type 3 configurations.

CONCLUSION

I recognize that the presence of a conclusion paragraph is not very common in technical publications. Perhaps this book will differ from the others rightly so due to its existence here! This book guided you through the meanders of the dc–dc river where rocks and obstacles are now hopefully well identified. The thorough description of some particular block shows how you can either use the obstacle as a support or try to oppose it. A perfect example is the TL431 network where the fast lane obviously hampers the configuration at first glance. Either you try to suppress its effect via an added circuitry or you think of a way to work with it. The final type 2 and type 3 implementations demonstrate actually how to use it to the benefit of the design. The flyback converter in Chap. 7 also identified the leakage inductor as a real offender. You either oppose it via a clamping network and it contributes to a loss, or you utilize it in an active clamp circuit improve the converter's overall performance. Both of the above examples show the difference between confronting the obstacle, with all the associated drawbacks, or thinking of a solution to actually benefit from the presence of the block and improve the operations.

As you have noticed, the author strove to maintain a fair balance between theoretical content and practical examples. Some people might think that, thanks to the power and ease of SPICE, theoretical foundations are unnecessary since trial and error looks like child's play on a computer. Even if this approach can sometimes work on low-volume projects, where individual trims are always possible, it turns into an impossible and dangerous exercise when quantities above 1000 per month are reached. A line down event will cruelly bring the designer back to the reality of parameter dispersion and teach him or her a lesson: always go through the equations that reflect the impact of hidden parameters so you can efficiently fight them once they move in production. Failure to do this, due to the pressure of time, will lead to irreparable failure.

I want to end this conclusion by quoting the French philosopher René Descartes who encouraged the readers of his "Discourse on Method" to avoid taking everything for granted. I would modestly suggest you do the same with this book: go through the text and derive the equations yourself. This is the best way to understand and gain insight into the circuit operation.

INDEX

ABM. *See* analog behavioral modeling (ABM)
absolute current error tolerance (ABSTOL), 286, 390
ac
  5 V, 10 A buck converter, 426–429
  circuits for multioutput flyback converter, 719–720
  current-mode 5 V, 1 A boost, 454–458
  DCM buck-boost, 479–480
  discontinuous current-mode 12 V, 2 A buck-boost, 479–483
  large-signal simulations, 106
  loop gain, 337–338
  output impedance, 13, 329
  small-signal models
    in model encapsulation, 137
    responses, 168–170
    terms, 107–108
  stacks, flyback converter secondary windings, 713–714
  sweeps
    boost converters, 445–447, 455–456
    CCM current mode, 700, 823
    closed-loop systems, 337, 339
    input filters, 463–464
    input ripple, 424
    low-frequency gain and peaking variations, 119
    operational amplifiers, 344
    optocouplers, 307–308, 310–312
    PSIM, 317–318
    shunt regulators, 314
    small-signal models, 128
  voltage-mode 12 V, 2 A buck-boost, 468–474
  voltage-mode 12 V, 4 A buck converter, 410–413
  voltage-mode 48 V, 2 A boost, 444–449

active clamps
  dead time generation, 387–388
  flyback converters
    design example, 622–625
    overview, 616–622
    simulation circuit, 625–627
  forward converter core reset solution, 778–795
  PWM switch model, 133
  TL431 feedback, 300
active integrator amplifiers. *See* type 1 amplifiers
active power factor correction, 527–528
Ampere's law, 396–397
ampere-seconds (coulombs), 26
ampere-turns (A-T) continuity
  boost converters, 44
  field probes, 368
  flyback converters, 581, 592
  forward converters, 743
  inductors, 353
  Lenz formula, 14
  switches, 21
  terminology, 393
amplifiers. *See also* error amplifiers; operational amplifiers (op amps); operational transconductance amplifiers (OTAs); type 2 amplifiers; type 3 amplifiers
  magnetic, 799–804
  selecting, 262
  type 1, 255–256, 262, 264, 288
  type 2a, 258–259, 262
  type 2b, 259–260, 262
analog behavioral modeling (ABM), 341–342, 348–350, 356, 374, 542
analog circuits, 3
apparent power, 512–513
area product, 732, 859, 866
astable generators, 372–377

871

INDEX

A-T continuity. *See* ampere-turns (A-T) continuity
attenuation
  calculating, 79–83
  input filters, 79–80, 460–462
  input ripple, 421–422
  type 2 amplifiers, 291–292
  type 3 amplifiers, 295
  voltage-mode buck converters, 267, 271
ATX power supplies, 93–94, 526, 739, 755–756, 849
audio susceptibility, 10, 12, 278, 282, 284, 348, 381–382
automated pole-zero placement, 323–326
auxiliary winding
  flyback converters, 642–656, 663, 677, 681, 685, 722–723, 725, 732–734
  forward converters, 846, 848, 856–857
avalanche, 526, 589, 612, 730
average current-mode, 535–543, 559
average inductor current, 49–50, 221, 441, 466–467, 487, 528, 539–540, 545, 553–555, 689
average power, 418–419
average simulations, 567–571, 715, 718
averaged function, 95–96
averaged models
  BCM boost, 532–535
  compensated DCM buck converters, 170
  current-mode boost converters, 200
  fixed-frequency peak current-mode PFC, 546–549
  generic switch inductor model, 221
  lossy model, 184–186
  small-signal, 95, 97, 213–223
  voltage-mode BCM, 193–194

B (flux density), 368, 393–397, 732–733, 744–745, 799–804, 859–860, 865–867
B elements, 341–343, 356, 391, 567, 570
bandgaps, 358, 645
bandwidth. *See* crossover frequency (f_c)
batteries. *See* car batteries, voltage-mode nonisolated boost from; lithium ion batteries, current-mode nonisolated boost from
BCM. *See* borderline conduction mode (BCM)
behavioral models, 288, 327. *See also* analog behavioral modeling (ABM)
Ben-Yaakov models, 220–223
B-H curves, 368, 764, 802
bias current
  optocouplers, 305–308
  TL431, 289, 291, 299–304, 326, 329
bias point analysis, 120, 129, 325

biasing, 106–107, 298–303
Bode plots
  automated pole-zero placement, 323–325
  boost converters, 444–446, 457–458
  buck converters, 244, 267–268, 271–272, 410, 412, 415, 429
  buck-boost converters, 469–470, 473, 481–482
  flyback converters, 219
  high-pass filters, 252
  low-pass filters, 251
  optocouplers, 309
  PFC, 535–537
  RHPZ, 254
  SEPIC, 220
  single zero networks, 252
  type 1 amplifiers, 256
  type 2 amplifiers, 258–259, 291–292
  type 3 amplifiers, 262, 294
body diodes, 377, 434, 597, 608, 621, 730, 780, 792, 798, 854
BOOLEAN functions, 341
boost converters. *See also* constant on-time borderline power factor correction; nonisolated boost converters
  BCM, 199, 528–532
  CCM, 44–47, 51–54, 545
  cutoff frequency, 82–85
  DCM waveforms, 47–50
  fixed-frequency DCM PFC, 550–555
  overview, 42–43
  PWM switch model, 113, 118, 177, 208
  RHPZ effect, 255
  ripple value, 54–55
  small-signal models, 124
  transfer equations, 229–231
  transition point DCM-CCM, 50–51
boost-buck converters, 211–212
bootstrap technique, 851–854
borderline conduction mode (BCM)
  constant on-time power factor correction, 529–535
  designing boost PFC circuit, 559–575
  inductor current in, 708
  input current in, 709
  overview, 21–25, 528–529
  PWM switch model, 187–202
borderline point, 37–38, 50
boundary point, 37–38, 50
B_r (remanent induction flux value), 395, 801–802
breakdown voltage (BV_{DSS})
  boost converters, 465
  buck converters, 439–440
  buck-boost converters, 486

INDEX **873**

breakdown voltage (BV_{DSS}) (*Cont.*):
  clamping networks, 600
  flyback converters, 726, 730–731
  forward converters, 849
  TL431, 327–328
bridge current, 303–304
B_{sat} (saturation flux density), 368–371, 395, 397, 800
buck converters. *See also* nonisolated buck converters
  CCM, 31–34, 39–41
  current-mode, 280–286
  DCM, 34–37, 130, 172
  generic current-mode controllers with, 380–381
  isolated, 739–745
  off-time events, 31
  on-time events, 30–31
  output impedance in open-loop configuration, 111
  PWM switch model, 113, 116, 153, 203
  ripple value, 41–42
  ripple voltage, 175–177
  SSA, 100–111
  transfer equations, 226–228
  transition point DCM-CCM, 37–39
  voltage-mode, 267–270
buck-boost converters. *See also* nonisolated buck-boost converters
  CCM, 57–59, 64–65
  DCM, 59–62
  isolated, 579–583
  off-time event, 56–57
  on-time event, 56
  pulsating voltage in, 177
  PWM switch model, 114, 206
  RHPZ location, 309
  ripple value, 65–66
  transfer equations, 231–235
  transition point DCM-CCM, 63–64
bulk capacitors, 493–494, 499, 506, 510–511, 526, 528–529, 563, 574
BV_{DSS}. *See* breakdown voltage (BV_{DSS})

capacitors
  boost CCM output ripple voltage, 51–54
  bootstrap, 851–854
  bulk, 493–494, 499, 506, 510–511, 526, 528–529, 563, 574
  charge balance, 26
  clamp, 620–622, 625–627, 775–778, 780–783, 789–795
  damping, 608–609
  dc block, 306, 330

capacitors (*Cont.*):
  drain-source, 724
  equivalent model, 89–93
  impedance, 249, 412, 606, 831, 856–858
  optocouplers, 305–306
  parallel damping, 76–79
  rectifier bridges, 493–498
  resonating, 158, 186, 286
  snubbing, 836
  voltage-adjustable passive elements, 351–353
CCM. *See* continuous conduction mode (CCM)
CD-ROM content, 2–3, 413, 868
charge and flux balance, 25–26
charge-balance law, 26–27
CL systems. *See* closed-loop (CL) systems
clamp capacitors, 620–622, 625–627, 775–778, 780–783, 789–795
clamping
  designing networks
    curing leakage ringing, 605–609
    diode selection, 609–610
    overview, 599–601
    RCD configuration, 601–603
    selecting k_c, 604–605
    TVS clamp, 612–614
    voltage variations, 610–612
  diodes, 343–344
  drain signal with, 591–597
  drain signal without, 588–591
  PWM switch model, 132–134
closed-loop (CL) systems
  buck converters, 110
  linear regulator, 7–9
  virtual ground in, 335–339
C_{lump} (lump capacitors), 588–589, 596–597, 620–621, 675, 762–765, 769, 784–785
CM. *See* common-mode (CM); current-mode (CM)
common-mode (CM), 404–405, 506, 687
comparators, 27–29, 138–141, 160, 194–195, 358–362, 372–373, 378, 549–551, 667–669
compensated gain, 269–270, 429, 482, 680, 700
compensation, independent pole-zero placement, 272–273
compensation loops
  amplifiers
    type 1, 255–256
    type 2, 256–258
    type 2a, 258–259
    type 2b, 259–260
    type 3, 261–262
  passive pole, 250–251
  passive zero, 251–252
  right half-plane zero, 253–255

874   INDEX

compensation ramp
  current-mode designs, 286
  DCM and CCM bucks, 172–175
  DCM current mode, 637–640
  fixed-frequency peak current-mode control, 544–546
  forward converters, 825
  lossy model in current mode, 183
  propagation delay and overpower protection, 655–660
  PWM switch model, 154–155, 159–161
  Ridley models, 213, 456
  voltage- and current-mode models, 381–383
compensator frequency response ($G(s)$), 9–10, 238, 247, 251, 278
component constraints
  flyback converter, 725–726
  forward converter, 849
  nonisolated boost converters, 465
  nonisolated buck converters, 439–441
  nonisolated buck-boost converters, 486
conditional stability, 247, 271–278, 471
conduction time, diode, 492–499, 505, 773, 775, 798
constant on-time borderline power factor correction
  averaged modeling of BCM boost, 532–535
  frequency variations in BCM, 531–532
  overview, 529–531
constant-current generators, 300, 661, 687, 834
continuous conduction mode (CCM)
  boost converters, 44–54, 445, 447–448, 451, 453, 457–458, 461–462, 465
  buck converters, 31–41, 409, 412, 418, 427, 437, 439–441
  buck-boost converters, 57–65, 468–469, 476–477
  feedback and control loops, 309, 316, 324
  leading edge blanking, 359
  mode transition, 143–145
  PFC, 520, 528–529, 535–537, 543, 557, 559, 565
  PWM switch model, 153–158, 172–175
  rectifier bridges, 497
  small-signal response of flyback topology, 635–636, 638–641
continuous equations, 102, 106
continuous functions, 95–96
control loops. *See* feedback and control loops
control variables, 242–243
controlled resistors, 179, 349
controller current-sense pin (CS), 431
controllers. *See also* generic controllers

controllers (*Cont.*):
  dedicated, 806
  high-voltage, 686
  NCP1216, 685
  voltage-mode BCM, 187–188
conversion. *See* power conversion
converter efficiency (η), 491
converter waveforms, 30
converters. *See also* boost converters; buck converters; buck-boost converters; flyback converters; forward converters; off-line converters
  CCM, 688
  input current signature, 79, 81
  input filtering, 66–67
  quickly stabilizing using shunt regulator, 314–316
  static input impedance, 73
  transfer equations, 225–231
  types of, 85
CoPEC (Colorado Power Electronic Center)
  small-signal models, 216–220
cores
  gapping, 767
  reset
    active clamp, 778–795
    detection, 722
    overview, 742–745
    RCD clamp, 767–778
    resonant demagnetization, 762–767
    third winding, 746–755
    two-switch configuration, 756–761
  saturable, 366–372
  terminology, 392, 397–398
coulombs (ampere-seconds), 26
coupled forward converters, 829
coupled inductors, 395, 806–817
coupling ratio, 364
critical conduction mode (CrM). *See* borderline conduction mode (BCM)
critical damping, 69
critical point, 37–38, 50
CrM (critical conduction mode). *See* borderline conduction mode (BCM)
cross-conduction, 387
crossover frequency (f_c), 92, 249–250, 267, 273–275, 719
cross-regulation, 372, 392
CS (controller current-sense pin), 431
CTR. *See* current transfer ratio (CTR)
Ćuk converters, 211–212
current
  cathode, 328–329

INDEX **875**

current (*Cont.*):
  closed-loop sources, 71–72
  input, 79, 487–489, 709
  modes, 224
  PWM switch model, 116
  rectifier bridges, 496–498, 506–508
  ripple, 24, 39, 41, 50, 54, 63, 65, 467
  shaping, 540–543
  sources, 71, 117, 165
  spikes, 91–92
  steps, 91
current transfer ratio (CTR)
  optocouplers, 303
  resistive dividers, 304–305
  variations, 300–301, 309, 314
current-mode (CM)
  5 V, 1 A boost from li-ion battery
    ac analysis, 454–458
    input filter, 460–464
    overview, 452–454
    transient analysis, 459–460
  5 V, 10 A buck from car battery
    ac analysis, 426–429
    overview, 425–426
    synchronous, 433–434
    transient analysis, 429–433
  buck converters, 215
  controllers, 30, 191–192
  discontinuous 12 V, 2 A buck-boost converter
    ac analysis, 479–483
    overview, 476–479
    transient analysis, 483–486
  forward converters, 749, 825, 849
  generic controllers, 378–382
  modulators, 29
  overview, 29–30
  PWM switch model. *See also* discontinuous conduction mode (DCM)
    borderline conduction mode, 194–202
    in CCM, 153–158
    current-mode instabilities, 146–151
    in DCM, 161–163
    deriving duty cycles d_1 and d_2, 163–165
    overview, 145–146
    parasitic elements effects, 182–186
    preventing instabilities, 151–153
    upgrading model, 158–161
  small-signal Ridley models, 213–214
  stabilizing buck converters with k factor, 280–286
current-to-voltage converters, 312
cutoff frequency, 82–85

cycle-by-cycle circuits, 683–684
cycle-by-cycle models, 142–144, 171–172, 842

D, d, or d_1. *See* duty cycles (D, d, or d_1)
D_0 (static duty cycle), 96, 110, 126, 159, 161
d_2 off duty cycle, 130–132, 163–166, 183
d_3 off duty cycle, 130–131
damping, 76–79, 160, 281–282, 381, 422–424, 455, 462, 546
damping factor (ζ), 68–70
Darlington output, 327
dc
  bias, 792
  block capacitors, 306, 330
  gain, 8–10, 278–279, 322, 412, 540
  output, 322, 564
  output resistance, 13, 61
  points, 123, 168, 245, 337, 463
  stacks, 713, 715
  sweep, 314, 463
  terms, 107–108
  transformers, 102–104
DCM. *See* discontinuous conduction mode (DCM)
dead time (DT), 377–379, 434
dead time generators, 387–390
dedicated controllers, 806
delay lines, 359–360
demagnetization, 34, 61, 803, 850
demonstration versions, simulation software programs, 3, 320, 868
differential mode, 687
digital signal processors (DSP), 3–4
ΔI_L (peak to peak ripple), 21–24, 40, 42, 54–55, 66, 148, 460, 564, 688, 842
diodes
  for 12-V, 250-W forward converters, 838–839
  blocking configuration, 695
  blocking point, 46
  boost converters, 451–452, 489–490, 564, 570–573
  buck converters, 417–419, 433
  buck-boost converters, 478
  for clamping networks, 609–610
  emission coefficient, 344
  leading edge blanking, 359
  rectifier bridges, 494–499
  transformers, 367
  VCO (Voltage-Controlled Oscillators), 374
direct duty cycle control, 28
discontinuity, 21, 51, 348, 533, 535, 562, 571
discontinuous conduction mode (DCM)
  boost converters, 47–51, 441, 446, 449, 453, 457, 465, 487–490, 550–558, 561

876　INDEX

discontinuous conduction mode (DCM) (*Cont.*):
　buck converters, 34–39, 168–170
　buck-boost converters, 59–64, 476–486
　current-mode designs, 322
　flyback converters, 314, 322, 398, 427,
　　597–599, 628–638
　operation, 25, 36–37, 43, 47, 49–50, 56
　overview, 21–25
　PWM switch model
　　buck instability, 172
　　building, 165–168
　　mode transition, 143–145
　　overview, 129–131
　　testing, 168–172
discrete Fourier analysis, 517
discrete silicon controlled rectifier, 663–664
discrete values, 95–97
drain
　clamping excursion, 591–597
　voltage at clamp level, 783
　without clamping action, 588–591
drain-source capacitors, 724
drain-source waveforms, 727
drive capability, 662–663
DSS (dynamic self-supply), 848
DT (dead time), 377–379, 434
dual transformers, 364
dual-output forward converters, 809
dummy sources, 351, 366
duty cycles (D, d, or d_1)
　boost converters, 441, 452, 489, 566, 575
　buck converters, 33, 427
　buck-boost converters, 467, 476, 479
　controlling, 789
　diode burdens, 797
　extremes, 408
　factory, 27–30
　generation, 382–384
　limit, 842
　modulation, 96
　overview, 18–19
　PFC, 539, 543, 545
　shunt regulators, 313–314, 317
dynamic self-supply (DSS), 848
dynamic variables, 104

Early effect, 10
Ebers–Moll models, 112–113
eddy currents, 367
EDF (Electricite de France), 506
efficiency, 4–7, 419, 432–433, 451, 460, 506,
　574–576, 604, 616, 842
Electricite de France (EDF), 506

electromagnetic force (emf), 396
electromagnetic interference (EMI) filters, 67,
　422, 499, 503, 506, 571
emf (electromagnetic force), 396
EMI (electromagnetic interference) filters, 67,
　422, 499, 503, 506, 571
energy
　battery, 452
　damping capacitor voltage, 608–609
　difference between peaks and valleys, 510
　leakage, and two-switch flyback, 614–616
　recycled, in active clamps, 616–618
　storage, 27, 42–44, 56–57, 518–520, 563,
　　584–585, 622
　supplied by capacitor to converter, 494–495
equivalent capacitors, 89–90, 305, 352
equivalent inductance, 13
equivalent inductor model, 398–400
equivalent series resistors (ESRs)
　boost converters, 54–55
　buck converters, 41
　buck-boost converters, 65
　effect, 65–66
　independent pole-zero placement, 273
　nonisolated boost converters, 444–449,
　　453–454
　nonisolated buck converters, 410–413, 426–427
　nonisolated buck-boost converters, 469–471,
　　478, 482–483
　output capacitor, 282
error amplifiers, 5–6, 110, 335, 345–348, 382,
　481, 539, 566–567
error voltage, 27–29, 145, 284, 382–383, 483,
　530, 540, 542, 545–546
ESRs. *See* equivalent series resistors (ESRs)
η (converter efficiency), 491
Excel spreadsheets, 323, 326

fan-out, 348–349
Faraday's law, 23, 396, 733, 860, 867
fast Fourier transform (FFT), 84–85, 381, 421,
　460–461, 464, 517–518, 525
fast lane, 288–294, 685
FB (feedback) pins, 288, 300–303, 312–313, 571,
　574, 654–655, 667, 685, 720, 821
f_c (crossover frequency), 92, 249–250, 267,
　273–275, 719
feedback (FB) pins, 288, 300–303, 312–313, 571,
　574, 654–655, 667, 685, 720, 821
feedback and control loops. *See also* amplifiers;
　TL431
　automated pole-zero placement, 323–326
　crossover frequency, 249–250

INDEX

feedback and control loops (*Cont.*):
  design, 2, 6
  k factor stabilization, 263–286
  observation points, 243–247
  optocouplers, 304–312
  overview, 241–243
  phase margin and transient response, 248–249
  shaping compensation loop, 250–262
  shunt regulators, 312–316
  small-signal responses with PSIM and SIMPLIS, 316–321
  stability criteria, 247–248
  virtual ground in, 335–339
feedback voltage, 151–152, 181, 187, 191, 198, 436, 667–669, 848
FFT (fast Fourier transform), 84–85, 381, 421, 460–461, 464, 517–518, 525
FHA (first harmonic approximation), 79–80
field probes, 368
filtering circuits, 66
filters. *See* input filtering; *LC* filters; *RLC* filters
first harmonic approximation (FHA), 79–80
first-order system, 187, 285–286, 628, 672, 719
fixed time step simulators, 95
fixed-frequency power factor correction
  average current-mode
    overview, 535–540
    shaping current, 540–543
  discontinuous boost, 550–555
  peak current-mode
    average modeling of, 546–549
    compensating, 544–546
    overview, 543–544
F_{line} (mains frequency), 570
flip-flops, 363
floating buck converters, 434–439, 739
flux (φ), 393
flux and charge balance, 25–26
flux circulation, 744
flux density (B), 368, 393–397, 732–733, 744–745, 799–804, 859–860, 865–867
flux excursion, 744
flux walk-away situation, 745
flyback, 93, 215, 234–235
flyback converters
  20-W, single-output power supply, 670–687
  35-W, multioutput power supply, 706–725
  90-W, single-output power supply, 687–706
  active clamp, 616–627
  clamping drain excursion, 591–597
  component constraints, 725–726
  considerations, 642–665
  CoPEC models, 218

flyback converters (*Cont.*):
  current spikes, 359
  DCM, 597–599
  designing clamping networks, 599–614
    curing leakage ringing, 605–609
    diode selection, 609–610
    overview, 599–601
    RCD configuration, 601–603
    selecting k_c, 604–605
    TVS clamp, 612–614
    voltage variations, 610–612
  drain signal without clamping, 588–591
  optocouplers, 309
  overview, 207, 579–583
  power factor correction, 555–559
  remanent induction flux level, 395
  small-signal response, 628–641
  standby power, 665–670
  two-switch, 614–616
  waveforms, 583–588
flyback current mode, 680
follow-boost technique, 574–575
forward converters
  auxiliary winding, 848
  component constraints, 849
  core reset
    active clamp, 778–795
    overview, 742–745
    RCD clamp, 767–778
    resonant demagnetization, 762–767
    third winding, 746–755
    two-switch configuration, 756–761
  dual-output, 808–809
  generic voltage-mode controllers, 384–386
  multioutput, 799–817
  overview, 205–206, 739–745
  single-output 12-V, 250-W
    diode selection, 838–839
    installing snubbers, 835–838
    MOSFET selection, 833–834
    overview, 828–833
    short-circuit protection, 846–848
    small-signal analysis, 839–841
    transient results, 841–846
  small-signal response
    current mode, 821–825
    multioutput, 825–828
    voltage mode, 817–821
  synchronous rectification, 796–798
Fourier. *See* fast Fourier transform (FFT)
fractional layers, 734, 861
freewheeling, 21, 31, 187, 396, 615–616, 739, 752–753, 765, 769, 796–797

878 INDEX

frequency. *See* mains frequency (F_{line}); switching frequency (F_{sw})
frequency evaluation, 80–82
frequency foldback, 670
frequency-dependent gain, 9–10
front-end acquisition board, 3
front-end filters, 2, 84, 422
front-end section. *See* off-line converters
F_{sw} (switching frequency), 97, 374, 532–535, 562, 633, 670, 710–711
full bridge topology, 94, 377–378, 388, 576
full-wave rectification, 10, 532

gain margins (GMs), 240, 247–248, 271–272, 310, 312, 322, 446, 458, 471
gapping, 725, 767, 861
gate charge, 434, 645, 651, 690, 692, 731, 834, 851
gate-source zeners, 731
generators, 372–377, 387
generic controllers
  current-mode, 378–382
  overview, 377–378
  start-up of, 642–644
  voltage-mode model, 382–386
generic simulation models, 341–343, 387–388
generic switch inductor models (GSIM), 221–225
GMIN, 286, 390
GMs (gain margins), 240, 247–248, 271–272, 310, 312, 322, 446, 458, 471
Graetz bridge, 491
$G(s)$ compensator frequency response, 9–10, 238, 247, 251, 278
GSIM (generic switch inductor models), 221–225
guard bands, 730

half-bridge drivers, 760–761, 851–854
half-bridge topology, 94, 228, 387–389
half-wave connections, 646–648
harmonics, 20, 79–85, 513–518, 520, 524–526
heat sink, 419, 566, 674, 677–679, 690–695, 713–715, 731, 838–839
high-pass filters, 251–252, 833
high-voltage controllers, 686
hold-up time, 501–502
HP4195 network analyzer, 322
hysteresis, 355–359, 361–362, 366, 369–373, 394–395
hysteretic, 437–439, 549–552, 650, 667–669

IF-THEN-ELSE functions, 341–342
IMAX, 382–386, 413, 416, 449–450, 474

impedance. *See also* input impedance
  line, 502–506
  output, 329
  overlapping, 72–76
  sweep of equivalent capacitor models, 89–90
incremental resistance, 67, 76
indirect energy transfer converters, 58–59
inductance, 396–397
inductive current
  average, 688–689
  in borderline mode operation, 708
  calculating attenuation, 79
  at steady states, 146
inductors
  downslope, 281
  flux, 26
  ohmic losses, 109
  operating modes, 21–25
  series resistance, 31, 129
  voltage excursion, 25, 37
  voltage-adjustable passive elements, 353–354
infinite magnetizing inductance, 826
in-line equations, 71–72, 341–343
input current (I_{in}), 79, 487–489, 709
input filtering
  calculating required attenuation, 79–80
  comprehensive representation, 70–71
  creating closed-loop current source with SPICE, 71–72
  damping filters, 76–79
  fundamental frequency evaluation, 80–82
  overlapping impedances, 72–76
  overview, 66–67
  RLC filter, 67–69
  selecting right cutoff frequency, 82–85
input impedance, 11, 67, 69–70, 72–76, 345, 422–423, 462–463, 521
input load responses, 275
input power factor, 498–499
input rejection, 124, 285
input ripple, 10, 17, 58–59, 82–83, 421–424
in-rush current, 501, 506–508, 539, 565, 573
instabilities
  current-mode, 146–151
  generic current-mode controllers with, 381–382
  prevention of, 151–153
  PWM current-mode DCM model buck, 172
installing snubbers, 835–838
instantaneous inductor current, 29, 32, 46
intersections, impedance, 76
Intusoft, 276, 365, 371, 386, 419, 503. *See also* IsSpice

INDEX

invariant internal architecture, 113–114
invariant operating slopes, 155
ISEC, 368
isolated Ćuk converters, 212
isolated SEPICs, 210–211

Jiles-Atherton model, 366
junction temperature (T_j), 694

k factor
  automated pole-zero placement, 323
  boost converters, 457
  buck converters, 427
  buck-boost converters, 482
  conditional stability, 270–272
  crossing over at selected frequency, 273–275
  current-mode buck converter, 280–286
  current-mode model and transient steps, 286
  independent pole-zero placement, 272–273
  versus manual pole-zero placement, 275–279
  optocouplers, 309, 312
  overview, 263
  small-signal responses, 317
  TL431, 291, 297
  type 1 derivation, 264
  type 2 derivation, 264–266
  type 3 derivation, 266–267
  voltage-mode buck converter, 267–270
k_c, 604–605
Kirchhoff's law, 288, 303–304

Laplace equation, 68–69, 308
large gain variations, 278
large-signal modeling
  averaged models, 224
  buck converters, 102
  CC-PWM switches, 156
  ESR-related parasitic effects, 178
  operations, 107
  PWM switch model, 118–121
  SSA, 105–106
LC filters
  boost converters, 444–445, 461
  buck CCM output ripple voltage calculation, 40–41
  buck-boost converters, 474–475
  fast lane, 290
  flyback converters, 684–685
  forward converter small-signal response, 817–818
  harmonic reduction, 524
  input ripple, 421
  oscillator creation, 71

LC filters (*Cont.*):
  output ripple, 419
  overview, 13
  passive PFC, 521
  and power conversion with switches, 19–21
  PWM switch model, 158–159, 175
  voltage-mode buck converters, 408
LDO (low-dropout) regulators, 7, 13
leading edge blanking (LEB), 171, 359–360, 378, 438
leading edge modulation, 799, 805, 808
leakage inductance (L_{leak}), 365, 372, 386, 400, 405, 726, 786, 813
leakage ringing, 605–609
LEB (leading edge blanking), 171, 359–360, 378, 438
LED current, 288, 300–301, 304, 310
LED resistors, 296–305, 310, 314
left half-plane zero (LHPZ), 127, 252
Lenz formula, 14
Lenz, Heinrich Friedrich Emil, 353
Lenz' law, 27, 353, 396–397
LHPZ (left half-plane zero), 127, 252
light load variation, 709
line impedance, 502–506
line ripple rejection, 348
linear equations, 106
linear regulators, 5–10, 14–18, 67
linearization, 106–108
lithium ion batteries, current-mode nonisolated boost from
  ac analysis, 454–458
  input filter, 460–464
  overview, 452–454
  transient analysis, 459–460
LLC resonant converters, 94, 320, 374, 376
L_{leak} (leakage inductance), 365, 372, 386, 400, 405, 726, 786, 813
load resistors, 13, 37, 186, 214, 225, 800
load step reaction, 278
logic gates, 168, 362–364
loop compensation, 310, 790, 806
loop gain, 9, 13, 273–274, 289, 312, 322, 336–337, 381, 542
loops. *See* compensation loops; feedback and control loops
losses, 417–418, 449–452, 506, 565–566, 597, 666–667, 674–675, 711. *See also* ohmic losses
lossy boost converters, 222–223
lossy model, 183–186, 203, 455
low standby power conversion techniques, 727
low-cost floating buck converters, 434–439

low-dropout (LDO) regulators, 7, 13
low-impedance switching pattern, 18
low-pass filters, 250–251, 313
LRC meters, 398
luminous flux intensity, 304
lump capacitors (C_{lump}), 588–589, 596–597, 620–621, 675, 762–765, 769, 784–785

magnetic amplifiers, 799–804
magnetic designs
  field definition, 393
  inductance, 396–397
  Maxwell's equations, 396
  overview, 392–393
  permeability, 393–395
  saturation, 397–398
magnetic flux, 21, 393
Magnetics Ferrite Core catalog, 732
magnetization cycle, 21
magnetizing current, 364, 398, 741, 744
magnetizing inductance, 103, 309, 364, 368, 397, 401–402
magnetomotive force (mmf), 393
manual pole-zero placement
  type 2 amplifiers, 332–335
  versus k factor stabilization, 275–279
Mathcad program, 127, 237, 555
matrix algebra, 86–89
maximum output current, 348
Maxwell, James Clerck, 396
mean magnetic path length (MPL), 393
mesh/loop analysis, 126
midband gain, 291–292, 303, 310, 333–334
mmf (magnetomotive force), 393
modulator gain, 138–142
mold compound, 694, 731
monolithic circuits, 413, 608, 730, 846
MOSFETs
  CCM PFCs, 565
  current-mode boost, 454, 460
  cycle-by-cycle simulation, 571
  dead time, 387, 390
  and diode losses, 449–451
  power switch waveforms, 417
  selecting, 833–834
  series-pass elements, 5
  synchronous, 433–434
  voltage drops, 180–182
MPL (mean magnetic path length), 393
μ_r, μ_i, μ_o (permeability), 366, 368–369, 393–395, 397
MUL (multiplier input), 562
multioutput converters, 850

multioutput flyback converters, 720–721
multioutput forward converters
  coupled inductors, 806–817
  magnetic amplifiers, 799–804
  synchronous postregulation, 804–806
multioutput transformers, 372
multiple winding-based circuits, 810
multiplier input (MUL), 562

NCP1200 controllers, 288, 291, 309–310, 312, 436–437, 645, 650, 655, 680, 699
NCP1216 controllers, 306, 650, 655, 685–686, 846, 848
negative feedback, 247
negative impedance loads, 72–73
negative resistor loading, 71
negative roots, 236
network analyzers, 2
no-load conditions, 16
nonisolated Ćuk converters, 211
nonisolated boost converters, 441–465
  component constraints for, 465
  current-mode 5 V, 1 from li-ion battery
    ac analysis, 454–458
    input filter, 460–464
    overview, 452–454
    transient analysis, 459–460
  in discontinuous mode, 487–490
  voltage-mode 48 V, 2 from car battery
    ac analysis, 444–449
    overview, 441–444
    transient analysis, 449–452
nonisolated buck converters, 407–441
  12 V, 4 voltage-mode from 28 V source
    ac analysis, 410–413
    diode, 418–419
    input ripple, 421–424
    output ripple and transient response, 419–420
    overview, 407–409
    power switch, 417–418
    transient analysis, 413–417
  5 V, 10 A current-mode from car battery
    ac analysis, 426–429
    overview, 425–426
    synchronous, 433–434
    transient analysis, 429–433
  component constraints for, 439–441
  low-cost floating, 434–439
nonisolated buck-boost converters, 465–486
  component constraints for, 486
  discontinuous current-mode 12 V, 2 A
    ac analysis, 479–483

INDEX

nonisolated buck-boost converters, discontinuous current-mode 12 V, 2 A (*Cont.*):
  overview, 476–479
  transient analysis, 483–486
  voltage-mode 12 V, 2 powered from car battery
    ac analysis, 468–474
    overview, 465–468
    transient analysis, 474–475
nonisolated SEPICs, 209–210, 219
nonisolated switch-mode power converters, 407
nonisolated topologies, 2
nonlinear controlled source, 341
nonlinear models, 102
non-pulsating output current, 33–34, 46–47
nonsinusoidal signals, 512–514
null average inductor voltage, 32

observation points, feedback and control loop, 243–247
off-line converters. *See also* power factor correction (PFC)
  overview, 491
  PFC
    designing BCM boost, 559–575
    fixed-frequency DCM boost, 550–555
    flyback converter, 555–559
    harmonic limits, 517–518
    hysteretic power factor correction, 549–550
    need for, 515–517
    need for energy storage, 518–520
    nonsinusoidal signals, 512–514
    overview, 510–512
    passive, 520–527
    total harmonic distortion, 514–515
  rectifier bridges
    100-W, operated on universal mains, 499–501
    capacitor selection, 493–495
    current in diodes, 498
    diode conduction time, 495–496
    hold-up time, 501–502
    input power factor, 498–499
    in-rush current, 506–508
    overview, 491–493
    rms current in capacitor, 496–498
    voltage doubler, 508–510
    waveforms and line impedance, 502–506
off-line forward converters, 817
Ohm, Georg Simon, 350
ohmic losses, 20, 33, 46–47, 59, 68–70, 109, 121–123, 180–183, 398, 460
op amps (operational amplifiers), 5, 286, 292–293, 317–318, 335–337, 341, 343–348

opening loops, 243–244
open-loop audio, 10, 67
open-loop Bode plot generation, 267, 323
open-loop boost converters, 84
open-loop buck converters, 111
open-loop gain, 298–299, 314–315, 330, 336–337, 341, 344, 346, 423, 481
open-loop small-signal response, 721
operating points, 106–108, 135–136
operating signals, 187–188
operational amplifiers (op amps), 5, 286, 292–293, 317–318, 335–337, 341, 343–348
operational transconductance amplifiers (OTAs), 323, 327, 339, 343–344, 530, 542, 566–567
options, 390–391
optocouplers, 2, 287–288, 291–292, 300–312, 322, 681
opto-isolated feedback system, 287
origin pole, 256, 288, 290, 341, 540
origin pole plus two coincident zero-pole pairs amplifier. *See* type 3 amplifiers
origin pole plus zero amplifier. *See* type 2a amplifiers
OTA (operational transconductance amplifiers), 323, 327, 339, 343–344, 530, 542, 566–567
output impedance, 9, 14, 72–76, 78, 86, 299, 327, 329, 422–423, 462, 503
output ripple, 39, 52, 58–59, 419–420, 489–490, 559, 684
output ripple voltage calculations
  boost converters, 51–54
  buck converters, 39–41
  buck-boost converters, 64–65
output transient steps, 278
output variables, 86–87
overdamping, 69, 160
overlapping impedances, 72–76
overshoots, 14–15, 202, 249, 262, 286, 300, 302, 378, 600, 611–612, 654
OVP (overvoltage protection), 564–565, 663–665

parallel damping, 76
parameters keyword, 275–276
parasitic capacitance, 33, 34, 47, 59–60
parasitic elements
  buck converters, 422
  buck-boost converters, 462
  capacitors, 89–90
  flyback waveforms with, 586–588
  flyback waveforms without, 583–586
  PWM switch model
    convergence issues with CM model, 186
    ohmic losses and voltage drops, 180–183

882 INDEX

parasitic elements, PWM switch model (*Cont.*):
   overview, 175–179
   testing lossy model in current mode, 183–186
   variable resistor, 179–180
passive poles, 250–251, 254
passive power factor correction
   improving harmonic content, 524–526
   overview, 520–524
   valley-fill passive corrector, 526–527
peak current-mode power factor correction
   average modeling of, 546–549
   compensating, 544–546
   overview, 543–544
peak inverse voltage (PIV), 838
peak to peak ripple (ΔI_L), 21–24, 40, 42, 54–55, 66, 148, 460, 564, 688, 842
periodic modulation, 148–149
permeability (μ_r, μ_i, μ_o), 366, 368–369, 393–395, 397
perturbation, 147–148, 285
PFC. *See* power factor correction (PFC)
phase boosts, 250
phase margin (PM), 247–249, 267, 275, 310, 320, 322–323, 457–458, 471, 482, 535, 567
phase rotation, 243, 323
pin count, 363–364
PIV (peak inverse voltage), 838
PM (phase margin), 247–249, 267, 275, 310, 320, 322–323, 457–458, 471, 482, 535, 567
point-of-load (POL) regulators, 407
poles. *See also* amplifiers
   automated placement, 323–326
   compensation loop passive, 250–251
   independent placement and k factor, 272–273
   manual placement of, 275–279, 332–335
   optocouplers, 306–312
   overview, 235–240
positive feedback oscillators, 247
postregulation technique, 799
power conversion. *See also* linear regulators
   boost converters
      CCM output ripple voltage calculations, 51–54
      CCM waveforms, 44–47
      DCM waveforms, 47–50
      ESR, 54
      off-time event, 44
      on-time event, 43–44
      overview, 42–43
      ripple value, 54–55
      transition point DCM-CCM, 50–51

power conversion (*Cont.*):
   buck converters
      CCM output ripple voltage calculation, 39–41
      CCM waveforms, 31–34
      DCM waveforms, 34–37
      ESR, 41
      off-time events, 31
      on-time events, 30–31
      overview, 30
      ripple value, 41–42
      transition point DCM-CCM, 37–39
   buck-boost converters
      CCM output ripple voltage calculation, 64–65
      CCM waveforms, 57–59
      DCM waveforms, 59–62
      ESR, 65
      off-time event, 56–57
      on-time event, 56
      overview, 55–56
      ripple value, 65–66
      transition point DCM-CCM, 63–64
   capacitor equivalent model, 89–93
   duty cycle factory, 27–30
   input filtering
      calculating required attenuation, 79–80
      comprehensive representation, 70–71
      creating closed-loop current source with SPICE, 71–72
      damping filters, 76–79
      fundamental frequency evaluation, 80–82
      overlapping impedances, 72–76
      overview, 66–67
      RLC filter, 67–69
      selecting right cutoff frequency, 82–85
   power supply classification by topologies, 93–94
   with resistors
      closed-loop system, 5–7
      generic linear regulators, 14–17
      overview, 3
      practical working example, 10–14
      resistive dividers, 3–5
   RLC transfer equations, 86–89
   with switches
      charge and flux balance, 25–26
      energy storage, 27
      filters, 19–21
      inductor operating modes, 21–25
      overview, 18–19

INDEX

power factor correction (PFC). *See also* active power factor correction
   designing BCM boost PFC
      average simulations, 567–570
      cycle-by-cycle simulation, 571–573
      follow-boost technique, 574–575
      overview, 559–567
      reducing simulation time, 570–571
   fixed-frequency DCM boost, 550–555
   flyback converter, 555–559
   harmonic limits, 517–518
   hysteretic power factor correction, 549–550
   need for, 515–517
   need for energy storage, 518–520
   nonsinusoidal signals, 512–514
   overview, 510–512
   passive
      improving harmonic content, 524–526
      overview, 520–524
      valley-fill passive corrector, 526–527
   total harmonic distortion, 514–515
power libraries, 2
power supply
   classification, 93–94
   test fixtures, 137–139
power switches (t_{on} and t_{off}), 417–418
preregulators, 527
primary inductance, 309–310, 366, 373, 400
propagation delay, 378, 438, 655–660
properness, 235
proportional plus pole amplifier. *See* type 2b amplifiers
proprietary syntax, 2
PSIM, *3, 95, 316–323, 791–792, 868*
P_{SW} (switching losses), 711, 837
pull-up resistors, 291, 300, 303, 305, 308, 312, 358
pulsating input current, 33–34, 46, 59, 66
pulsating output current, 59
pulse width modulator (PWM) switch model. *See also* current-mode (CM)
   in borderline conduction
      current-mode case, 194–198
      overview, 187
      testing current-mode BCM model, 198–202
      testing voltage-mode BCM model, 191–194
      voltage-mode case, 187–191
   circuits
      boost, 208
      buck, 203
      buck-boost, 206
      flyback, 207
      forward, 205–206

pulse width modulator (PWM) switch model, circuits (*Cont.*):
      isolated Ćuk converters, 212
      isolated SEPIC, 210–211
      nonisolated Ćuk converters, 211
      nonisolated SEPIC, 209–210
      overview, 202–203
      tapped boost, 208–209
      tapped buck, 204
   parasitic elements effects
      convergence issues with CM model, 186
      ohmic losses and voltage drops, 180–183
      overview, 175–179
      testing lossy model in current mode, 183–186
      variable resistor, 179–180
   PWMCM model, 280, 286
   PWMCM2 model, 380, 459
   PWMCMS model, 434
   voltage-mode case
      bipolar transistors, 112–113
      clamping sources, 132–134
      complex representation, 121–123
      deriving d_2 variable, 132
      discontinuous mode model, 129–131
      encapsulating model, 134–138
      invariant internal architecture, 113–114
      large-signal simulations, 118–121
      mode transition, 143–145
      overview, 111–112
      PWM modulator gain, 138–142
      small-signal model, 123–128
      terminal currents, 116
      terminal voltages, 117
      testing model, 142–143
      transformer representation, 117–118
      waveform averaging, 114–116
pulse width modulators (PWM), 19, 28–30, 142, 377, 379, 468
pulsed current sources, 79
pulsewise linear (PWL) statement, 12, 419
push-pull topology, 94
PWL statement. *See* pulsewise linear (PWL) statement
PWM. *See* pulse width modulators (PWM)
PWM switch model. *See* pulse width modulator (PWM) switch model
PWMBCMCM circuits, 203
PWMBCMVM circuits, 202
PWMCCMVM circuits, 202
PWMCM circuits, 202
PWMCM model, 280, 286
PWMCM_L circuits, 203

PWMCM2 model, 380, 459
PWMCMS generic controllers, 793
PWMCMS model, 434
PWMDCMCM circuits, 202
PWMDCMVM circuits, 202
PWMVM circuits, 202
PWMVM_L circuits, 203

QR (quasi-resonant) topologies, 187–188
QR mode (quasi-square wave resonance), 706–707, 723
quality coefficients (Q), 76
quasi square wave topologies, 187
quasi-resonant (QR) topologies, 187–188
quasi-square wave resonance (QR mode), 706–707, 723

ramp compensation
 buck DCM and CCM, 172–175
 CCM boost, 459
 CC-PWM switches, 159–161
 current-mode buck converters, 215–216, 281–282, 381–383, 427, 429, 431–432
 current-mode PWM switch model, 151–153
 flyback converters, 700–701
 peak current-mode control, 529, 543–549
RC filters, 20, 308
RC networks, 292, 296, 343, 355, 359, 361, 378, 387, 391, 449, 463
RCD clamps, 601–603, 767–778
RCD snubbers, 835
rectifier bridges
 100-W, 499–501
 capacitor selection, 493–495
 current in diodes, 498
 diode conduction time, 495–496
 hold-up time, 501–502
 input power factor, 498–499
 in-rush current, 506–508
 overview, 491–493
 rms current in capacitor, 496–498
 voltage doubler, 508–510
 waveforms and line impedance, 502–506
regulation loops, 299, 803
regulator output, 13
relative tolerance (RELTOL), 286, 390
remanent induction flux value (B_r), 395, 801–802
reset ratios, 748
reset volt-seconds, 803
reset winding technique, 755
resistive converters, 5
resistive dividers, 3–7, 303–304, 351, 539, 554

resistors. *See also* equivalent series resistors (ESRs); power conversion; sense resistors
 closed-loop system, 5–7
 controlled, 179, 349
 emitter, 308
 LED, 296–305, 310, 314
 linear regulators
  building simple generic, 14–17
  conclusion on, 17–18
  deriving useful equations with, 7–10
 load, 13, 37, 186, 214, 225, 800
 low collector, 308
 overview, 3
 practical working example, 10–14
 pull-up, 291, 300, 303, 305, 308, 312, 358
 resistive dividers, 3–5
 series, 30–31
 start-up, 644–649
 upper, 304
 variable, 179–180
 voltage-adjustable passive elements, 350–351
resonance, 13–14, 35, 48, 60, 75–76, 515
resonant demagnetization, 762–767
resonant frequency, 277, 317, 325, 374, 409, 445, 487
resonant reset, 768, 798
resonant topologies, 377
resonating capacitors, 158, 186, 286
resonating tank, 34, 47, 68, 751
reverse recovery time, diode (t_{rr}), 45, 58, 417, 691, 696, 794
RF emitters, 452
RHPZ (right half-plane zero), 127, 253–255, 309–310, 412, 444–447, 454, 468
Ridley models, 168–169, 174–175, 213–216, 344, 455–456, 577
right half-plane zero (RHPZ), 127, 253–255, 309–310, 412, 444–447, 454, 468
ringing-choke converters, 320–321
ripple. *See also* input ripple; output ripple
 amplitude, 789
 BCM boost PFC, 563–564
 boost converters, 51–55, 442–444, 449, 452–453
 borderline PFC, 530–533
 buck converters, 39–42
 buck-boost converters, 64–66, 466–468, 474–475, 477
 current, 22, 25
 current-mode buck, 425–426
 floating buck converter, 436–438
 PWM switch model, 175–176
 steering, 816
 voltage-mode buck, 407–409

RLC filters, 67–73, 75–77, 79, 124, 422–423, 464
RLC meters, 399, 422–423
RLC networks, 68–69
RLC transfer equations, 86–89
root-locus analysis, 238–240
Routh–Hurwitz criterion, 100
RS latches, 363

sag, 503–506
sampling, 115, 213, 381
saturable core models, 366–372
saturable inductors, 799
saturation flux density (B_{sat}), 368–371, 395, 397, 800
saturation transformers, 366–367, 397–398
sawtooth, 27–28, 138, 140, 312–313, 384, 538–539, 552, 557
Schottky diodes, 33, 417, 434, 451
secondary diode currents, 772
secondary inductance, 309–310
secondary *LC* filters, 290
secondary side current, 660–661
secondary side diode anode signal, 729
secondary windings, 714
self-oscillating circuits, 851
sense resistors, 198, 454, 479, 481, 483, 540–542, 561
SEPICs (single-ended primary inductance converters), 209–211, 528, 559
series resistors, 30–31
series voltage sources, 181
series-pass elements, 5–6
short-circuit protection, 653–654, 846–848
shunt regulators, 5, 287–288, 301, 312–316, 322, 326, 425–427
signal pulsing, 148–149
signatures
  averaged models, 535–536
  boost converter, 43
  buck, 421
  buck-boost converter, 56, 63
  CCM boost, 460–462
  converter, 79–82
  EMI, 611, 663
  PFC, 564
  TVS, 614
  valley-fill input, 527
silver boxes, 739
simple generic low-power regulators, 14–15
simple resistive dividers, 9
simplified switch-mode converters, 242
SIMPLIS software, 316–321, 791

SIMs (switched inductor models), 220
simulated ripple, 52–53
Simulation Program with Integrated Circuit Emphasis. *See* SPICE
simulation techniques
  astable generators, 372–377
  BCM boost PFC
    average, 567–570
    cycle-by-cycle, 571–573
    reducing time of, 570–571
  comparator with hysteresis, 361–362
  convergence options, 388–391
  dead time generation, 387
  feeding transformer models with physical values
    equivalent inductor model, 398–400
    three-winding T model, 401–405
    two-winding T model, 400–401
  generic controllers
    current-mode, 378–382
    overview, 377–378
    voltage-mode, 382–386
  generic models, 341–343, 387–388
  hysteresis switches, 355–357
  leading edge blanking, 359–360
  logic gates, 362–364
  operational amplifiers, 343–348
  sources with given fan-out, 348–349
  terminology of magnetic designs
    field definition, 393
    inductance, 396–397
    Maxwell's equations, 396
    overview, 392–393
    permeability, 393–395
    saturation, 397–398
  transformers
    multioutput, 372
    overview, 364–365
    simple saturable core model, 366–372
  undervoltage lockout block, 358–359
  voltage-adjustable passive elements, 349–354
single zero networks, 252
singled-ended configuration. *See* single-switch configuration
single-ended primary inductance converters (SEPICs), 209–211, 528, 559
single-output flyback converters, 207
single-pole double-throw configuration, 114
single-pole *RC* networks, 251
single-switch configuration, 756
single-switch forward converters, 739, 744, 750, 755
single-switch forward topology, 93

sinusoidal voltage, 491–492, 511–513, 528–532, 539–540, 544, 549–550, 562
sinusoidal waveform, 47, 141
skip-cycle operation, 28, 846
slope, 21, 38, 51, 63, 95, 154–155, 273–274, 291, 348, 394, 574–575
slope compensation, 381–382
slow lane, 288
small-signal analysis, 839–841
small-signal modeling. *See also* buck converters; pulse width modulator (PWM) switch model
  averaged models
    Ben-Yaakov models, 220–223
    CoPEC models, 216–220
    Ridley models, 213–216
  converter transfer equations
    boost, 229–231
    buck, 226–228
    buck-boost, 231–235
    overview, 225–226
  overview, 95–98
  poles, zeros, and complex plane, 235–240
  state space averaging
    dc transformer, 102–104
    large-signal simulations, 105–106
    overview, 98–100
small-signal PWM switch, 123–128
small-signal response
  of flyback topology
    CCM, 635–636, 638–641
    DCM, 628–635, 636–638
  of forward converters
    current-mode, 821–825
    multioutput, 825–828
    voltage mode, 817–821
  with PSIM and SIMPLIS, 316–321
small-signal switch-mode converters, 243
SMPS (switch-mode power supply), 1–2, 382, 422
S_n, 226, 228, 230–231, 233–234, 637–640, 699, 825
snubbers, 529, 835–838
snubbing capacitors, 836
spectral analysis, 84
SPICE (Simulation Program with Integrated Circuit Emphasis)
  convergence options, 388–391
  creating simple closed-loop current source, 71–72
  current-mode model, 286
  cycle-by-cycle models, 316
  Dc transformers, 103–104
  diode selection, 418–419

SPICE (Simulation Program with Integrated Circuit Emphasis) (*Cont.*):
  FFT capabilities, 517
  in-line equations, 341–342
  model of shunt regulator, 313–314
  optocouplers, 307
  series resistance, 403
  in time domain, 95–96
  TL431 model, 326–332
  transformer handling, 364
  voltage-adjustable passive elements, 349–350
spike amplitude, 33
SSA. *See* state-space averaging (SSA)
stability, feedback and control loop, 247–248
stabilization ramps, 153
standby power
  frequency foldback, 670
  origins of losses, 666–667
  overview, 665–666
  skipping unwanted cycles
    overview, 667–669
    with UC384X, 669–670
start-up resistors, 644–646, 648–649
start-up sequences, 151, 183, 356, 379, 438–439, 567, 612–613, 651, 806–807
state coefficient matrices (A), 87, 99
state equations, 98–99
state variables, 86–87, 98
state-space averaging (SSA)
  for buck converters
    linearization, 106–108
    overview, 100–102
    small-signal model, 108–111
  dc transformer, 102–104
  large-signal simulations, 105–106
  overview, 98–100
static duty cycle (D_0), 110, 126, 159
static errors, 8, 14, 242
static output impedance, 12
static output voltage definition, 8
status output impedance, 348
steady state, 24, 146, 150, 304, 378, 386, 439, 456, 500, 530–531
steady-state duty cycle (D_0), 96, 161
step line test, 283
step loading, 269, 271, 275, 282–283, 311, 322, 448, 569
stepping input voltage, 201
storage, energy. *See* energy
stress, 325, 479, 505, 570
subcircuits, 14, 341, 348–349, 351, 353–354, 359, 363–366, 387, 391

subharmonic oscillations, 146, 150–151, 280–281, 286, 312, 381–382, 455, 459, 543–544
subharmonic poles, 282, 286
subharmonic steady state, 150
superposition theorem, 336
sweeps
   of equivalent capacitor models, 89–90
   frequency, 13
   output impedance, 78
   transmittance, 77–78
SWhyste switch model (Hysteresis switch for PSpice), 358
switch network equivalent models, 216
switched inductor models (SIMs), 220
switched models, 95, 341–343. *See also* averaged models; pulse width modulator (PWM) switch model
switched-mode converters, 66
switchers, 312–313, 413, 437, 608, 730
switches
   charge and flux balance, 25–26
   energy storage, 27
   filters, 19–21
   hysteresis, 355–357
   inductor operating modes, 21–25
   overview, 18–19
switching converters, 90
switching losses (P_{SW}), 711, 837
switching regulators, 85
switching ripple, 14
switch-mode power supply (SMPS), 1–2, 382, 422
synchronous buck converters, 407, 433–434
synchronous postregulation, 804–806
synchronous rectification, 796–798

T (teslas), 393
T model, 400, 404, 583, 716, 812, 814–815
tapped boost converters, 208–209
tapped buck converters, 204
terminal currents, 116, 161–162
teslas (T), 393
THD (total harmonic distortion), 514–515, 517, 524–526, 543, 549, 555, 561, 569–570
thermal fatigue, 731
Thevenin generators, 7–9
third winding, 746–755
three-winding T model, 401–405
thyristor, 665
time steps, 95–96, 316–317
time-dependence, 156, 690, 741
T_j (junction temperature), 694

TL431
   biasing, 298–303
   overview, 286–291
   resistive divider, 303–304
   SPICE model, 326–332
      behavioral model, 326–328
      cathode current versus cathode voltage, 328–329
      netlist, 331–332
      open-loop gain, 330
      output impedance, 329
      overview, 326
      transient test, 331
   type 2 amplifier design example, 291–292
   type 3 amplifier with, 292–298
TLV431, 287–288, 303
toggling points, 143, 145
topologies, 3, 10, 93–94
TOPSwitch, 312
toroids, 742–743, 865–866
total harmonic distortion (THD), 514–515, 517, 524–526, 543, 549, 555, 561, 569–570
total loop gain, 13
trailing edge modulation, 799
transconductance amplifiers. *See* operational transconductance amplifiers (OTAs)
transfer equations
   converter
      boost, 229–231
      buck, 226–228
      buck-boost, 231–235
      overview, 225–226
   RLC, 86–89
transfer functions
   boost dc, 45–46, 51
   buck dc, 32, 38
   buck DCM, 34
   buck-boost dc, 64
   closed-loop, 471
   control-to-inductor current, 546–547
   current loop, 542
   deriving by matrix analysis, 70
   large-signal simulations, 118–119
   open-loop gain, 308
   RLC network, 68, 86
   small-signal models, 127, 129
   superposition theorem, 336
   TL431, 290, 292
   type 2 amplifiers, 333
   type 3 amplifiers, 273
transformers
   dc, 102–104
   feeding models of with physical values

888 INDEX

transformers, feeding models of with physical values (*Cont.*):
    equivalent inductor model, 398–400
    three-winding T model, 401–405
    two-winding T model, 400–401
  manufacturers, 687
  multioutput, 372–373
  overview, 364–365
  saturable models, 371
  simple saturable core model, 366–372
  voltage-mode PWM switch model, 117–118
transient analysis
  12 V, 2 A buck-boost converters, 474–475
  12 V, 4 A voltage-mode buck converters, 413–417
  5 V, 10 A current-mode buck converters, 429–433
  discontinuous current-mode 12 V, 2 A buck-boost converters, 483–486
transient iteration limit, 286
transient models, 97, 317
transient response, 248–249, 279, 284, 286, 310, 320, 330–331, 419–420, 422, 464, 567
transient simulation template, 702
transient steps, 11, 286
transient sweeps, 137–140
transient voltage suppressor (TVS), 612–614
transient waveforms, 275
transistors, 34, 56, 112–113. *See also* bipolar transistors; PNP transistors
transition point DCM-CCM, 37–39, 50–51, 63–64, 143–145
transmittance sweeps, 77–78
t_{rr} (reverse recovery time, diode), 45, 58, 417, 691, 696, 794
true-rms voltmeter, 18
turn-on sequences, 691–692
turn-on switching losses, 834
turns ratio, 24, 103, 131, 364–365, 386, 400, 404, 707–708, 816
TVS (transient voltage suppressor), 612–614
two-output forward converters, 808
two-switch configuration, 756–761
two-switch flyback converters, 614–616
two-switch forward converters, 94, 758, 760, 850
two-winding T model, 400–401
type 1 amplifiers, 255–256, 262, 264, 288
type 2 amplifiers, 256–258, 262, 264–266, 286, 290–293, 332–335
type 2a amplifiers, 258–259, 262
type 2b amplifiers, 259–260, 262
type 3 amplifiers, 261–262, 266–267, 273, 280, 292–298, 308, 316–317, 323, 412, 427

UC348X-based designs, 288, 300, 309, 344–348
undershoots, 14, 285–286, 322, 419, 439–440, 447
undervoltage lockout (UVLO), 358–359, 379, 391
universal input, 436, 508
universal mains, 499–501
upper resistors, 304
UVLO (undervoltage lockout), 358–359, 379, 391

valley switching, 187, 587, 591
valley-fill passive corrector, 526–527
variable resistors, 179–180
variable-frequency system, 22
VCOs (voltage-controlled oscillators), 341, 374–378
version syntax, 2
very high voltage integrated circuit (VHVIC), 649–651
VHVIC (very high voltage integrated circuit), 649–651
virtual grounds, 335–339
VM. *See* voltage mode (VM)
VM control (voltage-mode control), 27, 30
VNTOL (voltage error tolerance), 286, 390
voltage doublers, 508–510
voltage drops, 90–91, 175, 178, 186
voltage error tolerance (VNTOL), 286, 390
voltage limiters, 342
voltage mode (VM)
  12 V, 2 A buck-boost converters
    ac analysis, 468–474
    overview, 465–468
    transient analysis, 474–475
  12 V, 4 A buck converters
    ac analysis, 410–413
    diode, 418–419
    input ripple, 421–424
    output ripple and transient response, 419–420
    overview, 407–409
    power switch, 417–418
    transient analysis, 413–417
  48 V, 2 A boost converters
    ac analysis, 444–449
    overview, 441–444
    transient analysis, 449–452
  BCM controllers, 187–188
  buck converter, stabilizing with k factor, 267–270
  buck simulation results, 92
  converters, 273, 283, 285, 412
  forward converters, 818, 821, 849

voltage mode (VM) (*Cont.*):
  generic controllers, 382–386
  overview, 27–29
  PWM switch model, 134, 180–182, 187–194
  type of converters, 136
voltage reference, 5, 348
voltage runaway, 665, 846
voltage sources, 117
voltage transfer functions, 124
voltage-adjustable passive elements, 349–354
voltage-controlled oscillators (VCOs), 341, 374–378
voltage-controlled voltage sources, 5, 128, 342, 349
voltage-mode control (VM control), 27, 30
voltage-type feedback input, 312
volt-inductor balance, 47, 60
voltmeters, 18
volt-seconds, 24–25, 27, 45, 367–368, 397, 773–774
VSEC, 368
VSWITCH, 356

waveforms
  averaging, 114–116
  flyback, 583–588
  rectifier bridge, 502–506
  sinusoidal, 47, 141
  transient, 275
  two-switch forward converters, 760

weighted feedback, 718
WMCMX circuits, 202

x_1, 86–89, 98–102
x_2, 86–89, 98–102, 106
XFMR, 308, 311, 315, 364–365, 371, 373, 376, 385, 389, 403–404

ζ (damping factor), 68–70
zener diodes, 288, 292, 294, 300, 310, 313, 327–328, 330, 687
zero. *See also* amplifiers
  automated placement, 323–326
  coincident with origin pole, 291
  compensation loop passive, 251–252
  displacements, 426–427
  ESR, 273–275, 412
  independent placement and *k* factor, 272–273
  manual placement of, 275–279, 286, 332–335, 470
  overview, 235–240
  pole-zero placement and distortion, 540
  type 3 amplifiers with TL431, 297–298
  voltage, and change, 530
zero voltage switching (ZVS), 377, 780
zero-pole pair amplifiers. *See* type 2 amplifiers
Z_{in}, 423–424, 463
ZL capacitor type, 413
Z_{out}, 11, 13–14, 70, 329, 422–424, 462–463
ZVS (zero voltage switching), 377, 780

CD-ROM WARRANTY

This software is protected by both United States copyright law and international copyright treaty provision. You must treat this software just like a book. By saying "just like a book," McGraw-Hill means, for example, that this software may be used by any number of people and may be freely moved from one computer location to another, so long as there is no possibility of its being used at one location or on one computer while it also is being used at another. Just as a book cannot be read by two different people in two different places at the same time, neither can the software be used by two different people in two different places at the same time (unless, of course, McGraw-Hill's copyright is being violated).

LIMITED WARRANTY

Customers who have problems installing or running a McGraw-Hill CD should consult our online technical support site at http://books.mcgraw-hill.com/techsupport. McGraw-Hill takes great care to provide you with topquality software, thoroughly checked to prevent virus infections. McGraw-Hill warrants the physical CD-ROM contained herein to be free of defects in materials and workmanship for a period of sixty days from the purchase date. If McGraw-Hill receives written notification within the warranty period of defects in materials or workmanship, and such notification is determined by McGraw-Hill to be correct, McGraw-Hill will replace the defective CD-ROM. Send requests to:

> McGraw-Hill
> Customer Services
> P.O. Box 545
> Blacklick, OH 43004-0545

The entire and exclusive liability and remedy for breach of this Limited Warranty shall be limited to replacement of a defective CD-ROM and shall not include or extend to any claim for or right to cover any other damages, including, but not limited to, loss of profit, data, or use of the software, or special, incidental, or consequential damages or other similar claims, even if McGraw-Hill has been specifically advised of the possibility of such damages. In no event will McGraw-Hill's liability for any damages to you or any other person ever exceed the lower of suggested list price or actual price paid for the license to use the software, regardless of any form of the claim.

McGRAW-HILL SPECIFICALLY DISCLAIMS ALL OTHER WARRANTIES, EXPRESS OR IMPLIED, INCLUDING, BUT NOT LIMITED TO, ANY IMPLIED WARRANTY OF MERCHANTABILITY OR FITNESS FOR A PARTICULAR PURPOSE.

Specifically, McGraw-Hill makes no representation or warranty that the software is fit for any particular purpose and any implied warranty of merchantability is limited to the sixty-day duration of the Limited Warranty covering the physical CD-ROM only (and not the software) and is otherwise expressly and specifically disclaimed.

This limited warranty gives you specific legal rights; you may have others which may vary from state to state. Some states do not allow the exclusion of incidental or consequential damages, or the limitation on how long an implied warranty lasts, so some of the above may not apply to you.